Fundamental Processes in Atomic Collision Physics

NATO ASI Series

Advanced Science Institutes Series

A series presenting the results of activities sponsored by the NATO Science Committee, which aims at the dissemination of advanced scientific and technological knowledge, with a view to strengthening links between scientific communities.

The series is published by an international board of publishers in conjunction with the NATO Scientific Affairs Division

A Life Sciences **B Physics**	Plenum Publishing Corporation New York and London
C Mathematical and Physical Sciences	D. Reidel Publishing Company Dordrecht, Boston, and Lancaster
D Behavioral and Social Sciences **E Engineering and Materials Sciences**	Martinus Nijhoff Publishers The Hague, Boston, and Lancaster
F Computer and Systems Sciences **G Ecological Sciences**	Springer-Verlag Berlin, Heidelberg, New York, and Tokyo

Recent Volumes in this Series

Volume 127—Phenomena Induced by Intermolecular Interactions
edited by G. Birnbaum

Volume 128—Techniques and Concepts of High-Energy Physics III
edited by Thomas Ferbel

Volume 129—Transport in Nonstoichiometric Compounds
edited by George Simkovich and Vladimir S. Stubican

Volume 130—Heavy Ion Collisions: *Cargèse* 1984
edited by P. Bonche, Maurice Lévy, Philippe Quentin, and Dominique Vautherin

Volume 131—Physics of Plasma-Wall Interactions in Controlled Fusion
edited by D. E. Post and R. Behrisch

Volume 132—Physics of New Laser Sources
edited by Neal B. Abraham, F. T. Arecchi, Aram Mooradian, and Alberto Sona

Volume 133—Scaling Phenomena in Disordered Systems
edited by Roger Pynn and Arne Skjeltorp

Volume 134—Fundamental Processes in Atomic Collision Physics
edited by H. Kleinpoppen, J. S. Briggs, and H. O. Lutz

Series B: Physics

Fundamental Processes in Atomic Collision Physics

Edited by

H. Kleinpoppen
University of Stirling
Stirling, United Kingdom

J. S. Briggs
University of Freiburg
Freiburg, Federal Republic of Germany

and

H. O. Lutz
University of Bielefeld
Bielefeld, Federal Republic of Germany

Plenum Press
New York and London
Published in cooperation with NATO Scientific Affairs Division

Proceedings of a NATO Advanced Study Institute on
Fundamental Processes in Atomic Collision Physics,
held September 10–21, 1984, in Santa Flavia, Italy

Library of Congress Cataloging in Publication Data

Main entry under title:

Fundamental processes in atomic collision physics.

(NATO ASI series. Series B, Physics; v. 134)
"Proceedings of a NATO Advanced Study Institute on Fundamental Processes in Atomic Collision Physics, held September 10–21, 1984 in Santa Flavia, Italy."
Bibliography: p.
Includes index.
1. Collisions (Nuclear physics)—Congresses. 2. Massey, Harrie Stewart Wilson, Sir. I. Kleinpoppen, H. (Hans) II. Briggs, J. S. III. Lutz, H. O. IV. NATO Advanced Study Institute on Fundamental Processes in Atomic Collision Physics (1984: Santa Flavia, Sicily) V. Series.
QC794.6.C6F86 1985 539.7'54 85-24407
ISBN-13:978-1-4612-9256-2 e-ISBN-13:978-1-4613-2125-5
DOI: 10.1007/978-1-4613-2125-5

Softcover reprint of the hardcover 1st edition 1985

A Division of Plenum Publishing Corporation
233 Spring Street, New York, N.Y. 10013

Sir Harrie Massey at the NATO ASI on Fundamental Processes in Energetic Atomic Collisions, held September 20-October 1, 1982, in Maratea, Italy

SCIENTIFIC ADVISORY COMMITTEE

T. Anderson, Denmark
G. Ferrante, Italy
J.C. Lehmann, France
Sir Harrie Massey, United Kingdom
E. Merzbacher, United States of America
A. Niehaus, Netherlands
A. Scharmann, Germany

LOCAL COMMITTEE (University of Palermo)

P. Cavaliere (Chairman)
G. Ferrante
C. Leone
M. Zarcone
R. Daniele

ORGANIZING COMMITTEE (University of Stirling)

H.J. Beyer
A.J. Duncan
H.S. Hamdy
H. Kleinpoppen
H.A.G. Silim
W. Perrie

SECRETARY

H.J. Beyer

DIRECTORS

H. Kleinpoppen
J.S. Briggs
H.O. Lutz

PREFACE

The Proceedings of the Advanced Study Institute on Fundamental Processes in Atomic Collision Physics (Santa Flavia, Italy, September 10-21, 1984) are dedicated to the memory of Sir Harrie Massey, whose scientific achievements and life are reviewed herein by Sir David Bates.

At the first School on the above topic (Maratea, September 1983, Volume 103 in this series), Harrie Massey presented the introductory lectures, summarized the entire lecture program, and presented an outlook on future developments in atomic collision physics. In an after-dinner speech, Massey recalled personal reminiscences and historical events with regard to atomic collision physics, to which he had contributed by initiating pioneering work and by stimulating and surveying this branch of physics over a period of almost six decades. Participants in the Maratea School will always remember Harrie Massey as a charming and wonderful person who was most pleased to discuss with everyone--students, postdoctorals, and senior scientists--any topic in atomic collision physics.

Harrie Massey was a member of the Scientific Advisory Committee of the 1984 Santa Flavia School. Before his death he expressed his interest in attending this second School devoted to the presentation of recent developments and highlights in atomic collision physics. It is the desire of all authors to honor Harrie Massey with their contributions in these Proceedings.

As one of the highlights of the meeting, a very enjoyable and stimulating symposium on New Trends in Atomic Collisions was organized by Honorary Chairman Ugo Fano. The manuscripts of the lectures presented on this occasion, as well as short papers resulting from a poster session, are included in this volume.

The fact that the two Schools at Santa Flavia and Maratea attracted so many distinguished lecturers, junior and senior scientists, was very gratifying and evidence for the importance of atomic collision physics as a fundamental branch of physics. The success of the present School was the result of considerable efforts and cooperation from members of the Scientific Advisory Committee, the Local Committees of the Universities of Palermo and Stirling, the lecturers, invited speakers, and authors of contributed papers. The stimulating atmosphere of the meeting created by fruitful discussion between lecturers and participants was perhaps the most valuable asset of the School.

The Scientific Secretary, Dr. H.J. Beyer, deserves our deepest gratitude for all the work he carried out patiently and effectively during the meeting. We gratefully acknowledge the efforts of our secretary, Mrs. K. Murray, who carried the burden of organization through all stages of preparation of the School. We are most thankful for the generous support of the following sponsors:

Bundesministerium für Forschung und Technologie, Bonn
Deutsche Forschungsgemeinschaft, Bonn
Italian Research Councils
Lambda Physik GmbH, Göttingen
National Science Foundation, USA
Royal Society, London
Science and Engineering Research Council, London
Spectra-Physics GmbH, Darmstadt-Kranichstein
University of Palermo
University of Stirling

These sponsors enabled many scientists to attend the School.

H. Kleinpoppen
J.S. Briggs
H.O. Lutz
(Directors of the School)

Stirling, December 1984

CONTENTS

Ion-Atom Collisions

Atomic and Molecular Processes

H. S. W. MASSEY

LIFE, WORK, PERSONALITY AND CHARACTERISTICS

D. R. Bates

Department of Applied Mathematics & Theoretical Physics
Queen's University
Belfast, Northern Ireland

Harrie Stewart Wilson Massey was born in Melbourne on the 16 May 1908 and died in his home in Esher, Surrey on the 27 November 1983. He had an unequalled influence on atomic physics and its applications, directly through his research and monographs, and indirectly through his students, his students' students, their students, the line continuing.

I shall sketch his life and his individuality, and attempt an appreciation of his work concentrating on those extraordinarily fruitful four years 1929 to 1933 when he was a research student in Cambridge.

LIFE AND WORK

Australia

Massey was of mixed English, Irish and Swedish descent but above all was an Australian. He had an abiding love for his native country. He returned on over 20 occasions the first two times by sea, the next four times by slow propeller aircraft - and even to-day's journey by fast wide-body jet is tiresome to most people.

His father was also called Harrie Stewart, which is how I shall refer to him. The odd spelling of the first name is because of an aunt Harriet.

Harrie Stewart was a hunter, prospector and self-taught engineer. Having discovered gold and antimony in the little community of Hoddles Creek, in the bush 50 miles east of Melbourne, he settled there as a manager, first of a mine, then of a saw-mill. In 1907 he married Eleanor Wilson (whence Massey's third given name).

At the time there was no school in Hoddles Creek. Like his son, Harrie Stewart was farsighted and persuasive. Largely through his endeavours a State School, which had been closed since 1897, was reopened in 1912 in a tiny hall which had been built by a working-bee. In 1913 his only child entered this school, which had 20 to 30 boys and girls distributed through 8 classes and a single teacher, a Mr. Roadknight. Massey was a child prodigy, gifted with an amazing memory and with an amazing ability to master any subject quickly. Whereas pupils in State Schools were expected to progress at a rate of one grade per year Massey progressed seemingly effortlessly at a rate of two grades per year, attaining the Merit Certificate of the State School system at an age of only 9, instead of the more usual 13 to 14.

Massey remained nominally a pupil of Mr. Roadknight but was actually taking a correspondence course with the aim of winning a scholarship to University High School, Melbourne. He achieved his aim in 1920. The Principal of the High School interviewing him asked, "What do you wish to become?" He was somewhat taken aback by the 12 year old child's reply, "A university professor of science". In 1920 university professors of science were very much rarer than they are today.

Order had a strong appeal to Massey. At University High School he was captivated by chemistry, fascinated to learn that the endless variety of different substances are made up of relatively few elements, and that these elements had been arranged in a logical fashion in the Periodic Table. He spent most of his pocket money on chemicals for experiments he did in the kitchen at home.

In 1925 Massey won a scholarship which took him to Melbourne University. While there as usual he did more, far more, than was normal. By 1927 he had a First Class Honours B.Sc., in both Chemistry and Physics. Continuing with Mathematics he obtained a B.A. with First Class Honours in 1929, and at the same time as he did this course he worked for his M.Sc. in Physics. The work entailed an experimental project and a theoretical dissertation. The project for which he had C.B.O. Mohr as a collaborator, was on the reflection of soft X-rays from a metal surface. At the suggestion of the Professor of Natural Philosophy, T. H. Laby, he took as his dissertation topic "Wave mechanics". Massey was therefore at the frontier of research, Schrodinger's classic papers having been published as recently as 1926. Most of the literature was in German which Massey could not read easily. Yet his dissertation came to over 400 pages. Referring to it his old friend Mohr, now in retirement in Melbourne wrote last year "It was an outstanding one for all time in the Physics Department and still holds an honoured place in the Departmental Library for all to see". Having need for extra money Massey also undertook a heavy burden of tutoring at the University, and of coaching at a local school. Yet he made time to indulge in his passion for ball games: billiards, tennis, base-ball and of course cricket. And this was not all. While an undergraduate he met, wooed and married a beautiful girl, Jessica Bruce who gave him a daugher, and a happy marriage lasting almost 60 years.

Cambridge

In 1929 he won another scholarship which enabled him to go to the Cambridge of Rutherford. An accompanying perk was a free passage, First Class, to Tilbury, England on a liner of the Orient Company which stopped at least two days at each of Colombo, Bombay, Aden, Port Said, Naples and Gibraltar. It whetted a relish for seeing the world.

When Massey arrived in Cambridge he at once went to see Rutherford, an awe-inspiring figure. He said he wanted to do both theoretical and experimental research on collisions. Rutherford advised strongly against this on the grounds that the strain would be too great. With a confidence and self-knowledge remarkable in a young man of 21 Massey disregarded the advice of the great man who had attracted him to Cambridge and who he revered.

I will return to Massey's own work at Cambridge later. He was blithely happy during the four years he spent there, the only period during which his zest for science was untrammelled by teaching, administrative or policy making duties. He regarded himself as highly privileged that this fell to his lot. In particular he was deeply appreciative of being in the Cavendish Laboratory during the miracle year of 1932 when Cockcroft and Walton split the atom, when Chadwick discovered the neutron and when Blackett and Occhialini confirmed Anderson's identification of Dirac's positive electrons in the secondary products of cosmic rays using their counter controlled cloud-chamber. When he was dying he gave me a vivid account of how he heard of the disintegration of nuclei by energetic protons over half a century ago: "The discovery was kept confidential for some time, and I first learnt of it in the Cavendish Laboratory after tea one day. Mott came in and said to me, 'Have you heard that Cockcroft and Walton have done this?' Forbearing to say it in words he wrote down 'Li + H $\rightarrow$ 2He!'" His interest in science undiminished Massey recalled with satisfaction the wide range of work being done in the Cavendish Laboratory. Rutherford himself was concerned with α particle levels in nuclei; C. D. Ellis was determining the internal conversion coefficients of γ-rays; L. H. Gray and G.T.P. Tarrant were examining the absorption of γ-rays by heavy elements finding it to be curiously strong at quantum energy much in excess of 500 keV - a result that was to prove significant in connection with pair formation; C. D. Crowther was interested in the electromagnetic separation of isotopes; E. S. Shire was attempting to observe Mott's indistinguishability effects in the scattering of He nuclei at kev energies;P. Kapitza was investigating the effects of strong magnetic fields on the properties of materials; J. J. Thomson was carrying out experiments on what would now be called plasma oscillations; R. M. Chaudhri was studying the interactions of positive ions with surfaces; C.B.O. Mohr and F.H.Nicoll were making measurements on the angular distributions of inelastically scattered electrons; and J. A. Radcliffe was probing the ionosphere with radio waves. Note he dwelt on experiment, not theory.

Massey often spoke appreciatively of how he had been influenced by Rutherford, Chadwick and Mott. The influence of the first of these may have been responsible for his being rather averse to speculation in science. Referring to Rutherford's strong aversion to such speculation Wilson (1983) quotes him as warning, "Don't let me catch anyone talking about the universe in my laboratory".

Taking part in ball-games continued to be an important activity of Massey. He played for the Cavendish Cricket Club and was Captain in 1933 when the Club won the Inter-Laboratory Cup. He was a very good player. The Australian Test wicket-keeper, B. A. Barnett, a fellow Victorian who knew him well, advised that he could if he wished become a professional. Massey told me that his success as a player was mainly due to his having an exceptionally fast reaction time. He was indeed exceptionally fast at anything dependent on his mental processes: reading, understanding, writing doing mathematical analysis.

Returning to Massey's Cambridge period,the 1851 Senior Scholarship which he held was coming to its end. The job situation in science was appalling. Describing for me how he came to switch from atomic physics to geophysics after his brilliantly successful collaboration with Massey on the diffraction of slow electrons by atoms (§3) E. C. Bullard wrote, "I learnt of a job I might have helping Sir Gerald Lenox-Conyngham who was teaching surveying and geodesy to colonial service probationers. I asked Rutherford for his advice. He said, 'There are no jobs and there are a lot of people just in front of you, if I were you I would take any job I could get'. So I took the job and left atomic physics". Happily a new post, an Independent Lectureship in Mathematical Physics, was created in Queen's University,Belfast at the opportune time for Massey (October 1933). He applied with Fowler (who was for official purposes, his Supervisor) Mott and Chadwick as referees. A copy of Fowler's letter of recommendation has come into my possession and is of interest in showing how Massey was regarded half a century ago. It reads:

"Dr. Massey has done much first class work in theoretical physics since he came to the Cavendish Laboratory. He has worked mainly on the theory of collisions particularly between atoms or molecules and electrons at slow or fairly slow speeds of the electrons. In this field he occupies a quite outstanding position, for it is largely due to his work (much of it exceedingly laborious) that we are today confident that quantum mechanics is entirely adequate to account for all the effects in such collisions - though naturally owing to the complications of the calculations it is not always possible to confirm quantitatively all the details. He has also successfully applied his great knowledge of collision theory to other fields, notably the theory of gas viscosity and the theory of interaction of protons and neutrons. He must be classed as one of the most successful appliers of quantum mechanics to atomic physics among men of his standing in this country

R. H. Fowler"

It is odd that Fowler did not mention Theory of Atomic Collisions. Massey's application was successful - fortunately for me.

Belfast and pre-War London

Massey was initially the sole member of the Department of Mathematical Physics so that his teaching duties were onerous. They remained onerous even after an Assistant, R. A. Buckingham, was appointed in 1934 because Massey then added an extra Honours course and introduced a Post-graduate course. I became an undergraduate student at Queen's University in 1934 and testify that Massey was already a superb lecturer.

The numerical work which was entailed in much of Massey's research was done using a cylindrical slide rule and woefully inadequate tables of functions. It was time-consuming and Massey determined to improve the position shortly after he had settled in Belfast. For the immediate future he applied on March 20 1934 to the Royal Society for "£50 to be expended in paying an assistant to perform arithmetical calculations" on an investigation on nuclear forces in particular"to determine how the internal (anomalous) nuclear field depends on the charge of the interacting particle, the experimental phenomena considered to include anomalous scattering of α-particles, collisions of neutrons with nuclei and artificial disintegration by various nuclear projectiles". This must have seemed good value for money and Massey whose application received the support of Rutherford,then President of the Royal Society, got the grant. A letter from Rutherford dated May 15, 1934 is not without interest:

"Dear Massey,

I have received a box of apples and I think I received some time ago a statement from the Shipping people that you were sending one to me. If this be so my wife and I would like to thank you for this admirable gift. The apples arrived in good order and I am keeping the doctor away in the orthodox way. It is very good of you to send us such a useful gift.

I have sent a note to the R.S. supporting your claim for a grant in aid for a computer. I hope you be able to get the money alright.

Yours sincerely,
Rutherford"

Massey sought a different solution to the computing problem in the more distant future. At his instigation Mr. John Wylie the Physics Workshop Superintendent, constructed a small-scale differential analyser, the cost of the materials being born by another grant of £50 (which Massey obtained from the Queen's Better Equipment Fund) [41] . The differential analyser enabled the radial wave equation (including exchange) to be solved. However it was not easy to operate (cf Bates 1983). Its output was in the form of a light pencil line drawn on graph paper from which readings had to be taken. Massey and his associates made extensive use of the differential analyser in investigations on, for example, the properties of helium at low temperatures [44] and the photo-ionization of atomic oxygen [47] .

His stay in Belfast was crowned by the publication of the slim (105pp) First (1938) Edition of Negative Ions - in my opinion a minor masterpiece. Incidentally the interest he had in negative ions was indirectly responsible for his beginning, in 1937, the well-known series of investigations on recombination in the ionosphere [38, 47, 60, 65, 84, 88].

Massey moved to the Goldsmid Chair of Mathematics at University College in 1938. The Munich Agreement in September of that year convinced him that war was coming but his sense of imminent disaster did not decrease his enthusiasm for science. He had the differential analyser, brought from Queen's University to University College where it was later destroyed during an air-raid - which saved it from the ignominy of becoming obsolescent.

War

When the War began Massey was assigned to the Admiralty Research Laboratory, Teddington, to help in assessing defence measures against German aircraft laid magnetic mines which in September and October of 1939 caused the loss of nearly a dozen merchant ships at the entrance to harbours. Their effect would have been devastating if instead of beginning on a small scale, with limited stock and manufacturing capacity, Hitler and his General Staff had waited until they could suddenly start a large scale campaign. The defence measures consisted on the one hand of the degaussing of ships, and on the other hand of sweeping the mines by various means. Good use was made of the consummate skill at solving problems in electricity and magnetism which Massey had acquired as a university teacher.

A happy event of 1940 was Massey's election as Fellow of the Royal Society at the unusually young age of 32. He told me he learnt the good news through a note from C. G. Darwin who then was Director of the National Physics Laboratory which adjoined ARL. The note in a grubby envelope marked 'Confidential', was handed to him by a messenger as he left work for home on February 29. It was amongst his papers:

"My dear Massey,

My best congratulations to you on your nomination for the Royal Society. This was announced today but is to be regarded as confidential for the next fortnight until formal election. Still you may regard it as settled - always supposing you don't do something like deliberate murder in the interim.

I need not say how delighted I am about it, as indeed I expect you will be.

Best of luck.

Yrs. sinc.
C. G. Darwin"

In 1941 the emphasis changed from defence to attack, and Massey was appointed Deputy Chief Scientist at Mine Design Department (attached to H.M.S. Vernon) in Havant, near Portsmouth. He took with him 4 young scientists (R.A. Buckingham, F.H.C. Crick, J.C. Gunn and me) who had been associated with him at ARL. The Chief Scientist was approaching retirement and Massey as his Deputy, was in a key position which was not without awkwardness, and which needed the exercise of considerable tact. From the onset Massey had the respect of the scientists at MDD; his complete integrity and outstanding ability soon won him the respect of the production engineers and of the naval officers of the technical, operational and intelligence branches. He succeeded in bringing about the proper co-operation between these groups which hitherto had been sadly lacking. Undoubtedly this greatly increased the effectiveness of the British mining effort - an effort which sank about 1050 enemy vessels and damaged about another 540.

After the Quebec Agreement of August 19, 1943 there were demands that Massey should be released to join the Manhattan Project. The Admiralty resisted tenaciously but lost the conflict, the final decision being taken, Massey understood, at Cabinet level. In consequence Massey went to Berkeley, California, to lead the theorists working on the large scale separation of ^{235}U from natural uranium by the electro-magnetic method. Although this method was not ultimately utilized it found some favour despite its expense because of being sure. By the time Massey arrived the plant parameters had in fact been largely finalized by empirical methods [189] . Perhaps the covert reason for forcing the conflict with the Admiralty to a successful conclusion was to ensure as strong a British presence as feasible on the Manhattan Project for possible long-term benefits. The immediate consequence was that Massey provided with an opportunity of observing the organization of a huge research and development programme and with an opportunity, which he seized eagerly, of doing basic research on many problems which had been left unresolved. Moreover he found he was free to engage on experimental work with the magnificent resources of the 37-inch cyclotron laboratory at his disposal. He has acknowledged that he was enriched by his experiences in Berkeley. It may be said that Massey's coefficient of adsorption for experience was almost unity.

Post-war London

When Massey returned to University College in October 1945 he changed the applied mathematics section of the Mathematics Department into a largely theoretical physics section by appointing E. H. S. Burhop, R. A. Buckingham, J. C. Gunn and me to the staff.

He began to devote a considerable part of his time to the writing of those monographs which have so benefited our field: the 388 pp Second (1949) Edition and 858 pp Third (1965) Edition of Theory of Atomic Collisions; the 669 pp First (1952) Edition (with E. H. S. Burhop) and 3806 pp Second (1969-74) Edition (with also H. B. Gilbody) of Electronic and Ionic Impact Phenomena; and the 136 pp Second (1950) Edition and 741 pp Third (1976) Edition of Negative Ions. Referring to monographs Massey once said in an interview (Robinson 1981) "They are not worth writing unless you get them into a proper coherent and systematic form and this takes time and effort. The papers you work from are very diffuse and you have to find a relationship between them, so you end up with more knowledge than when you started out". Later in the same interview he commented, "I'm not so much an analyst as a synthesist in my mental processes. So I look for relations between things and have a pattern of linkages that helps me keep in touch with a very wide area of things. That's fairly unusual; most people are analytical in their approach". His popular books and textbooks cover much of physics but he was rather dismissive of the work they involved saying "You already know what you are going to write about".

Massey initiated experimental work being done in the Mathematics Department by encouraging Burhop to do particle physics using the nuclear emulsion technique, and by getting grants which enabled him to appoint R.L.F. Boyd and J. B. Hasted as research assistants working on atomic collision physics. In his capacity as Chairman of the Gassiot Committee of the Royal Society he started along the interconnected paths of making science policy and of facilitating international collaboration in research.

In 1950 Massey accepted the invitation to transfer to the Quain Chair of Physics at University College. At last he had a post of the type he had long wanted*. He gave top priority to ensuring

*Massey told me that Burhop (1968) was mistaken in thinking he "spent a week-end of agonizing indecision" regarding the invitation.

that the technical facilities available in his new department were as good as possible and to this end arranged for the appointment of H. S. Tomlinson ("a superlative engineer") who he had met at Berkeley. Under Massey's leadership the Physics Department was transformed into one of the largest in Britain with around 100 postgraduate students, and 30 postdoctoral fellows. In the field of atomic collision physics mention must be made of J. B. Hasted (slow heavy particle collisions), H. B. Gilbody (fast heavy particle collisions using a locally built 500 kv Van der Graaff generator) R.L.F. Boyd, J. W. Fox and E. J. Smith (atomic hydrogen experiments), D.W.O. Heddle (optical excitation functions), F. F. Heyman and G. R. Heyland (positron and positronium experiments) M. J. Seaton (theoretical atomic physics with applications to astrophysics). The width of the interests of the Department is illustrated by the variety of its research groups after a union with the Astronomy Department had been effected shortly before Massey's retirement in 1975. Their names (with in brackets, an indication of some of the topics covered) are as follows: (1) Theoretical Positron, Molecular and Nuclear Physics (nuclear three-body problem; calculation of cross sections for excitation of molecular rotation by slow positron impact.) (2) Theoretical Atomic Physics and Astrophysics (studies of atomic structures, transition probabilities and line-broadening parameters; interpretation of spectra of gaseous nebulae and of solar corona.) (3) Theoretical Elementary Particle Physics (K-meson reactions; quark model of primary interaction.) (4) Image Processing (application of cellular logic image processor; automatic road traffic census taking using infrared optical techniques.) (5) Experimental Atomic and Molecular Physics (studies of collisions of metastable atoms with free atoms and surfaces; elastic and inelastic differential scattering of low energy electrons by atoms.) (6) Infrared Astronomy (ground based observations at 10 and 20 microns and balloon-borne observations at 40 - 350 microns.) (7) Space Science and Atmospheric Structure Studies (measurement of winds and temperatures and analyses of results for global circulation patterns.) (8) Bubble Chamber and Emulsion Group (neutrino and antineutrino interactions; interaction of slow K^- mesons in hydrogen and helium.) (9) Astronomy (geological nature of surfaces of Moon and planets; physical nature of chromosphere and corona; collapse of interstellar gas clouds.) (10) Ultraviolet and Optical Astronomy Group (use of recently developed instrumentation, the Image Photon Counting System and the Speckle Interferometer; study of the interstellar gas at resonance lines not accessible from ground using star-stabilized platform.) (11) Mullard Space Science Laboratory (use of rocket and satellite born instruments for the study of role of magnetic field and particle motion in lower ionosphere, of electron temperature and density in plasmasphere, magnetopause and magnetospheric boundary, of solar X-ray events and of cosmic X-ray sources.) (12) Spark Chamber Group (use of facilities at CERN for study of possible exitence of Intermediate Vector Boson.) Bates, Boyd and Davis (1984) have given a brief account of the development of the Department.

As well as being engaged with his scientific endeavours proper Massey was a busy man of affairs. Internally at University College he was a member of the important Academic Staff Appointments and Promotion Committee 1947-75 and was Vice-Provost 1969-75. Externally he was one of the most influencial scientists in the United Kingdom. He served on the Council of the Royal Society 1949-51 and 1959-60, and was Physical Secretary 1969-78 (which entailed ex officio membership of about 50 committees and inter alia Chairmanship of the Ordnance Survey Committee). He was a Member of the Governing Board of the National Institute for Research in Nuclear Science 1957-65, was Chairman of the Council for Scientific Policy 1965-69 (which advised the Secretary of State for Education and Science on the exercise of his responsibilities for civil science policy and which inter alia led to the setting up of Flowers Committee and thence to the Computer Board) and was a Member of the Central Advisory Council for Science and Technology 1967-69. Because of his activities in various capacities (Chairman of the British National Committee on Space Research since 1959, Bureau Member of COSPAR 1959-78, President of the European Preparatory Committee for Space Research 1960-64, President of Council of ESRO 1964), he was the chief architect of the British and indeed European space research effort. He was a Member of the Anglo-Australian Telescope Board since 1974, Chairman since 1980, averting a serious financial crisis on one occasion by flying to Australia, and in a discussion with Government officials there presenting the Board's case unyieldingly. It is fitting that the base laboratory building of the Anglo-Australian Observatory in Epping, Sydney was on March 30 1984 named the Massey Building. At the ceremony H. H. Atkinson gave another illustration of Massey's steely resolve that right for science be done. He said "I was representing the United Kingdom on a committee of European governments drafting the convention for the proposed new European Space Agency. At one meeting it was proposed that science should not have a special place in the new agency which was to be devoted mostly to applied projects. Only the United Kingdom disagreed and I found myself in a minority of one. On my return from Paris I went to see Sir Harrie in London finding him behind one of many piles of books in his room at University College. I told him the position. Over the next few weeks he got in touch with scientists in each of the other European countries. By the time of the committee's next meeting every delegation had been lobbied by its own scientists and a miracle occurred: the previous decision was reversed - and unanimously. The result is that science still stands high in ESA".

With so much committee work it was inevitable that huge stacks of papers reached Massey's desk. I asked him how he managed to cope with these papers in view of his heavy other commitments. He smiled and said, "I don't read most of the papers. Nearly always they are banal, obvious. I've got a rather brutal technique. When the minutes arrive I merely turn the pages over to see if there's anything I don't already know. At a meeting I can pick up the important issues very quickly and then deal with any problems".

WORK 1929-33

In order to emphasise that Massey's achievements were the equivalent of the combined achievements of several very able men I will now return to his work during his Cambridge period. I do so because most of us have already passed through a corresponding period and therefore are in the position of being able to make a private comparison, whereas most of us have not experienced what is entailed in writing advanced monographs, in leading and administering a large science department, and in influencing science policy at national and international levels. These activities are vital. Not being a man to drift into any course of action Massey must have taken a deliberate decision to devote much of his time to them at the cost of sacrificing some of his potential for personal research. The decision cannot have been easy because his research capability remained very high throughout most of his life. He was superb technically, and had a superb knowledge of the relevant literature. However as he once remarked (Robinson 1981) this last is not always an advantage in that "often you fail to see a new way of doing things because you know too much about what is being done". Perhaps too, the speed and confidence with which he formulated a problem was not conducive to reflection.

Although the University's regulations required him to have a supervisor (R. H. Fowler) Massey was an independent research worker from the time of his arrival. As mentioned earlier he had already decided that his field would be collisions. He was exceptionally mature for a research student and except in two instances (which I will indicate later) studied problems which he had found for himself. My thesis is that working in largely uncharted territory, and with wretched computing facilities, he did enough for six Ph.D's and one D.Sc during those Cambridge years. I shall now categorize his researches in accord with my thesis.

1. Experiments on electron diffraction by atoms. In October 1929 Bullard, at the suggestion of Blackett, began an investigation on the angular distribution of electrons scattered by atoms to see if there was any pecularity associated with the Ramsauer effect in the total scattering cross section. He decided to start with argon. After some abortive initial work he was joined by Massey in February 1930, when he had got a little way in setting up the apparatus. Great care was needed, spurious angular distribution data having been published recently. Bullard and Massey [9] covered scattering angles up to about 120°, and took special trouble to show that the currents they were measuring really came from single scattering of their electron beam in the gas. They discovered that the scattered intensity after falling off monotonically went through a minimum and increased to a maximum. Massey interpreted this correctly. By

integrating the angular distribution curve at different energies the total cross section curve was obtained. It was in good agreement with that of Ramsauer. Bullard and Massey [11] then carried out similar measurements in helium, neon, nitrogen, hydrogen and methane and related their results perceptively to theory. Later, with Childs, Massey [21,23] constructed an apparatus which enabled measurements to be made on the angular distribution of electrons scattered by cadmium and zinc atoms as they emerged from a heated oven. Once again the results were related preceptively to theory.

2. First Born approximation. With Bullard, Massey [2] applied the Born approximation to calculate the scattered intensity $I(\theta)$ and total cross section Q for fast electrons in a Thomas-Fermi field. They showed that $I(\theta)Z^{-2/3}$ is a function of $vZ^{-1/3}\sin\frac{1}{2}\theta$ only and that $QZ^{-2/3}$ is a function of $vZ^{-1/3}$ only, v being the velocity and Z the atomic number, so that all their results could be represented by two curves. Massey and Mohr [10] carried through long complicated analysis to obtain closed expressions (in terms of Gegenbauer Polynomials) for the Born approximation for the excitation of the $1s \rightarrow n\ell m$ transitions of hydrogen by electron impact. They also treated the 1s-2s, 1s-2p, 1s-3p and 1s-3d singlet transitions of helium representing the ground state by the 3-parameter variational wave function of Hylleraas and Undheim and the excited states by the 2-parameter wave functions of Eckart. Massey and Mohr [19] further carried through the even longer and more complicated analysis arising for ionization. For both excitation and ionization they evaluated the expressions they obtained numerically, and made comparison with such experimental results as were available.

3. Electron exchange. Oppenheimer introduced electron exchange in 1928 as a possible explanation of the failure of Faxén and Holtsmark to reproduce the Ramsauer effect in their 1927 calculations. In the following year Holtsmark reproduced the Ramsauer effect on allowing for the polarizability of the target atom. However this did not cast doubt on the correctness of the concept of electron exchange. Massey and Mohr [10] were the first to treat exchange quantitatively. In the case of excitation by electron impact they succeeded in showing, after some formidable analysis, that exchange is responsible for strong multiplicity change transitions, a result of prime importance. They did detailed calculations on the excitation of the 2^3S, 2^3P, 3^3P and 3^3D terms of helium. As they noted, the success which Holtsmark had attained while ignoring exchange now seemed puzzling. Massey and Mohr [14] adduced evidence that it was due to the cancellation of the effects of exchange and of distortion in elastic scattering. The problem of including both exchange and distortion in calculations on inelastic scattering remained. Massey and Mohr [18] made progress at high energies, where the effect, of exchange is slight and at moderate energies, where the exchanged and directly scattering waves are of comparable intensity and interfere strongly. They even succeeded in obtaining numerical results for the cases of the excitation of the 2^1P and the 2^3P terms of helium. They also showed that the theory

is adequate to explain the maxima and minima that had been found experimentally in the angular distributions of electrons inelastically scattered by heavy atoms.

4. Electron-molecule collisions. Massey [5] used the Born approximation to treat elastic scattering by molecular hydrogen. He showed that the differential cross section, averaged over all molecular orientations is a function of v sin $\frac{1}{2}\theta$, where v is the velocity and θ is the scattering angle, and did computations which revealed the occurrence of diffraction effects. Massey [6] obtained similar diffraction effects in short X-rays scattered by molecular hydrogen. He and Mohr [13] established that there is a close relation between X-ray and electron scattering and gave a simple formula expressing this relation. They also greatly extended the range of Massey's Born approximation calculations on elastic scattering by molecular hydrogen. The final formula shows neatly the scattering by the two atoms considered seperately, the (relatively unimportant) scattering by the concentration of charge between the nuclei which produces the molecular binding and a straightforward diffraction factor due to the two scattering centres. It lead to satisfactory agreement with the measurements of Bullard and Massey [11] at energies above 80 eV. In the case of elastic scattering by N_2 Bullard and Massey [17] went further treating the scattering by the two atoms considered seperately by the Faxén-Holtsmark method (while neglecting the scattering by the concentration of charge between the nuclei). The agreement with their own experiments [11] was good (except at small scattering angles) above 60eV.

Massey [12] was the first to succeed in carrying out calculations on inelastic electron-molecule calculations. As he was quick to recognize there is one type of axially symmetrical field, that due to a dipole, for which a detailed theory is not required (provided the dipole moment is sufficiently small) because the Born approximation is valid even at very low energies. He proved the $\Delta J = \pm 1$ selection rule, and obtained a simple formula for the inelastic cross section (which is much greater than the elastic cross section). Turning to molecular hydrogen he and Mohr [13] calculated the Born approximation for the $X^1\Sigma \rightarrow B^1\Sigma$ transition showing that the diffraction factor is different from that for elastic scattering because of the initial and final bound states having opposite symmetries. Introducing exchange Massey and Mohr [13] also calculated the cross section for dissociation through the $X^1\Sigma \rightarrow b^3\Sigma$ transition.

5. Collisions between atomic systems and transport phenomena in gases
Massey and Smith [22] did extensive theoretical work on collisions between atomic systems using the wave version of Mott's perturbed stationary state method in its original form. The most lasting of this work is that on symmetrical resonance charge transfer. They obtained the correct formula for this process and were the first to treat the symmetry properties of the system properly. Using their formula they calculated the cross section for 1 keV

He^+ ions in helium. Massey and Mohr [25] recognized that the mobility of He^+ ions in helium is largely controlled by symmetrical resonance charge transfer and calculated this mobility.

The consideration of quantal effects in the elastic scattering of atoms and in transport phenomenon in gases was pioneered by Massey and Mohr [20,25] . In the first instance [20] they represented the atoms by rigid spheres for simplicity. They made a startling discovery. Letting r_o be the sum of the radii of the rigid spheres, they found that in the 'billiard ball limit' the quantal elastic cross section is $2\pi r_o^2$ which is twice the classical elastic cross section. As they demonstrated diffraction at unobservably small scattering angles is responsible for the effect. Using the rigid sphere model Massey and Mohr [20] calculated the viscosity of helium as a function of the temperature down to 15K. They found that their quantal treatment gave very much better agreement with experiment than the classical treatment. Massey and Mohr [25] next took the interaction between the atoms to be as proposed by Slater. They were no longer able to obtain the phase shifts analytically, and therefore resorted to the Born and Jeffreys approximations for them. Their calculations indicated that Slater's interaction does not yield a more accurate viscosity-temperature curve than the rigid sphere model. An interesting discovery was that the curve showing the total elastic cross section as a function of the velocity of relative motion has definite maxima and minima, both in the case of the rigid sphere model, and in the case of the Slater interaction. They explained that the phenomenon is due to the presence of a small number of terms only in the Faxén-Holtsmark series for the cross section when the velocity is low and the fact that, because of the identity of the colliding particles, only the even phases appear.

By a combination of the Jeffreys approximation and the random phase approximation Massey and Mohr [25] also succeeded in deriving a simple formula for the elastic scattering cross section corresponding to an inverse power law interaction.

6. Nuclear collisions. Using the Dirac wave equation, and applying the Born approximation, Massey [3] showed that the effect of a nuclear magnetic moment on the nuclear scattering of fast electrons is far too small to be invoked to explain why some calculations by Mott were not in good agreement with experiment. It was known that α-particle scattering by light nuclei do not obey the Rutherford formula if the velocities of relative motion be sufficiently large (of order 2×10^9 cm s^{-1}). Attempts had been made to account for the results classically by introducing various inverse power law, Fr^{-n}, interactions (with $n = 4$ or 5). From dimensional considerations Massey [4] showed that the scattering formula includes Planck's constant h unless

$$w = 2/n$$

w being the power of F on which the cross section depends. He concluded that the scattering formula includes h for $n > 2$ except in the extremely improbable event that w is fractional and therefore that significance could not be attached to the classical work that had been done on the anomalous scattering of α-particles. Massey obtained the Born approximation for the cross section in the cases n = 4, 5 and 6, and discussed the experimental data on scattering by magnesium and aluminium in the light of his results. However as he pointed out [15] the Born approximation is invalid because of the strong perturbation of the α-particle wave by the nuclear potential barrier. Massey included the effect of α-particle exchange finding it to be large at a resonance level.

Having shown that lithium is disintegrated by fast protons Cockcroft and Walton reported the similar disintegration of much heavier elements. In the course of our last meeting two months before his death Massey told me that he took up the problem at the request of R. H. Fowler. Using a formula due to Mott, which enabled a limit to be set to the contribution from a resonance level, he was able to avoid a tedious unrewarding calculation. In the case of heavy elements he found the limit to be less than the reported cross sections. Curiously he did not publish his important result in the open literature. Not being aware of this at the time I did not ask Massey the reason and can now only speculate. Perhaps he did not publish because he could not explain why the experimental results on the heavy elements were wrong. In 1933 Oliphant and Rutherford recognized the results were wrong because of contamination of the target by boron from the glass walls.

Chadwick's discovery of the neutron stimulated Massey to do further research at the frontier of nuclear physics. The then favoured model of the neutron was based on an idea Rutherford had put forward in 1920. According to the model the neutron consisted of an electron and a proton very tightly bound together, as conceivably might be allowed on a relativistic quantum theory (which was yet to be developed). By means of the model, and the Born approximation, Massey [16] discussed the passage of neutrons through matter, and the disintegration of neutrons by impact with protons and with heavy nuclei. Almost immediately after D.E. Lea discovered the γ-radiation emitted when neutrons were absorbed in a hydrogen-rich material, and interpreted it as arising from the association of a neutron and a proton to form a deuteron, Massey and Mohr [24] carried out two sets of calculations on the association rate. They assumed that the neutron is as envisaged by Rutherford and that there are exchange forces between it and the proton; and alternatively they assumed that the neutron behaves as a fundamental charge-free particle throughout the collision the radiation being due to the acceleration of the proton in the field of force. They showed that neither assumption could account for the observed rate.

Miscellaneous. For completeness I must mention that three of Massey's investigations while a research student do not fit my categorization. I will give only their titles: 'Theory of the extraction of electrons from metals by positive ions and metastable atoms' [1] , 'The triatomic hydrogen H_3^+'[7] and 'The theory of the extraction of electrons from metals by metastable atoms II' [8]. They more than balance the last item, [24] of (6) which although it carries the Cavendish Laboratory address probably represents research completed after Massey left.

The D.Sc of my fanciful categorization remains. It is of course the 283pp First (1933) Edition of Theory of Atomic Collisions, the first monograph on a rapidly growing subject, a monograph destined to become a classic. Mott invited Massey to be a co-author in 1931. Massey eagerly accepted the invitation and wrote chapters VII 'Scattering by a centre of force. Treatment by integral equation and miscellaneous theorems', VIII 'General theory of atomic collisions',IX 'The collisions of fast electrons with atoms - Born's approximation', X 'The elastic scattering of slow electrons by atoms', XI 'Inelastic collisions of electrons with atoms', XII 'The collisions of electrons with molecules' and XIII 'The collisions between massive particles'. It was an extraordinary achievement for so young a man, so busily engaged on research.

PERSONALITY AND CHARACTERISTICS

Massey's penetrating intelligence, sound judgement, energy and determination are sufficiently evident, from his career but I will illustrate his remarkable powers of concentration and his eidetic memory. Although he did much of his research and writing at home he did not have a room as a study: he could work at his desk in an alcove while his wife chatted to her friends in another part of the room, occasionally looking up from his work to interject a comment. He seemed to have powers of total recall: I have listened while he described events of long ago at Hoddles Creek, such as his finding a wombat in the school grounds, as vividly as if they had happened yesterday.

His own genius did not lead him to expect too much of students. He shared the opinion which Blackett [181] once expressed to him that "to award a Ph.D to a candidate who had worked satisfactorily on a subject chosen by his supervisor was almost an obligation".

His nature was kindly and warm. Biographers of some of the great report, approvingly, a habit of not suffering fools gladly. This meritorious habit, if meritorious it be, is not one I can claim for Massey who treated all people with the same consideration and courtesy.

He was patient and even tempered. I have seen him display irritation on only a few instances during the 50 years I have known him. When I was a research student of his before the War he once snapped at me for the abominable habit I had of misspelling 'nucleus' by inserting a gratuitous 'e' after the first letter. I now spell the word faultlessly. I have described another instance elsewhere (Bates 1983). The two instances are marked in my memory because of their being so untypical of him. He would however allow himself to become animated in political discussion. On one occasion during a train journey Massey, Blackett and Burhop so raised their voices in vehement argument that a fellow passenger felt it incumbent on him to rebuke them, and request that they moderate their tone. Although Massey did not lack a sense of humour it was rare for him to make a jesting remark. As an example of his appreciation of the ludicrous I recall his coming gleefully into my office at University College to tell me about a pile of luggage bound for an Eastern European city that he had seen at Victoria Station "Each bag had a label marked 'INTELLECTUALS' PEACE CONFERENCE'" he chuckled "It was like a scene in a book by Evelyn Waugh".

Naturally he had some quirks. While making it plain that he regarded time spent on morning coffee as an indefensible waste he was happy to idle away an hour or more over afternoon tea chatting about the news of the day. He had a cultured man's knowledge of modern literature but did not regard not having read a novel by Jane Austin as an omission which might with advantage be repaired.

As well as reading Massey's relaxations included watching T.V. and cricket, gardening and doing a little cooking. He loved travel.

Massey underwent serious surgery some years ago but made light of it and at the time was curiously anxious that news of it should not spread. After not having seen him for a year, because of my having been in the United States, I met him in London on July 12, 1983. He discussed a possible break with tradition in certain sections of the new edition of Theory of Atomic Collisions he was planning with Phil Burke as co-author. He was about ready to start writing and expected to have his part finished by Christmas. It was obvious to me that he had become gravely ill but he would not admit to this even, I think to himself. However in September he phoned me asking me to be responsible for the Royal Society's Biographical Memoir. I visited him in his home on September 27. He now accepted resignedly that he was dying of bone cancer. He faced his cruel fate with dignity, his main thoughts on those dear to him, most especially his wife. His mind, its powers unimpared, seemed an entity distinct from his stricken body. The human tragedy of the two being indissolubly linked weighed heavily on me as I sat by him while he talked at length about his life and work. One of the subjects raised was his religious beliefs. Did nebulae, planets and intelligent beings with self-awareness flow inevitably from the 'big bang'? Massey affirmed that he was a 'complete agnostic'.

Before he died Harrie Massey suffered much pain with that

fortitude which those who knew him would have expected. His last words may indicate that as death approached he changed from being a complete agnostic, or they may indicate that even when on the edge of the abyss he continued to try and give comfort. Make of them what you will. Addressing his wife Harrie said, "We shall meet again. I don't want to live here any longer."

REFERENCES

Bates, D. R., 1983, International Journal of Quantum Chemistry: Quantum Chemistry Symposium 17:5.
Bates, D. R., Boyd, R.L.F. and Davis, D. G., 1984, Biog. Mem of Fellows of Roy. Soc. 30 (in press).
Burhop, E.H.S., 1968, Adv. in Atom. Molec. Phys. 4:1.
Robinson, P., 1981, The Australian Physicist, 18:135.
Wilson, D., 1983, "Rutherford' Hodder and Stoughton, London.

NOTE

The following eleven pages give Massey's publications. A few slight or ephemeral pieces, such as brief introductory remarks at a Conference or Discussion Meeting have not been included.

H.S.W. MASSEY — JOURNAL PAPERS

1 1930 The theory of the extraction of electrons from metals by positive ions and metastable atoms.
Prof. Camb. Phil. Soc., 26, 386-401

2 (with E.C. Bullard) Remarks on the scattering of électrons by atomic fields.
Proc. Camb. Phil. Soc., 26, 556-563

3 Scattering of fast electrons and nuclear magnetic moments
Proc. Roy. Soc., A127, 666-670

4 Remarks on the anomalous scattering of alpha-particles from the quantum mechanical point of view
Proc.Roy.Soc., A127, 671-677

5 Theory of the elastic scattering of electrons in molecular hydrogen. Proc.Roy.Soc., A129, 616-627

6 1931 The theory of the scattering of short X-rays by molecular hydrogen. Prof.Camb.Phil.Soc., 27, 77-85

7 The triatomic hydrogen H_3^+. Proc.Camb.Phil.Soc., 27, 451-459

8 The theory of the extraction of electrons from metals by metastable atoms, II. Proc.Camb.Phil.Soc., 27, 460-468

9 (with E.C. Bullard) The elastic scattering of slow electrons in argon. Proc.Roy.Soc., A130, 579-590

10 (with C.B.O. Mohr) The collision of electrons with simple atomic systems and electron exchange. Proc.Roy.Soc., A132, 605-630

11 (with E.C. Bullard) The elastic scattering of slow electrons in gases, II. Proc.Roy.Soc., A133, 637-651

12 1932 The collision of electrons with rotating dipoles
Proc.Camb.Phil.Soc., 28, 99-105

13 (with C.B.O. Mohr) The collision of electrons with molecules.
Proc.Roy.Soc., A135, 258-275

14 (with C.B.O. Mohr) The collisions of slow electrons with atoms, I - General theory and elastic collisions.
Proc.Roy.Soc., A136, 289-311

15 The collision of alpha-particles with atomic nuclei
Proc.Roy.Soc., A137, 447-463

16 The passage of neutrons through matter
Proc.Roy.Soc., A138, 460-469

17 1933 (with E.C. Bullard) Scattering of electrons by nitrogen molecules. Proc.Camb.Phil.Soc., 29, 511-521

18 (with C.B.O. Mohr) The collision of slow electrons with atoms, II - General theory and inelastic collisions
Proc.Roy.Soc., A139, 187-201

19 (with C.B.O. Mohr) The collision of slow electrons with atoms, III - The excitation and ionization of helium by electrons of moderate velocity. Proc.Roy.Soc., A140, 613-636

20 (with C.B.O. Mohr) Free paths and transport phenomena in gases and the quantum theory of collisions, I - The rigid sphere model
Proc.Roy.Soc., A141, 434-453

21 (with E.C. Childs) Scattering of electrons by metal vapours, I - Cadmium. Proc.Roy.Soc., A141, 473-483

22 (with R.A. Smith) The passage of positive ions through gases. Proc.Roy.Soc., A142, 142-172

23 (with E.C. Childs) Scattering of electrons by metal vapours, II - Zinc. Proc.Roy.Soc., A142, 509 - 518

24 1934 (with C.B.O. Mohr) Radiative collisions of neutrons and protons. Nature, 133, 211

25 Free paths and transport phenomena in gases and the quantum theory of collisions, II - The determination of the laws of force between atoms and molecules. Proc.Roy.Soc., A144, 188-205

26 (with C.B.O. Mohr) The collision of slow electrons with atoms, IV. Proc.Roy.Soc., A146, 880-900

27 1935 (with C.B.O. Mohr) The double excitation of helium by electron impact. Proc.Camb.Phil.Soc., 31, 604-608

28 Excitation of molecular vibration by impact of slow electrons Farad.Soc., 31, 556-563

29 (with C.B.O. Mohr) The interaction of light nuclei, I. Proc.Roy.Soc., A148, 206-225

30 (with C.B.O. Mohr) The interaction of light nuclei, II. - The binding energies of nuclei H_1^3 and He_2^3 Proc.Roy.Soc., A152, 693-705

31 (with R.G.J. Fraser & C.B.O. Mohr) Scattering of molecular rays in gases. Zeit f Physik, 97, 11-12

32 1936 (with E.H.S. Burhop) The intensity of X-ray spectrum lines of heavy elements. Proc.Camb.Phil.Soc., 32, 461-470

33 (with E.H.S. Burhop) The relativistic theory of the Auger effect. Proc.Roy.Soc., A153, 661-682

34 (with R.A. Smith) Negative atomic ions. Proc.Roy.Soc., A155, 472-489

35 (with C.B.O. Mohr) The interaction of light nuclei, III - The binding energies of He^4, He^5, Li^6 and of nuclei type 4n Proc.Roy.Soc., A156, 634-654

36 1937 (with C.B.O. Mohr) Nuclear excitation and disintegration collisions involving strong interactions, I Proc.Roy.Soc., A163, 529-556

37 (with R.A. Buckingham) The nature of the interaction between neutrons and proton from scattering experiments. Proc.Roy.Soc., A163, 281-297

38 Dissociation, recombination and attachment process in the upper atmosphere, I. Proc.Roy.Soc., A163, 542-553

39 1938 (with C.B.O. Mohr) Anomalous scattering of particles and long-range nuclear forces. Proc.Camb.Phil.Soc., 34,498-501

40 The creation of electron pairs by nuclear capture of neutrons. Proc.Roy.Irish Acad., 44A, 77-85

41 (with J. Wylie, R.A. Buckingham & R. Sullivan) A small scale differential analyser: its construction and operation Proc.Roy.Irish Acad., 45A, 1-21

42 (with R.A. Buckingham) Long range forces between hydrogen molecules. Proc.Roy.Irish Acad., 45A, 31-45

43 (with E.H.S. Burhop) The probability of annihilation of positrons without emission of radiation
Proc.Roy.Soc., A167, 53-61

44 (with R.A. Buckingham) Low temperature properties of gaseous helium. Proc.Roy.Soc., A168, 378-389 Correction A169, 205

45 1939 (with H.C. Corben) The emission and absorption of heavy electrons. Proc.Camb.Phil.Soc., 35, 84-94

46 (with H.C. Corben) Elastic collisions of mesons with electrons and protons. Proc.Camb.Phil.Soc., 35, 463-473

47 (with D.R. Bates, R.A. Buckingham & J.J. Unwin) Dissociation, recombination and attachment processes in the upper atmosphere, II - The rate of recombination. Proc.Roy.Soc., A170, 322-340

48 1940 (with D.R. Bates) The continuous absorption of light by negative hydrogen ions. Astrophys. J., 91, 202-214

49 Appendix to article by E.H.S. Burhop: The inner shell ionization of atoms by electron impact. Proc.Camb.Phil.Soc., 36, 50-52

50 (with C.B.O. Mohr) Polarization of electrons by double scattering
Nature, 146, 264

51 (with R.A. Buckingham) Collisions of neutrons with deutrons and the nature of nuclear forces. Nature, 146, 776

52 1941 (with R.A. Buckingham) The scattering of neutrons by deutrons and the nature of nuclear forces.
Proc.Roy.Soc., A177, 123-151

53 (with D.R. Bates) Exchange effects in the theory of the continuous absorption of light, I - Ca and Ca^+
Proc.Roy.Soc., A177, 329-340

54 (with C.B.O. Mohr) The polarization of electrons by double scattering. Proc.Roy.Soc., A177, 341-357

55 (with R.A. Buskingham & S.L. Tibbs) A self-consistent field for methane and its applications. Proc.Roy.Soc., A178, 119-134

56 (with R.A. Buckingham & J. Hamilton) The low-temperature properties of gaseous helium, II. Proc.Roy.Soc., A179, 103-122

57 1942 The elastic scattering of fast positrons by heavy nuclei.
Proc.Roy.Soc., A181, 14-19

58 1943 (with D.R. Bates) The properties of neutral and ionized oxygen and their influence on the upper atmosphere
Reports on Prog. in Phys., 9, 62-74

59 (with D.R. Bates) The negative ions of atomic and molecular oxygen. Phil.Trans. Roy. Soc., A239, 269-304

60 (with D.R. Bates) The basic reactions in the upper atmosphere
Proc.Roy.Soc., A187, 261-296

61 1947 Semi-conductors. J. Sci., Instr., 24, 220-224

62 (with T-M. Hu) Nuclear forces and the magnetic moment of the deutrons. Nature, 160, 794-795

63 (with R.A. Buckingham) The collisions of neutrons with deutrons and the reality of exchange forces. Phys.Rev., 71, 558

64 Obituary Notice - Thomas Howell Laby. Proc.Phys.Soc., 59,506-508

65 (with D.R. Bates) The basic reactions in the upper atmosphere, II. - The theory of recombination in the ionized layers.
Proc.Roy. Soc., A192, 1-16

66 The peace-time application of atomic energy.
Year Book of World Affairs, 1, 265-278

67 1948 Atomic energy.
Journal of the Royal United Service Institution, 93, 390-402

68 (with R.A. Buckingham) The scattering of protons by deuterons.
Phys. Rev., 73, 260-261

69 (with E.H.S. Burhop & T-M. Hu) The effect of non-central forces on the collisions of high energy neutrons with protons.
Phys. Rev., 73, 1403-1404

70 (with D.R. Bates) A note on the ionic composition of the upper atmosphere.
Report of the Gassiot Committee of the Royal Society, published by the Physical Society, 48-51.

71 (with D.R. Bates & R.W.B. Pearse) The aurora. Ibid 97 - 105

72 (with R.A. Buckingham)Collisions of neutrons and protons with deuterons and the nature of nuclear forces
Rep.Internat.Conf., "Fundamental particles", (1946)
Phys.Soc., Lond., 1, 175-177

73 (with E.H.S. Burhop) The capture of neutrons by dueterons.
Proc.Roy.Soc., A192, 156-166

74 1949 (with E.H.S. Burhop & D. Bohm) The use of probes for plasma exploration in strong magnetic fields.
In The characteristics of electrical discharges in magnetic fields, (Eds.A. Guthrie & R.K. Wakerling, McGraw-Hill), 13-77

75 (with E.H.S. Burhop & G. Page) Experimental investigation of threshold pressure for stable operation of arcs Ibid 107-127

76 (with E.H.S. Burhop, W. Berkey, J.D. Craggs & J. Keene) Measurement of the absolute values of the cross sections for ionization of uranium tetra-chloride and uranium hexaflouride.
Ibid 127-145

77 (with E.H.S. Burhop & C. Watt) The ionization and dissociation of uranium tetra-chloride and uranium hexaflouride by electron impact. Ibid 145-166

78 (with E.H.S. Burhop, D. Bohm & R.W. Williams) A study of arc plasma. Ibid 173-334

79 Collisions between atoms and molecules at ordinary temperatures
Reports on Prog. in Phys., 12, 248-269

80 (with D.R. Bates, R.L.F. Boyd & E.H.S. Burhop) The atomic processes effective in the ionosphere.
Proc. International Scientific Radio Union, 7, 299-329

81 (with T-Hu) Non-central interactions between neutron and proton. Proc. Roy.Soc., A196, 135-159

82 1950 (part I with D.R. Bates & A. Fundaminsky and Part II with D.R. Bates, A. Fundaminsky & J.W. Leech) Excitation and ionization of atoms by electron impact - The Born & Oppenheimer approximations. Phil.Trans.A.Roy.Soc., Lon., 243, 93-143

83 (with B.L. Moiseiwitsch) The scattering of electrons by hydrogen atoms. Phys. Rev., 78, 180-181

84 1951 (with D.R. Bates) The negative ion concentration in the lower atmosphere. J.Atmos. & Terr.Phys., 2, 1-13

85 Formation of the ionosphere. Mixed Commission on Ionosphere, International Council of Scientific Unions, Proc.Second Meeting, Brussels, 1950, URSI, Brussels, 9-31

86 (with B.L. Moiseiwitsch) The application of variational methods to atomic scattering problems, I. - The elastic scattering of electrons by hydrogen atoms. Proc.Roy.Soc., A205, 483-496

87 1952 Gaseous ions and their reactions. Farad.Soc.Disc., 12, 24-33

88 (with D.R. Bates) On negative ions of molecular oxygen in the D-layer. J.Atmos. & Terr. Phys., 2, 253-254

89 Recombination of gaseous ions. Adv.in Phys., 1, 395-426

90 Electron scattering in solids. Adv. in Electronics, 4, 2-66

91 Recent laboratory measurements of recombination coefficients and the problems of ionospheric recombination. Proc. Mixed Commission on the Ionosphere, Canberra, 20-32

92 (with C.B.O. Mohr) Strong coupling in inelastic collisions of electrons with atoms. Proc. Phys. A65, 845-853

93 Leslie John Cromie, 1893-1950 Obituary Notices of Fellows of the Royal Society, 8, 97-105

94 (with R.A. Buckingham & S.J. Hubbard) The scattering of neutrons and of protons by deuterons. Proc.Roy.Soc., A211, 183-204

95 (with G.A. Erskine) The application of variational methods to atomic scattering problems, II - Impact excitation of the 2s level of atomic hydrogen - distorted wave treatment. Proc.Roy.Soc., A212, 521-530

96 Fundamental particles. Sci.Prog., 40, 193-212

97 1953 (with I. Abdelnabi) Inelastic collisions of electrons in helium and Townsend's ionization coefficient. Proc.Phys.Soc., A66, 288-296

98 (with B.L. Moiseiwitsch) Calculation of the 1s-2s electron excitation cross section of hydrogen by a variational method. Proc.Phys.Soc., A66, 406-408

99 (with D.R. Bates & A.L. Stewart) Inelastic collisions between atoms, 1. - General theoretical considerations. Proc.Roy.Soc., A216, 437-458

100 The collisions of deuterons with nucleons Prog.Nucl.Phys., 3, 235-270

101 1954 The nature of the upper atmosphere. Endeavour, 13, 81-85

102 Peaceful applications of atomic energy International Relations, 1, 10-16

103 (with M.J.M. Bernal)Metallic ammonium. Mon.Not.Roy.Astr.Soc., 114, 172-179

104 (with D.R. Bates)Slow inelastic collisions between atomic systems. Phil.Mag., 45, 111-122

105 Laboratory methods on investigating processes important in the high atmosphere. Proc. Conference on auroral physics, (University of Western Ontario, July, 1951), sect.3, 205-220 (Eds.N.C. Gerson, T.J. Kenshe & R.J. Donaldson Jr., Cambridge, Mass., Air Force Cambridge Research Center Geophysical research papers, 30).

106 (with C.B.O. Mohr) Gaseous reactions involving positronium. Proc.Phys.Soc., A67, 695-704

107 (with S. Hochberg & L.H. Underhill) The scattering of nucleons by alpha particles - the s-phases. Proc.Phys.Soc., A67, 957-966

108 (with B.L. Moiseiwitsch) The application of variational methods to atomic scattering problems, IV. - The excitation of the 2^1s and 2^3s states of helium by electron impact. Proc.Roy.Soc., A227, 38-51

109 1955 Fundamental primary processes in gaseous discharge. Appl.Sci. Res., (Hague), B5, 1-9

110 (with D.W. Sida) Collision processes in meteor trails Phil.Mag., 46, 190-198

111 Progress and problems in physics to-day. (Physical Society Presidential Address). Phys. Soc. Year book, 1, 1-12

112 (with S. Hochberg, H. Robertson & L.H. Underhill) The scattering of nucleons by alpha particles - determination of the spin-orbit interaction. Proc.Phys.Soc., A68, 746-753

113 (with A.H. deBorde) The collisions of nucleons with deuterons Proc.Phys.Soc., A68, 769-780

114 Recent laboratory experiments of interest for the physics of the ionosphere. Mixed Commission on the Ionosphere, International Council of Scientific Unions, Proc. Fourth Meeting, Brussels, Aug., 1954 URSI, Brussels, 32-40

115 Survey of nuclear forces In Proc. of 1954 Glasgow Conference on Nuclear and Meson Physics. (Eds. E.H. Bellamy & R.G. Moorhouse, Pergamon, London), 1-10

115 Atomic collision processes in astrophysics. Vistas in Astronomy, 1, 277-283

117 1956 Mesons, hyperons and antiprotons. Endeavour, 15, 117-127

118 Theory of atomic collisions. Hanbuch der Physik, Encyclopaedia of Physik, (Springer-Verlag, Berlin), 36, 232-306

119 Excitation and ionization of atoms by electron impact Ibid 307-408

120 (with R.O. Ridley) Application of variational methods to the theory of the scattering of slow electrons by hydrogen molecules Proc.Phys.Soc., A69, 659-667

121 Theory of the scattering of slow electrons. Rev.Mod.Phys., 28, 199-213

122 1957 Antiprotons and antineutrons. Sci.Prog., 45, 1-10

123 The earth satellite programme In Space research and Exploration, (Ed.D.R. Bates, Eyre and Spottiswoode, London), 100-124

124 (with A.H.A. Moussa) The elastic scattering of positrons by atoms and molecules. Proc.Phys.Soc., A71, 38-44

125 (with S. Khashaba) The excitation of the 2p state of hydrogen by slow electrons - distorted wave treatment. Ibid 574 - 584

126 1959 Space Research. Cont. Phys., 1, 81-95
127 (with J.D. Craggs) The collision of electrons with molecules
Hanbuch der Physik, Encyclopaedia of Physik, 37/1, 314-415
128 The excitation of molecular vibration and rotation by impact
of slow electrons. Phil.Mag., 4, 336-340
129 Chemistry of the upper atmosphere.
Proc.Chem. Soc., 1959, 377-383
130 Collisions (43rd Guthrie Lecture to the Physical Society)
Yearbook of the Physical Society, 15-28
131 Artificial satellites
Proc.Roy.Inst., G.B., 37, 347-354
132 1960 Scientific research with rockets and satellites
Bull. Inst.Phys., 11, 33-39
133 Theoretical aspects of the three-body problem at energies below
50MeV. IN Nuclear forces and the few-nucleon problem
(Eds. T.C. Griffiths and E.A. Power, Pergamon Press) 345-355
134 (with B.L. Moiseiwitsch) The excitation of the 2^3p state of
helium by electron impact. Proc.Roy.Soc.,A258, 147-158
135 1961 Space research. Endeavour, 20, 68-77
136 (with A.H.A. Moussa) Positronium formation in helium
Proc.Phys.Soc., 77, 811-816
137 (with A.E. Potter) Atmospheric photochemistry.
Roy.Inst. of Chemistry Lecture Series, 1, 1-27
138 1962 Engineering problems in space research.(The 49th Thomas
Hawksley Lecture, The Inst. of Mechanical Engineers).
The Engineer, 214, 920-923
139 1963 The International Geophysical Year in retrospect.
Endeavour, 22, 70-74
140 Problems raised by the growth of modern technology
International Relations, 2, 411-416
141 (with R.B. Bernstein, A. Delgarno & I.C. Percival) Thermal
scattering of atoms by homonuclear diatomic molecules.
Proc.Roy.Soc., A274, 427-442
142 1964 Space travel & exploration, 1-A general survey.
Introduction. Cont. Phys., 5, 241-254
143 Space travel & exploration, II. - Ways and means.
Cont. Phys., 5, 457-471
144 Space travel & exploration, III. - Scientific results - the
earth and the upper atmosphere. Cont. Phys., 6, 26-48
145 (with L. Hammett) The chemical bond.
International Science & Technology, 1964, 62-72
146 The present state of the study of atomic collisions
Proc.3rd. Int.Conf.on Physics of Electronic & Atomic
Collisions, 3-13
147 (with I.H. Sloan) The exchange polarization approximation for
the elastic scattering of slow electrons and positrons by
atoms and ions: Electron scattering by helium atoms Ibid 14-15
148 (with W.J. Cody, J. Lawson & K. Smith) The elastic scattering of
slow positrons by hydrogen atoms. Proc.Roy.Soc., A278, 479-489

149 Artificial satellites, I. - Introduction.
Vistas in Astronomy, 4, 29-110
150 1965 Space travel & exploration, IV - Scientific results:
The sun and solar-terrestrial relations: Short-wave astronomy
Cont.Phys., 6, 172-191
151 Space travel & exploration, V. - Lunar and planetary probes -
Meteorology and communications - Geodesy and navigation
Ibid 241-260
152 Space travel & exploration, VI - Exbiology - man in space.
Ibid 321-337
153 1966 Collisions of slow positrons with atoms
In Ouantum theory of atoms, molecules, solid state, (Academic
Press, New York), 203-215
154 The uncertainty principle and short-lived atomic and sub-atomic
states. Endeavour, 25, 59-64.
155 (with J. Lawson, J. Wallace & D. Wilkinson) Dispersion relations
and the elastic scattering of electrons by helium atoms
Proc.Roy.Soc., A294, 149-159
156 1967 Gaseous positronics
In Proceedings of Positron Annihilation Conference
(Academic Press, New York), 113-125
157 Problems of science policy
Proc.Roy.Inst. 41, 379-391
158 1968 Collisions between gas atoms
Endeavour, 27, 114-119
159 Report on space research
In The Royal Conference of Commonwealth Scientist, London:
The Royal Society, 184-187
160 1969 L'homme conquerant du cosmos
In L'Aventure Humaine, Encyclopaedia des Sciences de l'Homme,
(Ed. Paul Alexandre, Paris: Editions de la Grange-Bataliere),
177-189
161 The theory of the collisions of electrons with one- and two-
electron atoms and ions.
In Physics of the one- and two-electron atoms, Proc.
Sommerfeld Centennial Memorial Meeting, (Eds. F. Bopp &
H. Kleinpoppen. Amsterdam: North-Holland), 511-542
162 1971 David Forbes Martyn, 1906-1970
Biographical Memoirs of Fellows of the Royal Society, 17,
497-510
163 Atom-atom and atom-ion collisions, I. - Collisions between
neutral atoms in which no excitation occurs.
Cont. Phys., 12, 537-558
164 Slow collisions of positrons in gases.
In Atomic Physics, 2. Proceedings of the 2nd International
Conference on Atomic Physics (Ed. P.G.H. Sandars, Plenum Press:
London, New York), 307-343
165 1972 Atom-atom and atom-ion Collisions, II. - Collisions between
atoms and ions in which no excitation occurs - symmetrical
charge transfer. Cont. Phys., 13, 135-158

166 Atom-atom and atom-ion collisions, III - Collisions involving excitation transfer, spin exchange or depolarization, without internal energy change. Ibid 354-392

167 (with J. Lawson & S. Hara) Behaviour of positrons in molecular gases, II. - Dependence of annihilation rates of positrons in rare gases on the presence of molecules
J.Phys. B., 5, 599-608

168 Atomic and molecular reactions in space.
In From Plasma to Planet. Proceedings of the 21st Nobel Symposium (Ed. A. Elvius, Stockholm: Almquist & Wiskell), 17-37

169 Nuclear physics to-day and in Rutherford's day
Notes and Records of the Royal Society, 27, 25-44

170 1973 Atom-atom and atom-ion collisions, IV. - Inelastic collisions. Cont. Phys., 15, 497-512

171 Neutrinos. Endeavour, 32, 86-92

172 1974 Theories of the Ionosphere, 1930-1955
J.Atmos. & Terr. Phys., 36, 2141-2158

173 (with S. Hara) Annihilation of positrons in Ar-CO mixtures
J. Phys. B., 7, 262-263

174 1975 Negative ions, positive electrons (International Conference on the physics of electronic and atomic collisions - Invited lectures), Review papers and progress reports, Seattle: Univ. Washington Press, 3-26

175 Contributor to Memorial Meeting for Lord Blackett
Notes and Records of the Roy.Soc., 29, 138-140

176 Atomic energy and the development of large teams and organizations. Proc.Roy.Soc., A342, 491-497

177 1976 D.R. Bates - A sixtieth birthday tribute
In Atomic processes and applications, (Eds. P.G. Burke & B.L. Moiseiwitsch, North-Holland Publishing Co.), 3-12

178 (with N. Feather) James Chadwick, 1891-1974
Biographical Memoirs of Fellows of the Royal Society, 22, 11-70

179 Optical absorption by negative ions - photodetachment
Endeavour, 35, 58-65

180 Slow positrons in gases
Physics Today, 29, 42-46, 48-51

181 1977 Patrick Blackett - an appreciation. (First Blackett Memorial Lecture) Proceedings of the Indian National Science Academy, 43A, 1-17

182 1978 (with Sir Harold Thompson) David Christie Martin, 1914-1976
Biographical Memoirs of Fellows of the Royal Society, 24, 391-407

183 1979 Negative ions
Adv. in Atomic & Molecular Phys., 15, 2-36

184 Problems and possibilities in the study of electron-molecule collisions
In Electron-molecule scattering (Ed. S.C. Brown, Wily Interscience), 185-189

185 Ionic reactions in the laboratory and in planetary atmospheres.
Journal de Physique, 40, C7, 21-35

186 1980 Dissociative attachment
Endeavour, 4, 78-84
187 T.H. Laby, F.R.S. (The 1980 Laby Memorial Lecture)
The Australian Physicist, 17, 181-187
188 Space research (Australian Academy of Science Silver Jubilee Symposium). In Changing views of the Physical World, 1954 - 1979, vol. II. (Ed. Dr. G.K. White, Canberra, Australia), 87-122
189 1981 (with D.H. Davis) Eric Henry Stoneley Burhop, 1911-1980
Biographical Memoirs of Fellows of the Royal Society, 27, 131-152
190 (with Sir Kenneth Hutchison & J.A. Gray) Charles Drummond Ellis
Ibid 199-233
191 1982 Gaseous positronics - past, present and future
Can.J. Phys., 60,461-470
192 "Scattering experiments, (Atoms and Molecules)"
In 5 Edn. Encyclopaedia of Science and Technology, (Ed. Sybil P. Parker, McGraw-Hill Book Co., New York), 12, 81-88
193 Upper atmosphere physics
In Applied Atomic Collision Physics, (Eds. H.S.W. Massey, E.W. McDaniel & B. Bederson), Vol. I. Atmospheric Physics and Chemistry, (Eds. H.S.W. Massey & D.R. Bates, Academic Press, New York), 1-6
194 Structure of terrestrial atmosphere at middle latitudes
Ibid 7-21
195 Photochemistry of the midlatitude ionosphere
Ibid 22-77
196 Thermal balance in the thermosphere at middle latitude
Ibid 78-105
197 Atomic collisions and the lower ionosphere at midlatitudes
Ibid 106-151
198 High latitude ionosphere, exosphere and magnetosphere
Ibid 226-255
199 Ionospheres of planets and other bodies of solar system
Ibid 256-293
200 Atomic collisions in gaseous nebulae
Ibid 400-426
201 1983 Fundamental processes in atomic collision physics
In Fundamental processes in energetic atomic collisions. (Ed. H.O. Lutz, J.S. Briggs & H. Kleinpoppen, Plenum Press, New York), 1-38
202 Summary lecture
Ibid 659-668

BOOKS (AND LECTURES PUBLISHED OTHER THAN IN JOURNALS OR CONFERENCE PROCEEDINGS)

1933 (with N.F. Mott) The theory of atomic collisions, Oxford: Clarendon Press, (2nd Ed. 1949; 3rd Ed. 1965)
1938 Negative ions, Cambridge: Cambridge University Press, (2nd Ed. 1950; 3rd Ed. 1976)
1950 The atom and its nucleus, Brisbane: University of Queensland
1952 (with E.H.S. Burhop) Electronic and ionic impact phenomena, Oxford: Clarendon Press, (2nd Ed. with also H.B. Gilbody, 5 vols. 1969-1974)
1953 Atoms and energy, London: Elek Books, (wnd Ed. 1956)
1958 (with R.L.F. Boyd) The upper atmosphere, London: Hutchinson, (2nd Ed. 1960)
1959 (with H. Kestleman) Ancillary mathematics, London: Pitman, (2nd Ed. 1964)
1960 The new age in physics, London: Elek Books, (2nd Ed. 1967)
1961 (with A.R. Quinton) Basic laws of matter, Bronxville, N.Y.: Herald Books
1964 (with M.O. Robbins, R.L.F. Boyd, G.V. Groves & D.W.O. Heddle) Scientific Research in Space, London: Elek Books
Space physics, Cambridge: Cambridge University Press
1966 Space travel and exploration, London: Taylor & Francis
1979 Atomic and molecular collisions, London: Taylor & Francis
1984 (with M.O. Robbins) History of British Space Research, Cambridge, Cambridge University Press
1982 (Edited with E.W. McDaniel & B. Bederson) Applied atomic collision physics, New York: Academic Press, 5 volumes
Edited Negative Ions by B.M. Smirnov, translated by S. Chomet, London: McGraw-Hill
1971Chairman of working group. Report of a study on the support of scientific research in the universities (commissioned by the Council for Scientific Policy, London: HMSO-Cmnd.4798)

INTRODUCTION TO FUNDAMENTAL INTERACTIONS IN ATOMIC COLLISIONS

Eugen Merzbacher

Department of Physics and Astronomy
University of North Carolina at Chapel Hill
Chapel Hill, NC, 27514, U.S.A.

I. INTRODUCTION

In these introductory lectures an effort will be made to provide a review of background material which may be useful in following the other lectures to be given in this institute. We begin with a description of collision processes that is intentionally formulated in terms general enough to encompass the interactions of particles, photons, and composite systems, as they make their appearance in nonradiative and radiative atomic and molecular collision processes. The emphasis is on multichannel processes, and it is assumed that the reader has some familiarity with the less general formulations of the theory, such as those appropriate for elastic and inelastic scattering and ionization.[1] Decay processes will also be treated as "half-collisions", and the connection between the T matrix and decay rates will be established. The results will be expressed in terms of the density matrix as the most convenient tool for an analysis of experiments.

Once this general framework is outlined, it becomes profitable to study the specific interactions which govern the dynamical processes of atomic physics. Since much of the program of this institute deals with photons, some of the concepts of elementary quantum electrodynamics will be reviewed. The implications of gauge invariance will be discussed and a cursory review of relativistic quantum mechanics as it applies to atoms will be given.

In this chapter all formulas are written using units which imply that Planck's constant $\hbar = 1$, but it is well to remember that the conventional value of this constant is 0.66×10^{-15} eV-sec.

II. REVIEW OF COLLISION THEORY

In the context of atomic processes, a collision is a temporal chain of events that starts initially, as $t \to -\infty$, with the preparation of a beam of the noninteracting collision partners in what generically may be referred to as the "source". The beam is incident on a "target" and the collision partners interact in a common region of space after a suitable time span, around $t \sim 0$. Eventually, as $t \to +\infty$, a measurement and analysis of the final collision products is made in the "detectors". These three stages: $t \to -\infty$, finite $t \sim 0$, and $t \to +\infty$, characterize every collision process.

The description and theoretical analysis of a collision may be broken down into four steps:

1. The initial and final conditions must be formulated in terms of the specifications of the source, the target, and the detectors. The states specifying the experimentally defined initial and final conditions are characterized in terms of "arrangements" of the collision partners and the corresponding "channels". In the absence of the collision-inducing interaction, states develop in time according to separate channel Hamiltonians.

2. The connection between the asymptotic behavior of the channel states and the time development of the interacting systems must be established, either by proving or - if that is too difficult - by postulating certain mathematical relations and by defining the scattering operator or its representation, the S matrix, in a suitable basis.

3. The mathematical constructs (scattering and transition matrix elements) must be related to the measured quantities, such as transition probabilities, reaction yields, particle and radiation intensities, and cross sections. The density operator and the density matrix play an important role at this stage of the analysis.

4. The dynamics of the collision must be calculated from the equations of motion, assuming knowledge of the Hamiltonian $H(t)$, which may be explicitly time-dependent.

This four-step program requires that a strategy be devised to determine the unitary time development operator $U(t_1, t_2)$ which relates two states of the system at different times:

$$|\Psi(t_2)\rangle = U(t_2, t_1)\ |\Psi(t_1)\rangle \tag{1}$$

The time development operator must satisfy the condition $U(t, t) = 1$

and the "group" property

$$U(t_2, t_1) = U(t_2, t) \, U(t, t_1) \tag{2}$$

for all times t. For short time intervals the time development operator may be expressed in terms of the Hermitian Hamiltonian H(t) as

$$U(t + \varepsilon, t) = 1 - i\varepsilon \, H(t)$$

Relation (2) may be used to integrate this equation from an initial time t_1 to a final time t_2:

$$U(t_1, t_2) = T \exp[-i\int_{t_2}^{t_1} H(t') \, dt'], \tag{3}$$

where T is the time-ordering operator. To describe the progress of a collision from source to detection one eventually needs to know the time development operator $U(t_1, t_2)$ in the limit as $t_1 \to -\infty$ and $t_2 \to +\infty$. Although the expression (3) for the time development operator is formally appealing, it does not usually represent a practical means of obtaining explicit solutions to the dynamical problem posed by a collision experiment.

The implementation of the four-step program of collision theory will be briefly reviewed in the remainder of this section.[2]

1. The initial experimental conditions, generated by the source and the target, are generally described by a wave packet (state I) which can be traced back in time to $t \to -\infty$, with specified internal states for the collision partners. The incident channel is labelled by α, and $U^{(\alpha)}(t, 0)$ is the time development operator for the noninteracting channel states $|\Phi^{(\alpha)}\rangle$. Similarly, the outcome of the collision (state F) is described by wave packets which develop forward in time to $t \to +\infty$. The final states are $|\Phi^{(\beta)}\rangle$ in channels labelled by β. The channel Hamiltonians are H_α and H_β. They define the interactions V_α and V_β by the relation

$$H = H_\alpha + V_\alpha = H_\beta + V_\beta.$$

The channel dynamics of the system is described by the equations

$$\text{State I} = |\Phi_{in}^{(\alpha)}(-\infty)\rangle \quad \text{and} \quad |\Phi_{in}^{(\alpha)}(t)\rangle = U^{(\alpha)}(t, 0)\ |\Phi_{in}^{(\alpha)}\rangle$$

$$\text{State F} = |\Phi_{out}^{(\beta)}(+\infty)\rangle \quad \text{and} \quad |\Phi_{out}^{(\beta)}(t)\rangle = U^{(\beta)}(t, 0)\ |\Phi_{out}^{(\beta)}\rangle$$

2. In multichannel collision theory it is assumed that the pure channel "in" and "out" states are asymptotically related to states $|\Psi\rangle$ according to the following scheme:

$$U(t, 0)\ |\Psi\rangle \to |\Phi_{in}^{(\alpha)}(t)\rangle \quad \text{as } t \to -\infty$$

$$|\Psi\rangle = U^{\dagger}(-\infty, 0)\ U^{(\alpha)}(-\infty, 0)\ |\Phi_{in}^{(\alpha)}\rangle = \Omega_{+}^{(\alpha)}\ |\Phi_{in}^{(\alpha)}\rangle$$

$$U(t, 0)\ |\Psi\rangle \to |\Phi_{out}^{(\beta)}(t)\rangle \quad \text{as } t \to +\infty$$

$$|\Psi\rangle = U^{\dagger}(+\infty, 0)\ U^{(\beta)}(+\infty, 0)\ |\Phi_{out}^{(\beta)}\rangle = \Omega_{-}^{(\beta)}\ |\Phi_{out}^{(\beta)}\rangle$$

which defines the Møller "wave operators" $\Omega_{+}^{(\alpha)}$ and $\Omega_{-}^{(\beta)}$. The collision operator $S_{\beta,\alpha}$ links the initial and final channel states:

$$|\Phi_{out}^{(\beta)}\rangle = \Omega_{-}^{(\beta)\dagger}\ \Omega_{+}^{(\alpha)}\ |\Phi_{in}^{(\alpha)}\rangle = S_{\beta,\alpha}\ |\Phi_{in}^{(\alpha)}\rangle \tag{4}$$

The full collision operator S is defined as the direct sum of the $S_{\beta,\alpha}$ over all channels α and β. In the special case of a single-channel collision, the relations

$$S = U^{(\alpha)\dagger}(+\infty, 0)\ U(+\infty, 0)\ U(0, -\infty)\ U^{(\alpha)}(-\infty, 0)$$

$$= U^{(\alpha)\dagger}(+\infty, 0)\ U(+\infty, -\infty)\ U^{(\alpha)}(-\infty, 0) = U_{int}(+\infty, -\infty)$$

show that the S operator is the overall time development operator for the collision process in the interaction picture, $U_{int}(+\infty, -\infty)$.

The elements of the scattering matrix are simply related to the transition amplitudes for the collision. If the source and the target determine the initial channel state $|\Phi_{I}^{(\alpha)}\rangle$ and the detectors determine the final states $|\Phi_{F}^{(\beta)}\rangle$, the transition amplitude is given by

$$S_{FI} = \langle\Phi_F^{(\beta)}|S_{\beta,\alpha}|\Phi_I^{(\alpha)}\rangle = \langle\Phi_F^{(\beta)}|U^{(\beta)\dagger}(+\infty, 0)|\Psi_I^{(+)}(+\infty)\rangle$$
$$= \langle\Psi_F^{(-)}(-\infty)|U^{(\alpha)}(-\infty, 0)|\Phi_I^{(\alpha)}\rangle = \langle\Psi_F^{(-)}(t)|\Psi_I^{(+)}(t)\rangle \tag{5}$$

for all times t.

The choice of channels and channel Hamiltonians implies a degree of arbitrariness and discretion. In particular, channel Hamiltonians need not always correspond to "free" collision partners. If there are long-range (e.g. Coulomb) interactions between the collision partners, either before or after the collision, it may be desirable to include these interactions in the channel Hamiltonian. Asymptotically, the long-range forces introduce only slowly varying phases and thus do not disturb the usual connection between the transition amplitudes and the measured cross sections and intensities. Thus, such a choice does not necessarily conflict with the general objective of relating the channel states directly and simply to the states prepared by the source or analyzed by the detector. For example, if r is changed by one wavelength λ, the Coulomb wave function undergoes a phase change that, for macroscopically large r, amounts to λ/r, which is practically negligible.

One more remark is in order: The stable bound states of the system, defined by the Hamiltonian H, form a subspace of the full state vector space. It may sometimes be convenient to regard this subspace as a (trivial) channel by itself. The corresponding channel Hamiltonian is the bound-state projection of H, and there is no interaction (and no transition to any scattering state). Hence the Møller operators and the S operator for this channel are all equal to the identity.

3. So far it has been assumed that the initial and final states in a collision experiment are pure quantum mechanical states represented by well-defined state vectors. In actual experiments, the initial and final states are not precisely defined and only statistical distributions of certain physical quantities are known. For example, while the linear momentum of an incident wave packet may be rather sharply defined and known to reasonable accuracy, one is usually in the dark about its further detailed properties, such as the spatial extent of the wave packet, the presence of slow phase variations, and so forth. Put differently, there are infinitely many different wave packets which are compatible with the available experimental information about the incident beam and the detected particles. The density operator is then the appropriate tool for giving an account of the statistical nature of the prevailing experimental conditions.

The density (or statistical) operator[3] may be thought of as characterizing a collection of identical physical systems, each of which is assumed to be represented by a (pure) quantum state $|\Psi\rangle$. Applied to a collision, a (statistical) mixture is a "beam" of particles such that a fraction p_k is represented by $|\Psi_k\rangle$. In this interpretation, the particles are in their distinctive pure states owing to their different preparations.

In a conceptually somewhat different interpretation, the density operator represents a statistical ensemble, which may be associated with a single particle. Such an ensemble may be - but does not have to be - realized by a beam of particles. In this interpretation, depending on the nature of the preparation, a collection of identically prepared particles may nevertheless be represented by a pure state (corresponding to an idempotent density operator) or by a proper mixture (in which case the density operator is not idempotent). On the other hand, in the "beam interpretation" identical preparation implies a pure state for the beam. The distinction between these two views of the density operator has no discernible consequences in the application of the formalism to the analysis of collision experiments, but it assumes significance in other contexts.

The initial state of the incident beam can be described as a statistical ensemble of "in" states in channel α, represented according to statistical weights (probabilities) $p_{in}^{(\alpha)}$, and the final states are described by ensembles of "out" states in channels labelled by β, with statistical weights $p_{out}^{(\beta)}$. The transition probability from an initial to a final state becomes then

$$|S_{FI}|^2 = \sum |\langle\Phi_{out}^{(\beta)}|S_{\beta,\alpha}|\Phi_{in}^{(\alpha)}\rangle|^2 \; p_{out}^{(\beta)} \; p_{in}^{(\alpha)} \tag{6}$$

where the sum is taken over all "in" states in channel α and all "out" states in channel β that are represented in the statistical ensembles. If density operators for the source ("in") and the detector ("out") are defined by

$$\begin{aligned} \rho_{in}^{(\alpha)} &= \sum |\Phi_{in}^{(\alpha)}\rangle \; p_{in}^{(\alpha)} \; \langle\Phi_{in}^{(\alpha)}| \\ \rho_{out}^{(\beta)} &= \sum |\Phi_{out}^{(\beta)}\rangle \; p_{out}^{(\beta)} \; \langle\Phi_{out}^{(\beta)}| \end{aligned} \tag{7}$$

the transition probability may be expressed in compact form as

$$|S_{FI}|^2 = \text{Trace}\,(\rho_{out}^{(\beta)}\, S_{\beta,\alpha}\, \rho_{in}^{(\alpha)}\, S_{\beta,\alpha}^{\dagger}) \tag{8}$$

4. Finally, the elements of the dynamical aspects of multichannel collision theory will be outlined.[2] In their integral form, the equations of motion are

$$\begin{aligned} |\Psi_\alpha^{(+)}(t)\rangle &= |\Phi_{in}^{(\alpha)}(t)\rangle - i \int_{-\infty}^{t} G_\alpha(t,t')\, V_\alpha(t')\, |\Psi_\alpha^{(+)}(t')\rangle\, dt' \\ &= |\Phi_{in}^{(\alpha)}(t)\rangle - i \int_{-\infty}^{t} G(t,t')\, V_\alpha(t')\, |\Phi_{in}^{(\alpha)}(t')\rangle\, dt' \end{aligned} \tag{9}$$

and

$$\begin{aligned} |\Psi_\beta^{(-)}(t)\rangle &= |\Phi_{out}^{(\beta)}(t)\rangle - i \int_{t}^{+\infty} G_\beta(t,t')\, V_\beta(t')\, |\Psi_\beta^{(-)}(t')\rangle\, dt' \\ &= |\Phi_{out}^{(\beta)}(t)\rangle - i \int_{t}^{+\infty} G(t,t')\, V_\beta(t')\, |\Phi_{out}^{(\beta)}(t')\rangle\, dt' \end{aligned} \tag{10}$$

where the operators G_α and G_β are propagators or Green operators for the channels α and β, and $G(t,t')$ is the Green operator corresponding to the full Hamiltonian of the interacting system. The formalism summarized up to this point can accommodate Hamiltonians with time-dependent interactions, arising in semiclassical ion-atom collision theories, as well as the constant interactions of the full quantum mechanical collision theories.

If the channel Hamiltonian H_α and the interaction V_α are time-independent, it is convenient and conventional to consider stationary states which are limiting forms of the wave packets describing the collision process. In this case,

$$G(t,\, t') = e^{-i\, H\, (t-t')} \qquad \text{and} \qquad G_\alpha(t,\, t') = e^{-i\, H_\alpha\, (t-t')}$$

In the time-independent scattering formalism one then derives the Lippmann-Schwinger equation

$$\begin{aligned} |\Psi_\alpha^{(\pm)}\rangle &= |\Phi^{(\alpha)}\rangle + \frac{1}{E - H_\alpha \pm i\varepsilon}\, V_\alpha\, |\Psi_\alpha^{(\pm)}\rangle \\ &= |\Phi^{(\alpha)}\rangle + \frac{1}{E - H \pm i\varepsilon}\, V_\alpha\, |\Phi^{(\alpha)}\rangle \end{aligned} \tag{11}$$

The transition matrix becomes

$$S_{FI} = \langle\Psi_F^{(-)}|\Psi_I^{(+)}\rangle = \delta_{FI} - 2\pi i\ T_{FI}\ \delta(E_F - E_I) \tag{12}$$

where, in self-explanatory notation,

$$T_{FI} = \langle\Phi_F|\ V_I\ |\Psi_I^{(+)}\rangle = \langle\Psi_F^{(-)}|\ V_F\ |\Phi_I\rangle = \langle\Phi_F|\ T\ |\Phi_I\rangle \tag{13}$$

The triply differential cross section for a transition into the momentum range near $\mathbf{p}_F$ is

$$d^3\sigma_{FI} = \frac{2\pi\ |\langle\Phi_F|\ T\ |\Phi_I\rangle|^2}{v/(2\pi)^3}\ \delta(E_I - E_F)\ d^3p_F \tag{14}$$

For clarity it might be useful to write down the form which the Lippmann-Schwinger equation takes for single-channel potential scattering:

$$|\Psi_{\mathbf{k}}^{(+)}\rangle = |\mathbf{k}\rangle + \frac{1}{E - H_0 + i\varepsilon}\ V\ |\Psi_{\mathbf{k}}^{(+)}\rangle$$

$$= |\mathbf{k}\rangle + \int d^3k'\ \frac{2m}{k^2 - k'^2 + 2i\varepsilon m}\ |\mathbf{k}'\rangle\langle\mathbf{k}'|\ V\ |\Psi_{\mathbf{k}'}^{(+)}\rangle$$

In the coordinate representation, the last equation becomes, asymptotically for $r \to \infty$,

$$\phi_{\mathbf{k}}^{(+)} \to e^{i\mathbf{k}\cdot\mathbf{r}} - (2\pi)^3 m\ \langle\mathbf{k}'|\ T\ |\mathbf{k}\rangle\ \frac{e^{ikr}}{r}$$

leading to the identification of the scattering amplitude:

$$f_{\mathbf{k}}(\mathbf{k}') = -\ (2\pi)^3 m\ \langle\mathbf{k}'|\ T\ |\mathbf{k}\rangle \tag{15}$$

When the equations of time-independent multichannel collision theory, such as the generalized Lippmann-Schwinger equation, are formulated explicitly in terms of a suitably chosen basis, an infinite set of coupled equations for infinitely many amplitudes arises. To obtain solutions, for practical purposes, these equations are usually truncated. In principle, however, it is possible to replace such infinite sets of coupled equations by a finite set, or even by a single equation for just one selected amplitude, without making any approximations. This objective is accomplished by replacing the Hamiltonian of the system by an altogether different model Hamiltonian, in which a complicated "optical" potential simulates the effects of the eliminated open and closed collision channels.

If P is the projection operator that projects any state of the system unto the selected and specified channel subspaces of interest, and if $Q = 1 - P$ is the complement of P, the effective optical Hamiltonian is

$$H_{opt} = \frac{p^2}{2m} + E_\alpha + V_{opt} \tag{16}$$

The optical potential has a deceptively simple formal appearance[4]:

$$V_{opt} = P\, V^\alpha P + P\, V^\alpha\, Q \frac{Q}{E - QHQ + i\varepsilon} V^\alpha\, P \tag{17}$$

This potential is generally energy-dependent and nonlocal. As can be seen from the structure of the Green operator,

$$G_Q(E + i\varepsilon) = \frac{Q}{E - QHQ + i\varepsilon}$$

the optical potential may be non-Hermitian, corresponding to the absorption of probability (and, thus, particles) from the channel under consideration, if E is above the threshold for some excluded channel.

To conclude this section, it may be useful to indicate briefly the relation of formal scattering theory to the perturbation theoretical calculation of transition and decay amplitudes and decay rates. If the Hamiltonian can be split into a time-independent unperturbed Hamiltonian H_0 and a perturbation $V(t)$,

$$H = H_0 + V(t)$$

the integral form of the equation of motion is

$$|\Psi(t)\rangle = |\Phi(t)\rangle - i \int_{t_0}^{t} \exp[-iH_0(t - t')]\, V(t')|\Psi(t')\rangle\, dt'$$

where $|\Phi(t)\rangle$ is a solution of the equation

$$i\frac{d}{dt}|\Phi(t)\rangle = H_0|\Phi(t)\rangle ,$$

and the initial condition

$$|\Psi(t_0)\rangle = |\Phi(t_0)\rangle$$

is assumed. If the unperturbed state is selected to be an eigenstate $|s\rangle$ of H_0:

$$|\Phi(t)\rangle = e^{-iE_s t}|s\rangle,$$

the first-order transition amplitude to the stationary state

$$e^{-iE_k t}|k\rangle,$$

at time t, for $k \neq s$, is

$$\langle k|\Psi(t)\rangle = -i \int_{t_0}^{t} e^{i\omega_{ks}t'} \langle k|V|s\rangle \, dt' \tag{18}$$

If V is constant in time, and if the limits $t_0 \to -\infty$ and $t \to +\infty$ are taken, the transition amplitude becomes

$$\langle k|\Psi(+\infty)\rangle = -2\pi i \, \delta(\omega_{ks}) \, \langle k|V|s\rangle \tag{19}$$

corresponding to a transition rate (Fermi's Golden Rule),

$$\begin{aligned} w &= 2\pi \, \delta(\omega_{ks}) \, |\langle k|V|s\rangle|^2 \\ &= 2\pi \, \delta(\omega_{ks}) \, \langle k|V|s\rangle\langle s|V^+|k\rangle \end{aligned} \tag{20}$$

If the initial state is a statistical mixture of eigenstates $|s\rangle$ of H_0, with weights p_s, the transition rate can be expressed in terms of an initial state density operator

$$\rho_I = \sum_s |s\rangle \, p_s \, \langle s|$$

and the density operator $\rho_F = |k><k|$ for the final state as

$$W = 2\pi\ \delta(\omega_{FI})\ \mathrm{Trace}(\rho_F\ V\ \rho_I\ V^+) \tag{21}$$

Generally, this expression must be summed over the relevant final states.

If first-order perturbation theory is inadequate, equation (21) is replaced by

$$W = 2\pi\ \delta(\omega_{FI})\ \mathrm{Trace}(\rho_F\ T\ \rho_I\ T^+) \tag{22}$$

where the T operator is defined by equation (13). Equation (22) serves as a reminder that a collision or decay process transforms an initial "incoming" density operator ρ_{in} into an "outgoing" density operator[3]

$$\rho_{out} = \frac{T\ \rho_{in}\ T^+}{\mathrm{Trace}(T\ \rho_{in} T^+)} \tag{23}$$

III. INTERACTIONS IN ATOMIC COLLISION PHYSICS

Given the scale of the energies, spatial distances, and time intervals that are relevant in atomic collisions physics, the appropriate "fundamental" interactions for the description of atomic processes are provided by relativistic quantum electrodynamics (QED), the theory of interacting electron and photon fields in the presence of the nuclei. Here, only a very brief summary of the essentials of the theory will be given.

The relativistic Dirac theory of the electron of charge -e in an electromagnetic field $A^\mu(\phi, \mathbf{A})$ is

$$[\gamma_\mu(\frac{\partial}{\partial x_\mu} - \frac{ie}{c} A^\mu) + imc]\psi = 0 \tag{24}$$

where $\boldsymbol{\gamma} = \beta\boldsymbol{\alpha}$, $\gamma_0 = \beta$, and where

$$\boldsymbol{\alpha} = \begin{pmatrix} 0 & \boldsymbol{\sigma} \\ \boldsymbol{\sigma} & 0 \end{pmatrix}, \qquad \beta = \begin{pmatrix} 1 & 0 \\ 0 & -1 \end{pmatrix}, \qquad \boldsymbol{\sigma} = \begin{pmatrix} \sigma & 0 \\ 0 & \sigma \end{pmatrix}$$

are 4×4 matrices, the elements $\boldsymbol{\sigma}$ being the Pauli spin matrices. The Dirac spinor ψ has four components. By iteration of the first order Dirac equation one arrives at a second-order equation for ψ:

$$[(\nabla + \frac{ie}{c}\mathbf{A})^2 - \frac{1}{c^2}(\frac{\partial}{\partial t} - ie\phi)^2 - \kappa^2 - \frac{e}{c}\,\boldsymbol{\sigma}\cdot\mathbf{B} + \frac{ie}{c}\,\boldsymbol{\alpha}\cdot\mathbf{E}]\psi = 0$$

where $\kappa = 1/mc$. In this equation the term containing $\boldsymbol{\alpha}$ couples large and small components of the wave function.

The gauge invariance of the Dirac equation implies that if an arbitrary function g of the coordinates is introduced to define a gauge transformation,

$$A^{\mu} \rightarrow A^{\mu} + \frac{\partial g}{\partial x_{\mu}},$$

which keeps invariant the field

$$F_{\mu\nu} = \frac{\partial A^{\nu}}{\partial x_{\mu}} - \frac{\partial A^{\mu}}{\partial x_{\nu}},$$

a simultaneous transformation

$$\psi \rightarrow \psi\, e^{-ieg/c}$$

compensates for the change in the potentials and keeps the Dirac equation gauge invariant. Gauge invariance is a manifestation of the essential inseparability of electron and photon fields.

A simple example from nonrelativistic quantum mechanics serves to show how important it is to keep track at all times of the gauge that is being used. In ordinary quantum mechanics it is customarily asserted that

$$\psi = e^{i\mathbf{k}\cdot\mathbf{r}}$$

represents a particle moving with velocity $\mathbf{k}/m$. But this is true only if a gauge is adopted in which the vector potential $\mathbf{A}=0$, the usual tacit assumption. If, instead, the potential is transformed with a gauge function $g = (c/e)\,\mathbf{k}\cdot\mathbf{r}$, , giving the new vector potential $\mathbf{A} = (c/e)\,\mathbf{k}$, then the wave function is simply $\psi = 1$, but it still represents a particle moving with velocity $\mathbf{k}/m$. This remark shows that the local phase of the wave function, taken by itself without reference to the vector potential, has no independent physical meaning.

Even in classical mechanics, the choice of the gauge affects the ease with which problems can be formulated and solved. If an electron moves in an electrostatic field $\mathbf{E}(\mathbf{r})$, in the "Coulomb gauge", the potentials are

$$\mathbf{A} = 0, \quad \phi = \int \mathbf{E}(\mathbf{r})\cdot d\mathbf{r}$$

and the Hamiltonian is

$$H = \frac{p^2}{2m} - e\phi = \frac{p^2}{2m} + e \int \mathbf{E}(\mathbf{r})\cdot d\mathbf{r}$$

The corresponding Hamilton equations of motion are

$$\frac{d\mathbf{r}}{dt} = \frac{\mathbf{p}}{m} \quad \text{and} \quad \frac{d\mathbf{p}}{dt} = -e\mathbf{E}, \quad \text{giving} \quad m\frac{d\mathbf{v}}{dt} = -e\mathbf{E}$$

An entirely equivalent formulation is obtained if another gauge is used and the potentials are chosen to be

$$\mathbf{A} = -ct\,\mathbf{E}(\mathbf{r}), \quad \phi = 0$$

The Hamiltonian is now

$$H = \frac{1}{2m}(\mathbf{p} - e\mathbf{E}t)^2$$

but this expression does not represent the total energy and is not conserved. Hamilton's equations now become

$$\frac{d\mathbf{r}}{dt} = \frac{\mathbf{p} - e\mathbf{E}t}{m} = \mathbf{v}, \quad \frac{d\mathbf{p}}{dt} = e\mathbf{v}\cdot\nabla\mathbf{E}\ t$$

and the canonical momentum $\mathbf{p}$ can again be eliminated from these equations to yield Newton's second law just as before. The canonical momenta $\mathbf{p}$ are of course different for the two gauges, and it is amusing to note that for a uniform electric field, in the second gauge, $\mathbf{p}$ is a constant of the motion! According to the structure of the interaction terms in the two cases, the first choice of gauge may be termed a "length" gauge, whereas the second would then be referred to as a "velocity" gauge.

The theory of interacting electrons and photons is based on the Lagrangian density of QED,

$$\pounds = \overline{\psi}\,(ic\gamma_\mu \frac{\partial}{\partial x_\mu} - mc^2)\,\psi - \frac{1}{16\pi} F_{\mu\nu} F^{\mu\nu} + e\overline{\psi}\gamma_\mu A^\mu \psi \tag{25}$$

which is Lorentz and gauge invariant.

For applications to atomic physics, it is desirable to reduce the equations of motion to a form in which the instantaneous Coulomb interaction between the charged particles is made explicit. This is achieved[5] by casting the theory in the Coulomb gauge in which

$$\phi = A^0 = \int \frac{\rho(\mathbf{r}',t)}{|\mathbf{r} - \mathbf{r}'|}\, d^3r' \quad \text{and} \quad \nabla\cdot\mathbf{A} = 0 \tag{26}$$

The divergence condition expresses the transversality of electromagnetic radiation. The Hamitonian operator for the coupled fields can then be written in the form

$$H = \int \psi\,(-ic\gamma_k \frac{\partial}{\partial x_k} + mc^2)\,\psi\, d^3r + \frac{1}{8\pi}\int \{[\nabla\times\mathbf{A}]^2 + \frac{1}{c^2}(\frac{\partial\mathbf{A}}{\partial\mathbf{t}})^2\}\, d^3r$$
$$- \frac{1}{c}\int \mathbf{j}\cdot\mathbf{A}\, d^3r + \frac{1}{2}\iint \frac{\rho(\mathbf{r},t)\,\rho(\mathbf{r}',t)}{|\mathbf{r} - \mathbf{r}'|}\, d^3r\, d^3r' \tag{27}$$

The equations of motion in QED for the full interacting system of charged particles and photons must be simplified and approximated if concrete results are to be obtained from their solutions. To this end, for atomic structure and atomic collision calculations one resorts to perturbation treatments of relatively low order, circumventing the necessity of dealing with the infinite number of degrees of freedom of the electromagnetic field. Instead, an approximate Hamiltonian for the electrons is derived in terms of instantaneous, or retarded, interactions between the particles. For example, for the electron-electron interaction an effective retarded interaction can be derived, the Møller interaction. In the Lorentz gauge, defined by the vanishing of the four-divergence of A^μ, the interaction between electrons 1 and 2 has the form

$$V_{12} = \frac{(1 - \boldsymbol{\alpha}_1\cdot\boldsymbol{\alpha}_2)}{r_{12}} \exp(ikr_{12}) \tag{28}$$

where $ck = \omega$ is the energy transfer to one of the electrons (or, the energy of the virtual photon exchanged between the electrons).

In the Coulomb gauge ($\nabla \cdot \mathbf{A} = 0$), a different (Møller) electron-electron interaction is derived from QED, if the unperturbed electrons are assumed to be free:

$$V_{12} = \frac{1}{r_{12}} - \boldsymbol{\alpha}_1 \cdot \boldsymbol{\alpha}_2 \frac{\exp(ikr_{12})}{r_{12}} + \boldsymbol{\alpha}_1 \cdot \nabla \; \boldsymbol{\alpha}_2 \cdot \nabla \frac{\exp(ikr_{12}) - 1}{k^2 \, r_{12}}$$

These interactions both have the "defect" of being explictly dependent on the state of the electrons.

A different approach to the problem of eliminating the (virtual) photon field starts from an unperturbed Hamiltonian which includes the Coulomb interaction $1/r_{12}$ between the two particles:

$$H_0 = c\boldsymbol{\alpha}_1 \cdot [\mathbf{p}_1 + \frac{e}{c} \mathbf{A}(\mathbf{r}_1)] + \beta_1 mc^2 - e\phi(\mathbf{r}_1) + c\boldsymbol{\alpha}_2 \cdot [\mathbf{p}_2 + \frac{e}{c} \mathbf{A}(\mathbf{r}_2)] + \beta_2 mc^2 - e\phi(\mathbf{r}_2) + \frac{e^2}{r_{12}} \quad (29)$$

If first-order perturbation theory is applied and retardation neglected, the Breit interaction between two electrons is obtained[6]:

$$V_B = - \frac{e^2}{2r_{12}} [\alpha_1 \cdot \alpha_2 + \frac{\alpha_1 \cdot r_{12} \; \alpha_2 \cdot r_{12}}{r_{12}^2}] \quad (30)$$

The Breit interaction is not dependent on the state of the electrons. If in the Møller interaction retardation is neglected, the Breit and Møller interactions are equivalent in calculating the cross section for electron-electron scattering[7].

In recent times it has become practical in atomic physics to make calculations of form factors and related quantities directly in the 4×4 dimensional spinor space by using the full Breit or Møller interactions[8] between two atomic electrons, rather than employing a nonrelativistic reduction in terms of Pauli two-component wave functions[6].

IV. APPLICATIONS

The radiative decay of an atom provides an illustration of the concepts developed in the previous sections. For a one-electron atom, the effective perturbation term in the Hamiltonian, expressed in the electron coordinate representation, is

$$V(t) = -\frac{1}{c}\int \mathbf{j}\cdot\mathbf{A}\, d^3r \rightarrow \frac{e}{mc}\mathbf{A}\cdot\nabla \qquad (31)$$

If the emission of a photon with fixed momentum $\mathbf{k}$ and linear polarization (unit) vectors $\boldsymbol{\varepsilon}_1$ and $\boldsymbol{\varepsilon}_2$ is considered in the electric dipole approximation, the vector potential operator, apart from a constant, is effectively given by

$$\mathbf{A} \rightarrow a_1^+ \,\varepsilon_1 + a_2^+ \,\varepsilon_2 + a_1 \,\varepsilon_1 + a_2 \,\varepsilon_2$$

The operators a_1 and a_2 are annihilation operators for photons with the two perpendicular polarizations. The familiar results of time-dependent perturbation theory, reviewed in Section II, are then applied to a transition from an initial (I) zero-photon state $|0\rangle$ to a final (F) one-photon state $a_D{\dagger}|0\rangle$. The latter represents the detection of a photon that has passed through a polarization filter characterized by the (possibly complex) unit vector $\boldsymbol{\varepsilon}_D$. The transition matrix element $\langle F|(a_1{\dagger}\ \varepsilon_1 + a_2{\dagger}\ \varepsilon_2)\cdot\mathbf{p}|I\rangle$ is reduced by noting the relations

$$a_1\, a_D^+ \,|0\rangle = \boldsymbol{\varepsilon}_1\cdot\boldsymbol{\varepsilon}_D\, |0\rangle \quad \text{and} \quad a_2\, a_D^+ \,|0\rangle = \boldsymbol{\varepsilon}_2\cdot\boldsymbol{\varepsilon}_D\, |0\rangle$$

Hence, the transition matrix element can be written as

$$\boldsymbol{\varepsilon}_D^* \cdot (\varepsilon_1\, \varepsilon_1 + \varepsilon_2\, \varepsilon_2) \cdot \langle F|\,\mathbf{p}\,|I\rangle = \boldsymbol{\varepsilon}_D^* \cdot \langle F|\,\mathbf{p}\,|I\rangle \qquad (32)$$

and the decay rate, following equation (21), as

$$w \sim \boldsymbol{\varepsilon}_D^* \cdot \mathrm{Trace}(\rho_F\, \mathbf{p}\, \rho_I\, \mathbf{p}) \cdot \boldsymbol{\varepsilon}_D \qquad (33)$$

where the trace is taken over the state vector space of the atom. The tensor, $\mathrm{Trace}(\rho_F\, \mathbf{p}\, \rho_I\, \mathbf{p})$, fully characterizes the polarization properties of the emitted radiation[9,10].

Finally, the correlated emission of two photons, either in a two-step cascade or in direct two-photon decay, offers an example of the application of symmetry considerations. In the electric dipole approximation, the overall transition amplitude can be obtained in generalization of equation (32). It is a linear combination of expressions of the form

$$\boldsymbol{\varepsilon}_D^* \cdot \langle F|\ \mathbf{p}\ |M\rangle\langle M|\ \mathbf{p}\ |I\rangle \cdot \boldsymbol{\varepsilon}_{D'}^* + \boldsymbol{\varepsilon}_{D'}^* \cdot \langle F|\ \mathbf{p}\ |M\rangle\langle M|\ \mathbf{p}\ |I\rangle \cdot \boldsymbol{\varepsilon}_D^* \tag{34}$$

corresponding to a transition to the final two-photon state $a_D{}^\dagger a_{D'}{}^\dagger|0\rangle$ which represents the detection of two photons with polarizations $\boldsymbol{\varepsilon}_1$ and $\boldsymbol{\varepsilon}_2$. The symmetric form of expression (34) is dictated by the boson character of the photons. Effectively, expression (34) represents the matrix element of a Cartesian tensor operator which has rank two under rotations and which has even parity. If the transition takes place between atomic states which violate the triangular condition for angular momentum,

$$j_I + j_F > 2 > |j_I - j_F|,$$

as a $j_I=0 \rightarrow j_F=0$ transition does, the transition operator reduces to a scalar operator under rotations. If there is no parity change, it follows that for such transitions the transition amplitude is proportional to

$$\boldsymbol{\varepsilon}_D^* \cdot \boldsymbol{\varepsilon}_{D'}^* . \tag{35}$$

Under these conditions, the coincidence rate depends on the polarization vectors of the two detected photons according to

$$w \sim |\boldsymbol{\varepsilon}_D \cdot \boldsymbol{\varepsilon}_{D'}|^2 = \cos^2\phi = \tfrac{1}{2}(1 + \cos 2\phi) \tag{36}$$

where ϕ is the angle between the transmission directions of the two polarization filters. The quantum mechanical result expressed by equation (36) plays a central role in spectroscopic tests of Bell's inequality[11,12].

ACKNOWLEDGMENT

These lectures were prepared with partial support from the U.S. Department of Energy, under contract No.DE-AS05-76ER02408, and from the Kenan Trust Fund at the University of North Carolina at Chapel Hill. I thank Ms. Tricia Reeves for a critical reading of the manuscript.

REFERENCES

1. E. Merzbacher, "Theory of Coulomb Excitation and Ionization" in Fundamental Processes in Energetic Atomic Collisions, p. 319 - 348, H. O. Lutz, J. S. Briggs, and H. Kleinpoppen, eds., Plenum Press, New York, 1983.

2. J. R. Taylor, Scattering Theory, John Wiley and Sons, New York, 1972.

3. K. Blum, Density Matrix Theory and Applications, Plenum Press, New York, 1981.

4. R. G. Newton, Scattering Theory of Waves and Particles, Springer-Verlag, New York, 1982.

5. J. J. Sakurai, Advanced Quantum Mechanics, Addison-Wesley Publishing Co., Reading, 1967.

6. H. A. Bethe and E. E. Salpeter, "Quantum Mechanics of One- and Two-Electron Systems" in Handbuch der Physik, vol. XXXV, p. 88 - 436 , S. Flügge, ed., Springer-Verlag, Berlin, 1957.

7. M. E. Rose, Relativistic Electron Theory, John Wiley and Sons, New York, 1961.

8. M. H. Chen, "Relativistic Effects in Atomic Inner-Shell Transitions" in X-Ray and Atomic Inner-Shell Physics-1982, p. 331 - 345, AIP Conference Proceedings, B. Crasemann, ed., American Institute of Physics, New York, 1982.

9. G. Nienhuis, "The Possibility of a Complete Determination of Scattering Amplitudes for S→D Excitation by Electron-Photon Coincidence Measurements" in Coherence and Correlation in Atomic Collisions, H. Kleinpoppen and J. F. Williams, eds., Plenum Press, New York, 1980.

10. For a recent application see C. C. Havener, W. B. Westerveld, J. S. Risley, N. H. Tolk, and J. C. Tully, Phys. Rev. Lett. 48, 926 (1982); C. C. Havener, N. Rouze, W. B. Westerveld, and J. S. Risley, Phys. Rev. Lett. 53 1049 (1984); J. Burgdörfer and Louis J. Dubé, Phys. Rev. Lett. 52, 2225 (1984).

11. A. Aspect, P. Grangier, and G. Roger, Phys. Rev. Lett. 47, 460 (1981); A. Aspect, P. Grangier, and G. Roger, Phys, Rev. Lett. 49 91 (1982); A. Aspect, J. Dalibard, and G. Roger, Phys. Rev. Lett. 49, 1804 (1982).

12. W. Perrie, A. J. Duncan, H. J. Beyer, and H. Kleinpoppen, Phys. Rev. Lett. 54, 1790 (1985); see also the paper by A. J. Duncan, W. Perrie, H.J. Beyer and H. Kleinpoppen, in these Proceedings.

ELECTRON ATOM SCATTERING THEORY

P. G. Burke

Department of Applied Mathematics and Theoretical Physics
The Queen's University of Belfast
Belfast, BT7 1NN, Northern Ireland

ABSTRACT

The general theory of low energy scattering of electrons by atoms and ions is presented. First, general expressions for the scattering amplitudes and cross sections are obtained both when relativistic effects can be neglected and when, as in the case of scattering by heavy atoms, they must be included. Expansions of the total collision wave function is discussed in terms of independent electron coordinates and when hyperspherical coordinates are used and integrodifferential or differential equations are derived describing the motion of the scattered electron. A survey of the theory of electron impact ionization is presented with emphasis on excitation-autoionization and the threshold behaviour of the cross section. Then the role of resonances with a brief discussion of multichannel quantum defect theory is given. The lectures conclude with a summary of theories which enable electron impact processes at intermediate energies to be described. The lectures are illustrated throughout with examples of the comparison between theoretical prediction and experiment.

The following books are recommended as general references to atomic and molecular collision theory: Mott and Massey (1965), Smith (1971), Peterkop (1977), Nesbet (1980) and Bransden (1983). In addition the following review articles can be consulted with advantage: Bransden and McDowell (1977, 1978), Burke and Williams (1977), Byron and Joachain (1977), Burke and Eissner (1983), Seaton (1975, 1983), Henry (1981), Fano (1983) and Walters (1985).

1. BASIC PROCESSES

The basic processes that we will consider in these lectures are illustrated in the energy level diagram in figure 1.

We consider an electron incident upon a N-electron atom or ion of charge n+ in some state i. This state is usually the ground state but it may be an excited state prepared by photon absorption. After the collision the atom may be left in a discrete state f

$$e^- + A_i^{n+} \rightarrow e^- + A_f^{n+} \tag{1}$$

which corresponds to elastic scattering or excitation. If the final state lies in the continuum, represented by c in figure 1, then we have direct ionization

$$e^- + A_i^{n+} \rightarrow e^- + A_j^{(n+1)+} + e^- . \tag{2}$$

Near certain energies the excitation and ionization processes (1) and (2) may be dominated by the formation of intermediate resonant states. Hence in addition to the direct excitation process (1) we may have

$$e^- + A_i^{n+} \rightarrow A_k^{(n-1)+*} \rightarrow e^- + A_f^{n+} \tag{3}$$

where for an ion the levels k form a Rydberg series converging to a higher level of the target. Also, in addition to the direct

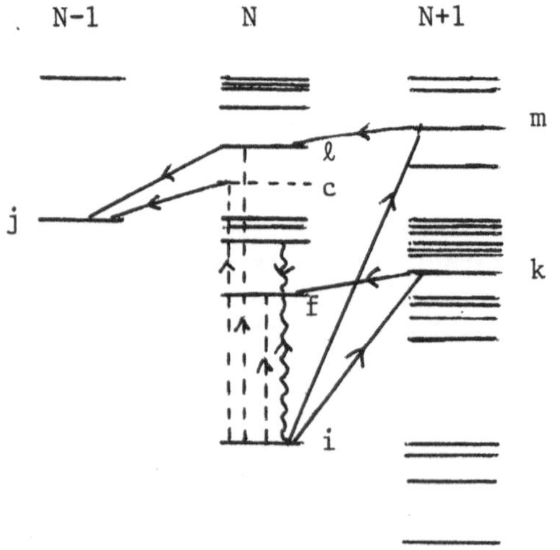

- - -➤- - - direct electron impact excitation or ionization
——➤—— electron absorption or emission
∿∿➤∿∿ photon absorption or emission

Figure 1. Energy level diagram of basic processes considered.

ionization process (2) we may have

$$e^- + A_i^{n+} \to e^- + A_\ell^{n+*} \to e^- + A_j^{(n+1)+} + e^- \qquad (4)$$

which is called excitation autoionization. Finally if the incident electron is close to threshold for exciting the state A_ℓ^{n+*} it may be captured giving rise to the ionization process.

$$e^- + A_i^{n+} \to A_m^{(n-1)+*} \to e^- + A_\ell^{n+*} \to e^- + A_j^{(n+1)+} + e^- \qquad (5)$$

which is often called electron capture ionization. Both processes (4) and (5) are important for the collision of electrons with complex ions (Crandall, 1981).

We will be interested in the energy dependence of the cross section particularly near to thresholds and the angular distribution of the scattered and/or emitted electrons in processes (1) to (5). In addition further information can be obtained, in many cases, by observing the radiative decay of the final atomic state where the photon may be observed in coincidence with the scattered electron. Finally further information about the collision process can often be obtained by using spin polarized electron beams and targets and analysing the polarization of the scattered electron and atom after the collision. The purpose of these lectures is to summarize the current status of the theory which will enable these observables to be predicted and compared with experiment. We will initially be mainly concerned with low energy collisions, where the velocity of the incident electron is of the same order or not much more than the velocity of the target electron taking an active role in the collision. Such collisions are of particular importance in many applications of electron atom and ion collisions in astrophysics, upper atmosphere physics and fusion research.

2. THE WAVE EQUATION AND CROSS SECTION

The Schrödinger equation describing the collision of an electron with a target atom or ion containing N electrons and with nuclear charge Z is

$$(H_{N+1} - E)\,\Psi = 0 \qquad (6)$$

where the (N+1)-electron Hamiltonian is defined by

$$H_{N+1} = \sum_{i=1}^{N+1} \left(-\tfrac{1}{2}\nabla_i^2 - \frac{Z}{r_i}\right) + \sum_{i>j=1}^{N+1} \frac{1}{r_{ij}} \qquad (7)$$

where $r_{ij} = |\underline{r}_i - \underline{r}_j|$ and $r_i = |\underline{r}_i|$ where $\underline{r}_i$ and $\underline{r}_j$ are the vector distances of the i th and j th electrons from the target nucleus which is assumed to have infinite mass and to be stationary at the origin of coordinates†. Initially we neglect all relativistic effects which restricts our discussion to light atoms and ions.

We also introduce the target eigenstates Φ_i by the equation

$$(H_N - E_i)\,\Phi_i = 0 \tag{8}$$

where H_N is the target Hamiltonian defined by eq. (7) with (N+1) replaced by N and where E_i are the target eigenenergies in atomic units.

We now look for a solution of eq. (6) corresponding to the process defined by eq. (1) where the electron is incident upon the target in some state Φ_i and is scattered leaving the target in some other state Φ_j. The asymptotic form of the wave function in the case of a neutral target is given by

$$\Psi_i \underset{r_{N+1}\to\infty}{\sim} \Phi_i\,\chi_{\frac{1}{2}m_i}(\sigma_{N+1})\,e^{ik_i z_{N+1}} + \sum_j \Phi_j\,\chi_{\frac{1}{2}m_j}(\sigma_{N+1})\,f_{ji}(\hat{\underline{r}}_{N+1})\,r_{N+1}^{-1}\,e^{ik_j r_{N+1}} \tag{9}$$

where $\chi_{\frac{1}{2}m_i}$ and $\chi_{\frac{1}{2}m_j}$ are the spin eigenfunctions for the incident and scattered electrons whose spin coordinate is σ_{N+1}, where the direction of spin quantization is usually taken to be the incident beam direction, and where f_{ji} is the scattering amplitude. The wave numbers k_i and k_j of the incident and scattered electrons are related to the total energy E of the electron-atom system by

$$E = E_i + \tfrac{1}{2}k_i^2 = E_j + \tfrac{1}{2}k_j^2 \;. \tag{10}$$

The second term on the right-hand-side of eq. (9) contains contributions at infinity from all atomic states which are energetically allowed i.e. for which $k_j^2 \geq 0$. If the incident energy is high enough, then continuum states of the target can be excited and will contribute to the asymptotic form. In this case, the asymptotic form defined by eq. (9) must be modified since the two or more outgoing electrons will partly screen each other depending on their relative energy and direction of ejection (Rudge and Seaton, 1964, Peterkop, 1977). We will ignore this complication now but will return to it later in these lectures when we consider electron impact ionization.

---

† Atomic units where $\hbar = m = e = 1$ are used where m and e are the mass and modulus of the charge of the electron. Hence the atomic unit of length $a_0 = \hbar^2/me^2 = 5.29.10^{-9}$cm and the atomic unit of energy $e^2/a_0 = 27.27$ eV which is twice the ionization energy of the hydrogen atom in its ground state.

A further modification of the asymptotic form (9) is required if the target is an ion. In this case the exponents must contain logarithmic phase factors which allow for the long range distortion by the Coulomb field. These terms however introduce no essential complication and we therefore omit them in order to maintain formal simplicity.

We can now obtain the cross section for a transition from the initial atomic state Φ_i to the final atomic state Φ_j where the scattered electron spin magnetic quantum changes from m_i to m_j. Calculating the incident and scattered flux due to the first and second terms in eq. (9) in the usual way gives

$$\frac{d\sigma_{ji}}{d\Omega} = \frac{k_j}{k_i} \left| f_{ji} \; (\theta,\phi) \right|^2 \tag{11}$$

for the differential cross section in units of a_o^2/steradian. The total cross section is obtained by integrating over all scattering angles.

Our basic problem in these lectures is to show how the scattering amplitude, and hence the cross section can be calculated by solving Schrödinger's equation (6) and to relate various features in the observed cross sections to the properties of the electron-electron interaction.

3. EXPANSION OF THE COLLISION WAVE FUNCTION AND CROSS SECTION

For the Hamiltonian defined by eq. (7), the total orbital angular momentum L, the total spin S, their components M_L and M_S in some preferred direction as well as the total parity π are conserved in the collision. This allows us to formally write down solutions of eq. (6) which are eigenstates of these quantum numbers as well as of the energy. Hence we adopt an expansion of the total wave function in the form

$$\begin{aligned} \Psi_i^\Gamma(X_{N+1}) = \mathcal{A} \sum_j \Phi_j^\Gamma \; (x_1 \ldots x_N; \hat{\underline{r}}_{N+1}\sigma_{N+1}) \; F_{ji}^\Gamma(r_{N+1}) \\ + \sum_j \chi_j^\Gamma \; (x_1 \ldots x_{N+1}) \; a_{ji}^\Gamma \end{aligned} \tag{12}$$

where $\Gamma \equiv L\,S\,M_L\,M_S\,\pi$ and will often be omitted for notational simplicity, X_{N+1} represents the space and spin coordination of all N+1 electrons, $x_i \equiv \underline{r}_i\sigma_i$ represents the space and spin coordinates of the i th electron, Φ_j^Γ are channel functions formed by coupling the target wave function $\Phi_j(x_1 \ldots x_N)$ with the angular and spin parts of the wave function for the (N+1)th electron whose reduced radial wave function is $F_{ji}^\Gamma\,(r_{N+1})$, $\mathcal{A}$ is the operator that antisymmetrises the first expansion with respect to interchange of any pair of electron coordinates and finally the χ_j^Γ are a set of quadratically integrable functions of the correct symmetry. Given the basis functions Φ_j^Γ and χ_i^Γ, we then have to derive and to solve equations for the functions F_{ii}^Γ and the coefficients a_{ii}^Γ corresponding to the boundary condition denoted by the subscript i. Expansion (12), without the quadratically integrable functions,was first written down by Seaton (1953) to describe collisions of electrons with complex atoms.

The antisymmetrization operator $\mathcal{A}$ has a dual purpose. It ensures that the wave function explicitly satisfies the requirements of the Pauli exclusion principle. Castellijo et al (1960) also pointed out that if it had not been included electron exchange effects would mean that there was a singularity in the integral over continuum states of H_N in the first expansion of eq. (12). The use of properly antisymmetric expansions removes this singularity and means that the correct boundary conditions can be satisfied provided only that the target states which can be energetically excited are retained in eq. (12). The energetically forbidden states can then be represented by the expansion over quadratically integrable functions.

In order to derive the equations satisfied by the functions F_{ii}^Γ and the coefficients a_{ii}^Γ we substitute eq. (12) into eq. (6) and project onto the channel functions Φ_i^Γ and onto the quadratically integrable functions χ_i^Γ. We obtain

$$\langle \Phi_j^\Gamma \mid H_{N+1} - E \mid \Psi_i^\Gamma \rangle = 0 \tag{13}$$

where the integral is carried out over all coordinates except for the radial coordinate of the (N+1)-electron and

$$\langle \chi_j^\Gamma \mid H_{N+1} - E \mid \Psi_i^\Gamma \rangle = 0 \tag{14}$$

where the integral is carried out over all coordinates. In simplifying these equations it is convenient to introduce the projection operator formalism of Feshbach (1962). We define P as the operator which projects onto the channel functions Φ_i^Γ and Q as the operator which projects onto the quadratically integrable functions χ_i^Γ. Eqs. (13) and (14) can then be rewritten as

$$P(H - E)\ (P + Q)\Psi = 0 \tag{15}$$

and

$$Q(H - E)\ (P + Q)\Psi = 0 \tag{16}$$

where we have omitted the superscript Γ and the subscript (N+1). We can then solve eq. (16) for $Q\Psi$ giving

$$Q\,\Psi = -\ Q\ \frac{1}{Q(H-E)Q}\ QHP\ \Psi \tag{17}$$

where we have assumed that we have chosen χ_i^Γ and Φ_i^Γ to be orthogonal so that PQ = QP = 0. Then substituting for $Q\Psi$ from (17) into (16) yields

$$P(H - PHQ \frac{1}{Q(H-E)Q} QHP - E) P \Psi = 0 \tag{18}$$

which are radial coupled integrodifferential equations for the functions $F_{ji}^\Gamma(r)$. We see that the close coupling equations

$$P(H - E)P \Psi^P = 0 \tag{19}$$

obtained by retaining just terms of the first expansion in eq. (12) are augmented by an additional "optical" potential

$$V_{opt} = - PHQ \frac{1}{Q(H-E)Q} QHP \quad . \tag{20}$$

This optical potential allows for short range correlation effects.

It is convenient to write out explicitly the coupled integro-differential equations (18). They have the form

$$\left(\frac{d^2}{dr^2} - \frac{\ell_i(\ell_i+1)}{r^2} + \frac{2Z}{r} + k_i^2\right) F_{ij}^\Gamma(r) = 2 \sum_\ell [V_{i\ell}^\Gamma (r) F_{\ell j}^\Gamma(r)$$

$$+ \int_0^\infty W_{i\ell}^\Gamma(r,r') F_{\ell j}^\Gamma(r')dr' + \int_0^\infty K_{i\ell}^\Gamma(r,r')F_{\ell j}^\Gamma(r') dr'] \tag{21}$$

where ℓ_i is the orbital angular momentum of the scattered electron, V and W are the direct and non-local exchange potentials arising in (19) and K is the non-local optical potential (20). Written out explicitly

$$V_{ij}^\Gamma(r_{N+1}) = \int \Phi_i^\Gamma(x_1 \ldots x_N; \hat{\underline{r}}_{N+1}\sigma_{N+1}) \sum_{i=1}^{N} \frac{1}{r_{i\,N+1}}$$

$$\Phi_j^\Gamma(x_1 \ldots x_N; \hat{\underline{r}}_{N+1}\sigma_{N+1}) \; d\,x_1 \ldots d\,x_N$$

$$d\hat{r}_{N+1} \; d\sigma_{N+1} \tag{22}$$

which has the asymptotic form

$$V_{ij}^\Gamma(r) = \frac{N}{r} \delta_{ij} + \sum_{\lambda=1}^{\lambda_{max}} a_{ij}^\lambda \, r^{-\lambda-1} \quad , \quad r \geq a \tag{23}$$

where a is the maximum range of the target states Φ_i and Φ_j. The first term in eq. (23) is the screening due to the N target electrons. When combined with the nuclear potential 2Z/r in eq. (21) it gives the long-range Coulomb potential seen by an electron incident upon an ion. The remaining terms in eq. (23) are the multipole potentials experienced by the electron. The leading $\lambda = 1$ term, which can easily be seen from eq. (22) to link channels coupled by the dipole operator, lead in second-order to the α/r^4 polarization potential.

The non-local potentials W and K vanish exponentially for large r, their ranges being determined respectively by the target states Φ_i and the quadratically integrable functions χ_i.

We look for solutions of eqs. (21) satisfying the boundary conditions

$$F^{\Gamma}_{ij}(0) = 0 \tag{24}$$

and

$$\begin{aligned} F^{\Gamma}_{ij}(r) &\underset{r\to\infty}{\sim} k_i^{-\frac{1}{2}}(\sin\theta_i\, \delta_{ij} + \cos\theta_i\, K^{\Gamma}_{ij}),\quad k_i^2 \geq 0 \\ &\underset{r\to\infty}{\sim} 0, \qquad k_i^2 < 0 \end{aligned} \tag{25}$$

where the index j which defines the linearly independent solutions goes over $j = 1,\ldots n_a$ where n_a is the number of open channels (where $k_i^2 \geq 0$) at the energy considered. Also

$$\begin{aligned} \theta_i &= k_i\, r - \tfrac{1}{2}\ell_i\pi - \eta_i\, \ell n\, 2\, k_i\, r + \sigma_i \\ \eta_i &= -(Z - N)/k_i \\ \sigma_i &= \arg\Gamma(\ell_i + 1 + i\,\eta_i)\ , \end{aligned} \tag{26}$$

These equations define the $n_a \times n_a$ dimensional K-matrix K^{Γ}_{ij}. It follows from unitarity and time reversal invariance that this matrix is real and symmetric.

In order to relate the asymptotic form of eq. (12) with eq. (9) it is convenient to recombine the independent solutions defined by eq. (25) giving in matrix notation n_a new solutions

$$\underline{G}^{\Gamma} = -2i\, \underline{F}^{\Gamma} .(\underline{1} - i\, \underline{K}^{\Gamma})^{-1} \quad . \tag{27}$$

It follows from eqs. (25) and (27) that the G^{Γ}_{ij} then have the asymptotic form

$$\begin{aligned} G^{\Gamma}_{ij} &\underset{r\to\infty}{\sim} k_i^{-\frac{1}{2}}\,(e^{-i\theta_i}\, \delta_{ij} - e^{i\theta_i}\, S^{\Gamma}_{ij}),\quad k_i^2 \geq 0 \\ &\underset{r\to\infty}{\sim} 0\quad ,\ k_i^2 < 0 \end{aligned} \tag{28}$$

where the $n_a \times n_a$ dimensional S-matrix is defined in terms of the K-matrix by the matrix equation

$$\underline{S}^{\Gamma} = \frac{\underline{1} + i\,\underline{K}^{\Gamma}}{\underline{1} - i\,\underline{K}^{\Gamma}} \quad . \tag{29}$$

The S-matrix is easily seen to be unitary and symmetric.

Following Blatt and Biedenharn (1952), the scattering amplitude is given by expanding the plane wave in eq. (9) in partial waves and equating the ingoing wave term in eqs. (9) and (12) for all Γ. We obtain

$$f_{ji}(\hat{\underline{r}}) = -i\left(\frac{\pi}{k_i k_j}\right)^{\frac{1}{2}} \sum_{\substack{L\,S\,\pi \\ \ell_i \ell_j m_{\ell_j}}} i^{\ell_i - \ell_j}(2\ell_i+1)^{\frac{1}{2}}(L_i M_{L_i}\ \ell_i\ 0|L\ M_L)$$
$$\times\ (L_j M_{L_j}\ \ell_j\ m_{\ell_j}|L\ M_L)\ (S_i\ M_{S_i}\ \tfrac{1}{2}m_i|S\ M_S)$$
$$\times\ (S_j\ M_{S_j}\ \tfrac{1}{2}\ m_j|S\ M_S)(S^{\Gamma}_{ji} - \delta_{ji})\ Y_{\ell_j m_{\ell_j}}(\hat{\underline{r}}) \tag{30}$$

corresponding to a transition from an initial state with angular, spin and parity quantum numbers $L_i\ S_i\ M_{L_i}\ M_{S_i}\ \pi_i$ to a final state with quantum numbers $L_j\ S_j\ M_{L_j}\ M_{S_j}\ \pi_j$. The quantities $(L_i M_{L_i}\ \ell_i\ 0|L\ M_L)$ etc are the usual Clebsch-Gordan coefficients.

The differential cross section is obtained by substituting eq. (30) into eq. (11), while the total cross section is obtained by averaging over initial spin states, summing over final spin states and integrating over all scattered electron directions giving

$$\sigma_{Tot}(\alpha_i L_i S_i \rightarrow \alpha_j L_j S_j) = \frac{\pi}{k_i^2} \sum_{\substack{LS\pi \\ \ell_i \ell_j}} \frac{(2L+1)(2S+1)}{2(2L_i+1)(2S_i+1)} |S^{\Gamma}_{ji} - \delta_{ji}|^2 \tag{31a}$$

in units of a_0^2 where α_i and α_j denote the other quantum numbers of the target. The corresponding collision strength is defined by

$$\Omega(i,j) = k_i^2\ (2L_i + 1)(2S_i + 1)\sigma_{Tot}(\alpha_i L_i S_i \rightarrow \alpha_j L_j S_j) \tag{31b}$$

which because it is dimensionless and symmetric is often considered rather than the cross section itself.

In the case of elastic scattering by ions, the Coulomb amplitude must also be added to the above result. This gives the usual Coulomb singularity in the angular distribution in the forward direction and causes the total elastic cross section to diverge.

4. ILLUSTRATIVE RESULTS FOR LOW ENERGY ELASTIC SCATTERING AND EXCITATION

In this section, we present a few illustrative results obtained using expansion (12) and solving the resultant coupled integrodifferential equations (21). Most of the recent results have been obtained using one of four very general computer programs. The first of these programs is based on the R-matrix method (Burke et al, 1971), the second is based on the reduction of eqs. (21) to a system of linear algebraic equations (Seaton, 1974), the third uses the non-iterative integral equations method (Smith and Henry, 1973a,b) and finally the fourth uses the matrix variational method (Nesbet, 1980). These methods and the associated programs have been reviewed by Burke and Eissner (1983). Many of the results, which now cover most light atoms and ions, are stored in a data bank at Queen's University Belfast (Berrington et al, 1984).

The simplest systems which can be studied are e^- - H and e^- - He^+ scattering. Schwartz (1961) obtained essentially exact results for the S-wave e^- - H phase shifts and Armstead (1968) for the P-wave phase shifts below the n = 2 threshold. They included the 1s state together with up to 50 Hylleraas-type quadratically integrable functions in eq. (12) and solved eqs. (21) using the Kohn variational method. Burke and Taylor (1966) and Taylor and Burke (1967) obtained accurate results for e^- - H and e^- - He^+ 1s-2s and 1s - 2p excitation close to threshold by including the 1s,2s and 2p states and up to 20 Hylleraas-type terms. Since these early results calculations have been carried out including the n=3 and recently the n=4 states.

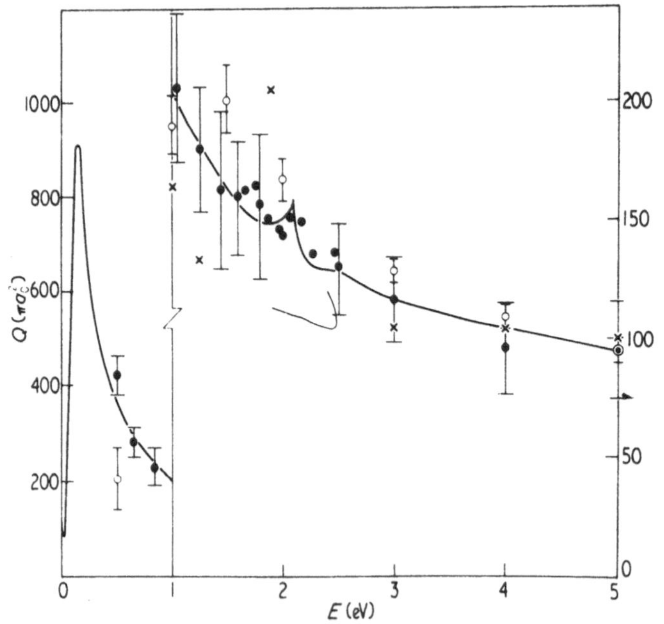

Figure 2. Total cross section for e^- - Na scattering (fig. 3 from Moores and Norcross, 1972)

Accurate results have also been obtained for electron scattering by light alkali metal atoms. In these atoms the closed shell core mainly provides a screening of the nuclear potential. The total cross section for e^- - Na scattering is shown in figure 2 where the calculation by Moores and Norcross (1972) includes the 3s and 3p states. It can be seen that good agreement is obtained with the experiments of Perel et al (1962) and Rubin et al (unpublished). The theoretical results exhibit a low energy ^3P resonance and a cusp and resonance in the ^1P state at the ^2P threshold. We note that many recent calculations and measurements on alkalis have used polarized beams and targets (Hanne, 1983, Kessler 1984).

In the case of e^- - He scattering accurate results can only be obtained if electron correlation effects are included in the target wave function Φ_i, defined by eq. (8), as well as in the collision wave function. These effects were included by O'Malley et al (1979) and Nesbet (1979) who represented the helium ground state by a configuration interaction expansion. Their results for the total and momentum transfer cross sections which are in excellent agreement with the experiments of Andrick and Bitsch(1975), Kauppila et al (1977) and Millroy and Crompton (1977) are shown in figure 3. Older calculations by Sinfailam and Nesbet (1972) are also shown.

Finally we mention a benchmark study of the 2s-2p excitation cross section of Be^+ carried out by Hayes et al (1977). The cross section for this transition is shown in figure 4.

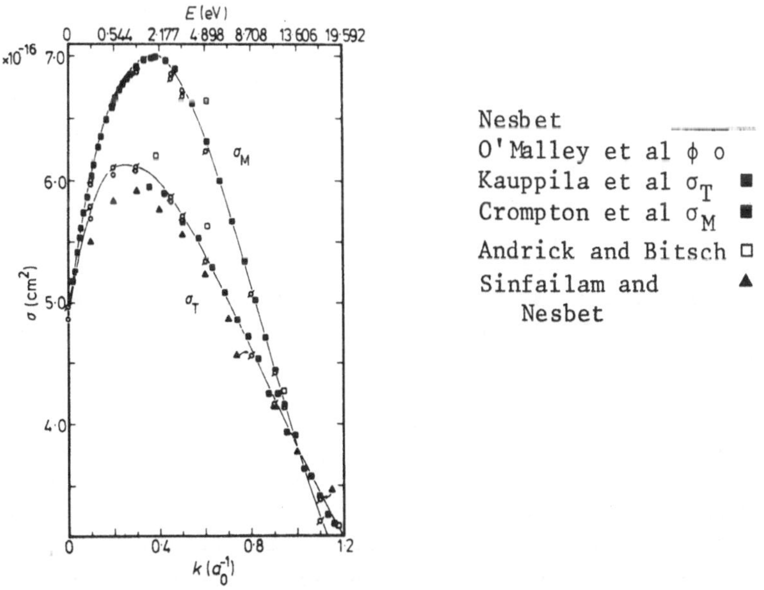

Figure 3. Total and momentum transfer cross section for e^- - He scattering (fig. 2 from Nesbet, 1979)

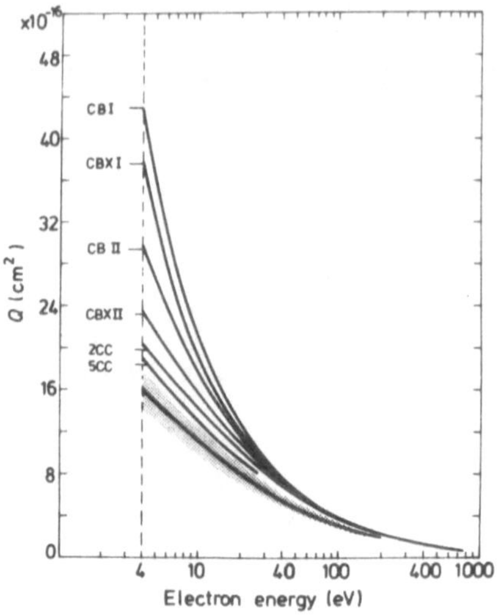

Figure 4. Cross section for $e^- - Be^+$ 2s - 2p excitation (fig. 1 from Hayes et al, 1977)

The experiments of Taylor et al (1980) are compared with 5 state 2s, 2p, 3s, 3p, 3d close coupling results of Henry et al (1978) and Hayes et al (1977), the 2 state 2s, 2p close coupling results of Hayes et al (1977) as well as a variety of Coulomb-Born calculations. The calculations clearly improve as the scattered wave is treated by solving eqs. (21) rather than by using a Coulomb wave and as more terms are included in the close coupling expansion. The final 5 state results are in reasonable accord with the experiment although it appears that convergence has not quite been achieved.

In conclusion, for light atoms and ions it is now possible to calculate low energy cross sections accurate to 10% or better except in very special circumstances. Such a circumstance arises for example in e^- - N scattering close to zero energy where the change in position of the $1s^2\ 2s^2\ 2p^4\ {}^3P^e$ resonance by a few meV strongly effects the absolute value of the cross section (Thomas et al, 1974, Burke et al 1974).

5. INCLUSION OF RELATIVISTIC EFFECTS

As the atomic number of the target increases, relativistic effects become important even for low incident electron energies. There are two main ways in which relativistic effects play a role. Firstly there is a direct effect which may be considered as a relativistic distortion by the strong nuclear potential of the wave function describing the motion of the scattered electron. Secondly there is an indirect effect caused by the change in the charge distribution of the target whose eigenstates are no longer described by the quantum numbers $\alpha_i\ L_i\ S_i\ M_{L_i}\ M_{S_i}\ \pi_i$ but by $\alpha_i\ J_i\ M_{J_i}\ \pi_i$.

There are several approaches for including relativistic effects. For light and intermediate weight atoms and ions, where the relativistic effects are small and the fine-structure intervals can be neglected in a first approximation the calculation is carried out in LS coupling using the non-relativistic Hamiltonian defined by eq. (7). The corresponding K-matrices are then recoupled to yield transitions between the fine-structure levels. This is most conveniently carried out using the pair coupling scheme of Racah (1942).

$$\underline{J}_i + \underline{\ell}_i = \underline{K} \qquad \underline{K} + \underline{s}_i = \underline{J} \tag{32}$$

where $\underline{J}_i$ is the total angular momentum of the target, and $\underline{\ell}_i$ and $\underline{s}_i$ are the orbital and spin angular momenta of the scattered electron. The corresponding recoupling coefficient is given by

$$\langle((L_iS_i)\ J_i,\ \ell_i)\ K,\ \tfrac{1}{2}\ ;\ J\ M_J|(L_i\ell_i)\ L,\ (S_i\ \tfrac{1}{2})\ S;\ J\ M_J\rangle$$

$$= [(2\ J_i + 1)(2L+1)(2K+1)(2S+1)]^{\frac{1}{2}}\ W(L\ell_iS_iJ_i;L_iK)W(L\ J\ S_i\tfrac{1}{2};S\ K) \tag{33}$$

where the W coefficients are the usual Racah coefficients. This procedure has been programmed by Saraph (1972, 1978) and works well provided that the incident electron energy is large compared with the fine-structure intervals.

For heavy atoms relativistic terms must be retained in the Hamiltonian. One method of doing this adopted by Jones (1975) Scott and Burke (1980) and Sinfailam (1980) is to use the Breit-Pauli Hamiltonian. This is obtained by reducing the Dirac equation and the Breit interaction to Pauli form (eg Bethe and Salpeter, 1972). The Hamiltonian for an electron incident on an N-electron target is then given by

$$H_{N+1}^{BP} = H_{N+1}^{NR} + H_{N+1}^{REL} \tag{34}$$

where the non-relativistic Hamiltonian is defined by eq. (7) and where H_{N+1}^{REL} consists of a sum of one- and two-body relativistic terms. The one-body terms are the spin-orbit interaction

$$H_{N+1}^{SO} = \tfrac{1}{2}\ \alpha^2\ Z\ \sum_{i=1}^{N+1} r_i^{-3}\ (\underline{\ell}_i\cdot\underline{s}_i)\ , \tag{35}$$

the mass correction term

$$H_{N+1}^{MASS} = -\frac{1}{8}\ \alpha^2\ \sum_{i=1}^{N+1} \nabla_i^4 \tag{36}$$

and the one-body Darwin term

$$H^D_{N+1} = -\frac{1}{8}\alpha^2 \; Z \sum_{i=1}^{N+1} \nabla_i^2 \left(\frac{1}{r_i}\right) \tag{37}$$

In addition, the two-body terms are the mutual-spin-orbit and spin-other orbit terms, the spin-spin term, the orbit-orbit term, the two-body Darwin term and the spin-contact term and are given by Scott and Burke (1980). The total wave function is then expanded in a form analogous to eq. (12) except that the target states must now be obtained using the Breit Pauli Hamiltonian and the wave function is an eigenstate of $J\, M_J$ and π rather than $LS\, M_L\, M_S$ and π. If we use the coupling scheme given by eq. (32) the channels are then denoted by $J_i\, K_i$ and ℓ_i. As a result the number of coupled equations analogous to (21) for the same number of target states increases very substantially. This added to the fact that the target states themselves are treated relativistically means that the solution of the equations is very much more lengthy than when relativistic effects are neglected. Nevertheless the formal methods for solving these equations to obtain the K-matrix and S-matrix are the same and the resultant expression for the cross section is

$$\sigma_{Tot}(\alpha_i J_i \to \alpha_j J_j) = \frac{\pi}{2\, k_i^2 (2\, J_i+1)} \sum_{\substack{J\pi \\ K_i K_j \ell_i \ell_j}} (2\, J+1) \, |S^{J\pi}_{ji} - \delta_{ji}|^2 \tag{38}$$

instead of eq. (31). This approach has been used to obtain cross sections for a number of heavy atoms and results will be presented later in these lectures.

An alternative approach is to base the scattering calculation on the Dirac Hamiltonian which has been widely used for atomic structure calculations (eg Grant, 1979). The Dirac Hamiltonian describing the collision of an electron by an N electron target is

$$H^{Dirac}_{N+1} = \sum_{i=1}^{N+1} \left(c\, \underline{\alpha}\cdot\underline{p}_i + \underline{\beta}'\, c^2 - \frac{Z}{r_i}\right) + \sum_{i>j=1}^{N+1} \frac{1}{r_{ij}} \tag{39}$$

where $\underline{\alpha}$ and $\underline{\beta}' = \beta - 1$ are the usual Dirac operators and c is the velocity of light. QED corrections to the electron-electron interaction considered by Breit, as well as other QED corrections are often included in atomic structure calculations but have not been considered so far for electron collisions.

Again, as in the case of the Breit-Pauli Hamiltonian the total wave function takes the general form of eq. (12) being diagonal in JM_J and π. However now both the orbitals describing the target and the scattered electron are defined by a four component spinor as

$$\phi(\underline{r},\sigma) = \frac{1}{r} \begin{bmatrix} P(r)\chi_{Km}(\hat{\underline{r}},\sigma) \\ Q(r)\chi_{-Km}(\hat{\underline{r}},\sigma) \end{bmatrix} \tag{40}$$

where the spin-angle function χ_{Km} is given by

$$\chi_{Km}(\hat{\underline{r}},\sigma) = \sum_{m_\ell m_i} (\ell\, m_\ell\, \tfrac{1}{2}\, m_i | j\, m) Y_{\ell m_\ell}(\hat{\underline{r}})\, \chi_{\frac{1}{2}m_i}(\sigma) \tag{41}$$

and where the subscript $K = j + \frac{1}{2}$ when $\ell = j + \frac{1}{2}$ and $K = - j - \frac{1}{2}$ when $\ell = j - \frac{1}{2}$. The resultant equations equivalent to eqs. (21) are now coupled first-order integrodifferential equation for the large and small components P(r) and Q(r) describing the continuum electron. Results have been obtained for electron scattering by hydrogen-like ions using this approach by Carse and Walker (1973) and by Walker (1974). In addition preliminary work has been carried out to develope a general code by Chang (1975) and by Norrington and Grant (1981).

In conclusion the use of the Dirac Hamiltonian is clearly more fundamental than the Breit-Pauli Hamiltonian. However technical difficulties have meant that so far the only significant results for heavy atoms have been with the latter Hamiltonian.

6. EXPANSION IN HYPERSPHERICAL COORDINATES

Expansions based on eq. (12) have been used for many years to describe electron collision with atomic hydrogen and hydrogen-like ions (eg Massey and Mohr, 1932, Percival and Seaton, 1957, Burke and Schey, 1962) and the most accurate elastic scattering and excitation cross sections have still been obtained by solving the corresponding integrodifferential equations (21) (see review by Callaway and McDowell, 1983). Nevertheless, in recent years it has been increasingly recognized that expansions in hyperspherical coordinates have a very important role to play in discussing doubly excited resonance states (Macek, 1968, Lin, 1974, 1983) as well as in the discussion of electron impact ionization (Wannier, 1953, Klar and Schlecht, 1976 and Peterkop, 1977). It should also be noted that hyperspherical coordinates have been used in the description of the bound states of helium (Fock, 1954, 1958, Demkov and Ermolaev, 1959), of the nuclear 3-body problem (eg Delves, 1960) and of the triatomic molecular system (eg Whitten and Smith, 1968).

In the case of two electrons interacting with a nucleus the hyperspherical coordinates are defined in terms of the electron polar coordinates (r_1,θ_1,ϕ_1) and (r_2,θ_2,ϕ_2) by

$$R = (r_1^2 + r_2^2)^{\frac{1}{2}}, \qquad \alpha = \tan^{-1} r_2/r_1 \tag{42}$$

while the remaining coordinates are usually chosen as before as

$(\theta_1\phi_1\theta_2\phi_2)$. The new hyperspherical coordinates (R,α) are illustrated in figure 5. Excitation corresponds to the situation where one electron stays close to the nucleus, i.e. r_1 or r_2 remains small, while ionization corresponds to the situation where both electrons tend to infinity simultaneously as indicated in this figure.

The Schrödinger equation (6) describing two electrons moving in the field of a proton can be expressed in terms of these coordinates as

$$\left(\frac{d^2}{dR^2} - \frac{\Lambda^2 + \frac{15}{4}}{R^2} + \frac{C}{R} + 2E\right)(R^{5/2}\,\Psi) = 0 \tag{43}$$

where the factor $R^{5/2}$ has been introduced to remove the first-order derivatives with respect to R. The potential energy is $-C/R$ where

$$C = R\left(\frac{2}{r_1} + \frac{2}{r_2} - \frac{2}{r_{12}}\right)$$

$$= \frac{2}{\cos\alpha} + \frac{2}{\sin\alpha} - \frac{2}{(1-\sin 2\alpha\cos\theta_{12})^{\frac{1}{2}}} \tag{44}$$

θ_{12} being the angle between the two vectors $\hat{\underline{r}}_1$ and $\hat{\underline{r}}_2$. The operator Λ^2 is defined by

$$\Lambda^2 = -\frac{1}{\sin^2\alpha\cos^2\alpha}\frac{d}{d\alpha}\left(\sin^2\alpha\cos^2\alpha\frac{d}{d\alpha}\right) + \frac{\underline{\ell}_1^2}{\cos^2\alpha} + \frac{\underline{\ell}_2^2}{\sin^2\alpha} \tag{45}$$

where $\underline{\ell}_1^2$ and $\underline{\ell}_2^2$ are the squared orbital angular momentum operators for the two electrons. Λ^2 is the square of the grand angular-momentum operator for six dimensions and is thus the Casimir operator for the O_6 group.

Figure 5. Illustration of the hyperspherical coordinates (R,α)

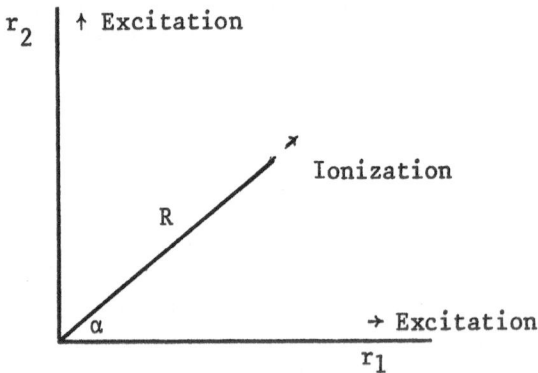

Eq. (43) resembles the Schrödinger equation for the radial motion of a particle moving in the potential $-C/R$ with the angular momentum given by $(\Lambda^2 + \frac{15}{4})/R^2$. However, unlike the similar equation for the hydrogen atom C depends on the angular coordinates α and θ_{12} and does not commute with Λ^2. The dynamics of the motion of the two electrons depends on the form of C and in figure 6 we present a three dimensional plot of $-C(\alpha,\theta_{12})$ for α and θ_{12} in the ranges $0 \leq \alpha \leq \pi/2$ and $0 \leq \theta_{12} \leq \pi$.

We see that at $\alpha = 0$ and $\pi/2$ the potential surface tends to $-\infty$ corresponding to the electron-nuclear attraction, while at $\alpha = \frac{1}{4}\pi$ and $\theta_{12} = 0$ there is a singularity due to the electron-nuclear repulsion. The saddle point in the potential energy surface at $\alpha = \frac{1}{4}\pi$ and $\theta_{12} = \pi$ corresponds to the situation where the two electrons are equidistant from and on opposite sides of the nucleus. It has been shown by Wannier (1953) to play a crucial role in the threshold behaviour of the ionization cross section. It is also important in determining the properties of doubly excited states of H^- and He. Both of these topics will be considered later in these lectures.

In order to solve eq. (43) we introduce the eigenfunctions of the operator Λ^2. As shown in Morse and Feshbach (1953) we can write

$$(\Lambda^2 - K(K+4))\, Z_{K\gamma}(\Omega) = 0 \tag{46}$$

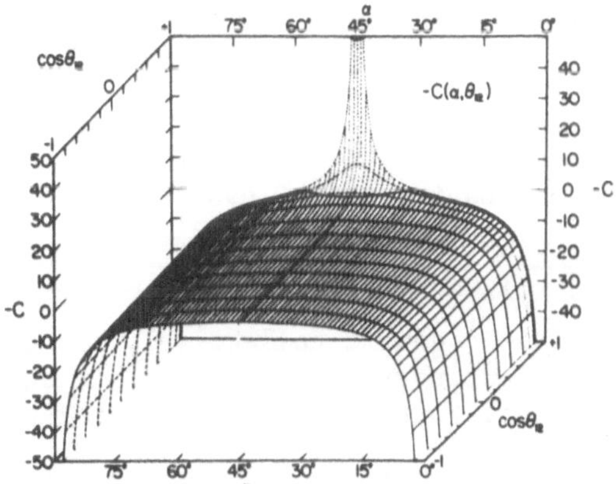

Figure 6. Three dimensional plot of $-C(\alpha,\theta_{12})$ for $Z = 1$ (fig. 1 from Lin, 1974)

where K is a non negative integer which can be written as

$$K = \ell_1 + \ell_2 + 2m , \tag{47}$$

ℓ_1 and ℓ_1 being the usual orbital angular momentum quantum numbers and m is a quantum number associated with the motion in α, and where γ represents the remaining quantum numbers required to specify the state. The functions $Z_{K\gamma}(\Omega)$, which depend on the variables $\Omega \equiv(\alpha,\theta_1,\phi_1,\theta_2,\phi_2)$, are called K harmonics. In order to eliminate the first derivative with respect to α in eq. (46) we introduce the functions

$$\phi_{\ell_1\ell_2 m}(\Omega) = \sin\alpha\cos\alpha Z_{K\gamma}(\Omega) \tag{48}$$

which satisfy the equation

$$\left(-\frac{\partial^2}{\partial\alpha^2} + \frac{\ell_1^2}{\cos^2\alpha} + \frac{\ell_1^2}{\sin^2\alpha} - (K+2)^2\right)\phi_{\ell_1\ell_2 m}(\Omega) = 0 \tag{49}$$

The eigenfunctions with symmetry corresponding to a given $\Gamma = L\ S\ M_L\ M_S\ \pi$ are given by (Macek, 1968)

$$\phi^{\Gamma}_{\ell_1\ell_2 m}(\Omega) = \frac{1}{\sqrt{2}}\left[f_{\ell_1\ell_2 m}(\alpha)\mathcal{Y}_{\ell_1\ell_2 LM_L}(\hat{\underline{r}}_1\hat{\underline{r}}_2)+(-1)^{\ell_1+\ell_2-L+S+m} f_{\ell_2\ell_1 m}(\alpha)\mathcal{Y}_{\ell_2\ell_1 L M_L}(\hat{\underline{r}}_1\hat{\underline{r}}_2)\right] \text{ if } \ell_1 \neq \ell_2 \tag{50}$$

$$\phi^{\Gamma}_{\ell_1\ell_2 m}(\Omega) = \tfrac{1}{2}\left[1+(-1)^{-L+S+m}\right] f_{\ell\ell m}(\alpha)\mathcal{Y}_{\ell\ell LM_L}(\hat{\underline{r}}_1\hat{\underline{r}}_2) \text{ if } \ell_1=\ell_2=\ell$$

where the $\mathcal{Y}_{\ell_1\ell_2 LM_L}(\hat{\underline{r}}_1\hat{\underline{r}}_2)$ are the usual total orbital angular momentum eigenfunctions obtained by coupling $Y_{\ell_1 m_{\ell_1}}(\hat{\underline{r}}_1)$ and $Y_{\ell_2 m_{\ell_2}}(\hat{\underline{r}}_2)$.

The total wave function for each Γ can then be expanded as

$$R^{5/2}\sin\alpha\cos\alpha\,\Psi^{\Gamma}_j = \sum_i \phi^{\Gamma}_i(\Omega)\,h^{\Gamma}_{ij}(R) \tag{51}$$

where the subscripts i and j denote collectively the quantum numbers $(\ell_1\ell_2 m)$. Substituting this expansion into eq. (43) and projecting onto the channel functions ϕ^{Γ}_i gives after using eq. (49)

$$\left(\frac{d^2}{dR^2} - \frac{(K+2)^2 - \frac{1}{4}}{R^2} + k^2\right) h^{\Gamma}_{ij}(R) = -\frac{1}{R}\sum_{\ell} V^{\Gamma}_{i\ell}\,h^{\Gamma}_{\ell j}(R) \tag{52}$$

where

$$V^{\Gamma}_{ij} = \langle \phi^{\Gamma}_i(\Omega)|C(\alpha,\theta_{12}|\phi^{\Gamma}_j(\Omega)\rangle \tag{53}$$

and $k^2 = 2E$. We see that eqs. (52) unlike eqs. (21) which they replace are coupled differential equations rather than coupled integro-differential equations. The Pauli exclusion principle is now contained in the form of the coupling potential $V^{\Gamma}_{i\,\ell\ell}$ and in the restrictions on the values of the summations over ℓ_1 ℓ_2 and m for each Γ. Eqs. (52) are an infinite set of coupled differential equations for each Γ and hence have to be approximated in some way in practical applications. What makes the hyperspherical coordinate approach particularly useful in describing doubly excited states of atoms, is that to a very good approximation the motion in the R variable can be separated from the motion in the other variables in a similar way to the Born-Oppenheimer separation of the electronic and nuclear motion in molecular physics. This suggests that the wave function can be approximately described by the Ansatz $\Phi^{\Gamma}_{\mu}(R,\Omega)\, H^{\Gamma}_{\mu}(R)$ where the function $\Phi^{\Gamma}_{\mu}(R,\Omega)$ is chosen to minimize the coupling at each R.

This consideration leads us to adopt the adiabatic expansion

$$R^{5/2}\,\Psi^{\Gamma}_{\nu} = \sum_{\mu} \Phi^{\Gamma}_{\mu}(R,\Omega)\, H^{\Gamma}_{\mu\nu}(R) \tag{54}$$

rather than expansion (51) where now the Φ^{Γ}_{μ} are chosen to diagonalize all terms in eq. (43) except the kinetic energy operator d^2/dR^2. Hence we define

$$\langle \Phi^{\Gamma}_{\mu}(R,\Omega) \,|\, \frac{\Lambda^2 + \frac{15}{4}}{R^2} - \frac{C}{R} \,|\, \Phi^{\Gamma}_{\nu}(R,\Omega)\rangle = U^{\Gamma}_{\mu}(R)\,\delta_{\mu\nu} \tag{55}$$

where the diagonalization in this equation has to be carried out separately for each value of R. The solution of this equation can be achieved by one of a number of methods as discussed by Lin (1981). Then substituting eq. (54) into eq. (43) and projecting onto the functions Φ^{Γ}_{μ} yields the coupled radial adiabatic equations

$$\left(\frac{d^2}{dR^2} - U^{\Gamma}_{\mu}(R) + k^2\right) H^{\Gamma}_{\mu\nu}(R) = \sum_{\mu'} W^{\Gamma}_{\mu\mu'}(R)\, H^{\Gamma}_{\mu'\nu}(R) \tag{56}$$

where the coupling potential $W^{\Gamma}_{\mu\mu'}$ is a differential matrix operator defined by

$$W^{\Gamma}_{\mu\mu'}(R) = 2\langle \Phi^{\Gamma}_{\mu}(R,\Omega) \,|\frac{d}{dR}|\, \Phi^{\Gamma}_{\mu'}(R,\Omega)\rangle \frac{d}{dR} + \langle \Phi^{\Gamma}_{\mu}(R,\Omega) \,|\frac{d^2}{dR^2}|\, \phi^{\Gamma}_{\mu'}(R,\Omega)\rangle \tag{57}$$

The extreme adiabatic approximation is obtained by neglecting all the terms on the right hand side of (56) while the adiabatic approximation is obtained by retaining in addition the diagonal term on the right hand side. If we retain all terms in the coupling potential then eqs. (52) and (56) are equivalent.

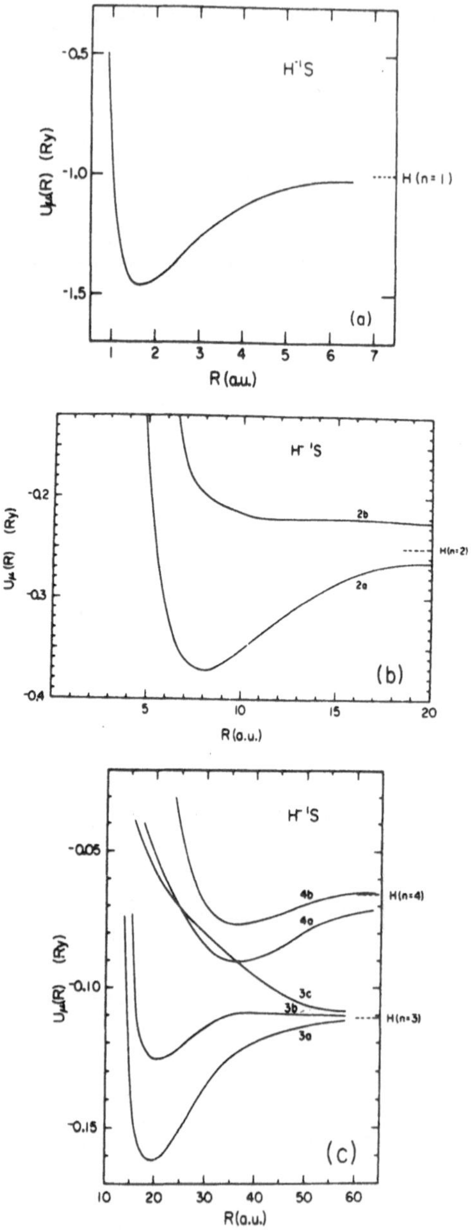

Figure 7. Adiabatic potential curves of $H^{-}\ {}^{1}S$ converging to the n = 1, 2, 3 and 4 states of H (fig. 4 from Lin, 1982a)

As an illustration of the adiabatic potential curves we show those for the $^1S^e$ state of H^- converging to the n = 1, 2, 3 and 4 states of hydrogen in figure 7.

We see that the potential curves exhibit a R^{-2} repulsion due to the $(\Lambda^2 + 15/4)$ term as $R \to 0$ and tend to $-1/n^2$ corresponding to the binding energy of the n th hydrogen level as $R \to \infty$. Accurate estimates for the energies of the ground $1s^2\ ^1S$ and double excited $2s^2\ ^1S$ states of H^- can be obtained by solving the extreme adiabatic equations with the appropriate potential $U_\mu^\Gamma(R)$. However calculation of resonance widths and excitation cross sections requires more than one channel in eq. (56) to be coupled.

Recently the hyperspherical expansion method has been used to calculate elastic phase shifts for e-H scattering by Christensen-Dalsgaard (1984a,b). In figure 8 we show her results compared with the accurate results of Schwartz (1961) and the 3 state 1s,2s,2p close coupling result. By using expansion (54) for small R and carefully matching onto expansion (12) for large R very accurate results could be obtained including a single state in each expansion.

The hyperspherical coordinate approach can be extended to more than two electrons in the field of a nucleus. For example, Clark and Greene (1980) have considered a hyperspherical analysis of three electron dynamics. However the method rapidly loses its simplicity as the number of electrons increases. An alternative way of extending the method, provided that not more than two electrons are in highly excited states or continuum states is to use a combination of expansions equivalent to (12) and (54). This is illustrated in figure 9.

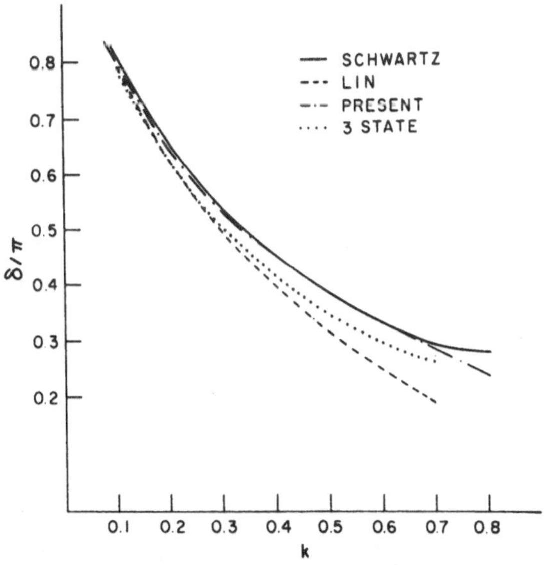

Figure 8. Elastic e-H phase shifts calculated using the hyperspherical method compared with other methods (fig. 2 from Christensen-Dalsgaard (1984b)

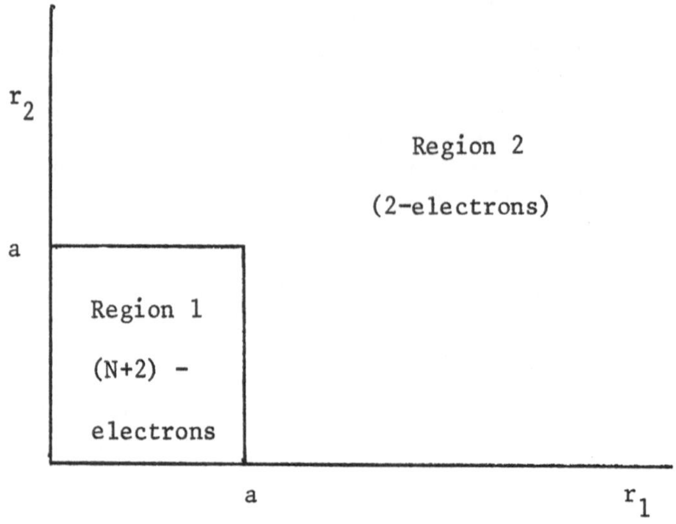

Figure 9. Extension of hyperspherical coordinates for an arbitrary number of electrons

In the inner region we adopt an expansion analogous to eq. (12) which now takes the form

$$\psi^{\Gamma}_{\ell}(X_{N+2}) = \mathcal{A}\sum_{ijk} \Phi^{\Gamma}_{i} (x_1 \ldots x_N; \hat{\underline{r}}_{N+1}\sigma_{N+1} \hat{\underline{r}}_{N+2}\sigma_{N+2}) u_j(r_{N+1}) v_k(r_{N+2}) a^{\Gamma}_{ijk\ell}$$

$$+ \sum_j \chi^{\Gamma}_{j} (x_1 \ldots x_{N+2}) b^{\Gamma}_{j\ell} \qquad (58)$$

In the first expansion N electrons are in orbitals which vanish by r = a while the remaining two electrons are represented by continuum orbitals u_j and v_k which are non-zero on the boundary. The second expansion consists of quadratically integrable functions which vanish by the boundary. The continuum orbitals are then matched on the boundary to an expansion involving just two electrons in the outer region. In this region an expansion in hyperspherical coordinates can be used. Such an approach is currently being developed to describe doubly excited states and ionization for a general atom or ion. A similar approach has also been used to describe doubly excited states of alkaline-earth atoms (Greene, 1981) and the helium negative ion (Watanabe, 1982) where the calculation in region 1 is replaced by a boundary condition on the surface obtained from a multichannel quantum defect analysis (Seaton, 1983) of the electron N-electron atom system.

7. ELECTRON IMPACT IONIZATION

We have already noted that the asymptotic form defined by eq. (9) must be modified when ionization is included. In order to avoid inessential complications we limit our discussion here to e - H scattering. We then have

$$\Psi(\underline{r}_1,\underline{r}_2) \underset{r_1\to\infty}{\sim} \phi_i(\underline{r}_2)\, e^{ik_i z_1} + \sum_{n\ell m} \phi_{n\ell m}(\underline{r}_2)\, f_{n\ell m}(\hat{\underline{r}}_1) r_1^{-1} e^{ik_n r_1}$$

$$+ \int_{k' \leq \sqrt{2E}} \phi(\underline{k}',\underline{r}_2)\, f(\underline{k},\underline{k}') r_1^{-1}\, e^{ikr_1 + i\eta(\underline{k},\underline{r}_1)}\, d\underline{k}' \qquad (59)$$

and

$$\Psi(\underline{r}_1,\underline{r}_2) \underset{r_2\to\infty}{\sim} \sum_{n\ell m} \phi_{n\ell m}(\underline{r}_1)\, g_{n\ell m}(\hat{\underline{r}}_2)\, r_2^{-1}\, e^{ik_n r_2}$$

$$+ \int_{k' \leq \sqrt{2E}} \phi(\underline{k}',\underline{r}_1)\, g(\underline{k}, \underline{k}')\, r_2^{-1}\, e^{ikr_2 + i\eta(\underline{k},\underline{r}_2)}\, d\underline{k}' \qquad (60)$$

where ϕ_i is the initial state of the atom. Since in the two electron case the space and spin coordinates are separable we have not included the spin functions in these equations.

The functions $\phi_{n\ell m}(\underline{r})$ and $\phi(\underline{k},\underline{r})$ are the discrete and continuous eigenfunctions of the hydrogen atom satisfying

$$(-\tfrac{1}{2}\nabla^2 + \tfrac{1}{r})\, \phi_{n\ell m}(\underline{r}) = E_n\, \phi_{n\ell m}(\underline{r}) \qquad (61a)$$

$$(-\tfrac{1}{2}\nabla^2 + \tfrac{1}{r})\, \phi(\underline{k},\underline{r}) = \tfrac{1}{2}k^2\, \phi(\underline{k},\underline{r}) \qquad (61b)$$

At each energy there are two linearly independent continuum solutions of eq. (61b). It can be shown (Peterkop, 1977) that in order for $f(\underline{k},\underline{k}')$ and $g(\underline{k},\underline{k}')$ to have the meaning of ionization amplitudes we must choose a solution $\phi^{(-)}$ corresponding to an "incident wave plus ingoing wave" boundary condition. We also normalise these solutions so that

$$\int \phi^{(-)*}(\underline{k},\underline{r})\, \phi^{(-)}(\underline{k}', \underline{r})\, d\underline{r} = \phi(\underline{k} - \underline{k}') \qquad (62)$$

The amplitude $f(\underline{k},\underline{k}')$ is the direct ionization amplitude where the electron at position $\underline{r}_1$ has momentum $\underline{k}$ and the electron at position $\underline{r}_2$ has momentum $\underline{k}'$, while the amplitude $g(\underline{k},\underline{k}')$ is the exchange ionization amplitude where the electron at position $\underline{r}_2$ has momentum $\underline{k}$ and the electron at position $\underline{r}_1$ has momentum $\underline{k}'$. From energy conservation

$$k^2 + k'^2 = 2E \qquad (63)$$

where E is the total energy so that the domain of integration in eqs.(59) and (60) is restricted to $0 \leq k' \leq \sqrt{2E}$. It follows from eqs. (59) and (60) that $f(\underline{k},\underline{k}')$ and $g(\underline{k}', \underline{k})$ describes the same physical process. Hence these amplitudes can differ only in a phase so that

$$g(\underline{k},\underline{k}') = e^{i\tau(\underline{k},\underline{k}')}\, f(\underline{k}',\underline{k}) \qquad (64)$$

This result was first given by Peterkop (1961). Finally we note that the phase factor $\eta(\underline{k},\underline{r})$ in eqs. (59) and (60) allows for the fact that at large distances there remains a long range Coulomb interaction between the two electrons and between the electrons and the nucleus giving rise to an additional logarithmic phase.

The ionization cross section can be obtained in the usual way from the asymptotic form given by eqs. (59) and (60). We obtain

$$\sigma = \frac{1}{k_i} \int_{k' \leq \sqrt{2E}} k|f(\underline{k},\underline{k}')|^2 \, d\underline{k}' \, d\hat{\underline{k}} \tag{65}$$

or

$$\sigma = \frac{1}{k_i} \int_{k' \leq \sqrt{2E}} k|g(\underline{k},\underline{k}')|^2 \, d\underline{k}' \, d\hat{\underline{k}} \tag{66}$$

It follows from eq. (64) that these two expressions are equal. Transforming to an integration over energy and using eq. (64), we may rewrite the cross section in the symmetrical form

$$\sigma = \int_0^{E/2} \frac{k\,k'}{k_i} [\int (|f(\underline{k},\underline{k}')|^2 + |g(\underline{k},\underline{k}')|^2) d\hat{\underline{k}} \, d\hat{\underline{k}}'] \, d\varepsilon' \tag{67}$$

where $\varepsilon' = k'^2/2$. We see that the integration in eq. (67) is over half the full energy range $0 \leq \varepsilon' \leq E/2$ where direct ionization is represented by the first term and exchange ionization by the second term.

Since the Hamiltonian describing the motion of two electrons in the field of a nucleus is symmetrical with respect to interchange of the two electron coordinates, the total wave function can be separated into symmetrical and antisymmetrical spatial parts

$$\Psi^{\pm}(\underline{r}_1,\underline{r}_2) = \Psi(\underline{r}_1, \underline{r}_2) \pm \Psi(\underline{r}_2,\underline{r}_1) \tag{68}$$

From the Pauli exclusion principle Ψ^+ must be multiplied by a singlet spin function and Ψ^- must be multiplied by a triplet spin function. The asymptotic form of $\Psi^{\pm}$ as $r_1 \to \infty$ is obtained from eqs. (59) (60) and (68) where the corresponding ionization scattering amplitude is

$$f^{\pm}(\underline{k},\underline{k}') = f(\underline{k},\underline{k}') \pm g(\underline{k},\underline{k}') \tag{69}$$

The ionization cross section for singlet and triplet scattering is

$$\sigma^{\pm} = \int_0^{E/2} \frac{k\,k'}{k_i} [\int |f(\underline{k},\underline{k}') \pm g(\underline{k},\underline{k}')|^2 \, d\hat{\underline{k}} \, d\hat{\underline{k}}'] \, d\varepsilon' \tag{70}$$

which can be written as

$$\sigma^{\pm} = \sigma_o \pm 2\sigma_{int} \tag{71}$$

where σ_o is given by eqs. (65) - (67) and σ_{int} is the interference term between the direct and exchange amplitudes. The cross section for unpolarized electrons and targets is then given in the usual way by

$$\sigma = \frac{3}{4}\sigma^{+} + \frac{1}{4}\sigma^{-} \quad . \tag{72}$$

We now have to determine the direct and exchange amplitudes. Peterkop (1962) and Rudge and Seaton (1964) derived integral expressions for these amplitudes which have formed the basis of most calculations. They found that

$$f(\underline{k},\underline{k}') = -(2\pi)^{-\frac{1}{2}} \left(\frac{k}{\sqrt{2E}}\right)^{2iz/k} \left(\frac{k'}{\sqrt{2E}}\right)^{2iz'/k'} \int \Psi (H_2 - E)\Phi^* d\underline{r}_1 d\underline{r}_2 \tag{73}$$

where z and z' must be chosen so that

$$\frac{z}{k} + \frac{z'}{k'} = \frac{1}{k} + \frac{1}{k'} - \frac{1}{|\underline{k}-\underline{k}'|} \tag{74}$$

If we take $\Phi = \Phi^{(-)}(z,\underline{k},\underline{r}_1)\ \Phi^{(-)}(z',\underline{k}',\underline{r}_2)$ where $\Phi^{(-)}(z,\underline{k},\underline{r})$ is a solution of eq. (61b) with the potential 1/r replaced by z/r with incident wave plus ingoing wave boundary conditions, then H_2-E may be replaced by

$$\frac{1}{|\underline{r}_1-\underline{r}_2|} - \frac{1-z}{r_1} - \frac{1-z'}{r_2} \tag{75}$$

With the choice of phase defined by eq. (73) then $\tau(\underline{k},\underline{k}')$ in eq. (64) is zero provided that Ψ is an exact solution of the Schrödinger equation. Hence eq. (73) defines both the direct and the exchange amplitudes.

In practice some opproximation must be made for the total wave function Ψ in eq. (73). The simplest approximation is to represent Ψ by the incident plane wave term in eq. (59). This gives a variety of Born type approximations depending on the choice of screening in the definition of Φ, the choice of phase $\tau(\underline{k},\underline{k}')$ in eq. (64), which can now no longer be assumed to be zero, and whether the cross section is defined by eq. (65), eq. (70) or by some other form. Some of these approximations have been discussed by Geltman et al (1963) and Rudge and Seaton (1964). Although they have been used to describe low energy collisions they are basically high energy approximations.

A more sophisticated choice of Ψ is to base it on expansion (12) where the unknown functions are obtained by solving the coupled integrodifferential equations (21). This approach was adopted by Burke and Taylor (1965) who calculated the ionization cross section of H and He^+ from the ground and from the 2s and 2p excited states by adopting a 3 state 1s, 2s, 2p close coupling expansion for Ψ.

Jacobs (1974) and Jacubowicz and Moores (1981) have extended the use of close coupling wave functions in electron impact ionization. In the latter work these functions were used in a symmetric way to represent both the initial and final states in the integral expression. They rewrote the ionization process as

$$e^- + A_i \rightarrow e^- + A_f \rightarrow e^- + A_j^+ + e^- \qquad (76)$$

where the initial state A_i is regarded as a bound state of the $(A_j^+ + e^-)$ system while the final state A_f is a continuum state of this system. These two states are then represented by close coupling wave functions while the incident and scattered electrons in eq. (76) are represented by Coulomb functions. The direct ionization amplitude for an electron incident upon an (N +1)-electron atom or ion is then defined by

$$f(\underline{k},\underline{k}') = -(2\pi)^{-\frac{1}{2}} \int \Psi_i^{\Gamma^*}(X_{N+1}) [\phi^{(-)^*}(Z\text{-}N\text{-}1,\underline{k}_i, x_{N+2}) \left(\sum_{i=1}^{N+1} \frac{1}{r_{i\,N+2}}\right)$$

$$\times \phi^{(-)}(Z\text{-}N\text{-}1,-\underline{k},x_{N+2})\, dx_{N+2}]\, \Psi_f^{\Gamma'}(X_{N+1})\, dX_{N+1} \qquad (77)$$

where Ψ_i^{Γ} and $\Psi_f^{\Gamma'}$ are expanded as in eq. (12) where in the case of Ψ_i^{Γ} the boundary condition defined by eq. (25) or eq. (28) must have all channels closed corresponding to the bound state of the (N+1)-electron system A_i.

This approximation is most appropriate when the incident electron and scattered electron are fast and the ejected electron is slow. In this case it describes excitation autoionization where A_f is a resonant state lying in the continuum which autoionizes giving $A_j^+ + e^-$. But it does not include strong interaction effects between the incident or the scattered electron and the target which occur at low impact energies. Hence it cannot give reliable results at or near to the ionization threshold. In addition the choice of the phase $\tau(\underline{k},\underline{k}')$ in eq. (64) is not well defined by this approximation and hence this quantity is treated on a parameter.

An alternative description of excitation autoionization adopted by (Henry, 1979, Griffin et al, 1982 and Burke et al, 1983) is to neglect autoionization during the excitation process. Hence the state A_f in eq. (76) can be treated as a bound state which can be excited in the usual way. This enables the excitation cross section for the transition $i \rightarrow f$ to be determined by solving eqs. (21) where the states i and f have been included in expansion (12). The ionization cross section is then determined as a second step by calculating the ratio of the autoionization rate to any stabilization rate of the state f. As an example the electron impact ionization cross section for Ca^+ is illustrated in figure 10.

In this figure, the calculations of Burke et al (1983) are compared with the measurements of Peart and Dolder (1975). Also shown are the Born results for the direct ionization cross section

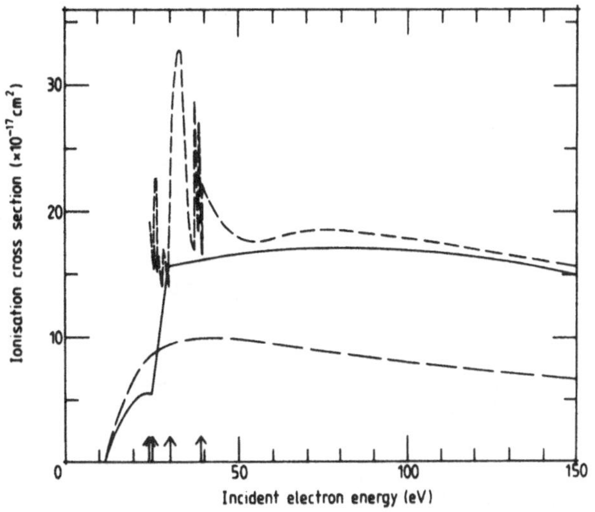

Figure 10. Electron impact ionization cross section of Ca^+ (fig. 2 from Burke et al, 1983)

calculated by McGuire (1977). Close to the ionization threshold only direct ionization $3p^6 4s \rightarrow 3p^6 k\ell$ is possible. However above about 25 eV excitation to the autoionizing states $3p^5 3d4s$ and $3p^5 4s^2$ becomes possible. This gives rise to a strongly enhanced ionization cross section. The structure close to threshold for excitation autoionization arises because the incident electron is captured into the states of the configuration $3p^5\ 3d^2 4s$, as indicated by eq. (5), which decays with the emission of two electrons. Although the experiment does not resolve this structure in Ca^+ it has been clearly seen in Ba^+ by Peart and Dolder (1975).

Unfortunately no theory at present properly includes all interactions arising in these processes. Thus the post-collision interaction effect seen by Smith et al (1974) where the scattered and ejected electrons in the final state in eq. (76) interact strongly together can still not be accurately predicted. However electron correlation at higher energies seen in recent (e, 2e) experiments (Ehrhardt, 1983) can now be largely understood and will be discussed in section 9.

We conclude this section with a brief discussion of the threshold dependence of the ionization cross section. We have already noted that the saddle point in the potential energy surface given in hyperspherical coordinates in figure 6 is important for threshold ionization. For two electrons moving in the field of a nucleus of charge Z the potential interaction can be expanded near this saddle point as

$$V(\alpha, \theta_{12}) = Z_o + \frac{1}{2} Z_1(\alpha - {}^{\pi}/4)^2 + \frac{1}{8} Z_2 (\theta_{12} - \pi)^2 + \ldots \quad (78)$$

where

$$Z_o = \frac{4Z-1}{\sqrt{2}} \quad , \quad Z_1 = \frac{12Z-1}{\sqrt{2}} \quad , \quad Z_2 = - \frac{1}{\sqrt{2}} \quad (79)$$

We see that the motion is stable in θ_{12} but unstable in α at constant R. Ionization close to threshold results from an increase in R accompanied by limited deviations of α from ${}^{\pi}/4$ and θ_{12} from π.

Wannier (1953), in an analysis of the threshold behaviour of the cross section, divided configuration space into three regions as illustrated in figure 11. It is not necessary to know the detailed behaviour of the electrons in the reaction zone, but only to make the quasi-ergodic assumption that the distribution of the boundary conditions at $R = R_o$ (i.e. the density of points in phase space) remains finite as $E \rightarrow 0$. Wannier then makes the assumption that for sufficiently large R_o the Coulomb potential varies slowly enough for classical mechanics to be valid in the Coulomb zone. Finally at very large $R \geq R_1$ the kinetic energy is larger than the potential energy and the particles move freely. As $E \rightarrow 0$ the free zone receeds and the Coulomb zone extends to infinity.

When the total orbital angular momentum L is zero then the motion of the two electrons can be described by the three variables R, α and θ_{12}. Making the substitution

$$\alpha = {}^{\pi}/2 + u_1 \quad , \quad \theta_{12} = \pi + u_2 \quad (80)$$

and retaining terms of equal order enables the classical equations of motion to be written as

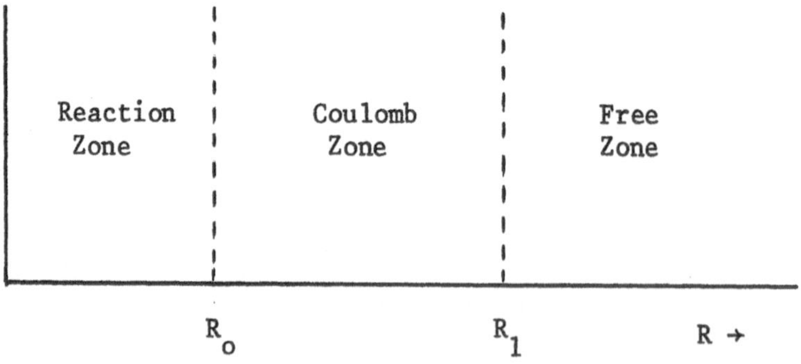

Figure 11. Division of configuration space for ionization

$$\frac{d^2R}{dt^2} = - \frac{Z_o}{R^2}$$

$$\frac{d}{dt}\left(R \frac{du_i}{dR}\right) = \frac{Z_i u_i}{R}, \quad i = 1, 2 \tag{81}$$

The solution of these equations as $E \to 0$, when combined with the quasi-ergodic assumption mentioned above, leads to the threshold behaviour for the cross section

$$\sigma = E^m \tag{82}$$

where

$$m = -\frac{1}{4} + \frac{1}{4}\sqrt{\frac{100Z-9}{4Z-1}} \tag{83}$$

When $Z = 1$ then $m = 1.127$ and as $Z \to \infty$ then $m \to 1$. A further important result is that the width of the angular distribution in θ_{12} is

$$\Delta\theta_{12} \sim E^{\frac{1}{4}} \tag{84}$$

ie the trajectories become more concentrated about $\theta_{12} = \pi$ as $E \to 0$.

This result has been confirmed using the quantal WKB approximation by Rau (1971) and by Peterkop (1971). In addition experiments by Cvejanovic and Read (1974) for e^- - He ionization in which the two outgoing electrons are observed in coincidence are consistent with both eqs. (83) and (84).

Klar and Schlecht (1976), Greene and Rau (1982) and Feagin (1984) have considered how the Wannier theory extends to other $LS\pi$ values. The initial expectation is that just as in the case of electron impact excitation of positive ions the threshold bahaviour is independent of these quantum numbers. However care has to be taken to ensure that the symmetry of the wave function does not force it to be zero at the saddle point. Greene and Rau showed that in all states except $^3S^e$ and $^1P^e$ the wave function can remain finite at the saddle point and hence the Wannier threshold law will pertain. However in these two states there will be a node at $\alpha = \pi/4$ because of the requirements of the Pauli exclusion principle. Peterkop (1983) has shown that the presence of this node will give rise to an exponent 3m where m is given eq. (83)

Evidence that the singlet and triplet ionization cross sections have the same threshold behaviour comes from measurements of the asymmetry parameter A defined by

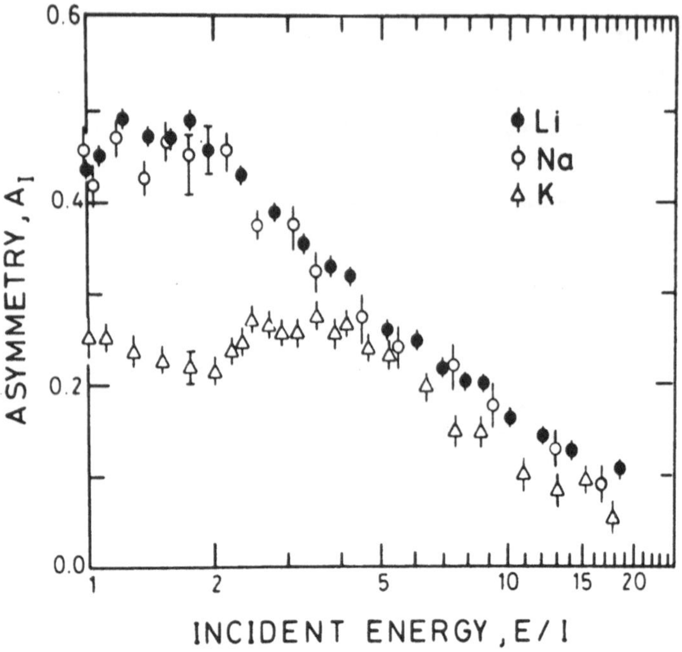

Figure 12. Ionization asymmetry for Li, Na and K (fig. 2 from Kessler, 1984)

$$A = \frac{\sigma^{\uparrow\downarrow} - \sigma^{\uparrow\uparrow}}{\sigma^{\uparrow\downarrow} + \sigma^{\uparrow\uparrow}} = \frac{1 - r}{1 + 3r} \qquad (85)$$

where r is the ratio of the triplet to singlet cross section. Alguard et al (1977) measured this quantity for atomic hydrogen while Hils and Kleinpoppen (1978), Hils et al (1982) and Baum et al (1983) measured it for Li Na and K. The results of Baum et al (1983) are shown in figure 12. This figure shows that the asymmetry tends to a value consistent with the above discussion at threshold.

The Wannier theory for the threshold behaviour of multiple ionization has also been considered by Klar and Schlecht (1976). They show that the cross section near threshold has the form given by eq. (82) with m = 2.270 for double ionization, m = 2.162 for triple ionization and in the limit where an infinite number of electrons are ionized m = 2.

In the Wannier theory discussed so far it is assumed that ionization proceeds via the configuration where $\alpha = \pi/4$ and $\theta_{12} = \pi$. Recently, Temkin (1982) has questioned this assumption and proposed that close to threshold the cross section may be dominated by configurations where the two electrons have unequal energies. He finds that the threshold law in this case is determined by the dipole field experienced by the outer electron due to the inner electron plus nucleus. This leads to a modulated quasi-linear threshold law

$$\sigma \sim E(\ell n\, E)^{-2}\, [1 + c\, \sin(\alpha\, \ell n\, E + \mu)]. \qquad (86)$$

Some evidence that such configurations play a role in ionization has come from a numerical solution of the time-dependent equation for electron impact ionization carried out by Bottcher (1982). However recent measurements by Kelley et al (1983) for e^--Na ionization show no significant oscillations and are consistent with eq. (82).

8. RESONANCES

There are basically two types of resonance which can occur when an electron is incident upon an atom or ion. The electron can be captured temporarily by the effective field of the target (usually in its ground state)

$$V_{eff}\,(r) = V(r) + \frac{\ell(\ell + 1)}{r^2} \qquad (87)$$

where V(r) is the static potential of the target modified by the usual polarization potential $-\alpha/r^4$ and $\ell(\ell + 1)/r^2$ is the centrifugal barrier.

A typical effective potential is illustrated in figure 13 which in this example supports two bound states B_1 and B_2 and one resonance state R. Examples of such resonances are in e^-- Ar scattering which has a broad d-wave resonance at about 10 eV and e-N_2 scattering which has a narrow $^2\Pi_g$ resonance, which is mainly d-wave, at about 2.4eV.

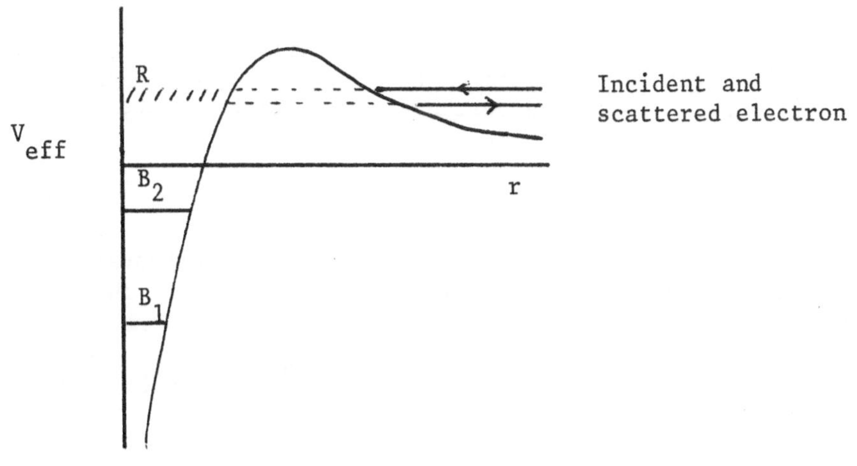

Figure 13. Effective Potential giving rise to Shape Resonances

The other type of resonance which dominates electron-ion collisions and plays an important role in electron collisions with many neutral atoms is the Feshbach resonance. This occurs, as illustrated in figure 1 when the incident electron is captured by the target which is in an excited state. As an example, in e-He^+ scattering below the n = 2 threshold of He^+ we have

$$e + He^+(1s) \rightarrow He^*(2s\, n\ell + 2p\, n\ell') \rightarrow e^- + He^+(1s). \tag{88}$$

Here the intermediate resonance state is a doubly excited state where one electron is in a 2s or 2p orbital and the other is in a Rydberg orbital where n = 2, 3, 4 etc. In a similar way we obtain Rydberg series of resonances converging to the excited thresholds of all ions.

In the case of a neutral atom the situation is more complicated since in the absence of a long-range Coulomb potential the resonances may lie above some of the thresholds of the target to which they are strongly coupled. For example in e - He scattering near the n = 2 threshold we have

$$e^- + He(1s^2) \rightarrow He^{-*}\ (1s\, 2\ell\, 2\ell') \rightarrow e^- + He(1s^2) \tag{89}$$

In this case, the $1s\, 2s^2\ {}^2S^e$ resonance at 19.3 eV lies below the $1s\, 2s\ 2^3S$ threshold at 19.82 eV but the $1s\, 2s\, 2p\ {}^2P^o$ resonance at 20.2 eV lies above this threshold while the $1s\, 2p^2\ {}^2D^e$ resonance at 20.8 eV also lies above the $1s\, 2s\ 2^1S$ threshold at 20.62 eV. Further, except in the case of atomic hydrogen, where the long-range degenerate dipole interaction causes an infinite series of resonances to exist below each excited threshold (Gailitis and Damburg, 1963), the infinite Rydberg series of resonances for an ion reduces to a finite, usually small, number of resonances at each threshold.

Calculations based on expansion (12) are a very convenient and appropriate way of describing these resonances provided the relevent open and closed channels are included. Following Feshbach (1962) we consider eq. (18) where P now projects onto the open channels retained in expansion (12) and Q projects onto the orthogonal space. We introduce the eigenfunctions of the operator QHQ by

$$QHQ|\xi_i\rangle = \varepsilon_i|\xi_i\rangle \tag{90}$$

where this operator may have a discrete spectrum of eigenvalues as well as a continuous spectrum. In the neighbourhood of an isolated discrete eigenvalue ε_s we may rewrite eq. (18) as

$$\left[PHP - \sum_{i\neq s} PHQ \frac{|\xi_i\rangle\langle\xi_i|}{\varepsilon_i - E} QHP - E\right] P\Psi = PHQ \frac{|\xi_s\rangle\langle\xi_s|}{\varepsilon_s - E} QHP\Psi \tag{91}$$

where the rapidly varying part of the optical potential has been separated out on the right hand side. This equation can be solved in terms of the Green's function G_s of the operator on the left hand side of eq. (91) and in terms of the solution Ψ_s of eq. (91) with the right hand side set equal to zero. We then find that the solution exhibits a resonance at $E_s = \varepsilon_s + \Delta_s$ where the resonance width is

$$\Gamma_s = 2| \langle\Psi_s|H|\xi_s\rangle|^2 \tag{92}$$

and the resonance shift is

$$\Delta_s = \langle\xi_s|QHP\ G_s\ PHQ|\xi_s\rangle\ . \tag{93}$$

This is valid provided that Γ_s and Δ_s are small and the resonances are not overlapping. These equations are equivalent to those obtained by Fano (1961) based on the configuration interaction between a discrete state and the continuum. They have formed the basis of a number of calculations of resonance positions and widths using various approximations for ξ_s and Ψ_s. However the real power of expansion (12) is that solution of the resultant eqs. (18) or (21) yield directly the effect of the resonances on the scattering amplitudes and cross sections independent of the width and separation of the resonances.

In the case of electron-ion scattering it is often more convenient to consider a Rydberg series of resonances rather than an isolated resonance as above. This is one of the motivations behind the development of multichannel quantum defect theory (Seaton, 1983). In this theory it can be shown that the K-matrix defined by eq. (25) can be written

$$\underline{K}^\Gamma = \underline{K}^\Gamma_{oo} - \underline{K}^\Gamma_{oc}\ (\underline{K}^\Gamma_{cc} + \tan \pi\underset{\sim}{\nu}_c)^{-1}\ \underline{K}^\Gamma_{co} \tag{94}$$

where the elements of the sub-matrices $\underline{K}^\Gamma_{oo}$, $\underline{K}^\Gamma_{oc}$, $\underline{K}^\Gamma_{co}$ and $\underline{K}^\Gamma_{cc}$ are obtained by analytically continuing the K-matrix from above the threshold where all channels are open to below the threshold where some channels are closed. The corresponding partitioning of the extrapolated K-matrix is

$$\begin{bmatrix} \underline{K}^\Gamma_{oo} & \underline{K}^\Gamma_{oc} \\ \underline{K}^\Gamma_{co} & \underline{K}^\Gamma_{cc} \end{bmatrix} \tag{95}$$

Also, if a channel i is closed ($k_i^2 < 0$) then ν_i is defined by

$$k_i^2 = -\frac{(Z-N)^2}{\nu_i^2}\ . \tag{96}$$

It follows that the physical K-matrix defined by eq. (94) has the dimension of the number of open channels but contains information about the closed channels through the second term on the right hand side.

The role of the diagonal matrix $\tan\pi\nu_i$ is to cause the determinant of the matrix $K^{\Gamma}_{cc} + \tan\pi\nu_c$ to vanish periodically in $1/k_i$ leading to the Rydberg series of resonances.

If K is calculated by solving eqs. (21) at a few energies above threshold then it can be extrapolated yielding an infinite series of resonances below threshold. This theory can also be used to fit experimental data above and below threshold with a few parameters.

Using a similar approach Gailitis (1963) has shown how the collision strength can be averaged over resonances. If the widths of the resonances are small compared with their separation then an averaged collision strength can be defined by

$$\bar{\Omega}(i,f) = \Omega^{>}(i,f) + \sum_{i'} \frac{\Omega^{>}(i,i')\Omega^{>}(i',f)}{\sum_{i''}\Omega^{>}(i',i'')} \quad . \tag{97}$$

Here $\Omega^{>}$ are collision strengths calculated above the new threshold and extrapolated to energies below this threshold, i' is summed over degenerate closed channels of the threshold and i'' is summed over all open channels.

As an example of QDT methods we show in figure 14 the collision strength for the $2p^2\ {}^3P - 2p^2\ {}^1D$ transition in $e^- - O^{2+}$ scattering near to the $2p^2\ {}^1S$ threshold in the $LS\pi = {}^2P^o$ state.

Accurate solutions of the coupled integrodifferential equations (21) were obtained by Eissner and Seaton (1974) with the inclusion

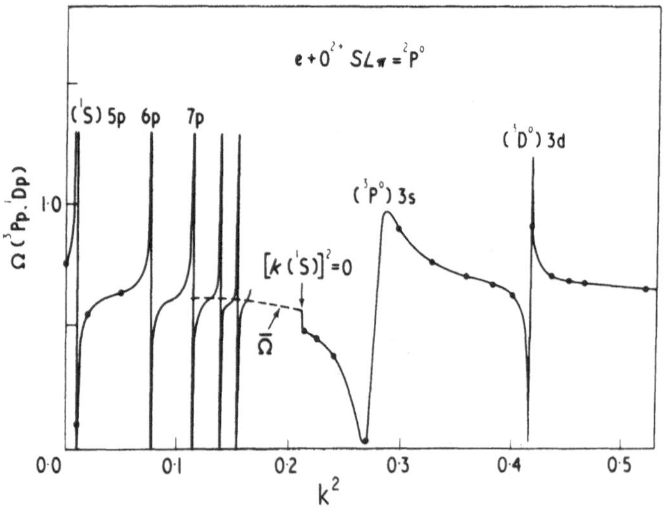

Figure 14. Collision strengths for $e^- - O^{2+}$ scattering (fig. 1a from Eissner and Seaton, 1974)

of nine channels belonging to the $1s^2\ 2s^2\ 2p^2$ and $1s^2\ 2s\ 2p^3$ configurations of $O^{2+}$and these are plotted as a solid line in the figure. There is a Rydberg series of resonances converging to the $2p^2\ {}^1S$ threshold and two series in the region above this threshold converging to higher thresholds. The Gailitis average extrapolated from above the $2p^2\ {}^1S$ threshold is shown as a broken curve.

We now discuss three examples of recent work on resonances in electron atom scattering. Reviews of earlier work have been given by Burke (1968) and Schulz (1973).

In the case of two electron atoms and ions such as H^- and He considerable effort has been devoted to understanding the radial and angular correlation of electrons in doubly excited states. This work is a natural extension of the program started by Cooper et al (1963) who first classified the He absorption experiments of Madden and Codling (1965). This work has been based mainly on the hyperspherical coordinate approach already discussed in these lectures or using group theoretical methods based on the O(4) group (Wulfman, 1973, Herrick and Kellman, 1980 and Kellman and Herrick, 1980).

We show in figure 15 the surface-charge density plots $|\Phi_\mu^\Gamma(r,\Omega)|^2$ defined by eq. (55) for ${}^1S\ H^-$. These plots are for the 3 adiabatic potential curves converging to the n = 3 threshold, where these curves have already been given in figure 7. In each case the plots are given for R = 20, 30 and 40 a.u. We see that the lowest channel $\mu = (3,1)$ shows a large charge density for $\theta_{12} \sim \pi$. This channel corresponds to the $3s^2\ {}^1S$ doubly excited resonance state. The highest channel $\mu = (3,3)$ has a large charge density for $\theta_{12} \sim 0$.

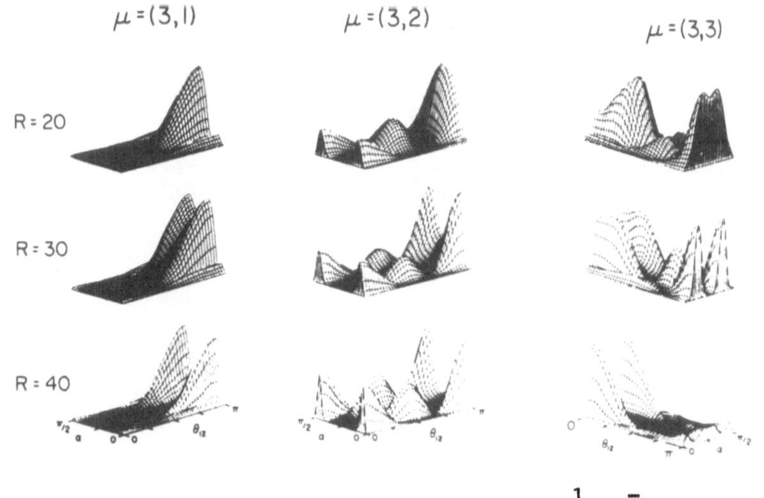

Figure 15. Surface-charge-density plots for ${}^1S\ H^-$ (fig. 2 from Lin 1982b)

Finally the middle channel $\mu = (3,2)$ has a distinct peak near $\theta_{12} = \pi/2$. For R = 20 the (3,1) channel shows a pronounced peak at $\alpha = \pi/4$, $\theta_{12} = \pi$ (the Wannier point). It was pointed out by Kellman and Herrick (1980) that this type of correlation is similar to a linear XYX molecule which can undergo rotations and bending vibrations. As R increases the charge density along $\alpha = \pi/4$ decreases and eventually the density corresponds to one electron in a bound state with the other far from the nucleus. Lin (1982a,b) has found that the lowest channel (n,1) below each n th excited threshold exhibits a large charge density at the Wannier point at a certain critical radius R_e near where the potential curve has a minimum. This critical radius of course increases as n increases. Hence ionization may be viewed as arising through successive "channel hopping" from one adiabatic curve to the next with both electrons remaining near the Wannier point as R increases.

Similar doubly excited states occur in e^- - He scattering except that now the nucleus is screened by the inert 1s electron core. In addition in this case very accurate experiments have been carried out by Brunt et al (1977) and Buckman et al (1983) which have observed groups of resonances associated with all excited thresholds up to n = 7. We show in figure 16 the metastable atom yield obtained by Buckman et al (1983) from 22eV to 25eV. The excited states of He, the doubly excited resonance states of He^- and the ionization threshold are marked in this figure. Recently a calculation based

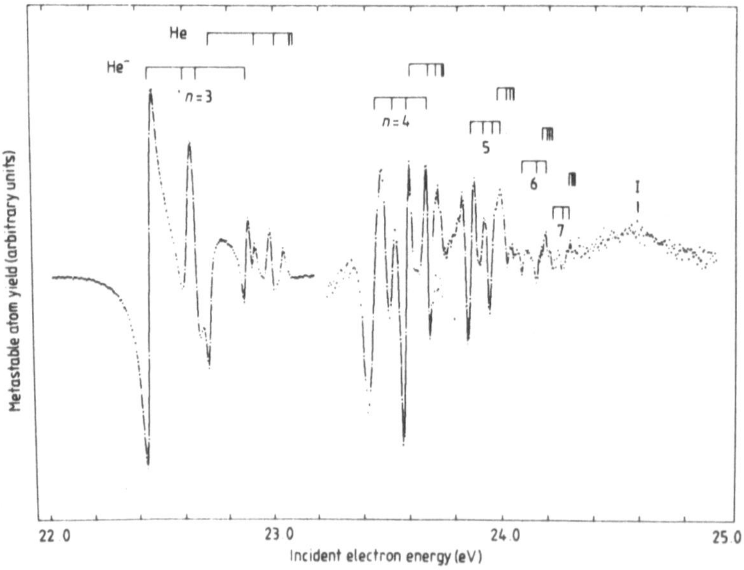

Figure 16. Metastable yield in e^- - He scattering (fig. 2 from Buckman et al 1983)

on expansion (12) has been carried out by Freitas et al (1984) where all 11 target states associated with the n = 1, 2 and 3 levels of He were included in the first expansion. Their results for the $1^1S - 2^3S$ and $1^1S - 2^1S$ cross sections from threshold to 23.5 eV are shown in figure 17.

Also shown in this figure are earlier 5 state results of Fon et al (1981) and a measurement by Johnston and Barrow (1983). The structure near the n = 3 threshold is in very good agreement with the experiments shown in figure 15. In addition a detailed analysis of the resonances show that they divide into two classes. The first class is of very narrow resonances which are closely associated

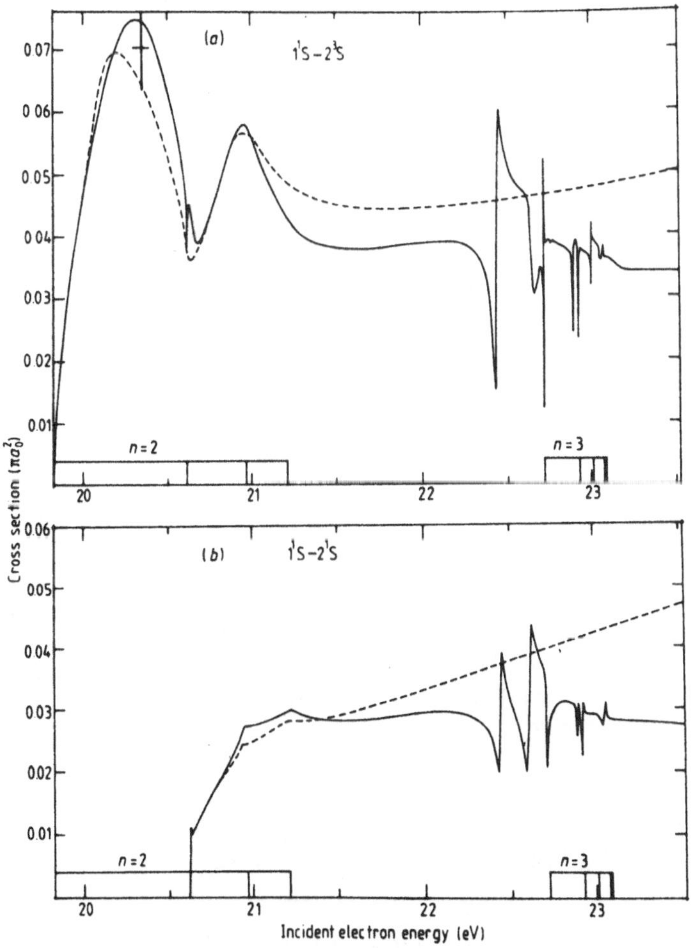

Figure 17. $1^1S - 2^3S$ and $1^1S - 2^1S$ excitation cross sections for e^- - He (fig. 2 of Freitas et al, 1984)

with a single threshold. Examples are a $^2S^e$ virtual state near the 2^1S threshold and a $^2S^e$ virtual state near the 3^3S threshold. These resonances are states where one electron is bound near the nucleus and the other is far away from the nucleus. The second class of resonance is broader, like the 1s $3s^2$ $^2S^e$ resonance at 22.44 eV, and are associated with the configuration where the electrons are equidistant from the nucleus near the Wannier point.

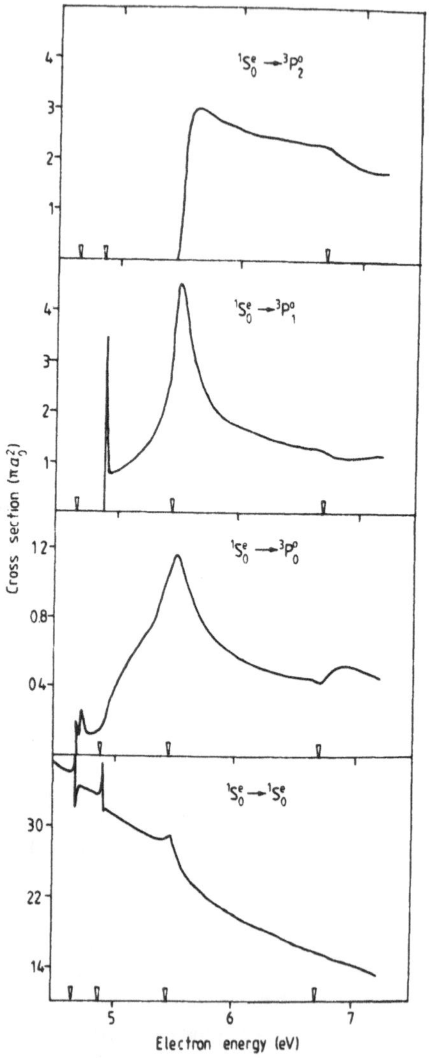

Figure 18. Calculated elastic and inelastic e^- - Hg cross sections (fig. 2 from Scott et al, 1983)

We conclude by discussing an example of resonances in electron scattering by heavy atoms where relativistic effects must be included. The case which has been studied in greatest detail experimentally is e^- - Hg scattering where polarized electron and atom beams have been used and where the polarization of the photons from the decay of excited states has been observed (Hanne, 1983). Recently the first excitation calculations have been carried out by Scott et al (1983) and Bartschat et al (1984) in which relativistic effects, strong coupling effects and resonances have been simultaneously included. They included 5 states $6s^2\ ^1S^e$, $6s\ 6p\ ^3P^o_{0,1,2}$ and $6s\ 6p\ ^1P^o_1$ in expansion (12) and based their calculations on the Breit-Pauli model Hamiltonian representing 3 electrons interacting in the field of an inert core of 78 electrons

$$H_3 = \sum_{i=1}^{3} [-\tfrac{1}{2} \nabla_i^2 + V(r_i) + \tfrac{1}{2} \alpha^2 Z r_i^{-3} (\underline{\ell}_i \cdot \underline{s}_i)] + \sum_{i>j=1}^{3} \frac{1}{r_{ij}} \qquad (98)$$

where V(r) is a model potential representing the core electrons calculated from Thomas-Fermi target orbitals. The target energy separations were adjusted to agree with experiment which partially represents other terms not included in eq. (98). The calculation which was carried out using the R-matrix method involved up to 13 coupled channels for each J and π values.

The results show resonances associated with the $6s^2 6p\ ^2P^o_{1/2,3/2}$, $6s\ 6p^2\ ^4P^e_{1/2\ 3/2\ 5/2}$, $6s\ 6p^2\ ^2P^e_{1/2\ 3/2}$, $6s\ 6p^2\ ^2D^e_{3/2\ 5/2}$ and $6s\ 6p^2\ ^2S^e_{1/2}$ states. The $6s\ 6p^2$ resonances dominate the excitation cross section to the $6s\ 6p\ ^3P_{0,1,2}$ states shown in figure 18. The estimated positions of the contributing resonances are given in Table 1.

Bartschat et al (1984) have calculated the integrated Stokes' parameters for the $6s\ 6p\ ^3P^o_1 \rightarrow 6s^2\ ^1S^e_0$ transition excited by polarized electrons and have found good agreement with experiments of Wolcke et al (1983). As an example, we show in figure 19 the comparison for η_1 defined as the degree of linear polarization of

Table 1. Calculated positions (eV) and assignments of the $6s\ 6p^2$ resonances of Hg^-.

$J\pi = \frac{1}{2}+$	$J\pi = \frac{3}{2}+$	$J\pi = \frac{5}{2}+$
4.7 (^{4}P)	4.7 (^{4}P)	4.9 (^{4}P)
5.8 (?)	5.0 (^{4}P)	5.5 (^{2}D)
6.7 (?)	5.5 (^{2}D)	
	6.7 (?)	

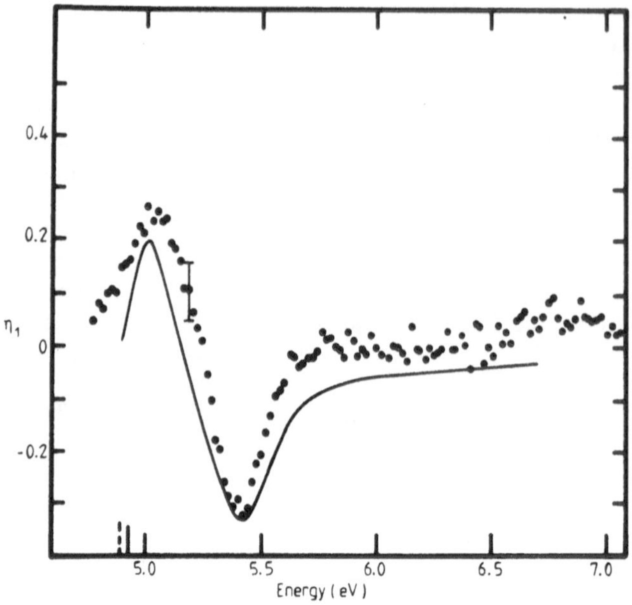

Figure 19. Integrated Stokes' parameter η_1 for 6s 6p $^3P^o_1 \rightarrow 6s^2\ ^1S^e_0$ in Hg (fig. 2 from Bartschat et al, 1984)

the light with respect to two orthogonal axis orientated at $\pm$ 45° to the incident electron beam direction. The incident electrons are traversely polarized and the light is observed in the direction of this polarization.

9. ELECTRON SCATTERING AT INTERMEDIATE ENERGIES

We conclude these lectures by considering electron atom scattering at intermediate energies. At low energies, where only a few channels are open, expansion (12) provides the most appropriate description of the collision. Here all open channels can be retained in the first expansion and the quadratically integrable functions allow for other short range correlation effects. We have seen that the resultant coupled integrodifferential equations (21) can then be solved to yield accurate results. At high energies an infinite number of channels are open, including ionizing channels, but these are weakly coupled so that the Born approximation is usually accurate. However at intermediate energies, which range from about the ionization threshold to an energy typically about ten times this threshold, there are clearly too many channels to include explicitly in the expansion of the wave function, and the coupling between these channels is too strong to be treated by the Born approximation.

There are basically two approaches to treat scattering at intermediate energies (Burke and Williams, 1977, Kingston and Walters,1982,

Walters, 1985). The first is to extend the low energy methods based on expansion (12) to high energies. The second is to push down the validity of the Born approximation by including second-order and higher order terms.

There have been a number of attempts to extend the low energy methods. The first is to represent the infinity of omitted states in the first expansion in eq. (12) by including a few pseudo-states. These are not eigenstates of the target Hamiltonian but instead are chosen to optimize the representation of the short range correlation effects or long range polarization effects (Burke and Schey, 1962, Damburg and Karule, 1967). In this way part of the continuum is included in expansion (12) since each pseudo-state can be expanded in terms of target eigenstates where the expansion coefficients include continuum contributions. In the case of atomic hydrogen the pseudo-state which gives the complete long range polarization can be written down exactly and was first considered by Temkin and Lamkin (1961) in their polarized orbital treatment of $e^- - H$ scattering. As an example we show in figure 20 calculations carried out including 1s and 2s eigenstates and $\overline{3s}$, $\overline{4s}$ and $\overline{5s}$ pseudo-states for ^1S $e^- - H$ elastic scattering in the radial limit by Burke and Mitchell (1973). This model in which only s states are included in expansion (12) was first considered by Temkin (1962). Also shown are calculations using square integrable functions alone to calculate the Fredholm determinant from which the phase shift can be extracted by analytic continuation (Rescigno and Reinhardt, 1974). The main difficulty with the pseudo-state method is obvious from this figure. It introduces unphysical thresholds and pseudo-resonances in the energy region of interest which have to be averaged in some way. Nevertheless the method

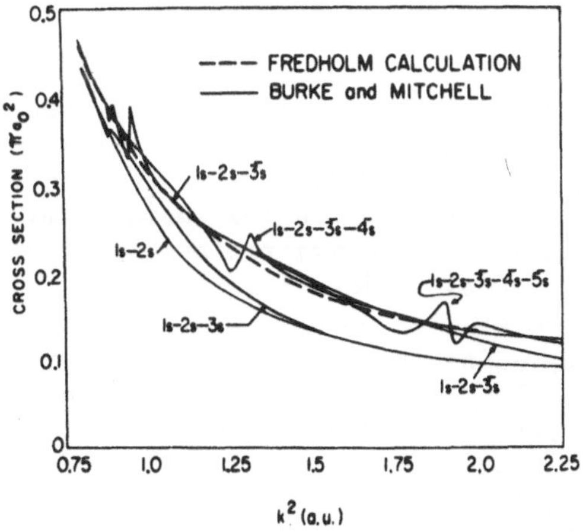

Figure 20. ^1S $e^- - H$ elastic cross section in the radial limit (fig. 2 from Rescigno and Reinhardt, 1974)

has been widely used to calculate elastic scattering and excitation cross sections at energies away from these singularities. In addition, Callaway and Oza (1984) have shown that the method can give reliable cross sections in the radial limit for e^- - H ionization by an appropriate projection of the pseudo-states onto the continuum.

Two other related low energy approaches to intermediate energies have been developed by Burke et al (1981) and Bransden and Stelbovics (1984). They note that the quadratically integrable functions in eq. (12) provide a discrete representation for the optical potential defined by eq. (20) which is given by

$$V_{opt} = - \sum_i \frac{PHQ\chi_i><\chi_i QHP}{\varepsilon_i - E} \tag{99}$$

where $\varepsilon_i = <\chi_i \, QHQ \, \chi_i>$. If the exact optical potential could be included in eq. (21) then of course the transitions between the states included in the first expansion in eq. (12) would be calculated exactly. However the representation (99) gives rise to pseudo-resonances close to the poles at ε_i if used without modification in the solution of eq. (21). In fact, as shown by Feshbach (1962), the exact optical potential has a branch point at the lowest threshold in Q space and is complex for higher energies. Hence the discrete set of poles in eq. (99) can be regarded as an approximate representation of the corresponding branch cut. In the work of Burke et al the "physical T-matrix" is extracted from the T-matrix calculated using eq. (99) by carrying out an appropriate average in the complex energy plane. In the work of Bransden and Stelbovics the "physical optical potential" is obtained by eliminating the poles in eq. (99) using the equivalent weight method of Heller et al (1973a,b). Both approaches have given good agreement with exact calculations for a model two channel problem and the method of Burke et al has been used to obtain cross sections at intermediate energies for many ions and molecules.

We conclude this brief review of low energy methods by mentioning the calculations of Poet (1978, 1980) who obtained essentially exact intermediate energy results for elastic scattering and 1s-2s excitation for e^- - H scattering in the radial limit. In this case the Schrödinger equation can be written as a partial differential equation in the two radial variables r_1 and r_2. Exact separable solutions can then be written down in the two regions $r_1 > r_2$ and $r_2 > r_1$ and matching along the boundary $r_1 = r_2$ gives the required physical solution. This work has provided benchmark results for checking methods which have wider application. Recently ionization cross sections have also been obtained for this model by Callaway and Oza (1984).

Approaching intermediate energies from the high energy region it seems reasonable to try an approach based on perturbation theory.

Ignoring exchange for the moment the Born series for the direct scattering amplitude can be written as (Joachain 1975, 1980).

$$f = \sum_{n=1}^{\infty} \bar{f}_{Bn} \tag{100}$$

where the n th Born term $\bar{f}_{Bn}$ contains the interaction between the incident electron and the target atom V n times and the Green's function $G_o^{+} = (E - H_N + i\varepsilon)^{-1}$ n - 1 times. Also we have written

$$H_{N+1} = H_N + V \tag{101}$$

where H_{N+1} and H_N are defined by eq. (7). It is also convenient to define the quantity f_{Bj} as the sum of the first j terms in eq. (100). The first term $\bar{f}_{B1} = f_{B1}$ is the familiar first Born amplitude which has been widely used to calculate elastic scattering, excitation and ionization (Bell and Kingston, 1974). The second Born term $\bar{f}_{B2}$ for a transition from state $|\underline{k}_i, i\rangle$ to $|\underline{k}_j, j\rangle$ is

$$\bar{f}_{B2} = 8\pi^2 \int d\underline{q} \sum_n \frac{\langle \underline{k}_j, j|V|\underline{q}, n\rangle\langle \underline{q}, n|V|\underline{k}_i, i\rangle}{q^2 - k_i^2 + 2(E_n - E_i) - i\varepsilon} \tag{102}$$

where the sum over intermediate states includes an integration over the continuum.

A useful approximation for $\bar{f}_{B2}$ is to replace the energy differences $E_n - E_i$ in the denominator by an average excitation energy $\bar{E}$ (Massey and Mohr, 1934). The sum over intermediate states can then be performed by closure. An improvement which has been widely used is to treat the first few terms in the sum exactly while treating the remainder by closure (Holt and Moiseiwitsch, 1968).

An important development in recent years has been the recognition of the importance of retaining consistently all terms in the Born series with similar energy dependence. We show in table 2 this dependence for elastic and inelastic scattering taken from the work of Byron and Joachain (1974, 1975)

We see that for elastic scattering the amplitude converges to the first Born approximation for all momentum transfer Δ but at lower energies it is necessary to include Re $\bar{f}_{B3}$ as well as the second Born terms to obtain a result correct to k^{-2}. In the forward direction convergence to the first Born approximation is slow because of contributions to Im $\bar{f}_{B2}$ from intermediate states. For inelastic scattering we see that the first Born approximation does not even give the correct high energy limit for large Δ. Instead this comes from Im $\bar{f}_{B2}$. Physically we can understand this by noting that at large scattering angles inelastic scattering involves a collision with the nucleus, to give the large scattering

Table 2. Energy dependence of terms in the Born series

Elastic	$\bar{f}_{B1}$	Re $\bar{f}_{B2}$	Im$\bar{f}_{B2}$	Re $\bar{f}_{B3}$	Im$\bar{f}_{B3}$
$\Delta<k^{-1}$	1	k^{-1}	$k^{-1}\ln k$	k^{-2}	k^{-3}
$k^{-1}<\Delta<k$	1	k^{-2}	k^{-1}	k^{-2}	k^{-3}
$k<\Delta$	Δ^{-2}	$k^{-2}\Delta^{-2}$	$k^{-1}\Delta^{-2}\ln\Delta$	$k^{-2}\Delta^{-2}\ln^2\Delta$	$k^{-3}\Delta^{-2}\ln\Delta$
Inelastic (s-s transitions)					
$\Delta<k^{-1}$	1	k^{-1}	$k^{-1}\ln k$	k^{-2}	k^{-3}
$k^{-1}<\Delta<k$	1	k^{-2}	k^{-1}	k^{-2}	k^{-3}
$k<\Delta$	Δ^{-6}	$k^{-2}\Delta^{-2}$	$k^{-1}\Delta^{-2}$	$k^{-2}\Delta^{-2}\ln\Delta$	$k^{-3}\Delta^{-2}$

angle, followed or preceded by an inelastic collision with the bound electron. Again in the forward direction Re $\bar{f}_{B3}$ must be retained to give the correct high energy limit to $O(k^{-2})$.

Since the third Born term is very difficult to evaluate Byron and Joachain suggested that to third order the scattering amplitude should be calculated using the "eikonal Born series"

$$f_{EBS} = \bar{f}_{B1} + \bar{f}_{B2} + \bar{f}_{G3} + g_{Och} \tag{103}$$

where f_{G3} is the third-order term in the expansion of the Glauber amplitude (Glauber, 1959, Byron and Joachain, 1977) in power of V. This is much easier to calculate and,on the basis of an analysis of potential scattering, has been conjectured to give the same result as Re $\bar{f}_{B3}$ in the limit of large k for all Δ. Lastly in eq. (103) the exchange effect is taken into account by the Ochkur (1964) amplitude g_{Och}.

An important application of the second Born approximation has been made recently by Byron et al (1982), to the (e, 2e) process in He

$$e^- + He(1^1S) \rightarrow He^+ (1s) + e^- + e^- \tag{104}$$

and a detailed comparison was made with the experimental measurements of Ehrhardt et al (1982). We show in figure 21 the triple differential cross section for an incident electron energy of 500 eV where the scattered electron is observed at 3.5° and the ejected electron has an energy of 5 eV. The dashed curve corresponds to the first Born approximation (multiplied by 0.87), the solid curve to the second Born approximation (the third-order Glauber and the exchange terms are small in this case) and the dots are

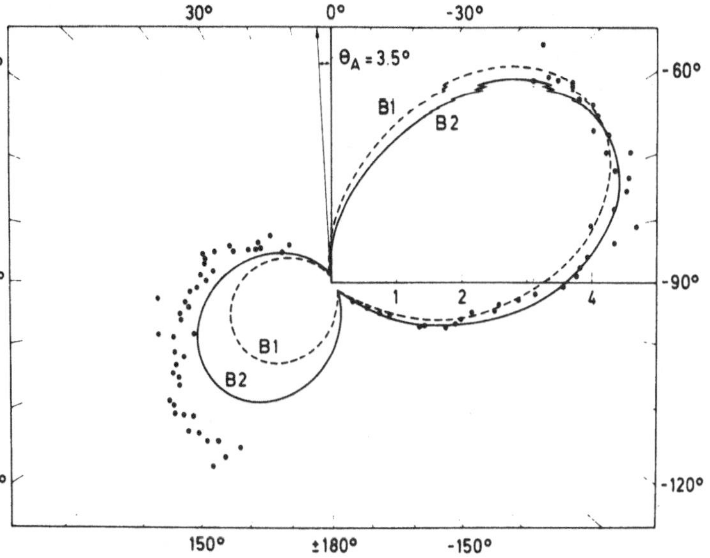

Figure 21. Triple differential cross section for e^- - He ionization (fig. 1 from Ehrhardt et al, 1982)

the experimental data. All results are normalized to the same value at an ejected electron angle of -60° which is near the maximum of this binary peak. It can be seen that the second Born approximation calculations yield very significant improvements over the first Born approximation. Indeed the second Born results exhibit (i) a shift of the binary peak to larger angles (ii) a shift of the recoil peak to larger angles (iii) a major enhancement of the magnitude of the recoil peak with respect to the value predicted by the first Born approximation. However there is still clearly a discrepancy which probably requires a better treatment of the slow ejected electron in the field of the ion as in the methods of Jacobs (1974) and Jacubowicz and Moores (1981) discussed earlier. A quantum defect fit of the data by Klar et al (1984) indicates that this is the case.

Bryon et al (1983) have recently also studied large-angle coplanar symmetric (e, 2e) collisions. They find, as expected, that at large scattering angles the second Born term is more important than the first Born term.

At lower energies or for atoms heavier than H and He perturbation series in powers of the full electron-atom interaction V is no longer appropriate. In this case, it is important to take into account the static potentials in the initial and final states exactly. This leads to the distorted-wave second Born approximation (DWSBA). One approach considered by Kingston and Walters (1980) is to correct the scattering amplitude calculated in the close coupling approximation by terms calculated in second-order allowing for intermediate transitions to all other states not included in the close coupling expansion. Hence we have

$$f_{ij}^{DWSBA} = f_{ij}^{cc} + \langle i|V_2 G_o^+ V_2|j\rangle \tag{105}$$

where the interaction potential V has been split into two parts

$$V = V_1 + V_2 \tag{106}$$

where V_1 is treated exactly in f_{ij}^{cc}.

An alternative approach considered by Bransden and Coleman (1972) and McCarthy and Stelbovics (1983a,b,c) is to approximate the optical potential V_{opt} in eq. (18) or eq. (21) by evaluating it by allowing for second-order transitions to intermediate states. Eqs.(21) are then in principle solved exactly including this second-order potential. A major problem with this method is that these second-order potentials are non-local and complex. This has lead to the study of local energy dependent approximations.

We also mention the well known distorted wave approximation (Mott and Massey, 1965) which has been widely used in electron atom scattering (Madison and Shelton, 1973). This has been successful for studying weak transitions at intermediate energies. It has also been extended by Eissner and Seaton (1972) to enable transitions for highly ionized ions and for transitions involving high total orbital angular momenta to be accurately calculated at all energies. The first-order many body theory of Thomas et al (1974) is similar in principle to the distorted wave approximation except that the scattered electron is assumed to move in the same field in the initial and final states.

Finally as an indication of the difficulties which are still experienced in obtaining good agreement between theory and experiment at intermediate energies we present in figure 22 a comparison between various theories and experiments for the λ and χ parameters for the $1^1S - 2^1P$ transitions in e^- - He scattering. If σ_m is the differential cross section for exciting the m th magnetic sublevel and

$$\sigma_o = |b_o|^2 , \quad \sigma_1 = \sigma_{-1} = |b_1|^2 \tag{107}$$

then the total differential cross section $\sigma = \sigma_o + 2\sigma_1$ and

$$\lambda = \sigma_o/\sigma \quad , \quad \chi = \arg(b_1) - \arg(b_o) \tag{108}$$

These quantities can be measured by observing the decay photon in coincidence with the scattered electron. In this figure ——— is the 5 state 1^1S, 2^3S, 2^1S, 2^3P, 2^1P close coupling calculation of Fon et al (1980); - - - - is the distorted wave calculation of Madison and Calhoun (1978); - · - · is the first-Born approximation;

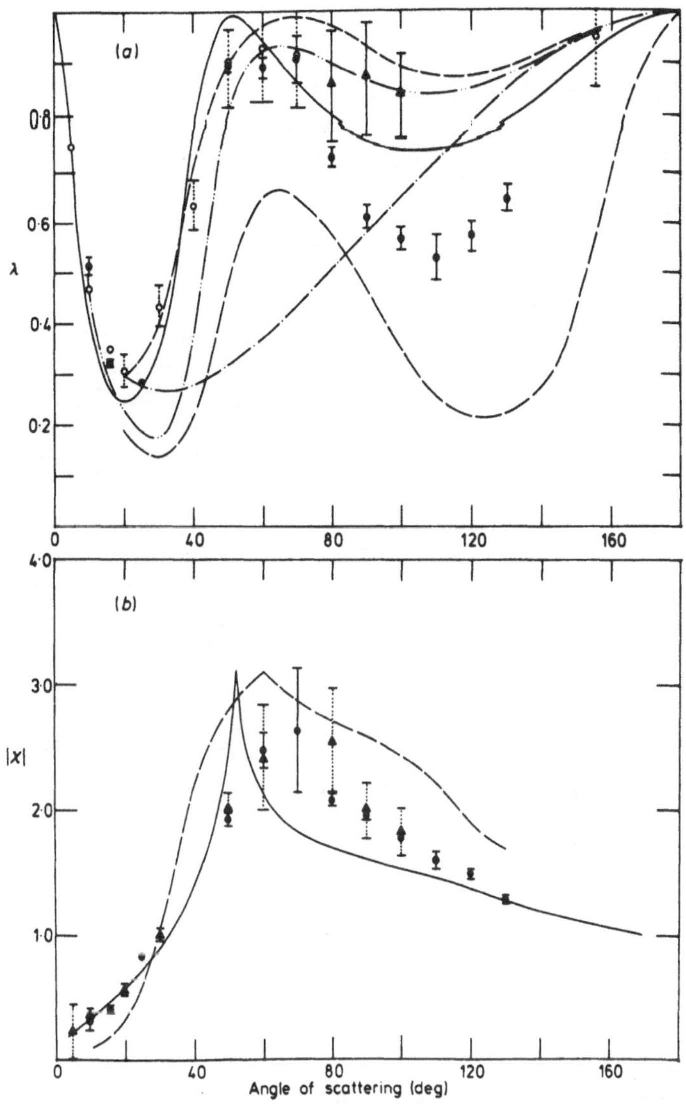

Figure 22. λ and χ parameters for 2^1P excitation in e^- - He scattering (fig. 7 from Fon et al, 1980)

- ·· - is the first order many body theory of Meneses et al (1978); -- is the distorted wave theory of Baluja and McDowell (1979). The experiments are, open circles, Sutcliffe et al (1978); triangles, Steph and Golden (1980), closed circles Hollywood et al (1979). We see that while there is good agreement between theory and experiment at small angles there are still unexplained discrepancies for large angle scattering.

In conclusion although considerable progress has been made in the theory of electron atom scattering at intermediate energies, a fully satisfactory theory for an arbitrary atom has yet to be developed.

REFERENCES

Alguard, M. J., Hughes, V. W., Lubell, M. S. and Wainwright, P.F., 1977, Phys. Rev. Letters, 39, 334-338.

Andrick, D. and Bitsch, A., 1975, J. Phys. B. 8, 393-406.

Armstead, R. L., 1968, Phys. Rev. 171, 91-93.

Baluja, K. L. and McDowell, M.R.C., 1979, J. Phys. B. 12, 835-846.

Bartschat, K., Scott, N. S., Blum, K. and Burke, P. G., 1984, J. Phys. B. 17, 269-277.

Baum, G., Moede, M., Raith, W. and Schroder, 1983, Int. Symp. on Polariz.and Correl. in elec.-atom coll., Munster.

Bell, K. L. and Kingston, A. E., 1974, Adv. Atom. Molec. Phys.10,53-130.

Berrington, K. A., Aggarwal, K. M., Hughes, J. G., Smith, F. J. and Elder, M., 1984 in "Information Quarterly for Atomic Processes and Applications, No.25" (SERC, Daresbury Laboratory, England), 2-5.

Bethe, H. A. and Salpeter, E. E., 1972, Quantum Mechanics of One- and Two-Electron Atoms (Springer-Verlag, Berlin).

Blatt, J. M. and Biedenharn, L. C., 1952, Rev. Mod. Phys. 24, 258-272.

Bottcher, C., 1982, J. Phys. B. 15 L463-L469.

Bransden, B. H., 1983, Atomic Collision Theory (Benjamin, New York, 2nd Edition).

Bransden, B. H. and Coleman, J. P., 1972, J. Phys. B. 5, 537-45.

Bransden, B. H. and McDowell, M.R.C., 1977, Physics Reports 30C,207 -303.

Bransden, B. H. and McDowell, M.R.C., 1978, Physics Reports, 46C, 249-394.

Bransden, B. H. and Stelbovics, A. T., 1984, J. Phys. B. 17, 1877-1888.

Brunt, J.N.M., King, G. C. and Read, F. H., 1977, J. Phys. B. 10, 433-448.

Buckman, S. J., Hammond, P., Read, F. H. and King, G. C., 1983, J. Phys. B. 16, 4039-4047.

Burke, P. G., 1968, Adv. Atom. Molec. Phys. 4, 173-219.

Burke, P. G., Berrington, K. A., LeDourneuf, M. and Vo Ky Lan, 1974, J. Phys. B. 7, L531-L535.

Burke, P. G., Berrington, K. A. and Sukumar, C. V., 1981, J. Phys. B. 14, 289-305.

Burke, P. G. and Eissner, W., 1983, in "Atoms in Astrophysics" edited by P. G. Burke, W. B. Eissner, D. G. Hummer and I. C. Percival (Plenum Press, New York), 1-54.

Burke, P. G., Hibbert, A. and Robb, D. W., 1971, J. Phys. B. 4,153-161.

Burke, P. G., Kingston, A. E. and Thompson, A., 1983, J. Phys. B. 16, L385-L389.

Burke, P. G. and Mitchell, J.F.B., 1973, J. Phys. B. 6, 320-328.

Burke, P. G. and Schey, H. M., 1962, Phys. Rev. 126, 147-162.

Burke, P. G. and Taylor, A. J., 1965, Proc. Roy. Soc. A. 287,105-122.

Burke, P. G. and Taylor, A. J., 1966, Proc. Phys. Soc. 88, 549-562.

Burke, P. G. and Williams, J. F., 1977, Physics Reports 34C, 325-369.

Byron, F. W. Jr., and Joachain, C. J., 1972, Physics Reports, 34C 233-324.

Byron, F. W. Jr., and Joachain, C. J., 1974, J. Phys. B. 7 L212-L215.
Byron, F. W. Jr., and Joachain, C. J., 1975, J. Phys. B. 8 L284-L288.
Byron, F. W. Jr., and Joachain, C. J. and Piraux, B., 1982, J. Phys. B. 15, L293-L296.
Byron, F. W. Jr., Joachain, C. J. and Piraux, B., 1983, J. Phys. B. 16, L769-L774.
Callaway, J. and McDowell, M.R.C., 1983, Comm. Atom. Molec. Phys. 13, 19-35.
Callaway, J. and Oza, D. H., 1984, Phys. Rev. A. 29, 2416-2420.
Carse, G. D. and Walker, D. W., 1973, J. Phys. B. 6, 2529-2544.
Castillejo, L., Percival, I. C. and Seaton, M. J., 1960, Proc. Roy. Soc. A. 254, 259-272.
Chang, J. J., 1975, J. Phys. B. 8, 2327-2335.
Christensen-Dalsgaard, B. L., 1984a, Phys. Rev. A29, 470-487.
Christensen-Dalsgaard, B. L., 1984b, Phys. Rev. A29, 2242-2244.
Clark, C. W. and Greene, C. H., 1980, Phys. Rev. A21, 1786-1797.
Cooper, J. W., Fano, U. and Prats, F., 1963, Phys. Rev. Letters 10, 518-521.
Crandall, D. H., 1981, Phys. Scr. 23, 153-162.
Cvejanovic, S. and Read, F. H., 1974, J. Phys. B. 7, 1841-1852.
Damburg, R. and Karule, E., 1967, Proc. Phys. Soc. 90, 637-640.
Delves, L. M., 1960, Nuc. Phys. 20, 275-308.
Demkov, Yu. N. and Ermolaev, A. M., 1959, Soviet Phys. JETP 9, 633.
Ehrhardt, H., 1983, Comm. Atom. Molec. Phys. 13, 115-125.
Ehrhardt, H., Fischer, M., Jung, K., Byron, F. W. Jr., Joachain, C.J. and Piraux, B., 1982, Phys. Rev. Letters 48, 1807-1810.
Eissner, W. B. and Seaton, M. J., 1972, J. Phys. B. 5; 2187-2198.
Eissner, W. B. and Seaton, M. J., 1974, J. Phys. B. 7, 2533-2548.
Fano, U., 1961, Phys. Rev. 124, 1866-1878.
Fano, U., 1983, Rep. Prog. Phys. 46, 97-165.
Feagin, J. M., 1984, J. Phys. B. 17, 2433-2451.
Feshbach, H., 1962, Ann. Phys. 19, 287.
Fock, V. A., 1954, Izv. Akad. Nauk. SSR Ser. Fiz. 18, 161.
Fock, V. A., 1958, K. Norske Vidensk. Selsk. Forh. 31, 138.
Fon, W. C., Berrington, K. A., Burke, P. G. and Kingston,A. E., 1978, J. Phys. B. 14, 2921-2934.
Fon, W. C., Berrington, K. A. and Kingston, A. E., 1980, J. Phys. B. 13, 2309-2325.
Freitas, L.C.G., Berrington, K. A., Burke, P. G., Hibbert, A., Kingston, A. E. and Sinfailam, A. L., 1984, J. Phys. B. 17, L303-L309.
Gailitis, M., 1963, Sov. Phys. JETP 17, 1328-1332.
Gailitis, M. and Damburg, R., 1963, Proc. Phys. Soc. 82, 192-200.
Geltman, S., Rudge, M.R.H. and Seaton, M. J., 1963, Proc. Phys. Soc. 81, 375-377.
Glauber, R. J., 1959, in "Lectures in Theoretical Physics" Vol. 1 edited by W. E. Brittin (Interscience, New York) 315.
Grant, I. P., 1979, Comput. Phys. Commun. 17, 149-161.
Greene, C. H., 1981, Phys. Rev. A23, 661-678.
Greene, C. H. and Rau, A.R.P., 1982, Phys. Rev. Lett. 48, 533-537.

Griffin, D. C., Bottcher, C. and Pindzola, M. S., 1982, Phys. Rev. A25, 1374-82.
Hayes, M. A., Norcross, D. W., Mann, J. B. and Robb, W. D., 1977, J. Phys. B. 10, L429-L434.
Hanne, G. F., 1983, Physics Reports, 95, 95-165.
Heller, E. J., Reinhardt, W. P. and Yamani, H. A., 1973a, J. Comput. Phys. 13, 536.
Heller, E. J., Rescigno, J. N. and Reinhardt, W. P., 1973b, Phys. Rev. A 8, 2946-2951.
Henry, R.J.W., 1979, J. Phys. B. 12, L309-L313.
Henry, R.J.W., 1981, Physics Reports 68, 1-91.
Henry, R.J.W., van Wyngaarden, W. L. and Matese, J. J., 1978, Phys. Rev. A 17, 798-800.
Herrick. D. R. and Kellman, M. E., 1980, Phys. Rev. A 21, 418-425.
Hils, D. and Kleinpoppen, H., 1978, J. Phys. B. 11, L283-L287.
Hils, D., Jitschin, W. and Kleinpoppen, H., 1982, J. Phys. B. 15, 3347-3357.
Hollywood, M. T., Crowe, A. and Williams, J. F., 1979, J. Phys. B. 12, 819-834.
Holt, A. R. and Moiseiwitsch, B. L., 1968, J. Phys. B. 1, 36-47.
Jacobs, V. L., 1974, Phys. Rev. A 10, 499-507.
Jacubowicz, H. and Moores, D. L., 1981, J. Phys. B. 14, 3733-3760.
Joachain, C. J., 1975, Quantum Collision Theory (North Holland Publ. Co.)
Joachain, C. J., 1980, in "Atomic and Molecular Processes in Controlled Thermonuclear Fusion" edited by M.R.C. McDowell and A. M. Ferendeci (Plenum Press, New York) 147-183.
Johnston, A. R. and Burrow, P. D., 1983, J. Phys. B. 16, 613-628.
Jones, M., 1975, Phil. Trans. Roy. Soc. A277, 587-622.
Kauppila, W. E., Stein, T. S., Jesion, G., Dababneh, M. S. and Pol, V., 1977, Rev. Sci. Inst. 48, 322.
Kelley, M. H., Rogers, W. T., Celotta, R. J. and Mielczarek, S. R., 1983, Phys. Rev. Lett. 51, 2191-2193.
Kellman, M. E. and Herrick, D. R., 1980, Phys. Rev. A22, 1536-1551.
Kessler, J., 1984, Comm. Atom. Molec. Phys. 14, 275-284.
Kingston, A. E. and Walters, H.R.J., 1980, J. Phys. B. 13, 4633-4662.
Kingston, A. E. and Walters, H.R.J., 1982, Comm. Atom. Molec. Phys. 11, 177-191.
Klar, H., Jung, K. and Ehrhardt, H., 1984, Phys. Rev. A29, 405-407.
Klar, H. and Schlecht, W., 1976, J. Phys. B. 9, 1699-1711.
Lin, C. D., 1974, Phys. Rev. A10, 1986-2001.
Lin, C. D., 1981, Phys. Rev. A23, 1585-1590.
Lin, C. D., 1982a, Phys. Rev. A25, 76-87.
Lin, C. D., 1982b, Phys. Rev. A26, 2305-2314.
Lin, C. D., 1983, Phys. Rev. Letters 51, 1348-51.
Macek, J. H., 1968, J. Phys. B. 1, 831-843.
Madden, R. P. and Codling, K., 1965, Astrophys. J. 141, 364-375.
Madison, D. H. and Calhoun, R. V., 1978, private communication to Sutcliffe et al, 1978.
Madison, D. H. and Shelton, W. N., 1973, Phys. Rev. A7, 499-513.
Massey, H.S.W. and Mohr, C.B.O., 1932, Proc. Roy. Soc. A136,289-311.

Massey, H.S.W. and Mohr, C.B.O., 1934, Proc. Roy. Soc. A146,880-900.
Meneses, G. D., Padial, N. T. and Csanak, Gy., 1978, J. Phys. B. 11, L237-L242.
Milloy, H. B. and Crompton, R. W., 1977, Phys. Rev. A15, 1847-1850.
Moores, D. L. and Norcross, D. W., 1972, J. Phys. B. 5, 1482-1505.
Morse, P. M. and Feschbach, H., 1953, Methods of Theoretical Physics (McGraw-Hill Book Company, New York).
Mott, N. F. and Massey, H.S.W., 1965, The Theory of Atomic Collisions (Oxford University Press, 3rd Edition).
McCarthy, I. E. and Stelbovics, A. T., 1983a, Phys. Rev. A28, 2693-2707.
McCarthy, I. E. and Stelbovics, A. T., 1983b, J. Phys. B. 16, 1233-1245.
McCarthy, I. E. and Stelbovics, A. T., 1983c, J. Phys. B. 16, 1611-1617.
McGuire, E. J., 1977, Phys. Rev. A16, 73-79.
Nesbet, N. F., 1980, Variational Methods in Electron-Atom Scattering Theory (Plenum Press, New York).
Nesbet, R. K., 1979, J. Phys. B. 12, L243-L248.
Norrington, P. H. and Grant, I. P., 1981, J. Phys. B. 14, L261-L267.
Ochkur, V. I., 1964, Sov. Phys. JETP 18, 503-508.
O'Malley, T. F., Burke, P. G. and Berrington, K. A., 1979, J. Phys. B. 12, 953-965.
Peart, B. and Dolder, K., 1975, J. Phys. B. 8, 56-62.
Percival, I. C. and Seaton, M. J., 1957, Proc. Camb. Phil. Soc. 53, 654-662.
Perel, J., Englander, P. and Bederson, B., 1962, Phys. Rev. 128, 1148-1154.
Peterkop, R. K., 1961, Proc. Phys. Soc. 77, 1220-1222.
Peterkop, R. K., 1962, Optika i Spectrosc. 13, 153.
Peterkop, R. K., 1971, J. Phys. B. 4, 513-521.
Peterkop, R. K., 1977, Theory of Ionization of Atoms by Electron Impact (Colorado Associated University Press).
Peterkop, R. K., 1983, J. Phys. B. 16, L587-L593.
Poet, R., 1978, J. Phys. B. 11, 3081-3094.
Poet, R., 1980, J. Phys. B. 13, 2995-3008.
Racah, G., 1942, Phys. Rev. 61, 537-539.
Rau, A.R.P., 1971, Phys. Rev. A4, 207-220.
Rescigno, T. N. and Reinhardt, W. D., 1974, Phys. Rev. A10, 158-167.
Rudge, M.R.H., and Seaton, M. J., 1964, Proc. Roy. Soc. A283, 262-290.
Saraph, H. E., 1972, Comp. Phys. Commun. 3, 256-268.
Saraph, H. E., 1978, Comp. Phys. Commun. 15, 247-258.
Schulz, G. J., 1973, Rev. Mod. Phys. 45, 378-422.
Schwartz, C., 1961, Phys. Rev. 124, 1468-1471.
Scott, N. S. and Burke. P. G., 1980, J. Phys. B. 13, 4299-4314.
Scott, N. S., Burke, P. G. and Bartschat, K., 1983, J. Phys. B. 16, L361-L366.
Seaton, M. J., 1953, Phil. Trans. Roy. Soc. A245, 469-499.
Seaton, M. J., 1974, J. Phys. B., 7, 1817-1840.
Seaton, M. J., 1975, Adv. in Atom. Molec. Phys. 11, 83-142.

Seaton, M. J., 1983, Rep. Prog. Phys. 46, 167-257.
Sinfailam, A. L., 1980, Aust. J. Phys. B. 33, 261-281.
Sinfailam, A. L. and Nesbet, R. K., 1972, Phys. Rev. A6, 2118-2125.
Smith, A. J., Hicks, P. J., Read, F. H., Cvejanovic, S., King, G.C.M., Comer, J. and Sharp, J. M., 1974, J. Phys. B. 7, L496-L502.
Smith, E. R. and Henry, R.J.W., 1973a, Phys. Rev. A7, 1585-1590.
Smith, E. R. and Henry, R.J.W., 1973b, Phys. Rev. A8, 572-575.
Smith, K., 1971, The Calculation of Atomic Collision Processes (Wiley, New York).
Steph, N. C. and Golden, D. E., 1980, Phys. Rev. A21, 759-770.
Sucliffe, V. C., Haddad, G. N., Steph, N. C. and Golden, D. E., 1978, Phys. Rev. A17, 100-7.
Taylor, A. J. and Burke, P. G., 1967, Proc. Phys. Soc. 92, 336-344.
Taylor, P. O., Phaneuf, R. A. and Dunn, G. H., 1980, Phys. Rev. A22, 435-444.
Temkin, A., 1962, Phys. Rev. 126, 130-142.
Temkin, A., 1982, Phys. Rev. Letters 49, 365-368.
Temkin, A. and Lamkin, J. C., 1961, Phys. Rev. 121, 788-794.
Thomas, L. D., Csanak, Gy., Taylor, H. S. and Yarlagadda, B. S., 1974, J. Phys. B. 7, 1719-1733.
Thomas, L. D., Oberoi, R. S. and Nesbet, R. K., 1974, Phys. Rev. A10, 1605-1611.
Walker, D. W., 1974, J. Phys. B. 7, 97-116.
Walters, H.R.J., 1985, Physics Reports, to be published.
Wannier, G. H., 1953, Phys. Rev. 90, 817-825.
Watanabe, S., 1982, Phys. Rev. A25, 2074-98.
Whitten, R. C. and Smith, F. T., 1968, J. Math. Phys. 9, 1103.
Wolcke, A., Bartschat, K., Blum, K., Borgmann, H., Hanne, G. F. and Kessler, J., 1983, J. Phys. B. 16, 639-655.
Wulfman, C., 1973, Chem. Phys. Lett. 23, 370-372.

DENSITY MATRICES, STATE MULTIPOLES, AND THEIR PHYSICAL INTERPRETATION

K. Blum

Institut für Theoretische Physik I
Universität Münster
D-4400 Münster

1. Introduction

In recent years, there has been a considerable increase in the level of sophistication of experimental techniques in atomic physics. These methods include coincidence measurements, collisions with spin-polarized particles, excitation by lasers, and scattering processes in strong electromagnetic fields. As a result of these investigations one has obtained data on many observables characterizing excitation processes. In order to interprete these data theoretically, to develop physical pictures of the excitation process, and to relate experimental observables with theoretical parameters, the application of density matrix techniques is required.

These methods and their application to various fields of physics have been discussed extensively in the literature (see for example Fano 1957, ter Haar 1961, Blum 1981, Macek 1983). In the present lectures the emphasis will be on the main concepts and their physical and geometrical interpretation.

In chapter 2 it will be discussed what can be learned about the state of excited atomic ensembles from general quantum mechanical arguments without using the mathematical formalism. The fundamental concepts are illustrated in terms of simple physical pictures. In chapters 3 and 4 reduced density matrices and state multipoles will be introduced and their physical importance discussed. A geometrical interpretation particularly of the alignment tensor will be given by relating this quantity to charge clouds of excited atoms and their internal symmetry. Some of the relations obtained will then be rederived in chapter 5, using the "natural frame" (Hermann and Hertel

1982) and pointing out some of its advantages. Finally, in chapter 6, some results are given on the decay of polarized atomic ensembles.

2. PHYSICAL INTERPRETATION OF SOME BASIC CONCEPTS

In this section we will introduce some basic concepts in a physical way. Consider a simple example, the excitation of ^1P-states of atoms from a ^1S -groundstate by protons (or by other projectiles which behave as spin- and structureless particles during the collision):

$$p + A(^1S) \rightarrow p' + A(^1P) \qquad (2.1)$$
$$\hookrightarrow A(^1S) + \gamma$$

After some time the excited atoms will decay to their groundstate by photon emission. We will assume that the collision time t_o is much shorter than the life time τ of the excited atoms

$$t_o < \tau \qquad (2.2)$$

If this assumption is satisfied then excitation and decay can be treated as independent processes. The atoms are excited during the collision, they remain a relatively long time (on the atomic scale) in the excited state and then they decay to the groundstate.

We will assume that the projectiles scattered in a given direction with momentum ($\vec{P}_1$) and the emitted photons are detected <u>in coincidence</u>. The important point is that the coincidence method allows to restrict the observation to light, emitted by a subensemble of excited atoms only, namely those atoms, which have been excited by the detected protons. The experiment allows to "isolate" this subensemble, so to speak, and to investigate this subensemble only. Our first task is to describe the quantum mechanical state of the atoms in this "isolated" subensemble between excitation and decay. That is, we are only interested in those atoms which scattered the detected projectiles. In section 3 we will give a formal derivation. Here, we will consider what can be learned from very general quantum mechanical arguments, without using the mathematical apparatus. This requires, however, some simplifying assumptions about the measuring process which will be relaxed later.

We will use a coordinate system where the z-axis is parallel to the incoming beam direction and where the x-z-plane is the collision plane (see fig. 1).

First of all, we have to consider the initial states. Before the collision all atoms are assumed to be in their groundstate which

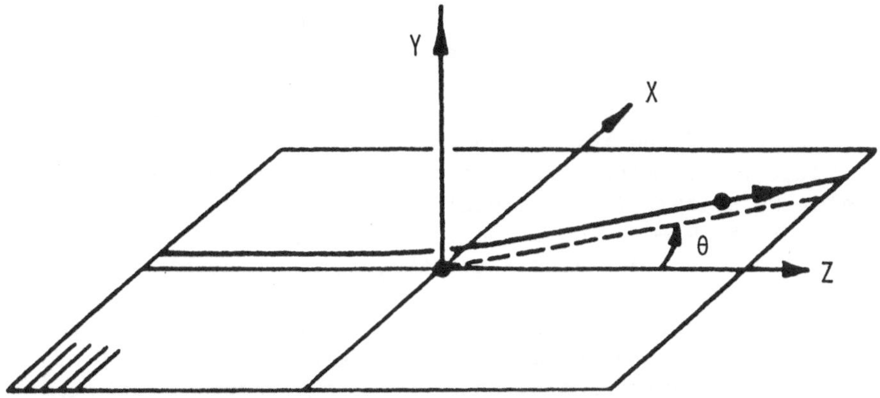

Fig. 1

Collision System

we will denote by $|0\rangle$. That is, all atoms are assumed to be in one and the same state and this common state of all atoms can be characterized by assigning a single state vector $|0\rangle$ to the whole ensemble. Ensembles of atoms in identical states are said to be in pure states.

Similarly, the state of the incoming protons is completely characterized, for example, by specifying the momentum (as long as all internal degrees of freedom can be neglected during the collision). Hence, by assuming that the initial protons have a well-defined momentum $\vec{p}_o$, we can assign a single state vector $|\vec{p}_o\rangle$ to the whole beam.

Finally let us assume that the scattered protons are detected in a given direction by a detector which accepts only protons with momentum $\vec{p}_1$. In this case the detected protons are in identical states with sharp momentum, characterized in terms of the single state vector $|\vec{p}_1\rangle$.

The initial states of both reaction partners and the state of the detected protons are therefore completely determined, that is, it is not possible - not even in principle - to obtain more knowledge on these states without destroying some of the information already obtained.

As a consequence, the atoms of the subensemble of interest have been excited under identical conditions and are therefore in identical states after the collision. We can characterize this common state of all the atoms of interest by assigning a single state vector $|\psi(p_1,p_o)\rangle = |\psi\rangle$ to the whole subensemble.

This result can also be obtained in a slightly different way. Since the states of the initial and of the detected particles have been completely determined the state of the corresponding atomic subensemble is necessarily completely determined, too. (The attempt to obtain more knowledge would require a more complete determination of initial and detected states - but that is not possible under the experimental conditions assumed here). The phrase "completely determined" means that one can be sure that the atoms are in states characterized by the eigenvalues of a complete, commuting set of observables. (The existence of this set is guaranteed by quantum mechanics for any state of "maximum knowledge"). Hence, since all the atoms of interest can be thought of as having the same values of these variables, they are in identical states and indistinguishable as far as quantum mechanics allows.

In conclusion we have arrived at the following result:
The quantum mechanical state of the experimentally "selected" atomic subensemble can be described in terms of a single state vector if (and only if) the initial states of both reaction partners and the state of the detected projectiles are completely determined under the given experimental conditions.

2.2 DETERMINATION OF THE STATE VECTOR

Our next task is to determine this state vector $|\psi\rangle$. A convenient way is to expand $|\psi\rangle$ in terms of a suitably chosen set of basis states. Since we are interested in ^{1}p-excitation we may choose the magnetic substates with magnetic quantum numbers $M = \pm 1, 0$ and expand $|\psi\rangle$ in terms of this basis set $|M\rangle$:

$$|\psi\rangle = \sum_M a(M)\,|M\rangle$$

where $a(M) = a(M\vec{p}_1,\vec{p}_o)$ is the amplitude of finding an atom with quantum number M in the subensemble. M is defined with respect to the incoming beam axis as quantization axis.

The amplitudes a(M) contain all information on the collision. Their determination constitutes a "complete experiment" in the sense defined by Bederson (1969).

It is important to have a clear understanding of superposition states. Expressions like eq. (2.3) does not mean that the atoms are in states with definite quantum numbers M, say, some atoms in the state $|+1\rangle$, others in $|-1\rangle$, others in $|0\rangle$. It would be absolutely wrong to associate such a picture with eq. (2.3). The choice of the quantization axis and of the corresponding basis set is arbitrary. Equally well we could have chosen a different axis Z' with different states $|M'\rangle$ (for example, following Hermann and Hertel (1982) the axis perpendicular to the scattering plane). The atoms do not "know" anything about this choice. Consequently, they are not in states with sharp magnetic quantum number M. Under the experimental conditions assumed here, the selected atoms are in one and the same state $|\psi\rangle$, this is a statement on the physics of the process. The right-hand side of eq. (2.3) is but a mathematical expression. Its usefulness stems from the fact that it represents a convenient characterization of $|\psi\rangle$ in terms of the amplitudes a(M), and the calculation of these amplitudes is the goal of scattering theory (see the lectures by P.G. Burke in this volume).

2.3 ILLUSTRATION

In order to visualize these results let us assume that the scattering energy is high enough so that the first Born approximation can be applied. It can be shown that in this case the atoms of the "selected" subensemble have sharp magnetic quantum number M' = 0 with respect to a praticular axis, namely the direction of the momentum transfer $\vec{q} = \vec{p}_1 - \vec{p}_0$ (see for example Macek 1983). Atoms in ^1P states can be represented by a dumb-bell orientated along the axis $\vec{q}$ (see fig. 2). In terms of a model, one may associate with these statements the following picture. All atoms in the subensemble of interest have the same shape of their charge clouds and these shapes have the same orientation in space.

The common state of all of the "selected" atoms is represented by the vector

$$|\psi\rangle = |M' = 0\rangle = \sum_M a(M)|M\rangle \qquad (2.4)$$

All atoms in the subensemble of interest are in one and the same state $|M' = 0\rangle$. All these atoms have made the same transition $|0\rangle \rightarrow |M' = 0\rangle$ during the collision, there are no transitions to states with definite M.

2.4 MIXED STATES

The discussion given in the preceding sections shows that to any pair of initial and final momenta $\vec{p}_0$ and $\vec{p}_1$ there is associated a subensemble of excited atoms, described by the state vector $|\psi(\vec{p}_1,\vec{p}_0)\rangle$ (provided the projectiles can be considered as spin- and structureless particles and the target atoms have spin 0). In terms

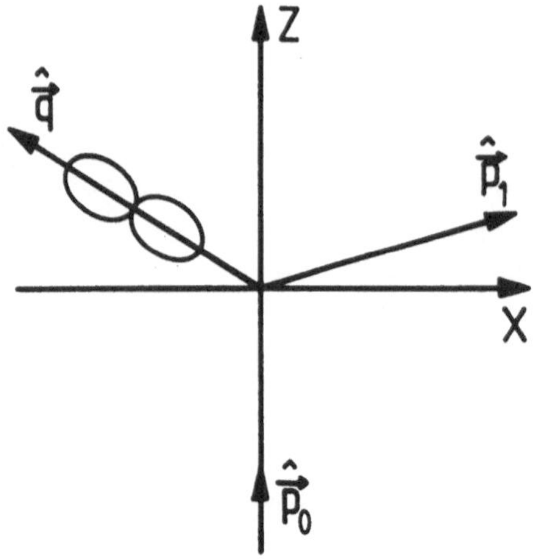

Fig. 2

Charge Cloud in Born Approximation (^ denotes unit vectors)

of a model, all atoms in a given subensemble may be visualized as having in particular the same shape orientated along the same direction. Atoms in other subensembles, excited by projectiles scattered in other directions, can be thought of as having different shapes and different orientations in space (in Born approximation only the orientation of the dumbbells differs corresponding to the various directions of $\vec{q}$).

Assume now that the scattered protons are not observed. In this case we have to consider the ensemble of <u>all</u> excited atoms, regardless of the direction of the scattered projectiles. This total ensemble may be visualized as consisting of atoms with different shapes and different orientations of these shapes. Clearly it is not possible to describe this ensemble in terms of a single state vector (for example, in terms of a linear superposition of the various states $|\psi(\vec{p}_1,\vec{p}_o)\rangle$). This would necessarily mean that all atoms are in identical states and that is not the case. Such an ensemble is said to be a <u>mixture</u> of the various states $|\psi(\vec{p}_1,\vec{p}_o)\rangle$ in contrast to an ensemble in a pure state characterized in terms of a single state vector.

More generally, <u>mixtures are obtained in scattering experiments where the initial and/or the final states are not completely determined</u>. This is the case, for example, if one or both of the reaction partners have non-vanishing spin and are not completely polarized in the initial state, or if the projectiles have been excited during the collision but experimentally no state selection is being performed (see for example section 3.1 in Blum (1981) for a more complete discussion). To describe such situations we have to generalize the usual formalism in terms of state vectors and linear superpositions of state vectors. In order to do this it is convenient to use the density matrix description.

2.5 DENSITY OPERATOR AND DENSITY MATRIX

Before continuing our discussion of scattering processes we will introduce the density matrix and briefly consider its main properties. Assume that the only information we have obtained about a given system is the following: We know the probability W_1 of finding an atom of the system in state $|\psi_1\rangle$, the probability W_2 of finding an atom in state $|\psi_2\rangle$ and so on. The expectation value $\langle Q\rangle$ of any observable Q for any state $|\psi_n\rangle$, multiplying $|Q_n\rangle$ by the corresponding statistical weight W_n, and sum over all these contributions:

$$\begin{aligned}\langle Q\rangle &= \sum_n W_n \langle Q_n\rangle \\ &= \sum_n W_n \langle\psi_n|Q|\psi_n\rangle\end{aligned} \tag{2.5}$$

Since

$$\langle\psi_n|Q|\psi_n\rangle = \mathrm{tr}|\psi_n\rangle\langle\psi_n|Q$$

expression (2.5) can be written as the trace over the product of two operators:

$$\begin{aligned}\langle Q\rangle &= \sum_n W_n\, \mathrm{tr}|\psi_n\rangle\langle\psi_n|Q \\ &= \mathrm{tr}\Big[\sum_n W_n|\psi_n\rangle\langle\psi_n|Q\Big]\end{aligned} \tag{2.6}$$

Defining the density operator ρ by the equation

$$\rho = \sum_n W_n|\psi_n\rangle\langle\psi_n| \tag{2.7}$$

eq. (2.6) is given by

$$\langle Q\rangle = \mathrm{tr}\,\rho\, Q \tag{2.8}$$

if we normalize according to

$$\mathrm{tr}\,\rho = 1 \tag{2.9}$$

As shown by the definition (2.7) ρ contains exactly the information we have obtained about the given system. It is therefore convenient to use this operator in order to characterize systems which are not in pure states.

As an example, consider an ensemble of unpolarized spin-S particles. This system can be thought of as being a mixture of states $|S\ M\rangle$ with equal probabilities

$$W_M = 1/(2S + 1)$$

According to the definition (2.8) this system is described by the density operator

$$\rho = 1/(2S + 1) \sum_M |SM\rangle\langle SM| \tag{2.10}$$

In general, the states $|\psi_n\rangle$ are not orthogonal to each other. Expanding any state $|\psi_n\rangle$ in terms of a conveniently chosen orthonormal set of basis states $|\phi_m\rangle$

$$|\psi_n\rangle = \sum_m a_m^{(n)} |\phi_m\rangle$$

we obtain

$$\rho = \sum_{nmm'} W_n\, a_{m'}^{(n)}\, a_m^{(n)} |\phi_{m'}\rangle\langle\phi_m| \tag{2.11}$$

Taking matrix elements we obtain an explicit matrix representation of the operator ρ, the so-called density matrix, with elements

$$\langle m'|\rho|m\rangle = \sum_n W_n\, a_{m'}^{(n)}\, a_m^{(n)*} \tag{2.12}$$

where the asterisk denotes the complex conjugate amplitude.

The density matrix is hermitian, that is,

$$\langle m'|\rho|m\rangle = \langle m|\rho|m'\rangle^* \tag{2.13}$$

as can be seen from eq. (2.12).

Knowledge of ρ allows the calculation of any expectation value $\langle Q\rangle$ of interest by applying eq. (2.8). This is an important result. We recall from quantum mechanics that all information on the behaviour of a given system can be expressed in terms of expectation values of suitably chosen observables. Thus, the basic problem is to calculate the expectation values. Since these can be obtained by using eq. (2.8) it follows that the denisty matrix contains all significant information on the system.

If a given system is in a pure state $|\psi\rangle$ then eq. (2.7) reduces to

$$\rho = |\psi\rangle\langle\psi| \tag{2.14}$$

It can be shown that a system, described by the density matrix ρ is in a pure state if, and only if, the following relation holds (see for example section 2.2 in Blum (1981)):

$$\mathrm{tr}(\rho^2) = (\mathrm{tr}\rho)^2 \tag{2.15}$$

3. THE DENSITY MATRIX OF ATOMS EXCITED IN COLLISIONS

3.1 THE REDUCED DENSITY MATRIX

In section 2.4 we have discussed that the state of excited target atoms must in general be characterized by a density matrix if the initial states of the particles and/or the final state of the detected projectiles is not completely determined experimentally. We will now consider how the relevant density matrix can be constructed and the information on the collision which is contained in its elements.

We will introduce all relevant concepts by discussing the following example:

$$B(^2S) + A(^1S) \rightarrow B^* + A(^1P) \tag{3.1}$$

We will allow for excitation of the projectiles B during the collision. It will be assumed that the momenta $\vec{p}_o$ and $\vec{p}_1$ of the initial and the detected projectiles are fixed by the experimental conditions and that no state-selection of the scattered projectiles B^* is being performed. For simplicity we will assume for the target atoms a ^1S-^1P-transition and the initial projectiles shall be unpolarized.

The analysis proceeds in several steps. The first one is obvious: Specify the initial state of the total system. The target atoms A are in their groundstate $|0\rangle$, the projectiles have sharp momentum p_o and are unpolarized. Applying eq. (2.10) and (2.14) the relevant density matrices are $|0\rangle\langle 0|$ for the target atoms and

$$|\vec{p}_o\rangle\langle\vec{p}_o| \times \frac{1}{2} \sum_{m_o} |m_o\rangle\langle m_o|$$

for the incoming projectiles, where $m_o = \pm\frac{1}{2}$ denotes the Z-component of the spin. The state of the total system is then described by a density matrix ρ_{in} which is given by the direct product of the density matrices of the reaction partners (since we may safely assume that target atoms and projectiles are uncorrelated before the interaction):

$$\rho_{in} = |0\rangle\langle 0| \times |\vec{p}_o\rangle\langle p_o| \times \frac{1}{2}\sum_{m_o} |m_o\rangle\langle m_o|$$

which can be written in the form

$$\rho_{in} = \frac{1}{2}\sum_{m_o} |0\vec{p}_o m_o\rangle\langle 0\vec{p}_o m_o| \tag{3.2}$$

The next step is the description of the final states. After the collision the total system is characterized by a density matrix ρ_{out}, which is related to ρ_{in} by the relation (see Blum, 1981, eq.(E5)):

$$\rho_{out} = T\,\rho_{in}\,T^{+} \tag{3.3}$$

where T is the transition operator. This operator contains all the information on the scattering dynamics. It is one of the main tasks of collision physics to determine T, that is, all its physically significant matrix elements.

In order to take matrix elements we have to choose a set of basis states. As in chapter 2 we will take the substates $|M\rangle$ for the target atoms in the ^1P-state. We characterize the final projectile states by the momentum p_1, the total angular momentum J and its z-component M_J. In this representation the matrix elements of eq.(3.3) are given by the expression:

$$\begin{aligned}\langle M',\vec{p}_1 J'M_J'|\rho_{out}|M,\vec{p}_1 JM_J\rangle &= \langle M',J'M_J'|\rho_{out}(\vec{p}_1)|M,JM_J\rangle \\ &= \langle M'J'M_J'|T\rho_{in}T^{+}|MJM\rangle\end{aligned} \tag{3.4}$$

where eq.(3.3) has been used.

Substitution of eq. (3.2) yields:

$$\langle M'J'M_J'|\rho_{out}(\vec{p}_1)|MJM_J\rangle = \frac{1}{2}\sum_{m_o} \langle M'J'M_J'\vec{p}_1 m_o\rangle\langle MJM_J\vec{p}_1|T|0p_o m_o\rangle^{*} \tag{3.5}$$

We introduce the scattering amplitudes as elements of the T-operator:

$$f(MJM_J\,m_o) = \langle MJM_J\vec{p}_1|T|0p_o m_o\rangle \tag{3.6}$$

where we have supressed the dependence on the fixed momenta $\vec{p}_o$ and $\vec{p}_1$. The scattering amplitudes will be normalized in such a way that their absolute square gives the differential cross section for the corresponding transition $|0\vec{p}_o m_o\rangle \to |MJM_J\vec{p}_1\rangle$:

$$|f(MJM_J m_o)|^2 = \sigma(MJM_J m_o) \tag{3.7a}$$

and the sum over all final quantum numbers and the average over initial spins gives the differential cross section:

$$\frac{1}{2}\sum_{\substack{MJ\\ M_J m_o}} |f(MJM_M m_o)|^2 = \sigma \tag{3.7b}$$

The elements (3.5) describe the total system, the detected projectiles and the target atoms of the corresponding subensemble. Suppose that we are only interested in the state of the excited atoms A but not in the scattered projectiles. That is, we assume that the scattered projectiles with momentum $\vec{p}_1$ are detected but that no state selection of B is being performed. We are interested in the state of those target atoms excited by the detected projectiles. Hence, what we really want is the density matrix $\langle M'|\rho(\vec{p}_1)|M\rangle$ which contains only the information on these states. The dependence on p_1 indicates that we consider that subensemble of atoms only excited by the detected projectiles as discussed in chapter 2. So our nect task is to project the elements of interest out of those of the total density matrix (3.5).

It can be shown that the density matrix $\rho(\vec{p}_1)$ is obtained from ρ_{out} by taking those elements of (3.5) which are diagonal in all unobserved variables J and M_J and by summing over the unobserved variables

$$\langle M'|\rho(\vec{p}_1)|M\rangle = \sum_{JM_J} \langle M'JM|\rho_{out}(\vec{p}_1)|MJM\rangle \tag{3.8}$$

The proof of this statement can be found in literature (see for example Fano (1957) or section 3.2 in Blum (1981)).

In general, a density matrix ρ, describing only one of two or more interacting particles, is called the <u>reduced density matrix</u>. It is one of the most fundamental concepts of the theory, not only for the description of scattering experiments but for quantum statistics in general, many-body theory, quantum optics and so on.

Substitution of eqs. (3.5) and (3.6) into eq. (3.8) yields:

$$\langle M'|\rho(\vec{p}_1)|M\rangle = \frac{1}{2}\sum_{JM_J m_o} f(M'JM_J)f(MJM_J)^* \tag{3.9}$$

It should be noted that no interference terms with $J' \neq J$ and $M_J' \neq M_J$ occur in eq. (3.9) because of the non-observation of these variables.

3.2 THE NUMBER OF INDEPENDENT PARAMETERS

Expression (3.9) shows what information on the scattering process can be obtained from an experiment under the given conditions: $\vec{p}_0$ and $\vec{p}_1$ fixed, no state selection of initial and detected projectiles. The elements of the reduced density matrix can be determined by measuring angular distribution and polarization of the emitted light (in coincidence with the scattered projectiles). The scattering amplitudes are the theoretical quantities which can be calculated numerically. Eq. (3.9) gives therefore the relation between the experimental observables and the theoretical parameters. It shows in particular what combinations of the scattering amplitudes must be taken in order to describe the experimental situation correctly.

An important point is the determination of the number of independent parameters which characterizes the reduced density matrix and, hence, the number of independent experiments necessary in order to determine ρ completely. First of all, it follows from eq. (3.9) that ρ is a 3×3 matrix with nine elements which may be complex. This corresponds to 18 real parameters. By applying the hermiticity condition (2.13) this number is reduced to nine.

Using the normalization (3.7) the diagonal elements are given by the partial differential cross sections:

$$\langle M'|\rho(p_1)|M\rangle = \sigma(M) \tag{3.10}$$

The off-diagonal elements are complex in general. They are bilinear combinations of the scattering amplitudes for different magnetic quantum numbers M' and M. These elements describe therefore the interference which exists between the corresponding magnetic substates of A under the given experimental conditions. Explicitly we have

$$\rho = \begin{pmatrix} \sigma(1) & \langle 1|\rho|0\rangle & \langle 1|\rho|-1\rangle \\ \langle 1|\rho|0\rangle^* & \sigma(0) & \langle 0|\rho|-1\rangle \\ \langle 1|\rho|-1\rangle^* & \langle 0|\rho|-1\rangle^* & \sigma(-1) \end{pmatrix} \tag{3.11}$$

where condition (2.13) has been used.

The number of independent parameters may be further reduced by symmetry requirements. Under the experimental conditions assumed here the scattering plane (X-Z-plane in fig. 1) is a plane of symmetry. Reflection invariance in this plane gives the relation:

$$<M'|\rho(\vec{p}_1)|M> = (-1)^{M'+M} <-M'|\rho(\vec{p}_1)|-M> \qquad (3.12)$$

(which can be derived similar to eq. (5.2)). This gives the relations:

$$\sigma(1) = \sigma(-1) \qquad (3.12a)$$

$$<1|\rho|0> = -<-1|\rho|0> \qquad (3.12b)$$

$$<1|\rho|-1> = <-1|\rho|1>^* \qquad (3.12c)$$

$$= <1|\rho|-1>^*$$

which corresponds to four conditions for the elements of the matric (3.11) (note that eq. (3.12b) corresponds to two conditions: one for the real part and one for the imaginary part of the complex element $<1|\rho|0>$). Hence, the matrix (3.11) is completely determined by five real parameters. Explicit parametrizations can be found in the literature (Nienhuis (1983), or section 3 in Blum and Kleinpoppen (1983). See also Hippler et al. (1982)).

3.3 EXCITATION OF PURE STATES

Finally let us assume that the projectiles are acting as spinless and excitationless particles. In this case no average over unobserved variables needs to be taken and eq. (3.9) reduces to the simple expression

$$<M'|\rho(\vec{p}_1)|M> = f(M')f(M)^* \qquad (3.13)$$

representing ρ by the projector

$$\rho = |\psi><\psi|$$

with

$$|\psi> = \sum_M f(M)|M> \qquad (3.14)$$

we obtain eq. (3.13). It follows therefore that the atoms of the subensemble of interest are in a pure state. Hence, we have obtained the result derived in chapter 2 in a different way. (Note, however, that the normalization in eq. (3.14) is different from that one used in chapter 2).

Reflection invariance gives the requirements

$$f(M) = (-1)^M f(-M) \quad . \tag{3.15}$$

It follows therefore that the matrix (3.13) is completely characterized by three independent real parameters, for example the magnitudes $|f(1)|$ and $|f(0)|$ and their relative phase χ, or the parameters σ $\lambda = \sigma(0)/\sigma$ and χ introduced by Eminyan et al. (1974).

We have obtained the result: If the experimental and dynamical conditions are such that atoms in pure states are isolated then the number of independent density matrix elements is reduced. In the case under discussion no state selection of the scattered projectiles is performed. Nevertheless, the fact whether the projectiles are excited during the collision or not is reflected in the number of independent parameters characterizing the state of the excited atoms of interest.

In conclusion, we have seen that only under very special conditions can an atomic system be "selected" in a pure state. This, however, is the case most textbooks on quantum mechanics concentrate on. In general, one has to construct the relevant reduced density matrix in order to describe more complex situations. Hence, the reduced density matrix can be considered as the natural generalization of the state vector or wave function of elementary quantum mechanics.

4. STATE MULTIPOLES AND THEIR PHYSICAL AND GEOMETRICAL INTERPRETATION

4.1 DEFINITIONS AND SOME FORMAL RESULTS

Any given ensemble of atoms can be characterized in terms of the relevant density matrix. Often, however, it is more convenient to use a description in terms of the so-called state multipoles. As we will see this method has several advantages. It provides an efficient way of expressing the inherent symmetries of the system and enables one to separate geometrical and dynamical factors. Furthermore, it is often possible to give a direct geometrical or physical interpretation of these quantities in contrast to the density matrix elements themselves.

State multipoles have been applied extensively in literature (see for example the reviews by Steffen and Alder 1975, Fano and Macek 1973, Omont 1977, Andrä 1979, Blum 1981 and references therein). Here, we will not develop the theory in detail but concentrate on a few examples of experimental interest. The emphasis will be placed on the discussion of the main concepts and on the physical interpretation of the multipoles.

We will restrict our discussion to atomic systems in states with well-defined orbital angular momentum L. The ensemble is characterized in terms of its density matrix ρ with elements $\langle LM'|\rho|LM\rangle$ in the angular momentum representation.

Consider the mean value $\langle\hat{\vec{L}}\rangle$ of angular momentum contained in the atomic system. This is equal to the angular momentum transfered to the atoms during the excitation process, assuming $\langle\hat{\vec{L}}\rangle = 0$ for the initial state. The components $\langle\hat{L}_i\rangle$ (i = x,y,z) of the angular momentum vector are related to the relevant density matrix by eq. (2.8) (assuming $tr\rho = 1$):

$$\langle\hat{L}_i\rangle = tr\rho L_i \tag{4.1}$$

For most calculations it is more convenient to use spherical components of the angular momentum vector, defined by the expressions

$$\langle\hat{L}_{\pm 1}\rangle = \mp(1/\sqrt{2})(\langle\hat{L}_x\rangle \pm i\langle\hat{L}_y\rangle) \ , \qquad \langle\hat{L}_o\rangle = \langle\hat{L}_z\rangle \tag{4.2}$$

Expressing these components in terms of the density matrix elements we obtain ($Q = \pm 1, 0$):

$$\begin{aligned}\langle L_Q\rangle &= tr\rho\hat{L}_Q \\ &= \sum_{M'M} \langle LM'|\rho|LM\rangle\langle LM|\hat{L}_Q|LM'\rangle \\ &= \sqrt{\frac{(2L+1)(L+1)L}{3}} \sum_{M'M} \langle LM'|\rho|LM\rangle \ (-1)^{L-M'}(LM,L-M'|1Q)\end{aligned} \tag{4.3}$$

where we have used

$$\langle LM|\hat{L}_q|LM'\rangle = (-1)^{L-M'}(LM,L-M'|1Q)\sqrt{\frac{(21+1)(L+1)L}{3}}$$

for the matrix elements of the angular momentum operators. $(LM,L-M'|1Q)$ denotes a standard Clebsch-Gordan coefficient.

We now define new parameters by omitting the normalization factor in eq. (4.3)

$$\langle T(L)_{1Q}\rangle = \sum_{M'M} \langle LM'|\rho|LM\rangle \ (-1)^{L-M'}(LM,L-M'|1Q) \tag{4.4}$$

for $Q = \pm 1,0$. These parameters are called the components of the orientation vector. This vector is proportional to the angular momentum vector:

$$\langle T(L)_{1Q}\rangle = \sqrt{\frac{3}{(2L+1)(L+1)L}} \langle \hat{L}_Q\rangle \qquad (4.4a)$$

The reason for this change of normalization is that eq. (4.4) can immediately be generalized to more complex quantities. Following Fano (1953) we define the state multipoles by the expression:

$$\langle T(L)_{KQ}\rangle = \sum_{M'M} \langle LM'|\rho|LM\rangle (-1)^{L-M'} (LM, L-M'|KQ) \qquad (4.5)$$

The Clebsch-Gordan coefficient vanishes automatically if the usual angular momentum coupling rules are not satisfied. This condition restricts the allowed values of K and Q for a given L:

$$K \leq 2L \quad , \quad -K \leq Q \leq K \qquad (4.6)$$

By considering their transformation properties under rotations it can be shown that the state multipoles (4.5) are tensors of rank K and component Q. The monopole $\langle T(2)_{00}\rangle$ is essentially a normalization constant as discussed below. For K=1 we have a vector, the orientation vector (4.4). The five parameters with K=2 and Q= 2,±1,0 constiute a second rank tensor, the so-called alignment tensor. The quantities with K=3 are the components of a third rank tensor and so on. Orientation vector and alignment tensor are the most important ones.

Applying the definition (4.5) and the hermicity condition (2.13) we obtain:

$$\langle T(L)_{KQ}\rangle^{*} = (-1)^{Q} \langle T(L)_{K-Q}\rangle \qquad (4.7)$$

which relates the multipoles with components Q and -Q to each other (the star denotes the complex conjugate multipole).

In particular, if ^{1}P-states have been excited, we have $K \leq 2$ from condition (4.6). The system is therefore described by a monopole, two independent components of the orientation vector $\langle T_{11}\rangle$ and $\langle T_{10}\rangle$, and the three independent components of the alignment tensor $\langle T_{22}\rangle$, $\langle T_{22}\rangle$ and $\langle T_{20}\rangle$. Since $\langle T_{00}\rangle$, $\langle T_{10}\rangle$and $\langle T_{20}\rangle$ are real because of condition (4.7), the system is completely characterized by nine real parameters. This is the same result as obtained in chapter 3.

The relation (4.6) can be inverted by applying the orthogonality relation of the Clebsch-Gordan coefficients:

$$\langle LM'|\rho|LM\rangle = \sum_{KQ} (-1)^{L-M'} (LM,L-M'|KQ) \langle T(L)_{KQ}\rangle \qquad (4.5a)$$

Hence, the two descriptions in terms of density matrix elements and of state multipoles are equivalent. Both sets of parameters contain exactly the same information.

It is often more convenient to use the language of state multipoles. One of the reasons for this choice is that symmetry properties of systems can usually more easily be expressed in terms of the parameters (4.5) than by using the density matrix elements. In the next section we will briefly consider some of the most common types of symmetry.

It is of great interest to study the dynamical mechanism by which orientation and alignment are produced. These problems will be discussed in the lectures of J. Briggs in this volume (see also J. Macek 1983).

4.2 SYMMETRY CONDITIONS FOR STATE MULTIPOLES

Let us start by considering excitation processes which possess a plane of symmetry, defined by the geometry of the excitation process. Throughout this chapter we will always assume that this plane is the x-z-plane of our coordinate system. It can be shown that atomic ensembles, invariant under reflections in this plane, are described by state multipoles which satisfy the following symmetry condition:

$$\langle T(L)_{KQ}\rangle = (-1)^{K+Q} \langle T(L)_{K-Q}\rangle \qquad (4.8)$$

(which can be derived by applying the methods in chapter 5). In particular

$$\langle T(L)_{KQ}\rangle = 0 \quad \text{if } K \text{ odd.} \qquad (4.8a)$$

Combining relation (4.8) with the hermiticity condition (4.7) we obtain in particular that the real part of the components $T_{1\pm1}$ and the vanishing parts of the components of the alignment tensor vanish.

As an example, consider ^1P-excitation in collisions where the scattered projectiles have been detected. The momenta $\vec{p}_o$ and $\vec{p}_1$ of initial and detected projectiles span a plane and, if no other direction is being defined, the atoms of the ensemble of interest cannot "distinguish" between "up" and "down" with respect to this plane, Consequently, the "selected" atomic subensemble must be invariant against reflection in this plane and is completely characterized in terms of five real parameters, one monopole, one independent component of the orientation vector and the three real components $\langle T_{22}\rangle$, $\langle T_{21}\rangle$, $\langle T_{20}\rangle$ of the alignment tensor (compare this result with the discussion given in chapter 3).

As another example, consider excitation of atoms by electron impact where the scattered electrons are not detected, but where the initial electrons are transversely polarized. The incoming beam direction (z-axis) and the polarization vector $\vec{P}$ (y-axis) define again a plane and, in general, the excited atoms must be invariant under refelections in this plane. However, the atoms "see" this plane only when the excitation is influenced by the initial spin, either by electron exchange processes, or by explicit spin-dependent forces, or a combination of both. Hence, by investigating experimentally whether the atomic subensemble has planar or axial symmetry, something can be learned about these spin-dependent interactions (for details see Bartschat and Blum 1982, Bartschat et al 1982, 1984, Jitschin et al 1984).

Let us now consider excitation processes which are axially symmetric with respect to a given axis which we will choose as z-axis of our coordinate system. Examples are excitations of atoms in collisions where the scattered projectiles are detected in forward direction or not detected at all. In both cases only a single axis parallel to the incoming beam axis is defined by the experimental conditions. As a consequence, the system must be invariant under rotations around z.

This axial symmetry can be expressed mathematically by the condition:

$$\langle T_{KQ}\rangle = 0 \quad \text{for} \quad Q \neq 0 \qquad (4.9)$$

Axially symmetric systems are therefore characterized by state multipoles with component Q = 0 only: $\langle T_{00}\rangle$, $\langle T_{10}\rangle$, $\langle T_{20}\rangle$, ...

A further classification is possible by considering whether the excitation process is also invariant against reflection in any plane through the symmetry axis. If the symmetry axis is defined by a <u>polar vector</u> (the momentum vector p_0 of the initial projectiles for example, or by an electric field) this vector does not change under reflections in planes through z. This holds in particular for the z-z-plane. The atomic ensemble of interest is therefore characterized by state multipoles which have to satisfy eq. (4.9) and, in addition, condition (4.8a). All multipoles with K odd are therefore necessarily zero. In particular, the orientation vector vanishes which means that no net angular momentum has been transferred to the atom during the collision.

It follows that, in axially symmetric excitations, orientated atomic systems can only be produced if the symmetry axis is defined by an axial vector, (for example, in excitations by longitudinally polarized projectiles, or circularly polarized photons, or magnetic fields). Under reflections in planes through z an axial vector will change its sign and the excitation process is therefore not invariant under reflections. The corresponding atomic ensemble is then characterized by state multipoles which have to satisfy eq. (4.9) but not the condition (4.8a). In particular, $\langle T_{10}\rangle$ will be different from zero in general.

As an example consider excitation of ^1P-states in a collision where the scattered particles are not observed. From the discussion above, it follows that the atomic ensemble is completely characterized by two parameters only, the monopole $\langle T_{00}\rangle$ and one alignment parameter $\langle T_{20}\rangle$. These tensors can be expressed in terms of the corresponding denstiy matrix elements by applying the definition (4.5), substituting numerical values for the Clebsch-Gordan coefficients, and by remembering that in our normalization the diagonal elements of the density matrix are given by the partial total cross sections Q(M) for excitation of the sublevel $|M\rangle$ (this follows from eq. (3.10) by integrating over all scattering directions). We obtain that the monopole is proportional to the total cross section $Q = 2Q(1) + Q(0)$:

$$\langle T_{00}\rangle = \frac{Q}{\sqrt{3}} \tag{4.10a}$$

and the alignment parameter is given by

$$\langle T_{20}\rangle = \sqrt{\frac{2}{3}}\,(Q(1) - Q(0)) \tag{4.10b}$$

$\langle T_{20}\rangle$ characterizes therefore the difference in the population numbers of the magnetic substates.

4.3 PHYSICAL INTERPRETATION OF THE ALIGNMENT TENSOR. RELATION TO THE CHARGE DISTRIBUTIONS

In this section we will give a physical interpretation of the components of the alignment tensor, and of some of the results derived above. We will do this by considering the shape of the charge cloud of excited atoms, and relating the parameters of this shape to density matrix elements and state multipoles. With regard to recent collision studies ^1P- and ^1D-excitation are of particular interest (Andersen and Neitzke 1983 and references therein). We will therefore concentrate on the discussion of singlet states with definite orbital angular momentum L.

ρ may be the reduced density matrix describing the atomic system of interest (where we allow, for example, for excitation of the projectiles). For simplicity we will assume throughout this chapter that the atoms are spinless. Otherwise one has first to construct the relevant reduced density matrix which is independent of the spin. In this case additional factors will occur in the formulas below.

Any density matrix can be transformed to diagonal form and represented by:

$$\rho = \sum_n W_n |\psi_n\rangle\langle\psi_n| \tag{4.11}$$

where $|\psi_n\rangle$ are atomic states. Atoms in a state $|\psi_n\rangle$ have a charge density given by $e|\psi_n(\vec{r})|^2$. The total charge distribution of the incoherent superposition (4.11) is then obtained by summing over the individual contributions:

$$\rho(\vec{r}) = e \sum_n W_n |\psi_n(\vec{r})|^2 \tag{4.12}$$

Since $\psi_n(\vec{r}) = \langle\vec{r}|\psi_n\rangle$ we can write eq. (4.12) in the form:

$$\begin{aligned}\rho(\vec{r}) &= e \sum_n W_n \langle\vec{r}|\psi_n\rangle\langle\psi_n|\vec{r}\rangle \\ &= e \langle\vec{r}|\Big[\sum_n W_n |\psi\rangle\langle\psi_n|\Big]\vec{r}\rangle \\ &= e \langle\vec{r}|\rho|\vec{r}\rangle\end{aligned} \tag{4.13}$$

Hence, for a given density matrix ρ, the charge distribution is obtained by transforming to the coordinate representation of ρ.

We will apply eq. (4.13) to the case where the system of interest is in states with sharp angular momentum L. The corresponding density operator is given by

$$\rho = \sum_{M'M} |LM'\rangle\langle LM'|\rho|LM\rangle\langle LM| \tag{4.14}$$

(ρ is the reduced density matrix of interest, that is, ρ contains the sum over all unobserved variables as discussed in chapter 3). Transforming to the coordinate representation we have

$$e\,\langle\vec{r}|\rho|\vec{r}\rangle = e\sum_{M'M} \langle LM'|\rho|LM\rangle\langle\vec{r}|LM'\rangle\langle LM|\vec{r}\rangle \tag{4.15}$$

The wave function $\langle\vec{r}|LM\rangle \equiv \psi(\vec{r})_{LM}$ can be separated into radial and angular part:

$$\psi(\vec{r})_{LM} = R(r)_L\; Y(\beta\alpha)_{LM} \tag{4.16}$$

where Y_{LM} is the corresponding spherical harmonic. β is the angle between the radius vector $\vec{r}$ and the z-axis of the chosen coordinate system (which we will not specify here) and α is the azimuth of $\vec{r}$ (that is, the angle between the x-axis and the projection of $\vec{r}$ into the x-y-plane). The x-z-plane is then specified by $\alpha = 0$, the x-y-plane by $\beta = \frac{\pi}{2}$.

Substitution of eq. (4.16) into eq. (4.15) yields:

$$e\,\langle\vec{r}|\rho|\vec{r}\rangle = e|R(r)_L|^2 \sum_{M'M} \langle LM'|\rho|LM\rangle\; Y(\beta\alpha)_{LM'} Y(\beta\alpha)_{LM} \tag{4.17a}$$

$$= e|R(r)_L|^2\; W(\beta\alpha) \tag{4.17}$$

where the angular part of $W(\beta\alpha)$ of the electric charge distribution is defined by the sum over M' and M in eq. (4.17a). By drawing $W(\beta\alpha)$ as a function of β and α a polar diagram of the charge cloud is obtained.

In order to see the relation between the shape $W(\beta\alpha)$ of the charge distribution and the state multipoles we expand ρ in terms of the $T(L)_{KQ}$ by substituting eq. (4.5a) into the expression for $W(\beta\alpha)$ which yields:

$$W(\beta\alpha) = \sum_{\substack{KQ\\M'M}} (-1)^{L-M}(LM',L-M|KQ)\;\langle T(L)_{KQ}\rangle\; Y(\beta\alpha)_{LM'}\; Y(\beta\alpha)^{*}_{LM} \tag{4.18}$$

This expression can be simplified by applying the addition theorem:

$$Y(\beta\alpha)_{LM'}\; Y(\beta\alpha)^{*}_{LM} \tag{4.19}$$

$$= \frac{(2L+1)}{\sqrt{4\pi}} \sum_{K'Q'} (-1)^M\; (2K'+1)^{-1/2}(L0,L0|K'0)\,(LM',L-M|K'Q')\; Y(\beta\alpha)_{K'Q'}$$

by sustituting eq. (4.19) into eq. (4.18) and performing the sum over the magnetic quantum numbers M' and M by applying the orthogonality relation of the Clebsch-Gordan coefficients we obtain finally

$$W(\beta\alpha) = (2L+1)(-1)^L \sum_{KQ} (4\pi(2K+1))^{-1/2} (L0,L0|K0) \langle T(L)_{KQ} \rangle Y(\beta\alpha)_{KQ} \quad (4.20)$$

It is important to note that only multipoles with even K contribute to eq. (4.20) (otherwise the Clebsch-Gordan coefficient vanishes). This result was to be expected since the tensors $T(L)_{KQ}$ with K odd are proportional to the corresponding magnetic multipoles and can therefore not contribute to the electric charge distribution.

We will now assume that a plane of symmetry exists and that the atomic system under discussion is invariant under reflection in this plane. We will now choose a specific coordinate system by requiring that the x-z-plane is the symmetry plane (for another choice see chapter 5). The charge cloud is then mirror-symmetric with respect to the x-z-plane. Since under reflection in this plane $\alpha \to -\alpha$ it follows that $W(\beta\alpha)$ must be independent of the sign of α, that is, $W(\beta\alpha)$ can only depend on the cosinus of α but not on the sinus. Formally, this result follows from eq. (4.20) by using the symmetry condition (4.8) (with K even) and using symmetry properties of the spherical harmonics.

Let us now specialize to systems with L = 1. Applying the condition (4.8) and substituting the explicit expressions for the spherical harmonics into eq. (4.20) we obtain the result:

$$W(\beta\alpha) = \quad (4.21$$

$$\frac{\sqrt{3}}{4\pi} \langle T_{00} \rangle - \frac{3}{4\pi}\left[\langle T_{22} \rangle \sin^2\beta \cos 2\alpha - \langle T_{21} \rangle \sin 2\beta \cos\alpha + \frac{1}{\sqrt{6}} \langle T_{20} \rangle (3\cos^2\beta - 1)\right]$$

Note that in cases where the system of interest is a pure ^{1}P-state, only three independent parameters exist according to the discussion in chapter 3. By expressing the multipoles in terms of the independent parameters eq. (4.21) simplifies. The case of pure ^{1}D-states has been discussed by Andersen and Neitzke (1983) and by Andersen et al (1983).

The shape $W(\beta\alpha)$ of the charge cloud for ^{1}P-excitation will usually look as shown in fig. 3. According to eq. (4.21) this charge distribution can be thought of as being composed of two incoherent parts, an isotropic contribution (due to $\langle T_{00} \rangle$) and one ellipsoid (corresponding to the second-rank tensor $\langle T_{2Q} \rangle$). The ellipsoid is characterized by its three principal axes and, since the contribution

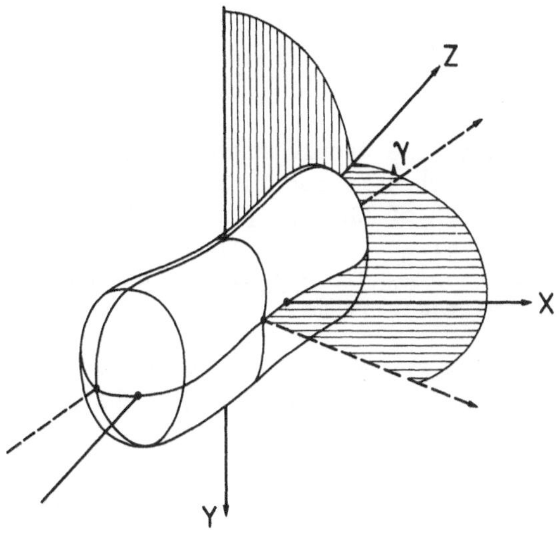

Fig. 3

Typical Charge Cloud for ^{1}P-excitation (from Herman and Hertel, 1977)

from the monopole is isotropic, the shape W(βα) must be characterized by the same three principal axes, denoted by x_{at} , y_{at} , z_{at} in fig.3.

Because of the mirror symmetry of the charge cloud two of its principal axes must lie in the x-z-plane and the third one is perpendicular to it. The shape is therefore determined by three parameters (apart from the normalization), the ratios of its principal axes and the angle γ between z_{at} and z (see fig. 3).

On the other hand, according to eq. (4.21), the shape is described by the three components $\langle T_{22}\rangle$, $\langle T_{21}\rangle$, $\langle T_{20}\rangle$ of the alignment tensor (the monopole fixes only the overall normalization. We therefore have to consider the relation between these two sets of parameters.

It has been shown (see for example Hertel and Stoll 1977) that the angle γ defined in fig. 3 is given by the following relation:

$$\mathrm{tg}\ \gamma = \frac{2\langle T_{21}\rangle}{\langle T_{22}\rangle - \sqrt{2}\langle T_{20}\rangle} \tag{4.22}$$

This gives a physical interpretation of the component $\langle T_{21}\rangle$: $\langle T_{21}\rangle$ characterizes the deviation between the z-axes and the principal axis z_{at} in the following sense: if $\gamma = 0$ then $\langle T_{21}\rangle = 0$ and vice versa.

It is always possible to transform to a coordinate system, called the atomic frame, in which $\langle T_{21}\rangle = 0$. In this frame the coordinate axes coincide with x_{at}, y_{at} and z_{at}.

$\langle T_{22}\rangle$ is most easily discussed in the atomic frame. Assume that we have transformed to this system. Then it can be shown that $\langle T_{22}\rangle$ is related to the difference in length of the two principal axes x_{at} and y_{at} in the following sense: If $\langle T_{22}\rangle = 0$ then these two axes have equal length and vice versa. Hence, $\langle T_{22}\rangle$ measures the difference between these two axes in the atomic frame.

The ratios of the principal axes are determined by $\langle T_{22}\rangle$ and $\langle T_{20}\rangle$ (when transformed to the atomic frame). $\langle T_{00}\rangle$ fixes the normalization.

If $\langle T_{22}\rangle = \langle T_{21}\rangle = 0$ then the atomic charge cloud is clearly axially symmetric with respect to z. The charge distribution is disc-like if $\langle T_{20}\rangle > 0$, and cigar-shaped if $\langle T_{20}\rangle < 0$.

In conclusion, we have shown that the components of the alignment tensor characterize the internal symmetry of atomic shapes and its orientation in space.

4.4 EXAMPLE: AXIALLY SYMMETRIC SYSTEMS

Finally, we will consider a simple example. Assume that we are dealing with an atomic ensemble, excited in a collision where the scattered projectiles have not been detected. The shape of the atomic charge cloud is then axially symmetric with respect to z (where z is defined by the incoming beam direction). Since $\langle T_{21}\rangle = \langle T_{22}\rangle = 0$, eq. (4.21) reduces to

$$W(\beta) = \frac{\sqrt{3}}{4\pi} \langle T_{00}\rangle - \frac{3}{4\pi\sqrt{6}} \langle T_{20}\rangle (3\cos^2\beta - 1) \qquad (4.23)$$

which is independent of α because of the symmetry.

The charge cloud is then orientated in z-direction (which coincides with z_{at}). The length A of its principal axis z_{at} is given by $2W(0)$. Form eq. (4.23) follows

$$W(0) = \frac{\sqrt{3}}{4\pi} (\langle T_{00}\rangle - \sqrt{2} \langle T_{20}\rangle)$$

Substitution of eq. (4.10) yields

$$W(0) = 3Q(0)/4\pi$$

Hence, the length A is determined by the total cross section Q(0) for excitation of the substate with magnetic quantum number M = 0:

$$A = 2W(0) = 6Q(0)/4\pi \qquad (4.24)$$

Similarly the length B of the principal axes perpendicular to z_{at} is proportional to $2W(\frac{\pi}{2})$. Eq. (4.21) gives:

$$W(\frac{\pi}{2}) = \frac{\sqrt{3}}{4\pi}\left(<T_{00}> + \frac{1}{\sqrt{2}}<T_{20}>\right)$$

and by substituting eq. (4.10) we obtain

$$B = 2W(\frac{\pi}{2}) = 6Q(1)/4 \qquad (4.25)$$

This discussion shows the importance of the determination of the total cross sections (the population numbers of the magnetic substates). Q(1) and Q(O) determine the shape of the electric charge distribution of atomic systems.

5. DESCRIPTION OF ATOMIC SYSTEMS IN THE "NATURAL FRAME"

5.1 STATE MULTIPOLES AND THEIR SYMMETRY RELATIONS

The discussion of processes possessing a plane of symmetry is often simplified when the so-called "natural frame" is used with axes x_N, y_N, z_N (Hermann and Hertel 1982). Here, the coordinate axes are chosen in such a way that the plane of symmetry is the x_N-y_N-plane, and the z_N-axis is perpendicular to the symmetry plane. In this frame the symmetry relations for the state multipoles have a different form (since the tensor component Q is defined with respect to a different quantization axis). In this section we will derive the symmetry relations for the multipoles in some detail. In the following section the relation between the charge clouds and tensor components in the natrural frame will be considered.

We will again concentrate on systems in states with sharp orbital angular momentum L. Throughout this chapter all third components (M,Q) are assumed to be defined with respect to z_N as quantization axis.

Reflection in the x_N-y_N-plane can be thought of as a combined operation: an inversion at the origin of the coordinate frame, followed by a rotation around z_N about an angle π.

Let us first consider the transformation properties of (orbital) angular momentum states under reflections. Denoting the inversion operator S_o the states $|LM>$ transform according to

$$S_o|LM> = (-1)^L|LM>$$

Application of the rotation operator $R(\pi)_z$ gives

$$R(\pi)_z |LM\rangle = e^{iM\pi} |LM\rangle$$
$$= (-1)^M |LM\rangle$$

Hence, denoting the reflection operator in the x_N-y_N-plane by $R = =(\pi)_z S_o$, we have

$$R|LM\rangle = (-1)^{L+M} |LM\rangle \tag{5.1}$$

We will now require that the x_N-y_N-plane is a symmetry plane, that is, the atomic system of interest shall remain invariant under reflections in this plane. Mathematically, this is expressed by the symmetry requirement

$$R^+ \rho R = \rho \tag{5.2}$$

where ρ is the density matrix of the atomic system under discussion.

Applying the relation (5.2) and the transformation property (5.1) we obtain for the density matrix elements

$$\langle LM'|\rho|LM\rangle = \langle LM'|R^+ \rho R|LM\rangle$$
$$= (-1)^{M'+M} \langle LM'|\rho|LM\rangle$$
$$= (-1)^{M'-M} \langle LM'|\rho|LM\rangle \tag{5.3}$$

(since L and M are the integers the sign of M in the phase factors can be changed. Eq. (5.3) shows that, as a consequence of the symmetry, all density matrix elements with M'+M odd are zero.

We define the state multipoles $\langle T(L)_{KQ}\rangle_N$ in the natural frame by eq. (4.5) which gives

$$\langle T(L)_{KQ}\rangle_N = \sum_{M'M} (-1)^{L-M'} (LM,L-M'|KQ) \langle LM'|\rho|LM\rangle$$
$$= \sum_{M'M} (-1)^{L-M'} (LM,L-M'|KQ) (-1)^{M'-M} \langle LM'|\rho|LM\rangle$$
$$= (-1)^Q \langle T(L)_{KQ}\rangle_N \tag{5.4}$$

where M'-M = Q because of the properties of the Clebsch-Gordan coefficient.

Eq. (5.4) is the symmetry condition for state multipoles in the natural frame which replaces eq. (4.8). Eq. (5.4) shows that all tensor components with Q odd vanish if the x_N-y_N-plane is a plane of symmetry. Consider for example ^1P-excitation. If the atomic ensemble is invariant under reflection in this plane it then can be characterized by the parameters $\langle T_{00}\rangle_N$, $\langle T_{10}\rangle_N$, $\langle T_{22}\rangle_N$ (which is complex in general), and $\langle T_{20}\rangle_N$. The components $\langle T_{11}\rangle_N$ and $\langle T_{2\pm1}\rangle_N$ vanish because of the symmetry of the system. It should be noted that the conditions (5.3) and (5.4) for the density matrix elements and the state multipoles, respectively, are simpler than the corresponding equations (3.12) and (4.8).

5.2 SHAPES OF CHARGE CLOUDS

In section 4.3 we have derived relations between the shape $W(\beta\alpha)$ of atomic charge distributions and the state multipoles. Up to eq. (4.20) the discussion is valid for any coordinate system since no reference to a particular frame has been made. We will therefore start with eq. (4.20) and specialize to the natural frame. β is then the angle between the radius vector $\vec{r}$ and z_N, and α the angle between x_N and the projection of $\vec{r}$ into the x_N-y_N-plane.

We will assume that the x_N-y_N-plane is a <u>plane of symmetry</u> for the system under discussion. Taking into account the symmetry relation (5.4) we obtain from eq. (4.20):

$$W(\beta\alpha) = \sum_{KQ} \frac{(-1)^L(2L+1)}{\sqrt{4\pi}\ (2K+1)} (L0,L0|K0)\ \langle T(L)_{KQ}\rangle\ Y(\beta\alpha)_{KQ} \tag{5.5}$$

with <u>K even</u>. Because of condition (5.4) only multipoles with <u>Q even</u> contribute to eq. (5.5).

Since the tensor components with $Q \neq 0$ are complex in general, we write

$$\langle T(L)_{KQ}\rangle = |\langle T(L)_{KQ}\rangle|\ e^{i\gamma(KQ)} \tag{5.6}$$

Consider now the terms with Q and (-Q) in eq. (5.5), apply the hermiticity condition (4.7), and use the relation

$$Y(\beta\alpha)_{KQ} = \sqrt{\frac{2K+1}{4\pi}}\ d(\alpha)^K_{0Q}\ e^{iQ\alpha} \tag{5.7}$$

where d^K_{0Q} are elements of the "small" rotation matrix which is real. We obtain

$$\langle T_{KQ}\rangle Y_{KQ} + \langle T_{K-Q}\rangle = 2d(\beta)^{(K)}_{0Q}\ \mathrm{Re}\left[\langle T_{KQ}\rangle e^{iQ\alpha}\right] \qquad (5.8)$$

$$= 2d(\beta)^{(K)}_{0Q}\ |\langle T_{KQ}\rangle|\ \cos(\gamma(KQ) + Q\alpha) \quad .$$

Keep now β fixed and consider the shape $W(\beta\alpha)$ as a function of α only (for example, consider the projection of the shape into the symmetry plane which corresponds to $\beta = \frac{\pi}{2}$). Substitution of the result (5.8) into eq. (5.5) shows that W has the following general form:

$$W = A + B\cos(\gamma(22)+2\alpha) + D\cos(\gamma(42)+2\alpha) + E\cos(\gamma(44)+4\alpha) + \ldots \qquad (5.9)$$

where the highest term depends on 2L . If the atomic system under discussion is in a ^1P-state (not necessarily a pure one) then eq. (5.9) reduces to

$$W = A + B\cos(\gamma + 2\alpha) \qquad (5.10)$$

Transformation to the atomic frame corresponds to $\gamma = 0$. If the system of interest is in a pure ^1D-state (that is, if in particular the scattered projectiles have not been excited and the collision was spin-independent) then it can be shown that $\gamma(22) = \gamma(42)$. This follows by expressing the state multipoles in terms of the scattering amplitudes. Eq. (5.9) reduces then to the general form:

$$W = A + \cos(\gamma+2\alpha) + \cos(\gamma'+4\alpha) \qquad (5.11)$$

with $\gamma' = \gamma(44)$. The relation (5.11) for pure ^1D-states has been derived by Andersen et al (1983) and we refer to this papaer for a further discussion (see also Andersen and Neitzke 1983 and references therein).

6. RADIATION FROM POLARIZED ENSEMBLES

In this section we will briefly consider how to determine density matrices and state multipoles by observing the radiation, emitted from the excited atoms. Experimental observables are the angular distribution and the polarization of the emitted light. These can be expressed conveniently in terms of the multipole parameters, characterizing the excited atomic ensembles. The relevant derivations have been discussed extensively in literature and will not be repeated here (see for example: Fano and Macek 1973, Blum 1981, Macek 1983).

As an example we will only give the result for the angular distribution $I(\theta,\phi)$ where θ and ϕ are the polar angles of the direction of observation. Assuming that the excited atomic ensemble of interest is invariant against reflections in the x-z-plane we obtain for a transition between states $L \rightarrow L_1$:

$$I(\theta,\phi) =$$

$$A(-1)^{L+L_1} \left[\frac{2(-1)^{L+L_1}}{3\sqrt{(2L+1)}} \langle T(L)_{00} \rangle - \begin{Bmatrix} 1 & 1 & 2 \\ L & L & L_1 \end{Bmatrix} \left(\langle T(L)_{22} \rangle \sin^2\theta \cos 2\phi - \langle T(L)_{21} \rangle \sin 2\theta \cos\phi + \frac{1}{\sqrt{6}} \langle T(L)_{20} \rangle (3\cos^2\theta - 1) \right) \right] \quad (6.1)$$

where A contains numerical factors and the reduced matrix elements describing the dynamics of the decay. It has been shown that, when dipole radiation is observed, angular distribution and polarization depend only on tensor components with $K < 2$. Information on the higher multipoles can be obtained by observing cascades, by studying superelastic scattering, or Auger emission.

The equations for the Stokes parameters of the emitted light have a similar structure to eq. (6.1). Any state multipole, occuring in these relations, gives rise to a specific angular dependence of the observables. Consequently, by studying the dependence of $I(\theta,\phi)$ and/or the Stokes parameters on the angles θ and ϕ the state multipoles can be determined.

Eq. (6.1) shows that $I(\theta,\phi)$ depends only on monopole and the components of the alignment tensor. In order to determine the atomic orientation one has to measure the circular polarization.

For further details on actual measurements and for experimental results we refer to the lectures of R. Hippler in this volume.

Acknowledgement: This work has been supported by the Deutsche Forschungsgemeinschaft in Sonderforschungsbereich 216 "Polarization and Correlation in Atomic Physics".

References

Anderson, N., Andersen, T., Dahler, J.S., Nielsen, S.E., Nienhuis, G., Refsgaard, K., 1983, J. Phys. B16, 817

Andersen, N., Neitzke, 1983, in: "Electronic and Atomic Collisions" J. Eichler, I.V. Hertel, N. Stolterfoht, eds., North-Holland, Amsterdam.

Andrä, H.J., 1979, in: "Progress in Atomic Spectroscopy B", W. Hanle, H. Kleinpoppen, eds., Plenum Press, New York.

Bartschat, K., Blum, K., 1982, Z.Phys. A304, 85

Bartschat, K., Hanne, G.G., Wolcke, A., 1982, Z. Phys. A304, 89

Bartschat, K., Scott, N.S., Blum, K., Burke, P.G., 1984, J. Phys. B17, 269
Bederson, B., 1969, Comments At. Mol. Phys. 1, 71
Blum, K., 1981, "Density Matrix Theory and Applications" Plenum Press, New York
Blum, K., Kleinpoppen, H., 1983, Phys. Reports 96, 251
Eminyan, M., McAdam, K., Slevin, J., Kleinpoppen, H., 1974, J. Phys. B7, 1519
Fano, U., 1953, Phys. Rev. 90, 577
Fano, U., 1957, Rev. Mod. Phys. 29, 74
Fano, U., Macek, J., 1973, Rev. Mod. Phys. 45, 553
Hermann, H., Hertel, I.V., 1982, Comments At. Mol. Phys. 12, 61
Hertel, I.V., Stoll, W., 1978, Adv. At. Mol. Phys. 13, 113, Academic Press, New York
Hippler, R., Malumet, G., Faust, M., Kleinpoppen, H., Lutz, H.O., 1982, Z. Phys. A304, 63
Jitschin, N., Osimitsch, S., Reihl, H., Kleinpoppen, H., Lutz, H.O., 1984, J. Phys. B17, 1899
Macek, J., 1983, in: "Fundamental Processes in Energetic Atomic Processes", H.O. Lutz, J.S. Briggs, H. Kleinpoppen, Plenum Press, New York
Nienhuis, G., 1984, J. Phys. B17, 587
Omont, A., 1977, Progr. Quantum Electronics 5, 69
Steffen, R., Alder, K., 1975, in: "Electromagnetic Interactions in Nuclear Spectroscopy", W. Hamilton, ed., North Holland, Amsterdam.
ter Haar, D., 1961, Rep. Prog. Phys. 24, 304

POLARIZATION EFFECTS IN ELECTRON-ATOM COLLISIONS

Benjamin Bederson

Physics Department
New York University
New York, New York 10003

INTRODUCTION

The three talks upon which this article is based are concerned with certain aspects of the electron-atom collision problem, particularly those which concern angular momentum orientation in both selection and analysis of collision partners in electron-atom collisions. I will not try to make this a comprehensive review of the subject. There are by now a large number of review articles and portions of books that do this;[1] some of these are listed in the references at appropriate places. Rather, I thought that this might be a good opportunity to give some interesting background, early history, and some pedagogic material and, of course, some prejudicial viewpoints from the work of myself and colleagues at New York University.

The first point to be emphasized about studies of polarization effects in collisions is the fact that in such selective experiments one observes what are in effect "partial" cross sections. These form a body of information which presents a unified picture of the electron-atom collision problem. From this viewpoint, all aspects of a particular collision problem should be taken together. These include so-called "grand total", total, and differential elastic and inelastic collisions, as well as spin- and angular-momentum selected collisions.

While all atomic systems can be explored using state selective techniques, the principal features and simplest theoretical analyses are best illustrated using "single-electron" atoms, e.g., atomic hydrogen and alkali metal vapors, and most, though by no means all of this article will be concerned with these.

Before presenting the historic background, I would like to clarify what, because of semantic ambiguity, can cause some confusion when discussing polarized beams. There are in fact a number of different ways in which the prefix "polar" appears in atomic collisions. The interaction energy of each of them exhibits a characteristic spatial dependence, roughly speaking, of an inverse r^n form, excepting (a) below. The main ones are:

(a) Spin-polarized, referring exclusively to free electrons or the ground-state of effective one-electron atoms (neglecting hyperfine structure). The polarization can be a "label" that describes dynamics of the excitation process (see (b) below). For elastic scattering, target and/or projectile spin polarization makes it possible to distinguish the "direct" and "exchange" interactions, or equivalently, singlet and triplet interactions.[2,3] These distinct (though coherent) scattering amplitudes result because of the symmetrization requirements on the total wave functions of fermions. The interaction is non-local and therefore cannot be characterized by an inverse r^n interaction (except in rough approximations, such as the Ochkur-Bonham type[4,5]). Usually the direct spin-spin interaction is negligible. At low energies, the spin-orbit is mostly (but definitely not entirely) negligible.

(b) Orbital angular momentum polarized. The interaction is describable in terms of state multipoles, e.g., quadrupolar, etc., with spherical harmonic angular variation of the interaction, depending on orientation and/or alignment ("orientation" is a synonym for polarized, i.e., the atom possesses net non-vanishing multipole moment; "alignment" describes a distribution of fine- or hyperfine-states whose weighted average is zero but is not equally distributed in accordance with their statistical weights[6,7]).

(c) Polarizable. The interaction is due to electric moments induced in the atom by the charged projectile, of the form

$$\bar{\alpha} \cdot \hat{r}/r^4 ,$$

with $\bar{\alpha}$ the polarization tensor.[8] This interaction also depends on orientation and alignment unless atomic state has $J \leq \frac{1}{2}$.

(d) Permanent electric dipole interaction. There are two cases here: first, polar molecules, and second, special cases of either degenerate atomic states of opposite parity, e.g., n=2 states of atomic hydrogen or two-level atoms in very strong resonant radiation fields, with Rabi frequencies comparable to inverse of interaction times. (This is not generally realizable in collision experiments; for such strong fields other interactions would likely mask simple dipole interaction.) The interaction is extremely long-range ($\sim r^{-2}$); (in fact, almost "pathological" in the sense that the total cross sections may diverge.[9])

Except for (d), all "polar" interaction described above will play a role in the discussion that follows.

The scattering formalism that set the stage for all electron-atom collision theory is contained in the several editions of Mott and Massey.[10] The seminal theoretical papers that initiated the "modern" era in electron-atom collisions are the articles "The Polarization of Atomic Line Radiation Excited by Electron Impact" by Percival and Seaton (1958)[11] and "Theory of Excitation and Ionization by Electron Impact" by Seaton (1966).[12] These articles explicitly show how particular observables, e.g., the polarization of resonance radiation, or singlet and triplet differential elastic cross sections, are related to various combinations of scattering matrix elements. Both papers rely on what is now dignified by the appellation "Seaton-Percival hypothesis", which is the assumption that spin and orbital angular momenta are separately conserved during a collision. This assumption has generally withstood the test of time well, under the limited conditions (light atoms, relatively crude electron energy resolution) in which most earlier experiments were performed. Recently experiments have readily observed departures from such "spin-conservation" experiments. A more complete analogous theory that includes the spin-orbit interaction term, with corresponding increase in complexity, is given by Burke and Mitchell.[13]

The essence of the Seaton analysis derives from the hierarchy of time sequences that govern the electron-atom collision. I write these here (although not all of them are necessary for the discussion that follows):

electron orbitting time in atom	$\sim 10^{-15}$sec
collision time	$\sim 10^{-15}$sec
atomic spin-orbit relaxation time	$\sim 10^{-12}$sec
collisional spin-orbit relaxation time	$< 10^{-12}$sec
nuclear hfs relaxation time	10^{-9}sec
inverse Rabi frequency	10^{-8}sec(1mW)-10^{-11}sec(1W)
radiative lifetime	$\sim 10^{-8}$sec

The above list tells us that for light atoms at low (up to the Kilovolt range) energies, the appropriate representation for the collision process is (L,S,M_L,M_S); the appropriate representation for observation of the collision (either by the atom or the resonant photon) is (F, M_F) where $\overline{F} = \overline{I} + \overline{J}$, and I is the nuclear spin; using a single-mode cw laser for atomic excitation does not change these.

Accordingly, taking these relaxation processes into account we can express the "master" statement that relates a partial cross section $\sigma(\Gamma\tau \rightarrow \Gamma'\tau')$ to elements of the so-called "transition" or "reactance" T-matrix as

$$\sigma(\Gamma\tau\to\Gamma'\tau') = \frac{\pi^2}{kk'} \times \Big| \text{sum over all permitted total spin and angular momenta and their projections not observed in going from } \Gamma\tau\to\Gamma'\tau' \times \text{appropriate vector coupling coefficients} \times \text{associated Legendre polynomials} \times T_{\Gamma,\Gamma'} \Big|^2 \qquad (1)$$

In this relation, Γ,Γ' represent all atomic quantum numbers before and after the collision and τ,τ' are the corresponding electron spin states. Elements T_{ij} are related to the corresponding scattering matrix elements S_{ij} by

$$T_{ij} = \delta_{ij} - S_{ij} \ .$$

The vector-coupling coefficients appear because in general quantization axes that are appropriate for the collision and state preparation and analysis may be, and in fact usually are, different.

The above expression reveals an important and essential feature of electron-atom collision theory: while the angular and vector coupling factors may be gruesomely complicated they are in principle analytic and tractable; the elements of the T matrix contain all the possible observables of a given collision process. We will later show how to specifically relate this to observables in the most fundamental of excitation processes: The S→P excitation in 1-electron atoms.

There is a final point I would like to make in this introduction. Because the fundamental interactions are known, collision physics is characterized by a very "close coupling" between theory and experiment. To a lesser extent such intimacy pervades all of physics. Purely experimental discoveries occur very infrequently in atomic physics: rather, progress in the field moves because of the "leapfrog" effect: experiment and theory continually jumping ahead of each other, with neither getting too far ahead, or behind. In fact, when a true experimental discovery is claimed, it will have a difficult time being accepted if existing theory did not predict it or cannot explain it. I will start the next section by describing an early example of such a case, and end this article by a more recent one.

EARLY HISTORY

The book Polarized Beams by J. Kessler[14] offers a comprehensive discussion of the production, use and detection of polarized beams of electrons; there is no need to duplicate that material here. See also an interesting and very readable review article by P. S. Farago.[15] In the section below we will briefly touch on modern source techniques.

The first "polarized electron" experiment of which I am aware was a remarkable study by Cox, MacIlwrath and Kurrelmayer,[16] published in 1928 and performed at New York University. It is worth quoting from the introduction of the original article:

"The already classic experiment of Davisson and Germer in which the diffraction of electrons by a crystal shows the immediate experimental reality of the phase-waves of de Broglie and Schrödinger suggested that it might be of interest to carry out with a beam of electrons experiments analogous to optical experiments in polarization. It was anticipated that the electron spin, postulated by A. H. Compton to explain the systematic curvature of the fog-tracks of β-rays, and recently so happily introduced in the theory of spectra by Uhlenbeck and Goudsmit might appear in such an experiment as the analogue of a transverse vector in the optical experiments."

It seems particularly remarkable, to me at least, that Cox et al decided to do a double scattering experiment, without exactly understanding why, arguing simply in analogy to the use of double scattering to observe polarized x-rays by Compton. Apparently, they had performed a double Mott-scattering experiment, although they were apparently not aware of the relativistic Mott theory that was to precede later, more specifically motivated polarized electron experiments. The Cox et al paper reports an essentially unambiguous left-right asymmetry produced by double-scattering. Years later, this was identified by Wu and Ambler as the definitive demonstration of nonconservation of parity (PNC) in weak interactions.[17] Aside from the remarkably intuitive sensible motivation of the early experiment two additional important features of this work are worth noting. First, it was in fact the double-scattering aspect of the experiment that allowed for observation of the asymmetry. This is because β-decay electrons are longitudinally polarized, while Mott-scattering can only distinguish transverse polarization. The first scattering (at $\pm 90^{o}$) simply transformed the polarization from longitudinal to transverse, thereby causing it to become distinguishable in the second scattering.

Perhaps even more important, the experiment apparently made no significant impact on polarized beam research (not to mention PNC), for the reason that there was no accompanying theory that could explain the observations. As already noted, this is a common thread throughout the modern development of atomic physics. At the end of this article we will see another interesting, more up-to-date example of the same phenomenon.

Shortly after the Cox et al experiment was published,

"Mott-scattering" appeared in proper theoretical context, to become one of atomic and nuclear collision physics' most important tools.[10] It remains so to this day. From the viewpoint of these lectures, "Mott-scattering" represents two important features: it refers to the spin-orbit interaction, a subject of fundamental interest in atomic physics, and of course remains the spin-analysis workhorse of polarized beam research.

If one is concerned with what could be characterized as "low energy" electron-atom collision physics (defined as the energy regions from thermal up to the kilovolt range), polarized beam research actually started its history in a somewhat different way. The first experiments involving polarization were of optical polarization, and the first significant experiments involved the polarization of resonant radiation from mercury excited by electron bombardment (Skinner,[18] Skinner and Appleyard[19]). Because angular momentum is transferred in direct excitation of optically allowed transitions by electrons it follows that the optical polarization of subsequent decay photons will reflect the orientation, that is, the values of the magnetic quantum numbers, of the excited atoms. At threshold, application of momentum conservation demands particularly simple polarization relations. These were already understood by Skinner and Appleyard, who were in fact surprised to find that their observed threshold polarization did not conform to these necessary predictions (Fig. 1).

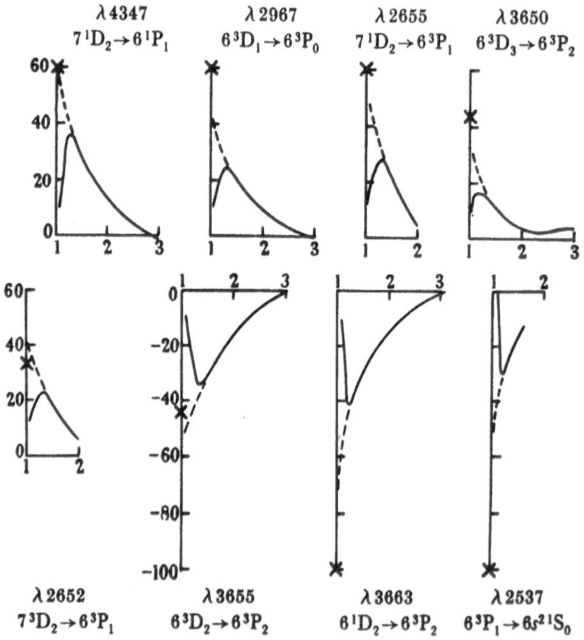

Fig. 1. Polarization of resonance radiation, for electron impact on mercury, Skinner and Appleyard.[19] Dashed lines are extrapolation to the expected threshold values.

In 1934 an extraordinary research monograph appeared, covering this subject in a comprehensive and skillful fashion. The book was (and is, since it still appears from time to time in current citations) Resonance Radiation and Excited Atoms, by Mitchell and Zemansky.[20]

The other, very different, exploitation of polarization in atomic collision physics concerned the case of spin exchange. Spin exchange was first observed in the laboratory in optical pumping experiments -- this was years after the theory of spin exchange in electron-atom collisions had been properly formulated by Oppenheimer,[21] in 1928, and, in detail, by Mott and Massey in the first edition of The Theory of Atomic Collisions, in 1933. However, I believe the first practical paper on spin exchange was in astrophysics. Purcell and Field[22] were interested in the relative populations of the F = 1,0 hyperfine levels of ground state atomic hydrogen, which determines the intensity of the 1420 MHz transition. These in turn are partially governed by thermal atom collisions, through spin-exchange. True, this is not a pure electron-atom collision situation, but even so, Purcell and Field were able to make reasonable estimates of the spin-exchange cross sections, which in turn made it possible to estimate absolute atomic hydrogen number densities.

Shortly thereafter optical pumping appeared on the scene,[23] and shortly after that the first true electron-atom spin exchange experiment was performed by Dehmelt.[24] Dehmelt used resonance radiation to optically align a sodium vapor. Indication of alignment is an increase of transmitted radiation through the cell, once the sodium alignment becomes saturated. Dehmelt then applied a weak rf discharge, to produce some free electrons. The free electrons became partially polarized by exchange collisions with aligned sodium atoms. Subsequent application of an rf magnetic field, in the presence of an external dc magnetic field, will cause depolarization of the free electrons, followed by a decrease in sodium alignment caused by exchange collisions, followed by an increased absorption of the resonant optical radiation. This is shown in Fig. 2, the first direct demonstration of spin-exchange. Fig. B is a background signal, no resonances, of transmitted optical intensity vs applied magnetic field. Fig. C shows the atomic sodium resonance (at 15.9 MHz), and Fig. A shows the free electron resonance (at 62.1 MHz). This observation was of particular importance because Dehmelt was able to directly determine (g-2) for the free electron, g being the gyromagnetic ratio. A rough estimate on the lower bound of the total elastic spin-exchange cross section was obtained by estimating line-widths, although this was not the principal goal of the experiment.

Shortly thereafter, in two independent experiments, Franken et al[25] used optical pumping, monitoring transmitted resonant radiation,

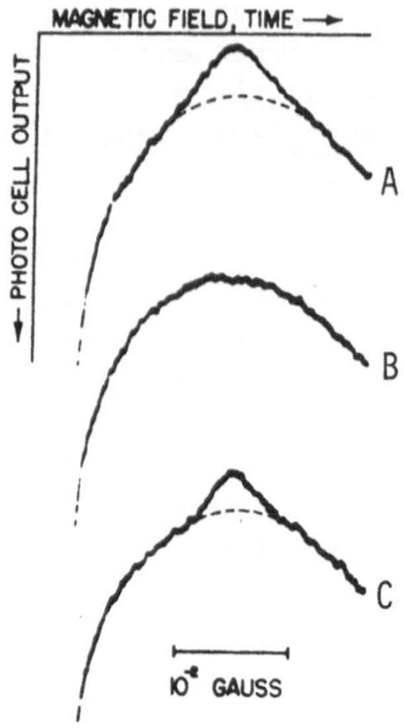

Fig. 2. The first "free electron" exchange signal, using optical pumping (Dehmelt[24]). See text for description of the observation.

to observe transfer of polarization from sodium to potassium (Franken) and from sodium to rubidium (Novick) via free electron exchange. The last such experiments we will mention were quantitative studies of spin exchange in rubidium and cesium, by Balling and Pipkin[27] in 1969, who exploited measurements of linewidths and frequency shifts to obtain estimates of the spin-exchange cross sections at thermal energies. I do not believe that there have been significantly improved experimental determinations of thermal spin-exchange cross sections since that time.

The first observation of "single-collision" spin exchange was made by Rubin et al[28] at New York University. They argued by analogy with magnetic resonance ("Rabi flop"): if one could affect the trajectory of an atomic beam using inhomogeneous magnetic fields by changing spin orientation by rf spectroscopy, why not by electron exchange? Indeed, this idea worked well, although there was one significant difference between the two means of changing hyperfine states: rf fields transmit little linear momentum ("recoil") and so the spin analyzer, called the "B" magnet by atomic beamists, must be capable of rotation around the interaction center. As an extra dividend, this discrimination in the atom scattering angles can be related to polar and azimuthal electron scattering angles, with surprisingly good resolution. The result is that differential exchange cross sections can be measured, and, by integration, total

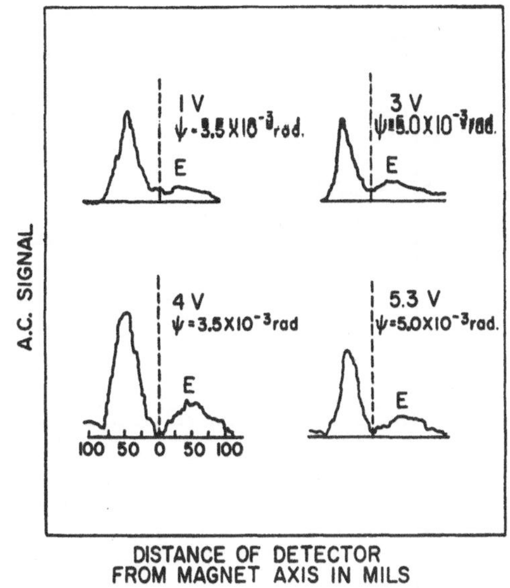

Fig. 3. First crossed-beam direct observation of elastic exchange scattering, Rubin et al.[28] The peaks marked E are due to exchange T.

exchange cross sections as well. Fig. 3 shows the first directly observed exchange cross sections. In this experiment the atom beam is first polarized and velocity-selected by an offset Stern-Gerlach magnet. The "B" magnet is actually an E-H gradient analyzer.[29] This device consists of electrically isolated, inhomogeneous magnet pole faces, which can simultaneously sustain congruent electric and magnetic fields. Atoms possessing negative effective magnetic moments can experience a null combined induced electric dipole moment force (i.e., due to the atomic polarizability) and magnetic moment force. In the exchange experiment the E-H gradient analyzer is set for extinction, so that it could only transmit atoms with spins antiparallel to the polarized beam, and accordingly any signal observed at E must be due to exchange (assuming no spin-orbit effects). Differential exchange cross section data for potassium at four energies is shown in Fig. 4, compared to an early close-coupling calculation of Karule.[30] These data were among the first to demonstrate the singular effectiveness of few-state close-coupling expansions for effective "single-electron" atoms, i.e., the alkalis. In fact, by a peculiarity of nature, the small energy denominator of the resonant ns-np transition in the alkalis results in much faster convergence than for the seemingly simpler case of atomic hydrogen!

Extensions of this technique to excitation cross sections can readily be made. Here one measures the so-called "spin-flip" cross sections -- the reaction that results in a reversal of ground-state polarization caused by a combination of direct and exchange

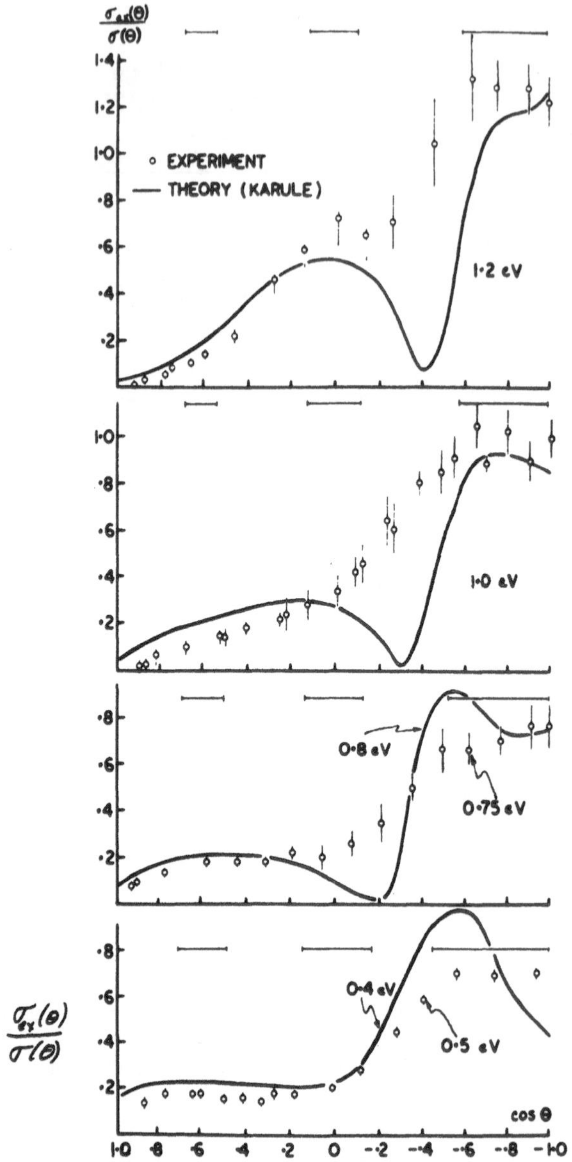

Fig. 4. $\sigma_{ex}(\Theta)/\sigma(\Theta)$ vs $\cos\Theta$, for elastic scattering of electrons by potassium, Collins et al.[28] Solid curves are results of close-coupling calculation of Karule.[30]

excitations to specific excited fine-structure states, which is followed by a mixing of these states when the atom relaxes into a $^2P_{1/2,3/2}$ multifold, in the Jm_J representation, finally followed by decay to ground states in accordance with the specific transition probabilities. These experiments are a subclass of more general ones which involve use of polarized electron beams as well as polarized atomic beams, and also, in a complete experiment, observation of the emitted photon, including its polarization, in coincidence.

A full analysis of this one electron S-P excitation problem was first performed by Rubin et al,[31] expressed directly in terms of scattering amplitudes, rather than in terms of coherence parameters and state multipoles, as was done later by Fano and Macek,[6] Kleinpoppen,[1,3] Macek and Hertel,[7] etc. A summary of some of the results is presented here. The analysis essentially covered all possibilities for determination of collision parameters for this simple excitation process, which, neglecting spin-orbit effects, consists of four scattering amplitudes ($f^{\pm 1}$, $g^{\pm 1}$, f^{0}, g^{0}) and therefore seven observable parameters (one arbitrary phase factor is unobservable). The superscripts refer to excitation with a change of M_L of ±1,0, the f,g refers to direct and exchange cross sections. An early example of spin flip data is shown in Fig. 5, which plots R (the ratio of spin flip to full cross section for small-angle scattering) as a function of energy above threshold.

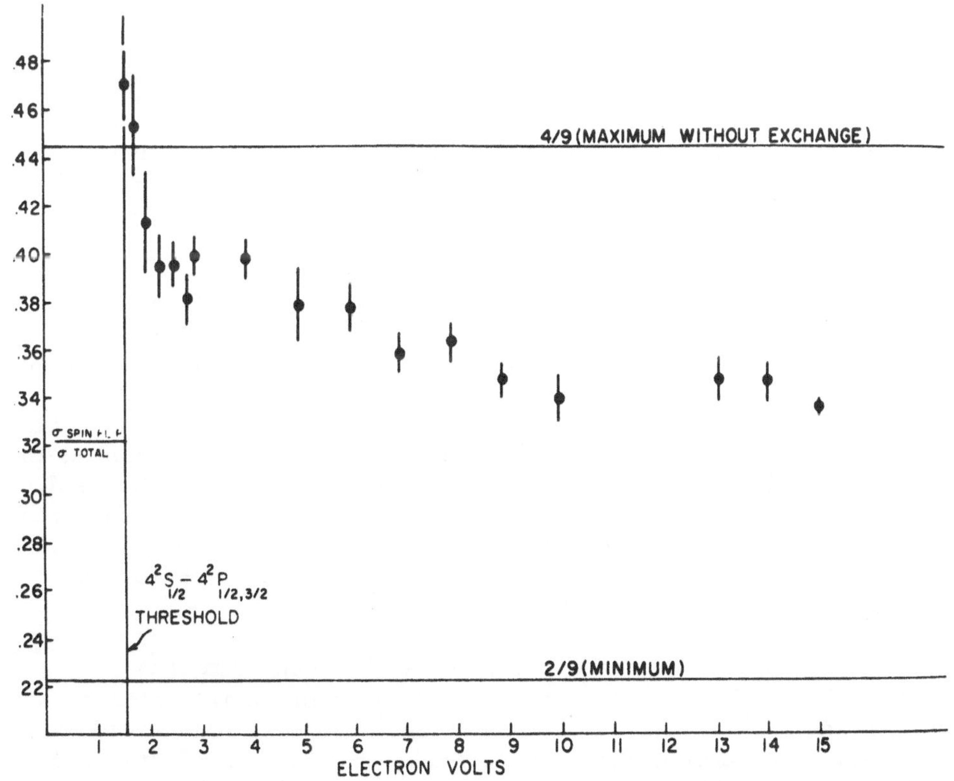

Fig. 5. Ratio of "spin-flip" to "total" cross section, 4^2P excitation in potassium. Rubin et al.[31]

Rubin et al[31] show that

$$R = \frac{4}{9} + \frac{1}{9}\,\frac{6|g^1|^2+|g^0|^2-4|f^1-g^1|^2-4|f^1|^2}{|f^0-g^0|^2+|g^0|^2+|f^0|^2+2(|f^1-g^1|^2+|g^1|^2+|f^1|^2)} \tag{2}$$

If the exchange amplitudes can be neglected then

$$R(\text{no exchange}) = \frac{4}{9} - \frac{4}{9}|f^1|^2/(2|f^1|^2+|f^0|^2) \; . \tag{3}$$

Alternatively, Eqs. (2) and (3) can be written as

$$R = \frac{4}{9} + \frac{1}{18}(|g^0|^2+10|g^1|^2-8\sigma_1)/\sigma \tag{4}$$

and

$$R(\text{no exchange}) = (\sigma_0 + \sigma_1)/\sigma \; , \tag{5}$$

where $\sigma=\sigma_0+2\sigma_1$ and σ_0,σ_1 are the differential cross sections for excitation with $\Delta m=0,\pm1$, respectively, i.e.,

$$\sigma_{0,1} = \frac{3}{4}|f^{0,1}-g^{0,1}|^2 + \frac{1}{4}|f^{0,1}+g^{0,1}|^2 \tag{6}$$

At threshold, both f^1,g^1 are zero, hence $\frac{4}{9} \leqslant R \leqslant \frac{5}{9}$.

The lower limit obtains if $g^0=0$, and the upper if $g^0=f^0$ (in both magnitude and phase). The data in Fig. 5 clearly show a small but non-negligible contribution to exchange at the threshold for S-P excitation in potassium.

More generally Rubin et al[31] derive a "complete" relation for all combinations of polarized atoms, electrons and photons. This expression includes all possible cases, including interference terms between the fine structure levels, which need to be resolved in order to measure f^0,f^1 and g^0,g^1 phase differences. The other five parameters can be obtained in coincidence experiments, as listed in Table I. It is interesting to observe that the laser can be used, in the time-reversed "superelastic" experiment, to prepare the atom in specific fine-structure states (this will be discussed later), so that the requirement on one or two meV energy resolution for the scattered electron is not required!

It should also be noted that, as can be seen from this discussion, in the present example where spin-orbit interactions are not considered, the use of polarized electrons will only yield information concerning the role played by exchange in elastic and inelastic interactions. The more general role played by collision

Table I. Relative probabilities that the atom (originally in $M_S=+\frac{1}{2}$ state) decays to a particular ground state M_S, with correlated photon including polarization, and polarized electron. From Ref. 31.

Inital spin state of incoming electron	Final spin state of atom M_s	Final spin state of outgoing electron	Probability of achieving specified final state with the emission of photon is proportional to: (photon parallel to final spin state)	(photon perpendicular to final spin state)
α	$+\frac{1}{2}$	α	$\frac{4}{9}\lvert f^0-g^0\rvert^2\sin^2 A+\frac{1}{9}\lvert f^1-g^1\rvert^2\cos^2 A$	$\frac{1}{9}\lvert f^1-g^1\rvert^2$
α	$+\frac{1}{2}$	β	0	0
α	$-\frac{1}{2}$	α	$\frac{2}{9}\lvert f^0-g^0\rvert^2\cos^2 A+\frac{4}{9}\lvert f^1-g^1\rvert^2\sin^2 A$	$\frac{2}{9}\lvert f^0-g^0\rvert^2$
α	$-\frac{1}{2}$	β	0	0
β	$+\frac{1}{2}$	α	$\frac{4}{9}\lvert g^1\rvert^2\sin^2 A+\frac{2}{9}\lvert g^0\rvert^2\cos^2 A$	$\frac{2}{9}\lvert g^0\rvert^2$
β	$+\frac{1}{2}$	β	$\frac{4}{9}\lvert f^0\rvert^2\sin^2 A+\frac{1}{9}\lvert f^1\rvert^2\cos^2 A$	$\frac{1}{9}\lvert g^1\rvert^2$
β	$-\frac{1}{2}$	α	$\frac{4}{9}\lvert g^0\rvert^2\sin^2 A+\frac{1}{9}\lvert g^1\rvert^2\cos^2 A$	$\frac{1}{9}\lvert g^1\rvert^2$
β	$-\frac{1}{2}$	β	$\frac{2}{9}\lvert f^0\rvert^2\cos^2 A+\frac{4}{9}\lvert f^1\rvert^2\sin^2 A$	$\frac{2}{9}\lvert f^0\rvert^2$

Tabulation of final electron and atom spin-states, assuming initial atomic spin state $m_s=+\frac{1}{2}$ and initial electron spin-states parallel (α) and antiparallel (β) to external magnetic field. $S\rightarrow P$ excitation in alkali (nuclear spin neglected).

dynamics can only be probed by studying the atomic and radiation polarization.

POLARIZED ELECTRONS

In recent years the technology of production of polarized electrons has taken great strides. The story is well-documented in the literature, so there is little need to include a detailed discussion here. As mentioned already, the history and status of the field up to about 1976 is well described in the book Polarized Electrons by J. Kessler.[14] The book contains thorough discussions of various early methods used to produce polarized electrons. Over the years a number of techniques have been used, with some success (and occasional failures). Most noteworthy among these methods were photoionization from polarized alkali atoms,[32] photoionization with circularly polarized light ("Fano effect"),[33] photoemission[34] and field emission[35] from magnetized surfaces, optical pumping sources[36] (using exchange between polarized atoms and free electrons as the early spin exchange experiments described in the Introduction), and differential electron scattering by heavy atoms, especially mercury.[37] This last method was probably the most fruitful technique, yielding much important scattering data by Kessler and others, prior to the introduction of the GaAs source. Just at the time of publication of the Kessler book, the GaAs polarized electron source appeared on the scene, and has by now become the standard source for polarized electrons in collision experiments. I believe that the first suggestions that electrons photoemitted from a GaAs surface using circularly polarized resonant light could be polarized were by Garwin, Pierce, and Siegmann,[38a] and Lampel and Weisbuch.[38b] The crucial trick that converted this interesting idea into a useful device could only have been conjured up by "surface", i.e., solid-state, physicists[39], introducing a $Cs-O_2$ surface that resulted in an effective negative electron affinity for conduction-band electrons, so that the polarized electrons as they migrated to the surface were ejected into the vacuum (incidentally, resulting also in good energy resolution). A detailed "nuts and bolts" description is given by Pierce et al.[40] As an example of the recent utility of this method, I show the near threshold measurements at the National Bureau of Standards by Kelley et al,[41] for the ionization of sodium. This experiment was motivated by a recent re-examination of the ionization threshold law by Temkin,[42] who predicted oscillations in the cross section above threshold, to modulate the Wannier $E^{1.127}$ power law, where E is the excess electron energy above the threshold value, if polarized electrons are used. The data clearly puts stringent limits on these oscillations (although of course not totally ruling them out).

I will mention only one more polarized electron experiment, out of the many that are now or have been recently performed. This is the experiment by Mollenkamp et al,[43] on mercury (using Fano effect polarized electrons) and xenon (using a GaAs source) to obtain, for elastic scattering and at several fixed energies and angles, what can be characterized as "complete" information. Here, because the atoms possess zero net electron spin, but with spin-orbit interaction not negligible, there are again only two scattering amplitudes, the direct f and spinflip g. There are four observables, σ, S, T, U, with σ $(= |f^2| + |g^2|)$ the differential cross section, S the "Sherman function",

$$S = \frac{2|f||g|\sin\gamma}{\sigma} \tag{7}$$

and T, U the polarization parameters,

$$T = \frac{|f|^2|g|^2}{\sigma} \tag{8}$$

and

$$U = 2|f||g|\cos\gamma \tag{9}$$

where γ is the phase difference between f,g. One constraint $(S^2+T^2+U^2 = 1)$ reduces the number of independent measurements to three. Results, for mercury, compared to calculations by Walker[44] and several earlier measurements, where available, are shown in Fig. 6. This is but one example, albeit a singularly interesting one, of a thriving activity that shows no sign of abatement.

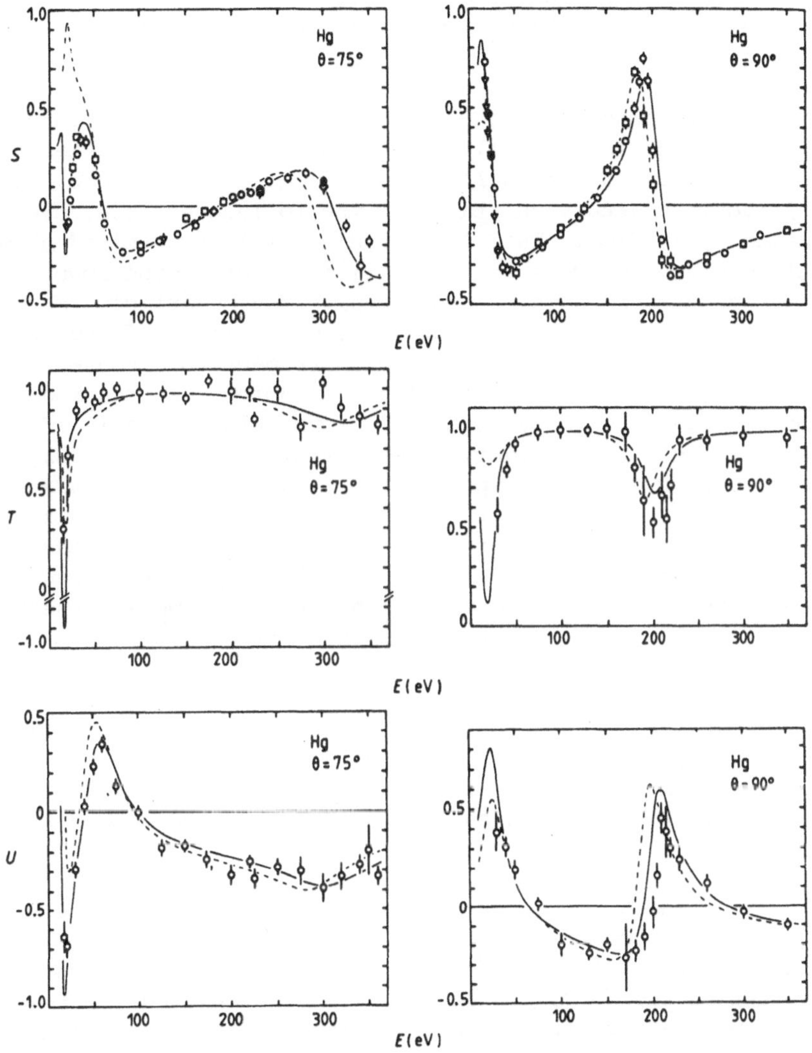

Fig. 6. From Mollenkamp et al.[43] Energy dependence of the parameters S, J, and U at $\Theta=75^{\circ}$ and 90° for mercury. See text. Circles are data of Ref. 43, other data from earlier work of Kessler and co-workers. Theoretical curves are those of Walker,[44] solid line with, and dashed line without exchange, convoluted with $\pm 2^{\circ}$ in S and $\pm 4^{\circ}$ in T,U diagrams.

The role of the laser in atomic collision physics is not simply stated, because it is now so ubiquitous, but for the purposes of this article it has a well-defined function: to prepare an atom in a specific fine -- or hyperfine -- excited state (see the Lambropoulos article in this volume for a broader interpretation of the laser's role). Viewed from the point of view of state-preparation (and, occasionally, analysis), the laser is then simply another means of producing and analyzing polarized beams. And, of course, and not so incidentally, at the same time it elevates the atom under a study to a (usually) short-lived excited state, so that interactions with such states, hitherto inaccessible, can be systematically studied. The first studies of laser-excited atoms interacting with electrons were performed by Hertel and coworkers;[45] this work is summarized in a comprehensive review article (Hertel and Stoll[46]). In these experiments, the orientation of the charge cloud of the laser-excited P-states with respect to a convenient quantization axis (the direction of the incoming electron beam), could be altered at will by rotating the polarization axis of the linearly polarized laser light. The de-excitation, or "superelastic" cross section, itself then experiences a modulation which tracks the polarization axis, and which is a measure of the relative strengths of the scattering amplitudes f^0, f^1. More recently Hertel has extended these studies to atom-atom collisions,[47] where orientation-dependent interactions are equally interesting; however, these experiments are not relevant to the present discussion.

In what remains of this paper I will concentrate mostly on some recent work at NYU related to laser excited scattering, partly because it is work with which I am most familiar, and partly because it illustrates the unique way that lasers can be used as a remarkably selective tool in performing polarization beam collision experiments.

The experiments to be described here use a modified version of what is known as the atomic beam recoil technique, as discussed above in connection with spin exchange and spin flip experiments. The relevant angles (ψ,χ) of the atomic scattering and (Θ,Φ) using MV, MV' as parameters (the linear atom momenta before and after collision). A vertical cross section of the apparatus is shown in Fig. 8. The long field-free drift space, between interaction and detection regions (3m in length) results in relatively high angular resolution. The entire drift-tube and detector assembly can be rotated in two dimensions about the interaction volume. In the laser excited experiments, the single-mode ring dye laser beam enters the interaction volume through a window in the top of the collision chamber.

A schematic of the experiment is shown in Fig. 9. The atomic

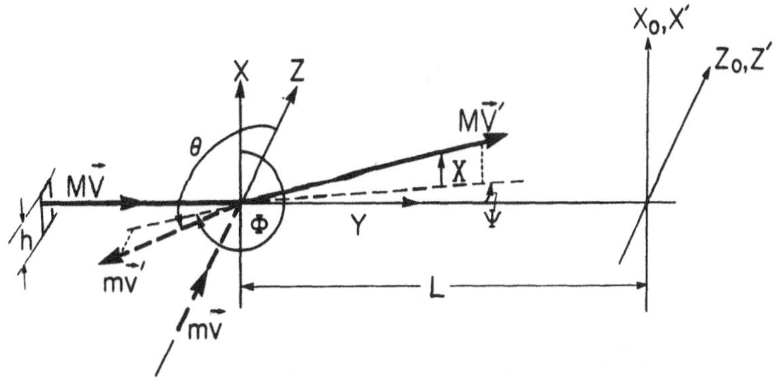

Fig. 7. Atomic-recoil and electron scattering angles. $M\vec{V}, M\vec{V}'$, atomic momenta before and after the collision; $m\vec{v}, m\vec{v}'$, electron momenta before and after the collision. h is the height of the interaction volume, L the distance between interaction volume and detector plane. In the absence of collisions an atom crosses the detector plane at X_0, Z_0; after a collision, the coordinates are X', Z'.

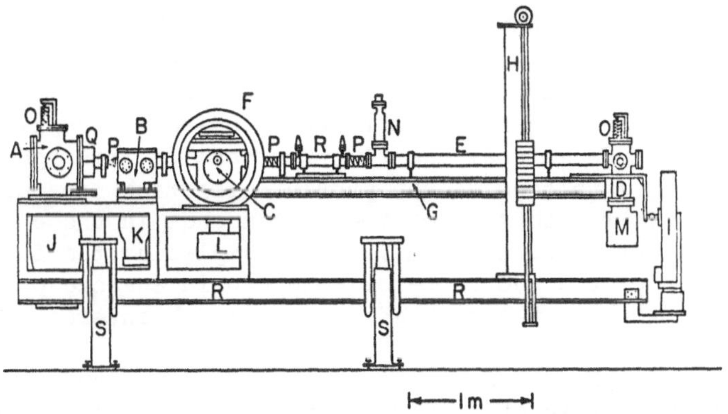

Fig. 8. General apparatus layout of electron-photon-atom double-recoil experiment. A, source chamber; B, intermediate chamber; C, collision chamber; D, detector chamber; E, drift tube; F, mount providing two-dimensional scanning motion for the detector; G, movable detector boom; H, counter-weight system; I, detector positioner; J, 10-in. diffusion pump; K, 4-in. diffusion pump; L, 300-l/sec ion pump; M, 150-l/sec ion pump; N, Ti sublimation pump; O, internal component positioner; P, flexible bellows; O, hexapole magnetic coils and yoke; R, aluminum mounting structure; S, vibration isolators (Ref. 49).

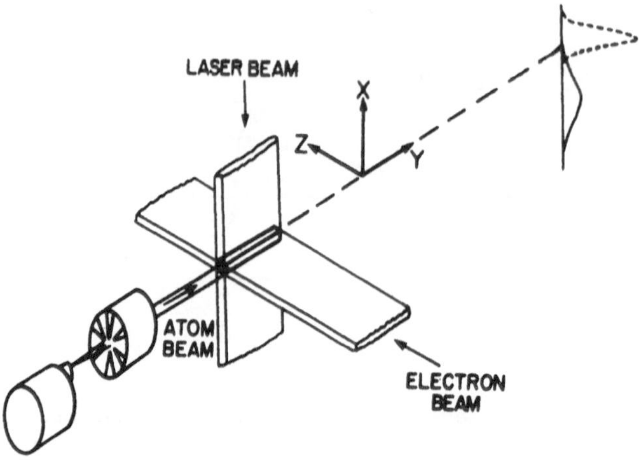

Fig. 9. Experimental arrangement of electron-photon-atom double-recoil experiment. The effusive atomic sodium beam is incident along the Y axis; the electron beam along the Z axis, and the laser beam along the X axis. The atomic beam is focused and state and velocity selected by a hexapole electromagnet. The curves at right show the shape of the undeflected (dash line) and photon-deflected (full line) atomic beams. (Ref. 49)

beam is focussed and partially state-selected by the six-pole magnet. An important distinction between ground- and excited-state experiments is that in the latter case the laser beam itself introduces significant recoil into the atom beam (roughly the same order of magnitude as does the electron beam). This "photon recoil" can be used to advantage -- the photon deflection is approximately proportional to the total time the atom spends in the excited state while in the interaction volume, so by displacing the detector in the direction of the photon recoil one can in essence preselect the fraction of atoms in the excited state, in the interaction volume.

In Fig. 9, photon recoil is in the -X direction, electron recoil (except for superelastic scattering) is in the +Z direction.

To get an idea of the actual photon recoil angles obtained first we note that the recoil angle α_0 per resonantly absorbed photon is

$$\alpha_0 = h\nu/MVc \qquad (10)$$

which is about 3×10^{-5} rads for a thermal sodium beam. This corresponds to a displacement of 9×10^{-3} cm at the detector. For circularly polarized, $F=2 \rightarrow F=3$ sodium light a typical atom emits between 250 and 500 spontaneous photons in traversing the interaction volume, so that the net photon recoil produces atom

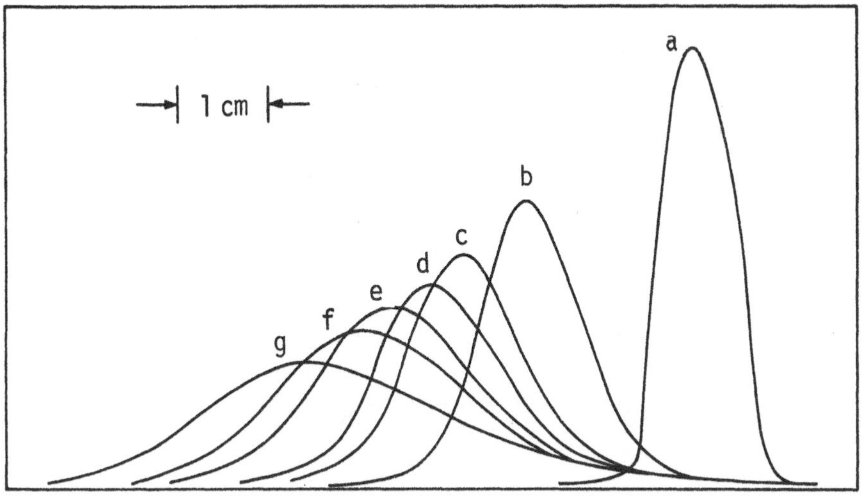

Fig. 10. (a) Electron-photon-atom double-recoil experiment atomic beam vertical profile, and laser-deflected profiles with nominal laser power, (b) 25 mW, (c) 50 mW, (d) 100 mW, (e) 200 mW, (f) 300 mW and (g) 500 mW. Detector is downstream of interaction region. (G. F. Shen, B. Jaduszliwer and B. Bederson, to be published.)

deflections of 3 to 6 cm, which is of the same order of magnitude as recoil caused by electrons. For linearly polarized $F=2 \rightarrow F=3$ sodium light, where optical pumping occurs because of residual $F=2 \rightarrow F=3$ transitions, the deflections are somewhat smaller ($\sim$1-2 cm), reflecting the fact that the excited state fractions are somewhat smaller. The photon recoil effect is shown in Fig. 10, which plots atom beam profile data taken at several values of laser power, for circularly polarized $F=3 \rightarrow 3$ light.

Scattering experiments can generally be performed in either "scattering out" or "scattering in" modes. The former is analogous to a transmission experiment, where normally one observes attenuation of an electron beam passing through a target gas; the latter is analogous to a differential measurement. Displacing the detector to $(-X_0')$, it is easy to show that the fraction f of atoms in the excited state in the interaction volume is

$$f = (X_0'/L\alpha_0)(\tau_a/\tau) \tag{11}$$

where τ, τ_a are the times spent by an atom in the interaction volume, and the atom mean lifetime, respectively. L is the distance from interaction volume to detector. A scattering-out experiment performed with the detector at X_0' gives an "effective" cross section

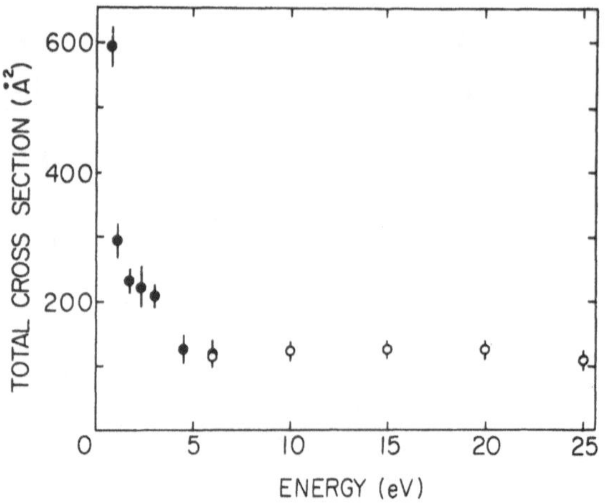

Fig. 11. Total cross sections for electron scattering on 3^2P sodium atoms. Black dots: Jaduszliwer et al (Ref. 48). Open circles: Ref. 49. Error bars denote statistical errors and a maximum estimate of possible systematic errors. (Ref. 49)

$$\sigma_{eff} = (1-f)\sigma_g - f\sigma_{ex} \qquad (12)$$

where σ_g, σ_{ex} are the ground and excited state cross sections, respectively. Fig. 11 shows results of two experiments performed in transmission for the "grand total" $^2P_{3/2}$ cross section. One of these was performed in a "high" magnetic field, in sodium (785 G); at this field the excited $^2P_{3/2}$ magnetic substates are sufficiently separated in frequency so that only the $M_J=3/2$, $M_I=3/2$ can be excited. In this circumstance we are dealing with an effective two-level system. The electron beam travels along the magnetic field, i.e., axis of quantization, direction, which is also the direction of the laser exectronic field, i.e., the photon and collision frames of reference are the same; the scattering was therefore from a pure $|1,1\rangle$ state in L, M_L. The other experiment, performed more recently, was in weak field. In the case of circularly polarized light, the atoms are excited to the $^2P_{3/2}$, F=3 M=3 state in the photon frame; the quantization axis is -X (Fig. 7), for right-circularly polarized light incident from the -X direction. The electrons are incident from the Z direction. The atomic state in the collision frame can be determined by using the appropriate rotation matrices to be

$$|\Psi_{coll}> = \frac{1}{8}\Big\{|3/2,3/2> + \sqrt{3}\,|3/2,1/2> + \sqrt{3}\,|3/2,-1/2>$$
$$+ |3/2,-3/2>\Big\}\Big\{|1/2,1/2> + |1,1> + \sqrt{2}\,|1,0> + |1,-1>\Big\} \quad (13)$$

where the first bracket refers to nuclear spin quantum numbers, $|I,M_I>$; the second bracket, to the electron spin, $|S,M_S>$, and the third one, to the orbital angular momentum $|L,M_L>$; this last bracket is the only one entering explicitly into the collisional dynamics.

At the one energy (6eV) for which data were obtained in both the early and the present experiment, the cross sections are in very good agreement; $(120 \pm 24) \times 10^{-16}$ cm^2, respectively. This implies that the anisotropy contribution at 6eV is at most of the order of the combined errors, perhaps 30%.

To conclude this summary I would like to mention, as promised earlier, a recent experiment which offers a certain symmetry, or rather asymmetry, to the first experiment discussed (Cox et al). Register, Trajmar[50] et al have performed some differential superelastic measurements of electron scattering by laser-excited $^1P_1$ barium. In this work a linearly polarized laser beam was incident on a barium beam, in the plane of scattering. The superelastic scattering signal was measured as a function of polarization direction Ψ of the laser light. The surprising result was obtained that at small scattering angles (Θ_e), the superelastic scattering signal was asymmetric to reflection of Θ_e in the scattering plane (with respect to the incoming electron's direction). The asymmetry depended upon the incident laser angle (Θ_ν). It could be characterized by an asymmetry phase factor α which depended upon the electron energy, Θ_e, and Θ_ν.

Both incoming barium and electron beams were unpolarized, and accordingly there does not appear to be any mechanism at work that would distinguish positive Θ_e from negative Θ_e. Therefore there should be no asymmetry, regardless of details of the interaction. A schematic of the experimental setup is shown in Fig. 12. Representative scattering data (E = 100 eV, Θ_e = $-2^o, 0^o, +2^o$) is shown in Fig. 13. In this Figure the abscissa is the polarization orientation Ψ, the ordinate is scattering signal intensity in arbitrary units. The oscillation in each curve represents the effect of the orientation of the P-wave with respect to the incoming electron beam, similar to that explored in detail by Hertel and Stoll. The mystery is in the phase shift of the say -2° curve with respect to the $+2^o$ one. For a thorough discussion of the experiment, and for detailed results, please refer to the original article.

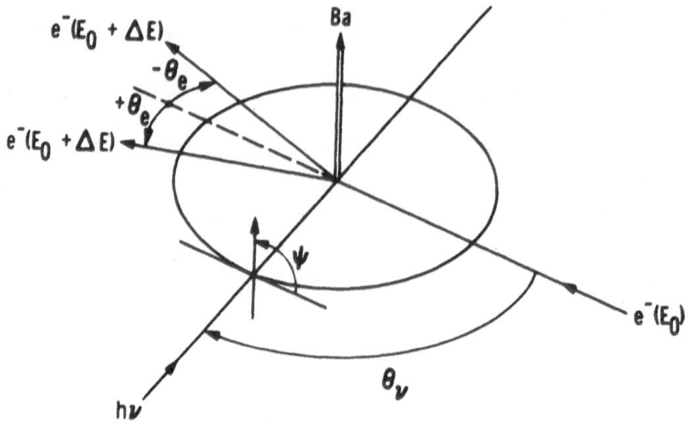

Fig. 12. Schematic representation of the scattering geometry, Register et al, Ref. 50. The Ba beam is perpendicular to the scattering plane. Laser beam is in the scattering plane at angle Θ_ν with respect to the incoming electron beam. Scattering to the left is associated with positive and to the right with negative scattering angles. Polarization of the light is rotated clockwise from the Ba target point of view and in the present experiments we start our measurements always at $\Psi=90^\circ$.

The main point I would like to make here is similar to that I made with respect to the Cox et al experiment. Lacking a theoretical explanation, the experimental result remains suspended, so to speak. One doesn't quite know what to do with it. Doubtless, attempts will be made to duplicate it (I don't know of any yet, however). But no one is ready to make any fundamental change in collision theory at this time; the attitude is, one might say, a "wait and see" one.

The authors themselves summarize the situation properly. They state,[50] "either (1) the results are an artifact of the experimental conditions, which we cannot recognize at present, or (2) an effect has been observed which is inconsistent with our present understanding of collision physics for heavy atoms." It will be interesting to await the verdict.

CONCLUSION

I have attempted in the talks upon which this article is based to present a short summary of some very selective work in polarized atomic collisions. Early history was emphasized, to show the strong continuity with past research that helps characterize this still young and developing field. I have naturally emphasized work performed at New York University mainly because it is work with which I am most familiar. I have also chosen a few selected examples from the enormous range of polarized beam research currently being pursued, particularly in Germany and the US.

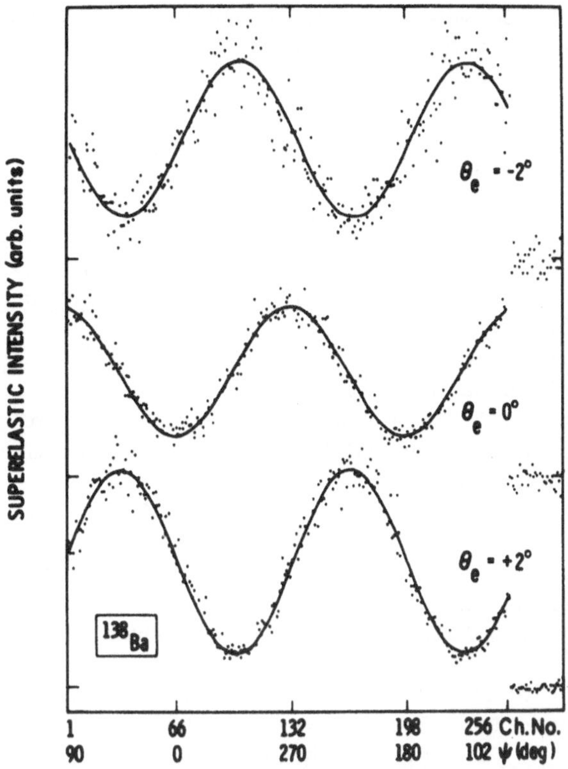

Fig. 13. Superelastic scattering intensities as a function of laser light linear polarization angle (Ψ) and channel number of the multichannel scalar (ch. No.) at E_0=100eV and Θ, ranging from -2° to +2°. Laser position corresponded to Θ_ν=45°. Experimental points and a fit to them are indicated. Background signal is shown on the right side for each scattering angle. (Register et al, Ref. 50.)

ACKNOWLEDGMENTS

The work at New York University was performed in collaboration with my colleagues Kenneth Rubin, Tom M. Miller, and Bernardo Jaduszliwer, as well as some excellent students including R. E. Collins, R. Dang, P. Weiss, A. Tino, and G. F. Shen.

Some of the NYU work described herein was, and is, supported by the National Science Foundation.

REFERENCES

1. see e.g. H. Kleinpoppen, Advances in Atomic and Molecular Physics, Academic Press, NY, 423 (1979); M. S. Lubell, Atomic Physics 5 325 (1977); B. Bederson, Atomic Physics 3 401 (1973).

2. B. Bederson, Comments in Atomic and Molecular Physics, 1, 41, 65 (1969); 2, 160 (1971).
3. K. Blum and H. Kleinpoppen, Phys. Rep. 52, 204 (1979); H. Kleinpoppen, Phys. Rev. A 3, 2015 (1971).
4. V. I. Ochkur; Zh.Tksp.Teor.Fiz. 45, 734 (1963) (Soviet Physics JETP 18, 503 (1964).
5. R. Bonham, J. Chem. Phys. 36, 3260 (1962).
6. U. Fano and J. Macek, Rev. Mod. Phys. 45, 553 (1973).
7. J. Macek and I. V. Hertel, J. Phys. B, 7, 2173 (1974).
8. T. M. Miller and B. Bederson, Advances in Atomic and Molecular Physics 13, 1 (1977).
9. D. W. Norcross and L. A. Collins, Advances in Atomic and Molecular Physics 18, 241 (1982).
10. M. F. Mott and H. S. W. Massey, The Theory of Atomic Collisions Oxford U. Press, First Edition 1933; Second Edition 1949; Third Edition 1965.
11. I. C. Percival and M. J. Seaton, Proc. Camb. Phil. Soc. 53, 654 (1957).
12. M. J. Seaton, Atomic and Molecular Processes, Ed. D. R. Bates, Academic Press, NY (1966).
13. P. G. Burke and J. F. B. Mitchell, J. Phys. B 6, L 161; 7, 214 (1974).
14. Polarized Electrons, J. Kessler, Springer-Verlag (1976).
15. P. S. Farago, Rep. Prog. Phys. 34, 1055 (1971
16. R. T. Cox, G. G. MacIlwraith and B. Kurrelmayer, Proc. Nat. Acad. Sci. USA 14, 544 (1928). Note that the classic Mott paper appeared after the Cox experiment; N. F. Mott, Proc. Roy. Soc. A124, 425 (1929).
17. C. S. Wu, E. Ambler, R. W. Hayward, D. D. Hoppes, and R. P. Hudson, Phys. Rev. 105, 1413 (1957).
18. H. W. B. Skinner, Proc. Roy. Soc. A112, 642 (1926).
19. H. W. B. Skinner and E. T. S. Appleyard, Proc. Roy. Soc. A117, 224 (1927). The first "modern" polarization of resonance radiation experiment was performed by W. E. Lamb and T. H. Maiman, Phys. Rev. 105, 573 (1957).
20. Resonance Radiation and Excited Atoms, A. C. G. Mitchell and M. W. Zemansky, Cambridge U. Press (1934).
21. J. R. Oppenheimer, Phys. Rev. 32, 361 (1928).
22. E. M. Purcell and G. B. Field, Astrophys. J. 124, 542 (1956).
23. A. Kastler, J. Opt. Soc. Am. 47, 460 (1957).
24. H. G. Dehmelt, Phys. Rev. 109, 381 (1958).
25. P. Franken, R. Sands, and J. Hobart, Phys. Rev. Lett. 1, 52 (1958).
26. R. Novick and H. E. Peters, Phys. Rev. Lett. 4, 54 (1958).
27. L. C. Balling, R. J. Hanson, and F. M. Pipkin, Phys. Rev. 133, A607 (1964).
28. a) K. Rubin, J. Perel, and B. Bederson, Phys. Rev. 117, 151 (1960); b) R. E. Collins, B. Bederson, M. Goldstein, and K. Rubin, Phys. Lett. 27A 440 (1968).

29. B. Bederson, J. Eisinger, K. Rubin, and A. Salop, Rev. Sci. Inst. 31, 852 (1960).
30. E. M. Karule, in Atomic Collisions III, ed. V. Ya. Veldre (Latvian Academy of Sciences, Riga, 1965); Transl: J.I.L.A. Information Center Report No. 3, Univ. of Colorado, Boulder, Colorado (unpublished) pp 29-48.
31. K. Rubin, B. Bederson, M. Goldstein, and R. E. Collins, Phys. Rev. 182, 201 (1969).
32. W. W. Hughes, R. L. Long, Jr., M. S. Lubell, M. Posner, and W. Raith, Phys. Rev. A 5, 195 (1972).
33. U. Fano, Phys. Rev. 178, 131 (1969).
34. G. Busch, M. Campagna, and H. C. Siegman, J. Appl. Phys. 41, 1044 (1970).
35. E. Kisker, G. Baum, A. H. Mahan, and W. Raith, Phys. Rev. Lett. 29, 1651 (1976).
36. M. V. McCulker, L. L. Hatfield, and G. K. Walters, Phys. Rev. Lett. 22, 817 (1969).
37. G. F. Hanne and J. Kessler, J. Phys. B 9, 805 (1976).
38. a) E. Garwin, D. T. Pierce, and H. C. Siegmann, Helv. Phys. Acta 47, 343 (1974); b) G. Lampel and C. Weisbuch, Solid State Commun. 16, 877 (1975).
39. For a detailed discussion, see D. T. Pierce and F. Meier, Phys. Rev. B 13, 5484 (1976).
40. D. T. Pierce, R. J. Celotta, G-C. Wang, W. N. Unertl, A. Galejs, C. E. Kuyatt, and S. R. Mielczarek, Rev. Sci. Inst. 51, 478 (1980).
41. M. H. Kelley, W. T. Rogers, R. J. Celotta, and S. R. Mielczarek, Phys. Rev. Lett. 51, 2191 (1983).
42. A. Temkin, Phys. Rev. Lett. 49, 365 (1982).
43. R. Mollenkamp, W. Wubker, O. Berger, K. Jost, and J. Kessler, J. Phys. B 17, 1107 (1984).
44. D. W. Walker, Adv. Phys. 20, 257 (1971).
45. I. V. Hertel and W. Stoll, J. Phys. B 7, 583 (1974); H. W. Hermann and I. V. Hertel, Comm. At. Mol. Phys. 12, 61; 127 (1982); H. W. Hermann and I. V. Hertel, Z. Phys. A 307, 89 (1982); H. W. Hermann, I. V. Hertel and M. H. Kelley, J. Phys. B 13, 3465 (1980).
46. I. W. Hertel and W. Stoll, Advances in Atomic and Molecular Physics, 13, (1977).
47. I. V. Hertel, Advances in Chemical Physics, J. Wiley, New York, K. Lawley, ed. 475 (1982); Fundamental Processes in Energetic Atomic Collisions, H. O. Lutz, J. S. Briggs, and H. Kleinpoppen, eds., Plenum Press (1983).
48. B. Jaduszliwer, R. Dang, P. Weiss, and B. Bederson, Phys. Rev. A 21, 808 (1980).
49. B. Jaduszliwer, G. F. Shen, J-L. Cai and B. Bederson, Phys. Rev. A 31, 1157 (1985).
50. D. F. Register, S. Trajmar, G. Csanak. S. W. Jensen, M. A. Fineman, and R. T. Poe, Phys. Rev. A 28, 151 (1983).

THEORY OF HEAVY-PARTICLE COLLISIONS

J.S. Briggs

Fakultät für Physik/Universität Freiburg
Hermann-Herder-Straße 3
D-7800 Freiburg i. Br.

1. FROM THE MOLECULAR PICTURE TO THE BORN APPROXIMATION

The basic theoretical methods applicable to the description of electron excitation, ionisation and capture in atomic collisions have been discussed in detail in the lectures of Barat, Briggs, Merzbacher and Taulbjerg at the Maratea A.S.I. in 1982[1]. Here only a brief summary of these fundamental ideas will be presented to allow emphasis to be laid on more recent developments in this field. Particular attention will be paid to the production of coherent, aligned or oriented states of projectile or target atoms as a result of the collision. It will emerge that the choice of co-ordinate frame and of quantisation axis has a strong effect upon the interpretation of the dynamics of the production of such states.

The general problem to be solved is that of the motion of a number of electrons in the combined field of two nuclei in close collision (here interest will be concentrated on those collisions in which the centre-of-mass energy exceeds several keV). Removal of the overall centre-of-mass motion leads to a problem of 3(N+1) dimensions where N is the number of electrons. Although it may seem trivial, the choice of independent co-ordinates in which to write down the Schrödinger equation for such a few-body system has a profound effect upon the "appearance" of the mathematical problem and the ease with which one or other physical approximation may be introduced to render the problem tractable. The appropriate choice of co-ordinates depends principally upon the velocity of the collision and upon the initial and final configurations of the (N+2)-particle system under study. For example, four possible choices are shown in fig. 1 for the one-electron, two-nuclei problem. There are others of more complicated form, such as hyperspherical co-ordinates in which the ratios of inter-particle separations appear.

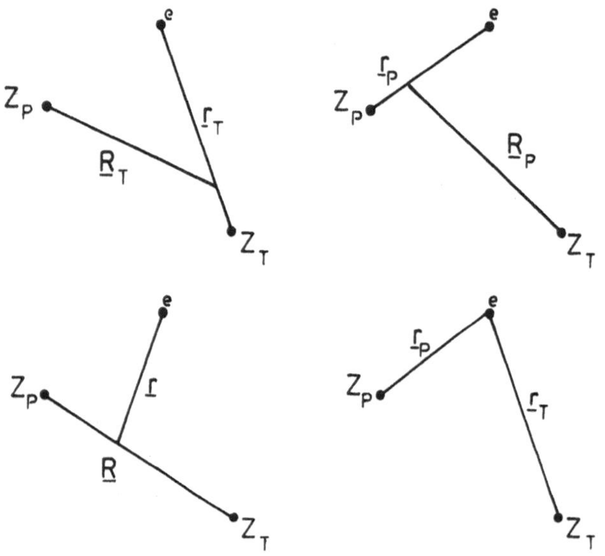

Fig. 1. Four possible choices of the six independent co-ordinates required for the motion of two nuclei and one electron.

Having chosen a set of independent co-ordinates the Schrödinger equation may be written down. For example when the internuclear separation $\underline{R}$ and the co-ordinates of each electron with respect to the centre-of-mass of the nuclei plus all other electrons are taken as independent co-ordinates the Schrödinger equation is (in units $e = m = \hbar = 1$)

$$\Big[-\frac{1}{2\mu}\nabla^2_R - \sum_i^N \frac{1}{2\nu_i}\nabla^2_{r_i} + \sum_i^N \{V_T(\underline{r}_{Ti}) + V_P(\underline{r}_{Pi})\}$$

$$+ V_{PT}(R) + \sum_{i,j}^N \frac{1}{|\underline{r}_i - \underline{r}_j|} - E \Big] \Psi(\underline{r},\underline{R}) = 0 \quad .. \qquad (1)$$

where V_T, V_P, V_{PT} represent the electron-target nucleus, electron-projectile nucleus and nucleus-nucleus Coulomb interactions respectively, and μ and ν_i are appropriate reduced masses. A hierarchy of approximations has been developed in the solution of this problem. These generally rely upon the smallness of one or other of the three ratios m/M, v/v_e and Z_P/Z_T. The first is the ratio of electron mass to nuclear mass, the second the ratio of collision velocity v to an effective electron orbital velocity v_e and the third the relative magnitude of charges of projectile and target. The character of the theoretical method used depends mostly upon the ratio v/v_e, which itself depends strongly upon whether outer-shell (low v_e) or inner-shell (high v_e) electrons are under consideration. At lower velocities the electrons occupy a limited portion of the available Hilbert space since transitions involving large energy and momentum changes are improbable. However there is sufficient time for multiple interactions to occur and this leads to an expansion of the wavefunction in a limited basis i.e. an infinite number of interactions take place but between a limited number of states. By contrast, at higher collision velocities, all electronic states including those in the continuum are readily accessible but the collision time is usually so short as to allow only a limited number of interactions. Then the wavefunction is expanded to a certain order of interaction but all intermediate virtual states are possible i.e. a limited number of interactions takes place between an infinity of states. All theoretical methods considered here fall into one or other of these two categories.

The equation (1) is written in terms of co-ordinates which essentially refer each electron's motion to the centre of the diatomic molecule composed of two nuclei plus electrons. This particular choice of co-ordinates is most appropriate for slow collisions ($v/v_e \lesssim 1$) where the wavefunction is expanded in a basis set of the form,

$$\Psi(\underline{r},\underline{R}) = \sum_n F_n(\underline{R})\, \chi_n(\underline{r},\underline{R}) \qquad (2)$$

Substitution into eq. (1) leads to the close-coupled set of equations

$$\sum_n \langle\chi_m|\chi_n\rangle \left[-\frac{1}{2\mu}\nabla_R^2 + V_{PT}(R) - E \right] F_n$$

$$+ \sum_n \left[-\frac{1}{2\mu} \langle\chi_m|\nabla_R^2|\chi_n\rangle F_n - \frac{1}{\mu} \langle\chi_m|\nabla_R|\chi_n\rangle \cdot \nabla_R F_n \right]$$

$$+ \sum_n \langle\chi_m|H_e|\chi_n\rangle F_n = 0 \qquad (3)$$

where H_e is the electronic Hamiltonian

$$H_e = \sum_i -\frac{1}{2\nu_i} \nabla_{r_i}^2 + V_P(\underline{r}_i,\underline{R}) + V_T(\underline{r}_i,\underline{R}) + \sum_{ij} \frac{1}{|\underline{r}_i-\underline{r}_j|} \qquad (4)$$

The most popular expansion at low velocities is in terms of Born-Oppenheimer molecular states which diagonalise the electronic Hamiltonian for fixed $\underline{R}$ (the so-called adiabatic states). These represent an exact solution in the limit $v/v_e = 0$ and the coupling between them comes from the finite nuclear motion represented by the terms involving ∇_R. The solution of these coupled equations usually proceeds by an expansion in partial waves which rapidly leads to a basis of very high order as the collision velocity increases. To simplify this problem recognition can be made of the smallness of the ratio m/M and the fact that the inelastic energy of the collision is normally only a small fraction of the available kinetic energy, to replace the quantum-mechanical internuclear motion by a classical one. That is, the nuclear motion is described by a fixed trajectory determined by $\underline{R}(t)$, independent of the electronic state. Then the time-independent Schrödinger equation is replaced by a time-dependent one

$$(H_e(t) - i \sum_i \frac{\partial}{\partial t}\Big|_{\underline{r}_i}) \Psi (\underline{r},\underline{R}(t)) = 0 , \qquad (5)$$

where the subscript on the $\partial/\partial t$ operator indicates the electronic co-ordinate to be held fixed whilst time derivatives are taken. Expansion in a basis set

$$\Psi (\underline{r},t) = \sum_k C_k(t) \chi_k(\underline{r},t) \qquad (6)$$

leads to a set of coupled equations

$$i \sum_k \langle\chi_\ell|\chi_k\rangle \dot{C}_k(t) = \sum_k \langle\chi_\ell|(H_e - i \sum_i \frac{\partial}{\partial t}\Big|_{\underline{r}_i})|\chi_k\rangle C_k(t) \qquad (7)$$

for the occupation amplitudes $C_k(t)$.

For few-electron systems, good approximations to the Born-Oppenheimer molecular states (e.g. Hartree-Fock or better) may be obtained, although it is never possible to fully diagonalise the electron-electron interaction. As the number of electrons increases the treatment of electron-electron interaction becomes more and more difficult. However, in certain cases, for example the excitation of inner-shell electrons, it is possible to reduce the many-electron problem to an effective one-electron problem. If many-electron states are built from a fixed set of one-electron Hartree-Fock molecular orbitals, if the residual electron-electron interaction is ignored and if transitions between fully-closed and fully-open shells only are considered then the problem reduces to that of the motion of one hole in the spin-independent, one-electron, electron - nuclear potential. Henceforth, such a one-hole or one-electron picture will be considered for simplicity. The departures from the one-electron picture are currently under study and it is a particularly challenging problem to sort out the effects of spatial electron correlations present in initial or final bound states from the temporal correlations induced by the perturbation from the moving nuclear charges. Since in collisions involving highly-charged ions multiple electron excitation is often the rule rather than the exception the proper treatment of the many-electron collision problem is likely to become more necessary as the deficiencies of the independent-electron model are exposed.

The essential consequences of the molecular orbital (MO) model of atomic collisions for alignment of states produced by impact are very clearly and directly seen in the classical trajectory picture. The basic Born-Oppenheimer MO have well-defined symmetry with respect to the internuclear axis $\underline{R}(t)$. As $|\underline{R}|$ changes, the energy of each MO changes but the symmetry is invariant. Whether or not the electronic motion is adiabatic depends upon the ability of the electron to follow faithfully the contraction (or expansion) of the internuclear distance and the rotation of the internuclear axis in space. Mathematically this is expressed by the decomposition of the time differential operator as

$$\frac{\partial}{\partial t} = \dot{R} \frac{\partial}{\partial R} + \dot{\theta} \frac{\partial}{\partial \theta} \tag{8}$$

where $\theta(t)$ is the instantaneous orientation of the internuclear axis with respect to some laboratory-fixed direction (usually the beam direction). The first term, the radial coupling, represents the rate of contraction or expansion of the internuclear axis. It preserves the symmetry of the MO but changes their radial nodal structure i.e. it couples different MO of σ symmetry, of π symmetry and so on. Transitions between MO of different symmetry are caused by the second term in (8), the rotational coupling. Transitions occur when electrons cannot adjust to the rapid rotation of the internuclear axis and therefore must change their symmetry with respect to this axis, involving concomitant changes in energy. This is the origin of inelastic transitions in slow ion-atom collisions. Since a given MO has definite symmetry along R, then predominant population of one MO during a collision implies automatically a strong alignment and/ or orientation with respect to the final direction of the internuclear axis. Except for collisions at very small scattering angles, the description of the corresponding alignment with respect to the beam axis will not be so simple. For this reason, when a differential experiment is performed, relation of the alignment and orientation parameter to the final direction of motion is to be preferred.

The use of MO in the description of slow atomic collisions is well-documented for both outer and inner shells and only a simple example will be given here, which typifies a large class of recent experiments where capture of electrons from hydrogen atoms by incident ions of high charge is observed. The MO correlation diagram is shown in fig. 2 for a typical case of a bare carbon ion incident on atomic hydrogen. The electron occupies a σ orbital initially which at large distances looks like an H(1s) orbital. At certain internuclear distances this MO comes close to other σ MO in a sequence of avoided crossings. The radial motion couples these MO such that at each avoided crossing, as the electron is squeezed by the radial motion, it has the choice of more or less retaining its size (and thereby its energy) but changing its nodal structure (to look like a high-n orbital) or, it can try to keep its 1s structure but

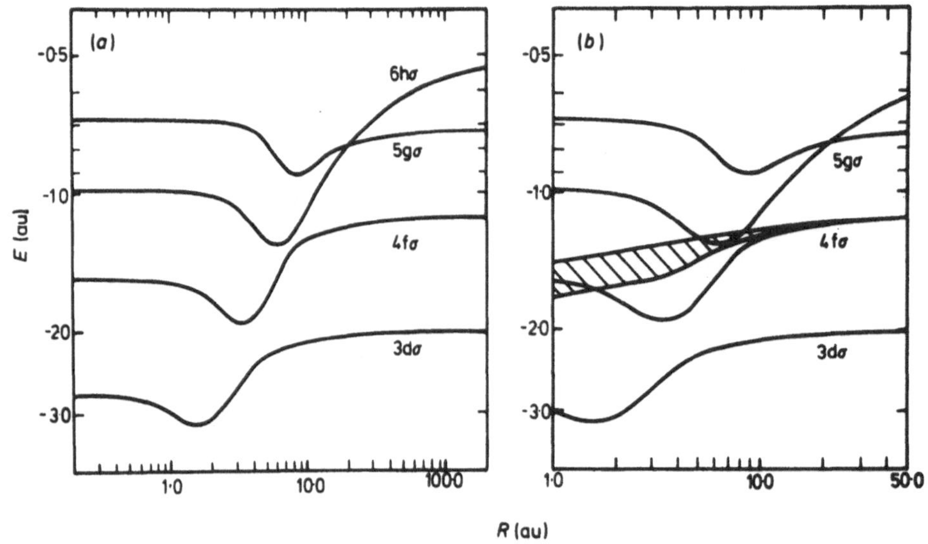

Fig. 2. The MO correlation diagram of the $(CH)^{6+}$ molecule. Only σ levels are shown explicitly. The hatched area illustrates the location of levels of other symmetry correlating to the n=4 level of the C^{5+} ion. From ref. 2.

then must shrink and thereby increase its binding energy. The former choice is represented by adiabatic motion along the curves of fig. 2, the latter by non-adiabatic motion, (jumping across the avoided crossings) whose limit at R=0 is the 1s MO of the united atom. Associated with each σ-state is a manifold of states correlating to the same separated-atom or united-atom levels but with higher λ-quantum number i.e. π, δ etc. MO. The rotational motion causes the electron to "slip off" the internuclear axis and thereby acquire angular momentum about this axis and to change its symmetry with respect to this axis. Quantum-mechanically the consequence of this radial and rotational coupling is that the original σ MO develops during the course of the collision into a coherent linear superposition, (expressed by eq. (6)), of all other MO with which it becomes near-degenerate at some stage.

As the collision velocity is increased, the Hilbert space must be expanded to allow coupling to other energetically accessible levels, including ultimately those in the continuum. For this purpose a basis set must be chosen, ideally on grounds of physical reasonableness, but often with the help of those two great principles of theoretical physics, Expediency and Hope. In the intermediate velocity region $v/v_e \approx 1$, the principal components of the basis are usually certain of the undistorted eigenfunctions of the separated target and projectile. These may be augmented by other functions e.g. pseudo-state wavefunctions, Sturmian functions, united-atom eigenfunctions, two-centre Hylleraas functions or square-integrable functions designed to represent the continuum. In all cases however, when the time-dependent Schrödinger equation (5) is solved in a basis which diagonalises some part of the coupling matrix on the r.h.s. of eq. (7), the symmetry of the residual perturbation may be related to the internuclear axis $\underline{R}(t)$, which forms a natural symmetry axis for the problem.

In the limit in which the probability of coupling of a given initial state to the final state of interest is small, as for example in the case of ionisation from an inner-shell, the coupled equations (7) may be solved approximately in terms of first-order perturbation theory. Such an approach can be valid under a variety of conditions. For example if $Z_P/Z_T \ll 1$ the incident projectile always represents a small perturbation of the target inner-shell electrons even when the collision is slow ($v/v_e<1$) and 'a fortiori' when $v/v_e>1$. In fast collisions when $Z_P \approx Z_T$ the first order perturbation theory (first Born approximation) may also be used. However when $Z_P \approx Z_T$ the experimental conditions are usually such that the collision <u>is not fast</u> and more elaborate multiple-scattering methods are needed. Curiously enough, when $Z_P \approx Z_T$ and the collision is sufficiently slow it has been established that another type of first order approximation is valid.[3] This is applicable to the ionisation of electrons out of deep-lying MO in slow collisions, where ionisation takes place very near to the united atom R=0 limit. Then the first-order solution of the coupled MO can be re-written in terms of the perturbation of an electron, temporarily occupying a united atom orbital, by the coherent Coulomb field of both nuclei. This first-order amplitude is,

$$C_f(\infty) = -i \int_{-\infty}^{\infty} \langle \phi_f^{uA}(\underline{r},t) | (V_P(t)+V_T(t)) | \phi_i^{uA}(\underline{r},t) \rangle \, dt \tag{9}$$

Here the electron is considered to be attached to the molecular centre of charge and to make a transition between initial and final united atom orbitals. The coherence of the nuclear fields leads to the result that dipole transitions are identically zero for $Z_P=Z_T$ and strongly suppressed in near-symmetric collisions. This has implications for the velocity distribution of electrons ejected from inner-shells into the continuum in slow heavy-ion collisions. However, it must be remembered that this distribution refers to electrons ejected in the moving frame of the centre of charge. To compare with any measured distribution the calculated velocity distribution must be suitably transformed to a laboratory-fixed frame.

In the more usual case of $Z_P \ll Z_T$ or fast collisions, the target electrons are considered to remain attached to the target nucleus and to make a transition from initial separated atom orbital $\phi_i(\underline{r}_T,t)$ to final $\phi_f(\underline{r}_T,t)$. In first Born approximation the transition amplitude is

$$C_f(\infty) = -i \int_{-\infty}^{\infty} \langle \phi_f(\underline{r}_T,t) | V_P(t) | \phi_i(\underline{r}_T,t) \rangle \, dt \tag{10}$$

Here the target nucleus is considered as the origin of co-ordinates and its recoil is ignored. Although not apparently connected with the molecular picture, even in the Coulomb perturbation theory the internuclear axis appears as an axis of symmetry. This is seen by writing the Coulomb matrix element explicitly as

$$\int d\underline{r}_T \, \phi_f^*(\underline{r}_T) \frac{Z_P}{|\underline{r}_T-\underline{R}|} \phi_i(\underline{r}_T) \tag{11}$$

and expanding the Coulomb operator in partial waves. The lth partial wave contains an angular factor

$$P_\ell(\hat{\underline{r}}_T \cdot \hat{\underline{R}}) = \sum_m Y_{\ell m}^*(\underline{r}_T \cdot \underline{v}) \, Y_{\ell m}(\underline{R} \cdot \underline{v}) \tag{12}$$

Hence, when states are quantised along the instantaneous internuclear axis $\hat{\underline{R}}(t)$, only $\Delta m=0$ transitions are allowed. These correspond to a (time-dependent) linear combination of all possible Δm transitions amongst states quantised in a laboratory-fixed frame, as shown by (12). In fact, if the perturbation theory is properly developed from separated atom states quantised in the molecular frame, it is found that the co-efficients $Y_{\ell m}(\underline{R}.\underline{v})$ are just those required to diagonalise the $\Theta\, \partial/\partial\Theta$ rotational coupling operator.

In high velocity collisions, where usually a fully quantum-mechanical treatment is made with the internuclear motion described by plane waves, a further axis appears as an axis of symmetry. This is the axis of momentum transfer during the collision[4], which in a differential experiment can be well-defined. This way of looking at the collision is complementary to that which considers the internuclear axis as axis of quantisation. The latter, although very descriptive, is not a fixed axis of course. If a Fourier transform of the Coulomb potential is made then the matrix element (11) reads

$$\frac{1}{(2\pi)^{3/2}} \int d\underline{r}_T \int d\underline{k}\, \phi_f^* (\underline{r}_T)\, e^{i\underline{k}\cdot\underline{r}_T}\, \phi_i(\underline{r}_T)\, \tilde{V}(\underline{k})\, e^{-i\underline{k}\cdot\underline{R}} \qquad (13)$$

Expansion of the operator $\exp(i\underline{k}\cdot\underline{r}_T)$ in partial waves gives an angular factor

$$P_\ell(\hat{\underline{r}}_T\cdot\hat{\underline{k}}) = \sum_m Y^*_{lm}(\hat{\underline{r}}_T\cdot\hat{\underline{v}})\, Y_{lm}(\hat{\underline{k}}\cdot\hat{\underline{v}}) \qquad (14)$$

It can be shown that to order m/M the momentum $\underline{k}$ is equal to the momentum transfer vector obtained in a plane-wave treatment of nuclear motion. Hence, when states are quantised along $\hat{\underline{k}}$ only $\Delta m=0$ transitions with respect to this axis are allowed and these correspond to a definite linear combination of transitions for all possible Δm amongst states quantised along the beam axis.

The conjugate relation between $\underline{k}$ and $\underline{R}$ is apparent from the $\exp(-i\underline{k}\cdot\underline{R})$ factor in (13) and a connection between the two pictures can often be made. For example at high velocities, the momentum transfer is nearly perpendicular to the beam axis. This corresponds in the trajectory picture to transitions only occurring in large impact parameter collisions near to the distance of closest approach where the internuclear axis is also nearly perpendicular to the beam axis.

The coherent population of excited states, either via excitation or capture processes, is an emerging theme of heavy-ion studies. The most obvious example of coherent excitation is the process of ionisation i.e. excitation of electrons to an unbound state. States of the same energy but different ℓ are degenerate and the velocity distribution of ejected electrons provides a very sensitive test of scattering theories. However the most interesting example is ionisation in a pure Coulomb field where the coherence in the continuum extrapolates below the threshold to appear as a coherent excitation of degenerate lm substates belonging to a different n principal quantum number. Unfortunately the hydrogen atom is a difficult target with which to work and the observation of near-threshold electrons (energies less than a few electron volts) or cascade photon emission from high Rydberg states is not an easy technical task. The method of electron loss spectroscopy[5], however, offers an elegant solution to these problems. A hydrogen atom or one-electron hydrogenic ion is used as projectile and the distribution of continuum electrons emerging from collision with a neutral target is observed. The electrons moving with velocities close to that of the projectile nucleus after the collision are those ionised from the projectile into states of very low momentum. This distribution peaks near to the beam velocity, usually with a characteristic cusp-shaped distribution. However these electrons are easily measured since they are fast in the laboratory frame. For example electrons ionised with zero energy in the frame of a 1 MeV He^+ beam have an energy of 140 eV in the laboratory frame. The angular distributions measured in the laboratory frame can readily be transformed back to those in the projectile frame or alternatively calculated velocity distributions can be transformed to the laboratory frame to compare with measurement. Usually the direction of the outgoing projectile nucleus is not measured. Then the cross-section for production of electrons with momentum $\underline{q}$ in the projectile frame can be written as,

$$\frac{d\sigma}{d\underline{q}} = \frac{1}{4\pi} \sum_{k=0}^{\infty} (2k+1)\, P_k(\cos\theta) \sum_{ll'm} \{(2l+1)(2l'+1)\}^{\frac{1}{2}} (-1)^{l+l'-m} \begin{pmatrix} l & k & l' \\ -m & 0 & m \end{pmatrix} \begin{pmatrix} l & k & l' \\ 0 & 0 & 0 \end{pmatrix} \langle qlm|\sigma|ql'm\rangle \tag{15}$$

where $\cos\theta = \hat{\underline{q}}\cdot\hat{\underline{v}}$ and where the density matrix element is defined by

$$\langle qlm|\sigma|ql'm'\rangle = \delta_{mm'}\, \{4\pi^2 v\}^{-1} \int d\underline{k}\; t_{qlm}(\underline{k})\, t^*_{ql'm'}(\underline{k})\; \delta(\underline{k}\cdot\underline{v}+\Delta\varepsilon) \tag{16}$$

Here, $\Delta\varepsilon$ is the inelastic energy of the collision and the transition amplitude

$$t_{qlm}(\underline{k}) = F(k)\, \langle qlm|e^{i\underline{k}\cdot\underline{r}}|i\rangle \tag{17}$$

involves the matrix element for transfer of momentum $\underline{k}$ to the projectile in initial state $|i\rangle$ and a factor F(k) dependent only on the target (essentially the Fourier transform of the effective target potential).

The coherent nature of the ionisation into different ℓ states is obvious from eq. (15). The expression becomes considerably simpler in the limit of threshold ionisation $q\to 0$. Then it may be shown that for an initial state nlm the angular distribution of electrons at threshold can be represented in the form[6]

$$\frac{d\sigma}{d\underline{q}} \rightarrow \frac{\sigma}{q} \sum_{k=0}^{2n} \beta_k \, P_k(\cos\theta) \tag{18}$$

where σ is the isotropic part of the cross-section and the β_k parameters are independent of q. In particular is to be noted that for an initial ground state only $P_2(\cos\theta)$ appears. A recent experiment[7] with atomic hydrogen (fig. 3) has indicated the absence of odd multipoles in the measured distribution. It must be remembered that the result (18) depends a) upon ionisation in a pure Coulomb field and b) the validity of the first Born approximation. Already in the data of fig. (3) the presence of higher even multipoles than k=2 indicate the influence of higher Born terms. For non-hydrogenic targets the unravelling of the effect of departures from a Coulomb potential from the departures from the first Born approximation remains a task for the future. Nevertheless the technique of electron-loss spectroscopy has already shown itself to be of great value. The ability to study near - threshold electrons could be applied to multiple ionisation e.g. the double-ionisation of two-electron ions would throw considerable light on the interesting problem of three-body Coulomb break-up near threshold.

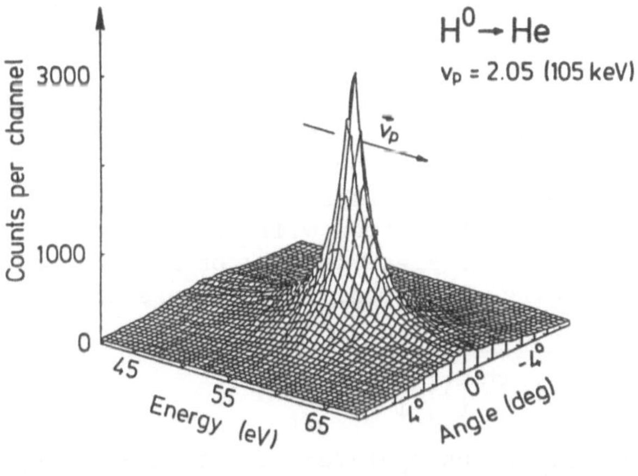

Fig. 3. Distribution of electrons ejected into the continuum from hydrogen atoms in collision with He atoms. From ref. 7.

2. ELECTRON CAPTURE AT HIGH VELOCITY

The essential differences between the transfer of electrons between two colliding nuclei (electron capture) at low and high velocity is illustrated in figure 4. When the translational electron kinetic energy difference between the two frames is negligible compared with the electron binding energies then the molecular picture is appropriate and transfer into the 1s, 2s or 2p states of the projectile takes place by coupling of the $1s\sigma_g$, $2s\sigma_g$, $2p\sigma_g$, $2p\sigma_u$ and $2p\pi_u$ MO as discussed earlier. In this region the probability of capture is identical to that of excitation in a symmetric collision. As the collision velocity increases such that $\frac{1}{2}mv^2$ becomes of the order of the binding energy, the momentum and energy changes necessary for the electron to change frames result in an effective splitting of the two sets of levels (fig. 4) and a reduction of the capture cross-section with respect to the corresponding excitation cross-section. In the limit that $\frac{1}{2}mv^2$ greatly exceeds the binding energies, the large energy and momentum changes necessary to transfer to the moving frame of the projectile become decisive and the capture cross-section falls very much more rapidly than the excitation cross-section. In addition, because of the effective degeneracy of bound states on one nucleus with continuum states of the other, it becomes essential to include continuum states in the representation of the scattering wavefunction. Since the collision is also fast one is automatically led to consider expansion of the T operator for transfer, rather than diagonalisation in some basis set of square-integrable functions.

As emphasised by Shakeshaft and Spruch[8] the capture process can occur via a real three-body encounter, where the high-momentum components of the bound-state wavefunctions play a crucial role, via a sequence of essentially two-body collisions or by emission of a photon. The latter process, radiative electron capture, will not be considered here. The non-radiative capture proceeds either through the electron-nucleus, the nucleus-nucleus or the electron-electron interactions. In the simplest case to be considered initially the explicit electron-electron interaction will be ignored, as appropriate to a collision involving one electron only or an independent particle model of a real atomic collision. In this case the T-matrix element reads

$$T_{fi} = \langle \Phi_f | V_f (1 + G^+ V_i) | \Phi_i \rangle \qquad (19)$$

where G^+ is the full Green operator of the interacting two-nuclei, one-electron system and

$$V_f = V_T + V_{PT}$$
$$V_i = V_P + V_{PT} \qquad (20)$$

are the perturbations in final and initial channel respectively. The initial and final wavefunctions are, with reference to the co-ordinate systems of fig. 1,

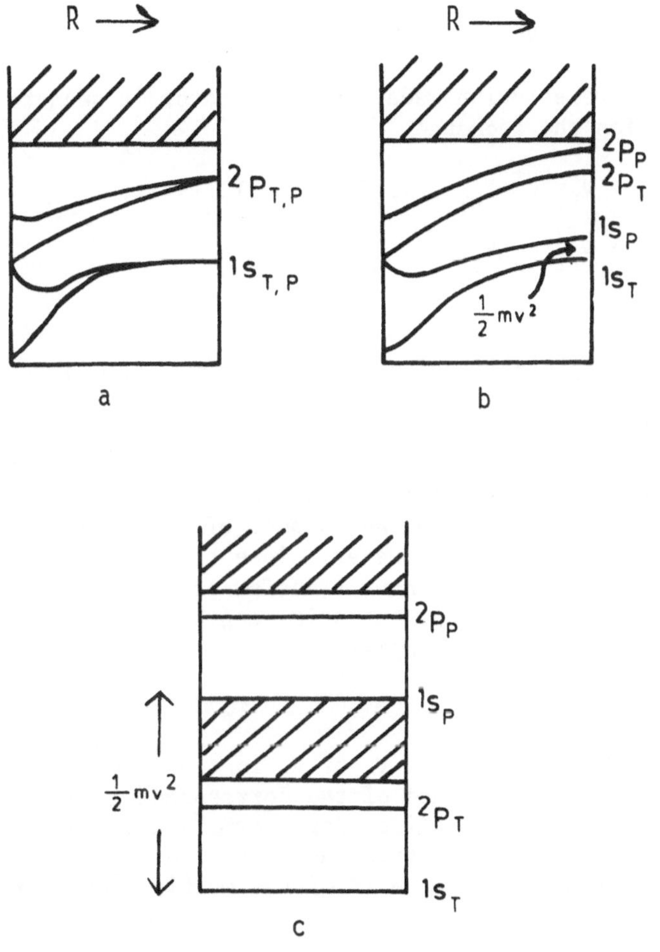

Fig. 4. Schematic ordering of projectile and target levels in a proton-hydrogen atom collisions for a) $\frac{1}{2}$ $mv^2 = 0$, b) $\frac{1}{2}$ $mv^2 < 1$ a.u., c) $\frac{1}{2}$ $mv^2 > 1$ a.u.

$$\Phi_i = (2\pi)^{-3/2} \exp\{i\, \underline{K}_i \cdot \underline{R}_T\}\, \phi_i\, (\underline{r}_T)$$
$$\Phi_f = (2\pi)^{-3/2} \exp\{i\, \underline{K}_f \cdot \underline{R}_P\}\, \phi_f\, (\underline{r}_P) \qquad (21)$$

It is significant that the interaction between the two particles of like charge, the two nuclei, appears in both initial and final perturbation. In fact, when the full Green operator G^+ is written in terms of the free-particle Green operator G_o as

$$G^+ = G_o^+ + G_o^+ (V_P + V_T + V_{PT})\, G^+ \qquad (22)$$

and an iterative solution made, the T operator is seen to contain as a sub-series the sequence

$$T \simeq V_{PT} + V_{PT}\, G_o^+\, V_{PT} + V_{PT} G_o^+\, V_{PT} G_o^+\, V_{PT} + \dots$$
$$= V_{PT} + V_{PT}\, G_{PT}^+ V_{PT}$$
$$= T_{PT} \qquad (23)$$

The corresponding contribution to the T-matrix element $T_{fi} \simeq \langle\Phi_f|T_{PT}|\Phi_i\rangle$ represents the lowest-order description of "knock-on" capture in which the target nucleus recoils by (off-shell) elastic scattering from the incident projectile nucleus. The amplitude for capture is then the "shake-over" amplitude decided by the overlap of initial and final bound states. A further aspect of this knock-on capture process will appear later. However, close to the forward direction (corresponding to small momentum transfer or large impact-parameter collisions) the internuclear potential can be neglected and the T operator written in the simplified form

$$T = V_{T,P} + V_T G^+ V_P \qquad (24)$$

Here the appearance of either V_T or V_P in the leading (first Born) term reflects the equivalence of the corresponding matrix elements. As is now appreciated, the first Born term is never an adequate approximate form of the T matrix, even at high energies. Rather it is necessary to include some part of the multiple-scattering operator

$$F \equiv V_T G^+ V_P$$
$$= V_T \sum_{n=0}^{\infty} G_o^+ \{(V_T + V_P) G_o^+\}^n\, V_P \qquad (25)$$

and various approximations have been suggested[9].

An iterative equation for the exact operator F can be obtained in the following way[10]. Use is made of the identities

$$G^+ = G_T^+ (1+V_P G^+)$$
$$G^+ = G_P^+ (1+V_T G^+) \tag{26}$$

and $V_P G_P^+ = T_P G_o^+$ where $T_P \equiv V_P + V_P G_P^+ V_P$

$V_T G_T^+ = T_T G_o^+$ where $T_T \equiv V_T + V_T G_T^+ V_T$

to write the F operator as

$$\begin{aligned} F = V_T G^+ V_P &= T_T G_o^+ (1+V_P G^+) V_P \\ &= T_T G_o^+ \{V_P + V_P G_P^+ (1+V_T G^+) V_P\} \\ &= T_T G_o^+ T_P + T_T G_o^+ T_P G_o^+ V_T G^+ V_P \end{aligned} \tag{27}$$

i.e. $F = F_o + F_o G_o^+ F$ (28)

where $F_o \equiv T_T G_o^+ T_P$.

It will emerge that the particular expansion obtained by an iterative solution of eq. (28) orders the F operator in terms of the (odd) number of the switches of frame which the electron makes in transferring from target to projectile. The lowest order term F_o collects all those terms of the Born series in which a single switch of frame is made and the corresponding form of the T-matrix

$$T_{fi} \approx \langle \Phi_f | V_{T,P} + T_T G_o^+ T_P | \Phi_i \rangle \tag{29}$$

has been called the distorted wave Born (DWB) approximation for electron capture. (In view of the use of this term where distortion occurs due to the internuclear potential V_{PT} , it may be more appropriate to call this approximation the 'one-switch' approximation). In its distorted wave form the T-matrix element of eq. (29) may be written

$$T_{fi} \approx \langle \Phi_f | V_{T,P} | \Phi_i \rangle + \langle \chi_f^- | V_T G_o^+ V_P | \chi_i^+ \rangle$$

where, $\chi_f^- = (1+G_T^- V_T)\Phi_f$

$$\chi_i^+ = (1+G_P^+ V_P)\Phi_i \tag{30}$$

The one-switch approximation is completely symmetric in the operation of the potentials V_T and V_P. Where one or other of these is dominant in the interaction region i.e. $Z_T \gg Z_P$ or $Z_P \gg Z_T$, the infinite order interaction represented by T_T or T_P may be replaced by its first order approximation V_T or V_P as appropriate. The T-matrix elements resulting from eq. (29) are the two forms of the impulse approximation (IA), also called the strong potential Born (SPB) approximation by Macek and co-workers[11],

$$T_{fi} \simeq \langle \Phi_f | V_P + T_T G_o^+ V_P | \Phi_i \rangle$$
$$= \langle \chi_f^- | V_P | \Phi_i \rangle \qquad (31)$$

when $Z_T >> Z_P$, or

$$T_{fi} \simeq \langle \Phi_f | V_T + V_T G_o^+ T_P | \Phi_i \rangle$$
$$= \langle \Phi_f | V_T | \chi_i^+ \rangle \qquad (32)$$

when $Z_P >> Z_T$.

The high-velocity limit of the one-switch approximation is clearly the second Born approximation in which the electron collides once in each potential

$$T_{fi} = \langle \Phi_f | V_{T,P} | \Phi_i \rangle + \langle \Phi_f | V_T G_o^+ V_P | \Phi_i \rangle \qquad (33)$$

The double-scattering term, also obtained from eq. (30) by replacing distorted by undistorted waves, is known to provide the leading term in an expansion of the total cross-section in terms of inverse powers of the collision velocity and leads to a structure known as the 'Thomas peak' in the differential cross-section for capture (see fig. 5).

An approximation similar to the one-switch approximation, also derived from a distorted wave approach and known as the continuum distorted wave (CDW) approximation[12], has found wide use mainly because of the ease with which it may be calculated. Crothers[13] has shown that the CDW may be obtained by writing the Hamiltonian in the non-orthogonal co-ordinates $\underline{r}_P$, $\underline{r}_T$ (see fig. 1) in which the term $\nabla_{\underline{r}_P} \cdot \nabla_{\underline{r}_T}$ appears as part of the kinetic energy operator. This form acts as the perturbing term in a distorted wave expansion whose lowest-order term reads, in the impact parameter formalism

$$f = -i\, N(\nu_T)\, N(\nu_P) \int_{-\infty}^{\infty} dt \, \langle \phi_f \; {}_1F_1\{-i\nu_T, 1, -i(\underline{v}\cdot\underline{r}_T + vr_T)\} \\ | \nabla_{\underline{r}_T} \cdot \nabla_{\underline{r}_P} | \phi_i \; {}_1F_1\{i\nu_P, 1, i(\underline{v}\cdot\underline{r}_P - vr_P)\} \rangle \qquad (34)$$

where $\nu_P = Z_P/v$, $\nu_T = Z_T/v$ and N is a Coulomb wave normalisation factor. The distorted waves appearing in the CDW matrix element have been shown[9] to be derivable from the distorted waves χ_f^-, χ_i^+ of the IA (eq. (30)) by use of a further peaking approximation. The CDW approximation gives rise to structure in the differential cross-section similar to that shown by the second Born approximation but predicts additional structure which is probably not physical. Its contribution to the total cross-section to leading order in 1/v does not agree with the second Born result. However, recent work[13] indicates that the asymptotic differences between second Born and lowest order CDW may be removed when CDW is expanded to second order.

The cross-section for electron capture in proton-helium atom collisions at 7.4 MeV is shown in fig. 5 as a function of the proton scattering angle. The Thomas peak is clearly seen and of the various multiple-scattering approaches the DWB or 'one-switch' approximation appears to give the better overall agreement[14]. It is also seen that at this high velocity the capture process is confined to extremely small angles near the forward direction which justifies the neglect of the internuclear potential in the calculations. However, the probability of capture in large-angle scattering, though small, is finite and has recently been measured.[15] As already mentioned, the 'knock-on' process, described by the T-matrix element of eq. (23), provides one mechanism for capture solely by operation of the internuclear potential. More generally scattering by the internuclear potential combines with scattering of the electron by one or other nucleus to contribute to the full T-matrix element. Of particular interest in this connection are the second Born terms in the full T

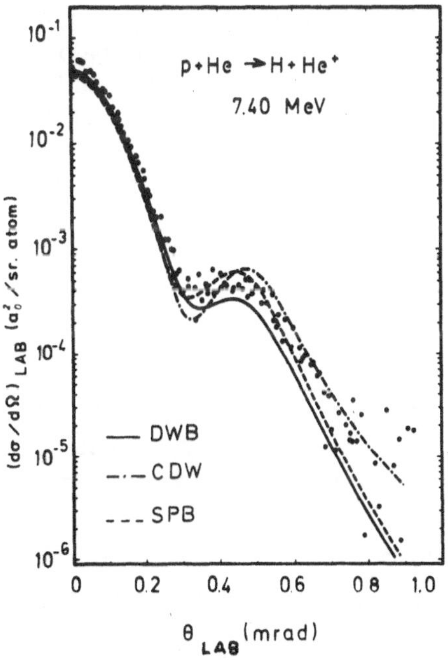

Fig. 5. The differential cross-section for electron capture in p-He collisions at 7.4 MeV. The theoretical curves are for 1s-1s transitions only and have been convoluted with the detector resolution. From ref. 14.

operator of eq. (19), of which there are four;

$$\begin{aligned}
&\text{a) } V_T G_o^+ V_P \\
&\text{b) } V_{PT} G_o^+ V_P \\
&\text{c) } V_T G_o^+ V_{PT} \\
&\text{d) } V_{PT} G_o^+ V_{PT}
\end{aligned} \tag{35}$$

Of these, the last is just part of the T operator for internuclear scattering which has already been discussed. In the other three, Dettmann and Leibfried[16] have pointed out the possible existence of critical angles for the incident nucleus scattering at which the T operator diverges. If binding is ignored, the condition for capture is essentially that the electron and incident nucleus emerge with zero relative momentum after the collision. When the kinematic conditions for this to occur via two binary collisions (as described by the second Born terms) are satisfied a pole appears in the free-particle Green function G_o^+. This is precisely the origin of the Thomas peak arising from term a), one scattering V_P followed by one scattering V_T. The kinematic conditions are shown in fig. 6a. When $M_P < M_T$, term b) also contains critical angle scattering corresponding to initial scattering of the electron by the projectile followed by scattering of the projectile by the target nucleus. In the particular limit $M_P/M_T \to 0$ whose kinematic conditions are shown in fig. 6b, the critical angle for capture by this mechanism is 60°. Some evidence for the presence of this type of critical angle has been seen recently[15]. When $M_P < M_T$, the term c) shows no critical angle since the second collision is of the electron with the recoiling target nucleus which has insufficient velocity. For $M_P > M$ however the argument is

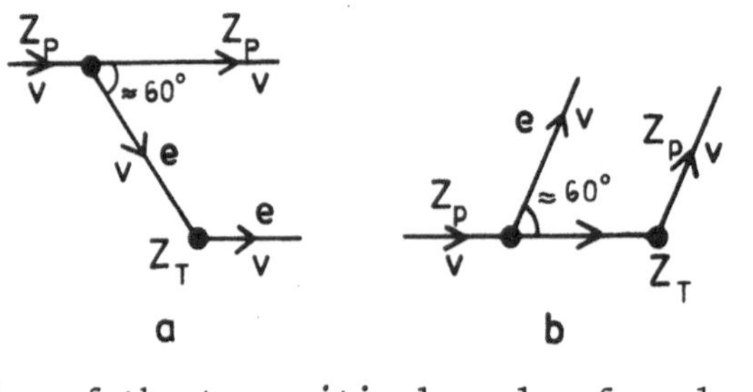

Fig. 6. Kinematics of the two critical angles for electron capture in the case $M_T >> M_P$.

reversed and term c) shows critical angle scattering whilst term b) does not.

Although outside the direct scope of this lecture, it is interesting to note that for positron impact the two critical angles of fig. 6, one at small and one at large angle, co-alesce at a positron scattering angle of 45°. As has been noted[17] the terms a) and b) are then identical except for a change of sign at the second scattering, so that the terms interfere constructively or destructively according as the parity of the final positronium state is odd or even.

That the process of electron capture even in high-velocity collisions is a rather delicate business is illustrated by the necessity to include higher-order scattering and the dominance of the correlated binary collisions leading to the Thomas mechanism. Recent work[18] has shown that the delicacy of the process increases as one considers capture from or into the multiplicity of states of higher and higher quantum number. A particularly interesting feature of such capture events, although unfortunately restricted to capture by bare nuclei, is the coherent population of different lm sub-states of a given n manifold. This has been studied experimentally for states of low m and the corresponding elements of the capture density matrix extracted from the experimental results[19]. In a pure Coulomb field not only the orbital angular momentum vector is a constant of the motion but also the Runge-Lenz vector. It has been shown how to relate the components of these vectors to various combinations of the density matrix elements. Since the two vectors are also constants of the classical motion in a Coulomb field it is possible to relate the shape of the coherent quantum-mechanical state produced by capture to the corresponding classical orbit of an electron around the projectile nucleus.[18] This has provided further evidence for the necessity to include multiple-scattering terms in high-velocity capture theories. For example the expectation value of the component of the Runge-Lenz vector along the beam axis is zero in the first Born approximation, for all n manifolds. Experiment and higher order theory show this not to be the case. As shown in fig. 7 for capture into the n = 3 shell of hydrogen, the corresponding classical orbit is such that the electron 'lags behind' the proton and endows the atom with a dipole moment.

It is clear that in the Coulomb field, such gross features of the electron distributions resulting from coherent population of different lm states belonging to a different n value (or energy) extrapolate smoothly through the ionisation continuum. The resulting

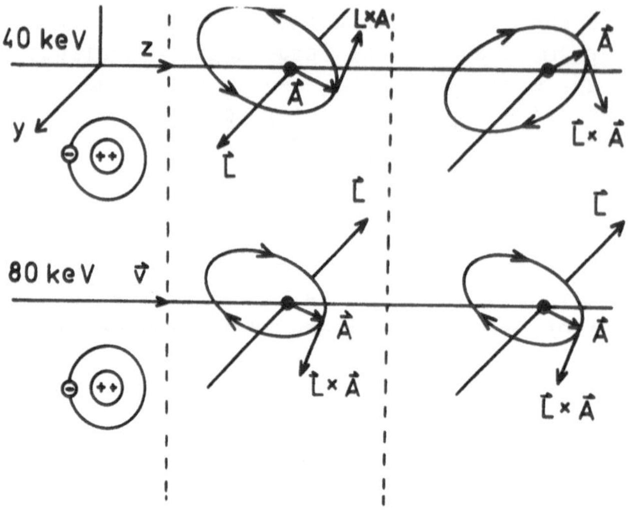

Fig. 7. Possible directions of the angular momentum vector L and the Runge-Lenz vector A of the state produced by capture into the n=3 hydrogenic level. From ref. 18.

continuum electrons arising from 'electron capture to the continuum' (ECC) have a cusp-shaped distribution in the laboratory frame which is not in general symmetric around the beam velocity. The origin of this forward-backward asymmetry is similar to that for capture to the n=3 state discussed above and has been shown to arise already in the second Born term for capture to the continuum[20].

Nothing has been said so far about the explicit role of the electron-electron interaction in high-velocity capture and indeed very little is known. In the very first classical treatment of capture via two correlated binary collisions Thomas recognised that in the second collision the electron to be captured could scatter off one of the other target electrons rather than off the target nucleus. A simplified treatment of the corresponding second Born term[21] indicates that this process also gives rise to the Thomas peak structure, but with lower amplitude. Its presence in the experimentally measured p-He differential cross-section has not yet been identified. A specific capture process has been recognised however when both projectile and target carry electrons into the collision. In the projectile frame, the target electrons appear almost as a "beam" of free electrons in the high-velocity collision limit. Under these circumstances the well-known features of electron-ion collisions should manifest themselves in the ion-atom collision. A good example of such a phenomenon is radiative electron capture which can be viewed as radiative recombination of the (quasi-free) target electrons with the projectile ion. The process of dielectronic recombination is a resonant process in electron-ion scattering in which the incoming electron has just sufficient energy to excite an electron of the ion and itself become temporarily bound. This doubly-excited electronic state can then stabilise by emission of a photon leading to capture of the incident electron. In ion-atom collisions, the target electrons can play the same role as the incident electrons in the electron-ion collision. Recently the dielectronic recombination process leading to capture of target electrons has been identified in ion-atom collisions and called resonant transfer and excitation (RTE)[22]. However the sharp resonant character of dielectronic recombination in electron-ion collisions is considerably broadened by the momentum distribution of the bound target electrons in ion-atom collisions. The experiments performed to date have used highly-charged ions (incident charge $\gtrsim$ 10) which necessitates the incorporation of the strong projectile field in the theory i.e. a multiple-scattering approach is again necessary. To lowest order in the electron-electron interaction the T-matrix element for the RTE process[23] appears as

$$T_{fi} = \langle\Phi_f(12)|V_{ee}(1+G^+_{1P}V_{1P})|\Phi_i(12)\rangle$$

where $\Phi_f(12)$ is a product of a doubly excited projectile state and a plane wave describing relative motion of the final fragments (the internuclear potential is again neglected), $\Phi_i(12)$ describes a product of a bound state of electron 1 on the target, electron 2 on the projectile and the plane-wave relative motion of target and projectile. The operator V_{ee} is the electron-electron interaction and the operator $G^+_{1P}V_{1P}$ describes the distortion of the target bound-state electron in the strong Coulomb field of the projectile ion. This T-matrix element has been evaluated in some cases and the resulting theoretical RTE cross-section reproduces the overall features of the measured cross-section[22], although detailed agreement particularly in the absolute magnitude is lacking.

All of the above serves to indicate that the last word on high-velocity electron capture is far from spoken. Although much has been accomplished in the last few years, a comprehensive picture of the role played by each type of interaction and each type of Born term is only just beginning to emerge. The important question of the convergence of the Born series for capture via pure Coulomb forces is still open.

REFERENCES

1. Fundamental Processes in Energetic Atomic Collisions eds. H.O. Lutz, J.S. Briggs and H. Kleinpoppen, NATO ASI Series B, Vol. 103, (1983) (New York: Plenum).
2. J. Vaaben and J.S. Briggs, J. Phys. B10, L521, (1977).
3. J.S. Briggs, J. Phys. B8 L485 (1975).
4. M.R.C. Mc Dowell and J.P. Coleman "Introduction to the theory of Ion-Atom Collisions" (Amsterdam: N. Holland). (1970)
5. "Forward Electron Ejection in Ion Collisions" ed. K.-O. Groeneveld, W. Meckbach and I.A. Sellin (Springer: Heidelberg) LNP 213 (1984).
6. a) J. Burgdörfer in ref. 5.
   b) J.S. Briggs and M.H. Day, J. Phys. B13 4717, (1980).
7. W. Meckbach, R. Vidal, P. Focke, I.B. Nemirovsky and E. Gonzales Lepera Phys. Rev. Lett. 52 621 (1984).
8. R. Shakeshaft and L. Spruch Rev. Mod. Phys. 51 369, (1979).
9. a) Dz. Belkic, R. Gayet and A. Salin, Phys. Reports 56, 279, (1979).
   b) J.S. Briggs, J. Macek and K. Taulbjerg Comm. At. Mol. Phys. 12, 1, (1982).
   c) K. Taulbjerg and J.S. Briggs, J. Phys. B16, 3811, (1983).
10. J.S. Briggs (to be published).
11. a) J. Macek and R. Shakeshaft, Phys. Rev. A22, 1441, (1980).
    b) J. Macek and K. Taulbjerg, Phys. Rev. Lett. 46, 170, (1981).
    c) J. Macek and S. Alston, Phys. Rev. 26, 250, (1982).
12. I.M. Cheshire, Proc. Phys. Soc. 84, 89, (1964).
13. D.S.F. Crothers, J. Phys. B15 2061, (1982).
    J. Phys. B17 L177, (1984).
14. a) S. Alston, Phys. Rev. A27, 2342, (1983) and to be published.
    b) R.D. Rivarola, A. Salin, M.P. Stöckli, J. Physique Lett. 45, L259, (1984).
    c) E. Horsdal-Pedersen, C.L. Cocke and M. Stöckli, Phys. Rev. Lett. 50, 1910, (1983).
15. E. Horsdal Pedersen, P. Loftager and J.L. Rasmussen, J. Phys. B15 2461, (1982).
16. K. Dettmann and G. Leibfried, Zeits. Phys., 218, 1, (1969).
17. R. Shakeshaft and J. Wadehra, Phys. Rev. A22, 968, (1980).
18. J. Burgdörfer and L.J. Dubé, Phys. Rev. Lett. 52, 2225, (1984).
    Phys. Rev. A (to be published).
19. C. Havener, W. Westerveld, J. Risley, N. Tolk and J.C. Tully, Phys. Rev. Lett. 48, 926, (1982).
20. R. Shakeshaft and L. Spruch, Phys. Rev. Lett., 41, 1037, (1978).
21. J.S. Briggs and K. Taulbjerg, J. Phys. B12, 2565, (1979).
22. a) J.A. Tanis, E.M. Bernstein, W.G. Graham, M. Clark, S.M. Shafroth, B.M. Johnston, K. Jones and M. Meran, Phys. Rev. Lett. 49, 1325, (1982).
    b) D. Brandt, Phys. Rev. A27, 1314, (1983).
23. J.M. Feagin, J.S. Briggs and T.M. Reeves, J. Phys. B17 1057 (1984).

ANGULAR CORRELATION AND POLARIZATION STUDIES OF ATOMIC COLLISIONS

Rainer Hippler

Fakultät für Physik, Universität Bielefeld

D-4800 Bielefeld 1

1. INTRODUCTION

In recent years an increasing interest in detailed information about atomic collision processes has led to the development of angular correlation experiments. In a well-defined collision experiment the direction of the incident projectile is usually fixed (z-axis) and provides a kind of quantization axis for the spatial distribution of atomic angular momenta produced during the collision. This may result in a collisionally induced anisotropy reflecting non-uniform population of magnetic substates $|m\rangle$. It is obvious that such an anisotropy is only possible for atomic states with total angular momentum $J > 0$. Let us now consider, for example, a $^1P_1$-state whose wavefunction we may express as a linear combination of substates $|m\rangle$

$$|^1P_1\rangle = \sum_m f_m \, |m\rangle \tag{1}$$

where f_m is the corresponding excitation amplitude. For a moment we may use a set of substates $|+1\rangle$, $|0\rangle$, and $|-1\rangle$, where m is the projection of the orbital angular momentum L=1 on a given axis. In terms of substate population an isotropic state requires uniform population of all substates (i.e. $|f_m|=|f_{m'}|$ for all m,m'). Anisotropic states are commonly referred to as oriented and/or aligned (Fano and Macek 1973, Blum and Kleinpoppen 1979, 1983). The term orientation is used if there is a net angular momentum in a given m-direction, and requires $|f_m|\neq|f_{-m}|$. Alignment refers to an unequal population of substates aligned in different directions, i.e. $|f_m|\neq|f_{m'}|$ (Fig.1.).

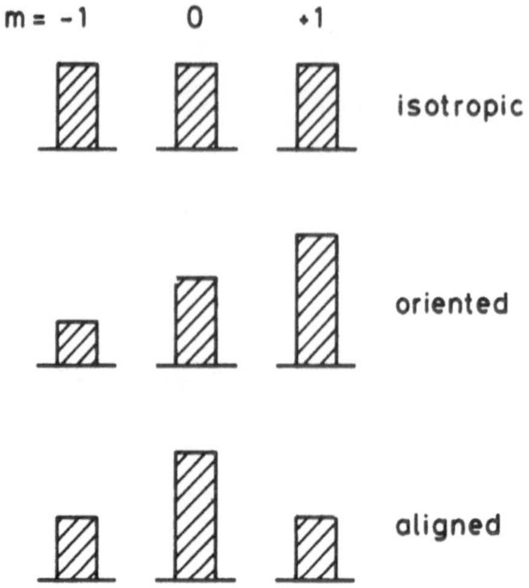

Fig.1: Schematic representation of different substate populations.

The collision-induced alignment and orientation (Fano and Macek 1973) may be investigated in angular correlation or polarization studies of characteristic photons or Auger electrons. The simplest experiments of this kind involve a measurement of the angular distribution of the emitted photon or Auger electron, and may yield the spatial distribution of the atomic angular momenta in z-direction and averaged over all possible scattering angles of the projectile. Since in such an experiment the collision system possesses rotational symmetry around the z-axis, no information about the distribution of the angular momenta perpendicular to the z-axis may be extracted.

Polarization of line radiation in a spherical symmetric collision process (for instance, a total cross section measurement) was first described in the independent dipole model (see, e.g. Oppenheimer 1927). With the incident projectile along the z-axis, and observing the emitted light at an angle θ_γ with respect to $\underline{z}$, the light intensity $I(\theta_\gamma)$ may be written as

$$I(\theta_\gamma) = I_\parallel \sin^2(\theta_\gamma) + I_\perp\{1 + \cos^2(\theta_\gamma)\} \tag{2}$$

with $I_\parallel$ and $I_\perp$ the light intensities with electric vector parallel and perpendicular to the z-axis, respectively. Defining the degree of linear polarization P_1 as

$$P_1 = (I_\parallel - I_\perp) / (I_\parallel + I_\perp) \tag{3}$$

we notice that P_1 may be determined either from a polarization measurement or, alternatively, from an angular distribution measurement using

$$I(\theta_\gamma) \;/\; I(90^o) = 1 - P_1 \cos^2(\theta_\gamma) \qquad (4)$$

Thus, identical information may be obtained either from angular distribution or from linear polarization measurements.

2. EXCITATION BY ELECTRON IMPACT

In the following we shall start with a discussion of atomic collision processes involving excited $^1P_1$-states. As an example we shall consider excitation of helium atoms by electron impact

$$e^- + He(1^1S_0) \rightarrow e^- + He^*(n^1P_1).$$

Let us first consider the angular shape of a $^1P_1$ substate aligned, for example, in z-direction. The angular shape of this particular substate $|z\rangle$ resembles that of a dumb-bell. Similarly, we may have substates $|x\rangle$ and $|y\rangle$ aligned in x- and y-direction, respectively. Altogether, these three substates form a useful basis-set for a $^1P_1$ wavefunction. In Fig.2 we show this basis-set which we will use in the following discussion. [An alternative basis set frequently used is composed of substates $|+1\rangle$, $|0\rangle$, $|-1\rangle$, where we have $|0\rangle = |x\rangle$, and $|\pm1\rangle = \mp\ 1/\sqrt{2}\ \{|x\rangle \pm i|y\rangle\}$.] Also indicated is the behaviour of the wavefunction upon reflection at the scattering (x-z) plane. We notice that the $|y\rangle$ substate changes sign when reflected at the scattering plane. Thus, whereas only $|y\rangle$ has so-called negative reflection symmetry, both $|x\rangle$ and $|z\rangle$ have positive reflection symmetry. Since during a collision reflection symmetry is conserved, only states with either positive or negative reflection symmetry can be excited. Starting from an inital $^1S_0$ ground state with positive reflection symmetry, the $|y\rangle$ substate cannot be excited. The excited-state wavefunction $^1P_1$ thus becomes particular simple, as it only contains contributions from substates aligned along x and z,

$$|^1P_1\rangle = f_x|x\rangle + f_z|z\rangle \qquad (5)$$

where f_x and f_z are the complex excitation amplitudes.

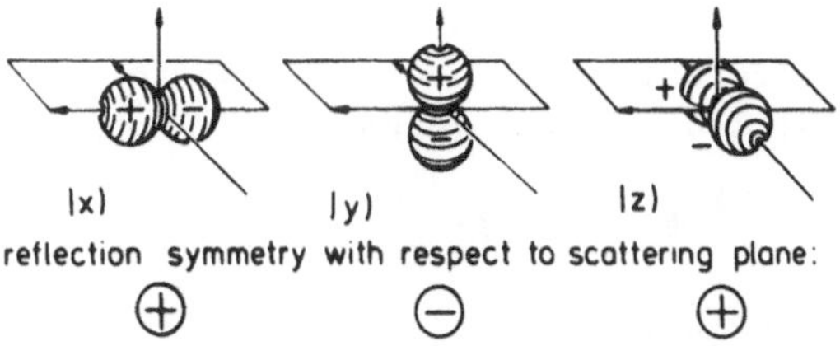

Fig.2: Real $^1P_1$ basis functions (as per Hermann and Hertel 1982).

Next we shall consider the decay of the excited $^1P_1$ state to a $^1S_0$ state by emission of one photon. A substate $|m\rangle$ then corresponds to an electric dipole oscillating in m-direction. The emission characteristics of such a dipole is that the amplitude of the photon field is proportional to $\sin(\theta_\gamma)$, if θ_γ is the angle of photon emission with respect to the m-axis. In our particular example of a $^1P_1$ excitation followed by a $^1P_1 \rightarrow {}^1S_0$ photon transition we obtain for the (normalized) photon intensity $I(\theta_\gamma)$ in the x-z plane, where θ_γ now refers to the z-axis,

$$I(\theta_\gamma) = |f_x \cos(\theta_\gamma) + f_z \sin(\theta_\gamma)|^2 \qquad (6)$$

The first coincident angular correlation experiment was performed by Eminyan et al. (1973, 1974) for excitation of He($2\,^1P_1$) by electron impact. They looked for photons emitted during the decay of the excited helium state to the He($1\,^1S_0$) ground state. To select a particular scattering process the emitted photons were detected in coincidence with electrons scattered through a given scattering angle θ_γ. The angle of photon detection was varied during the experiments. Results of such a coincident angular distribution measurement are displayed in Fig.3 for an incident electron energy of about 80 eV and two scattering angles of 16° and 100°. Note that, particularly for small scattering angles, a pronounced angular dependence of the photon emission is observed. The solid lines are a least-squares fit to the experimental data points, assuming an angular dependence as predicted by Eq. 6. Also given in Fig.3 is a theoretical prediction based on plane wave Born approximation (PWBA). In PWBA the transition amplitude for a collision induced transition between an initial and a final state, $|i\rangle$ and $|f\rangle$, respectively, may be expressed within a normalization constant as

$$f(K) = \langle f|\exp(iKr)|i\rangle \qquad (7)$$

where $\underline{K} = \underline{k}_i - \underline{k}_f$ is the momentum transfer. This matrix element is invariant to a rotation around $\underline{K}$ (Macek 1983), which implies that no net angular momentum may be transferred perpendicular to $\underline{K}$. Moreover, $f(K)$ obeys a selection rule $\Delta m = 0$ with respect to $\underline{K}$. Thus, choosing for a moment a coordinate frame where the z-axis is parallel to $\underline{K}$ we have for the substate excitation amplitudes $f_0(K) = f(K)$ and $f_1(K) = 0$. Then only one particular excitation amplitude is different from zero, and the normalized photon intensity as predicted by PWBA reads

$$I_{PWBA}(\theta_\gamma) = \sin^2(\theta_\gamma - \gamma(K)) \; .$$

Thus, the PWBA predicts that the shape of the excited state resembles a dumb-bell aligned along the K-axis, which is inclined by an angle $\gamma(K)$ with respect to the z-axis. Comparing the predictions of PWBA with the experimental results for the two scattering angles shown in Fig.3 we note that PWBA reproduces the experimental data fairly well for a scattering angle of 16°, whereas the experimental data for 100° differ widely from the PWBA prediction.

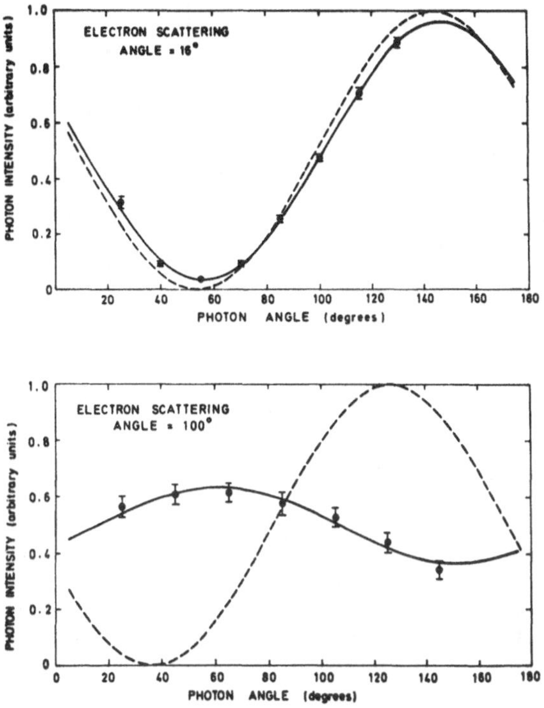

Fig.3: Photon angular distribution following excitation of $He^*(2^1P)$ by electron impact for scattering angles of 16° and 100°. The incident electron energy was 81.2 eV (Hollywood et al. 1979).

2.1 λ and χ Parameters

The data displayed in Fig.3 may be described in terms of so-called λ and χ parameters introduced by Eminyan et al. (1974). They are defined as

$$\lambda = |f_0|^2/\sigma, \text{ and } \chi = \arg(f_1 - f_0),$$

with $\sigma = \Sigma_m |f_m|^2$, the (differential) cross section summed over all substates. λ gives the relative population of the m = 0 substate, and χ the phase difference between the two complex excitation amplitudes f_1 and f_0 for the |+1> and |0> substates, respectively. In Fig.4 we give experimental and theoretical results for the e + He → e + $He(2^1P_1)$ collision system at an incident electron energy of about 80 eV. The experimental results are compared with theoretical calculations based on various approximations. For λ

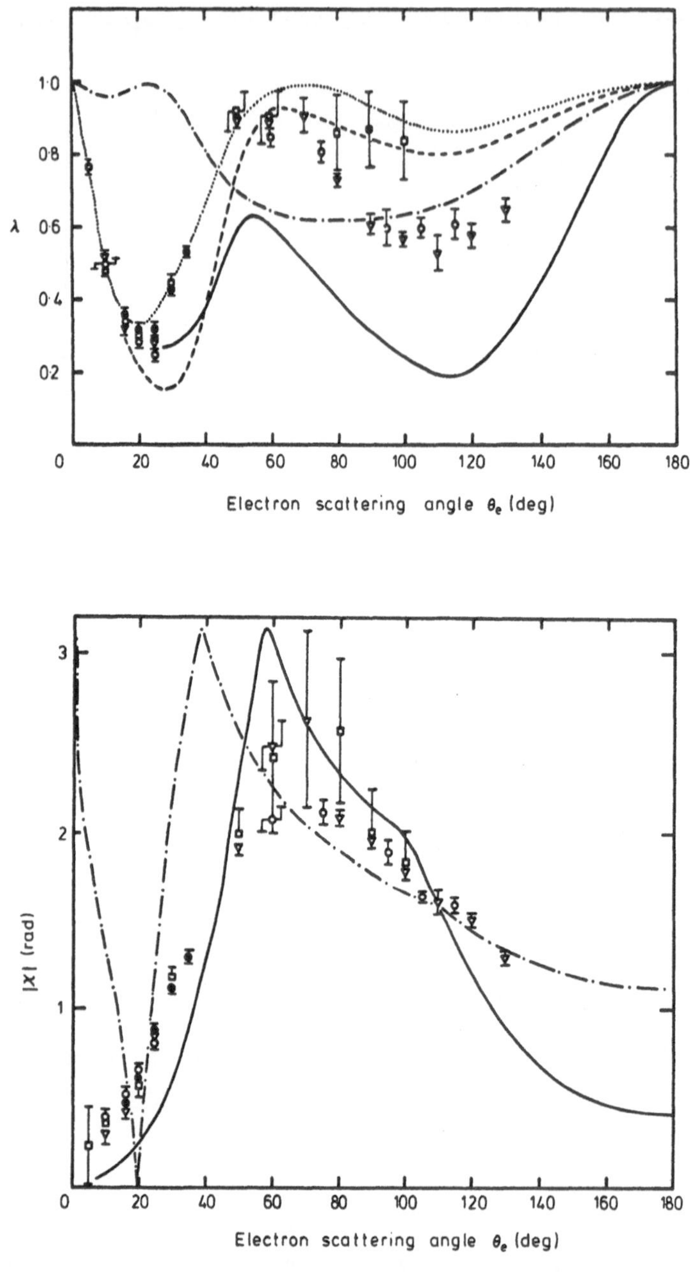

Fig.4: Variations of λ and χ with electron scattering angle for an incident electron energy of 80 eV. O, Slevin et al. (1980a,b); ●, Eminyan et al. (1974); ∇, Hollywood et al. (1979); □, Steph and Golden (1980); ......, Madison and Calhoun (from Sutcliffe et al. 1978); -----, Thomas et al. (1974); -.-.-.-., Catalan and Roberts (1979); ______, Scott and McDowell (1976).

calculations in PWBA give a satisfactory agreement with experimental data only at small scattering angles. Better agreement is here obtained with more sophisticated approximations, for instance distorted-wave calculations. For χ PWBA predicts $\chi = 0$, which clearly deviates almost anywhere from the experimental data. The reason for this striking discrepancy may be seen in the fact that no net angular momentum transfer is possible in PWBA, while the experimental data show that just this happens.

2.2 Density Matrix Description

In general, in a collision experiment the final state must not necessary be a pure one. Then, a description by two parameters, for example λ and χ no longer applies. Let us again consider a $^1P_1$ state. A so-called mixed state may be characterized by a density matrix $\underline{\rho}$ (see, e.g. Blum 1981). This density matrix has the form

$$\underline{\rho} = \begin{pmatrix} \rho_{xx} & 0 & \rho_{zx} \\ 0 & \rho_{yy} & 0 \\ \rho_{xz} & 0 & \rho_{zz} \end{pmatrix} \tag{8}$$

where $\rho_{ab} = f_a^* f_b$, f_a, and f_b are the corresponding excitation amplitudes. Hermiticity further requires $\rho_{xz} = \rho_{zx}^*$. We note that since $|y\rangle$ is the only substate with negative reflection symmetry all density matrix elements ρ_{yk} and ρ_{ky} with $k = x,z$ vanish. A pure state requires that only density matrix elements for substates with either positive or negative reflection symmetry are different from zero. Returning to our example of excitation from a $^1S_0$ state we have for the density matrix element $\rho_{yy} = 0$. In addition we have the requirement that so-called degree of coherence $\mu = |\rho_{xz}|/(\rho_{xx} + \rho_{zz})^{1/2}$ is equal to unity.

2.3 Polarization Studies

If a collision process produces a pure $^1P_1$ state, the light emitted during the decay of that state to a $^1S_0$ state will be 100% polarized. The experimental set-up for a polarization study is shown in Fig.5. In general, the collision process may prepare the three substates $|x\rangle$, $|y\rangle$, and $|z\rangle$, which, for a subsequent $^1P_1 \rightarrow {}^1S_0$ transition correspond to electrical dipoles oscillating in x, y, and z-direction, respectively. Thus, light emitted by each individual oscillator in m-direction will be linearly polarized with respect to that direction. The polarization properties of the emitted light may be characterized by so-called Stokes parameters P_i (see, e.g. Born and Wolf 1970). Three Stokes parameters are required to describe the polarization properties of the light emitted into a given direction. They are related to the density matrix elements ρ_{ab} and are defined as

$$P_1 = [I(0) - I(90)]/I \sim \varrho_{zz} - \varrho_{xx}$$

$$P_2 = [I(45) - I(135)]/I \sim \varrho_{xz} + \varrho_{zx} = 2\,\mathrm{Re}(\varrho_{xz}) \qquad (9)$$

$$P_3 = [I(+) - I(-)]/I \sim \varrho_{xz} - \varrho_{zx} = 2\,\mathrm{Im}(\varrho_{xz})$$

where $I = I(\alpha)+I(\alpha+90) = I(+)+I(-)$, with $I(\alpha)$ the light intensity transmitting a linear polarizer oriented at an angle α with respect to the z-axis, and $I(+)$ and $I(-)$ the intensity of circular polarized light with positive and negative helicity, respectively. The Stokes parameters measured perpendicular to the scattering plane are proportional to linear combinations of density matrix elements for |x> and |z> substates. To determine ϱ_{yy} an additional measurement is required. From an in-plane linear polarization measurement we obtain

$$P_4 = [I_{\parallel}(0) - I_{\parallel}(90)]/I \sim \varrho_{zz} - \varrho_{yy} \qquad (10)$$

If the collision process produces a pure state, the emitted light will be fully polarized and the total degree of polarization $P = (P_1^2 + P_2^2 + P_3^2)^{1/2} = P_4$ will be equal to unity. A measurement of the total polarization P thus provides an essential test for the coherence of the collision process. Following the early angular distribution measurements of Eminyan et al. (1974) Standage and Kleinpoppen (1975) could confirm that the total polarization of the light emitted in e - He collisions is in fact equal to unity.

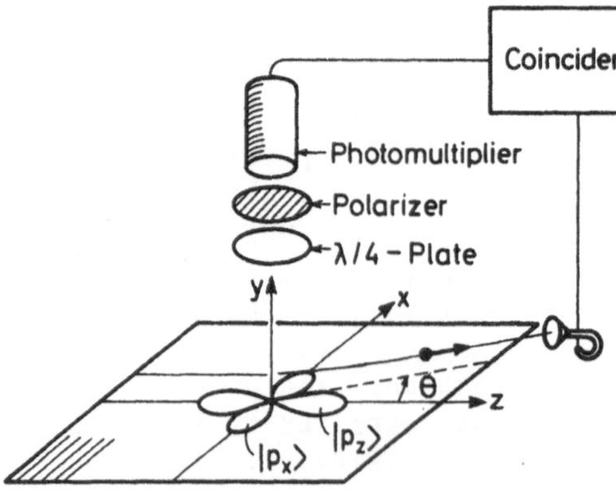

Fig.5: Experimental set-up for polarization studies. Only the |x> and |z> substates are shown (Andersen 1980).

2.4 Depolarization and Incoherence

In many experiments, the total polarization is less than unity. This happens to be the case, when either the collision process itself does not produce a pure state (incoherence), and/or if depolarization occurs in the time between the state was produced by the collision and its decay (depolarization). Examples for incoherence during the collision is if initial and/or final states are not fully specified. For example, in electron-hydrogen collisions we have two spin-channels, with the incident electron's spin either parallel or anti-parallel to the target electron's spin. In principle one may distinguish between these two channels by performing spin preparation and spin analysis before and after the collision, respectively. Another example is averaging over the direction of the outgoing projectile, i.e. in all those experiments where the scattering angle of the projectile is not determined.

Depolarization may take place in the time interval between collisional excitation and the subsequent deexcitation. It is caused by the interaction of the orbital angular momentum with external or internal fields. Here we consider the interaction with internal magnetic fields caused by electronic and nuclear spins. Assuming that during the collision a well-defined angular momentum state characterized by an orbital angular momentum quantum number L is produced, this L-state couples, for example, with the electronic spin S to a state characterized by a total angular momentum J. Usually it is save to assume that during the collision the projectile does not interact with the target electronic or nuclear spin. This hyphothesis is generally justified, since the dominant interaction is due to the Coulomb fields of projectile and target electrons, with the spin-interaction typically some orders of magnitude smaller. Thus, both electronic and nuclear spin remain in good approximation unaffected by the collision. Even, if the collision prepares a pure L-state, the post-collisional coupling of L with (the randomly distributed spin) S to J causes a depolarization since now L performs a precession around J. The depolarization depends on the precession time (typically about 10^{-12} s) and the lifetime of the excited state, and can be calculated from angular momentum algebra (Flower and Seaton 1967, see also Blum 1981).

The depolarization caused by angular momentum coupling has been investigated by Hafner and Kleinpoppen (1967). Performing a measurment of the integrated (i.e. integrated over the scattering of the projectile) linear polarization P_1 of some resonance transitions in alkali atoms excited by electron impact at incident energies close to threshold, they investigated the influence of different angular momentum couplings. The threshold polarization values, as derived from their measurements, are given in Table 1.

For threshold excitation by electron impact, a selection rule $\Delta m_L=0$ holds which is a consequence that the scattered electron can only escape at threshold if its final angular momentum is equal to zero. Thus, in absence of depolarization effects the threshold polarization should approach 100%, as was observed, for example, for electron impact excitation of Ca($4\,^1P_1$) (Ehlers and Gallagher 1973). The observed threshold polarization values for the (unresolved) $^2P_{3/2;1/2} \rightarrow {}^2S_{1/2}$ transitions in Li-6, Li-7, and Na-23 are all considerably smaller. Also given in Table 1 are theoretical values for the threshold polarization of these transitions, taking into account that depolarization is caused by interaction of an 100% polarized L=1 state with unpolarized electronic and nuclear spins (Percival and Seaton 1958, Flower and Seaton 1967). The good agreement between the experimental observation and the theoretical calculations essentially proves that angular momentum coupling may result in a considerable depolarization, and may be accounted for on the basis of angular momentum algebra.

Table 1: Threshold polarization for the first resonance lines of Li-6, Li-7, Na-23. Experimental results of Hafner and Kleinpoppen (1967) are compared with theoretical predictions of Flower and Seaton (1967). The nuclear spin quantum number is I=1 for Li-6, and I = 3/2 for Li-7 and Na-23, respectively.

	Threshold Polarization P_1	
	Theory	Experiment
Li-6	37.9%	(39.7±3.8)%
Li-7	21.3%	(20.6±3.0)%
Na-23	14.1%	(14.8±1.8)%

In collisions of electrons with heavy atoms, such as xenon or mercury, there might also be some collisional interaction between and with the electronic spins. In all these cases and without spin-analysis the final state in general is a mixed one, and a two-parameter description by, for example, λ and χ parameters might be insufficient (Zaidi et al. 1978, McGregor et al. 1982, see also Hanne et al. 1981).

3. INNER SHELL IONIZATION

Vacancies in atomic shells created by photon or particle impact may be also oriented and aligned (see, e.g., Mehlhorn 1983, Hippler 1984). This orientation and alignment may show up in the angular distribution of emitted characteristic photons (photon energies typically a few keV) or of ejected Auger electrons following the decay of inner-shell vacancies.

3.1 Electron Impact

First evidence for a non-isotropic emission of Auger electrons following the decay of argon L3 ($2p_{3/2}$) vacancies produced by electron impact was observed by Cleff and Mehlhorn (1971, 1974). Measurements for the L3-shell alignment of magnesium by electron impact are shown in Fig.6. The experimentally observed (integrated) alignment

$$A_{20} = (\sigma_{3/2} - \sigma_{1/2}) / (\sigma_{3/2} + \sigma_{1/2})$$

where σ_m is the cross section for ionizing the m-substate, as a function of the relative velocity $V=v_p/v_e$, where v_p is the projectile velocity and v_e the target electron velocity, is negative close to threshold and approaches $A_{20} \approx 0$ for $V \approx 7$.

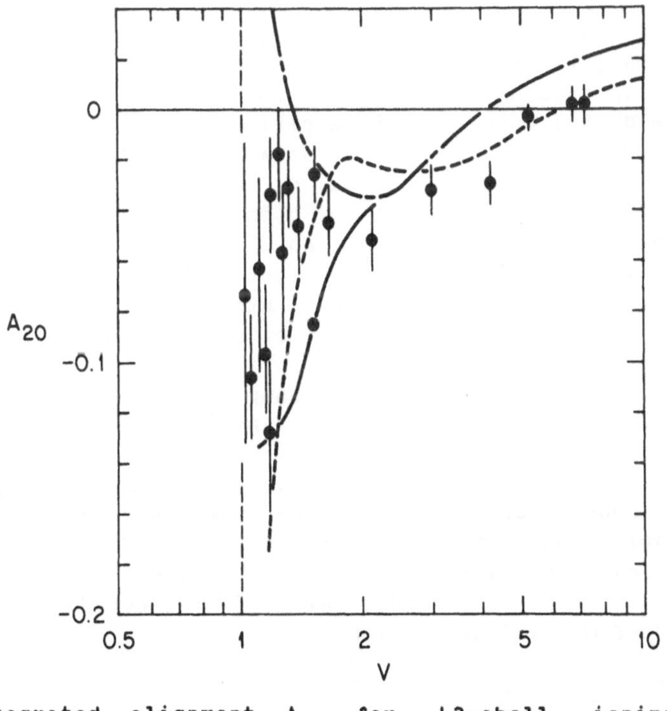

Fig.6: Integrated alignment A_{20} for L3-shell ionization in electron-magnesium collisions. Experimental data of Dubois et al. (1981) are compared with PWBA calculations using -.- hydrogenic or --- Hermann-Skillman wavefunctions, and with —— DWBA calculations (Berezhko and Kabachnik 1982).

Of particular interest is the threshold value of A_{20}. Previous measurements for argon (Cleff and Mehlhorn 1974, Sandner and Schmitt 1978) gave a small A_{20} only, in contrast to measurements on xenon (Aydinol et al. 1980), which indicated a large threshold alignment. The magnesium measurements show that the alignment rapidly decreases close to threshold, but do not provide evidence that A_{20} may become as small as -1. In fact, as was shown by Sandner and Schmitt (1978), threshold ionization in contrast to threshold excitation does not obey a selection rule $\Delta m = 0$ and thus we may have $A_{20} > -1$ at threshold. Of further interest is a comparison of the experimental data with theoretical calculations. This comparison shows that in the velocity regime under consideration the alignment is very sensitive to the wavefunctions used in the calculations, showing drastic differences if either hydrogen-like (HL) or Hartree-Slater (HS) wavefunctions are used in PWBA calculations. More sophisticated calculations making use of distorted wave Born approximation (DWBA) qualitatively predict the decrease of the alignment close to threshold, but a quantitative agreement is still lacking.

3.2 Proton Impact

In Fig.7 the (integrated) alignment A_{20} for L3-shell ionization of medium to heavy target atoms (atomic number $Z = 47$ to 79) by proton impact is displayed as a function of incident relative velocity V. Within the quoted accuracy, the experimental data closely follow a universal curve. The experimental data agree well for $V > 0.2$ with theoretical calculations based on PWBA. At lower incident velocities PWBA predicts a large negative alignment, which is clearly not seen in the experimental data. This deviation from PWBA is explained by the Coulomb-deflection of the projectile in the nuclear field of the target atom. As was pointed out by Jitschin et al. (1981) and Palinkas et al. (1980) the large A_{20} predicted by PWBA may be observed in a direction which correlates somehow with the direction of the outgoing projectile. If the outgoing projectile direction is not specified, as is the case for integrated alignment experiments, this results in some averaging over scattering angle and, most important, azimuthal angle ϕ, and hence in a considerable decrease of the observable alignment. Simple models, which account for this Coulomb-deflection, as well as theoretical calculations in which Coulomb-deflection is explicitely taken into account, give in fact reasonable agreement with the experimental observation.

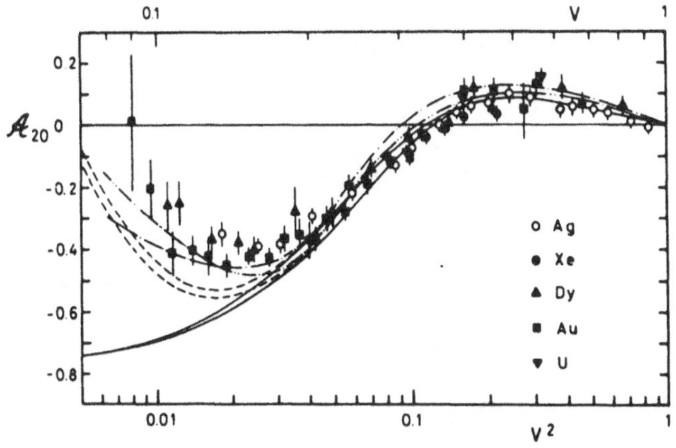

Fig.7: Integrated L3-alignment A_{20} for proton impact on different target atoms (Z = 47-92). Experimental data for O, Ag; ●, Xe, ▲, Dy; ■, Au; and ▼, U (Jitschin et al. 1979, 1982, Richter et al. 1981, Palinkas et al. 1980, 1981) are compared with ——, PWBA calculations (Sizov and Kabachnik 1980); and --- including corrections for Coulomb-deflection (Jitschin et al. 1981) for Ag (upper curves) and Au (lower curves), and RSCA calculations for (-..-), Ag and (-.-) Au (Rösel et al. 1982). (From Jitschin et al. 1982).

3.3 Differential Alignment

The first differential angular correlation experiment of an inner atomic shell ionized by electron impact was reported by Sewell and Crowe (1982). They measured the angular distribution of Ar-L Auger electrons in coincidence with electrons scattered by 15° and 21°. Since then, a number of experiments have been carried out for this collision system, and the effects of Auger-line broadening due to (long-range) post-collisional interaction (PCI) and interference effects have been studied in some detail (see, e.g., Crowe 1984, Sandner 1985).

Zehendner et al. (1985) have investigated the impact parameter dependence of L3-shell alignment in 4 MeV proton-samarium collisions. They have measured polar and azimuthal angular distributions of characteristic Ll x-rays in coincidence with scattered protons. This coincident photon angular distribution $I(\theta_{\gamma},\phi)$ may be expressed in terms of (differential) alignment parameters A_{2Q} (Q = 0,1,2) as

$$I(\theta_\gamma,\phi) = I_0 \{1 + \alpha_2 [(1/2) A_{20} (3 \cos^2(\theta_\gamma) - 1)$$
$$- \sqrt{(3/2)}\, A_{21} \sin(2\theta_\gamma) \cos(\phi)$$
$$+ \sqrt{(3/2)}\, A_{22} \sin^2(\theta_\gamma) \cos(2\phi)]\} \quad (11)$$

where α_2 depends on the photon transition, and A_{2Q} given by (with $\varrho_{xx} + \varrho_{yy} + \varrho_{zz} = 1$)

$$A_{20} = (1/2) [1 - 3\varrho_{zz}]$$
$$A_{21} = Re(\varrho_{zx}) \quad (12)$$
$$A_{22} = (1/2) [\varrho_{yy} - \varrho_{xx}]$$
$$O_1 = Im(\varrho_{zx})$$

A measurement of the orientation $O_1 = \langle L_y \rangle/[L(L+1)]$ giving the excess (spin-averaged) angular momentum L_y perpendicular to the scattering plane requires a circular polarization analysis of the emitted light which is prevented in the x-ray regime due to the low efficiency of both linear and circular polarizers. In Fig.8 the experimental alignment data versus impact parameter b are shown together with theoretical calculations using the semi-classical approximation with non-relativistic (SCA) and relativistic (SCAR) hydrogenic wavefunctions. In general, the agreement between experiment and theoretical calculations is satisfactory. The data for A_{20} strongly support the intuitive picture that at small impact parameters preferentially $|z\rangle$ becomes ionized, whereas at larger impact parameters also ionization of $|x\rangle$ and $|y\rangle$ occurs. The data also show that there is an appreciable incoherence during the collision, indicated by a small A_{21} and an either small or positive A_{22}. A positive A_{22} requires $\varrho_{yy} > 0$ and is probably caused by averaging over kinetic energy and direction of the ionized electron.

3.4 Light and Heavy Ion Impact

In Fig.9 experimental data for the L3-alignment of gold and uranium induced by ion impact at a fixed relative velocity ($V^2 \approx 0.045$) are

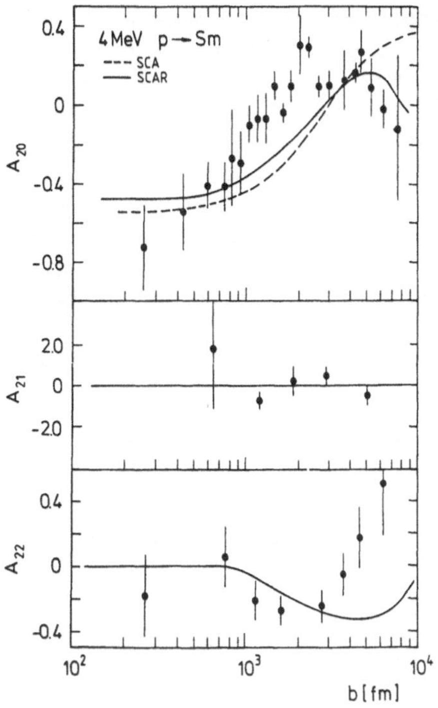

Fig.8: Differential L3-shell alignment A_{2Q} (Q = 0,1,2) versus impact parameter for 4 MeV proton-samarium collisions. The experimental data are compared with calculations based on semi-classical approximation with non-relativistic (SCA) or relativistic (SCAR) hydrogenic wavefunctions (Zehendner et al. 1985).

given as a function of the atomic number of the lighter collision partner. For proton impact, this corresponds to a velocity where a large negative alignment is experimentally observed. For heavier collision partner A_{20} is generally smaller (Jitschin et al. 1983). Of particular interest is the data point for uranium on gold, for which on the basis of a model of symmetric collisions (quasi-molecular model) a large negative alignment of A_{20} = -1 might be expected. Although this molecular orbital (MO) model of atomic collisions was found to give excellent results when predicting the impact parameter dependence of inner-shell vacancy production in slow ion-atom collisions (see, e.g., Lutz 1983, Shanker et al. 1982, 1984), the predictions of this model are not in agreement with the experimental observation. The most likely reason for this may be seen in the way in which the experimental data were extracted. Already in alignment studies by light ion impact, and the more so in symmetric or near-symmetric collisions, in addition to L3-shell vacancy creation a large number of outer shell vacancies are produced in the same collision. Thus, the observed anisotropy no longer directly reflects the alignment of the L3-vacancy, but also depends on the outer-shell configuration. This leads to a considerable reduction of the observable anisotropy (see, e.g. Jitschin 1984). Therefore, such heavy collision system are at present not very useful for a study of alignment and orientation in symmetric collision systems. A probably better choice for a study of this collision regime are few-electron systems, which will be discussed in the following.

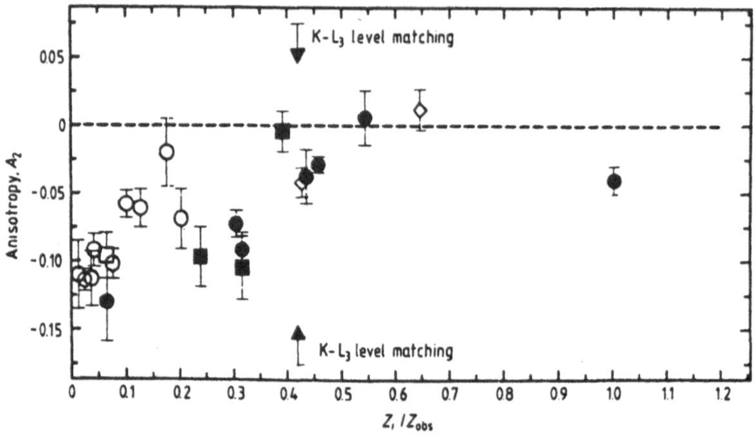

Fig.9: Anisotropy A_2 of the L1 transition versus Z_1/Z_{obs} for fixed relative velocity V≈0.212; open symbols: Z_1 → Au, Pb, U (Jitschin et al. 1979, 1982, 1983); closed symbols: U → Z_1, Z_1 → U (Stachura et al. 1984)

4. SYMMETRIC COLLISIONS

In symmetric or near-symmetric collisions an intuitive picture is that of two slowly approaching nuclei surrounded by electrons and transiently forming a quasi-molecule (Fano and Lichten 1965). This approach is well justified as long as the velocity of the two approaching nuclei is small compared to the (classical) electron velocity of the accompanying electrons under consideration for a particular collision. Usually, within this model one calculates the energies of a single electron moving in the field of the two nuclei separated by an internuclear distance R, with other electrons accounting for some shielding of the nuclear charges. Fig.10 gives a so-called correlation diagram for the H^+-H collision system. For clarity, the l-degeneracy for states of the same main quantum number n has been removed. In this picture, excitation of hydrogen to the 2p-state takes place when the two nuclei approach each other. An electron in the $2p\sigma$ molecular orbital (MO) may be transferred at small internuclear distances by so-called rotational coupling (i.e. coupling induced by the rotation of the internuclear axis, see, e.g. Macek 1983, Briggs 1985) to the $2p\pi$ state. This molecular state correlates with the atomic $H(2p\pm1)$ state.

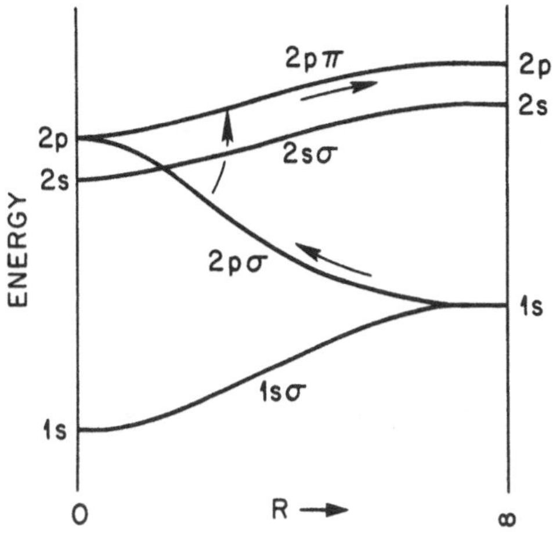

Fig.10: Schematic correlation diagram for H^{+}+H (see text).

4.1 Quasi-One Electron Systems

Since no (differential) angular correlation or polarization studies have been performed yet for one-electron systems, we shall discuss some results for quasi-one electron systems (Menner et al. 1981, Andersen et al. 1980) like Li-He. One may think of this collision system as a single electron outside two (inactive) core states $Li^{+}(1s^{2})$ and $He(1s^{2})$. The collision process Li(2s) + He → Li(2p) + He was studied by Menner et al. (1981) performing measurements at a center-of-mass-energy of 366 eV together with theoretical calculations. Despite its suggested simplicity, the Li-He collision system shows some complexity. Basically, Li(2p) excitation may proceed via two different excitation mechanism, being thought of as non-localized direct excitation taking place at relatively large and rotational coupling operating at small internuclear separations.

Another quasi-one electron system studied recently by Bähring et al. (1983, 1984) is Na^{+}+Na(3s) → Na^{+}+Na^{*}(3p). Making use of an experimental technique in which sodium ions are superelastically scattered on laser-excited Na^{*}(3p) atoms in well defined angular momentum substates, they could demonstrate that the (de-)excitation process performs via a Σ_u-Π_u rotational coupling (Fig.11).

However, at large internuclear separations the electron cloud decouples from the rotation of the internuclear axis. This decoupling (Fig.12) happens where the potential energy curves correlating with Na(3p) merge with each other and was found to agree reasonably with predictions of Grosser (1981). A full theoretical account of this collision system was given recently by Allan et al. (1985).

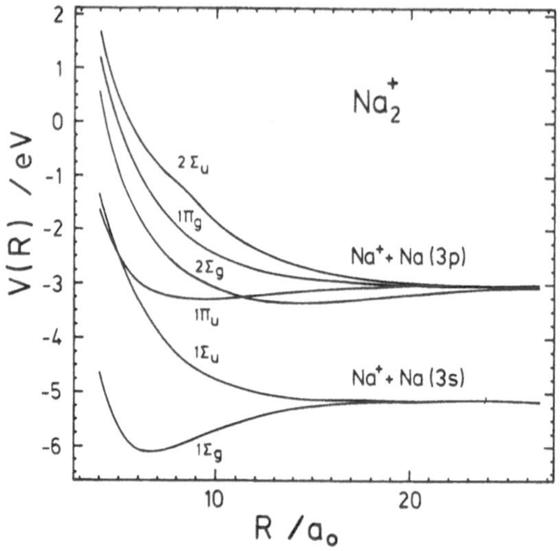

Fig.11: Potental energy curves for Na_2^+ (Habitz 1975)

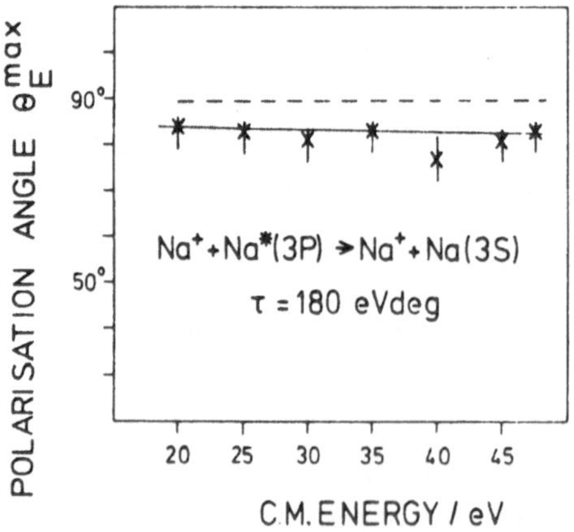

Fig.12: Decoupling of the electron cloud from the molecular axis (given by the deviation of the experimental polarization angle from 90°) versus center-of-mass-energy in Na^++Na^*(3p) → Na^++Na(3s) superelastic collisions (from Bähring et al. 1983).

4.2 Two-Electron Systems

Two-electron systems have received some attention so far as they allow to study electron-correlation effects in collisions. Prototype for such a system is $(H-He)^+$. For this system the excitation of H(2p) was studied in charge changing H^+-He collisions (Mueller and Jaecks 1982, Hippler et al. 1984a). Fig.13 shows the corresponding correlation diagram as obtained by Macias et al. (1981, 1983). In this quasi-molecular picture H(2p) excitation is thought to perform in two steps, the first of which takes place at large internuclear separations and couples the 1sσ ground state with the 2pσ state. At small internuclear separations the electron is transferred by rotational coupling from the 2pσ to the 2pπ state. This should result in $H(2p_{\pm 1})$ excitation only.

Fig.14 shows the relative population of $H(2p_0)$ as a function of impact parameter for 1-4 keV proton-helium collisions (Hippler et al. 1984). It is seen that only at large internuclear distances $H(2p_0)$ excitation is small (and vice-versa $H(2p_{\pm 1})$ excitation large). For small internuclear separations $H(2p_0)$ excitation strongly increases, indicating the importance of other excitation mechanism. This increase of $H(2p_0)$ is in reasonable agreement with model calculations of Fritsch (1984), using a one-electron picture.

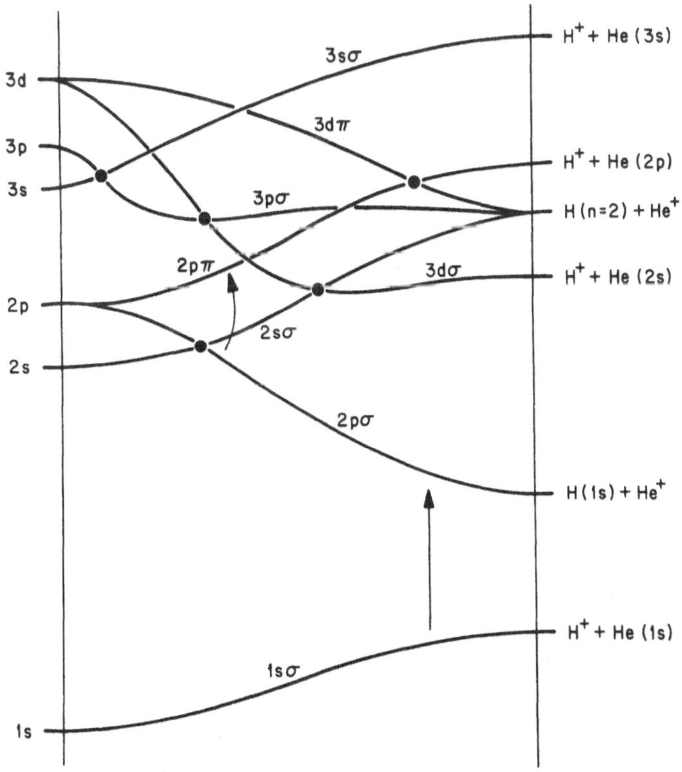

Fig.13: Schematic correlation diagram for $(H-He)^+$ (after Macias et al. 1981, 1983)

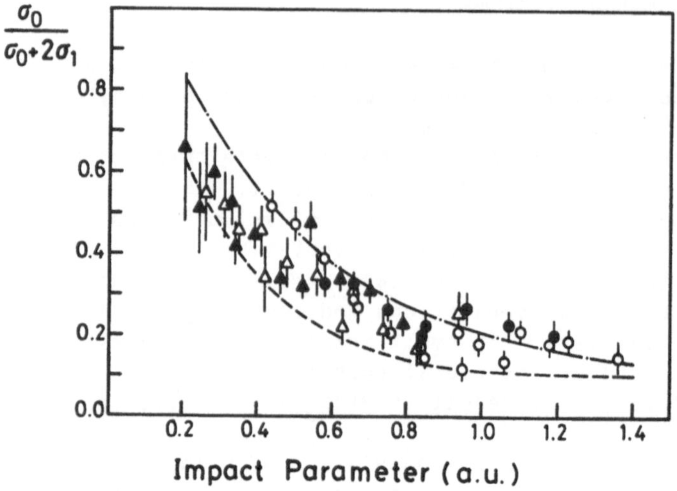

Fig.14: Relative $H(2p_0)$ excitation as a function of impact parameter for o 1 keV, ● 1.5 keV, Δ 3 keV, and ▲ 4 keV H^+-He collisions (Hippler et al. 1984) in comparison with model calculations for --- 1 keV and -.-.- 4 keV (Fritsch 1984).

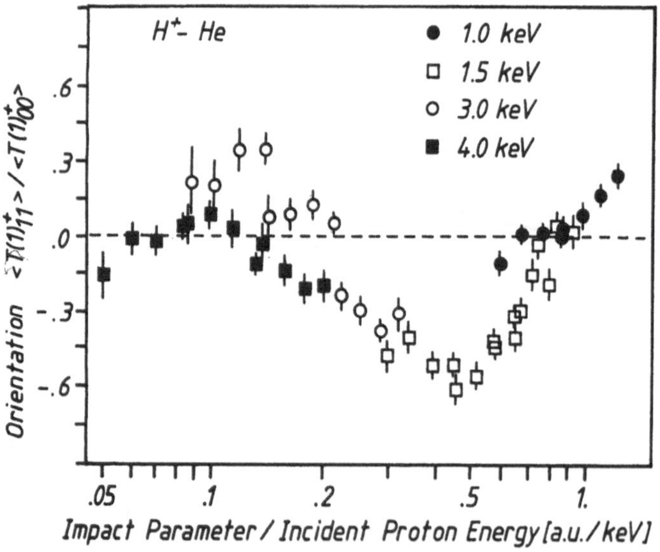

Fig.15: Orientation of H(2p) as a function of impact parameter divided by incident energy in ● 1 keV, □ 1.5 keV, o 3 keV, and ■ 4 keV H^+-He collisions (Hippler et al. 1984).

While the relative population of the H(2p) substates was found to scale reasonably with impact parameter, such a scaling was not observed for the orientation. Fig.15 shows the relative orientation observed for this collision system, which shows a pronounced dependence of both impact parameter and incident electron energy. Experimentally it was observed that an approximate scaling may be achieved by plotting the data versus impact parameter divided by the incident collision energy. So far, this scaling is not understood, nor does it agree with the model calculations of Fritsch (1984).

5. TWO-ELECTRON EXCITATION

A number of investiagations have dealt with excitation processes involving two electrons in a single collision. For example, using the photon-scattered projectile coincidence method Fayeton et al. (1981) studied the collision system

$$He + He \rightarrow He^*(2^1P) + He^*(2^1P)$$

where both helium atoms are excited from the ground to the 2^1P_1 state. The same collision system was also investigated by De Vlieger et al. (1981, 1982) measuring the angular correlation between two emitted photons in a photon-photon coincidence experiment. Similar to previous studies for the $H^+ + Ar \rightarrow H(2p) + Ar^+(^2p)$ collision system (Hippler et al. 1982, 1985a) such investigations are incomplete if not, as has not been done yet, the final states of both collision partners and the scattering angle of the projectile are specified.

Two-electron excitation has also been observed for collision systems like

$$He^+ + He \rightarrow \begin{cases} He^+ + He^{**}(2p^2\ {}^1D) \\ He^{**}(2p^2\ {}^1D) + He^+ \end{cases}$$

for double excitation in both target or projectile. In this particular case, both electrons are excited in the same atom, forming an autoionizing state which decays by ejection of an electron with characteristic kinetic energy. Boskamp et al. (1982, 1984a) have performed some experiments for this collision system (see also Morgenstern 1983); the results for the angular distribution of characteristic electrons ejected from both projectile and target in coincidence with scattered projectiles are displayed in Fig.16. During the collision two electrons are promoted via the $(2p\sigma)^2\ {}^2\Sigma_g$ molecular state and rotationally coupled to the $(2p\pi)^2\ {}^2\Sigma_g$ or ${}^2\Delta_g$ states, which both correlate with $He^{**}(2p^2\ {}^1D)$. Although the lifetime of the excited He^{**} is long compared to the collision time, it is sufficiently short due to the long-range nature of the Coulomb-force to decay while it still feels the ionic charge of the second collision partner. This so-called post-collision interaction (PCI) causes an appreciable broadening and shift of the kinetic energy of the ejected electron compared to the field-free value. Due to the collision dynamics the kinetic energy of ejected electrons is generally different for ejection from projectile and target. Thus, it is possible to distinguish between projectile and target excitation. However, this does not hold for certain ejection angles, where the collision dynamics together with the line broadening from the PCI just prevents this. For these particular angles sharp structures are observed in the angular distributions of the ejected electrons, which result from interference between projectile and target excitation. Taking into acccount (incoherent) contributions from two decay channels resulting in gerade (g) and ungerade (u) final molecular states (and vice-versa for the ejected electron, the total, electron plus two helium ions, final state has gerade parity), the ejected electron intensity may be written as

$$\begin{aligned} I &= (1/4)\ |A_g|^2 + (3/4)\ |A_u|^2 \\ &= (1/4)\ |A_t+A_p|^2 + (3/4)\ |A_t-A_p|^2 \end{aligned} \tag{13}$$

where $A = A(E-E_D)$ is the transition amplitude, g and u refer to gerade and ungerade final molecular states, t and p to target and projectile, respectively. E is the kinetic electron energy in the moving frame and E_D is the Doppler-shift. The solid lines in Fig.16 are calculated using Eq. (13) with the proper angular and energy dependence for the phases of A and reproduce the sharp interference structures without any additional assumption.

6. ELECTRIC AND MAGNETIC FIELDS

6.1 Electric Fields

We have already seen how a long-range (post-collisional) interaction may modify the energy and angular distribution of

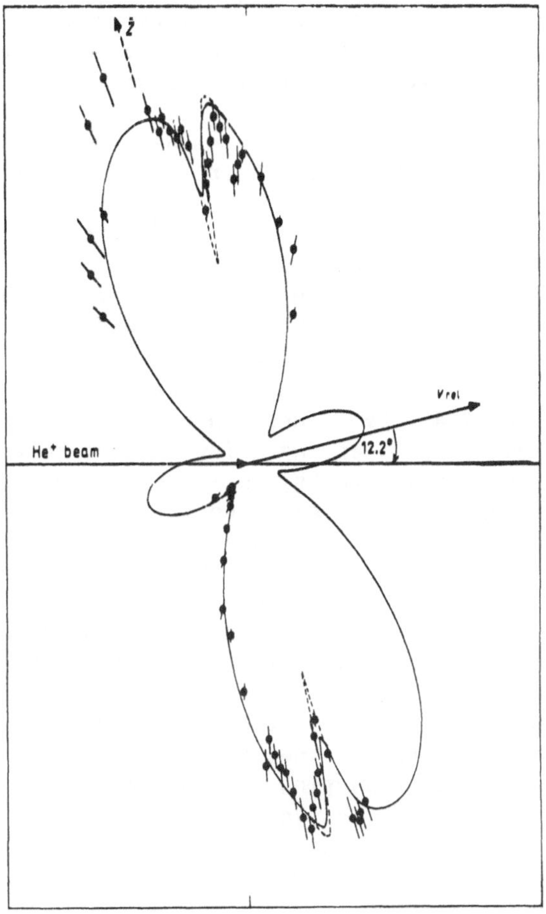

Fig.16: Angular distribution of autoionisation electrons from $He^{**}(2p^2\ ^1D)$ ejected in the scattering plane for 2 keV He^+-He collisions. The scattering angle was 12.2^0 (Morgenstern 1983).

ejected electrons. Recently, an even more striking feature of PCI was observed for double excitation in $Li^+ + He \rightarrow Li^+ + He^{**}$ collisions. While in He^+ + He collisions the dominant double excitation mechanism at few keV incident energies is formation of $He^{**}(2p^2\ ^1D)$, Li^++ He collisions in addition produce a large fraction of $He^{**}(2s2p\ ^1P)$ states. The electric field of the Li^+ ion causes a mixing of doubly excited states with opposite parity. The resuling "Stark-beats", though of a frequency which is not accessible by time-resolved electron spectroscopy, are observed as intensity variations of the angular distributions of ejected electrons. As before in He^+ + He collisions, the PCI causes the kinetic energy of the ejected electrons to be a function of the internuclear separation and thus on the time elapsed since excitation took place (Boskamp et al. 1984b).

PCI effects also play a role in $H^+ + He \rightarrow H(n) + He^+$ charge changing collisions. Due to their near-degeneracy hydrogenic states with identical main quantum number n are coherently excited during the collision, and, due to PCI, states of opposite parity remain mixed for internuclear separations as large as 100 a.u. This PCI is usually neglected in many calculations, although it may result in a significant excitation transfer between different l-states. In Fig.17 the integrated alignment A_{20} for H(2p) excitation in $H^+ + He \rightarrow H(2p) + He^+$ collisions is displayed. A strong variation of A_{20} as a function of incident proton energy is observed. At low energies around a few keV the total alignment is small, in sharp contrast to this differential measurements show a large alignment here (see above). The integrated alignment shows a maximum around 8 keV, it reaches a deep minimum of about -45% around 40 keV. Afterwards, up to the largest incident energy investigated for this collision system so far, A_{20} remains negative, with a small tendency to increase. Also shown in Fig.17 are theoretical calculations using the continuum-distorted wave (CDW) approximation. The two CDW calculations differ whether PCI is included in the calculation or not (Burgdörfer and Dubé 1984). It is noted that for incident energies larger than 50 keV the PCI causes a significant reduction of the calculated A_{20}. However, none of the theoretical curves is in full accord with the experimental data. At present it is not clear whether this points to a failure of the CDW approximation or of the way in which PCI is included in the calculation.

So far, coherent excitation of H(n) states has been observed by Sellin et al. (1979), Krotkov and Stone (1979), and Back et al. (1984) for H(n=2), and by Mahan and Smith (1977) and Havener et al. (1982, 1983) for H(n=3). In these measurements the coherent excitation was observed by applying external electric fields, thereby mixing l-states of opposite parity. Both Sellin et al. (1979) and Back et al. (1984) made use of time-resolved spectroscopy, which provides a method to directly observe the corresponding Stark-beats. For the $H^+ + He$ collision system studied by Sellin et al. (1979) this time-resolution was provided by the time-of-flight of the excited fast H(n=2) atom and the corresponding density matrix was determined. Back et al. (1984) investigated H(n=2) excitation by electron impact.

In Fig.17 the coincidence signal for decay of H(n=2) states in an electric field of 250 V/cm is shown for 350 eV electron impact on atomic hydrogen. The signal shows the decay of the Stark-mixed states, which may be expressed as linear combinations of field-free wavefunctions of the n=2 states. The decay of these Stark states

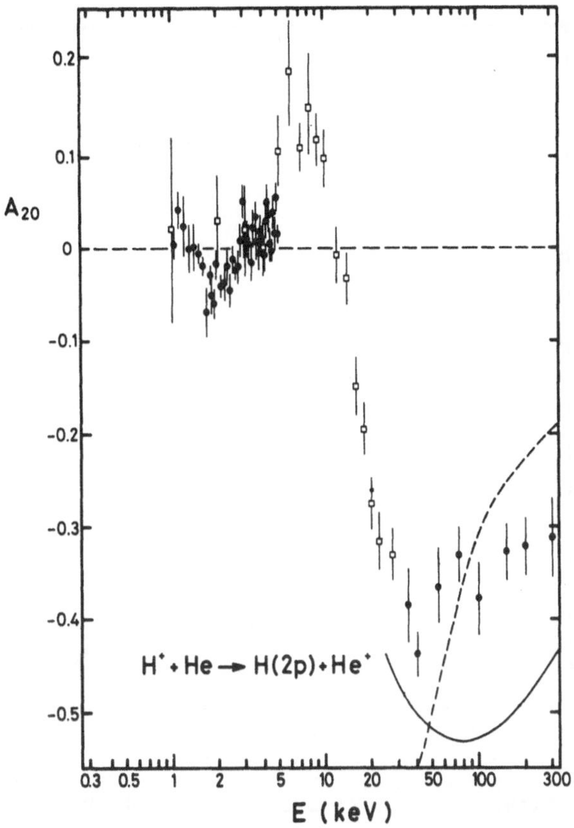

Fig.17: Integrated alignment A_{20} for charge exchange excitation of H(2p) in H^{+}+He collisions: ● Hippler et al. (1985b), □ Teubner et al. (1970); —— CDW calculations without PCI; --- CDW calculations including PCI (Hippler et al. 1985b).

is governed by a sum of three exponentials and allows, at least in principle, to determine the relative population of individual (field-free) substates.

If the H(n=2) states are coherently excited, the interference between s- and p-states could give rise to Stark-beats. Thus, at a field strength of 250 V/cm the exponential decay should be modulated with periods of 0.1 and 0.6 ns. In the experiment of Back et al. (1984) the time resolution was only sufficient to observe the longer beat-frequency, which corresponds to interference between states which asymptotically reduce to $2s_{1/2}$ and $2p_{1/2}$ in the field-free limit. Fig.18 clearly shows evidence of the expected beat structure, although the data were not of sufficient quality to analyze quantitatively the relative phase of the beat-amplitude.

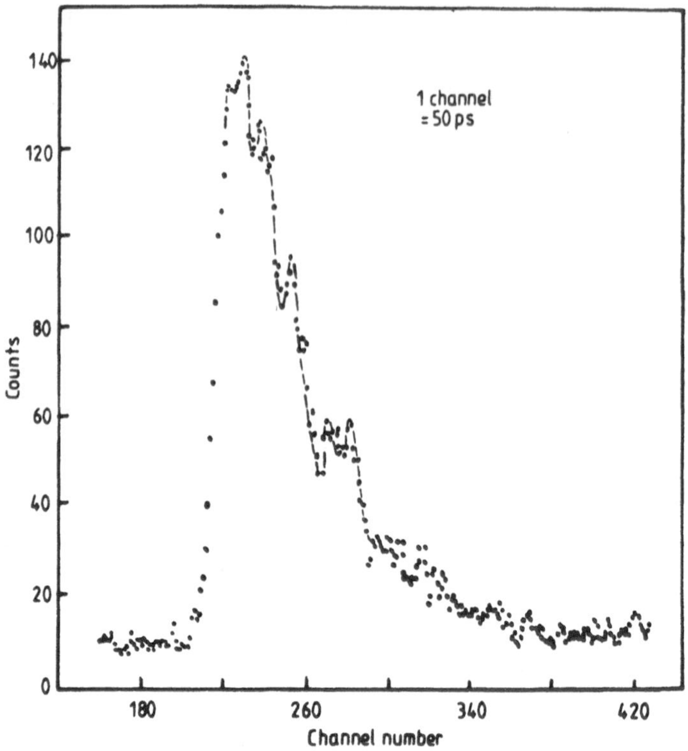

Fig.18: Decay curve for H(n=2) excited by 350 eV electron impact on atomic hydrogen recorded with an electric field of 250 V/cm. The electron scattering angle was 5° (Back et al. 1984).

Havener et al. (1982, 1983), by applying perpendicular and parallel (with respect to the z-axis) electric fields and performing a polarization analysis of emitted light, have determined the n=3 density matrix formed by electron capture in 40 to 80 keV $H^+ + He \rightarrow H(n=3) + He^+$ collisions. Fig.19 gives the electron probability distribution for an incident proton energy of 40 keV as obtained from their data. As a most striking feature Havener et al. (1982) observed a large s-p coherence (corresponding to a large and positive dipole moment), indicated by the forward-backward asymmetry of the electron distribution of Fig.19. The large center-peak reflects the fact that at the chosen experimental conditions dominantly the 3s state is populated. It should be noted that Burgdörfer (1981) has recently introduced an alternative description of alignment and orientation in terms of pseudo-spin operators constructed from orbital angular momentum operator and from the Runge-Lenz vector which might prove to be particular suitable for high-n states. For example, within this formalism the real part of the s-p_0 coherence is expressed as the z-component of the Runge-Lenz vector. As was emphasized by Burgdörfer (1984), this formalism is also useful to parametrize the asymmetries of continuum states such as the well-known cusp-asymmetry in electron capture to the continuum (ECC) processes.

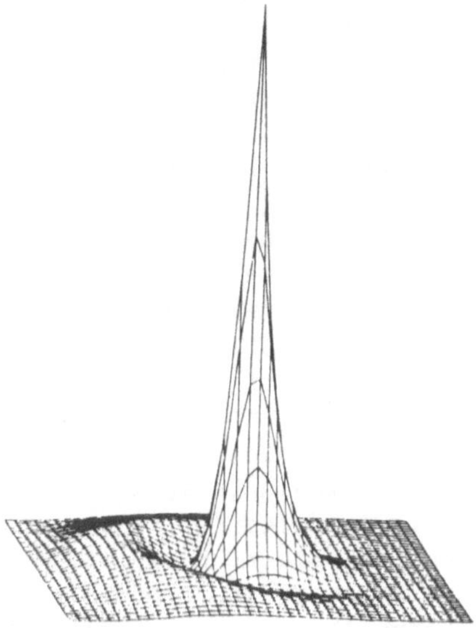

Fig.19: H(n=3) electron probability distribution for 40 keV H^+ + He charge changing collisions (Havener et al. 1983).

6.2 Magnetic Fields

A number of experiments thus far have utilized the use of external magnetic fields in angular correlation or polarization studies. As in the case of internal magnetic (for example, spin-orbit) interaction, the external magnetic field modifies or even may destroy the observable anisotropy. Consider, for instance, a magnetic field H_y in y-direction and interacting with initially prepared $|x\rangle$ and $|z\rangle$ substates having them rotate around $\underline{y}$ with Larmor frequency $\omega_L = g_j \mu_0 m_j H_y/h$, where g_j is the g-factor, and μ_0 the Bohr magneton. This Hanle-effect is now known for more than 60 years (Hanle 1924). If the lifetime τ of the excited state is small compared to the precession time $1/\omega_L$ ($\omega_L \tau \ll 1$), for instance if $H_y = 0$, then little or no precession occurs and the angular distribution remains unaffected. In contrast, if the precession time is fast enough that the initially produced x- and z-substates perform many precessions during their lifetime, all anisotropy will be smeared out.

The interaction of the magnetic field with the magnetic substates thus provides a means to distinguish between excited states having different lifetimes. This method was used by Grosser and Neitzke

(1982) to separate contributions from $Ne^*(^1P_1)$ and $Ne^*(^3P_1)$ excitation in He^+ + Ne collisions. Similarly, Aynacioglu et al. (1982) used the method to separate direct $He^*(3^1D)$ excitation from excitation via cascading transitions. Using a level-crossing technique utilizing the combined effect of electric and magnetic fields, Schilling et al. (1982) determined the density matrix for excitation of helium D-states, whereas Neitzke and Andersen (1984) used the combined action of internal and external magnetic fields to determine ϱ for Li + He → $Li^*(3^2D)$ + He collisions. Recently, Richard et al. (1985) observed Zeeman-beats during the decay of $F^{7+}(1s2p\ ^3P_2)$ states and used it for a determination of g-factors. Finally, Winter et al. (1984) used the Zeeman-beat technique to investigate the orientation of angular momenta of ground state nitrogen atoms formed by the interaction of N^+ ions with a solid copper surface at grazing incidence. This ion-beam surface interaction at grazing incidence is known to produce a large orientation of the orbital angular momentum of reflected atoms. Using a newly developed technique, Winter et al. (1984) could demonstrate that the ion-solid interaction leaves the (neutralized) nitrogen atom in a state-selected ($2p^3\ ^2D$) and highly polarized core. No population of other core states was observed. This result was interpreted by proposing a model of selective electron capture of an oriented electron by an already oriented ion core. Within this model, population of the nitrogen $2p^3\ ^2P$ state is forbidden by the Pauli-principle.

6.3 Electromagnetic Fields

Angular distribution studies are also of interest for collision processes taking place in intense electromagnetic fields such provided by powerful lasers. During a collision in such a field the atomic species not only interact with themselves but simultaneously or successively with the electromagnetic field, thereby absorbing or emitting photons. An example for such a process is simultaneous electron-photon excitation of atoms,

$$e + \omega + A \rightarrow e + A^*$$

where the atom A is excited from its ground state to an excited state A^*. Both, photon ω and electron e jointly contribute to the atomic excitation energy E_{if}; the energy loss ΔE of the electron is therefore given by $\Delta E = E_{if} - \hbar\omega$. For this particular excitation process the interaction V may be written as $V = V_1 + V_2$, where $V_1 = -1/r_1 + 1/r_{12}$ is the well-known Coulomb interaction of the incident electron with target nucleus and target electron. $V_2 = -(\underline{A}/c)(\underline{p}_1+\underline{p}_2)$ is the interaction of the electromagnetic field $\underline{A}$ with projectile and target electron.

In Fig.20 we display total cross sections for H(2s) excitation calculated in second-order perturbation theory as a function of the kinetic energy of the incident electron. The photon energy is 1.17 eV, which corresponds to the fundamental frequency of a Nd:YAG laser. The calculated cross sections are of the order of πa_0^2, for a laser intensity of 10^{13} W/cm^2. This tremendous intensity is either difficult to achieve with nowadays lasers or to maintain longer than few nanoseconds. This is the reason why this process has escaped detection so far, although there is some hope that it might be possible in the near future. The calculations presented in Fig.20 show that the cross section for this process depends on the polarization of the incident light. In dipole approximation, the atomic transition amplitude for this process is proportional to $\underline{K} \cdot \underline{\varepsilon}$, where $\underline{K}$ is the momentum transfer and $\underline{\varepsilon}$ the polarization vector. Since at small incident electron energies we have $\underline{K} \parallel \underline{z}$, the total cross section becomes large for $\underline{\varepsilon} \parallel \underline{z}$ (i.e. $\theta_\gamma = 0^\circ$). For large incident electron energies $\underline{K} \perp \underline{z}$, and a large cross section should be observable for $\underline{\varepsilon} \perp \underline{z}$.

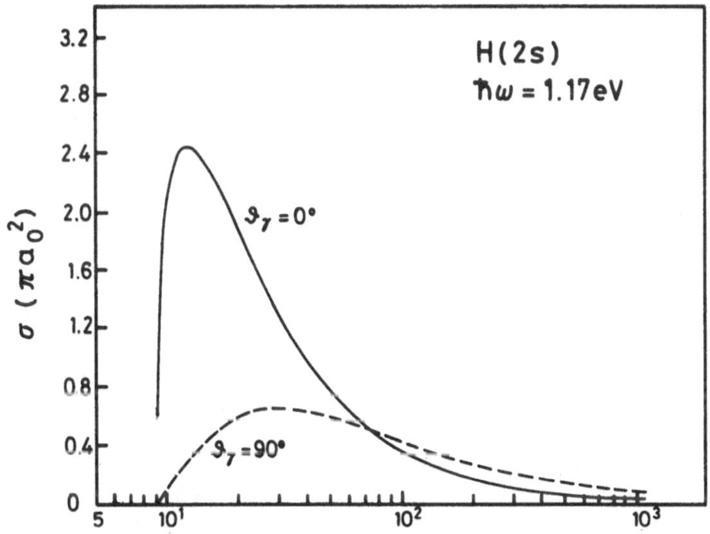

Fig.20: Total cross section for simultaneous electron-photon excitation of H(2s) as a function of incident electron energy for two polarization angles $\theta_\gamma = 0^\circ$ and 90°. The photon energy is 1.17 eV; the photon flux corresponds to 10^{13} W/cm^2 (Jetzke et al. 1984).

Acknowledgements:

The author acknowledges helpful discussions with Professor H. O. Lutz, Professor H. Kleinpoppen, Dr. W. Jitschin, Dr. U. Wille, M. Faust, R. Wolf, W. Harbich, and S. Jetzke. Part of this work was supported by the Deutsche Forschungsgemeinschaft in SFB 216 "Polarisation und Korrelation in atomaren Stoßkomplexen".

References

Abignoli, M., Barat, M., Baudon, J., Fayeton, J., Houver, J. C., 1972, J. Phys. 5:1533.

Allan, R. J., Bähring, A., Haussen, J., 1985, to be published.

Andersen, N., 1980, Physikalische Blätter 36:219.

Andersen, N., Andersen, T., Østgaard Olsen., J., Horsdal Pedersen, E., 1980, J. Phys. B 13:2421.

Aydinol, M., Hippler, R., McGregor, I., Kleinpoppen, H., 1980, J. Phys. B 13:989; 1981, in "Coherence and Correlation in Atomic Collisions", H. Kleinpoppen and J. F. Williams, eds., Plenum Press, New York, p. 205.

Aynacioglu, A. S., Oppen, G. von, Perschmann, W.-D., Szostak, D., 1981, J. Phys. 14:2611.

Back, C. G., Watkin, S., Eminyan, M., Rubin, K., Slevin, J., Woolsey, J. M., 1984, J. Phys. B. 17:2695.

Bähring, A., Hertel, I. V., Meyer, E., Schmidt, H., 1983, Z. Physik A 312:293

Bähring, A., Hertel, I. V., Meyer, E., Meyer, W., Spies, N., Schmidt, H., 1984, J. Phys. B 17:2859.

Berezhko, E. G., Kabachnik, N. M., 1982, J. Phys. B 15:2075.

Blum, K., 1981,

Blum, K., Kleinpoppen, H., 1979, Phys. Reports 52:203; 1983, Phys. Reports 96:251.

Briggs, J. S., 1985, this volume.

Born, M., Wolf, E., 1970, Principles of Optics, Pergamon Press.

Boskamp, E., Griebling, O., Morgenstern, R., Nienhuis, G., 1982, J. Phys. B 15:3745.

Boskamp, E., Morgenstern, R., Straten, P. van der, Niehaus, A., 1984a, J. Phys. B 17:2823.

Boskamp, E., Morgenstern, R., Verlinde, H. L., Niehaus, A., 1984b, 9th Int. Conf. Atomic Physics, Seattle, Book of Abstracts, p. A 98.

Burgdörfer, J., 1981, Phys. Rev. A 24:1756; 1983, Z. Phys. A 309:285; 1984, in: "Forward Electron Ejection in Ion Collisions", K. O. Groeneveld, W. Meckbach, I. A. Sellin, eds., Springer-Verlag, Berlin, p. 32.

Burgdörfer, J., Dubé, L., 1984, Phys. Rev. Lett. 52:2225.

Catalan, G., Roberts, M. J., 1979, J. Phys. B 12:3947.

Cleff, B., Mehlhorn, W., 1974, J. Phys. B 7:605; 1971, Phys. Lett. 37A:3.

Crowe, A., 1984, in: "Electronic and Atomic Collisions", Book of Invited Papers, J. Eichler, I. V. Hertel, N. Stolterfoht, eds., North-Holland, Amsterdam, p. 97

Dubois, R. D., Mortensen, L., Rødbro, M., 1981, J. Phys. B 14: 1613.

Ehlers, V. J., Gallagher, A. C., 1973, Phys. Rev. A 7:1573.

Eminyan, M., MacAdam, K. B., Slevin, J., Kleinpoppen, H., 1973, Phys. Rev. Lett., 31:576; 1974, J. Phys. B 7:1519.

Fano, U., Macek, J. H., 1973, Rev. Mod. Phys. 45:553.

Fayeton, J. A., Houver, J. C., Brenot, J. C., Barat, M., 1981, J. Phys. B 14:2599.
Flower, D. R., Seaton, M. J., 1967, Proc. Phys. Soc., London, 91:59.
Fritsch, W., 1984, private communication.
Grosser, J., 1981, J. Phys. B 14:1449.
Grosser, J., Neitzke, H.-P., 1982, Z. Phys. A 304:49.
Habitz, P., 1975, Ph.D. Thesis, University of Bonn.
Hafner, H., Kleinpoppen, H., 1967, Z. Phys. 198:315.
Hanle, W., 1924, Z. Physik 30:93.
Hanne, G. F., Wemhoff, K., Wolcke, A., Kessler, J., 1981, J. Phys. 14:L507
Havener, C., Westerveld, W., Risley, J., Tolk, N., Tully, J. C., 1982, Phys. Rev. Lett. 48:926.
Havener, C. C., Rouze, N., Westerveld, W. B., Risley, J. S., 1983, XIIIth ICPEAC Electronic and Atomic Collisions, Book of Abstracts, p. 479.
Hermann, H. W., Hertel, I. V., 1982, Comments At. Mol. Phys. 12: 61 and 121.
Hippler, R., Malunat, G., Faust, M., Kleinpoppen, H., Lutz, H. O., 1982, Z. Phys. A 304:63.
Hippler, R., 1983, in: "Fundamental Processes in Energetic Atomic Collisions", H. O. Lutz, J. S. Briggs, H. Kleinpoppen, eds., Plenum, New York, p. 551.
Hippler, R., 1984, in: "Progress in Atomic Spectroscopy, Part C", H. J. Beyer, H. Kleinpoppen, eds., Plenum Press, p. 511.
Hippler, R., Faust, M., Wolf, R., Kleinpoppen, H., Lutz, H. O., 1984, 9th Int. Conf. Atomic Physics, Seattle, Book of Abstracts, p. 103; 1985a, Phys. Rev. A.
Hippler, R., Harbich, W., Faust, M., Lutz, H. O., Dubé, L., 1985b, to be published.
Hollywood, M. T., Crowe, A., Williams, J. F., 1979, J. Phys. B: 12:819.
Jetzke, S., Faisal, F. H. M., Hippler, R., Lutz, H. O., 1984, Z. Phys. A 315:271.
Jitschin, W., Kleinpoppen, H., Hippler, R., Lutz, H. O., 1979, J. Phys. B 12:4077.
Jitschin, W., Lutz, H. O., Kleinpoppen, H., 1981, in: "Inner-Shell and X-ray Physics of Atoms and Solids", D. J. Fabian, H. Kleinpoppen, L. M. Watson, eds., Plenum Press, New York, p. 89.
Jitschin, W., Kaschuba, A., Kleinpoppen, H., Lutz, H. O., 1982, Z. Phys. A 304:69.
Jitschin, W., Hippler, R., Shanker, R., Kleinpoppen, H., Schuch, R., Lutz, H. O. 1983, J. Phys. B 16:1417.
Jitschin, W., 1984, J. Phys. B 17:4179.
Krotkov, R., Stone, J., 1980, Phys. Rev. A 22:473.
Lutz, H. O., 1983, S. Afr. J. Phys. 6:83.
Macek, J., Jaecks, D. H., 1971, Phys. Rev. A 4:2288.

Macek, J., 1983, in: "Fundamental Processes in Energetic Atomic Collisions", H. O. Lutz, J. S. Briggs, H. Kleinpoppen, eds., Plenum, New York, p. 39.
Macias, A., Riera, A., and Yañez, 1981, Phys. Rev. A 23:2941.
Macias, A., Riera, A., Yañez, M., 1983, Phys. Rev. A 27:206 and 213.
McGregor, I., Hils, D., Hippler, R., Malik, N. A., Williams, J. F., Zaidi, A. A., Kleinpoppen, H., 1982, J. Phys. B 15:L411.
Mahan, A. H., Smith, S. J., 1977, Phys. Rev. A 16:1789.
Mehlhorn, W., 1983, in: "Proc. Int. Seminar High-Energy Ion-Atom Collision Processes", Debrecen 1981, Akademiai Kiado, Budapest, p. 83.
Menner, B., Hall, Th., Zenle, L., Kempter, V., 1981, J. Phys. B 14:3693.
Morgenstern, R., 1983, in: "Fundamental Processes in Energetic Atomic Collisions", H. O. Lutz, J. S. Briggs, H. Kleinpoppen, eds., Plenum, New York, p. 567.
Mueller, D. W., Jaecks, D. H., 1982, in: "Proc. 8th Int. Conf. Atomic Physics", Book of Abstracts, Göteborg, p.B 55.
Neitzke, H.-P., Andersen, T., 1984, J. Phys. B 17:1559.
Oppenheimer, J. R., 1927, Z. Physik 43:27.
Palinkas, J., Sarkadi, L., Schlenk, B., 1980, J. Phys. B 13:3829.
Palinkas, J., Schlenk, B., Valek, A., 1981, J. Phys. B 14:1157.
Percival, I. C., Seaton, M. J., 1958, Phil. Trans. Roy. Soc., London, A 251:113.
Richard, P., Sanders, J. M., Stöckli, M. P., Brenn, R., 1985, to be published.
Richter, G., Brüssermann, M., Ost, S., Wigger, J., Cleff, B., Santo, R., 1981, Phys. Lett. 82A:412.
Rösel, F., Trautmann, D., Baur, G., 1982, Z. Phys. A 304:75
Sandner, W., Schmitt, W., 1978, J. Phys. B 11:1833.
Sandner, W., Völkel, M., 1984, J. Phys. B 17:L597.
Sandner, W., 1985, this volume.
Schilling, W., Oppen, G. von, Perschmann, W.-D., Szostak, D., 1981, J. Phys. B 14:2617.
Scott, T., McDowell, M. R. C., 1976, J. Phys. B: Atom. Molec. Phys. 9:2235.
Sellin, I., Liljeby, L., Mannervik, S., Hultberg, S., 1979, Phys. Rev. Lett. 42:570.
Sewell, E. C., Crowe, A., 1982, J. Phys. B 15:L357.
Shanker, R., Hippler, R., Wille, U., Lutz, H. O., 1982, J. Phys. B 15:2041.
Shanker, R., Wille, U., Bilau, R., Hippler, R., McMurray, W. R., Lutz, H. O., 1984, J. .Phys. B 17:1353.
Sizov, V. V., Kabachnik, N. M., 1980, J. Phys. B 13:1601.
Slevin, J., Porter, H. Q., Eminyan, M., Defrance, A., Vassilev, G., 1980, J. Phys. B. 13: L23-25.
Slevin, J., Porter, H. Q., Eminyan, M., Defrance, A., Vassilev, G., 1980, J. Phys. B 13:3009.

Stachura, Z., Bosch, F., Hambsch, F. J., Bing Liu, Maor, D., Mokler, P. H., Schönfeldt, W. A., Wahl, H., Cleff, B., Brüssermann, M., Wigger, J., 1984, J. Phys. B 17:835.
Standage, M., Kleinpoppen, H., 1975, Phys. Rev. Lett. 36:577.
Steph, N. C., Golden, D. E., 1980, Phys. Rev. A 21:759 and 1848.
Sutcliffe, V. C., Haddad, G. N., Steph, N. C., and Golden, D. E., 1978, Phys. Rev. A 17:100.
Teubner, P. J. O., Kauppila, W. E., Fite, W. L., Girnius, R. J., 1970, Phys. Rev. A 2:1763.
Thomas, L. D., Csanak, G., Taylor, H. S., Yarlagadda, B. S., 1974, J. Phys. B 7:1719.
Vlieger, G. J. N. E. de, Eck, J. van, Pijkeren, D. van, Heideman, H. G. M., 1981, J. Phys. B 14:3943.
Vlieger, G. I. N. E. de, Heideman, H. G. M., Eck, J. van, Nienhuis, G., 1982, J. Phys. B 15:L345.
Winter, H., Langheim, M., Schirmacher, A., Zimny, R., Andrä, H. J., 1984, Phys. Rev. Lett., 52:1211.
Zaidi, A. A., McGregor, I., Kleinpoppen, H., 1978, J. Phys. B 11: L151.
Zehendner, S., Baptista, G. B., Justiniano, E., Konrad, J., Schmidt-Böcking, H., Schuch, R., 1985, to be published.

ION-MOLECULE ASSOCIATION

D. R. Bates

Department of Applied Mathematics & Theoretical Physics
Queen's University
Belfast BT7 1NN, Northern Ireland

1. INTRODUCTION

In ion-molecule association on ion X^+ and a molecule Y are drawn together by their mutual long-range attraction to form an activated complex

$$X^+ + Y \rightarrow XY^{+*} \tag{1}$$

which is prevented from dissociating

$$XY^{+*} \rightarrow X^+ + Y \tag{2}$$

by being stabilized through energy being removed by a collision with an ambient gas molecule

$$XY^{+*} + M \rightarrow XY^+ + M \tag{3}$$

(ter-molecular ion-molecule association) or by photon emission

$$XY^{+*} \rightarrow XY^+ + h\nu \tag{4}$$

(radiative ion-molecule association).

Ion-molecule association is an important means of building up complicated ions beginning with simple ions.

Silicon, iron and magnesium are introduced into the Earth's atmosphere by the ablation of meteors. Their atoms are readily ionized by charge transfer with the O_2^+ and NO^+ ions that are present near the 100 km level. Composition measurements by

Zbinden et al (1975) show that Si^+ is depleted relative to Fe^+ and Mg^+ below 100 km presumably due to the formation of molecular ions followed by dissociative recombination. Goldberg (1975) has found evidence for SiO_2^+ ions and Fahey et al (1981) have suggested that one of the processes involved is

$$Si^+ + O_2 + M \rightarrow SiO_2^+ + M \quad (5)$$

which is much faster than the corresponding Fe^+ and Mg^+ ter-molecular association (See §3.1). Ferguson et al (1981) have discussed possible reaction schemes.

After Williams (1972) had drawn attention to the possibility of neutral-neutral radiative association as a source of molecules in intersteller clouds Black and Dalgarno (1973) introduced

$$C^+ + H_2 \rightarrow CH_2^+ + h\nu \quad \text{(when energy} \geq 0.4\ \text{eV)} \quad (6)$$

and Herbst and Klemperer (1973) pioneered the inclusion of ion-molecular radiative association in modelling work on synthesis in these clouds. One of the key reactions is

$$CH_3^+ + H_2 \rightarrow CH_5^+ + h\nu \quad . \quad (7)$$

While most interest relates to radiative ion-molecule association in cool (10K to 100K) clouds Mitchell (1984) has suggested that process (7) may be the precursor of the formation of methane by

$$CH_5^+ + e \rightarrow CH_4 + H \quad (8)$$

in hot (10,000K) shock-heater clouds. The temperature range that must be covered in therefore wide.

Much data on the rates of ter-molecular ion-molecule association has been obtained from flow or drift tube measurements (cf Smith and Adams 1979, Adams and Smith 1981, Smith et al 1983, Bohringer and Arnold 1983, Bohringer et al 1983, van Koppen et al 1984) and more recently (Rowe et al 1984) from a supersonic jet apparatus. Radiative ion-molecule association does not lend itself to study in the laboratory. Nevertheless McEwan et al (1980) and Bass et al (1981) have obtained results on

$$CH_3^+ + HCN \rightarrow CH_3^+ . HCN + h\nu \quad (9)$$

from an ion-cyclotron resonance investigation and Barlow et al (1984) have obtained results on process (7) using the trapped ion technique. By comparison the results of theorists have been modest. Our approach will be that of the loose collision complex of Klots (1971) developed from first principles. We will concentrate on theory and will only cite enough experimental data to illustrate the position.

2. STATISTICAL THEORY

Denote number densities and internal partition functions by [ ] and Q with the species indicated by its chemical symbol. We have that in equilibrium

$$\frac{[XY^{+*}]}{[X^+][Y]} = \frac{h^3}{(2\pi\mu kT)^{3/2}} \frac{Q(XY^{+*})\psi}{Q(X^+)\,Q(Y)} \tag{10}$$

where μ is the X^+ - Y reduced mass, T is the temperature and ψ is the ratio of the electronic statistical weight of XY^+ to the product of the electronic statistical weights of X^+ and Y. The Q's are taken to include symmetry numbers and nuclear spin factors. These last cancel except in special circumstances (§2.5).

Assuming for the present that the rate A_S at which the stabilization step (3) or (4) takes place is much less than the mean frequency $\bar{k}_b$ of dissociation (2) we see that the rate coefficient of ion-molecule association expressed in two-body form is

$$k_2 = \frac{h^3}{(2\pi\mu kT)^{3/2}} \frac{Q(XY^{+*})\psi A_S}{Q(X^+)\,Q(Y)} \; . \tag{11}$$

In the case of the ter-molecular process

$$A_S = \beta_c\, Z\, [M] \tag{12}$$

where Z is the rate coefficient for XY^+ - M hitting collisions and β_c is an efficiency factor (Troe 1977a). At low ambient gas densities it is convenient to extract [M] from the right of Eq. (11) to get the ter-molecular rate coefficient

$$k_3 = \frac{h^3}{(2\pi\mu kT)^{3/2}} \frac{Q(XY^{+*})\psi\beta_c Z}{Q(X^+)\,Q(Y)} \; . \tag{13}$$

For radiative ion-molecule association A_S is normally the probability A_R of the transition concerned (See however §2.9).

Calculation of the partition functions $Q(X^+)$ and Q(Y) is straightforward. As noted by Herbst (1980) the classical formulae are invalid at the low temperatures of interstellar clouds where they must be replaced by direct summations. This was not done in some of the early investigations the oversight leading to erroneous extrapolations of experimental results on ter-molecular ion-molecule association.

Because the activated complex carries the association energy E_o and because angular momentum is conserved in the encounter calculation of $Q(XY^{+*})$ is more troublesome.

First we will recall a formula given by Troe (1977b) for the energy density of vibrational states of a neutral complex. Let E_Z be the zero point energy and a be the factor which Whitten and Rabinovich (1963) devised to improve the semi-classical approximation. Let the number of oscillators be increased by m to s when the complex is formed from the reactants and let ν_i be their frequencies. Assuming for simplicity that there is no internal rotation we have following Troe (1977b), that the density of vibrational states at the dissociation threshold is

$$\rho_{vib}(E_o+aE_Z) = \{(s-1)/(s-1.5)\}^m (E_o+aE_Z)^{s-1}/(s-1)! \prod_{i=1}^{s} h\nu_i \quad (14)$$

The density distance E above threshold may be written

$$\rho_{vib}(E_o+aE_Z+E) = F_E \, \rho_{vib}(E_o+aE_Z) \quad (15)$$

with

$$F_E = \left\{1 + \frac{E}{E_o+aE_Z}\right\}^{s-1} . \quad (16)$$

Next consider the effect of rotation. Let J be the rotational quantum number and let ζ, which depends on J and in the case of a non-linear complex on another quantum number K or index τ be the rotational energy. The rotational influences the density of vibrational states: thus it reduces by ζ the energy available to the vibrational modes; and again, if the complex is non-linear, it increases the number of levels between which energy may flow consistent with the angular momentum being conserved. To take account of this Eq. (15) is replaced by

$$\rho_{vib}(E_o+aE_Z+E,J) = F_J \, F_E \, \rho_{vib}(E_o+aE_Z) \quad (17)$$

where F_J is as given in table 1. In addition we must introduce the factor $N_{rot}(J,E)$ arising directly from rotation to get

$$Q(XY^{+*}) = \frac{\rho_{vib}(E_o+aE_Z)}{\sigma} \iint F_E F_J \, N_{rot}(J,E) \exp(-E/kT) \, dJ dE \quad (18)$$

where σ is the symmetry number. The determination of $N_{rot}(J,E)$ is our immediate objective.

Denote the statistical weights, internal momentum quantum

Table 1.

Form of complex	Rotational energy ζ	F_J
linear	BJ(J + 1)	$\{1 - \frac{\zeta(J)}{E_o+aE_Z}\}^{s-1}$
spherical top	BJ(J + 1)	$2J\{1 - \frac{\zeta(J)}{E_o+aE_Z}\}^{s-1}$
symmetrical top	$BJ(J+1)+(A-B)K^2$	$\sum_{K=-J}^{K=+J} \{1 - \frac{\zeta(J,K)}{E_o+aE_Z}\}^{s-1}$
asymmetrical top	dependent on J and on index τ which can have 2J+1 values	$\sum_{\tau} \{1 - \frac{\zeta(J,\tau)}{E_o+aE_Z}\}^{s-1}$

numbers and energies of level i of X^+ and j of Y by g(X,i), J(X,j) and E(X,i) and by g(Y,j), J(Y,j) and E(Y,j) respectively and set

$$E(i,j) \equiv E(X,i) + E(Y,j) \quad . \tag{19}$$

We shall express $N_{rot}(J,E)$ in terms of these entities and of a long range interaction parameter allowing for the important collecting effect of this interaction by incorporating the conservation of angular momentum in the treatment of the kinetics of the encounter.

We are concerned with ion-molecule association when all the levels of the reactants are thermally populated and also when only some of the levels are thermally populated and the remainder are unoccupied; and we have need to be able to cover the low temperature region where nuclear spin factors do not cancel. We shall begin with a model which is simple and instructive.

2.1 Model in which each reactant has only one active level

By 'active' level (which is an artificiality) we have mean a level from which association may take place and which may be entered on dissociation.

We will suppose that the long-range interaction is that between the charge on the ion and the induced dipole on the neutral of polarizability α so that a hitting collision between the reactants only ensues if the orbital angular momentum quantum number J_o satisfies the condition

$$J_o \leq J_L \tag{20}$$

where

$$J_L = (8\mu^2 \alpha e^2 \varepsilon)^{1/4}/\hbar \tag{21}$$

ε being the energy of relative motion. The hitting collision cross section is

$$q = \pi\hbar^2 (J_o + 1/2)/\mu\varepsilon , \quad J_o \leq J_L \tag{22}$$

$$= 0 \qquad J > J_L \tag{23}$$

and we may identify this cross section with the association cross section. Since the reactant's energy of relative motion has the normalized distribution

$$u(\varepsilon)d\varepsilon = 2\varepsilon^{1/2} \exp(-\varepsilon/kT)\, d\varepsilon/\pi^{1/2}(kT)^{3/2} \tag{24}$$

we see that the association rate into the energy interval dE around E is

$$A(i,j,E) = C \sum_{J_o=0}^{J_o=J_L(i,j,E)} (J_o + 1/2) \tag{25}$$

with

$$C = 2(2\pi)^{1/2} \hbar^2 g(X,i) g(Y,j) \exp(-E/kT)/(\mu kT)^{3/2} Q(X^+) Q(Y) \tag{26}$$

$$J_L(i,j,E) = \{8\mu^2 \alpha e^2 [E - E(i,j)]\}^{1/4}/\hbar \tag{27}$$

i and j being the active levels. We require that part A (i,j,J,E) of A(i,j,E) that pertains to association into the rotational levels J of the complex. This necessitates considering the internal angular momenta (Bates 1985). Vector coupling of J(X,i) and J(Y,j) leads to a set of states with angular momenta $P\hbar$ where

$$P = |J(X,i)-J(Y,j)|, |J(X,i)-J(Y,j)| + 1, \ldots, J(X,i)+J(Y,j). \tag{28}$$

The probability of a particular P arising from J(X,i) and J(Y,j) is

$$p_1(P) = (2P+1)/\{2J(X,i)+1\}\{2J(Y,j) + 1\} \tag{29}$$

and the probability of P and J_o combining to form a state J is $p_2(J)$ where

$$p_2(J) = (2J+1)/\{(2P+1)(2J_o+1)\} \tag{30}$$

if

$$|J - J_o| \leq P \tag{31}$$

and

$$p_2(J) = 0 \tag{32}$$

if

$$|J - J_o| > P. \tag{33}$$

Hence the contribution to A(i,j,J,E) from a J_o satisfying condition (31) is

$$C(J_o + \frac{1}{2})\ p_1(P)p_2(J) = \frac{C(J + \frac{1}{2})}{\{2J(X,i)+ 1\}\{2J(Y,j) + 1\}} \tag{34}$$

and from any other J_o is zero. If f(i,j,J,E) be such that the number of values of J_o that satisfy conditions (20) and (31) is

$$\{2J(X,i) + 1\}\{2J(Y,j) + 1\}\ f(i,j,J,E) \tag{35}$$

we see from formula (34) that

$$A(i,j,J,E) = C(J + \tfrac{1}{2})\ f(i,j,J,E) \tag{36}$$

(Light, 1967). A straightforward count gives

$$f(i,j,J,E) = \frac{1}{\{2J(X,i)+1\}\{2J(Y,j)+1\}} \sum_{\text{all } P} g(J,P,E) \tag{37}$$

where g(J,P,E) is as in table 2.

Consider next the dissociation rate into dJdE given by

$$D(J,E)dJdE = \frac{h^3(\psi/\sigma)\rho_{vib}(E_o+aE_Z)F_EF_JN_{rot}(J,E)k_b(J,E)\exp(-E/kT)dJdE}{(2\pi\mu kT)^{3/2}\ Q(X^+)\ Q(Y)} \tag{38}$$

where $k_b(J,E)$ is the dissociation frequency. With only (X,i) and (Y,j) being involved in association and dissociation in our model it is obvious from phase space theory that

$$\left.\begin{aligned} N(J,E) &= 2J + 1 \quad , \quad J \leq J\ (\max) \\ &= 0 \qquad\quad\ , \quad J > J\ (\max) \end{aligned}\right\} \tag{39}$$

with

$$J(\max) = J_L(i,j,E) + J(X,i) + J(Y,j)\quad . \tag{40}$$

Equation the association and dissociation rates yields

$$k_b(J,E) = g(X,i)g(Y,k)f(i,j,J,E)/hF_EF_J(\psi/\sigma)\rho_{vib}(E_o+aE_Z) \tag{41}$$

Table 2. Function g(J, P, E) of equation (37)

J_L-range	J-range	g(J, P, E)
any	$J \geq P + J_L + 1$	0
$J_L \geq 2P$	$J_L - P \leq J \leq J_L + P + 1$	$J_L + P + 1 - J$
	$P \leq J \leq J_L - P$	$2P + 1$
	$J \leq P$	$2J + 1$
$2P \geq J_L \geq P$	$P \leq J \leq J_L + P + 1$	$J_L + P + 1 - J$
	$J_L - P \leq J \leq P$	$J_L + J + 1 - P$
	$J \leq J_L - P$	$2J + 1$
$P \geq J_L$	$P - J_L \leq J \leq P + J_L$	$2J_L + 1 - 2\|P-J\|$
	$J < P - J_L$	0

Note: the state indictors i and j have been omitted from J(X), J(Y), P and J_L.

The factor f(i,j,J,E) may be interpreted as the fraction of the g(X,i)g(Y,j) states which can be reached by combining the given J with a J_o satisfying condition (20) so that the numerator of the fraction on the right of Eq. (41) is the number of channels that are accessible when account is taken of angular momentum being conserved. Hence Eq. (20) is a form of standard unimolecular theory rule (cf Forst 1973) expressing the energy-time uncertainty relation.

2.2 Generalization from model

It is apparent from Eq. (20) that the dissociation frequency of the actual complex is

$$k_b(J,E) = W(J,E)/h\ F_E F_J(\psi/\sigma)\ \rho_{vib}(E_o + aE_Z) \tag{42}$$

where

$$W(J,E) = \sum_{i,j}^{*} g(X,i)g(Y,j)f(i,j,J,E) \tag{43}$$

the asterisk on Σ signalling that for given E the summation is over all pairs making

$$\varepsilon(i,j) \equiv E - E(i,j) \tag{44}$$

positive. Substitution for $k_h(J,E)$ from Eq. (42) into Eq. (38) gives the dissociation rate D(J,E)dJdE. The association rate A(J,E)dJdE may be obtained from Eq. (36) by summing over all contributing pairs of levels. From the equality of D(J,E) and A(J,E) we get

$$N_{rot}(J,E) = \sum^{*}_{h,k} \{g(X,h)g(Y,k)f(h,k,J,E)(2J+1)/W(J,E)\} \tag{45}$$

where the asterisk signals that the summation is over all (h,k) pairs making ε(h,k) positive. The dependence of the J-range on (h,k) prevents simplification of Eq. (45). Knowing $N_{rot}(J,E)$ the internal partition function $Q(XY^{+*})$ and hence the rate coefficient k_3 may be obtained (at least in principle) from Eqs. (18) and (13) respectively.

It is easy to make the modification required if only certain of the levels of the reactants have a thermal population and the remaining levels are unpopulated. All that need be done is to restrict summation (45) to the populated levels and to replace the partition functions $Q(X^+)$ and $Q(Y)$ in Eq. (13) by effective partition functions $Q'(X^+)$ and $Q'(Y)$ that differ from $Q(X^+)$ and $Q(Y)$ in having no contributions from the unpopulated levels of the reactants. Summation (43) is unchanged.

2.3 Effect of internal angular momenta

Calculations on the effect which the internal angular momenta have on k_3 are being done (Bates 1985) but results are not yet available. As far as can be judged from the simple model of §2.1 the effect is probably modest. Thus according to the model if the rotational energy ζ is small compared to $(E_o + aE_Z)$ inclusion of angular momenta increases

$$\eta \equiv \int_o^{J_{max}} N_{rot}(J,E)\,dJ \tag{46}$$

(to which k_3 is proportional) from

$$\eta(L) = J_L^2 \quad \text{to} \quad \eta(L+I) = \{J_L + J(X,i) + J(Y,j)\}^2 \tag{47}$$

if the complex is linear and from

$$\eta(L) = \frac{4}{3} J_L^3 \quad \text{to} \quad \eta(L+I) = \frac{4}{3} \{J_L + J(X,i) + J(Y,j)\}^3 \tag{48}$$

if the complex is non-linear. As a preliminary guide let us consider the value of J_L corresponding to an energy of relative motion of kT and the value of $\{J(X,i) + J(Y,j)\}$ corresponding to each of the reactants having this rotational energy. For

$(N_2^+ + N_2)$ association the complex is linear and

$$J_L = 21T^{1/4} \quad , \quad \{J(X,i) + J(Y,j)\} = 1.2T^{1/2} \; ; \qquad (49)$$

while for $CH_3^+ + O_2$ association the complex is non-linear and

$$J_L = 18T^{1/4} \quad , \quad \{J(X,i) + J(Y,j)\} = 1.0T^{1/2} \; . \qquad (50)$$

As T is raised from 100K to 400K the ratio $\eta(L + I)/\eta(L)$ therefore increases from 1.4 to 1.6 in the former case and from 1.6 to 1.9 in the latter case. Although quantitative significance should not be attached to these numbers it would seem likely from them that internal angular momenta are of only minor importance as far as the temperature dependence of the rate coefficient is concerned. It is hence worth considering this dependence when internal angular momenta are ignored which greatly simplifies the problem.

2.4 An integral formula and the temperature dependence

The expression for the internal partition function of the complex obtained on assuming that the factors $f(i,j,J,E)$ are unity if $J < J_L$ and vanish otherwise and substituting from Eq. (45) into Eq. (18) may be written

$$Q(XY^{+*}) = \frac{\rho_{vib}(E_o+aE_Z)}{\sigma} \int_o^\infty \frac{F_E \exp(-E/kT)}{W(E)} \sum_{h,k}^{*} g(h,X)g(k,Y) \int_o^{J_L(h,k,E)} F_J(2J+1)\,dJdE \qquad (51)$$

with

$$W(E) = \Sigma^* \; g(X,i) \; g(Y,j) \qquad (52)$$

which implies that channels allowed by the asterisk are fully effective.

As it stands Eq. (51) is troublesome to use. To remedy this defect we must replace the summations by integrations as is permissible if the separation between neighbouring energy levels be small compared to kT. Introducing

$$I(\varepsilon) \equiv \int_o^{J_L(h,k,E)} F_J(2J + 1)\,dJ \qquad (53)$$

and replacing F_E by

$$F_T = 1 + \frac{(s-1)kT}{E_o+aE_Z} + \ldots . \qquad (54)$$

outside the integral (Troe 1977b) we see that Eq. (51) may be recast as

$$Q(XY^{+*}) = \frac{1}{\sigma} F_T \rho_{vib}(E_o + aE_Z) \sum_{h,k}^{*} g(X,h) g(Y,k) \exp(-E(h,k)/kT) \int_o^\infty \frac{I(\varepsilon)\exp(-\varepsilon/kT)d\varepsilon}{W(E(h,k)+\varepsilon)} \tag{55}$$

$$= \frac{1}{\sigma} F_T \rho_{vib}(E_o + aE_Z) \int_o^\infty \frac{dW}{dE} \exp(-E/kT) \int_o^\infty \frac{I(\varepsilon)\exp(-\varepsilon/kT)d\varepsilon}{W(E(h,k)+\varepsilon)} \tag{56}$$

If only rotational levels need be taken into account W(E) has the simple form

$$W(E) = A\, E^{\ell/2} \tag{57}$$

where ℓ is the number of rotational degrees of the reactants and A is a constant which need not be specified because it cancels in Eq. (56) leaving

$$Q(XY^{+*}) = \frac{1}{\sigma} F_T \rho_{vib} (E_o + aE_Z) P_{rot} (chn) \tag{58}$$

with

$$P_{rot}(chn) = \frac{1}{2}\ell \int_o^\infty E^{\frac{1}{2}\ell - 1} \exp(-E/kT) \int_o^\infty \frac{I(\varepsilon)\exp(-\varepsilon/kT)d\varepsilon}{(E+\varepsilon)^{\frac{1}{2}\ell}} dE \tag{59}$$

in which chn is a reminder that allowance has been made for the dissociative channels. If the rotational energy ζ is small compared to $E_o + aE_Z$ we have that

$$I(\varepsilon) \propto \varepsilon^p \tag{60}$$

with

$$p = \frac{1}{2} \text{ for linear complex}, \quad p = \frac{3}{4} \text{ for non-linear complex} \tag{61}$$

and we see from Eq. (59) that

$$P_{rot}(chn) \propto T^{(1+p)} \quad . \tag{62}$$

If we may take F_T of Eq. (54) to be unity and the stabilization rate A_S to be independent of T we find from Eq. (13) that

$$k_3 \propto T^{p-\frac{1}{2}-\ell/2} \tag{63}$$

ℓ being the number of degrees of rotational freedom of the reactants.

It is of interest to use approximation (39) of simple phase space theory to calculate the factor P_{rot}(ph.sp) corresponding to

P_{rot}(chn) of Eq. (59) for comparison purposes. Rewriting Eq. (21) as

$$J_L = (\varepsilon/\varepsilon_o)^{1/4} \tag{64}$$

with

$$\varepsilon_o = \hbar^4/(8\mu^2 \alpha e^2) \tag{65}$$

we obtain

$$P_{rot}(ph,sp) = (kT/\varepsilon_o)^{3/2} \Gamma(3/2), \quad \text{linear complex} \tag{66}$$

and

$$P_{rot}(ph,sp) = (kT/\varepsilon_o)^{7/4} \Gamma(7/4), \quad \text{non-linear complex} . \tag{67}$$

Reference to Eq. (62) shows that the T-variation of P_{rot}(chn) and P_{rot}(ph,sp) are identical. The ratio P_{rot}(ph,sp)/P_{rot}(chn) may be regarded as the factor f(D) by which the association rate coefficient k_3 is depressed by the dissociative channels. This depression tends to become more marked as the number of rotational degrees of freedom of the reactants and of the complex are raised.

Approximation (39) is relatively easy to use. The correction which should be applied to the results to allow for the dissociative channels may be obtained from Table 3. However as will be recalled, this table ignores the internal angular momenta of the reactants (and entails the implicit assumption that an open channel is fully effective

Table 3. Factor f(D) by which association rate coefficient k_3 is depressed by the dissociative channels.

number of degrees of rotational freedom of reactants (ℓ)	f(D) linear complex	f(D) non-linear complex
2	0.67	0.57
3	0.59	0.48
4	0.53	0.41
5	0.49	0.37
6	0.46	0.33

2.5 Nuclear spin

Specific account must be taken of nuclear spin if T is so low that only a few states of one or both reactants are occupied. This must be done for instance when treating radiative association involving molecular hydrogen or hydrides in cool interstellar clouds.

The pair of reactants remains in a particular nuclear spin state with quantum number S which is conserved throughout the encounter. It will be our practice to affix S as a subscript to the symbol for entities pertaining to this nuclear spin state where such distinguishing is needed.

The modification that must be made to Eq. (42) to get the dissociation frequency $k_{b,S}(J,E)$ is straightforward: thus

$$k_{b,S}(J,E) = W_S(J,E)/h(1+2S)n(S)F_E F_J(\psi/\sigma)\rho_{vib}(E_o+aE_Z) \qquad (68)$$

where

$$W_S(J,E) = \sum_{i,j}^{*} g(X,i)g(Y,i)\ \Lambda_S(i,j)f(i,j,J,E) \qquad (69)$$

$\Lambda_S(i,j)$ being the probability that nuclear spin state S may be formed by reactants in states i and j and where $n(S)$ is the number of nuclear multiplets of spin S which occur. The weights $g(X,i)$ and $g(y,i)$ include the appropriate nuclear spin factors. The factor $(1 + 2S)n(S)$ is introduced into Eq. (68) to make the denominator the standard h times the energy density of vibrational states that may dissociate along the channels in the numerator. The semblance of increasing the denominator may seem odd. However if the numerator counted channels of all nuclear spin states the denominator would have to include the factor

$$\prod_{\text{all atoms}} (1 + 2I_a) \equiv \sum_S (1 + 2S)\ n(S) \qquad (70)$$

I_a being the nuclear spin of one of the atoms involved.

To take account of the difference between Eq. (42) and (68) we see that Eq. (45) must be replaced by

$$N_{rot}(J,E) = \sum_S (1 + 2S)n(S)\ N_{rot,S}(J,E) \qquad (71)$$

with

$$N_{rot,S}(J,E)=\sum_{h,k}^{*}\{g(X,h)g(Y,k)f(h,k,J,E)\Lambda_S(h,k)(2J+1)/W_S(J,E)\} \qquad (72)$$

Calculations on

$$CH_3^+ + H_2 \rightarrow CH_5^+ + h\nu \qquad (73)$$

using Eq. (70) are being done (Bates 1985). Although the final results are not yet available it is evident that

$$CH_3^+ + H_2 \;(J = 1) \rightarrow CH_5^+ + h\nu \tag{74}$$

is much slower than

$$CH_3^+ + H_2 \;(J = 0) \rightarrow CH_5^+ + h\nu \tag{75}$$

because the internal energy carried by the ortho hydrogen increases the number of dissociative channels accessible. Calculations in which the internal angular momenta are included give that the ratio of the rates is 0.079 at 10K.

2.6 Diffuse fringe to complex

Up until now we have treated the internal partititon function $Q(XY^+)$ as in standard unimolecular theory (cf Troe 1977b) disregarding the charge except that in obtaining the rotational contribution we took specific account of the long range interaction between the reactants. This long range interaction has another effect. Rotational excitation of either of the reactants when they collide may leave their energy of relative motion negative so that they move in a closed orbit. Because of the range of the interaction such orbits may be quite extended. They provide the complex with a diffuse fringe making it qualitatively different from a neutral complex. Eq. (14) leads to an underestimation of the partition function since it does not take account of the fringe. However a rough calculation indicates that the fractional extent of the underestimation is inappreciable except perhaps when $E_o + aE_Z$ is small and the association rate is low.

2.7 Finiteness of dissociation frequency

The simplest way of taking the finiteness of the dissociative frequency $k_b(J,E)$ into approximate account is to replace Eq. (11) which may conveniently be written in the form

$$k_2 = V(T)\, A_S \;, \tag{76}$$

by

$$k_2 = V(T) \left\{ \frac{\overline{k}_b(T)\, A_S}{\overline{k}_b(T) + A_S} \right\} \tag{77}$$

where $\overline{k}_b(T)$ is the appropriate mean value of this frequency. In the case of an ion-induced dipole interaction the mean dissociation frequency may be found from

$$\bar{k}_b(T) = Z_L/V(T) \qquad (78)^*$$

where Z_L is the Langevin collision coefficient given (cf Gioumousis and Stevenson 1958) by

$$Z_L = 2\pi(\alpha e^2/\mu)^{\frac{1}{2}} \qquad (79)$$

$$= 7.41 \times 10^{-10}\ \tilde{\alpha}^{\frac{1}{2}}\ (10/\tilde{\mu})^{\frac{1}{2}}\ cm^3\ s^{-1} \qquad (80)$$

the tilde symbol over α and μ in Eq. (80) indicating that polarizability of Y and the X^+ - Y reduced mass are in units of $10^{-24}cm^3$ and on the ^{12}C chemical amu scale. Because of the T-dependence of V(T), which is the same as that of k_3 of Eq. (63), the effect is greatest at low T where it tends to reduce the T-dependence of k_2 the expression for which becomes

$$k_2 = V(T)\ A_S/\{1 + V(T)\ A_S/Z_L\}\ . \qquad (81)$$

Clearly a false inference on the T-dependence of V(T) may be drawn from experimental data (which of course cannot be taken at the low density limit) if the finiteness of $\bar{k}_b(T)$ is not born in mind as pointed out by Mickens (1983) and others.

An obvious improvement to the theory would be to replace A_S outside the integral expression for the rate coefficient by

$$k_b(J,E)\ A_S/\{k_b(J,E) + A_S\} \qquad (82)$$

within this integral. The practical gain is limited by A_S being poorly determined.

2.8 Stabilization by collisions

According to the strong collision approximation stabilization ensues from every hitting collision between the activated complex and

---

* A few years ago it was common practice amongst physicists to express results in terms of a mean lifetime $\bar{\tau}(T)$ of the activated complex defined not as $\bar{k}_b(T)^{-1}$, as of course it should be, but instead obtained from $k_b(E)^{-1}$ by assigning to E some plausible value (straightforward averaging being avoided because the integral diverges): thus the rate coefficient for association was

$$k_2 = Z_L\ A_S\ \ \bar{\tau}(T) \qquad (a)$$

Bates (1979) pointed out the incorrectness of the practice. He was nevertheless criticised for it by Bass et al (1981) who through a misreading of his paper supposed that it was he who was responsible for Eq. (a) and for $\bar{\tau}(T)$ being erroneously defined.

an ambient gas molecule so that the stabilization rate coefficient Z_S is just the rate coefficient Z for such collisions. For a non-polar gas (the main type we are considering) Z may be taken to be the Langevin rate coefficient Z_{L3}, the subscript 3 being added to distinguish the coefficient from that occurring in Eq. (78).

Troe (1977a) has sought to improve upon the strong collision approximation by introducing the efficiency factor β_c of Eq. (12). He discussed β_c and its T-dependence by adopting a simple model describing the transfer of rotational and vibrational energy and setting up and solving the appropriate master equation. Practical utilization of his model is hampered by lack of knowledge on the several parameters introduced. Troe (1977b) was however able to rationalize data on unimolecular rates in ambient argon by the representation

$$\beta_c/(1 - \beta_c^{\frac{1}{2}}) = - \langle \Delta E \rangle /kT \tag{83}$$

where $\langle \Delta E \rangle$ is the mean energy charge which the complex experiences in a collision. He judged the dependence of $\langle \Delta E \rangle$ on T to be weak or negligible and found he could fit the data on reactions in ambient argon by taking $-\langle \Delta E \rangle$ to be 0.4 k cals mole^{-1} (0.017eV).

Cates and Bowers (1980) have made measurements on

$$(CH_3)_3 NH^+ + (CH_3)_3 N + M \rightarrow \{(CH_3)_3 N\}_2 H^+ + M \tag{84}$$

at 300K with M various inert species. Assuming that β_c is unity. when M is $(CH_3)_3N$ they deduced (amongst other results) that β_c is 0.31 when M is He and is 0.46 when M is N_2.

Bass et al (1981) have made a study of

$$CH_3^+ + HCN + He \rightarrow CH_3.HCN^+ + He \quad . \tag{85}$$

In an analysis of experimental data at 300K they took Z to be 3.4×10^{-9} cm^3 s^{-1}, which is much greater than the Langevin rate coefficient to allow for HCN being a polar molecule. They inferred that β_c is 0.1 which in so far as Eq. (83) is correct implies that $-\langle \Delta E \rangle$ is here only about 4×10^{-3}eV.

2.9 Radiative stabilization rate

Finding a reliable method of predicting the probability A_R of radiative stabilization is proving a difficulty problem.

Woodin and Beauchamp (1979) applied a quasi-thermodynamic method due to Dunbar (1975) and deduced A_R to be in the range 10s^{-1} to 100s^{-1} for internal energies 1 to 3 eV. In his early work on radiative association Herbst (1979, 1980a,b,c) took

$$A_R \triangleq 10^2 \bar{n} \ s^{-1} \quad (86)$$

where $\bar{n}$ is the average number (typically 10 to 100) of excited vibrational quanta in the complex. Later Herbst (1982) developed a statistical approach (incorporating experimental and theoretical information on the fundamental 1 → 0 transition of some polyatomic neutral molecules). He concluded that Eq. (86) overestimates A_r, his revised values being in the range 30 s^{-1} to 300 s^{-1} at an internal energy of around 3 eV. Some information on infra-red intensities of polyatomic molecules have recently been provided by quantal calculations on H_3O^+ by Colvin et al (1983) and on $HCNH^+$ by Lee and Schaefer (1984). On incorporating the information into his theory Herbst (1985) found A_R at 3 eV internal energy to be 2.3 x 10^3 s^{-1} and 1.1 x 10^3 s^{-1} for H_3O^+ and $HCNH^+$ respectively. He speculated that the true values might be even higher owing to the negelected overtone and combination transitions. It is to be noted that such transitions are of special importance in connection with radiative association at very high temperatures (§1) where a photon emitted in the fundamental transitions might not carry enough energy to cause stabilization.

From an analysis of laboratory data on process (9) obtained by McEwan et al (1980) and by themselves Bass et al (1981) found that the stabilization rate A_S is around 1.4 x $10^4 s^{-1}$ (correct to within a factor of 5). They suggested that this may not be the photon emission rate A_R but may instead be a curve-crossing rate to an excited electronic state which then radiates.

3. MEASUREMENTS ON TER-MOLECULAR ION-MOLECULE ASSOCIATION

3.1 Absolute magnitude of rate coefficient

The energy density of vibrational states is the most important factor determining the magnitude of the rate coefficient. It is evident from Eq. (14) that a high k_3 is favoured by the association energy E_o and by the number s of oscillators in the complex being large which entails the number $N(A)$ of atoms in the complex being large since

$$s = 3N(A) - 3 - \ell' \quad (87)$$

where ℓ' is the number of rotational degrees of freedom in the complex. The pattern is well known and is sufficiently illustrated by the examples in table 4.

The results of a series of measurements by Herbst et al (1983) on the rate at which association occurs between various hydrocarbon ion A^+ and H_2 in ambient helium

Table 4. Ion-molecule association in ambient helium at 80K

Reactants	Association energy E_o (eV)	Rate Coefficient k_3 ($cm^6 s^{-1}$)	Reference
$O_2^+ + H_2$	small	7.4×10^{-31}	Adams et al (1970)
$O_2^+ + O_2$	0.42	2.0×10^{-29}	Bohringer et al (1983)
$N_2^+ + N_2$	1.1	3.8×10^{-28}	Bohringer et al (1983)
$CH_3^+ + H_2$	1.7	1.5×10^{-27}	Adams & Smith (1981)
$C_2H_2^+ + H_2$	2.6	7×10^{-27}	Herbst et al (1983)

$$A^+ + H_2 + He \rightarrow H_2A^+ + He \qquad (88)$$

show that a large value of E_o does not prevent a process from being slow. Ferguson et al (1984) have rationalized the data neatly (including the data on the process (5) family). They pointed out that the rate coefficient for process (88) in all instances investigated is below the observable limit of 10^{-30} cm^6 s^{-1} unless the reaction energy ΔE for the related process

$$A^+ + H_2 \rightarrow HA^+ + H + \Delta E \qquad (89)$$

is in the range

$$-4 \text{ k cal mole}^{-1} \leq \Delta E \leq 0 \quad . \qquad (90)$$

Their suggestion is that endothermic dissociation (89) occurs (without the products being able to separate indefinitely) if the reactants gain energy of relative motion in range (90) as they approach. According to them this aids process (88) in two ways. Firstly it ensures that the hydrogen atoms of the molecule can separate (without crossing a potential hill) and find their way to the most stable positions of the complex so that the full association energy E_o is utilized. Secondly it increases the rate coefficient because of the lifetime of the complex being extended by the $AH^+ - H$ mode - so-called 'endothermic trapping' . The argument regarding the utilization of the full E_o is surely valid. However calculations (Bates 1985) have shown that the effect of endothermic trapping on the lifetime of the complex is inappreciable.

The absolute measurements have not yet been matched by theoretical work. In order to evaluate k_3 from the formulae given in §2 there must be accurate quantal calculations on the oscillator frequencies ν_i and the ambient gas must be such that the collision efficiency β_c can be taken to be unity with assurance. Unfortunately ν_i are known reliably in only a few cases and in these the measurements have usually been made only in ambient helium so that β_c is very uncertain.

3.2 Temperature variation

The measured variation of k_3 with T provides a partial test of the theory. We will confine our attention to two association processes.

Measurements on k_3 for

$$N_2^+ + N_2 + N_2 \rightarrow N_4^+ + N_2 + 1.1 \text{ eV} \tag{91}$$

have been published by four groups. Expressed in $cm^6 s^{-1}$ they may be represented as follows: $7.9 \times 10^{-29} (300/T)^{1.7}$, Moet-Ner and Field (1974); $6.8 \times 10^{-29} (300/T)^{1.64}$, Bohringer and Arnold (1983); $6.0 \times 10^{-29} (300/T)^{1.85}$ Rowe et al (1984); $5.5 \times 10^{-29} (300/T)^{1.67}$, van Koppen et al (1984). Quantal calculations by de Castro et al (1981) show that N_4^+ is linear with rotational constant B about 0.12 cm^{-1}. In this circumstance F_j of table 1 has only a slight influence on the T-variation. The influence of F_T of Eq. (54) is also slight and in the opposite sense. On the strong collision approximation that β_c is unity (which is not affected by the discovery by Smith et al (1984) that stabilization takes place via N_2 switching) we therefore see that theory gives the T-variation of k_3 to be close to that in Eq. (63). Detailed calculations (Bates 1985) which however disregard the effect of the internal angular momenta of the reactants yield

$$k(89) \propto (300/T)^{2.04} \quad . \tag{92}$$

Taking the internal angular momenta into account would be expected to decrease the index n of the (300/T) factor somewhat. Calculations on this are being performed. Any departure from the strong collision approximation would be expected to increase the index. One cannot yet be sure whether or not the apparent slight discrepancy with the index indicated by experiment is real.

Process (91) has a superficially similar companion

$$O_2^+ + O_2 + O_2 \rightarrow O_4^+ + O_2 + 0.42 \text{ eV} \quad . \tag{93}$$

Two sets of measurements on the rate coefficient have been published. The power law representations are: $2.6 \times 10^{-30} (300/T)^{3.2}$, Payzant et al (1973); 4.0×10^{-30} $(300/T)^{2.93}$, Bohringer and Arnold (1983). An index to the (300/T) factor so much greater than for process (91) had not been expected. It was recognized that the O_4^+ complex may differ from the N_4^+ complex by being non-linear (cf Conway and Janik 1970) but the initial supposition was that the main effect of this would merely be to decrease the index by an amount of 0.25 in accord with Eq. (63). The perception then was that the rotational constant B of O_4^+ is about 0.15 cm^{-1} and that the factor F_i would therefore introduce only a slight correction as in the case of process (91) so that the attractively simple spherical top approximation (cf table 1) would be adequate. This is incorrect. The rotational constant A is about 0.8 cm^{-1} or 1.6 cm^{-1} according as O_4^+ has a regular trapezoid or T-shaped structure and with B about 0.15 cm^{-1} the factor F_i of table 1 markedly depresses the contribution from the levels with the higher J values. Moreover if $\zeta(J,K)$ exceeds 1.5×10^3 cm^{-1} insufficient energy remains to excite the vibrational mode having the highest frequency and if it exceeds 1.8×10^3 cm^{-1} insufficient energy remains to excite the vibrational mode having the next highest frequency. Now when there is sufficient energy to excite a particular mode that mode should be excluded from Eq. (14) for the vibrational energy density of states. Exclusion of the two modes that have been specified reduces $\rho_{vib}(E_o + a E_z)$ greatly thereby making the decrease of $k_b(93)$ with increase in T much steeper in the 100K to 400K region (Bates 1985). Table 5 gives the calculated values of the index n for some selected values of A and B (including A = B in order to allow the effect of A being large to be demonstrated). The strong collision approximation is again assumed. As may be seen acceptable values of A and B yield fairly satisfactory agreement with experiment.

Table 5. Calculated T-variation of rate coefficient for process (93)

O_4^+ rotational constants A cm^{-1}	B cm^{-1}	index n of (300/T) factor
0.1	0.1	2.0
0.8 or 1.6	0.1	2.3
0.15	0.15	2.5
0.8 or 1.6	0.15	2.7
0.2	0.2	2.8
0.8 or 1.6	0.2	3.0

3.4 Measurements on radiative ion-molecule association

Association of CH_3^+ with HCN is unusual in that under the conditions of the ion-cyclotron resonance study of McEwan et al (1980) and Bass et al (1981) (temperature around 300K, pressure 1×10^{-3} to 4×10^{-3} torr) the radiative process

$$CH_3^+ + HCN \rightarrow CH_3 . HCN^+ + h\nu \tag{94}$$

and the ter-molecular process

$$CH_3^+ + HCN + He \rightarrow CH_3 . HCN^+ + He \tag{95}$$

proceed at comparable rates. The measurements yield that

$$k_2(94) = 1.0 \times 10^{-10} \ cm^3 \ s^{-1} \tag{96}$$

and

$$k_3(95) = 2.7 \times 10^{-25} \ cm^6 \ s^{-1} .$$

In their analysis of the data Bass et al (1981) took the finiteness of $\overline{k}_b$ into account.

The trapped ion measurements of Barlow et al (1984) on

$$CH_3^+ + H_2 \rightarrow CH_5^+ + h\nu \tag{97}$$

give that $k_2(97)$ is 1.1×10^{-13} $cm^3 \ s^{-1}$ for normal hydrogen at 13K. On taking the ratio of the association rate coefficient for ortho hydrogen to that for para hydrogen to be as calculated with allowance for the effect of the internal angular momenta of the reactants we hence obtain

$$k_2(97; \ ortho, \ 13K) = 2.8 \times 10^{-14} \ cm^3 \ s^{-1} \tag{98}$$

and

$$k_2(97; \ para, \ 13K) = 3.6 \times 10^{-13} \ cm^3 \ s^{-1} . \tag{99}$$

References

Adams, N. G., Bohme, D. K., Dunkin, D. B., Fehsenfeld, F. C. and Ferguson, E. E., 1970. J. Chem. Phys., 52:3133.

Adams, N. G. and Smith, D., 1981, Chem. Phys. Lett., 79:563.

Barlow, S. E., Dunn, G. H. and Schauer, M., 1984, Phys. Rev. Lett., 52:902.

Bass, L., Chesnavich, W. J. and Bowers, M. T., 1979, J. Am. Chem. Soc., 101;5493.

Bass, L. M., Kemper, P. R., Anicich, V. G. and Bowers, M. T., 1981, J. Am. Chem. Soc., 103:5283.

Bates, D. R., 1979, J. Phys. B. Atom. Molec. Phys., 12:4135.
Bates, D. R., 1983, Astrophys. J., 270:564.
Bates, D. R., 1984, J. Chem. Phys., 81:288.
Bates, D. R., 1985, (in preparation).
Black, J. H. and Dalgarno, A., 1973, Astrophys. Lett., 15:79.
Bohringer, H. and Arnold, F., 1983, J. Chem. Phys., 77:5534.
Bohringer, H., Arnold, F., Smith, D. and Adams, N. G., 1983, Int. J. Mass Spect. Ion. Phys., 52:25.
Cates, R. D. and Bowers, M. T., 1980, J. Am. Chem. Soc., 102:3994.
Colvin, M. E., Raine, G. P., Schaefer, III, H. F. and Dupois, M., 1983, J. Chem. Phys., 79:1551.
Conway, D. C., 1975, J. Chem. Phys., 63:2210.
Conway, D. C. and Janik, G. S., 1970, J. Chem. Phys., 53:1859.
de Castro, S. C., Schaefer III, H. F. and Pitzer, R. M., 1981, J. Chem. Phys., 74:2228.
Dunbar, R. C., 1975, Spectrochim. Acta, 31A:797.
Fahey, D. W., Fehsenfeld, F. C., Ferguson, F. C. and Viehland, L. A., 1981, J. Chem. Phys., 75:669.
Ferguson, E. E., Fahey, D. W., Fehsenfeld, F. C. and Aldbritton, D. L., 1981, Planet. Space Sci. 29:307.
Ferguson, E.E., Smith, D. and Adams, N.G., 1984, J.Chem.Phys. 81:742.
Forst, W., 1973, Theory of Unimolecular Reactions, p.61, Academic Press, New York.
Gioumousis, G. and Stevenson, D. P., 1958, J. Chem. Phys., 29:294.
Goldan, P. D., Schmeltekopf, A. L., Fehsenfeld, F. C., Schiff, H.I. and Ferguson, E. C., 1966, J. Chem. Phys., 44:4095.
Goldberg, R. A., 1975, Radio Science, 10:329.
Herbst, E., 1979, J. Chem. Phys., 71:2201.
Herbst, E., 1980a, Ap. J., 237:462.
Herbst, E., 1980b, J. Chem. Phys., 72:5284.
Herbst, E., 1980c, Ap. J., 241:197.
Herbst, E., 1982, Chem. Phys., 65:185.
Herbst, E., 1985, Ap. J. (in press).
Herbst, E. and Klemperer, W., 1973, Ap. J., 185:505.
Herbst, E., Adams, N. G. and Smith, D., 1983, Ap. J., 209:329.
Klots, C. E., 1971, J. Physic. Chem. 75:1526.
Lee, T. J. and Schaefer, III, H. F., 1984, J. Chem. Phys., 80:2977.
Light, J. C., 1967, Discuss. Farad. Soc. 44:14.
Meot-Ner, M. and Field, F. H., 1974, J. Chem. Phys., 61:3742.
Mickens, R. E. 1983, J. Chem. Phys., 79:1102.
Mitchell, G. F., 1984, Ap. J. Supplement, 54:81.
McEwan, M. J., Anicich, V. G., Huntress, W. T., Kemper, R. P. and Bowers, M. T., 1980, Chem. Phys. Lett., 75:278.
Payzant, J. D., Cunningham, A. J. and Kebarle, P., 1973, J. Chem. Phys., 59:5615.
Rowe, B. R., Dupeyrat, G., Marquette, J. B. and Gaucherel, P., 1984, J. Chem. Phys., 80:4915.
Smith, D. and Adams, N. G., 1979 in Gas Phase Ion Chemistry (ed. M. T. Bowers) Vol. 1, p. 2, Academic Press, New York.
Smith, D., Adams, N. G., Alge, E. and Herbst, E., 1983, Ap. J. 272:365.

Smith, D., Adams, N. G. and Alge, E., 1984. Chem. Phys. Lett., 105:317.
Troe, J., 1977a, J. Chem. Phys., 66:4745.
Troe, J., 1977b, J. Chem. Phys., 66:4758.
van Koppen, P.A.M., Jarrold, M. G., Bowers, M. T., Bass, L. M. and Jennings, K. R., 1984, J. Chem. Phys., 81:288.
Whitten, G. Z. and Rabinovitch, B. S., 1963, J. Chem. Phys., 38:2466.
Williams, D. A., 1972, Astrophys. Lett., 10:17.
Woodin, R. L. and Beauchamp, J. L., 1979, Chem. Phys., 41:1.
Yang, J.-H. and Conway, D. C., 1964, J. Chem. Phys., 40:1729.

MANY BODY CALCULATIONS IN ATOMIC PHYSICS

Hugh P. Kelly

Department of Physics
University of Virginia
Charlottesville, VA 22901

INTRODUCTION AND REVIEW OF RAYLEIGH SCHRÖDINGER PERTURBATION THEORY

In these lectures, the use of many-body perturbation theory for atomic calculations will be reviewed. The major emphasis will be on use of the linked-cluster many-body perturbation theory first derived by Brueckner[1] and Goldstone.[2] General reviews of many-body theory are given in books such as those by Thouless,[3] Fetter and Walecka,[4] March, Young, and Sampanthar,[5] and Lindgren and Morrison.[6] In understanding many-body theory, it is useful to begin with Rayleigh-Schrödinger (RS) theory for a one-particle system and to relate it to the many-body case. A similar discussion has been presented previously.[7]

Consider nondegenerate RS perturbation theory[8] with the Hamiltonian

$$H = H_o + H' \tag{1}$$

split into an unperturbed part H_o and a perturbation H'. In calculating eigenstates $|\psi_\alpha>$ and eigenvalues E_α for H, we use a complete set of unperturbed states $|n>$ satisfying

$$H_o|n> = E_n^{(o)}|n> \tag{2}$$

These states are used to calculate $|\psi_\alpha>$ and E_α in perturbation expansions

$$|\psi_\alpha> = \sum_{n=0}^{\infty} |\psi_\alpha^{(n)}> \quad , \tag{3}$$

and

$$E_\alpha = \sum_{n=0}^{\infty} E_\alpha^{(n)} \quad , \tag{4}$$

with $|\psi_\alpha^{(o)}> = |\alpha>$.

Using "intermediate normalization",

$$|\psi_\alpha^{(n)}> = (E_\alpha^{(o)}-H_o)^{-1}Q[(H'-E_\alpha^{(1)})|\psi_\alpha^{(n-1)}> - E_\alpha^{(2)}|\psi_\alpha^{(n-2)} \;\ldots.$$

$$E_\alpha^{(n-1)}|\psi_\alpha^{(1)}>] \quad , \tag{5}$$

$$E_\alpha^{(n)} = <\alpha|H'|\psi_\alpha^{(n-1)}> \quad , \tag{6}$$

where

$$Q = 1 - |\alpha><\alpha| \quad . \tag{7}$$

Diagrams analogous to those used in the many-body approach may be used to represent matrix elements of H' and of $E_\alpha^{(n)}$ as shown in Fig. 1. The state $|\alpha>$ is referred to as an unexcited state, and all other states $|n>$ are called excited states. An occupied excited state is called a particle and when $|\alpha>$ is not occupied, it is referred to as a hole state. Particle lines are drawn with an arrow directed upward, and hole lines are drawn with an arrow pointed down.

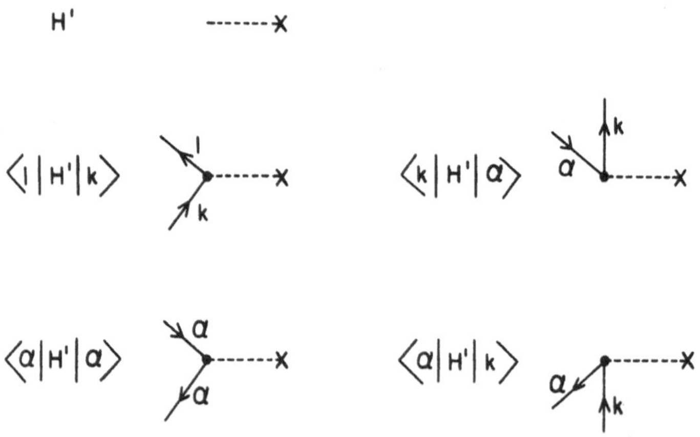

Fig. 1. Diagrams representing matrix elements of H'. Upward-going lines correspond to particle lines or occupied excited states. Lines labelled α represent a hole in state $|\alpha>$.

Diagrams corresponding to $E_\alpha^{(n)}$ are shown in Figs. 2 and 3. In the diagrams, going from bottom to top corresponds to right to left in Eq. (6). In Fig. 2(a), $E_\alpha^{(1)} = \langle\alpha|H'|\alpha\rangle$, and the lines labelled α in the matrix element $\langle\alpha|H'|\alpha\rangle$ have been connected into a closed loop. Fig. 2(b) is the diagram for

$$E_\alpha^{(2)} = \sum_{n\neq\alpha} \langle\alpha|H'|n\rangle\langle n|H'|\alpha\rangle[E_\alpha^{(o)}-E_n^{(o)}]^{-1} \quad . \tag{8}$$

In order to draw all diagrams for $E_\alpha^{(n)}$:

1) Draw n horizontal lines -----X.
2) Connect the lines in all possible ways so that there are no free lines and no disconnected parts. There is one arrow entering and one arrow leaving at each connection point.
3) Write the corresponding matrix elements.
4) Between each pair of horizontal lines write the energy denominator given by (number of lines labelled α)x$E_\alpha^{(o)}$ minus $\sum E_n^{(o)}$, where the sum includes the particle (upgoing) lines which are present between the pair of horizontal lines.
5) Attach the factor $(-1)^{h+1}$ where h is the number of hole lines (i.e., lines labelled α). Note that in each diagram the particle and hole lines make up a single closed loop.

Note that sums over excited states $|n\rangle$ are implied. Also, there can be more than one hole line labelled α as shown in Fig. 2(d). From these rules, there are (n-1)! diagrams for $E_\alpha^{(n)}$.

Diagrams for $E_\alpha^{(3)}$ are given in Fig. 2(c) and (d) and reproduce the expression

$$E_\alpha^{(3)} = \langle\alpha|H' \frac{Q}{E_\alpha^{(o)}-H_o} (H'-E_\alpha^{(1)}) \frac{Q}{E_\alpha^{(o)}-H_o} H'|\alpha\rangle \quad , \tag{9}$$

which is given in terms of matrix elements upon inserting the identity $I = \sum |n\rangle\langle n|$ between each pair of interactions. Note that the contribution in Eq. (9) involving $-E_\alpha^{(1)}$ is given by Fig. 2(d).

Diagrams for $E_\alpha^{(4)}$ are shown in Fig. 3. In diagrams (b), (c), (e), and (f) h=2, and h=3 in (d). Diagrams (e) and (f) sum to give the term proportional to $-E_\alpha^{(2)}$ in the mathematical expression for $E_\alpha^{(4)}$. The diagrams for $|\psi_\alpha^{(n)}\rangle$ are obtained from the diagrams for $E_\alpha^{(n+1)}$ by removing the top interaction.

The perturbation expansion for $|\psi_\alpha^{(n)}\rangle$ may also be obtained by considering the time-development operator $U(t,t_o)$ discussed in the lecture by Professor Merzbacher in this school. We define

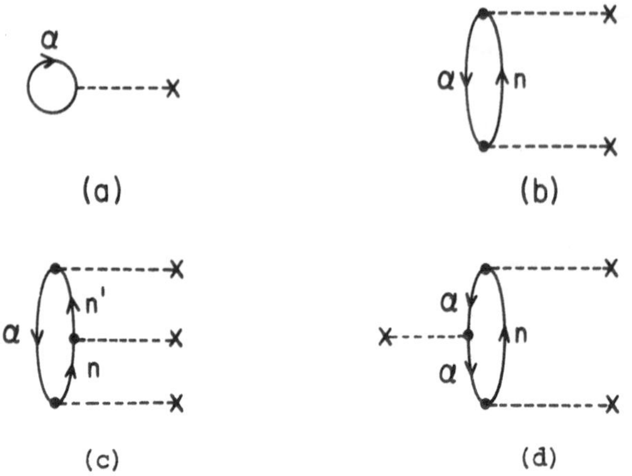

Fig. 2. Diagrams corresponding to (a) $E_\alpha^{(1)}$ (b) $E_\alpha^{(2)}$ (c) and (d) represent $E_\alpha^{(3)}$.

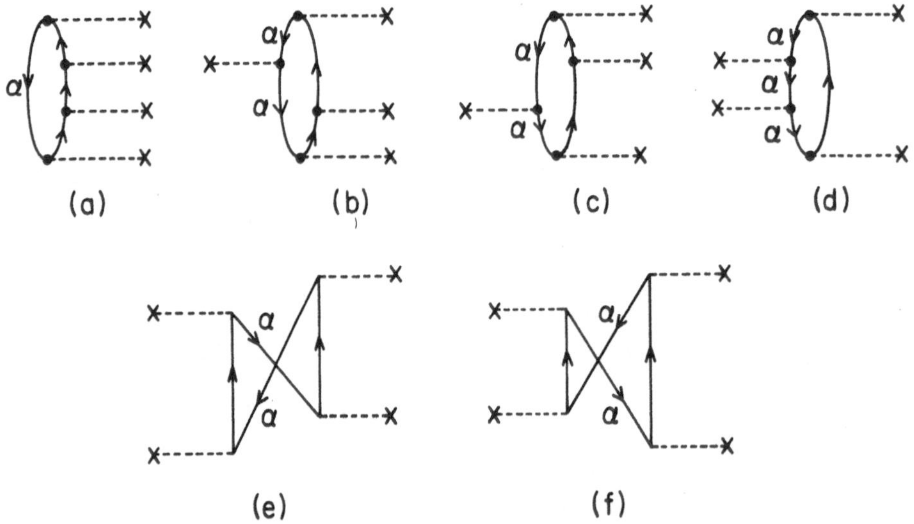

Fig. 3. Diagrams corresponding to $E_\alpha^{(4)}$.

$$\psi(t) = e^{-iH_0t}U(t,t_o)\psi(t_o) \quad , \tag{10}$$

with $U(t_o,t_o) = 1$. Substituting into the time-dependent Schrödinger equation,

$$i\frac{\partial\psi}{\partial t} = H\psi \quad , \tag{11}$$

we find

$$i\frac{\partial U}{\partial t} = H'(t)U \quad , \tag{12}$$

where

$$H'(t) = \lim_{\varepsilon\to o^+} e^{iH_0t}H'e^{-iH_0t}e^{\varepsilon t} \quad . \tag{13}$$

The factor $e^{\varepsilon t}$ as $\varepsilon\to o^+$ has been inserted for convergence. The solution of Eq. (12) may then be written

$$U_\varepsilon(t,-\infty) = \sum_{n=0}^{\infty}(-i)^n\int_{-\infty}^{t}H'(t_1)dt_1\ldots\int_{-\infty}^{t_{n-1}}H'(t_n)dt_n. \tag{14}$$

We then have $\psi(t=0) = U_\varepsilon(0,-\infty)\psi(-\infty)$ with $\psi(-\infty) = \psi^{(o)}$. However, we want $\langle\psi^{(o)}|\psi\rangle = 1$ so we write

$$\psi(t=0) = \lim_{\varepsilon\to 0}\frac{U_\varepsilon(0,-\infty)\psi^{(o)}}{\langle\psi^{(o)}|U_\varepsilon(0,-\infty)|\psi^{(o)}\rangle} \tag{15}$$

Carrying out the integrations in Eq. (14),

$$U_\varepsilon\psi^{(o)} = [1 + (E_\alpha^{(o)}-H_o+i\varepsilon)^{-1}H' + (E_\alpha^{(o)}-H_o+2i\varepsilon)^{-1}H'(E_\alpha^{(o)}-H_o+i\varepsilon)^{-1}H' + \ldots.]\psi^{(o)}. \tag{16}$$

According to the linked-cluster theorem,[2]

$$U_\varepsilon(0,\infty)\psi^{(o)} = [U_\varepsilon(o,-\infty)\psi^{(o)}]_L\langle\psi^{(o)}|U_\varepsilon(0,-\infty)|\psi^{(o)}) \tag{17}$$

where the subscript L indicates that only "linked" terms are included which means that there are no closed, disconnected parts.

A typical factorization is illustrated in Fig. 4. Note that

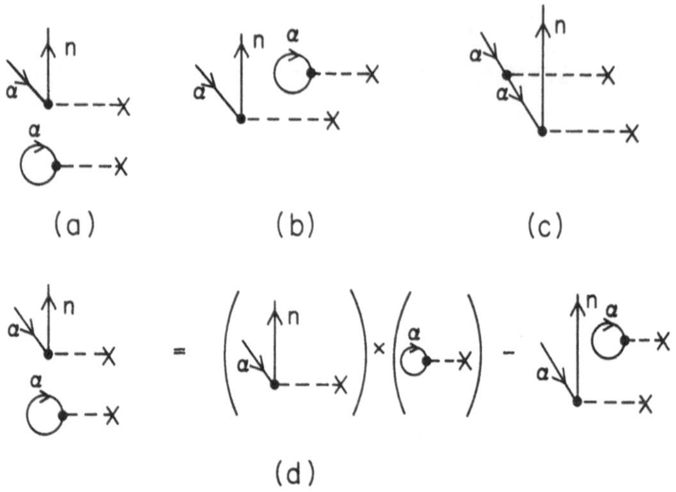

Fig. 4. (a) Diagram allowed in $U_\varepsilon(0,-\infty)\psi^{(o)}$. (b) Diagram not allowed in $U_\varepsilon\psi^{(o)}$ because when a hole is created in state $|\alpha\rangle$ as shown in bottom of diagram, upper interaction involving state $|\alpha\rangle$ cannot occur. (c) Diagram which is negative of diagram shown in (b). Diagram (c) is considered linked. (d) Factorization of $U_\varepsilon\psi^{(o)}$. Note that the subtracted piece at the right corresponds to diagram (c).

the factorization shown in (d) results in a remainder which is the negative of (b), which can be written as (c). The diagram for $|\psi^{(2)}\rangle$ shown in (c) corresponds to the energy diagram for $E^{(3)}$ shown in Fig. 2(d). In fact, the terms of the form $-E^{(j)}|\psi^{(i)}\rangle$ in the Rayleigh-Schrödinger perturbation expansion for $|\psi^{(i+j)}\rangle$ all arise in this way.

THE MANY-BODY PROBLEM

We wish to solve the problem of N identical fermions interacting through two-body potentials v_{ij}

$$H\psi = E\psi \quad , \tag{18}$$

where

$$H = \sum_{i=1}^{N} T_i + \sum_{i<j}^{N} v_{ij} \quad , \tag{19}$$

and T_i is the sum of the kinetic energy operator for the i-th particle and all one-body potentials. For atoms of atomic number Z,

$$T_i = -\frac{\nabla_i^2}{2} - \frac{Z}{r_i} \tag{20}$$

Atomic units are used throughout this article. We approximate the effects of the two body interactions on the i-th particle by an effective potential V_i which may be non-local but is Hermitian. The Hamiltonian is then divided into an unperturbed part

$$H_o = \sum_i T_i + V_i \quad , \tag{21}$$

and a perturbation

$$H' = \sum v_{ij} - \sum V_i \quad . \tag{22}$$

The unperturbed wave function $\psi^{(o)}$ is a solution of

$$H_o\psi^{(o)} = E^{(o)}\psi^{(o)} \quad , \tag{23}$$

where $\psi^{(o)}$ is a determinant or linear combination of determinants containing N single-particle solutions of the equation

$$(T+V)\phi_n = \varepsilon_n\phi \quad . \tag{24}$$

Eq. (24) is also used to calculate a complete set of single-particle states which are used to evaluate the linked-cluster perturbation expansions for ψ and E.

$$\psi = \sum_L \left(\frac{1}{E_o - H_o} H'\right)^n \psi^{(o)} \quad , \tag{25}$$

and

$$E = E^{(o)} + \langle\psi^{(o)}|H'|\psi\rangle \quad ,$$

where $\sum_L$ indicates that only "linked" terms are included. One can again write diagrams for ψ as shown in Fig. 5. The rules for writing diagrams are essentially those for the one-particle case except that there are now interactions with both v and -V, and the factor $(-1)^{h+1}$ becomes $(-1)^{h+\ell}$, where h is still the number of internal hole lines and ℓ is the number of closed loops, which can now exceed unity.

Diagram (b) of Fig. 5 does not contribute to Eq. (25) because

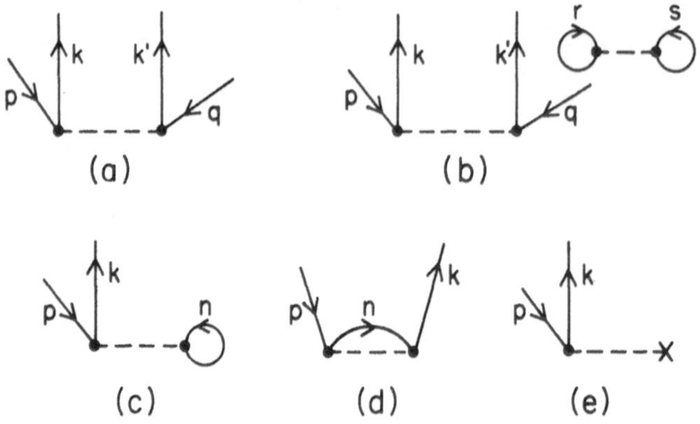

Fig. 5. Wave function diagrams. Dashed line between two solid lines represents interaction v. (a) Linked diagram contributing to $\psi^{(1)}$ involving double excitation. (b) Unlinked diagram involving unlinked part with interaction between states r and s. (c) Excitation involving coulomb matrix element $\langle kn|v|pn\rangle$. (d) Excitation involving exchange matrix element $-\langle nk|v|pn\rangle$. (e) Excitation due to $-\langle k|V|p\rangle$.

it has an unlinked part due to the interactions between states r and s. The expression for the diagram of Fig. 5(a) is

$$(\varepsilon_p+\varepsilon_q-\varepsilon_k-\varepsilon_{k'})^{-1}\langle kk'|v|pq\rangle\ \Phi_{pq}^{kk'}\quad , \tag{26}$$

where $\Phi_{pq}^{kk'}$ is a determinantal wave function with single-particle states p and q replaced by k and k'. In $\psi^{(1)}$, we also sum over excited states k and k' and also sum over pairs of unexcited states p and q.

Diagrams (c), (d), and (e) of Fig. 5 add to zero when V is chosen to be the Hartree-Fock potential defined by matrix elements

$$\langle k|V_{HF}|p\rangle = \sum_{n=1}^{N} (\langle kn|v|pn\rangle-\langle nk|v|pn\rangle). \tag{27}$$

However, if excited states are calculated in this potential, they interact with N other electrons rather than N-1 as expected physically. This difficulty is circumvented by noting that the potential of Eq. (27) can be generalized[9-13] to

$$V = V_{HF} + (1-P)\Omega(1-P)\quad , \tag{28}$$

where

$$P = \sum_{n=1}^{N} |n\rangle\langle n| \quad , \qquad (29)$$

and Ω is any Hermitian operator which may be chosen to represent a desired physical situation. Note that V of Eq. (28) when operating on one of the N Hartree-Fock orbitals of $\psi^{(o)}$ reduces to V_{HF} since $(1-P)|n\rangle$ equals zero.

In evaluating diagrams we sum over complete sets of single-particle states. Sums over bound states are evaluated individually and extrapolated to infinity by noting[14]

$$\lim_{n\to\infty} n^3\langle cn|v|ab\rangle = C \quad , \qquad (30)$$

where C is a constant. Continuum states are given by

$$\langle \underset{\sim}{r}|k,\ell,m_\ell,m_s\rangle = R(k,\ell,r)Y(\theta,\phi)\chi_{ms} \quad . \qquad (31)$$

$$\lim_{r\to\infty} R(k,\ell,r) = \cos[kr+\delta_\ell+(q/k)\ln 2kr-\tfrac{1}{2}(\ell+1)\pi]/r \qquad (32)$$

where $V \to q/r$. Then

$$\sum_k = \sum_n + \frac{2}{\pi}\int_0^\infty dk \quad . \qquad (33)$$

Integrals over the continuum are carried out numerically using a sufficient number of points to represent the integrand.

APPLICATIONS OF MANY-BODY THEORY

Correlation Energies

The first applications of many-body perturbation theory were to correlation energies E_{corr} defined as

$$E_{corr} = E - E_{HF} \quad , \qquad (34)$$

where E is the exact nonrelativistic energy and E_{HF} is the Hartree-Fock energy. When V_{HF} has been used for the ground state, E_{corr} is $\sum_{n=2}^{\infty} E^{(n)}$.

For an electron pair ab,

$$E_{ab}^{(2)} = \sum_{k,k'}' \frac{<ab|v|kk'>(<kk'|v|ab>-<k'k|v|ab>)}{\varepsilon_a+\varepsilon_b-\varepsilon_k-\varepsilon_{k'}} \quad . \tag{35}$$

Consider, for example, the neutral beryllium atom $1s^2 2s^2$. Calculations[14] using V_{HF} of Eq. (27) and including states with ℓ=0,1, and 2 showed that $E_{2s-2s}^{(2)}$ = -.0285 a.u., but approximate inculsion of higher-order terms changed this result to -0.0439 a.u. It was found that much faster convergence was obtained by calculating the excited orbitals in the field of $(1s)^2$ 2s. The final total result calculated for E_{corr} was -0.092 a.u. as compared with an experimental value -0.0943 a.u. and an extremely accurate configuration interaction calculation of -0.0943 a.u. by Bunge.[15] Another Be calculation of interest is a many-body treatment of Be as an open-shell system by Salomonson et al.[16] which gave E_{corr} = -.090 a.u. through second-order. Lindgren and Salomonson[17] have studied various coupled-electron-pair methods which include certain interactions to all orders of perturbation theory and have obtained very good results for E_{corr} for Be. Extensive discussions of these and other calculations for Be are given by Lindgren and Morrison.[6]

Calculations for oxygen[18] yielded -0.271 a.u. for pair correlations, 0.011 a.u. from three-body terms,[19] and -.003 from single excitations. The total calculated E_{corr} is -0.260 a.u. as compared with -0.258 from experiment. The close agreement is fortuitous since the accuracy of the calculated results is approximately 5%. The importance of three-body terms was also demonstrated in configuration interaction calculations by Viers et al.[20] and by Barr and Davidson[21] who found that for Ne the three-body effects reduced the contribution from independent pair correlations by approximately 20%. We note that for quite large systems, one must include quadruple excitations in configuration-interaction calculations to obtain the same accuracy as obtained in a many-body calculation with double excitations.

An alternative to evaluation of the direct sum over excited states is to solve a differential equation to obtain the perturbed wave function.[22-24] Consider, for example, the pair equation

$$[\varepsilon_a+\varepsilon_b-h(1)-h(2)]\eta_{ab}(1,2) = v_{12}|ab> + v_{12}\eta_{ab}(1,2)$$
$$-\sum_{cd} |cd><cd|(v_{12}|ab> + v_{12}\eta_{ab}(1,2)) \quad , \tag{36}$$

where the sum over cd runs over occupied states. The solution $\eta_{ab}(1,2)$ then accounts for certain pair-excitation effects to all orders. In Eq. (36), $h(1) = T_1 + V_1$. A similar equation may be written to account for sums over single excitations.[25]

Hyperfine Structure

There have been many applications[26-30] of many-body theory to calculations of hyperfine structure. In this case we wish to calculate $<\sigma_{hfs}>$, where σ_{hfs} is one of the usual hyperfine operators. If we add σ to the Hamiltonian, then $<\sigma>$ is given by all energy diagrams which have only one interaction with σ but any number of interactions with the perturbation v-V. An early calculation[27] evaluated the Fermi contact term given by the operator $\sigma_s = \sum s_{z_i}\delta(\underset{\sim}{r}_i)$ for atomic oxygen $2p^4\ ^3P$. In a restricted Hartree-Fock starting point, $<\sigma_s>_{HF} = 0$. However, exchange interactions with $2p^4$ differ according to whether $s_z = \pm 1/2$ for the 2s and 1s electrons, and so there are first-order corrections to ψ involving excitations of $2s^\pm$ and $1s^\pm$. These contributed .0126 to $<\sigma_s>$, with .224 from 2s excitations and -.211 from 1s excitations. Calculation and estimates of higher-order terms resulted in a value of .0601 as compared with the experimental value[31] 0.0569. Effects from the other hyperfine operators for atomic oxygen were also calculated successfully.[32]

A very careful calculation of higher-order effects in hyperfine structure has been given by Mårtensson[30] for the 4d ^2D states of Rb. Using the pair function of Eq. (36), she has calculated all corrections for ψ involving single and double excitations to all orders in perturbation theory, and has found in this case that high-order effects are very important. For example, the spin-dipolar parameter $<r^{-3}>_{sd}$ is 0.041 in the Hartree-Fock approximation, but with correlations was calculated to be -0.043 as compared with -0.047 from experiment. In these calculations, correlation corrections to the orbitals themselves were found to be important and modified the orbitals from Hartree-Fock to Brueckner orbitals.

Extensive discussions are given by Lindgren and Morrison[6] and by Lindgren.[29] Also, relativistic orbitals have been used successfully in many-body calculations of hyperfine structure by Andriessen and Das and coworkers.[28] Their papers also provide references to earlier extensive work on nonrelativistic many-body calculations of hyperfine structure by Das and coworkers.

Auger Rates

Consider an atom with an inner electron vacancy such as $Ne(1s^{-1})$. If we calculate the correlation energy for this system, we find[33,34] that energy denominators can vanish for transitions such as $p,q \rightarrow 1s,k$, where p,q are outer orbitals and k is in the continuum with energy $k^2/2$. In this case we add a small imaginary part $i\eta$ to the denominators and use the prescription

$$\lim_{\eta \to 0^+} (D+i\eta)^{-1} = PD^{-1} - i\pi\delta(D) \quad , \tag{37}$$

where P represents principal value integration and $\delta(D)$ is the Dirac delta function. Then

$$E^{(2)} = \sum_n \frac{|\langle 1sn|v|pq\rangle|^2}{\varepsilon_p+\varepsilon_q-\varepsilon_{1s}-\varepsilon} + \frac{2}{\pi} P \int_0^\infty dk \frac{|\langle 1sk|v|pq\rangle|^2}{\varepsilon_p+\varepsilon_q-\varepsilon_{1s}-k^2/2}$$

$$-i\,\frac{2}{k}\,|\langle 1sk|v|pq\rangle|^2 \quad , \tag{38}$$

where

$$\varepsilon_p + \varepsilon_q - \varepsilon_{1s} - k^2/2 = 0 \quad , \tag{39}$$

and exchange terms are understood to be included in the matrix elements. The term $2/k|\langle 1sk|v|pq\rangle|^2$ equals $\Gamma/2$, where Γ is the decay rate as is readily seen from the time dependence of ψ which is $\exp(-iEt)$.

Higher-order terms can be calculated, and were evaluated[34] for $Ne(1s^{-1})$ to correct the amplitude $\langle 1sk|v|pq\rangle$. In order to obtain contributions for specific final-state multiplets, the determinantal states corresponding to the diagrams were projected onto the multiplet states. For example, for the neon transition $1s2s^22p^6 \rightarrow 1s^22p^6 + e^-$, the calculated decay rate was 0.951×10^{-3} a.u. in the lowest order (Eq. (38)), and was reduced to 0.488×10^{-3} a.u. upon inclusion of correction diagrams in the next order. The corresponding experimental decay rate[35] is $(0.55 \pm .11) \times 10^{-3}$ a.u.

Many-body theory is also very useful in evaluating unusual processes such as the two-electron-one-photon transition[36] as illustrated for the iron atom.[37] In this case a heavy-particle collision has left an iron atom with two 1s vacancies which are filled by a $2p2s \rightarrow 1s^2$ transition emitting a single photon. An analogous process has also been calculated[38] for the Auger decay of Li $2s^2$ 2p in which two electrons fill the $1s^2$ shell and the continuum electron carries off the excess energy. The ratio of this double rate to the single Auger decay rate was calculated to be 0.80×10^{-3}. Another interesting process is the double Auger rate again illustrated by the decay of Li $2s^2$ 2p. In this case, one electron fills one of the vacant 1s orbitals and both of the remaining electrons are emitted. This process was calculated[39] to be 6.8% of the total $2s^2$ 2p Auger rate.

Parity Violation

There has been great interest in recent years in the measurement and calculation of effects of parity nonconservation (PNC) in atomic transitions due to the weak neutral currents.[40] The nuclear-spin independent part of the PNC interaction, which is expected to be dominant for heavy atoms, is given by[41]

$$V_{PN} = \sum_{i=1}^{N} - \frac{G_F}{2\sqrt{2}} Q_w \rho_N(r_i)(\gamma_5)_i \quad , \tag{40}$$

where G_F is the Fermi coupling constant (2.18×10^{-14} a.u.), Q_W is the weak charge, γ_5 is the usual Dirac operator, ρ_N is the nuclear density, and

$$Q_w = Z(1-4\sin^2\theta_w) - N \quad , \tag{41}$$

where Z and N are, respectively, the proton and neutron numbers, and θ_W is the Weinberg angle with $\sin^2\theta_W = 0.215$, the current best value when radiative corrections are taken into account. The nuclear density $\rho_N = (4/3\pi R_N^3)^{-1}$ with $R_N = (2.2677 \times 10^{-5})A^{1/3}$ and with A the atomic mass.

For a given transition between states of the same parity the PNC optical rotation is determined from the parameter

$$R = \mathrm{Im}(E1/M1) \quad , \tag{42}$$

where E1 and M1 are, respectively, the electric and magnetic dipole matrix elements. Treating V_{PN} as a perturbation, E1 in lowest order between states $|F\rangle$ and $|I\rangle$ is given by

$$E1 = \sum_k \left(\frac{\langle F|\underset{\sim}{D}|k\rangle\langle k|V_{PN}|I\rangle}{E_I - E_k} + \frac{\langle F|V_{PN}|k\rangle\langle k|\underset{\sim}{D}|I\rangle}{E_F - E_k} \right) \tag{43}$$

From the commutator,

$$[\underset{\sim}{r}, H] = ic\underset{\sim}{\alpha} \quad , \tag{44}$$

$$\langle b|\underset{\sim}{r}|a\rangle = (E_a - E_b)^{-1}\langle b|ic\underset{\sim}{\alpha}|a\rangle \quad , \tag{45}$$

and $\underset{\sim}{D} = -\underset{\sim}{r} = -\omega^{-1} ic\underset{\sim}{\alpha}$,

where $\omega = E_I - E_F$. The two forms for D only give equivalent results when the states $|I\rangle$, $|F\rangle$, and $|k\rangle$ are exact eigenstates of H in the absence of V_{PN}.

Many calculations have been carried out for the $6s^26p^3$ levels of Bi and measurements have been made for the $^4S_{3/2} \rightarrow {}^2D_{3/2}$, $^2D_{5/2}$ transitions.[42] For the $^2D_{3/2}$ transition R was calculated (corrected for $\sin^2\theta_W = 0.215$) with Hartree-Fock states to be[43] -17.74×10^{-8} (L) and -11.45×10^{-8} (V) where L refers to $\tilde{D} = -\tilde{r}$ and V refers to $\tilde{D} = -\omega^{-1}ic\tilde{\alpha}$. Harris, Loving, and Sandars[44] have included higher-order corrections and obtained the value -10.9×10^{-8} (L), corrected with $\sin^2\theta_W = .215$. Their calculation approximately included "shielding" correlation corrections to all orders and all remaining correlations to first order. A more recent calculation by Mårtensson, Henley, and Wilets[45] including correlation effects yielded the results -7.53×10^{-8} (corrected to $\sin^2\theta_W = 0.215$). The present experimental value[42] for the $^2D_{3/2}$ transition is $(-10.5\pm1.3) \times 10^{-8}$.

For the $^4S_{3/2} \rightarrow {}^2D_{5/2}$ transition in bismuth, Harris et al.[44] have calculated -12.7×10^{-8} and Mårtensson et al.[45] calculated -10.1×10^{-8} as compared with the experimental value[42] -10×10^{-8}. The theoretical values are corrected to $\sin^2\theta_W = 0.215$.

For the $^3P_o \rightarrow {}^3P_1$ transition in lead, the experimental result[42] is $(-9.9\pm2.5) \times 10^{-8}$ as compared with a theoretical value[42] -13×10^{-8}. For cesium ($6s \rightarrow 7s$) there is also reasonable agreement between calculations and experiment.[42]

For the $6p_{1/2} \rightarrow 7p_{1/2}$ transition in thallium, Im E1/β has been measured,[46] where β is the Stark amplitude parameter. The result is -1.73 ± 33 mV/cm as compared with a theoretical calculation[47] -1.31 ± 0.26 without correlations and -1.04 ± 0.16 including correlations.[48] At this point there is a need for additional calculations, especially for thallium.

Photoionization

The subject of photoionization has received much attention during the past decade, both experimentally[49] and theoretically.[50-53] The use of synchrotrons and techniques of photoelectron spectroscopy[54] have enabled the measurement of many partial cross sections including those for photoionization with excitation. On the theoretical side, sophisticated techniques have been developed to account for effects of electron correlations for closed shell atoms and more recently for open-shell atoms as discussed in the excellent review by Starace.[50]

The photoionization cross section $\sigma(\omega)$ may be expressed in terms of the frequency-dependent polarizability $\alpha(\omega)$ by[55]

$$\sigma(\omega) = 4\pi \frac{\omega}{c} \operatorname{Im} \alpha(\omega) \quad , \tag{46}$$

where

$$\alpha(\omega) = -\sum_k |\langle\psi_k|\sum_{i=1}^{N} z_i|\psi_o\rangle|^2 \left(\frac{1}{E_o-E_k-\omega} + \frac{1}{E_o-E_k+\omega+i\eta}\right) , \tag{47}$$

where $|\psi_o\rangle$ and $|\psi_k\rangle$ are exact many-particle states corresponding to the initial and final state, respectively. The $i\eta$ in the second denominator is treated according to Eq. (37) and then Im $\alpha(\omega)$ is proportional to $|\langle\psi_k|\sum z_i|\psi_o\rangle|^2$ upon using Eq. (33). Perturbation theory may also be used to calculate the many-particle length matrix element $\langle\psi_k|\sum_i^N z_i|\psi_o\rangle$. Alternatively, the velocity form $\langle\psi_k|\sum_i \frac{d}{dz_i}|\psi_o\rangle$ may also be used. When exact eigenstates ψ_o and ψ_k are used, the two expressions are related by

$$\langle\psi_k|\sum z_i|\psi_o\rangle = (E_o-E_k)^{-1}\langle\psi_k| \sum_i \frac{d}{dz_i}|\psi_o\rangle. \tag{48}$$

It is customary to calculate both length and velocity forms and to note that the results must be equal when the perturbation expansion has converged. However, agreement of length and velocity results is a necessary but not sufficient condition for convergence.

Diagrams for a transition $p \rightarrow k$ are shown in Fig. 6. The lowest-order contribution to the many-particle matrix element is given by $\langle k|z|p\rangle$ (or $\langle k|\frac{d}{dz}|p\rangle$ in the velocity case) where p and k are single-particle matrix elements. First-order correlations in the ground state are represented by the diagram of Fig. 6(b) and its exchange. This diagram represents the expression

$$\sum_{k'} \frac{\langle q|z|k'\rangle\langle kk'|v|pq\rangle}{\varepsilon_p+\varepsilon_q-\varepsilon_k-\varepsilon_{k'}}, \tag{49}$$

where the sum over k' includes all excited single-particle states. First-order correlations in the final state are represented by diagram (c) and its exchange (d). The expression for diagram (c) is given by

$$\sum_{k'} \frac{\langle kq|v|pk'\rangle\langle k'|z|q\rangle}{\varepsilon_q-\varepsilon_{k'}+\omega+i\eta}. \tag{50}$$

The expression for diagram (d) is obtained by replacing $\langle kq|v|pk'\rangle$ by $-\langle qk|v|pk'\rangle$ in Eq. (50). We note that the denominator of Eq. (50) can vanish when k' is a bound excited state and $\varepsilon_q-\varepsilon_{k'}+\omega = 0$. In this case there is a resonance due to the $q \rightarrow k'$ excitation. Summation of higher-order diagrams can be made which will shift the position of the resonance and provide an imaginary contribution to the denominator corresponding to the half-width of the resonance.

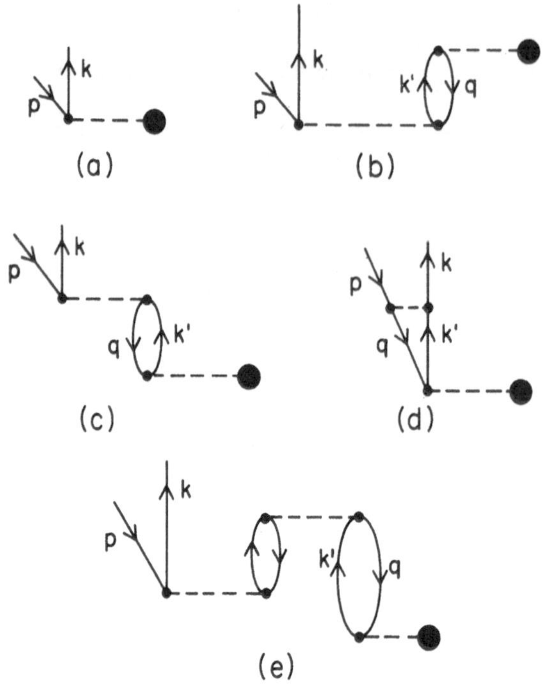

Fig. 6. Diagrams corresponding to the transition $p \to k$. Dashed line ending in large dot represents interaction with z (length) or $\frac{d}{dz}$ (velocity). (a) $\langle k|z|p\rangle$ (b) First-order correction due to ground state correlations. Exchange diagram should also be included. (c) First-order correlation in excited state. (d) Exchange diagram corresponding to (c). Diagrams with interactions with passive unexcited states are assumed to be cancelled by interactions with -V. (e) Higher-order RPA-type diagram.

For diagrams (c) and (d), when p and q are orbitals of the same closed subshell and when the excited states k and k' have the same orbital angular momentum, the potential may be chosen so that diagrams (c) and (d) are cancelled by interactions with the single-particle potential -V. Consider, for example, the neutral argon atom $3s^23p^6\ {}^1S$. After photon absorption, the final state is $3p^5kd\ {}^1P$ or $3p^5ks\ {}^1P$ and we denote it by Φ_k. We then define V by requiring

$$\langle\Phi_k|v-V|\Phi_k\rangle = 0 \quad . \tag{51}$$

Then interactions with -V will produce the desired cancellation as well as cancelling interactions in the excited state with the passive unexcited states. We then use this result along with Eq. (28) to define an appropriate Hartree-Fock potential for all orbitals.

Calculations using this single-particle potential have been carried out for argon and give good results.[56] For the $3p^6$ subshell, there is a considerable length-velocity discrepancy (approximately a factor of two), but the experimental results are bracketed above by the length calculation and below by the velocity calculation. Upon inclusion of the initial state correlation diagram of Fig. 6(b), the length and velocity results come into close agreement. A further improvement with experiment is provided by inclusion of diagram (e) which is a type of diagram included to all orders in the random phase approximation with exchange (RPAE).[57] Resonance structure due to $3s \rightarrow np$ resonances has also been calculated in good agreement with experiment for argon.[56] Extensive calculations for all the rare gases have been carried out using the RPAE by Amusia, Cherepkov, and Chernysheva.[57,58]

An interesting example of correlations is exhibited by the photoionization of the 3s subshell of argon as shown in Fig. 7. The Hartree-Fock length and velocity results are in good agreement (only length is shown). However, the experimental results are in qualitative disagreement with the Hartree-Fock calculations. The theoretical calculations including correlations are in good agreement with experiment. In a perturbation calculation it is found that the lowest order diagram of Fig. 6(a) is less important than the final state correlation diagram of Fig. 6(b) with p=3s and q=3p. We may think of this process as one in which the photon is absorbed (virtually) by the $3p^6$ subshell and the excitation is then transferred (by coulomb interaction) to the 3s subshell. The theoretical calculations in Fig. 7 include certain correlation effects to all orders. However, even a low-order calculation[64] including only diagrams (a)-(d) of Fig. 6 is in fairly good agreement with experiment.

Calculations for open-shell atoms present a severe challenge, but extensive calculations are now beginning to be carried out by various theoretical approaches such as RPAE, R-matrix theory, and many-body perturbation theory. The system most studied has been Cl $3p^5\ ^2P$. It has been found very important to include the coupling between final state channels involving single excitations such as $3p^4(LS)kd\ ^2L_f$ and $3p^4(LS)ks\ ^2L_f$, where L_f equals 0, 1, or 2. There is now rather good agreement among the recent calculations carried out by different methods.[65] An interesting feature is the appearance of resonance structure prior to the ^1D and ^1S edges due to $3P^4(^1D)nd$, ns and $3p^4(^1S)nd$,ns excitations. Calculations by Brown et al.[66] using MBPT and LS coupling predicted resonance structure prior to the ^1D edge to consist of one broad series and one narrow series, although the broad series was actually a superposition of $(^1D)nd\ ^2P$ and $(^1D)nd\ ^2D$. The narrow series was calculated to be $(^1D)ns\ ^2D$. A subsequent experiment by Ruščić and Berkowitz[67] verified the broad

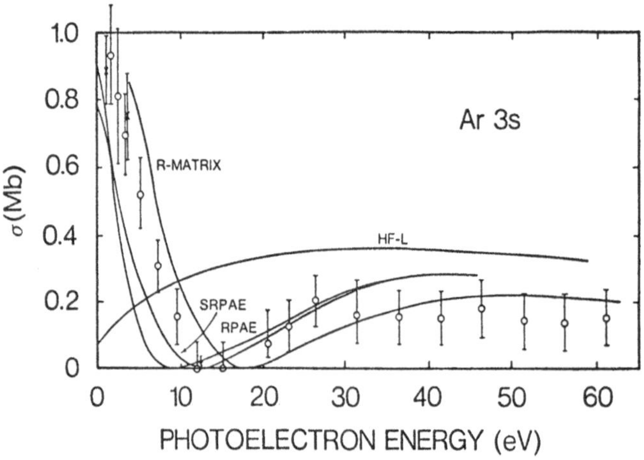

Fig. 7. Photoionization cross section for the $3s^2$ subshell of argon. R-matrix, Burke and Taylor, ref. 59; RPAE, Amusia and Cherepkov. ref. 58; SRPA, approximate RPAE calculation by Lin, ref. 60; HF-L, Kennedy and Manson, ref. 61; expt: o Houlgate et al., ref. 62; x Samson and Gardner, ref. 63

series but found two narrow series. Hansen et al.[68] have recently interpreted the second narrow series as $(^1D)nd\ ^2S$ with small admixtures of $(^1D)nd\ ^2P$ due to spin-orbit mixing.

Brown et al.[66] have also calculated the 3s subshell cross section for Cl. The results are qualitatively the same as for argon except for the two edges, and the deviation from the Hartree-Fock calculations is somewhat more pronounced.

For closed-shell atoms there have also been many calculations and measurements of the angular asymmetry parameter β defined[58] (for linearly polarized radiation) by

$$\frac{d\sigma}{d\Omega} = \frac{\sigma}{4\pi}\left(1+\beta\ P_2(\cos\theta)\right) \quad , \tag{52}$$

where θ is the angle between the polarization vector of the incident radiation and the direction of the ejected electron. In Fig. 8 is shown a calculation of β for the $5p^6$ subshell of xenon using the RPAE method by Amusia and Ivanov.[69] The importance of coupling of the $5p^6$ subshell with the $4d^{10}$ subshell is quite apparent.

Another aspect of photoionization calculations is the question of relaxation. That is, when the ejected electron energy is sufficiently low, the remaining ion has time to relax from the initial state with vacancy to the true ionic state. The outgoing electron will see a different field in the two cases. At high energies there is insufficient time for relaxation. An exact treatment of this time-dependent problem can (in principle) be made by calculating the appropriate relaxation diagrams in MBPT.[72] This is, however, not a simple task. An approximate treatment of relaxation effects near threshold may be made by calculating the continuum orbitals in the field of the relaxed ion. This has been done with considerable success for the $4d^{10}$ subshell of Ba where the relaxation effects are large.[73,74]

Resonance structure has been of much interest in recent calculations. Double-electron resonance structure such as $4s^2 \rightarrow 3dnp, 4pns$ has been calculated for the 4s cross section of calcium[75,76] and also for the photoionization with excitation cross section[76] leaving Ca^+3d. Similar resonance structure has also been recently calculated for the 3s cross section of Mg.[77]

Another type of resonance structure which has been studied recently[78-82] occurs in open-shell atoms such as the transition metal atoms with configurations $3p^63d^n4s^2$. There is very large resonance structure due to $3p^63d^n4s^2 \rightarrow 3p^53d^{n+1}4s^2$ resonances in the 4s and 3d subshell cross sections. In the diagrams of Fig. 6 the lowest-order contribution to this resonance is given by diagram (c)

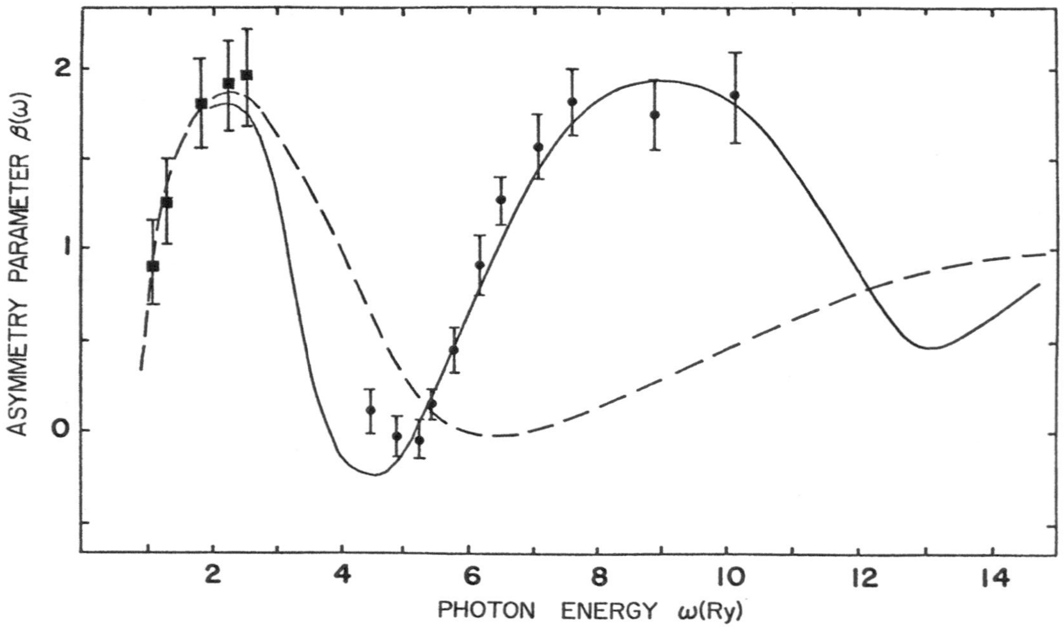

Fig. 8. Angular asymmetry parameter β for the $5p^6$ subshell of xenon. ——— and - - - - calculations with and without coupling with $4d^{10}$, Amusia and Ivanov, ref. 69; expt: ■ Lynch et al., ref. 70; Torop et al., ref. 71

with p=4s or 3d, q=3p, and k'=3d. Results of various calculations and experiments for Mn $3p^6 3d^5 4s^2$ are shown in Fig. 9. The MBPT calculations are given by solid curves. Although the calculations indicate $3p \rightarrow 3d$ resonance structure in the 4s cross section (lowest solid curve), it is not clear why there is so much discrepancy (more than a factor of two) with experiment. The $3p \rightarrow 3d$ resonance in the 3d cross section is represented rather well by the MBPT[83] and RPAE[85] calculations. The MBPT calculation by Garvin et al.[83] also calculated resonance structure due to $3p^6 3d^5 4s^2 \rightarrow 3p^5 3d^5 4s^2 (^7P)nd,ns$ Rydberg transitions. At higher energies (72-74 eV) their calculations showed an absence of resonance structure before the $3p^5 3d^5 4s^2 (^5P)$ edge where it would normally be expected. This was interpreted as due to the mixing of photoionization with excitation channels in which the resonance state $3p^5 3d^5 (^5P) 4s^2$ nd,ns decays to $3p^6 3d^3 4s^2 (nd,ns)\varepsilon f$. For the $3p^5 3d^5 4s^2 (^7P)nd,ns$ levels this decay is spin-forbidden. Nevertheless, the resonance structure has not been observed prior to the $3p^5 3d^5 4s^2 (^7P)$ edge. It is speculated[86] that this is due to mixing with $3p^6 3d^4 4s\ \varepsilon f, \varepsilon p$; i.e., a 4s electron fills the 3p hole and a 3d electron is ejected. Calculations are needed to check this conjecture.

The angular asymmetry parameter β was calculated[83] for the 3d subshell and showed a large resonance due to the $3p \rightarrow 3d$ absorption. The calculations are in generally good agreement with experimental results,[80,81] although the experimental resonance is less pronounced than the calculated one.

An aspect of photoionization which is presently of great interest is the process of photoionization with excitation in which the ion is left in an excited state. A very interesting calculation of this process was carried out by Scott et al.[76] in their calculation of the photoionization of neutral calcium. In the photon range 7.8 eV to 9.0 eV they found that the total cross section was dominated by the $3dkp(^1P)$ channel in which the calcium ion is left in the $3d(^2D)$ level. Furthermore, they predict strong resonance structure due to 4pns resonances with $n \geq 6$.

Another system which has been of considerable interest in recent years is $He(1s)^2$ with photoionization with excitation leaving He^+ in the n=2 levels. There has been controversy as to whether or not σ_{2p} dominates over σ_{2s} near threshold. Recent experiments and calculations agree that it does.[87-90] Another interesting aspect is the relatively large resonance structure in the n=2 cross section discovered by Woodruff and Samson[91] and confirmed by Shirley et al.[90] Very recent many-body calculations by Salomonson et al.[92] are in good agreement with the experimental results as shown in Fig. 10. Correlations in the initial state were included by radial pair functions with angular momenta up to $\ell=6$ and coupled to all orders as described by Mårtensson.[24] The final state correlations were included by the coupled equations approach, solving integral

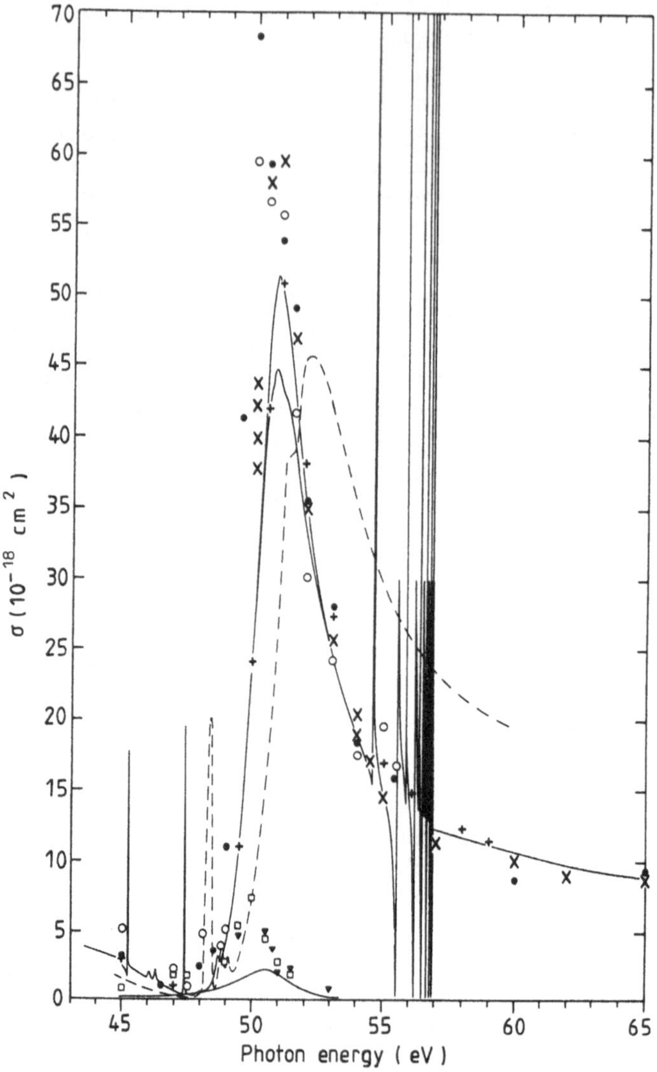

Fig. 9. Resonance structure in 3d and 4s subshell cross sections of atomic manganese. The highest full curve is the 3d cross section calculated in ref. 83 by the coupled equations method, ref. 66, from 50 eV to 65 eV. The solid curve just below is from ref. 83 including spin-orbit effects and extends from 43 eV to 53 eV. The broken curve is a calculation by Davis and Feldkamp, ref. 84. The plus signs are from an RPAE calculation, ref. 85. Experimental data for the 3d cross section are from ref. 78 (open circles), ref. 81 (full circles), and ref. 80 (crosses). Measurements for 4s (lowest full curve) are from ref. 78 (squares) and ref. 81 (triangles). Experimental data were normalized to fit the calculation of ref. 83 at energies above 53 eV.

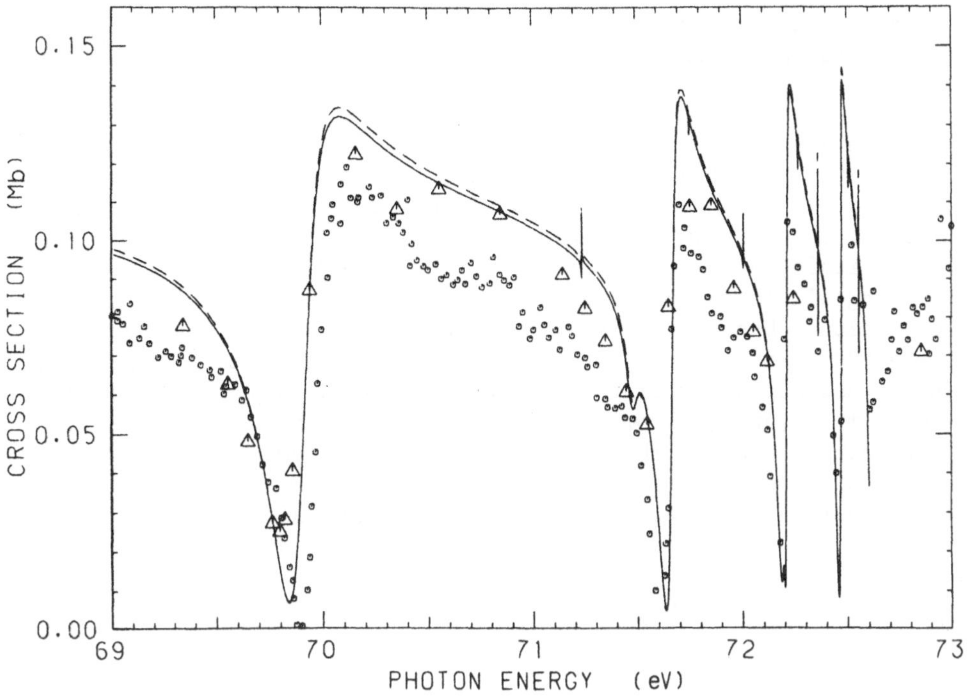

Fig. 10. The total n=2 photoionization cross section in the resonance region leading up to the n=3 threshold at 73.0 eV. Solid (dashed) line represents the calculation of ref. 92 using length (velocity) operator. The dots are experimental results from Woodruff and Samson, ref. 91, and triangles from Lindle et al., ref. 89.

equations for transition amplitudes as described by Brown et al.,[66] which is essentially a K-matrix approach.[50] The following nine final state channels were included: 1skp, 2skp, 2pks, 2pkd, 3skp, 3pks, 3pkd, 3dkp, and 3dkf. The symbol k refers to both bound and continuum orbitals. For each angular momentum, eight bound and forty-two continuum orbitals were used.

Double Photoionization

In this process one photon is absorbed and two electrons are ejected. It is a very interesting process since it cannot occur without electron correlations. although it can be roughly approximated by relaxation or shake-off effects. Another interesting aspect is that, in principle, an infinite number of partial waves contribute, although it is usually expected that a reasonable description of the process is obtained with a small number of pairs of partial waves. Almost all calculations thus far have used many-body perturbation theory, and results have been obtained for neon, argon, beryllium, carbon, and helium.[53]

The many-body calculations for helium[93] included both kskp and kpkd pairs of partial waves. It was found that certain higher-order many-body effects in the final state could be included by calculating $\ell=1$ orbitals in the field of $He^{+}1s$ and by calculating $\ell=0$ and $\ell=2$ orbitals in the field of He^{++}. The calculations involved calculating all lowest-order diagrams contributing to double photoionization. Dipole matrix elements of single excitations were corrected by including ground state correlations. Both final state and ground state correlations were found to be necessary, and it was found important to allow for higher-order diagrams of the ground state correlations. These were summed approximately by considering ratios of third order correlation energy diagrams to the second-order energy diagrams. Diagrams contributing to double photoionization are shown in Fig. 11, and their sum gives the many-body dipole matrix element $Z(pq \rightarrow k'k)$. Since the outgoing electrons share the energy, the cross section σ^{++} is given by

$$\sigma^{++}(\omega) = 16\,\frac{\omega}{c}\int_0^{k_{max}} dk\,\frac{|Z(pq\rightarrow k'k)|^2}{k'} \quad , \tag{53}$$

where

$$k' = [2(\varepsilon_p+\varepsilon_q-\frac{k^2}{2}+\omega)]^{1/2} \quad , \tag{54}$$

and

$$k_{max} = [2(\varepsilon_p+\varepsilon_q+\omega)]^{1/2} \quad , \tag{55}$$

Results of the calculations for helium are shown in Fig. 12. The agreement with experiment is seen to be excellent, especially for the velocity calculation. However, this agreement may be somewhat fortuitous, since the overall accuracy of the calculation probably doesn't exceed 20%. The calculation is dominated by the kskp partial waves, with the kpkd partial waves contributing approximately 10% of the total cross section near the maximum and approximately 25% at higher energies. Similar results, but with worse agreement with experiment, were obtained for the rare gases neon[94,95] and argon.[95] For argon it was found that ionization of $(3p)^2$ and $(3s3p)$ were both important, and similarly for neon.

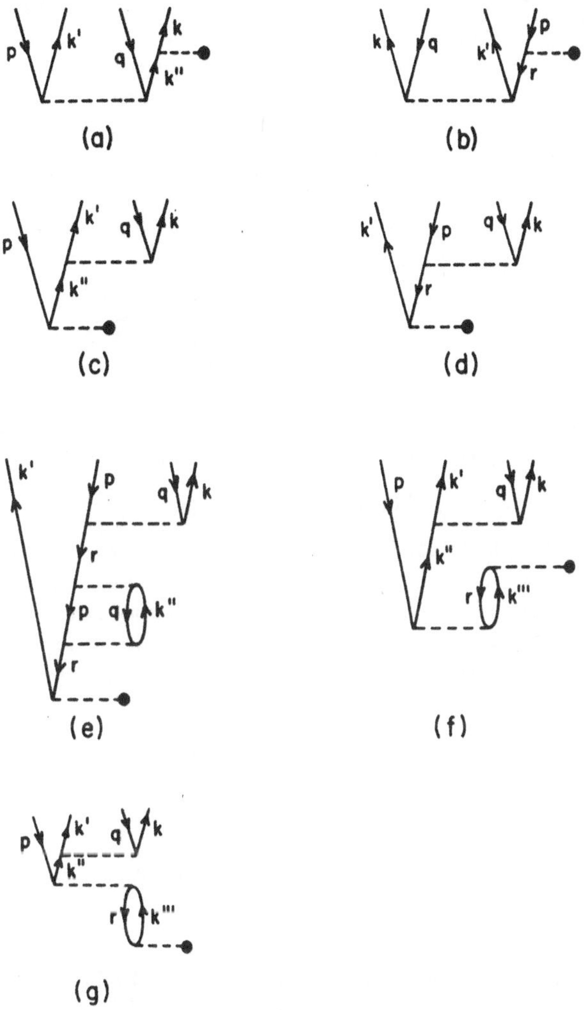

Fig. 11. Diagrams contributing to the double ionization matrix element $Z(pq \rightarrow k'k)$. Solid dots indicate interaction with the dipole operator Z. Diagrams (a)-(d) are lowest-order contributions. Diagram (f) illustrates ground state correlation corrections to the dipole matrix element $\langle k'|Z|p\rangle$. Diagrams (e) and (g) illustrate typical higher-order diagrams.

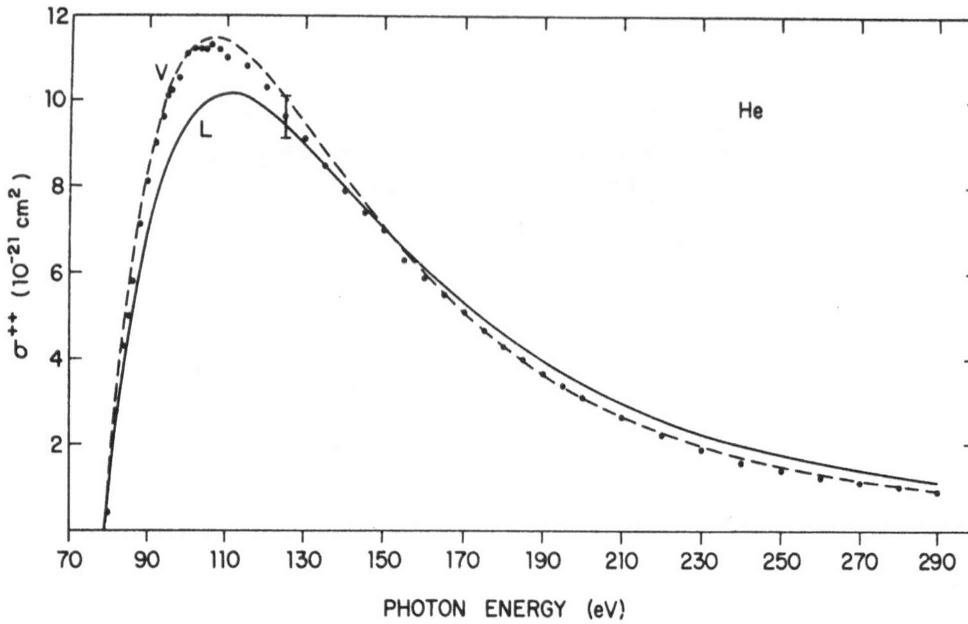

Fig. 12. Double photoionization cross section of helium. Curves for the correlated length (L) and velocity (V) cross sections include contributions from both kskp and kpkd channels, ref. 93. Experimental data from Bizau and Wuilleumier, ref. 96.

CONCLUDING REMARKS

In these lectures we have seen how many-body perturbation theory can be applied to a wide range of problems ranging from structural properties such as correlation energies and hyperfine structure to dynamical properties such as transitions induced by weak neutral currents and photoionization cross sections. In addition, many-body perturbation theory is particularly useful in studying complex processes such as multiple Auger transitions, photoionization with excitation, and double photoionization. One advantage of this method is that through the diagrams we can obtain a physical picture of complicated processes in terms of single-particle transitions. However, there remain many challenges such as methods to include higher-order terms and many applications where it would be desirable to use relativistic orbitals, although very impressive results have been obtained for closed-shell atoms using the relativistic random phase approximation.[51]

ACKNOWLEDGEMENTS

I wish to thank Dr. S. L. Carter and Dr. Sten Salomonson for very helpful discussions and the U.S. National Science Foundation for support of this work.

REFERENCES

1. K. A. Brueckner, Phys. Rev. 97:1353 (1955); 100:36 (1955).
2. J. Goldstone, Proc. Roy. Soc. A 239:267 (1957).
3. D. J. Thouless, "The Quantum Mechanics of Many-Body Systems," 2nd ed., Academic, New York (1972).
4. A. L. Fetter and J. D. Walecka, "Quantum Theory of Many-Particle Systems," McGraw Hill, New York (1971).
5. N. H. March, W. G. Young, and S. Sampanthar, "The Many-Body Problem in Quantum Mechanics," Cambridge University Press, Cambridge (1967).
6. I. Lindgren and J. Morrison, "Atomic Many-Body Theory," Springer-Verlag, Berlin (1982).
7. H. P. Kelly, 1976, in "Photoionization and Other Probes of Many-Electron Interactions," F. Wuilleumier, ed., Plenum, New York, p. 83.
8. A. Messiah, "Quantum Mechanics," North-Holland, New York (1961).
9. L. M. Frantz, R. L. Mills, R. G. Newton, and A. M. Sessler, Phys. Rev. Lett. 1:340 (1958).
10. B. A. Lippman, M. H. Mittleman, and K. M. Watson, Phys. Rev. 116:920 (1959).
11. R. T. Pu and E. S. Chang, Phys. Rev. 151:31 (1966).
12. H. J. Silverstone and M. L. Yin, J. Chem. Phys. 49:2026 (1968).
13. S. Huzinaga and C. Arnau, Phys. Rev. A1:1285 (1970).
14. H. P. Kelly, Phys. Rev. 136B:896 (1964).
15. C. F. Bunge, Phys. Rev. A14:1965 (1976).
16. S. Salomonson, I. Lindgren, and A.-M. Mårtensson, Phys. Scr. 21:351 (1980).
17. I. Lindgren and S. Salomonson, Phys. Scr. 21:335 (1980).
18. H. P. Kelly. Phys. Rev. 144:39 (1966).
19. H. P. Kelly, 1971, in "Atomic Physics 2," P. G. Sanders, ed., Plenum, London, p. 227.
20. J. W. Viers, F. E. Harris, and H. F. Schaefer III, Phys. Rev. A1:24 (1970).
21. T. L. Barr and E. R. Davidson, Phys. Rev. A1:644 (1970).
22. J. I. Musher and J. M. Schulman, Phys. Rev. 173:93 (1968).
23. V. McKoy and N. W. Winter, J. Chem. Phys. 48:5514 (1968).
24. A.-M. Mårtensson, J. Phys. B12:3995 (1979).
25. R. M. Sternheimer, Phys. Rev. 84:244 (1951).
26. E. S. Chang, R. T. Pu, and T. P. Das, Phys. Rev. 174:1 (1968).
27. H. P. Kelly, Phys. Rev. 173:142 (1968).
28. M. Vajed-Samii, S. N. Ray, T. P. Das, and J. Andriessen, Phys. Rev. A20:1787 (1979).

29. I. Lindgren, Repts. Prog. Phys. 47:345 (1984).
30. A.-M. Mårtensson, J. Phys. B12:3995 (1979).
31. J. S. M. Harvey, Proc. Roy. Soc. (London) A285:581 (1965).
32. H. P. Kelly, Phys. Rev. 180:55 (1969).
33. R. L. Chase, H. P. Kelly, and H. S. Köhler, Phys. Rev. A3:1550 (1971).
34. H. P. Kelly, Phys. Rev. A11:556 (1975).
35. W. Mehlhorn, D. Stalherm, and H. Verbeek, Z. Naturforsch A23: 287 (1968).
36. W. Wölfli, Ch. Stoller, G. Bonani, M. Suter, and M. Stöckli, Phys. Rev. Lett. 35:656 (1975).
37. H. P. Kelly, Phys. Rev. Lett. 37:386 (1976).
38. R. L. Simons, H. P. Kelly, and R. Bruch, Phys. Rev. A19:682 (1979).
39. R. L. Simons and H. P. Kelly, Phys. Rev. A22:625 (1980).
40. C. Bouchiat, 1981, in "Atomic Physics 8," D. Kleppner and F. M. Piplin, eds., Plenum, New York, p. 83.
41. M. A. Bouchiat and C. Bouchiat, J. Phys. B35:899 (1974).
42. T. P. Emmons, J. M. Reeves, and E. N. Fortson, Phys. Rev. Lett. 52:86 (1984).
43. S. L. Carter and H. P. Kelly, Phys. Rev. Lett. 15:966 (1979).
44. M. J. Harris, C. E. Loving, and P. G. H. Sanders, J. Phys. B11: L749 (1978).
45. A.-M. Mårtensson, E. M. Henley, and L. Wilets, Phys. Rev. A24: 308 (1981).
46. P. S. Drell and E. D. Commins, Phys. Rev. Lett. 53:968 (1984).
47. D. Neuffer and E. D. Commins, Phys. Rev. A16:844 (1977).
48. B. P. Das et al., Phys. Rev. Lett. 49:32 (1982).
49. J. A. R. Samson, 1982, in "Handbuch der Physik XXXI," S. Flügge and W. Mehlhorn, eds., Springer-Verlag, p. 123.
50. A. F. Starace, ibid., p. 1.
51. W. R. Johnson, 1983, in "Atomic Physics 8," I. Lindgren, A. Rosen, and S. Svanberg, eds., Plenum, New York, p. 149.
52. M. Ya. Amusia, ibid., p. 287.
53. H. P. Kelly, ibid., p. 305.
54. M. O. Krause, 1980, in "Synchrotron Radiation Research," H. Winick and S. Doniach, eds., Plenum, New York, p. 101.
55. U. Fano and J. W. Cooper, Rev. Mod. Phys. 40:441 (1968).
56. H. P. Kelly and R. L. Simons, Phys. Rev. Lett. 30:529 (1973).
57. M. Ya. Amusia, N. A. Cherepkov, and L. V. Chernysheva, Sov. Phys. JETP 33:90 (1971).
58. M. Ya. Amusia and N. A. Cherepkov, Case Studies in Atomic Physics 5:47 (1975).
59. P. G. Burke and K. T. Taylor, J. Phys. B8:2620 (1975).
60. C. D. Lin, Phys. Rev. A9:181 (1974).
61. D. J. Kennedy and S. T. Manson, Phys. Rev. A5:227 (1972).
62. R. G. Houlgate, J. B. West, K. Codling, and G. V. Marr, J. Elect. Spectrosc. 9:205 (1976).
63. J. A. R. Samson and J. L. Gardner, Phys. Rev. Lett. 33:671 (1974).

64. M. S. Pindzola and H. P. Kelly, Phys. Rev. A12:1419 (1975).
65. S. Shahabi, A. F. Starace, and T. N. Chang, Phys. Rev. A30: 1819 (1984), and references therein.
66. E. R. Brown, S. L. Carter, and H. P. Kelly, Phys. Rev. A21: 1237 (1980).
67. B. Ruščić and J. Berkowitz, Phys. Rev. Lett. 50:675 (1983).
68. J. E. Hansen, R. D. Cowan, S. L. Carter, and H. P. Kelly, Phys. Rev. A30:1540 (1984).
69. M. Ya Amusia and V. K. Ivanov, Phys. Lett. 59A:194 (1976).
70. M. J. Lynch, K. Codling, and A. B. Gardner, Phys. Lett. 43A: 213 (1973).
71. L. Torop, J. Morton, and J. B. West, J. Phys. B9:2035 (1976).
72. G. Wendin, Phys. Lett. 51A:291 (1975).
73. M. Ya. Amusia, V. K. Ivanov, and L. V. Chernysheva, Phys. Lett. 59A:191 (1976).
74. H. P. Kelly, S. L. Carter, and B. E. Norum, Phys. Rev. A25: 2052 (1982).
75. Z. Altun, S. L. Carter, and H. P. Kelly, Phys. Rev. A27:1943 (1983).
76. P. Scott, A. E. Kingston, and A. Hibbert, J. Phys. B16:3945 (1983).
77. C. H. Greene and P. O'Mahony, to be published.
78. R. Bruhn, E. Schmidt, H. Schröder, and B. Sonntag, Phys. Lett. 90A:41 (1982).
79. J. P. Connerade, M. W. D. Mansfield, and M. A. P. Martin, Proc. Roy. Soc. A350:405 (1976).
80. P. H. Kobrin, U. Becker, C. M. Truesdale, D. W. Lindle, H. G. Kerkhoff, and D. A. Shirley, J. Elect. Spectrosc. Relat. Phenom. 34:129 (1984).
81. M. O. Krause, T. A. Carlson, and A. Fahlmann, Phys. Rev. A30: 1316 (1984).
82. E. Schmidt, H. Schröder, B. Sonntage, H. Voss, and H. E. Wetzel, J. Phys. B18:79 (1985).
83. L. J. Garvin, E. R. Brown, S. L. Carter, and H. P. Kelly, J. Phys. B16:L269 (1983).
84. L. C. Davis and L. A. Feldkamp, Phys. Rev. A17:2012 (1978).
85. M. Ya. Amusia, V. K. Ivanov, and L. V. Chernysheva. J. Phys. B14:L19 (1981).
86. B. Sonntag, private communication.
87. K. A. Berrington, P. G. Burke, W. C. Fon, and K. T. Taylor, J. Phys. B15:L603 (1982).
88. V. Schmidt, H. Derenbach, and R. Malutzke, J. Phys. B15:L523 (1982).
89. D. W. Lindle et al., to be published.
90. D. A. Shirley et al., AIP Conf. Proc. No. 94:569 (1982).
91. P. R. Woodruff and J. A. R. Samson, Phys. Rev. Lett. 45:110 (1982); Phys. Rev. A25:848 (1982).
92. S. Salomonson, S. L. Carter, and H. P. Kelly, J. Phys. B, to be published.
93. S. L. Carter and H. P. Kelly, Phys. Rev. A24:170 (1981).

94. T. N. Chang and R. T. Poe, Phys. Rev. A12:1432 (1975).
95. S. L. Carter and H. P. Kelly, Phys. Rev. A16:1525 (1977).
96. J. M. Bizau, F. Wuilleumier, unpublished.

ANGLE- AND SPIN-RESOLVED PHOTOELECTRON SPECTROSCOPY

U. Heinzmann

Fakultät für Physik
Universität Bielefeld
D - 4800 Bielefeld, F.R.G.

INTRODUCTION AND GENERAL COMMENTS

Experimental analysis of the photoelectron-spin polarization vector in photoionization using circularly polarized light (Fano effect[1]) was up to 1984 restricted to angle-integrated measurements[2] without resolution of the kinetic energy of the photoelectrons corresponding to different ionic states. With the development of the new German dedicated electron storage ring for synchrotron radiation BESSY in Berlin, a light source of circularly polarized vacuum ultraviolet (vuv) radiation with sufficiently high intensity has become available, which makes angle- and energy-resolved spin-polarization transfer studies from circularly polarized radiation onto photoelectrons feasible. These measurements could be performed with free atoms,[3,4] atoms adsorbed on solid surfaces[5,6] as well as with a solid state system[7] even in a photon energy range ≧ 10 eV, where conventional methods for producing circularly polarized radiation break down because no transparent or even double refracting material exists. These studies using circularly polarized radiation complement recent photoelectron spectroscopy measurements with free randomly oriented[8] as well as free oriented molecules[9]. One of the reasons why these experiments have been done is to find a set of parameters measured in the experiments which characterize the photoemission process quantummechanically completely. It builds a bridge from the atoms via the molecules via the adsorbates up to the three dimensional solid state and makes this cross comparison not only in terms of intensities and polarizations but also by means of dipole matrix elements and phase shift differences of continuum wave functions for single channels, which are energy degenerate but have been isolated by the data-combination of different non-redundant experiments.

The reaction plane of symmetry for an angle- and spin-resolved photoionization process of an unpolarized atom or unoriented molecule using circularly polarized radiation is shown in Fig. 1. Because the momentum of the photon is negligibly small compared with the momentum of the photoelectron (valid in nonrelativistic approximation if photon energy $\leq$ 100 eV) there is a forward-backward symmetry in the reaction plane of Fig. 1. It also makes no difference whether right handed circularly polarized radiation comes from the left or left handed comes from the right. The rotational symmetry around the direction of the photon momentum causes, that both electron spin polarization components perpendicular to the photon spin have to vanish for photoelectron emission angles $\theta = 0, \pi/2, \pi$. This is shown in Fig. 2, where the angle dependences of intensity $I(\theta)$ and spin polarization components are shown for a certain atomic photoionization process, which has been simultaneously resolved with respect to all variables one has: radiation wavelength 80 nm, radiation polarization σ^+, electron emission angle θ, electron kinetic energy corresponding to the final ionic state $Xe^+ \ ^2P_{1/2}$, the 3 components of the electron spin polarization vector $\vec{P}(\theta)$: $P_\perp(\theta)$ perpendicular to the reaction plane, $A(\theta)$ parallel to the photon spin, $P_p(\theta)$ perpendicular to the photon spin but in the reaction plane.

The curves in Fig. 2 are fits to the experimental points[3,10] (the size of a typical error-bar cross is given in the middle part) and are in accordance with the theoretical predictions by Cherepkov[11] and Lee[12]:

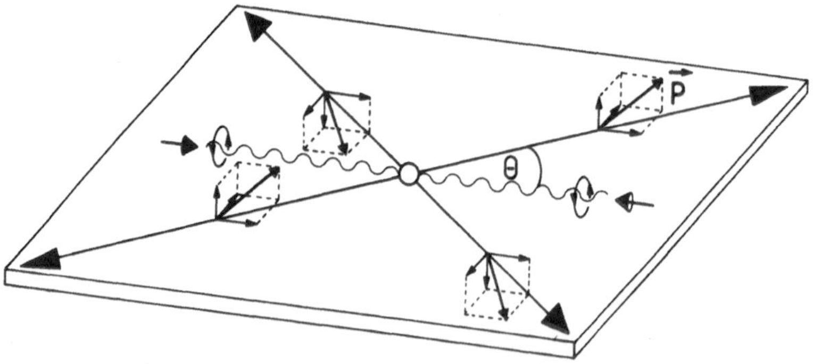

FIG. 1. Photoionization reaction plane using circularly polarized radiation. The results do not depend on whether right handed circularly polarized light comes from the left or left handed comes from the right.

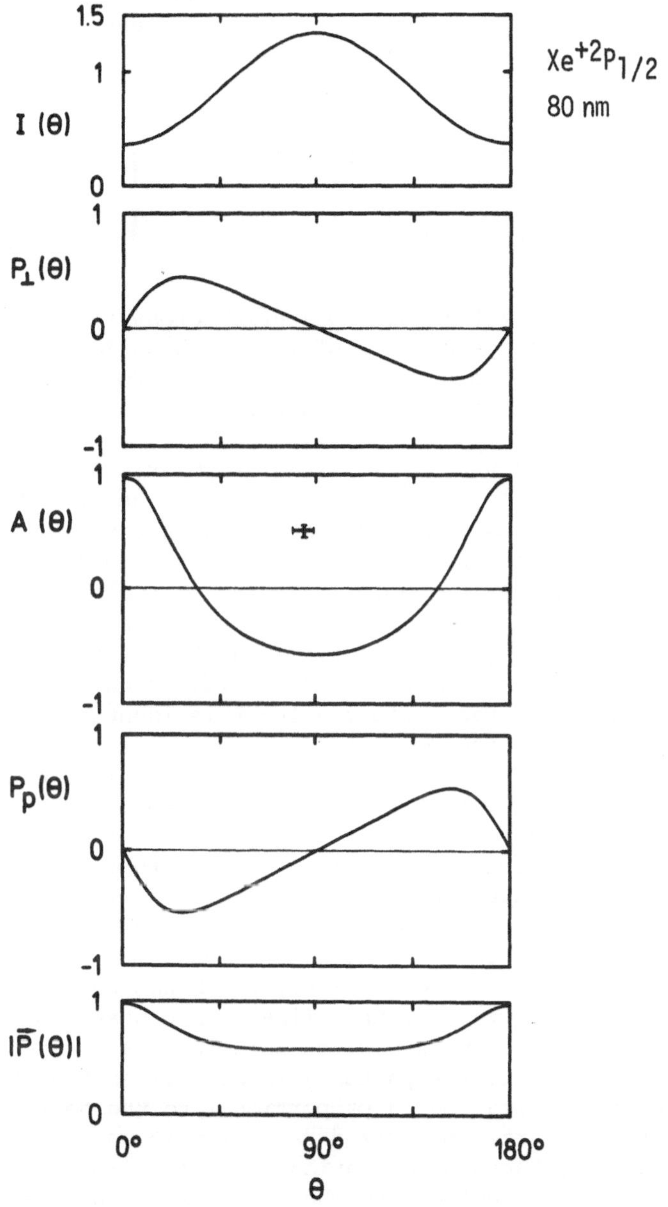

FIG. 2. Fit-curves of the experimental results (the size of a typical error-bar cross is given in the middle part) describing the angular dependences of the photoelectron intensity I(θ), of the 3 components and the length of the spin-polarization vector for photoionization of Xe atoms using radiation of 80 nm corresponding to photoelectrons leaving the ion in the $^2P_{1/2}$ state.[3]

$$I(\theta) = 1 - \frac{\beta}{2}\left(\frac{3}{2}\cos^2\theta - \frac{1}{2}\right)$$

$$P_\perp(\theta) = 2\,\xi \sin\theta\,\cos\theta\,/I(\theta)$$

independent on helicity of light

$$A(\theta) = \pm\left(A - \alpha\left(\frac{3}{2}\cos^2\theta - \frac{1}{2}\right)\right)\,/I(\theta)$$

$$P_p(\theta) = \pm\,\alpha \sin\theta\,\cos\theta\,/I(\theta)$$

+ for σ^+ light
- for σ^- light

β, ξ, A, α and the total photoionization-cross section Q are the so called dynamical parameters of the photoionization process, which are energy dependent and which are one possible set for a complete quantummechanical characterization.

$A(\theta)$ and $P_p(\theta)$ vanish, if linearly polarized or unpolarized instead of circularly polarized radiation is used.[13,14] All five curves in Fig. 2 show a mirror symmetry with respect $\theta = \pi/2$, but $P_\perp(\theta)$ and $P_p(\theta)$ with changing sign. Thus, the polarizations of opposite sign cancel one another, if the photoelectrons ejected are extracted by an electric field regardless of their direction of emission. The only non-vanishing component of the spin polarization in an angle-integrated measurement is $A(\theta)$ which yields A as the average value. This Fano-effect value A is identical with $A(\theta)$ for the so called magic angle $\theta = 54°$, where the second Legendre polynomial vanishs. To determine A in an angle resolved experiment yields the advantage, that it can be now also studied as function of the electron energies by use of an electron spectrometer in the experiment, which was impossible in the former original type of experiment to determine A angle integrated. It is also worth noting that within the error limits the photoelectrons emitted into forward direction $\theta = 0$ have been found[3] to be completely spin polarized (Fig. 2 middle part), which has been explicitly theoretically predicted for this final ionic state one and a half decades ago.[15] This complete electron-spin polarization in forward direction parallel to the photon spin as well as the fact, that the electron polarization is proportional to the degree of photon polarization if partly polarized radiation is used, allows to use the headline "spin-polarization transfer" from spin polarized photons onto photoelectrons to characterize the process.

The lowest part of Fig. 2 demonstrates that the length of the electron-polarization vector never vanishs as function of the emission angle θ. This can be generalized by the experimentally improved rule, that in an angular resolved photoemission experiment on atoms, molecules, adsorbates or solid states it is rather very common than exceptional to get spin polarized photoelectrons.

The main components of the two apparatus we have built up at the new German electron storage ring BESSY - one for the studies of atomic and molecular photoionization[3] and one for photoemission experiments with solid surfaces[7] and adsorbates[5] - are briefly discussed here; they are partly shown in Fig. 3. The synchrotron radiation is dispersed by a 6.5 m N.I. UHV monochromator of the Gillieson type[16], not shown in Fig. 1, with the electron beam in the storage ring being the virtual entrance slit. A spherical mirror and a plane holographic grating (1200 lines/mm) form a 1 : 1 image of the tangential point in the exit slit. With a slit width of 2 mm a bandwidth of 0.5 nm has been achieved. Apertures movable in vertical direction are used to select radiation emitted above and below the storage ring plane, which has positive or negative helicity, respectively. In the plane, the synchrotron radiation is linearly polarized.

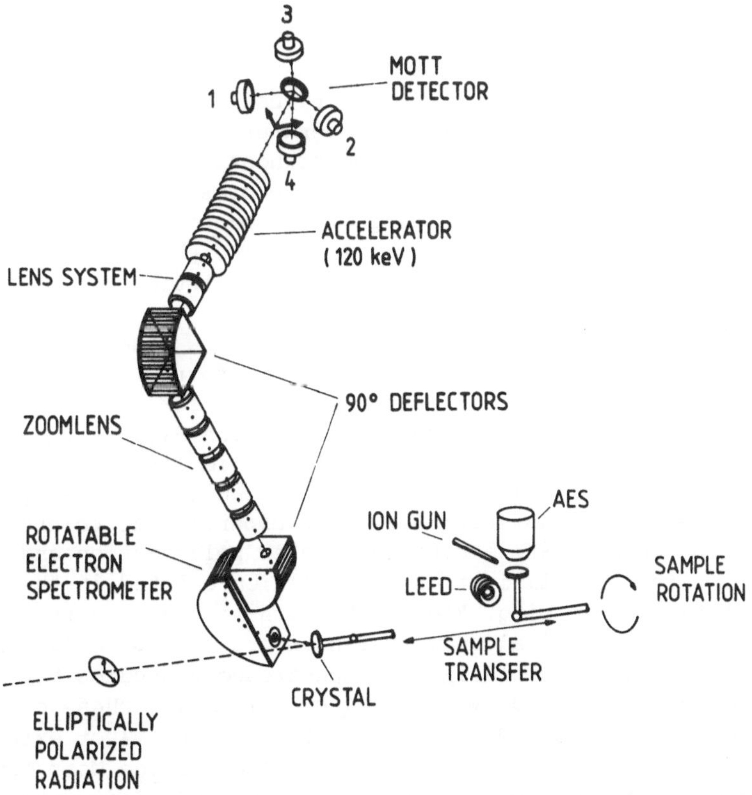

FIG. 3. Schematic diagram of the apparatus, built up at BESSY, shown for the general case of off-normal photoemission.[7]

The monochromatized and in general elliptically polarized light hits the target (crystal, atomic beam) producing photoelectrons in a region free of electric or magnetic fields. The photoelectrons emitted in the reaction plane at an angle θ are energy analyzed in a simulated hemispherical electron spectrometer[17], which is rotatable around the normal of the reaction plane. An electrostatic deflection by 90° directs the electron beam along the axis of rotation of the electron spectrometer. After a second deflection by 90° the electron beam is accelerated to 120 keV and scattered at the gold foil of the Mott detector[18]. $A(\theta)$ and $P_{\perp}(\theta)$ both being transverse components, are simultaneously determined from the left-right scattering asymmetry measured by two pairs of detectors as shown in Fig. 3. Instrumental asymmetries could be easily eliminated by taking advantage of the reversal of light helicity and of the change of the emission angle from θ to $-\theta$ as well as by use of 4 additional detectors in forward scattering directions in the Mott detector, not shown in Fig. 3.

In the solid state apparatus, the sample is cleaned by ion bombardment, heating in oxygen, and flashing; it is characterized by low energy electron diffraction (LEED) and scanning Auger electron spectroscopy in a separate preparation chamber. The crystal on top of a three-axes manipulator moveable between preparation and photoemission chamber, can be cooled by use of a temperature-controlled liquid He-Cryostat to temperatures of less than 40 K. The adsorbate is introduced via a doser nozzle which kept the background pressure below 10^{-9} mbar (base pressure $5 \cdot 10^{-11}$ mbar), allowing the continuous monitoring of the photoelectron spectra and LEED pattern as function of coverage. The photoelectrons emitted into a cone ±3° are energy analyzed at a resolution of 90 meV FWHM.

The optical degrees of polarization of the synchrotron radiation have been measured[3] by means of a rotatable four mirror analyzer[19] not shown in Fig. 3. Fig. 4 shows the results for the circular polarization P_{circ} and the linear polarization P_{lin} as functions of the vertical angle ψ (±0.1 mrad). The solid lines which represent the theoretical predictions according to Schwinger's theory and which show excellent agreement with the experimental points (error-bar crosses) demonstrate a complete linear polarization and a vanishing circular polarization of radiation emitted in the plane of the BESSY-storage ring. Fig. 5 shows the integrated results for the case the vertical angular ranges are from ψ to ±5 mrad. The photoelectron spinpolarization spectroscopy studies have been performed with P_{circ} = 95 % and P_{lin} = 31 %. Under these conditions a photon flux of a few $10^{11} s^{-1}$ passes the monochromator exit slit and hits the phototarget. Typical count rates in the Mott detector were a few s^{-1} for gas phase experiments and $10^{3} s^{-1}$ for studies with solids and adsorbates.

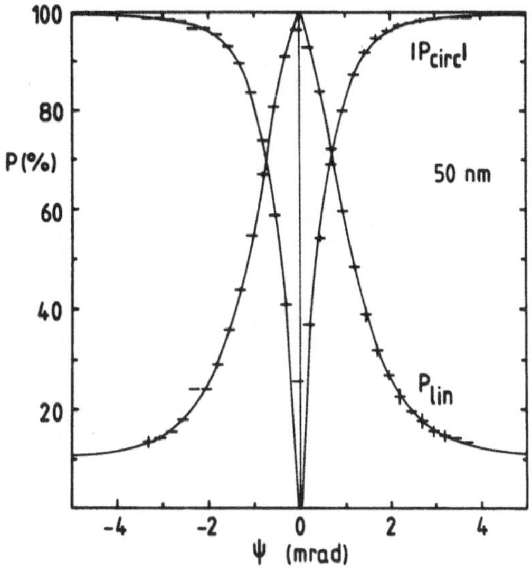

FIG. 4. Degree of circular and linear polarization P_{circ} and P_{lin}, respectively, of vuv synchrotron radiation emitted from the BESSY storage-ring plane as function of the vertical angle ψ (±0.1 mrad).[3]

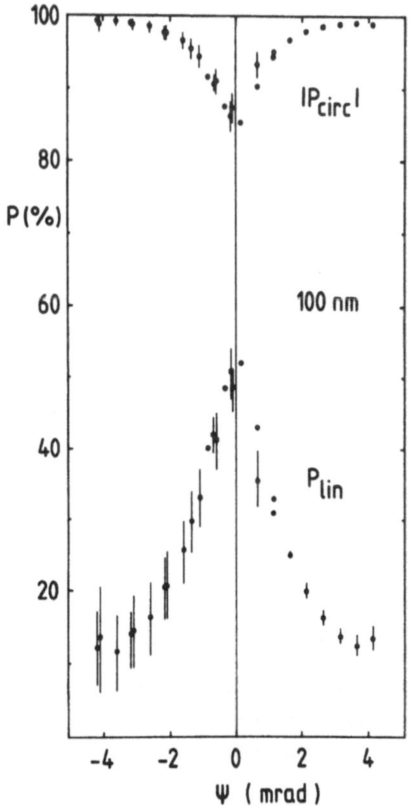

FIG. 5. As Fig. 4 but integral in a vertical angular range from ψ to ±5 mrad.[3] The wavelength dependence of P_{circ} and P_{lin} between 50 and 100 nm is the same within the experimental uncertainties.

All photoelectron spin polarization effects in atoms arise due to the existence of the spin-orbit interaction. Because of that the ℓ and m_ℓ quantum numbers are no longer good and thus the "spin momentum transfer" is no longer performed from the photon spin to the orbital angular momentum ℓ and m_ℓ but to the total angular momentum j and m_j of which the photoelectron spin is a part. Discussing this influence of the spin-orbit interaction quantitatively, however, one has to distinguish between two cases, which will be discussed in this chapter at certain examples in detail:

1. Photoionization of atoms, where the discrete atomic or ionic states involved show a fine structure splitting induced by the spin-orbit coupling.
2. Photoionization of an atomic s-subshell, where neither the ground-state nor the final ionic state shows a splitting.

Case 1 is fulfilled for photoionization of rare gase atoms; Fig. 6 shows the photoelectron spectrum of argon.[20] The two peaks correspond with the ionic states $^2P_{1/2}$ and $^2P_{3/2}$ of Ar^+, split by the existence of the spin-orbit interaction. Both peaks in the spectrum yield spin polarized photoelectrons but with a spin-polarization degree of opposite sign. Or otherwise, in the case

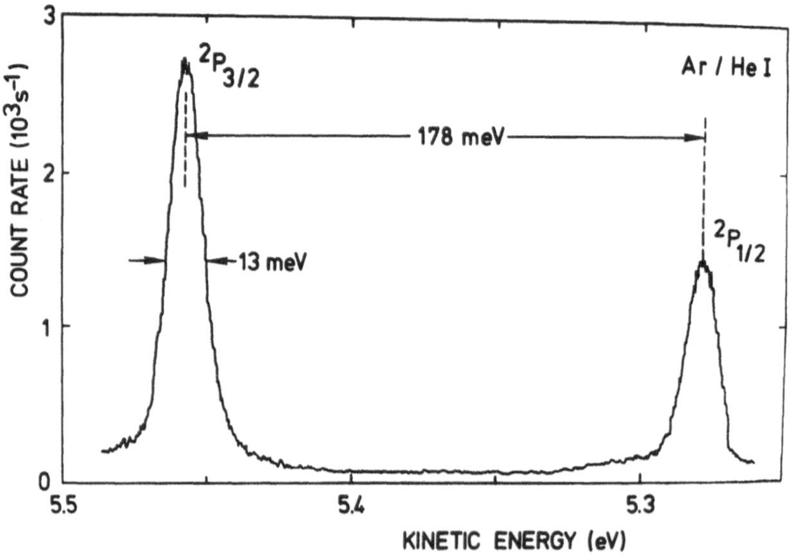

FIG. 6. Photoelectron spectrum of free argon atoms[20] using HeI vuv radiation (21.22 eV) and a simulated hemispherical electron spectrometer[17].

the spin-orbit interaction is not resolved by use of an appropriate electron spectrometer, the polarizations of opposite sign for both unresolved peaks would almost cancel one another. A quantitative example is shown in Fig. 7 as the wavelength dependences of the dynamical spin parameters α and A for photoionization of xenon. The agreement of the experimental data (error bars[3]) with the theoretical predictions (RRPA solid curves[21], RPAE dashed curve[22]) is good.

One needs the spin-orbit interaction and its resolved splitting in Fig. 6 in order to get polarized photoelectrons. It is, however, worth noting that the magnitude of the electron-spin polarization

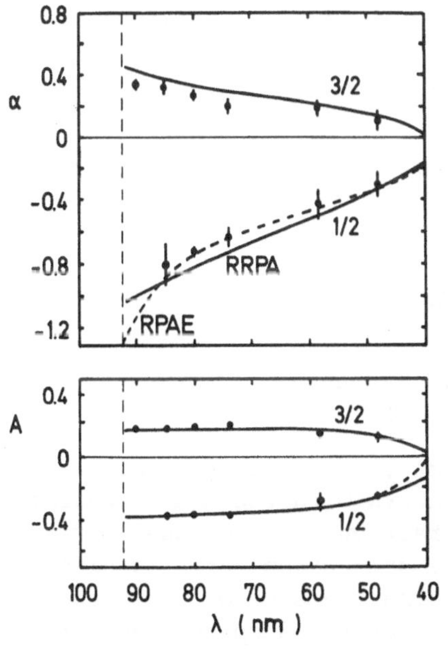

FIG. 7. Experimental results of the spin parameters α and A,[3] upper and lower part, respectively, as functions of the radiation wavelength for photoelectrons leaving the xenon ion in the $^2P_{3/2}$ and $^2P_{1/2}$ final states in comparison with theoretical predictions: RRPA[21], solid curve; and RPAE[22], dashed curve.

in both peaks does not depend on whether the spin-orbit interaction is strong or weak. While the fine-structure splitting in Xe^+ is seven times larger than in Ar^+, the magnitudes of the polarizations given by height and shape of the wavelength dependence of the dynamical spin parameter ξ shown in Fig. 8 are nearly the same for Ar, Kr and Xe. The only main difference in the three parts of Fig. 8 results in the different photoionization thresholds which shift the curves horizontally. Curves follow from calculations using RRPA (solid[23]), RPAE (dotted[22]) and MQDT (chained[19,24]).

Case 2 is discussed for the photoionization of mercury atoms as example:

$$\text{Hg } 6s^2\ (^1S_0) \longrightarrow \text{Hg } 6s\ (^2S_{1/2})\varepsilon p \left\{ \begin{array}{l} ^1P_1 \\ ^3P_1 \end{array} \right.$$

The photoionization transitions into the two energy degenerate continuum final states $^1P_1$ and $^3P_1$ are described by the singlet and triplet amplitudes D_S and D_T, respectively, as well as by the difference of the continuum-phase shifts $\delta_S-\delta_T$. In terms of the transition amplitudes and phases, the dynamical parameters read[12,25,26]:

$$Q = \tfrac{4}{3}\pi^2\alpha a_0^2\omega(D_S^2+D_T^2)$$

$$\beta = \frac{2D_S^2-D_T^2}{D_S^2 + D_T^2}$$

$$\xi = \frac{3\sqrt{2}D_S D_T \sin(\delta_S-\delta_T)}{4(D_S^2+D_T^2)}$$

$$A = \frac{D_T^2-2\sqrt{2}D_S D_T\cos(\delta_S-\delta_T)}{2(D_S^2+D_T^2)}$$

$$\alpha = \frac{-D_T^2-\sqrt{2}D_S D_T\cos(\delta_S-\delta_T)}{D_S^2 + D_T^2}$$

It is remarkable that the asymmetry parameter β depends incoherently upon the matrix elements with the consequence that neglecting the spin-orbit interaction ($D_T \equiv 0$) the "parity-favored" transition D_S yields $\beta = 2$. The spin parameter ξ is given by a single interference term containing the sine of the phase shift difference. It is worth noting, that all 3 spin-parameters ξ, A, α which are a measure for the 3 components of the spin-polarization vector are proportional to the "parity-unfavored" matrix element D_T. This means, that in this case the magnitude of the electron polarization is a measure of the strength of the spin-orbit interaction,

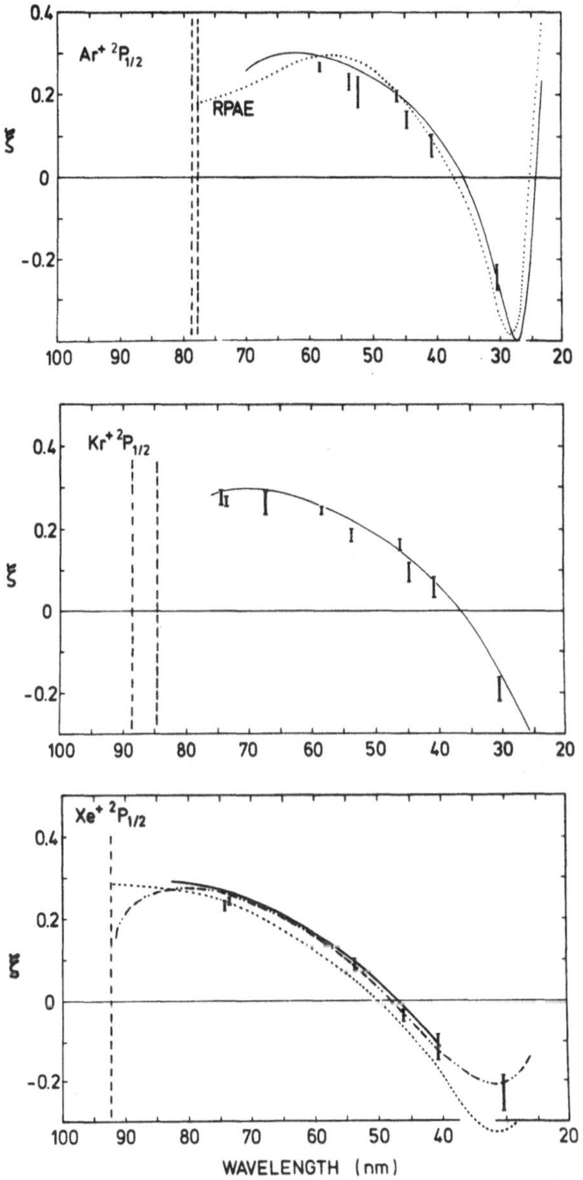

FIG. 8. Experimental results (error bars) of the spin parameter ξ for photoelectrons corresponding to the ionic state $^2P_{1/2}$ of Ar[13] (upper part), Kr[13] (middle part) and Xe[13,19] (lower part) in comparison with theoretical curves RRPA[23] (free), RPAE [22](dashed), MQDT[19,24] (chained). The vertical dashed lines represent the ionization thresholds.

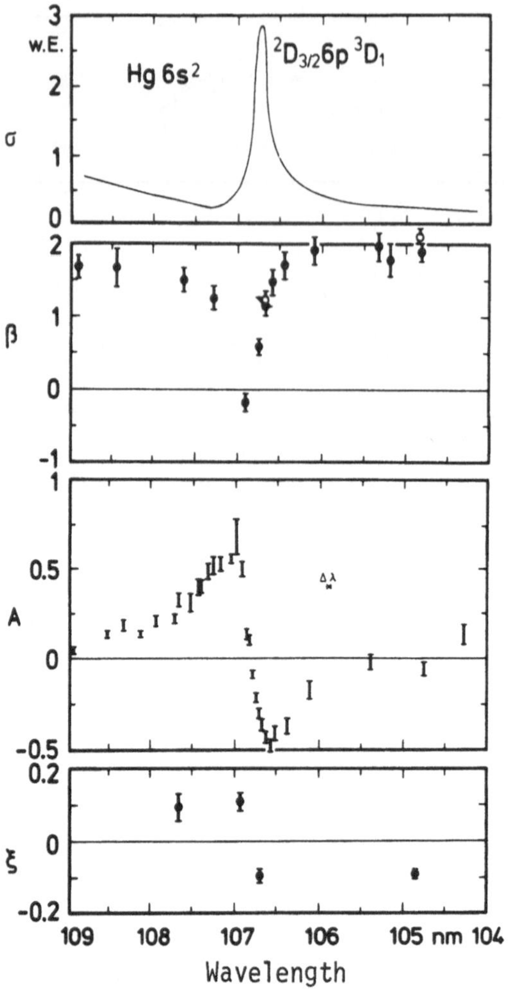

FIG. 9. Photoionization of mercury $6s^2$ in the autoionization region; cross section (upper part)[27]; asymmetry parameter β[28,29], spin parameters A and ξ[30].

which influences the photoionization process here in the final continuum state without a fine-structure splitting. Neglecting the spin-orbit coupling, all three spin parameters must vanish.

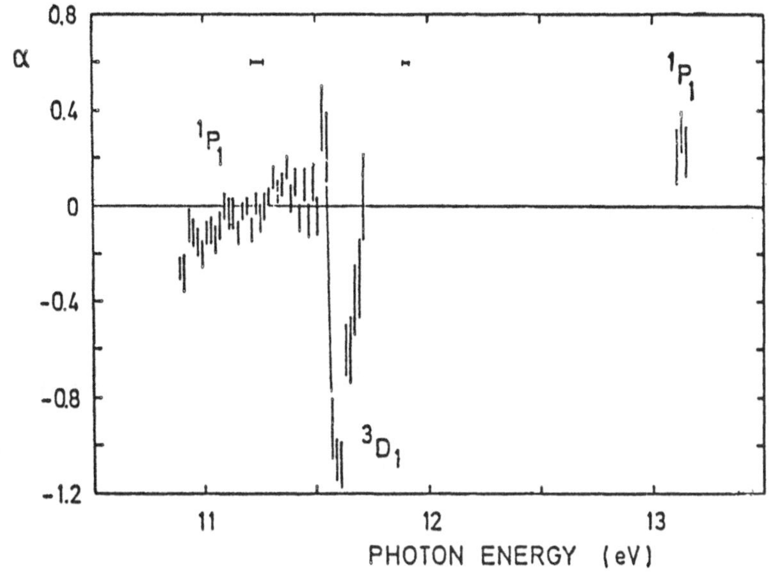

FIG. 10. The same as Fig. 9 but spin parameter α^{31}.

A strong enhancement of the influence of the spin-orbit interaction and thus a pronounced photoelectron-spin polarization has been seen in resonance regions, where effects of configuration interaction, channel mixing and many electron correlations play an important role. Fig. 9 and 10 give an example for the photoionization of mercury $6s^2$ in the autoionization region (via a virtual excitation of the $5d^{10}$ subshell into $5d^9 6p$) with respect to total photoionization cross section[27], the asymmetry parameter β[28,29], the spin parameters A[30], ξ[30] and α[31], respectively. All 5 dynamical parameters show a pronounced variation as function of the wavelength.

The combination of the data given in Figs. 9 and 10 allows to determine the matrix elements D_S and D_T as well as the phase-shift difference $\delta_S - \delta_T$ separately; the results[31] are shown in Fig. 11; the error bars contain the uncertainties of all experimental quantities involved. The singlet and triplet amplitudes show quite different behavior: the parity favored D_S follows the shape of the cross section and is always different from zero, whereas the unfavored D_T exhibits three changes of sign. In the two $^1P_1$ resonances, the phase difference (lower part of Fig. 11) varies only weakly across these resonances. For the $^3D_1$ resonance, however, we find completely different conditions. Here the triplet amplitude is negative and the phase shift difference between the singlet and triplet partial continuum waves shows a sudden change of sign which is typical for the variation of a relative phase across a resonance.

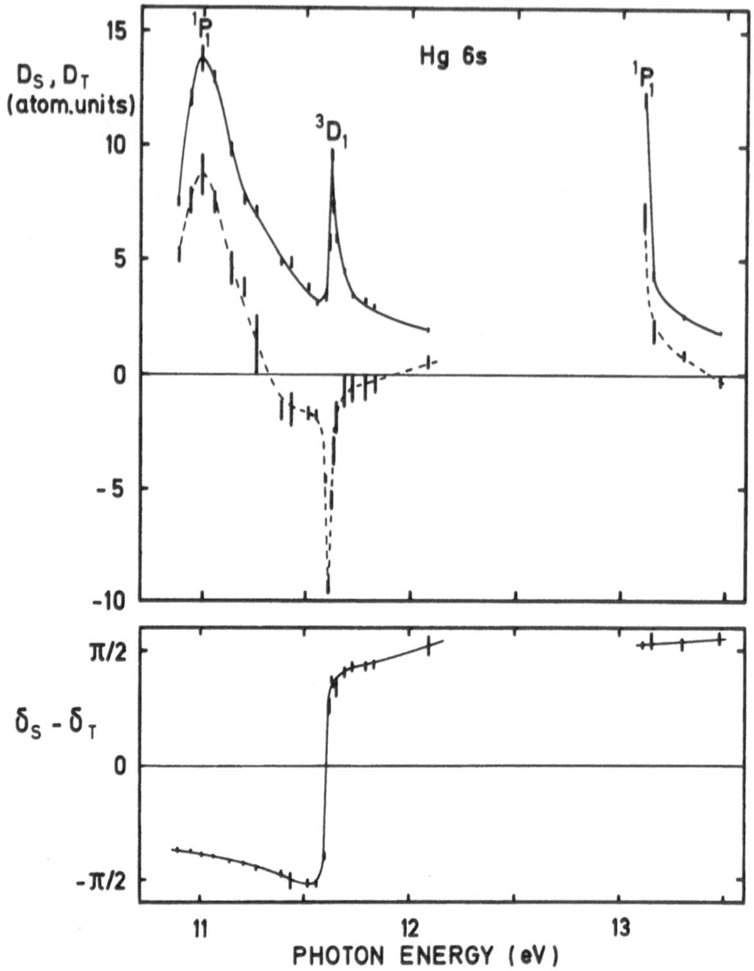

FIG. 11. Singlet and triplet matrix elements D_S and D_T (upper part) and corresponding phase shift difference (lower part)[31] obtained by use the data shown in Figs. 9 and 10. (curves: to guide the eyes).

In the same way as for the s-subshell of Hg the photoionization of atoms can also be characterized in terms of matrix elements and phase-shift differences in the case the ions show a fine structure splitting. But then, as discussed, the influence of the spin-orbit interaction onto the phases is in general a weak effect as for example seen in Fig. 12 showing the phase-shift difference between εf and εp partial waves in photoionization of the d subshell of Hg.[32] The full and open points with error bars correspond to the ionic states $^2D_{5/2}$ and $^2D_{3/2}$, respectively. The main contri-

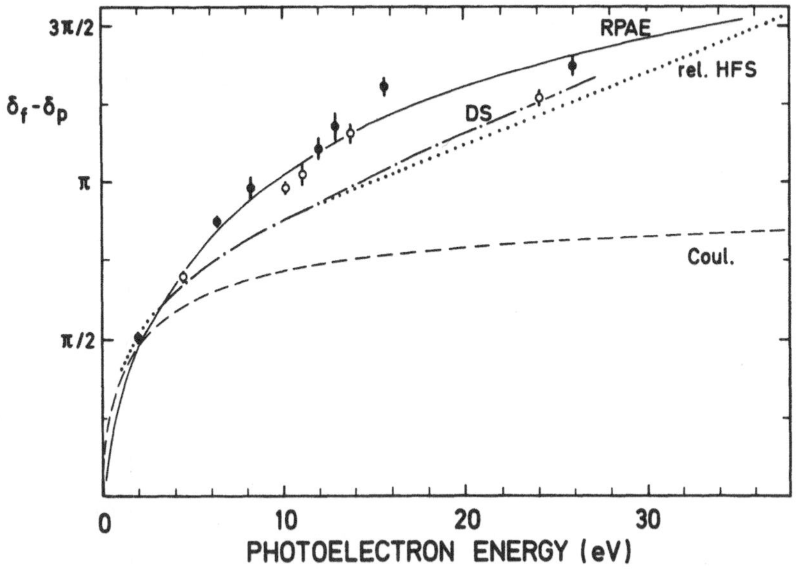

FIG. 12. Phase-shift difference between the εf and εp partial wave[32] in Hg 5d photoionization; full and open symbols correspond to the ionic states $^2D_{5/2}$ and $^2D_{3/2}$, respectively. Theoretical predictions Coulomb-phase difference $\sigma_f-\sigma_p-\pi$, relativistic Hartree Slater[33], Dirac Slater[34], RPAE with intertransition correlations between 5d → εp and 5d → εf.[35]

bution of the phase-shift difference especially close to the threshold comes from the Coulomb-phase shift, which varies by π/2 within 2.3 eV kinetic energy above the threshold. At higher energies effects of interchannel coupling play an important role as the comparison of the experimental data in Fig. 12 with the theoretical curves (rel. HFS[33], DS[34], RPAE[35]) shows.

The influence of parity-unfavored transitions is not only an important effect in autoionization resonances as discussed in Fig. 11 but also close to a Cooper minimum, where one matrix element changes its sign as function of the photon energy. This is the case for the $5p_{3/2}-\varepsilon d_{3/2}$ transition at xenon.[10,36] The corresponding parity unfavored matrix element[4], which is a measure for the influence of the spin-orbit interaction, shows a pronounced enhancement as Fig. 13 upper part demonstrates, which is not seen in any corresponding parity favored matrix element. The phase shift (lower part Fig. 13) of the unfavored transition with reference to the phase of the favoured $5p_{3/2}-\varepsilon s_{1/2}$ transition of xenon is very constant as function of the photon energy.

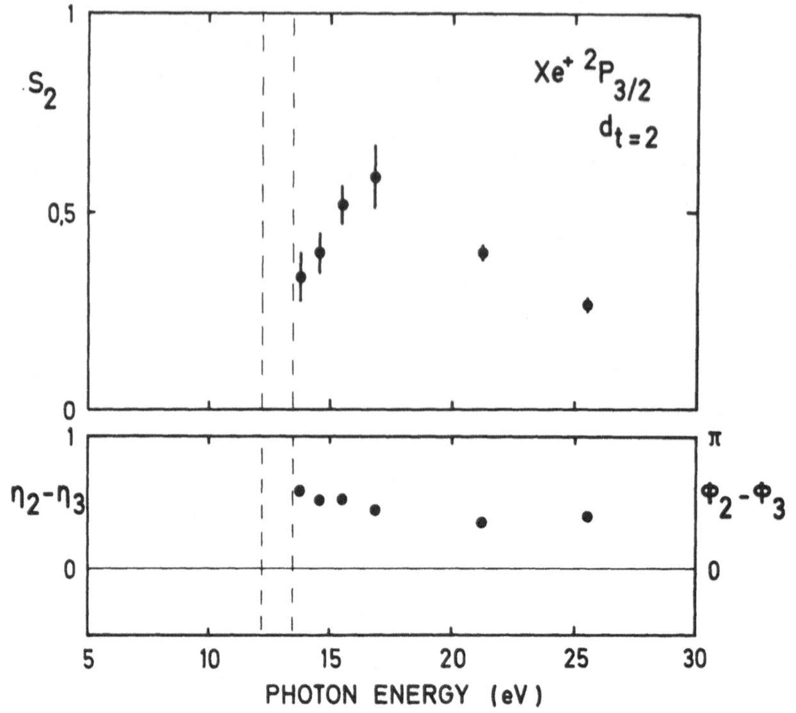

FIG. 13. Parity unfavored 5p → εd photoionization transition of Xe describing the strength of the influence of the spin-orbit interaction.[4] Upper part: reduced dipole-matrix element; Lower part: phase shift of the unfavored transition with reference to the phase of the favored s-transition. The dashed vertical lines represent the ionization thresholds.

MOLECULAR PHOTOIONIZATION

In molecular photoionization one has to take into account that the intramolecular Coulomb interaction is usually much stronger than the spin-orbit interaction. Therefore, it was believed over a period of several years that an electron polarization cannot occur in the photoionization of a randomly oriented molecular beam if one assumes the intramolecular axis as quantization axis the spin-polarization vector follows. But nevertheless, pronounced electron-polarization effects have been found[8] in the photoionization of randomly oriented halogen molecules by unpolarized radiation. Both cases discussed for atoms exist for

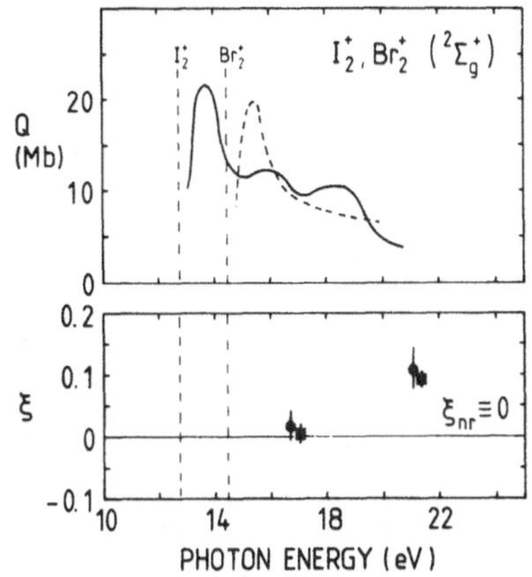

FIG. 14. Experimental results of the spin parameter ξ for photoelectrons leaving Br_2^+ (squares) and I_2^+ (circles) in their $^2\Sigma_g^+$ states[8] in comparison with the corresponding partial cross sections Q [38](Br_2 dashed, I_2 solid). The vertical lines indicate the adiabatic ionization thresholds.

molecular photoionization, too. Fig. 14 shows in the lower part the spin parameter ξ for photoelectrons leaving Br_2^+ (squares) and I_2^+ (circles) in their $^2\Sigma_g^+$ ionic state, where neither the ground neutral nor the final ionic state has any fine-structure splitting. The spin polarization, which occurs close to the photon energy where the cross section (Fig. 14 upper part) strongly decreases, is analogous to the well known Fano effect[1] in s-subshell ionization of alkali atoms. The dynamical spin parameters here are direct measure for the evidence of the spin-orbit interaction in the continuous spectrum.

Ionizing a π-orbital of halogens yields photoelectron spectra, which show a spin-orbit fine-structure splitting corresponding to the ionic total angular momentum 3/2 and 1/2 as in the rare-gas analogon. The behavior of spin polarizations and photoelectron intensities for the outermost orbitals of Br_2, I_2, CH_3Br and CH_3I is the most striking example studied in atomic and molecular photoionization with respect to the fact that photoelectron intensity data follow a certain theoretical prediction[37] - in our case the nonrelativistic model neglecting the influence of the spin-orbit

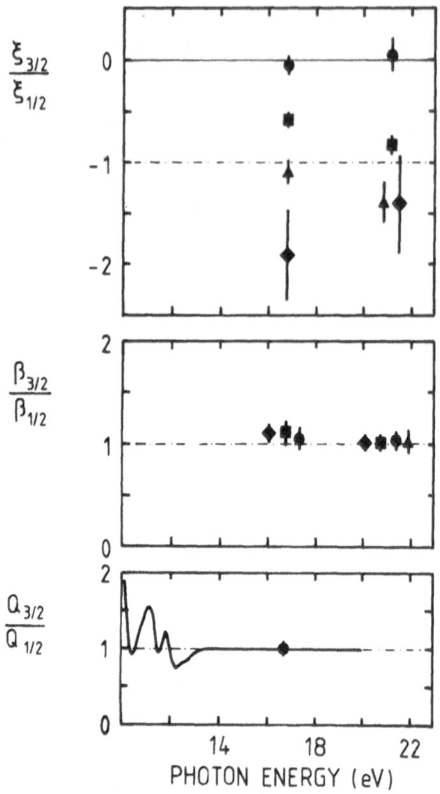

FIG. 15. Comparison of the ratios of the spin parameter ξ, asymmetry parameter β[8] and partial photoionization cross section Q[38] (for I_2 only) with the nonrelativistic predictions[37] (chain lines) for photoelectrons from the outermost orbitals of Br_2, I_2, CH_3Br, and CH_3I (squares, circles, triangles, and diamonds, respectively). The results correspond to photon energies of 16.85 and 21.22 eV.

interaction onto the molecular continuum states - whereas spin polarizations do not. Fig. 15 summarizes all experimental ratios of the spin parameter ξ, the asymmetry parameter β[8] and the partial cross section Q[38] for the spin-orbit components of these lone-pair orbitals. In all cases the ratio of β agrees with the theoretical prediction of +1 and the branching ratio $Q_{3/2}/Q_{1/2}$ (for I_2 only) is also identical to the statistical value over the energy range outside the threshold region. In contrast to this behavior of the differential cross section, the ratios of the spin parameters show a significant systematic deviation. While $\xi_{3/2}/\xi_{1/2}$ is close to -1 for CH_3Br (triangles) and not far from -1 for Br_2 (squares), it is zero for I_2 (circles), and tends to -2 for CH_3I (diamonds). In contrast to the cross sections, the spin polarizations are very sensitive to any phase shift of the continuum wave functions induced by the spin-orbit interaction. This, however, is stronger for heavier atoms in molecules than for lighter.

A very recent experiment of angular resolved photoelectron spectroscopy of free oriented CH_3I molecules has been performed for the first time.[9] CH_3I molecules in a supersonic beam have been oriented with respect to the molecular axis parallel to an external field by use of an electric hexapole in a "Stern-Gerlach" type analogous experiment. The photoelectrons ejected by vuv radiation in a region of very weak field (0.3 V/cm) from the lone-pair orbital at the iodine atom show a pronounced asymmetry in intensities depending on whether they are emitted parallel or antiparallel to the intramolecular axis. If the methyl group is directed toward the electron spectrometer, a photoelectron current I^+ is detected, if the iodine atom is directed, a current I^-. Fig. 16 shows the asymmetries I^-/I^+ measured for both spin-orbit components in the photoelectron spectrum and for two vuv photon energies (NeI and HeI light) as function of the focussing voltage in the hexapole.[9] The heights of the full points with error bars are roughly proportional to the degree of molecular orientation which has been estimated to be between 0.24 and 0.40. Comparing this degree of orientation with the asymmetry ratios found, the forward backward photoelectron-emission asymmetry parallel to the molecular axis must be a pronounced effect for a complete orientation of the molecules.

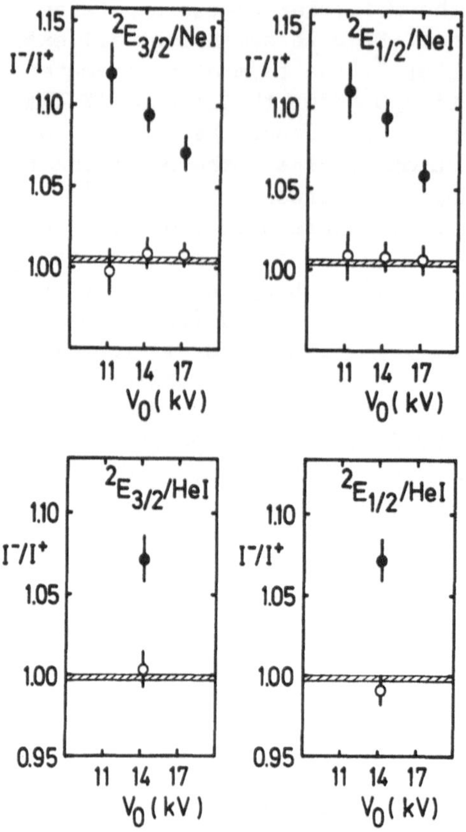

FIG. 16. Experimental results of the intensity asymmetry of photoelectrons emitted angular resolved parallel or antiparallel to the molecular axis of a free oriented CH_3I molecule (full points with error bars).[9] The open points and the dashed areas represent the corresponding results with a randomly oriented molecular beam showing the apparatus-related asymmetries.

PHOTOEMISSION FROM ATOMS ADSORBED ON SOLID SURFACES

Using circularly polarized synchrotron radiation at BESSY spin polarized photoemission from the valence orbitals of Xe and Kr atoms physisorbed on the Pt(111) single-crystal surface has been studied for normal light incidence and normal (angular resolved) emission. Two spin-resolved photoemission spectra[5,6] are shown in Fig. 17 for Kr and Xe monolayers adsorbed on Pt(111). The peak at lowest binding energy (1) has nearly complete negative spin polarization and corresponds to the $p_{3/2}|m_j| = 3/2$ hole state of the rare gas atoms, whereas peaks 2 and 3 are highly positive polarized ($|m_j| = 1/2$). These polarization values quantitatively correspond to the experimental results in the gas phase[3] (Fig. 2 middle part $\theta = 0$) except that for free atoms there is no ener-

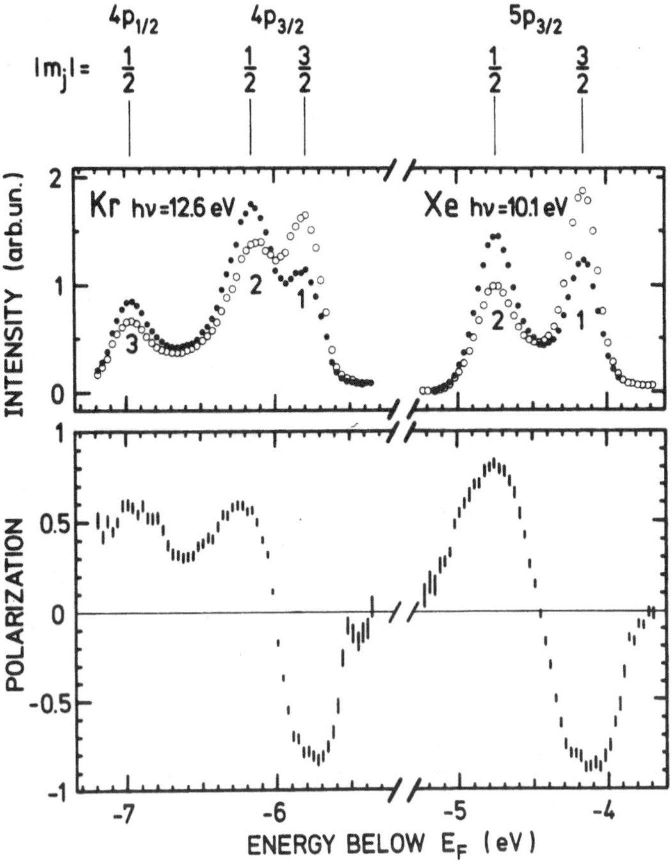

FIG. 17. Spin-resolved photoelectron spectra of Kr and Xe monolayers at full coverage on Pt(111) in normal photoemission.[5,6] Upper part, intensities scattered into the two counters 3 and 4 of the Mott detector (Fig. 3) as full and open circles. Lower part, photoelectron-spin polarization obtained from the count rates in the upper part, normalized to a complete circular photon polarization.

getic splitting of the m_j substates. These spin-polarization results confirm experimentally the peak assignment proposed in the literature[39] as shown in Fig. 17 which indicates that the m_j splitting is caused by lateral Xe-Xe interactions.

Fig. 18 gives two examples of spin-polarization data obtained for the different peaks in Fig. 17 plotted as function of the photon energy for an incommensurate hcp and a commensurate $\sqrt{3}$ layer of Xe. The polarization shows pronounced resonance structures which partly correspond with structures of the photoelectron intensities measured and shown in the upper part of Fig. 18. These structures which are discussed in more detail elsewhere[5,6] may partly be due

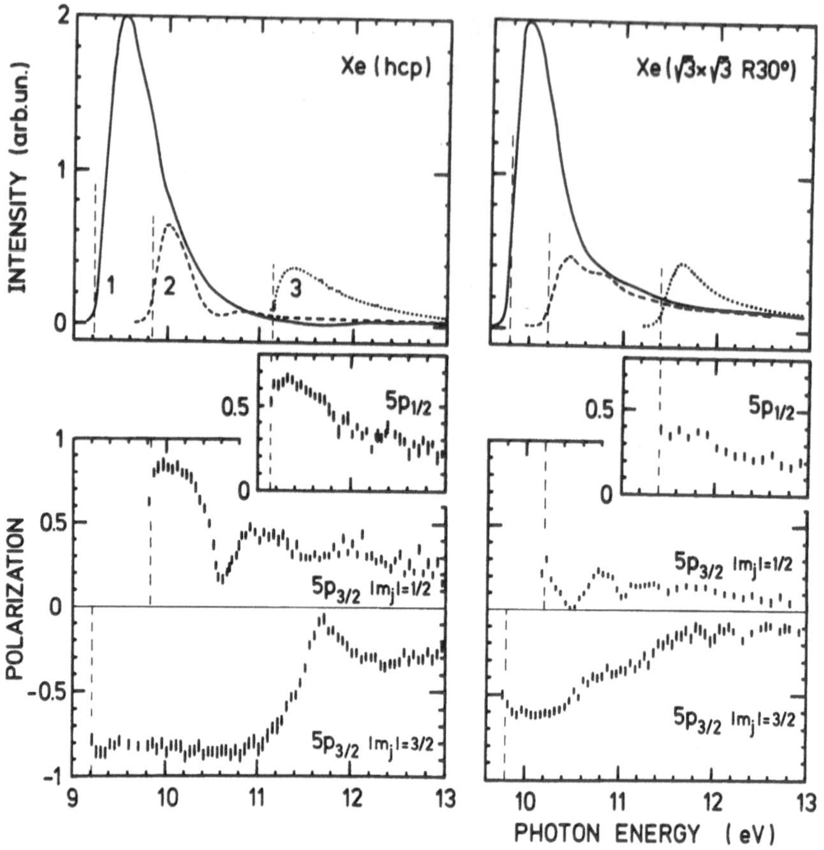

FIG. 18. Photoelectron intensities (upper part) and spin polarizations (lower part) of the Xe adsorbate photoemission peaks (normal emission) as function of the photon energies for hcp (incommensurate layer) and the $\sqrt{3}$ commensurate layer.[5,6] Peaks 1, 2, 3 are numbered within increasing binding energies (vertical dashed lines) corresponding to the rare gas hole states $5p_{3/2,1/2}$.

to atomic effects like autoionization resonances (the analogous spin parameter A of free xenon atoms measured[19] in comparison with theoretical curves[12,19,40] and the photoionization-cross section Q[19] are shown in Fig. 19) or Cooper minima or due to typical surface effects like electron diffraction patterns or resonances induced by the surface barrier. Further studies of different adsorbates on different substrates with different crystallographic structure and interatomic distances shall allow to answer these questions more quantitatively.

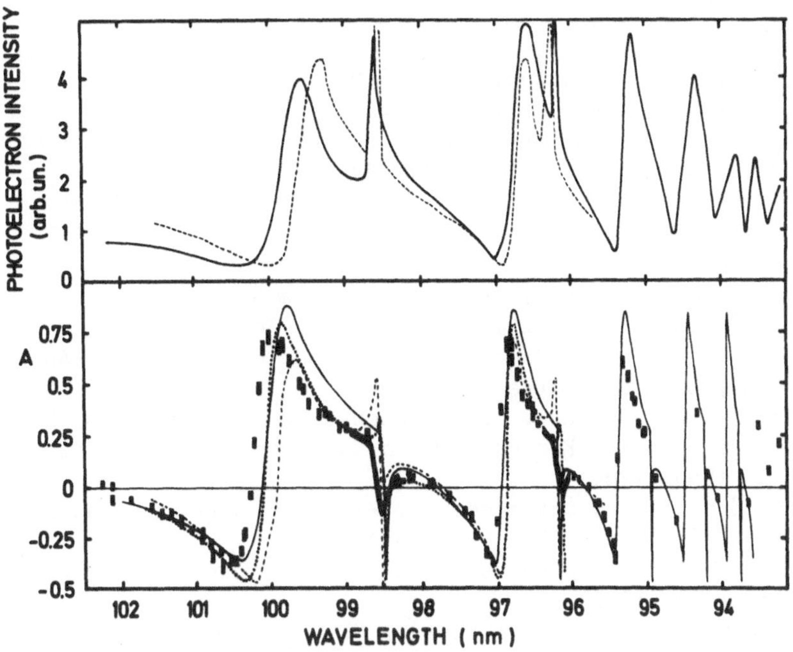

FIG. 19. Photoionization of Xe atoms in the autoionization range: (top) cross section, photoelectron intensity; (bottom) spin polarization parameter A. Experimental results (error bar rectangles lower part and full curve upper part[19]); theoretical curves: dashed[12], dotted[19,24], full[40].

OUTLOOK AND ACKNOWLEDGEMENT

It is the purpose of the angle- and spin-resolved photoelectron spectroscopy to find a set of non-redundant experimental data which characterize the photoeffect quantummechanically completely. This has been shown for atoms successfully. To build a quantitative bridge from the free atoms, via the free randomly oriented molecules up to the three dimensional crystal[7] will be the main topic of the angle- and spin-resolved photoemission studies in the future. Thus atomic physics can become an applied method to study and to understand more complicated systems like condensed matter. There is no doubt, that correlation effects studied in details for atoms play an important role there, too.

The author wishes to express his thanks to the coworkers Drs. A. Eyers, Ch. Heckenkamp, S. Kaesdorf, F. Schäfers, and G. Schönhense for the measurements performed and many intensive discussions. Support by the BMFT, DFG, and MPG is gratefully acknowledged.

REFERENCES

1. U. Fano, Phys. Rev. 178, 131 (1969)
2. U. Heinzmann, Appl. Opt. 19, 4087 (1980)
3. Ch. Heckenkamp, F. Schäfers, G. Schönhense, and U. Heinzmann, Phys. Rev. Lett. 52, 421 (1984)
4. Ch. Heckenkamp, F. Schäfers, and U. Heinzmann, 2. ECAMP Amsterdam 1985, book of abstracts
5. G. Schönhense, A. Eyers, U. Friess, F. Schäfers, and U. Heinzmann, Phys. Rev. Lett. in press (1985)
6. G. Schönhense, A. Eyers, U. Friess, F. Schäfers, and U. Heinzmann, Surface Science Symposium Obertraun (Austria) 1985, book of abstracts
7. A. Eyers, F. Schäfers, G. Schönhense, U. Heinzmann, H. P. Oepen, K. Hünlich, J. Kirschner, and G. Borstel, Phys. Rev. Lett. 52, 1559 (1984)
8. G. Schönhense, V. Dzidzonou, S. Kaesdorf, and U. Heinzmann, Phys. Rev. Lett. 52, 811 (1984)
9. S. Kaesdorf, G. Schönhense, and U. Heinzmann, Phys. Rev. Lett. in press (1985)
10. Ch. Heckenkamp, doctoral thesis, F. University of Berlin (1984), unpublished
11. N. A. Cherepkov, Zh. Eksp. Teor. Fiz. 65, 933 (1973) (Sov. Phys. JETP 38, 463 (1974))
12. C. M. Lee, Phys. Rev. A 10, 1598 (1974)
13. U. Heinzmann, G. Schönhense, and J. Kessler, Phys. Rev. Lett. 42, 1603 (1979) and J. Phys. B 13, L 153 (1980)
14. G. Schönhense, Phys. Rev. Lett. 44, 640 (1980)
15. B. Brehm, Z. Phys. 242, 195 (1971)
16. A. Eyers, Ch. Heckenkamp, F. Schäfers, G. Schönhense, and U. Heinzmann, Nucl. Instrum. Meth. 208, 303 (1983)
17. K. Jost, J. Phys. E 12, 1006 (1979)
18. J. Kessler, "Polarized Electrons", Springer, Berlin (1976)
19. U. Heinzmann, J. Phys. B 13, 4353 (1980)
20. G. Schönhense, doctoral thesis, University Münster (1981), unpublished
21. K. N. Huang, W. R. Johnson, and K. T. Cheng, At. Data Nucl. Data Tabl. 26, 33 (1981)
22. N. A. Cherepkov, J. Phys. B 12, 1279 (1979)
23. K. N. Huang, W. R. Johnson, and K. T. Cheng, Phys. Rev. Lett. 43, 1658 (1979)
24. J. Geiger, Z. Phys. A 282, 129 (1977) and private communication (1979)
25. H. Klar, J. Phys. B 13, 3117 (1980)
26. K. N. Huang and A. Starace, Phys. Rev. A 21, 697 (1980)
27. B. Brehm, Z. Naturforsch. 21a, 196 (1966)
28. B. Brehm and K. Höfler, Phys. Lett. 68A, 437 (1978)
29. K. Höfler, PhD-thesis, University Hannover (1979), unpublished
30. F. Schäfers, G. Schönhense, and U. Heinzmann, Z. Phys. A 304, 41 (1982)
31. G. Schönhense, F. Schäfers, Ch. Heckenkamp, U. Heinzmann, and A. M. Baig, J. Phys. B 17, L 771 (1984)

32. G. Schönhense and U. Heinzmann, Phys. Rev. A 29, 987 (1984)
33. Y. S. Kim, R. H. Pratt, A. Ron, and H. K. Tseng, Phys. Rev. A 22, 567 (1980)
34. F. Keller and F. Combet-Farnoux, J. Phys. B 12, 2821 (1979) and 15, 2657 (1982)
35. K. Ivanov, S. Yu. Medvedev, and V. A. Sosnivker, unpubl. Report No 615 (1979) A. F. Ioffe Phys. Techn. Inst. Leningrad
36. U. Heinzmann, J. Phys. B 13, 4367 (1980)
37. N. A. Cherepkov, J. Phys. B 14, 2165 (1981)
38. J. H. Carver and J. L. Gardner, J. Quant. Spectrosc. Radiat. Transfer 12, 207 (1972)
39. K. Horn, M. Scheffler, and A. M. Bradshaw, Phys. Rev. Lett. 41, 822 (1978)
40. W. R. Johnson, K. T. Cheng, K. N. Huang, and M. LeDourneuf, Phys. Rev. A 22, 989 (1980)

ATOMIC PROCESSES UNDER STRONG ELECTROMAGNETIC FIELDS*

P. Lambropoulos

University of Crete and Research Center of Crete
Iraklion, Crete, Greece

and

Physics Department, University of Southern California
Los Angeles, CA 90089-0484, USA

*Supported in part by the National Science Foundation
Grant No. PHY-8306263

PROLOGUE

My initial intention was to concentrate on multiphoton autoionization and the effects of laser intensity on electron correlation. Listening to the other lecturers, however, I was overcome by the irresistible temptation to show how many of the topics they discussed (photoelectron angular distributions, spin-polarization, double ionization, etc.) have their multiphoton counterparts with even larger variety and complexity. Thus, changing my original plan, I presented a general introduction to the field, followed by a collection of topics which illustrate their strong connection with traditional atomic physics as well as their novel character due to the strong and nonlinear coupling to the radiation. Having had the advantage of a week of listening and commenting, I included some topics upon the stimulation of spirited but friendly discussions. In addition, I devoted a substantial part of the lectures to three important current topics: Above threshold ionization, multiple ionization and multiphoton autoionization.

The material in this article has followed the lectures. It represents an overview but not an in depth review of the field. As a result, the references cover only a small part of the vast literature. They do, however, provide a more than adequate collection for a good first acquaintance with the field. In a way, this article is a guided tour through these references without the technical details. More technical and elaborate reviews can be found in two recent volumes which the interested reader will locate in the list of references.

I. MULTIPHOTON PHENOMENA AND THEIR THEORETICAL DESCRIPTION

The interaction of intense electromagnetic radiation with electrons bound in atoms, molecules or surfaces leads to qualitatively new behavior which in general can not be understood in terms of traditional weak-field experience. As a result, a new field has emerged which I will refer to as intense field physics. The intensities at which such behavior is to be expected range from about 10^8 W/cm² to 10^{17} W/cm² or more and are available at frequencies ranging from infrared to VUV, with rapid progress being made towards the XUV. The corresponding photon fluxes after focusing into the interaction region range from 10^{26} to 10^{36} photons/cm² sec. The laser sources producing such powers are pulsed with pulse durations from 10^{-8} to 10^{-12} sec, with even shorter pulses (subpicosecond to femtosecond) becoming more readily available. Intensities up to 10^{12} W/cm² can be obtained with commercially available lasers, while in many experiments, the intensity is routinely of the order of 10^{10} W/cm². To get a feeling for the size of this intensity, which roughly corresponds to a photon flux of $I=5\times10^{28}$ photons/cm² sec, let us take a rather small photoionization cross section $\sigma = 10^{-20}$ cm², and a pulse duration $\tau_L = 5\times10^{-9}$ sec. The probability of single-photon ionization $P = \sigma I \tau_L$ during the laser pulse under these conditions is 2.5 which is larger than 1. This means that all atoms in the interaction region will be ionized in a typical experiment. For the sake of comparison, it is worth keeping in mind that for a traditional source or even synchrotron, the photon flux is at most of the order of 10^{15} within a similar bandwidth and the interaction time may be as large as 10^{-5} sec. This leads to a transition probability $P \simeq 10^{-10}$.

To set the stage for some of the phenomena in this field, let us consider the sequence of transitions depicted in Fig. 1. First, we assume that the frequency ω_1 is such that $\hbar\omega_1 = E_1-E_0$, where the figure depicts the energy levels and ionization potential of some atom, and that $2\hbar\omega_1$ is larger than the ionization potential $E_\infty-E_0$. If σ_0 is the cross section for one-photon absorption from $|0\rangle$ to $|1\rangle$, σ_1 the photoionization cross section, τ_1 the lifetime of state $|1\rangle$, and I_1 the photon flux, the overall transition probability per unit time for absorption and subsequent ionization is

$$W_2 = (\sigma_0 I_1)\tau_1(\sigma_1 I_1) = (\sigma_0\tau_1\sigma_1)I_1^2 \equiv \hat{\sigma}_2 I_1^2 \qquad (1)$$

where we have lumped all atomic parameters into a new parameter $\hat{\sigma}_2$. If, for the sake of a qualitative estimate, we take $\sigma_0 \simeq 10^{-16}$ cm², $\sigma_1 \simeq 10^{-20}$ cm², and $\tau_1 = 10^{-7}$ sec, we obtain $\hat{\sigma}_2 = 10^{-43}$ cm⁴ sec. We have just derived the simplest case of a multiphoton process, namely resonant two-photon ionization. For a flux as low as $I_1 = 10^{25}$ and a pulse duration $\tau_L \simeq 5\text{x}10^{-9}$ sec, the probability of ionization during the pulse is $\tau_L W_2 = 5\text{x}10^{-2}$ which means 5% of the atoms in the interaction volume are ionized. Note that for a conventional source with flux at most 10^{15} and interaction time 10^{-5} sec the corresponding probability is 10^{-18}, which makes the process virtually unobservable.

To consider a case only slightly more general, assume a photon frequency ω which, as indicated in Fig. 1, does not coincide with any atomic transition frequency. Two-photon ionization is again possible, if the intensity is sufficiently large. The mathematical description of the process is now

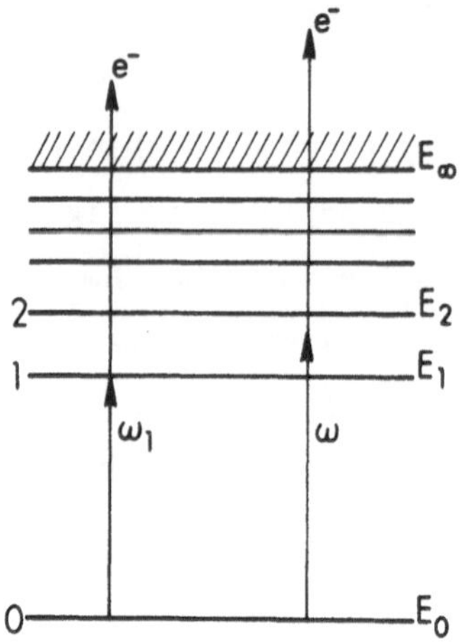

Fig. 1. Resonant and non-resonant 2-photon ionization

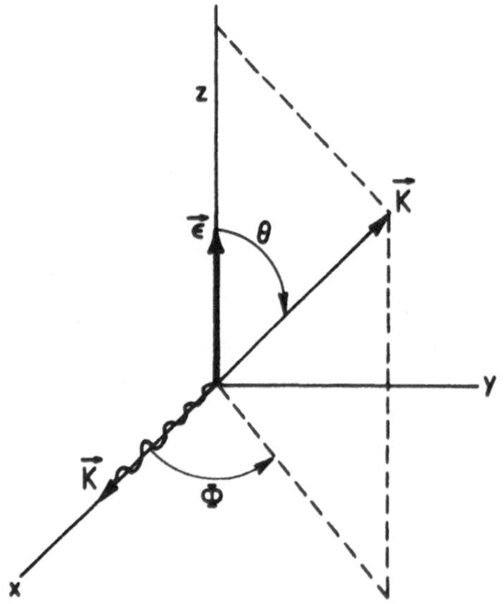

Fig. 2. System of coordinates relative to light propagation and polarization

considerably more complicated. A qualitative argument, however, can again give the flavor. The absorption of the first photon, in this case, does not lead to a real atomic state but to a virtual one $|i\rangle$ which, in general, is a linear superposition of all atomic states to which $|0\rangle$ is connected via an electric dipole (E1) transition. Let $|i\rangle = \sum_n \alpha_n |n\rangle$, where $|n\rangle$ indicates such states with the coefficients α_n determined through a mathematical procedure which we sidestep for the moment. If we view $|i\rangle$ as a short-lived state, whose lifetime τ_i, roughly speaking, is $1/\Delta$ where Δ is the detuning of ω from the nearest of the states $|n\rangle$, and we assign a cross section σ_0 for excitation from $|0\rangle$ to $|i\rangle$, and σ_i for ionization from there, we can write the equivalent of Eq. (1) as

$$W_2 \cong (\sigma_0 \frac{1}{\Delta} \sigma_i) I_1^2 = \hat{\sigma}_2 I_1^2 \qquad (2)$$

If ω is completely off-resonance, say half-way between states $|1\rangle$ and $|2\rangle$, then $1/\Delta$ is of the order of 10^{-14} sec. or even shorter. This change alone, would reduce $\hat{\sigma}_2$ by seven orders of magnitude to the value 10^{-50} $cm^4 sec.$, which in fact is not too far from the typical value of a non-resonant two-photon ionization generalized cross-section $\hat{\sigma}_2$.

The above argument, even though it has given a remarkably close approximation to an order of magnitude estimate can offer nothing else. Even the presence of σ_0 and σ_i as cross sections is extremely tenuous since no real atomic state is reached as an intermediate step. From the definition of $|i\rangle$, it is clear that a better way of looking at ionization is to calculate a bound-free cross-section through the

matrix element $\langle f(\vec{K})/e\vec{\varepsilon}\cdot\vec{r}/i\rangle$ where $e\vec{\varepsilon}\cdot\vec{r}$ is the electric dipole operator for radiation with polarization vector $\vec{\varepsilon}$ and $f(\vec{K})$ a continuum state of wave-vector $\vec{K}$ and energy $E = \hbar K^2/2m$. It remains to determine the coefficients a_n which must be related to the strength of dipole transitions from $|0\rangle$ to $|n\rangle$ and to the detuning $\hbar\omega-(E_n-E_0)$.

In fact, a systematic calculation[1] leads to the expression $a_n = \frac{\langle n|\vec{\varepsilon}\cdot\vec{r}|0\rangle}{E_n-E_0-\hbar\omega}$ which, with the inclusion of all appropriate constants, gives

$$\hat{\sigma}_2 = \frac{(2\pi\alpha)^2}{4\pi^2} \frac{mK}{\hbar} \omega^2 \int |M_{f0}^{(2)}|^2 \, d\Omega_{\vec{K}} \tag{3}$$

where

$$M_{f0}^{(2)} = \sum_n \frac{\langle f|\vec{\varepsilon}\cdot\vec{r}|n\rangle\langle n|\vec{\varepsilon}\cdot\vec{r}|0\rangle}{\frac{1}{\hbar}(E_n-E_0)-\omega} \tag{4}$$

with α being the fine structure constant and $d\Omega_{\vec{K}}$ the differential angle in the direction of propagation $\vec{K}$ of the outgoing photoelectron. It is assumed here that the one-electron bound states are written in the form $R_n(r)Y_{\ell m}(\delta,\phi)$ while the continuum state $f_{\vec{K}}(r)$ is as usual expanded in partial waves according to

$$f_{\vec{K}}(\vec{r}) = 4\pi \sum_{L=0}^{\infty} i^L e^{-i\delta_L} G_L(Kr) \sum_{M=-L}^{L} Y_{LM}^*(\theta,\Phi)Y_{LM}(\delta,\phi) \tag{5}$$

where δ_L is the total phase shift of the L partial wave and $G_L(Kr)$ the corresponding radial part. The vectors $\vec{r}$ and $\vec{K}$ are defined by the spherical coordinates (r,δ,ϕ) and (K,θ,Φ), respectively, in a system of coordinates shown in Fig. 2 for light linearly polarized. If the integration over angles is not carried out in Eq. (3), the angle dependent quantity $|M_{f0}^{(2)}|^2$ represents the angular distribution of the emitted photoelectrons, to which we return later on. It is worth pointing out at this point that if we define the Green's function

$$G(\vec{r},\vec{r}';E) = \sum_{m} \frac{|n(\vec{r})\rangle \langle n(\vec{r}')|}{(E_n - E)} \tag{6}$$

where E is an arbitrary energy parameter, then $M_{f0}^{(2)}$ is also written as

$$M_{f0}^{(2)} = \int d^3\vec{r} \int d^3\vec{r}' \; f^*_{\vec{K}}(r)(\vec{\varepsilon}\cdot\vec{r})G(\vec{r},\vec{r}';E_0 + \hbar\omega)(\vec{\varepsilon}\cdot\vec{r}')\psi_0(\vec{r}') \tag{7}$$

with the Green's function being evaluated at $E = E_0 + \hbar\omega$.

The calculation of a two-photon transition, therefore, involves an infinite summation over intermediate states including the continuum, as does the Green's function. Writing the expression in terms of $G(\vec{r},\vec{r}';E)$ allows for the possibility of evaluating $M^{(2)}$ by determining G first (for example, through the solution of a differential equation) and then performing a double integration, thus avoiding the explicit summation over n. The reader will notice that when the photon frequency is resonant with one internediate state, one of the denominators of the infinite summation vanishes and the corresponding term dominates the summation. The easiest way out of the resulting singularity is to replace E_n by $E_n + \frac{i}{2}\gamma_n$ where γ_n is the width of the excited state $|n\rangle$. After some rearrangement of constants, one can thus arrive at the equivalent of Eq. (1) as a special case of the more general equation. This derivation is far from rigorous and we return to this resonant case later on.

It should be evident now that 3- or N-photon processes can be obtained by a direct generalization of the arguments presented above. Before writing a general result for an N-photon process, we note that for 3-photon ionization the transition amplitude $M_{f0}^{(3)}$ has the form

$$M_{f0}^{(3)} = \sum_{n}\sum_{m} \frac{\langle f|\vec{\varepsilon}\cdot r|n\rangle \langle n|\vec{\varepsilon}\cdot r|m\rangle \langle m|\vec{\varepsilon}\cdot r|0\rangle}{(\hbar^{-1}(E_n - E_0) - 2\omega)\;(\hbar^{-1}(E_m - E_0) - \omega)} \tag{8}$$

which in terms of the Green's function can be written as

$$M_{fo}^{(3)} = \int d^3\vec{r}_3 \int d^3\vec{r}_2 \int d^3\vec{r}_1 \; f_{\vec{K}}^{*}(\vec{r}_3)(\vec{\varepsilon}\cdot\vec{r}_3)G(\vec{r}_3,\vec{r}_2;E_2)(\vec{\varepsilon}\cdot\vec{r}_2)$$

$$G(\vec{r}_2,\vec{r}_1;E_1)(\vec{\varepsilon}\cdot\vec{r}_1)\psi_0(\vec{r}_1) \; , \qquad (9)$$

from which one can easily guess the generalization to a process of any order. The generalized cross-section $\hat{\sigma}_N$ for N-photon ionization can thus be written[1] as

$$\hat{\sigma}_N = \frac{(2\pi\alpha)}{4\pi^2} \frac{mK}{\hbar}\omega^N \int |M_{fo}^{(N)}|^2 \, d\Omega_{\vec{K}} \qquad (10)$$

and the transition probability per unit time is given by

$$W_N = \hat{\sigma}_N I^N \, . \qquad (11)$$

In cgs units, $\hat{\sigma}_N$ is measured in $cm^{2N}\ sec^{N-1}$. For example, in such units, a typical 3-photon ionization cross section is 10^{-79} $cm^6\ sec^2$. Since it takes three photons to bridge the ionization potential in this case, there are two types of resonance with intermediate states possible: Single-photon resonance (from the initial to some intermediate state) or two-photon resonance (from the initial to some intermediate state which now has the same parity as the initial state). Obviously, with increasing N, the number of possibilities for resonances with intermediate states increase. In principle, if one could employ N-1 lasers with different frequencies, one could achieve N-1 simultaneous resonances. In practice, however, the simultaneous use of more than two lasers is technically extremely difficult. On the other hand, the successive excitation of molecular vibrational states with a single laser (to the extent that anharmonicity allows) represents a multiphoton transition involving several (near-) resonances. But as far as atoms are concerned, we only expect one m-photon resonance with m possibly changing as we scan a single laser in an N-photon ionization process.

The calculation of a generalized cross-section $\hat{\sigma}_N$ requires a method for the accurate performance of the infinite summations in addition to having good wave functions. A good potential does of course enable one to do both, but that is easier said than done. Any method that enables one to calculate a Green's function would be useful in these calculations. Single-channel quantum defect theory provides, for example, such a technique which has been employed rather widely in calculations of this type. The fundamental limitation of a quantum defect Green's function is the one-electron approximation on which it is based. There is, of course, multichannel quantum defect theory (MQDT), but its use in the construction of a Green's function for the performance of the multiple infinite summations is not practical. In that case, one can simply employ MQDT to calculate the matrix elements appearing in the infinite summations and limit the summation to a finite number of terms. This method of truncated summation usually yields results whose usefulness depends on how well the truncated summation can be shown to converge. Otherwise, it can only be viewed as a technique yielding results of only qualitative accuracy. This is not the place to review the techniques for such calculations. The interested reader is referred to an earlier article[1] where these methods have been reviewed. For most such methods a compromise has to be made between an exact infinite summation of terms with less than exact matrix elements and a truncated (or somehow not exact summation) of exact matrix elements. The type of the problem and the conditions of the experiment usually provide a guide as to which is preferable.

Considerable experience has accumulated by now on the calculation of 2- to 4-photon ionization generalized cross sections of atoms for which a one-electron model is expected to be reasonably valid. The alkali atoms are of course the ideal candidates and that is where most of the calculations are, although some calculations for rare gases have also been published. Absolute measurements of $\hat{\sigma}_N$ are extremely rare, mainly because it is technically difficult to determine the absolute number of atoms in the interaction region owing to the difficulty in determining the exact interaction region itself. The reader should remember that significant focusing is necessary, especially for processes of higher order. The existing few absolute measurements, as well as other evidence, have provided the opportunity to test some of the calculations and to build some intuition and confidence as to the size of the relevant parameters.[2-7] Even in He which is not expected to behave as a one-electron atom, the comparison between theory and experiment has been quite satisfactory.[6,7] The success is probably due to the fact that the relevant experiment involved ionization of the 2s excited state which is not too far from a one-electron state. For the alkali atoms, the theory has in fact been sufficiently good to provide the correct guidance[3] in cases where some experimental results have shown extremely large unexplainable discrepancies from theoretical predictions.[8,9] On the other hand, the quantitative comparison between theory and experiment is not always at the level of 20% but more like factors of 2 or 3. Again, the difficulties inherent in such experiments have prevented until now more quantitative comparisons.

Part of the other evidence mentioned above has to do with the ratio of total ionization for light linearly polarized to that for circularly polarized. This is an effect peculiar to multiphoton ionization, since for single-photon ionization, simply changing the light polarization does not affect the total ionization signal, as long as the initial state of the target atom is spherically symmetric. The simplest way of showing why it will be different for multiphoton ionization is to note that once the first photon is absorbed, the second photon "sees" an atom no longer spherically symmetric, because it already has the information of the polarization of the first photon; and so on. It does therefore make a difference whether the light is circularly, linearly or otherwise polarized. Another way of saying the same thing is through Fig. 3 where the channels of multiphoton transitions from an S-state are shown for linearly and circularly polarized radiation. The selection rule $\Delta m=0$ for linear polarization and $\Delta m=\pm 1$ for circular polarization is what causes the difference in the channels. Thus it is not too surprising that the total rates will be different for different polarizations. The exact value of the ratio does of course depend on the details of the atomic structure and that is why such measurements provide a valuable test of calculations, without requiring absolute cross section measurements. It is worth noting at this point that, although linearly polarized light can be written as a superposition of circularly polarized light, the total multiphoton rate for linear polarization can not be obtained from the linear superposition of rates for opposite circular polarizations. That is so because the process depends on the field non-linearly, which again is one more important difference between single- and multiphoton processes.

Let us now consider briefly the frequency dependence of multiphoton ionization. As the radiation frequency is varied, whenever an m-photon resonance with an intermediate state occurs, the generalized cross section peaks. Around such an intermediate resonance, the cross section exhibits a line-shape which results from the interference of the transition amplitude through that state with the non-resonant amplitude through all other states. This line-shape will usually be asymmetric of the typical form shown in the examples of Fig. 4. Often it will resemble that of an asymmetric autoionizing resonance. In fact, a formal similarity can be shown by considering the amplitude of the resonance as interfering with an energy-insensitive background due

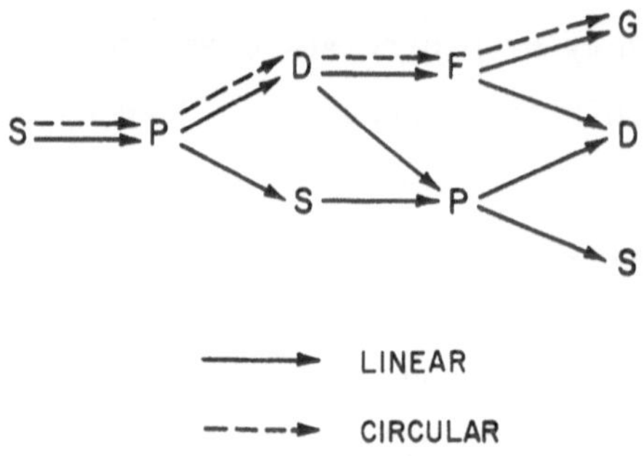

Fig. 3. Angular momentum channels for linear and circular light polarizations.

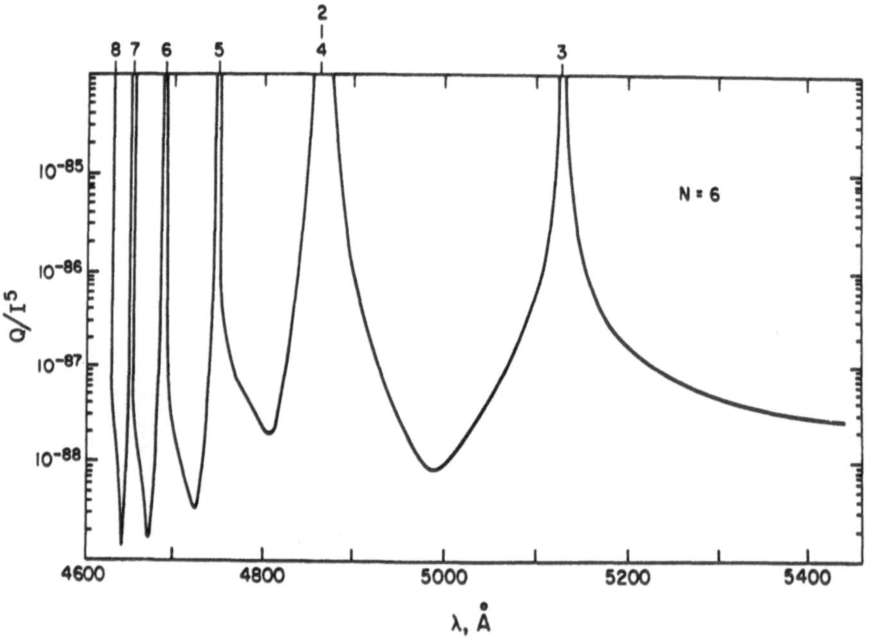

Fig. 4. 6-photon ionization cross section of H as a function of frequency (ref. 13).

to the amplitude of all other states. This similarity has been pointed out sometime ago by Beers and Armstrong,[10] as have the limitations of it by Dixit and Lambropoulos.[11] What must be underscored here, however, is that this formal similarity does not imply autoionization of any sort. It can be discussed even in the context of a hydrogen atom where autoionization in photoabsorption is impossible since it is truly a one-electron atom. This distinction must be kept in mind when we come to the discussion of multiphoton autoionization.

A paragraph should be devoted to multiphoton processes in hydrogen. Its Green's function is known exactly and the accuracy of generalized cross sections is limited only by numerical considerations. Gontier and Trahin,[12] as well as Karule,[13] have performed a considerable amount of work on this atom and have produced results for multiphoton ionization of various orders. More recently, Reinhardt and collaborators[14] have also calculated multiphoton ionization in hydrogen, using a technique that goes well beyond the lowest non-vanishing order of perturbation theory and should be valid for fairly large intensities. Unfortunately, there are virtually no experiments on hydrogen, because of the well-known difficulties in dealing with hydrogen. There are a couple of exceptions. Bayfield[15] has for a number of years now been concerned with ionization of high-n states of hydrogen employing a microwave field. His intention has been the investigation of multiphoton ionization versus tunneling ionization and most recently diffusive ionization which has received considerable attention in the last two years or so. An experiment on 3-photon ionization, via a two-photon intermediate resonance with the 2p state, was reported in 1984 by an MIT group,[16] but it is my understanding that the experiment is now being repeated. Thus one is left with some disappointment in that the atom which we know best theoretically we hardly know experimentally.

II. SOME SPECIAL FEATURES OF MULTIPHOTON PROCESSES

Having discussed the formalism of multiphoton processes and some of its direct consequences, we turn in the next few pages to the exploration of certain special features of these processes; features that are intimately related to the multiphoton character. We showed at the outset that even the simplest multiphoton process depends non-linearly on the field. It is really this non-linear character of the process that sets it apart from a single-photon process. If the field becomes strong (in the sense of saturating transitions) this non-linearity acquires additional complexity. It is worth noting here that, even the simplest single-photon process, resonance fluorescence for example, becomes non-linear in some sense if the resonant exciting radiation is so strong as to make the induced lifetime of the excited state comparable to or larger than its spontaneous lifetime. In that limit, resonance fluorescence acquires a multiphoton character in the sense that many induced emissions and absorptions take place before a single spontaneous emission can occur. Thus we see that, in almost any context of the interaction of radiation with matter, a multiphoton process implies non-linearity and vice-versa.

Although the emphasis in this article is on the atomic aspects of laser interactions, I should at least mention in passing one of the important field effects that set such nonlinear interactions apart from traditional singlephoton processes. The simplest consequence of this non-linearity was the dependence of an N-photon process on the Nth power of the laser intensity. But one may ask whether the way the photons arrive at the atom makes any difference in the rate of a multiphoton process. In other words, does the process depend on whether the photons arrive one at a time or in bunches of more than one, within the time it takes to complete the transition? For a completely non-resonant process, this characteristic time is of the order of 10^{-14} sec. Since N photons must be absorbed within this time for the N-photon absorption to be completed, on purely physical grounds one would expect that bunching would help the process. This is in fact borne out of the mathematical treatment[1] of this question. In more technical terms, the N-photon process turns out to be not simply proportional to the Nth power of the intensity, but to the Nth order intensity correlation function G_N. Thus a more precise expression for the rate of multiphoton ionization is

$$W_N = \hat{\sigma}_N G_N \tag{12}$$

where G_N depends on the state of the photon beam and that state is determined by the source. Two extreme types of sources are: A single-mode, well-stabilized laser (often identified with a purely coherent state of the field) and a multimode laser with a large number of statistically independent modes which can be considered equivalent to a thermal (chaotic or completely incoherent) source. I will use the terms coherent and chaotic to distinguish these two sources. A multimode laser operating under such conditions is indistinguishable from a thermal source as far as its intensity correlation properties are concerned. It is well-known that a chaotic source has substantial bunching compared to a purely coherent source which has none, except for the accidental bunching of a Poisson distribution. In other words, the probability of, say, three photons arriving within the above short time is much higher for a chaotic than a coherent source. In mathematical terms, G_N can be expressed in the form $G_N = \beta_N I^N$ where I is the photon flux per cm² and β_N a coefficient depending on the state of the source. It is equal to unity for a coherent source and equal to N! for a chaotic source. Thus N-photon ionization with a chaotic source is more efficient by a factor of N! than with a purely coherent source. The significance of this result is easily recognized if we note that, for example, in 11-photon ionization (which has repeatedly been observed in the laboratory) chaotic light is more efficient by seven orders of magnitude.[17]

If we go by the intuition based on traditional spectroscopy and optics, the above result is counter-intuitive since chaotic (incoherent) light turns out to be more efficient than coherent. This intuition, however, is based on (weak-field) linear processes and is not a safe guide for the regime of non-linear interactions. In weak-field linear transitions, it is the intensity that matters and the bandwidth whenever there is a resonance with an atomic state. In strong-field non-linear (multiphoton) transitions, higher order correlation functions of the field are equally important. It could be said that in a multiphoton process not only the radiation "sees" the internal structure of the atom (through the intermediate states that enter the description) but also the atom sees the internal structure of the radiation through its correlation functions. It is also relevant to recognize that an N-photon transition represents an N-fold photon coincidence experiment which otherwise would require N counters. In this case, the atom itself provides all the "electronics" through the characteristic time of absorption which, as we noted above, can be as short as 10^{-14} sec.

The problem of the interplay between field correlations and multiphoton processes has many more dimensions than one might suspect from the above brief discussion. In the presence of intermediate resonances under strong field excitation, not only intensity correlations but also field-amplitude correlations play an important role in the process. As noted above, even single-photon resonance fluorescence under strong field excitation becomes a multiphoton process because of the many absorptions and stimulated emissions (so-called Rabi oscillations) that may take place before its completion. The intensity and bandwidth of the field are not sufficient in that case to determine the total behavior, and correlations of all orders enter the picture. The interested reader will find a wealth of phenomena in the large body of literature[18-23] on these topics which have developed into a fascinating subfield of laser interactions (often called quantum optics) involving a combination of atomic, non-linear and stochastic physics.

Let us turn now to another aspect of multiphoton interactions having to do with the vector character of the electromagnetic field. We have seen already that total ionization of a spherically symmetric atom depends on the light polarization. But the phenomenon most sensitive to the light polarization is the photoelectron angular distribution and spin polarization which we now discuss in some detail.

The angular distribution of the emitted photoelectrons depends on the initial state, the light polarization, the order of the process and on whether there are resonances with intermediate states. The various combinations of these factors lead to an enormous variety of distributions from which a wealth of information about the underlying atomic structure can be extracted. If the light is linearly polarized, we will adhere to the definition of angles of Fig. 1, while for light circularly polarized the angle θ is defined with repect to the direction of propagation of the photon. One can show[1] quite generally that, for N-photon ionization of a spherically symmetric initial state, the angular distribution can be written as

$$f(\theta) = \sum_{n=0}^{N} \beta_{2n} \cos^{2n}\theta \qquad (13)$$

where the coefficients β_{2n} are the parameters containing the information on atomic structure and are given by complicated combinations of multiphoton matrix elements including the infinite summations. It is the value of these matrix elements that depends rather sensitively on the frequency. If the frequency is such that N-1 photons are in resonance with a real atomic state, the last photon causes ionization of that state and the angular distribution will reflect the properties of that state.[24] If on the other hand, there is no resonance with any intermediate state, the distribution will reflect the virtual intermediate states and will be rather sensitive to changes in the photon frequency.[6] Despite the long history of experimental studies of single-photon ionization angular distributions, it is only in the last three years or so that distributions from multiphoton ionization have begun attracting systematic attention.[25] It is the combination of angular with energy analysis[26] that promises to prove a very valuable tool in this field. Most of the experiments presently under way deal with situations of ionization with resonant intermediate states. The interplay between light polarization, bandwidth, pulse duration and intensity can reveal many aspects[25] of the intermediate state such as fine structure, hyperfine structure and even interaction of that state (during the laser pulse) with other atoms. It would take us well beyond the scope of this article to go into the details of this topic. I will simply give some examples of such distributions in Fig. 5 as well as an example of a distribution apparently from a hybrid resonance of two separating atoms.[26]

Closely related to the information obtained from a photoelectron angular distribution is the photoelectron spin-polarization when the light is circularly polarized. As is well-known,[27] photoelectrons ejected by circularly polarized light in single-photon ionization are spin-polarized, only when there is significant spin-orbit coupling in the continuum which requires rather special conditions. Thus the phenonemom is somewhat unusual in single-photon ionization and is sought at regions of the continuum where Cooper minima occur. It is assumed of course that we are talking about unpolarized atoms, because if the

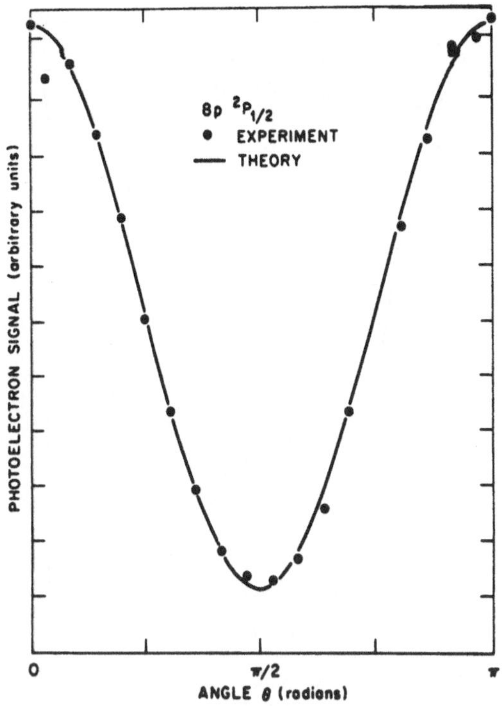

Fig. 5a. Photoelectron angular distribution from resonant 2-photon ionization of Cs via the $8P_{1/2}$ state (ref. 26).

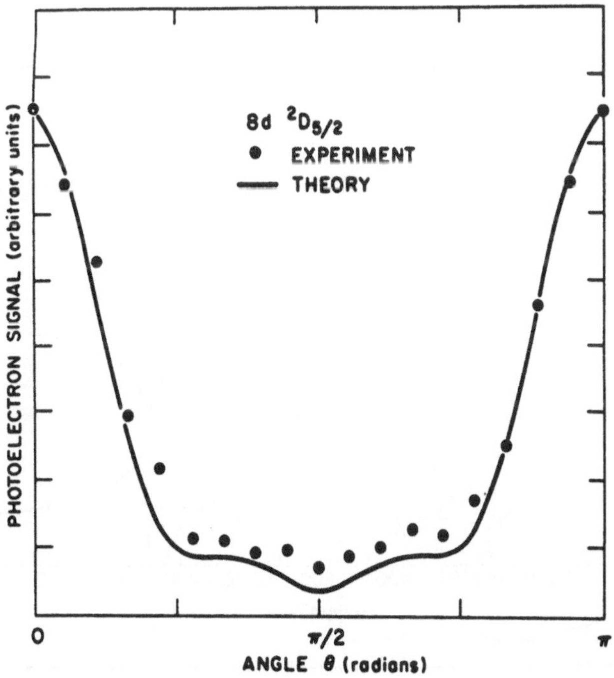

Fig. 5b. Photoelectron angular distribution from two-photon-resonant 3-photon ionization of Cs via the $8d(D_{5/2})$ state (ref. 26).

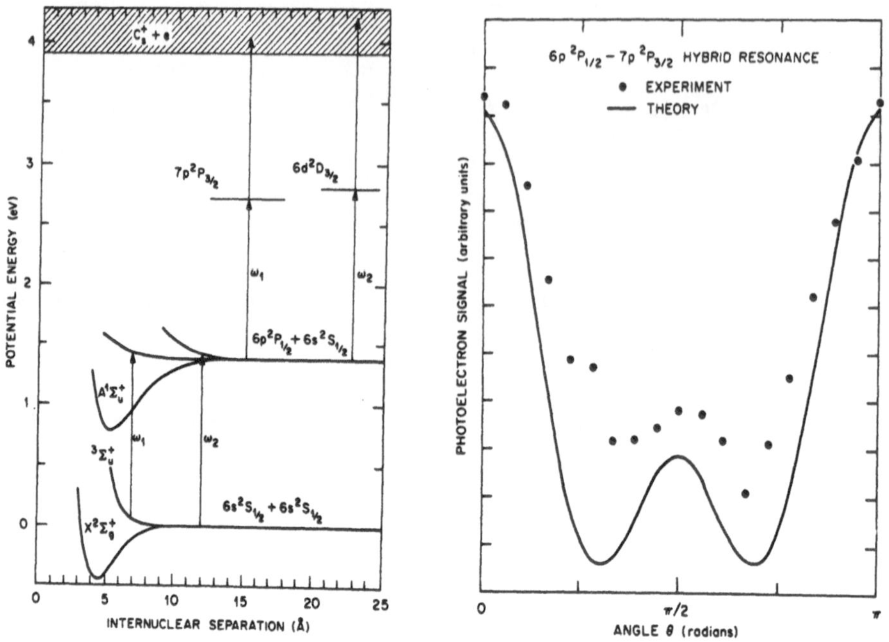

Fig. 5c,d. Probable hybrid resonance combined with quadrupole transitions in doubly resonant 3-photon ionization and the resulting photoelectron angular distribution (ref. 26).

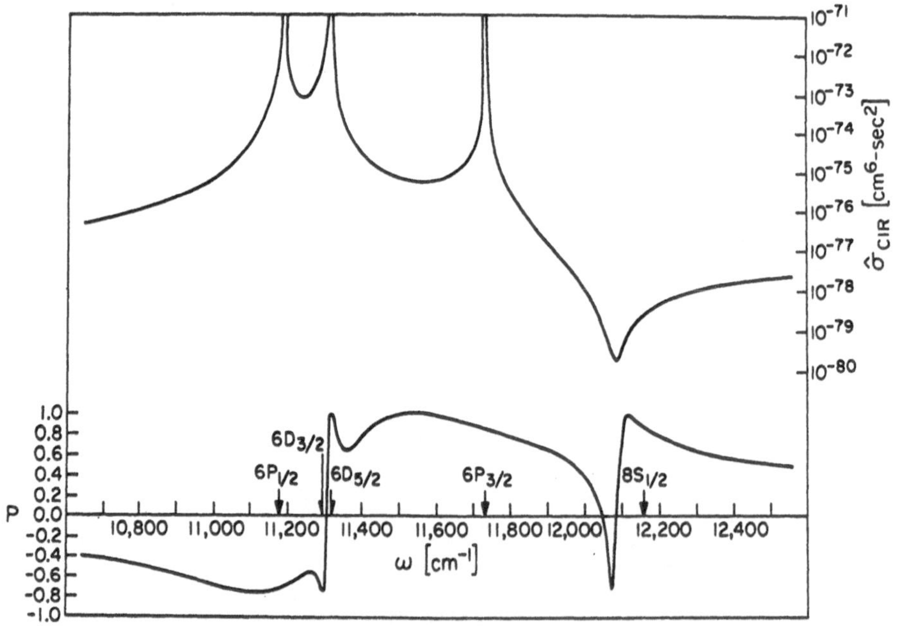

Fig. 6. Generalized cross-section and photoelectron spin-polarization in 3-photon ionization of Cs near the $6S_{1/2} \rightarrow 6P_j$ resonance.

initial atomic state is polarized, there will always be some degree of polarization in the ejected photoelectron. In multiphoton ionization with circularly polarized photons, the photoelectrons will have some spin polarization more often than not, independently of whether there is significant spin-orbit coupling in the continuum.[1] This polarization is due to the effect of spin orbit coupling of intermediate states and will be large whenever one of the photons is near a state split by the spin-orbit interaction. For example, the simplest situation[1] of this type occurs in two-photon ionization of an alkali atom when the photon frequency is in near resonance with a doublet state $nP(j = \frac{1}{2}, \frac{3}{2})$. As the photon is tuned from below the $P_{1/2}$ through the energy splitting and above the $P_{3/2}$, the spin-polarization of the photoelectron varies from about -50% to +100% and slowly falls off to zero as the photon is tuned away and the detuning becomes much larger than the spin-orbit separation $E_{nP_{3/2}} - E_{nP_{1/2}}$. Another example, from three-photon ionization of Cs, is shown in Fig. 6. Qualitatively one could say that the intermediate photons polarize the atom so that the ionization by the last photon ejects an electron from a polarized atom. In fact, there is no reason for all photons to have the same polarization. Combinations of different polarizations are of course experimentally more demanding but would lead to a much richer structure in the spin-polarization of the photoelectron. After an initial surge of interest in spin-polarized photoelectrons from multiphoton ionization, the subject has been dormant for more than eight years. One of the reasons must be the progress made in polarized electron sources by single-photon ionization. I do not elaborate further on this topic but, in view of the many discussions on spin-polarized electrons in this conference, I wanted to point out that the type of information obtained from the study of such effects in single-photon ionization can also be obtained from multiphoton ionization. Chances are we will be seeing more of this as the field matures and experimentalists begin looking into this type of more detailed information. It is also conceivable that the interest in multiphoton ionization as a source of spin-polarized electrons may revive.

I do not touch on the problem of electron scattering in the presence of laser radiation as it is the subject of the lectures by Prof. Ferrante. I do, however, present here a brief comment on one aspect of such processes which is not discussed by Prof. Ferrante but has been brought up by Dr. Hippler. My comment is also intended to show how our knowledge of multiphoton cross sections can be useful in other contexts. Dr. Hippler has discussed the possibility of an experiment on inelastic electron scattering in the presence of a strong laser. A hydrogen atom in its ground state is subject to the simultaneous action of an electron beam and a laser of photon energy $\hbar\omega$ = 1.17 eV which corresponds to a Nd laser. Let us assume, for generality, that the energy of the incoming electron is sufficient to excite the n=2 states of the atom which are about 10 eV above 1s. If the laser is present and sufficiently strong, the excitation of the n=2 states can also take place via a second order process in which part of the energy is given by the incoming electron and the rest by the photon. Thus if the kinetic energy of the incoming electron is E_0, the inelastically scattered electrons should show a peak at $E_0-(E_{2\ell}-E_{1s})$ without the laser, but also at $E_0-(E_{2\ell}-E_{1s})-\hbar\omega$ with the laser present. Higher order processes with peaks at $E_0-(E_{2\ell}-E_{1s})-N\hbar\omega$, where N>1, are also possible with lower probability.

Let us concentrate on the peak for N=1. A calculation by Jetzke et al.[28] shows that, for intensities of the order of 10^{12} to 10^{13} W/cm², an obvervable peak will appear at scattered electron energy $E_0-(E_2-E_1)-1.17$ eV. One may ask, however, about the possible effect of the laser intensity on the n=2 state and its influence on the scattering. For the above photon frequency, it takes three photons to ionize the n=2 state, whether it be 2s or 2p. The scattering-excitation process therefore leaves the system at a decaying state, the decay rate being the rate of 3-photon ionization of n=2. Another way of looking at it is to consider it as a final state interaction problem. In any case, the simplest way of describing the effect, without going into any equations, is to note that 3-photon ionization will broaden the n=2 states so that the electron-photon excitation leads to a broadened state. As a result, the peak for N=1 will be broadened by that amount. It remains to calculate that broadening which is given by $\hat{\sigma}_3 I^3$ where $\hat{\sigma}_3$ is the 3-photon ionization of the n=2 and I the photon flux per cm^2 per sec. I have calculated $\hat{\sigma}_3$ which turns out to be 1.5×10^{-78} cm^6 sec^2. For a laser intensity 10^{12} W/cm^2, the photon flux is about 0.6×10^{31} photons/cm^2 sec^2. Thus $\hat{\sigma}_3 I^3 \cong 3 \times 10^{14}$ sec^{-1} which is equivalent to about 1.24 eV while the photon energy is 1.17 eV. The peak therefore is wider than its distance from the main peak (for N=0) which means that it would not be observable under these conditions. It would simply merge with the background around the main peak. One could consider lowering the laser intensity, but then the height of the peak goes down as well, thus making its observation problematic for a different reason. Note that the intensity I used above is one order of magnitude smaller than the 10^{13} contemplated by Dr. Hippler and found as desirable in the paper by Jetzke et al.[28] Given that the 3-photon broadening increases with the cube of the intensity, the situation is experimentally much more unfavorable at that intensity. It is evident that the higher order peaks will also be equally broad thus merging with each other.

I should stress, before closing this subject, that the problem as discussed here is oversimplified. The rigorous theoretical treatment of this scattering-excitation to a state whose lifetime (because of multiphoton ionization) becomes possibly as short as the collision time itself is a very interesting problem in its own right.

III. RESONANT MULTIPHOTON PROCESSES

Although this article began with the discussion of resonant 2-photon ionization, the analysis and its conclusions were deceptively simple, having overlooked many subtle aspects which make this topic very rich in physical effects and applications. That simple analysis would be almost correct if we were interested only in weak field transitions (as in pre-laser resonance spectroscopy) with ideal monochromatic sources. On the contrary, it is mostly the opposite set of circumstances that is of interest from the standpoint of this article. Thus I should at least present a summary of the main features of such processes.

Returning to Fig. 1, let me first recall that in general the transition from $|0\rangle$ to $|1\rangle$ can be multiphoton, as can the transition from $|1\rangle$ to the continuum. Assume these to be m- and n-photon transitions, respectively, in an overall N-photon process (where $N=m+n$) and no other intermediate resonances for this particular frequency. One can show that all other atomic states can be eliminated from the problem giving rise to an m-photon matrix element coupling $|0\rangle$ to $|1\rangle$, an n-photon ionization matrix element coupling $|1\rangle$ to the continuum and AC Stark shifts for states $|0\rangle$ and $|1\rangle$. Since atoms can be excited to and populate state $|1\rangle$ and from there either be ionized or deexcited by stimulated m-photon emission, we can no longer calculate ionization in terms of a transition probability per unit time. Such a quantity, as a time-independent entity, may not exist because ionization may be a rather complicated function of time; although at certain limits of the parameters the transition probability per unit time may be meaningful. Thus we must consider the dynamics of the process through the equations for the density matrix representing levels $|0\rangle$ and $|1\rangle$, as well as the coupling to the continuum. If we denote by ρ_{00}, ρ_{11} and ρ_{01} the respective matrix elements, and represent by $\vec{E}(t) = \vec{\varepsilon}\,\mathcal{E}\,e^{i\omega t} + \vec{\varepsilon}^{\,*}\,\mathcal{E}^{*}\,e^{-i\omega t}$ the laser electric field (the dipole approximation is assumed), a long derivation[23] whose details can not be included here, leads to the following set of differential equations:

$$\left[\frac{d}{dt} + i(\Delta - S_{10}) + \frac{1}{2}(\Gamma_1^{ion} + \Gamma_1^{o})\right]\sigma_{01}^{(m)}(t) =$$

$$= \frac{i}{2}[\sigma_{11}(t) - \sigma_{00}(t)]\Omega_R \qquad (14a)$$

$$\frac{d}{dt}\,\sigma_{00}(t) = \Gamma_1^{0}\sigma_{11}(t) + \mathrm{Im}[\Omega_R^{*}\,\sigma_{01}^{(m)*}(t)] \qquad (14b)$$

$$\left[\frac{d}{dt} + \Gamma_1^{ion} + \Gamma_1^{0}\right]\sigma_{11}(t) = -\,\mathrm{Im}[\Omega_R^{*}\,\sigma_{01}^{(m)*}(t)] \quad . \qquad (14c)$$

Instead of the matrix elements of ρ, we have introduced here the slowly varying $\sigma(t)$ defined by $\sigma_{01}^{(m)}(t)=\rho_{01}(t)\,e^{-im\omega t}$, $\sigma_{00}(t)=\rho_{00}(t)$ and $\sigma_{11}(t)=\rho_{11}(t)$. The detuning Δ is defined by $\Delta=m\omega-\frac{1}{\hbar}(E_1-E_0)$ and it appears in the equations modified by the difference S_{10} of the laser-induced AC Stark shifts S_1 and S_0 which are proportional to the laser intensity. Because of these shifts, whether the process is resonant or not depends on the intensity, even if the frequency is such that $\Delta=0$ at low intensity. Γ_1^{ion} and Γ_1^{0} are, respectively, the (n-photon) ionization and spontaneous decay widths of state $|1\rangle$, with Γ_1^{ion} being proportional to the nth power of the laser intensity and of course Γ_1^{0} being independent of intensity. The quantity $\Omega_R=2\tilde{\mu}_{01}^{(m)}\mathcal{E}^{m}$

(where $\mu_{01}^{(m)}$ is the m-photon matrix element between $|0\rangle$ and $|1\rangle$) is the Rabi frequency coupling the two resonant states. Total ionization, as opposed to angle-resolved, is given by the quantity

$$P(t) = 1 - \sigma_{00}(t) - \sigma_{11}(t) \tag{15}$$

and is in general time-dependent. Under certain conditions, we may obtain the simple form $P(t) = 1-e^{-Wt}$ where W is independent of t. In that case, we are in the regime of applicability of Fermi's golden rule and W is just given by $\left(\frac{dP}{dt}\right)_{t=0}$. As indicated earlier, this is certainly true in the weak field limit $\Omega_R, \Gamma^{ion} \ll \Gamma_1^0$. On the other hand, a complicated damped oscillatory behavior may ensue if Ω_R is the dominant parameter. For a meaningful comparison with experiments, P(t) must be calculated for the duration of the laser pulse, preferably using a model temporal pulse shape that approximates the real pulse. Often a square pulse shape may be adequate, provided care is taken to ensure that artificial transcients due to the sudden turn on and off do not distort the predicted behavior. Since the shift S_{01} is intensity-dependent and the intensity is a function of time, so is the shift. The resonance condition, therefore, depends not only on intensity but on time as well. One of the effects of this dependence is that the line profile, as the frequency is tuned around the unperturbed resonance, exhibits an asymmetric broadened shape. The ionization width Γ_1^{ion} also follows the time evolution of the pulse as does the Rabi frequency Ω_R. Moreover, their sensitivity to the pulse evolution may be quite large since they are in general non-linear functions of intensity.

An additional aspect of field correlations enters the problem when we recognize that $\mathcal{E}$ appearing in Eqs. (14a-c) does not have to be constant but in fact undergoes stochastic fluctuations. In general, we will have amplitude and phase fluctuations, since $\mathcal{E}$ is complex. Given the presence of a resonance, the bandwidth will now play a role as will the intensity fluctuations discussed earlier. In fact, it is no longer possible to separate the effect of intensity from phase fluctuations; unless the laser happens to be of a perfectly amplitude-stabilized (single-mode) kind that has only phase fluctuations. Because of this stochastic behavior of the field, the parameters Ω_R, Γ_1^{ion}, S_{01} and the matrix elements $\sigma_{ij}(t)$ are stochastic quantities, as are the differential equations themselves. The experimentally observable quantity for ionization now is

$$\langle P(t)\rangle = 1 - \langle\sigma_{00}(t)\rangle - \langle\sigma_{11}(t)\rangle \tag{16}$$

where the angular brackets represent averages over the stochastic fluctuations of the field. Thus we must take the stochastic average of the differential equations. But in doing so, we encounter terms like $\langle\Omega_R^* \sigma_{01}^{(m)*}(t)\rangle$ which can not in general be separated into a product of averages; the two stochastic quantitites Ω_R and σ_{01} can not be decorrelated. Only in special cases can this be done, as in the case of weak field or in the case of a laser undergoing pure phase fluctuations.[18] The general treatment of these problems for strong fields with arbitrary amplitude fluctuations represents one of the most interesting examples of interplay between non-linear and stochastic physics.

This problem and its extentions has grown into a whole subfield[23] with rather sophisticated mathematical techniques. Extentions to multilevel systems, especially in connection with multiphoton transitions in molecules, have been attracting attention recently. A particular special case[18] of the two-level plus ionization problem, namely two-photon-resonant 3-photon ionization (corresponding to m=2, n=1) has been analyzed quite extensively in the literature because it is also of special spectroscopic interest as, for example, in measuring ionization cross-sections of excited states, isotope separation, four-wave mixing, etc.

IV. ABOVE THRESHOLD AND MULTIPLE IONIZATION

When it takes N photons of frequency ω to bridge the gap between the initial state and the first ionization threshold, the kinetic energy of the outgoing photoelectron is

$$E_K = N\hbar\omega - E_\infty \tag{17}$$

where E_∞ denotes the ionization potential to the lowest threshold. There is, however, no physical or mathematical reason for the electron to stop absorbing photons after having acquired the above minimum kinetic energy. This additional absorption of photons is called above threshold ionization (ATI) or more properly I think continuum-continuum absorption (CCA). If it is not usually observed in multiphoton or even single-photon ionization, it is because the relevant transition probabilities are relatively smaller, thus requiring higher intensity. Moreover, it can not be observed accidentally, as it is necessary to perform energy analysis of the outgoing photoelectrons. Thus in general Eq. (13) should be written as

$$E_K = (N + Q)\hbar\omega - E_\infty \tag{18}$$

where Q takes the values 0,1,2,3····· The experimental observation of ATI has been attracting considerable attention during the last few years.[29-32] At first, it was not clear whether the effect could be separated from the possible acceleration of electrons due to the gradient of the electric field and, in fact, early attempts were hampered by the rather broad background of accelerated electrons. But at this point, there are several experimental results on this effect. Its signature is a pattern of peaks in E_K spaced by intervals equal to the photon energy. Within a certain range of laser intensities, around 10^{12} W/cm², the peaks are expected to decrease in height as Q increases. What will happen at much higher intensities is still a rather unsettled question, but on the basis of general considerations, one would expect such peaks to tend to at first become equal in height.

There are two other related phenomena from which ATI is quite distinct, although it may be confused with. The phenomenon of free-free absorption of photons by electrons in the field of ions or atoms is one.[33] In that case, the electron is injected from outside and has no relation to the target which serves as third body for the absorption of the photons. In ATI on the other hand, the electron absorbs photons in the field of the ion from which it is emerging, and moreover this absorption is coherent. This means that after having absorbed N photons, the electron has particular values of angular momentum and as photons continue being absorbed the angular momentum changes with definite phase relations from one absorption to the next. This process would in principle be observable even if only one atom was present. This, of course, places a serious burden on the experimentalist who must ensure that the conditions of the observation are such that the influence of the field of other ions is negligible. If not, it is impossible to disentangle this effect from inverse Bremmstrahlung in the field of all ions present. The second phenomenon from which it must be disentangled experimentally is that of autoionization or, in general, the participation of a second electron through either real or virtual excitations. In atoms where either the frequency or the intensity or both conspire to make this possible, peaks above threshold may represent such double excitations. This is easier to separate from ATI since its importance is decidedly dependent on the structure of the atoms in question. Thus alkaline earths, with well-known doubly excited states relatively near the first threshold, would not be good candidates for ATI experiments. In the energy region in which the effect is genuine, the atom should behave like a one-electron atom.

The best studied system in this respect is Xe, which has two single-electron thresholds corresponding to the two angular momentum states of the core, namely $P_{3/2}$, and $P_{1/2}$. Data by the Saclay group[29-31] have resolved two series of peaks corresponding to the two different thresholds and spaced by the appropriate energy, namely 1.32 eV. Within each series, the peaks are spaced by the photon energy, as expected. These experiments have been performed with the Nd laser (λ - 1064 nm) as well as its second harmonic (λ = 532 nm). For λ=532 the lowest order ionization is 6, independently of the state of the residual ion. Most recently, Hippler et al.[34] have reported new data on the photoelectron angular distribution for the first and second peak, for each threshold. In the absence of quantitative calculations for such data, it is difficult to assess the meaning of these results.

A tendency for the distrubutions to become more anisotropic with increasing order can be expected on general grounds since each additional photon introduces a spherical harmonic of order higher by one than already present. Exceptions to this generally expected tendency do of course exist depending on the particular atom and frequency of the radiation. Thus the definitive interpretation of these results must await the appropriate theoretical input.

The peak heights observed by the Saclay group follow more or less what is expected on the basis of perturbation theory; they become lower with increasing order. On the other hand, data with λ=1064 nm (corresponding to 11- or 12-photon ionization depending on the state of the residual Xe^+ ion) by Kruit et al.[32] have shown discrepancies from the above picture. The peak heights, for example, were found to scale as I^N and not as I^{N+Q}. It has also been reported that the first peak (for Q=0) tended to vanish for intensities up to 70 TW/cm². This has been attributed to an upward shift of the ionization threshold and is supported by a theoretical calculation based, however, on a model one-dimensional atom.[35] Unfortunately these effects have not as yet been generally reproduced by others, and given the complexity of the undertaking, one must await further corroboration. It should also be kept in mind that Xe is not a model atom, but a real one, whose core is polarizable and excitable, which makes this writer rather skeptical as to the direct relevance of one-dimensional atomic models.

For reasons still not well understood, ATI had seemed difficult to observe in atoms other than rare gases. Two relatively recent observations, however, have shown the existence of at least one additional photon absorption above threshold in Cs. Petite et al.[30] have reported the observation of a 5-photon peak after 4-photon ionization while Dodhy et al.[36] have reported a 3-photon ionization peak above 2-photon ionization. In both cases, the additional peaks were small and no higher order peaks were detectable. Photoelectron angular distributions were reported in both experiments, with fairly good agreement with theory. In addition, Petite et al.[30] have measured and successfully compared with theory the ratio of the 5-photon to the 4-photon peak. Finally, as of the writing of this paper, Compton and collaborators[37] seem to have observed 3- and 4-photon peaks above 2-photon ionization of Rb.

It appears that the difference in saturation intensities between rare gases and alkalis may explain why ATI is not as easy to observe in the alkalis. For example, the saturation intensity of Xe at 532 nm is about twice that of Cs at 1064 nm. Saturation intensity here means the intensity at which the ground state is depleted. An up to date review of this topic can be found in a recent article by Agostini and Petite.[38]

Given that ATI has been shown to occur rather readily, there is no reason for a second electron not to absorb photons at these large intensities. If this does happen, it will lead to double ionization, and then triple ionization, and so on. This is now an established experimental fact and states of high ionization have been observed. Again the Saclay group (A. L' Huillier et al.[39-41]) were the first to observe multiply charged ions in rare gases, under excitation by the 2nd harmonic of the Nd laser with intensity 10^{12} to 10^{13} W/cm². Again Xe was the first target in this study with the radiation wavelength being 532 nm. As noted earlier, the lowest order process for the electron to be ionized is 6. Doubly ionized Xe, requires a 15-photon absorption, while the highest state of ionization, Xe^{5+} requires 74 photons. The same phenomenon was subsequently observed in other rare gases by the same group.[41] More recently, Rhodes and collaborators[42,43] have reported similar results on a larger sample of atomic species with radiation of larger intensity and frequency (10^{14}-10^{15} W/cm² and λ=193 nm). To this we must add a somewhat older series of experiments in which double and triple ionization was observed in alkaline earths first by Aleksakhin et al.[44] and later by Feldman et al.[45]

Thus the experimental evidence for multiple ionization is well established. Theoretical understanding on the other hand is at this point speculative at best. I will nevertheless try to summarize possibilities and put the various options in perspective.

The simplest mechanism for multiple ionization would be consecutive ionization, with the second electron being excited after the first has been ejected. In view of ATI, however, it is not clear what would be meant by "ejected", since the electron keeps absorbing photons into the continuum. But if we were to assume that, in any case, ions are produced in sequence, we would expect to observe a dependence on laser intensity reflecting this sequence. A log-log plot of ionization versus laser intensity should show the second ion production beginning when the first has saturated. In other words, the doubly charged ions, appearing when the ground state has been depleted, and so on. Moreover, each curve should exhibit a slope characteristic of the order of each process, e.g., 6 for the production of Xe^{+}, 10 for the production of Xe^{2+}, and so on, for radiation at λ=532 nm. Although this picture is in part compatible with the experimental evidence, there are other details which suggest that there is more to it than sequential ionization. For example, the Saclay group have measured the 10-photon ionization of Xe^{+} (to produce Xe^{2+}) and have found an unexpectedly large value. Also a comparison of the yields of successive ions has shown the rather surprising tendency to not decrease fast, even though the order of non-linearity from one ion to the next increases drastically. Thus the signal for Xe^{2+} is only 100 times smaller than for Xe^{+}, at an intensity of about 1.5×10^{13} W/cm² and wavelength λ=1064 nm. Note that it takes 11 photons of that frequency to ionize Xe and another 29 to ionize Xe^{+}. Thus it looks as if at that intensity, it is almost as easy to cause 29-photon as 11-photon ionization. At least on the basis of the up-to-now accumulated experience, this does not seem compatible with a picture of sequential one-electron transitions. One might suspect that the strong correlation between the electrons of the closed shell would lead to the simultaneous excitation of more than one electron. The experimental results of Rhodes[42,43] also lend support to this idea. His data having been obtained with higher intensities show higher charge multiplicities and make conceivable the possibility that a whole shell may be ionized collectively. Rhodes has in fact advanced such a hypothesis and has argued that the equivalent of a giant dipole resonance found in nuclear excitations may be at work. At this point, this can be nothing more than a picture as it lacks the theoretical foundation. In fact, some of his data provide evidence towards a somewhat subtler picture. In Kr, he has observed four-wave mixing[43] through an inner subshell excited state of the type $4s4p^6 4d$, which predominantly autoionizes but also decays radiatively. It takes four 193 nm photons to reach that state. Upon close examination of the spectrum of Kr and Kr^+ one can show that the most probable channel for this process is a two-photon excitation of one of the outer $4p^6$ electrons to a 6p state combined with a two-photon excitation of one of the $4s^2$ electrons to the 4p state while 6p falls back to 4d. It so happens that the first excited state of Kr^+ involves the excitation of an electron from the $4s^2$ subshell rather than one of the $4p^5$; a feature found in the other rare gases as well. This and other evidence, that space does not permit to go into, suggest that we are dealing with more than just blowing off the $4p^6$ shell or even part of it. Most probably, an intricate combination of multielectron excitations from more than the outer subshell is at play. It is of course known that in heavy ion collisions, whole shells may be "blown off". Surely a laser of intensity 10^{14} W/cm² has an enormous amount of energy capable of blowing off many shells. But this energy is in the form of photons which, depending on their frequency, can "see" the internal structure of the atom, as the experiment on Kr clearly suggests. Not only electrons from the shell nearest to the outer one, but inner electrons may undergo multiphoton excitations. And we know next to nothing about multiphoton excitation of inner shells. Most recent experimental results by Rhodes and collaborators[46] seem to point to the possibility of multiphoton-induced Auger transitions which should certainly be taking place since the energy is available. The burden for theory and experiment in the coming years is to assess the probabilities for all of these processes.

Turning now towards a completely different viewpoint, we should recognize that when it takes 30 photons to ionize an ion, these look like low frequency photons compared to the ionization potential. And this raises the possibility of tunneling, which should be the dominant mechanism of ionization in the limit of low frequency combined with high intensity. The question thus arises: Under what circumstances are some of these multiply charged species the result of tunneling ionization? For the uninitiated reader, I should simply mention that tunneling under an alternating field means that as the potential bends in alternating directions, the electrons have a probability of tunneling through, as in a DC field, except that it must happen within a time $1/\omega$. Roughly speaking, multiplying the probability of tunneling each time by the frequency of bending, gives the probability of ionization.[1] Unfortunately, tunneling under strong lasers is not well understood or even sufficiently tested. Most recently, the issue has become even more complicated by the discussions in connection with diffusive ionization[47] as opposed to both multiphoton and tunneling. What probably is happening could be viewed as multielectron excitation and ionization for the first two or perhaps three ionization species and then possibly some form of tunneling. Much, of course, depends on the photon frequency. Thus in the experiments of Rhodes which employed big uv photons it is more likely that multielectron excitations played a dominant role than in experiments with the 1064nm infrared photons for which tunneling might be expected.

If double or multielectron multiphoton excitation is a matter of conjecture and debate for the rare gases, it is an expected and proven occurrence in alkaline earths. These atoms have a valence shell with configuration ns^2 and relatively low ionization potentials. Many doubly excited states above the first ionization threshold have been known from single-photon ionization studies with UV sources. It is not surprising therefore that multiple ionization has been observed[44,45] in these atoms and at intensities as low as 10^{10} W/cm². Two-electron ionization would be expected to be enhanced considerably by the dense spectrum of doubly excited states between the two thresholds. A third electron on the other hand must come from a doubly charged ion with a rare gas structure; which brings us back to the situation discussed earlier, except for the apparently low intensity at which triple ionization of alkaline earths is reported to have been observed. I should note, however, that these studies have not been as extensive and systematic as for the rare gases and one must refrain from a final judgement until further evidence is in. It is nevertheless firmly established experimentally that multiphoton transitions to autoionizing states have been observed.[46] Transitions between autoionizing states have also been observed, although a clear identification of states will require additional work. Ions have been left in excited states and it is almost certain that CC transitions leading to an autoionizing state have also been observed. Finally, the lineshapes of the autoionizing states have been found to exhibit strong dependence on laser intensity. The most detailed, although by no means complete, experiments are those by Feldman and Welge[48] at intensities up to 10^{11} W/cm². I believe that understanding the behavior of such doubly excited states in strong fields is an important step towards a quantitative understanding of multielectron excitation under strong fields. In the remainder of this article, I review the main ideas and results on this topic which is still in a state of rapid evolution.

V. AUTOIONIZATION UNDER STRONG FIELDS

For the purposes of this brief summary of the topic, I use the simplest formulation[49] of transitions to autoionizing states (AS) in which an AS is described by a discrete configuration $|a\rangle$ coupled to a continuum $|c\rangle$ to which it decays by electron-electron interaction V. A single-photon transistion to an AS involves a dipole transition coupling the initial state $|g\rangle$ to $|a\rangle$ and $|c\rangle$ through the matrix elements D_{ga} and D_{gc} of the electric dipole operator D. The state $|a\rangle$ is coupled to $|c\rangle$ through the matrix element V_{ca} which is related to the autoionization width Γ_a through the equation

$$\pi|V_{ca}|^2 = \frac{1}{2}\Gamma_a . \tag{19}$$

The parameter q which is a measure of the relative strength of the transition $|g\rangle\to|a\rangle$ and $|g\rangle\to|c\rangle$ is defined by

$$q = \frac{\langle\Phi|D|g\rangle}{\pi V_{ca}\langle c|D|g\rangle} \tag{20}$$

where Φ_E is the state

$$|\Phi_E\rangle = |a\rangle + \int dE_c \frac{V_{ca}}{E-E_c} |c\rangle \tag{21}$$

containing a superposition of the discrete (doubly excited) configuration and the continuum. The value of q can range from $-\infty$ to $+\infty$ and characterizes the symmetry, or lack of it, of the autoionization line profile as the frequency of the exciting radiation is tuned around the resonance. If this detuning, measured in units of $\frac{1}{2}\Gamma_a$ is denoted by ε, the line-shape can be expressed as

$$R^2 = \frac{(q + \varepsilon)^2}{1 + \varepsilon^2} . \tag{22}$$

Thus the line profile can be described in terms of the two parameters q and Γ_a which are atomic parameters independent of the exciting field. For details on the derivation of these equations, the reader is referred to the relevant literature.[49]

Much changes when an autoionizing resonance is to be reached through a multiphoton transition.[50,51] In order to discuss the ideas in a specific context, I show in Fig. 7 a schematic representation of 3-photon autoionization including one additional transition to a higher AS. Note the different continua to which the two AS decay and how the radiation can induce transitions between the AS and the continua, as well as the continua themselves. Consider first the 3-photon transition to the AS. The most naive approach would be to simply calculate 3-photon matrix elements from $|g\rangle$ to $|a\rangle$ and from $|g\rangle$ to the continuum $|c\rangle$, denoted by $D_{ga}^{(3)}$ and $D_{gc}^{(3)}$, and use them in Eq. (16) thus obtaining an expression for q. In this approximation, Γ_a would retain the same value given by Eq. (15) which is independent of the field. Assuming for example that $|g\rangle$ has J=0 and $|a\rangle$ has J=1, this AS can be reached by a single-photon as well as by a 3-photon transition. The first point to be made is that, even in this most simple-minded approach (which can be shown to be valid in weak field), q in the 3-photon will not be the same as in the single-photon case; simply because $\langle\Phi|D^{(3)}|g\rangle$ and $\langle c|D^{(3)}|g\rangle$ do not have the same ratio as their single-photon counterparts. Thus when we generalize the spectroscopy by going into the multiphoton regime, q ceases to be independent of the process and values of q known from single-photon spectroscopy are not directly useful in multiphoton spectroscopy of AS. Of course if $|a\rangle$ has J=3, it is not accessible at all to single-photon transitions.

If the field is not weak, as must be the case in most multiphoton experiments, a different formulation is necessary leading to substantially different predictions. The field can be strong in more than one sense, reflecting different regimes of intensity. It can be sufficiently strong to saturate the 3-photon transition $|g\rangle\to|a\rangle$. More precisely, this is said to be the case when $|\langle\Phi|D^{(3)}|g\rangle|^2$ is larger than Γ_a. At that intensity, the coupling between $|g\rangle$ and $|a\rangle$ is stronger than autoionization and

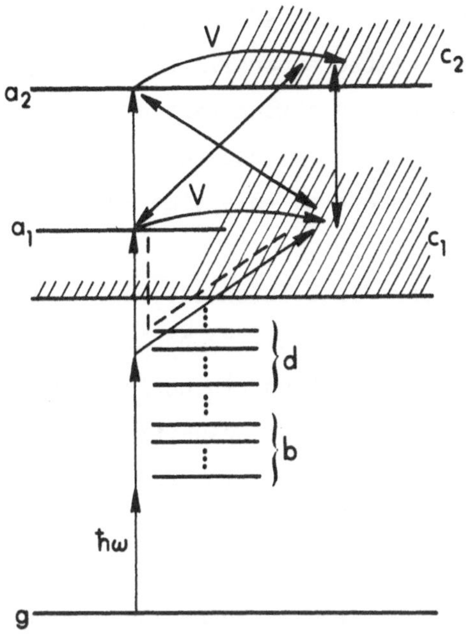

Fig. 7. Schematic representation of 3-photon autoionization simultaneously with a further transition to a higher autoionizing resonance. Arrows marked by V indicate configuration interaction. Unmarked arrows indicate electric dipole couplings.

the atom will oscillate between $|a\rangle$ and $|g\rangle$ many times before autoionizing. This requires rather substantial intensities and it does not appear to have occurred in experiments so far, although it is not too far outside the range of intensities of ongoing experiments. It can of course occur in 2-photon autoionization at lower intensities, or even in single-photon autoionization if an intense source at the appropriate frequency is available. The consequences on the line profile are dramatic. For sufficiently large intensity (but not unrealistic) the line may not show any maximum at all but simply a minimum. Moreover, the total ionization yield can no longer be calculated in terms of a transition probability per unit time, but must be calculated in terms of a probability which can exhibit strong time-dependence during the laser pulse. A number of these consequences have been discussed in theoretical papers[49] over the last three years. In the absence of any related experiments, however, theory seems to have run too far ahead and a good part of it is probably too idealized to have direct relevance to future experiments. Further elaboration on this aspect at this time would in my opinion be premature.

There is an intermediate intensity regime where equally dramatic changes may be and most probably have been observed.[48] Even if the intensity is not sufficiently strong to saturate (in the previous sense) the $|g\rangle \rightarrow |a\rangle$ transition, it can be sufficiently strong to make a 2-photon transition from $|a\rangle$ to the continuum (via intermediate states below and above threshold, but mostly below) larger than Γ_a. Let $D^{(2)}_{ac}$ be the 2-photon matrix element representing this transition. The related schematic path representing $D^{(2)}_{ac}$ is shown by dashed lines in Fig. 7. Although physically plausible, the origin of this contribution requires a rather lengthy derivation which can be found in the relevant papers.[50,51] The physical meaning of this contribution is evident from the figure. It represents an additional channel from $|a\rangle$ to $|c\rangle$ which is induced by the field and interferes with V_{ca}. As a result, the width of the AS now becomes

$$\frac{1}{2}\ \tilde{\Gamma}_a = \pi |V_{ca} + D^{(2)}_{ca}|^2 \tag{23}$$

while the q parameter also changes to

$$\tilde{q} = \frac{\mu^{(3)}_{ag} + P\int dE_c \dfrac{\tilde{V}_{ac}\ \mu^{(3)}_{cg}}{E-E_c}}{\pi\ [(V_{ac} + \mathcal{E}^2 \mu^{(2)}_{ac})]\ \mu^{(3)}_{cg}} \tag{24}$$

where

$$\tilde{V}_{ac} = V_{ac} + D^{(2)}_{ac} = V_{ac} + \mathcal{E}^2\ \mu^{(2)}_{ac} \tag{25}$$

with μ denoting the dipole operator $e\vec{\varepsilon}\cdot\vec{r}$ and $\mathcal{E}$ the amplitude of the laser electric field which is taken to have the form $\vec{\varepsilon}\,\mathcal{E}\,e^{i\omega t} + \vec{\varepsilon}^{*}\mathcal{E}^{*}\,e^{-i\omega t}$. The changes are much more profound now. First of all, $\tilde{\Gamma}_a$ is no longer an autoionization width because it is modified by the field. It will still be the observed width of the resonance, but it is due to the radiation as much as to configuration interaction. In fact, since it is the amplitudes that are added in Eq. (23), it is in principle possible for $\tilde{\Gamma}_a$ to be smaller than Γ_a if the two amplitudes have opposite signs. Since $D_{ca}^{(2)}$ is proportional to the square of the field amplitude, it will be negligible for small $\mathcal{E}$, becoming significant at larger values of $\mathcal{E}$ eventually overtaking V_{ca}. In that limit, it represents the dominant mechanism for the decay of $|a\rangle$ into the continuum. Although it is in principle possible for the width $\tilde{\Gamma}_a$ to vanish at some intensity, caution is recommended in taking this literally because other decay channels, even if small, will play a role. Not only is the parameter $\tilde{q}$ now different from q but is also intensity-dependent for the same reason that $\tilde{\Gamma}_a$ has become intensity-dependent. We expect therefore very substantial distortions of the line profile and ionization rate. A guide as to the size of intensities at which this will occur can be obtained from Eq. (23). If we have an AS of field-free width Γ_a, these effects will become significant at an intensity for which $|D_{ca}^{(2)}|^2 = |\mathcal{E}|^4|\mu_{ca}^{(2)}|^2 = I^2|\mu_{ca}^{(2)}|^2$ is comparable to Γ_a. If Γ_a is a few cm^{-1} and $\mu_{ca}^{(2)}$ is a typical two-photon (ionization) matrix element,[1] one finds that intensities of about 10^9 to 10^{10} W/cm² represent the region of these changes. Thus aside from the change of the line profile, we expect an eventual broadening of the resonance even if at first it goes through narrowing. More specific and quantitative estimates for Sr have given the same range of intensities.[51] The experimental results of Feldman and Welge[48] also show substantial broadening and change of the line profile within about that range of intensities. Although the theory seems to be in reasonable agreement with the experiment as far as this broadening is concerned, it is too early to declare the subject understood. From my point of view, the theory is not yet sufficiently quantitative and the experimental data not sufficiently extensive and detailed.

Should this be called power broadening? Not quite I think. In the usual context of transitions between two bound states, one of which decays in some way, the transition is said to be power broadened when the Rabi frequency is larger than the decay width. In the present context, this corresponds to the transition rate between $|g\rangle$ and $|\Phi\rangle$ becoming larger than Γ_a, which was not the case above. If one looks into the theoretical details[49] of this problem quite a bit more than can be done in this article, one finds that the concept of power broadening is more complex and subtle in autoionization than in usual bound-bound transitions. In addition, the length of the interaction time (usually determined by the laser pulse duration) also causes an apparent broadening[49] in any resonant transition and this must always be kept in mind when power broadening is invoked.

Since in the example we have been discussing, the transition $|g\rangle \rightarrow |a\rangle$ is 3-photon, while the process represented by $D_{ca}^{(2)}$ is 2-photon, it is reasonable to expect that there will be an intensity regime for which $D_{ca}^{(2)}$ is comparable to V_{ca} while $D_{ga}^{(3)}$ is much smaller. If on the other hand, $|g\rangle$ and $|a\rangle$ are coupled through a 2-photon transition, in which case both $D_{ga}^{(2)}$ and $D_{ca}^{(2)}$ are of second order, then there may not be an intensity regime where $D_{ga}^{(2)}$ is negligible compared to $D_{ca}^{(2)}$. In that case, we shall only have true power broadening, without being able to separate the contribution of the term $D_{ca}^{(2)}$. We must then calculate the coupling between $|g\rangle$, $|a\rangle$ and $|c\rangle$ by solving the complete set of differential equations for the amplitudes of these states. A rather extensive discussion of this and related situations has been presented by the author and collaborators in a series of articles.[49-51]

The calculations of these effects have been tested in Sr, mainly because the only experimental data available until recently were on Sr. It so happens and Sr has one doubly excited configuration ($5p^2$) below threshold. This gives rise to a few doubly excited states, singlet as well as triplet, of this configuration. Such states can serve as two-photon intermediate resonances in 3-photon ionization and autoionization. But these would be doubly excited intermediate states, and if the third photon were tuned to an AS, we would have the opportunity to study the simultaneous excitation of two electrons thus beginning to obtain some feeling about the essentials of multiphoton multielectron excitation. By contrast, a two-photon intermediate state of the type 5snd, could be thought of as involving the excitation of one electron. Experiments in which autoionizing doubly excited states are reached through a sequence of single excitation (one at a time) steps have been employed for a few years now by Gallagher, Cooke and others.[52] It would be of much interest to also have experiments involving excitations of two electrons at every step. Of course, in the general case of non-resonant multiphoton autoionization, all of these channels contribute at the same time.

In Fig. 7. I have also indicated a transition from the AS $|a\rangle$ to a higher one $|b\rangle$. The problem now becomes much more involved theoretically as well as conceptually. Because of the complexity, general solutions are not possible unless one oversimplifies the problem to the point of making it useless for the interpretation of experiments. From this type of problem we hope to assess the probability for transitions between AS leading to double ionization. At relatively low intensities, autoionization is so fast that transitions between AS would not have the time to occur with any substantial probability. For sufficiently high intensity on the other hand, one would think that the transition upwards can occur before the state autoionizes. That may be the case, but it will take careful and quantitative estimates to be convincing, because, as shown in the schematic representation of Fig. 7, the various combinations of coupling with the continua create a rather complex picture. Many of these couplings add coherently (interfere) and one can not be sure about the end result, unless calculations for specific and realistic atomic models are performed. We have begun obtaining the results of such calculations which will be published elsewhere.[53] I will simply mention here that such transitions appear quite possible at intensities in the region of 10^{10} to 10^{11} W/cm². The way and extent to which the various continua couple to the states appears to play a decisive role on the possibility of enhancing the upward transition compared to autoionization. We know yet nothing about the role of the coupling between the continua themselves. For the moment, it has been left out of the calculation on the basis of estimates that indicate it to be of secondary importance.

One very important aspect of resonant or near-resonant multiphoton transitions is the laser-induced AC Stark shifts of all states. This shift is proportional to the laser intensity and is particularly important for near-resonant states as they may shift closer to or farther from resonance with increasing intensity. For multiphoton autoionization, we have the AS itself which is on or near resonance as the frequency is tuned accross the line profile. We may also have an intermediate near resonant state at the same time. For example, as a 3-photon transition is tuned to the 5p6s (^{1}P) AS of Sr, two photons are not too far from resonance with an intermediate state of the type 5s5d(^{1}D). Although for sufficiently weak intensity, $2\hbar\omega$ is not on resonance as ω is tuned across the profile, it may in fact come into resonance for higher intensity. This has in fact been shown[51,53] to be the case in our calculations and the effect can be quite dramatic. Not only do width, line-profile, asymmetry etc. undergo large changes, but also the total ionization increases tremendously. As the intermediate state shifts into resonance at some intensity, we have a situation of double resonance with an obvious increase of ionization.

The calculations of these processes are quite involved. We have to cope not only with the atomic problem of configuration interaction, but also the field must be included on an equal footing because it does not enter as a weak probe; it can be as strong as or stronger than configuration interaction. It will be some time before we can accumulate the same expertise now existing for regular multiphoton ionization. But it is already clear that very interesting physics is contained in the existing data and that the general outline of the landscape can be guessed from the theory up to this point.

REFERENCES

1. P. Lambropoulos, Adv. At. Mol. Phys. 12, 87 (1976).
2. P. Lambropoulos and M. Teague, J. Phys. B 9, 587 (1976); also J. Phys. B 9, 1251 (1976).
3. M. Teague, P. Lambropoulos, D. Goodmanson and D. Norcross, Phys. Rev. A 14, 1057 (1976).
4. M. Crance and M. Aymar, J. Phys. B 12, 3665 (1979).
5. M. Crance and M. Aymar, J. Phys. B 13, 4129 (1980).
6. T. Olsen, P. Lambropoulos, S. E. Wheatley, and S. P. Roundtree, J. Phys. B 11, 4167 (1978).
7. L. A. Lompre, G. Mainfray, B. Mathieu, G. Watel, M. Aymar and M. Crance, J. Phys. B 13, 1799 (1980).
8. J. Morellec, D. Normand, G. Mainfray and C. Manus, Phys. Rev. Lett. 44, 1394 (1980).
9. M. Aymar and M. Crance, J. Phys. B 15, 719 (1982).
10. B. L. Beers and L. Armstrong Jr., Phys. Rev. A 12, 2447 (1975).
11. S. N. Dixit and P. Lambropoulos, Phys. Rev. A 19, 1576 (1979).
12. Y. Gontier and M. Trahin, Phys. Rev. A 4. 1896 (1971); also Phys. Rev. A 7, 2069 (1973).

13. E. Karule, J. Phys. B 4, L67 (1971); also in Atomic Processes, Report of the Latvian Academy of Sciences, Paper No. YΔK539.188 pp. 5-24 (1974).
14. S. I. Chu and W. P. Reinhardt, Phys. Rev. Lett. 39, 1195 (1977).
15. J. E. Bayfield, L. D. Gardner and P. M. Koch, Phys. Rev. Lett. 39, 76 (1977); J. E. Bayfield, Phys. Reports 51, 317 (1979).
16. L. R. Brewer, D. Kleppner and D. Kelleher, Bull. Am. Phys. Soc. 29, 824 (1984).
17. C. Lecompte, G. Mainfray, C. Manus and F. Sanchéz, Phys. Rev. A 11, 1009 (1975).
18. P. Agostini, A. T. Georges, S. E. Wheatley, P. Lambropoulos and M. D. Levenson, J. Phys. B 11, 1733 (1978).
19. S. N. Dixit and P. Lambropoulos, Phys. Rev. A 21, 168 (1980).
20. A. T. Georges and P. Lambropoulos, Phys. Rev. A 18, 587 (1978).
21. P. Zoller and P. Lambropoulos, J. Phys. B 13, 69 (1980).
22. L. A. Lompre', G. Mainfray, C. Manns and J. P. Marinier, J. Phys. B 14, 4307 (1981).
23. A. T. Georges and P. Lambropoulos in Adv. in Electronics and Electron Physics 54, 191 (1980).
24. S. N. Dixit and P. Lambropoulos, Phys. Rev. A 27, 861 (1983).
25. G. Leuchs and H. Walther in "Multiphoton Ionization of Atoms", edited by S. L. Chin and P. Lambropoulos, Academic Press, N.Y. (1984).
26. R. N. Compton, J.A.D. Stockdale, C. D. Cooper, X. Tang and P. Lambropoulos, Phys. Rev. A 30, 1766 (1984); J. C. Miller, C. D. Cooper, X. Tang and P. Lambropoulos, Phys. Rev. A 30, 1775 (1984).
27. J. Kessler, "Polarized Electrons", Springer-Verlag, Berling-Heidelberg-New York (1976).
28. S. Jetzke, F.H.M. Faisal, R. Hippler and H. O. Lutz, Z. Physik A 315, 271 (1984); see also, F.H.M. Faisal, in Coherence and Correlation in Atomic Physics, edited by H. Kleinpoppen and U. F. Williams, Plenum, New York (1980).
29. P. Agostini, F. Fabre, G. Mainfray, G. Petite and H. K. Rahman, Phys. Rev. Lett. 42, 1127 (1979).
30. G. Petite, F. Fabre, P. Agostini, M. Crance and M. Aymar, Phys. Rev. A 29, 2677 (1984).
31. F. Fabre, P. Agostini and G. Petite, Phys. Rev. A 27, 1682 (1983).
32. P. Kruit, J. Kimman, H. G. Muller and M. J. van der Wiel, Phys. Rev. A 28, 248 (1983).
33. This phenomenon constitutes the main theme of the lectures by Prof. Ferrante in this volume.

34. R. Hippler, H-J Humpert, H. Schwier, S. Jetzke and H. O. Lutz, J. Phys. B 16, L713 (1983).
35. H. G. Muller, A. Tip and M. J. van der Wiel, J. Phys. B 16, L679 (1983).
36. A. Dodhy, R. N. Compton and J.A.D. Stockdale, Phys. Rev. Lett. 54, 422 (1985).
37. R. N. Compton, private communication.
38. P. Agostini and G. Petite in "Multiphoton Processes", edited by P. Lambropoulos and S. J. Smith, Springer-Verlag, Berlin (1984).
39. A. L' Huillier, L-A. Lompre', G. Mainfray and C. Manus, Phys. Rev. Lett. 48, 1814 (1982).
40. A. L' Huillier, L-A. Lompre', G. Mainfray and C. Manus, J. Phys. B 16, 1363 (1983).
41. A. L' Huillier, L-A. Lompre', G. Mainfray and C. Manus, Phys. Rev. A 27, 2503 (1983).
42. T. Luk, H. Pummer, K. Boyer, M. Shahidi, H. Egger and C. K. Rhodes, Phys. Rev. Lett. 51, 110 (1983).
43. K. Boyer, H. Egger, T. Luk, H. Pummer and C. K. Rhodes, J. Opt. Soc. Am. B 1, 3 (1984).
44. I. Aleksakhin, I. Zapesochnyi and V. Suran, JETP Lett. 26, 11, (1977).
45. D. Feldman, J. Krautwald, S. L. Chin, A. von Hellfeld and K. Welge, J. Phys. B 15, 1663 (1982).
46. C. K. Rhodes, private communication.
47. G. Casati, B. V. Chirikov and D. L. Shepelyansky, Phys. Rev. Lett. 53,2525 (1984).
48. D. Feldman and K. Welge, J. Phys. B 15, 1651 (1982).
49. P. Lambropoulos and P. Zoller, Phys. Rev. A 24, 379 (1981).
50. K. Rai Dastidar and P. Lambropoulos, Phys. Rev. A 29, 183 (1984).
51. Y. S. Kim and P. Lambropoulos, Phys. Rev. A 29, 3159 (1984).
52. R. R. Freeman, L. A. Bloomfield, W. E. Cooke, J. Bokor and R. M. Jopson in "Multiphoton Processes", edited by P. Lambropoulos and S. J. Smith, Springer-Verlag (1984).
53. P. Lambropoulos and X. Tang, to be published.

PARTICLE-ATOM COLLISIONS IN STRONG LASER FIELDS

Gaetano Ferrante

Istituto di Fisica della Facoltà di Scienze

Via Archirafi 36, 90123 Palermo, Italy

1. INTRODUCTION

Atomic collisions in the presence of strong laser fields represent a new research area of growing interest. The basic peculiarity of this new class of elementary collision processes is that a third body enters the collision event, besides the projectile and the target. In the simplest cases, this third body (the field) is characterized by parameters such as the frequency ω , the wave number $\underline{K}_L$, the intensity and polarization vector $\hat{e}_L$. The field may interfere with the collision event (i) exchanging energy and momentum in quantized form ($l\hbar\omega$, $l\hbar\underline{K}_L$ with l an integer positive or negative); (ii) introducing a new physical axis (the direction $\hat{e}_L$ of polarization of its electric field); and (iii) making possible transitions which otherwise are simply forbidden. The presence of new parameters in the collision problem justify the expectation that a new collision physics may evolve from this new kind of three-body interactions, and that new possibilities and effects will come into reality. Historically, investigations of this new kind of collision processes were started in connection with important applications, such as laser gas breakdown, laser-plasma interaction, and plasma heating to high temperature by means of electromagnetic radiation(1-3). The particular elementary process addressed was that of electron-ion collisions in the presence of a laser field. To this same process will be largely devoted the present lecture. In more general terms, this lecture deals in detail with the theory of laser-assisted potential scattering, and it amounts, in practice, to restrict the treatment to cases when the atom internal structure and the laser modification of it play no significant role. Instead, the coupling of the scattered particle with the field is included in the theory exactly. This choice is dictated by pedagogic reasons, because laser-assisted potential scattering is the simplest example to treat and to introduce most of the basic concepts and models, which are common to more intricate processes, in which the internal structure of the colliding objects plays a role; second, by scientific reasons, because the existing theory is sufficiently well established and experiments too are available (the same is not true for field-assisted particle-atom collisions). Within the theory of laser-assisted potential scattering, special emphasis is placed on low-frequency treatments, which have received a lot of attention for the exact results they are apt to yield. In the last part of the lecture, it is shown how the theory is generalized to include information concerning the field modifications of the atom internal structure, and different cases are considered. Finally, the present status and some future directions in this subject are briefly discussed.

2. POTENTIAL SCATTERING IN A LASER FIELD: BASIC APPROXIMATIONS

We start by considering the problem of two interacting nonrelativistic, structureless charged particles (m_1 , e_1) and (m_2 , e_2), in the presence of a strong radiation field represented by its vector potential $\underline{A}(r,t)$. In the laboratory reference frame let the corresponding Schrödinger equation be

$$\left\{\sum_{\alpha=1}^{2}\frac{1}{2m_\alpha}\left[\frac{\hbar}{i}\underline{\nabla}_\alpha-\frac{e_\alpha}{c}\underline{A}(\underline{r},t)\right]^2+V(|\underline{r}_1-\underline{r}_2|)\right\}\psi(\underline{r}_1\,\underline{r}_2\,,t)$$

$$=i\hbar\,\dot{\psi}(\underline{r}_1,\underline{r}_2,t) \qquad (2.1)$$

2.1 Classical description of the assisting field

In writing the Schrödinger equation in the form (2.1) we have implicitly assumed that the radiation field may be treated not as a collection of photons, within the second quantization formalism, but simply as a prescribed function of coordinates and time. It amounts to the so-called "classical" or "external field" approximation. It is a fairly good approximation for intense fields, when stimulated processes dominate largely over spontaneous ones. More precisely, we can state the condition for treating the field classically by requiring that the occupation numbers of the field oscillators of any wave number K and the polarization $\hat{e}$ be much larger than unity:

$$n_{Ke} \gg 1 \tag{2.2}$$

This condition is easily converted into a condition on the value of amplitude of the electric field[4]

$$E_0 \gg (\Delta\lambda/\lambda)^{1/2} \lambda^{-2} \times 10^{10} \, (V/cm), \tag{2.2'}$$

where λ and $\Delta\lambda$ are, respectively, the field wavelength and bandwidth. In the visible, taking typically $\lambda = 5 \times 10^3$ (Å) and $\Delta\lambda = 10^{-1}$ (Å) one finds that E_0 must exceed few Volt/cm, which is easily satisfied by any real laser. It must be pointed out that the problem we are considering may be dealt with also within a fully quantum mechanical treatment[5]; however, the simpler semiclassical theory presented here is fully adequate to the scope of this lecture[6]. From Table 1, where are reported the wavelengths and photon energies of three frequently used lasers (Ruby, Nd:glass and CO_2 lasers), it is immediately concluded that typical laser wavelengths are much larger than atomic dimensions a_0 and effective ranges of projectile-target interactions R_0:

$$\lambda \gg (a_0, R_0) \tag{2.3}$$

TABLE 1

Laser	Wavelength (Å)	Photon Energy (eV)
Ruby	6,900	1.8
Nd:glass	10,600	1.17
CO_2	106,000	0.117

From Table 1 and inequality (2.3) we may easily assume that over distances of the order of magnitude of atomic dimensions a_0 or of the effective range of interaction R_0 the vector potential $\underline{A}(\underline{r},t)$ does not vary appreciably as a function of $\underline{r}$, so that $\underline{A}(\underline{r},t) \approx \underline{A}(t)$ (Dipole Approximation). The inequality (2.3) is sufficient to justify the dipole approximation only for weak fields. For intense fields, to inequality (2.3) it is wise to add some other condition, in which enter explicitly intensity-dependent quantities. For the dipole approximation to be valid, we must additionally require, for instance, that

$$v_{osc} = (eE_0/m\omega) \ll c, \tag{2.4}$$

v_{osc} and c being, respectively, the particle oscillatory velocity imparted by the field and the velocity of light. Alternatively, we may ask that the amplitude of the classical excursion

$$\alpha_0 = (eE_0/m\omega^2)$$

be not much larger than a_0 and R_0. Recent investigations[7,8] on problems partially similar to those discussed in this lecture have shown that the dipole approximation is likely to become suspect or even wrong at the following intensities: 10^{15} - 10^{16} Watt/cm^2 for the Nd:glass laser, and 10^{13} - 10^{14} Watt/cm^2 for CO_2 laser. Finally, the dipole approximation (which amounts to neglect of the photon momentum) is likely to become inadequate when the scattering parameters vary very rapidly as a function of the transferred momentum[6].

To complete our model to represent the radiation field, we assume that it is homogeneous, single mode, filling all space. As to the polarizations, both linear and circular polarization may be considered without much change both in the mathematics and the physics (as far as we do not consider, as is largely the case here, the atom internal structure). Thus we will have

$$\underline{A}_L(t) = \underline{A}_0 \cos\omega t = (c\underline{E}_0/\omega)\cos\omega t \tag{2.5}$$

for a linearly polarized field, and

$$\underline{A}_c(t) = (cE_0/\omega)(\hat{x}\cos\omega t + \hat{y}\sin\omega t) \tag{2.6}$$

for a circularly polarized field.

Actually, real lasers may be poorly described by the simple models (2.5) and (2.6) which do not account, for instance, for the pulsed regime, the focalization (the finite extent), the multimode character, the statistical fluctuations, and so on. Luckily, in several cases some of the above features of the real lasers are easily incorporated in the theory just by taking the appropriate averaging over the result obtained with the use of the simple models (2.5) and (2.6).

2.2 The initial and final states

Before we start to discuss how to construct the unperturbed initial and final states, we transform eq.(2.1) to the cm reference frame. The change of coordinates in (2.1),

$$(\underline{r}_1, \underline{r}_2) \rightarrow [\underline{r} = \underline{r}_1 - \underline{r}_2 \ , \ \underline{R} = (\underline{r}_1 m_1 + \underline{r}_2 m_2)/(m_1 + m_2)]$$

and the dipole approximation for the field give

$$\left[\frac{1}{2M_R}\left(\frac{\hbar}{i}\underline{\nabla}_R - \frac{Q}{c}\underline{A}\right)^2 + \frac{1}{2m}\left(\frac{\hbar}{i}\underline{\nabla}_r - \frac{e}{c}\underline{A}\right)^2 + V(r)\right]\psi(\underline{R},\underline{r},t)$$

$$= i\hbar\dot{\psi}(\underline{R},\underline{r},t) \qquad (2.7)$$

where

$$Q = e_1 + e_2 \ ; \qquad M_R = m_1 + m_2 \ ;$$

$$e = (m_1 e_2 - m_2 e_1)/(m_1 + m_2) \ ; \qquad m = m_1 m_2/(m_1 + m_2)$$

The change of variables has yielded the familiar result of separation of the cm motion from the internal, relative motion. This point is reminded here, first to draw attention to the circumstance that when an external field is present, generally, the cm system is no more an inertial system. Second, to stress that the separable form (2.7) is made possible only by the dipole approximation. Luckily, in most cases of interest (collisions of electrons and positrons with massive target), M_R is much larger than the reduced masses, so that the non-inertiality of the cm of the colliding particles may be neglected, compared with the effect of the field on the relative motion. This simplification will be adopted in the following. Thus we are left with the Schrödinger equation of the relative motion

$$\left[\frac{1}{2m}\left(\frac{\hbar}{i}\nabla - \frac{e}{c}\underline{A}\right)^2 + V(r)\right]\phi(\underline{r},t) = i\hbar\dot{\phi}(\underline{r},t) \qquad (2.8)$$

From (2.8) the initial and final states are constructed by assuming the static potential V(r) to be the perturbation responsible for the transition. In doing so, we are just following the conventional scattering theory. Thus the unperturbed initial and final states are given by (2.8) with V(r) = 0,

$$\frac{1}{2m}\left(\frac{\hbar}{i}\nabla - \frac{e}{c}\underline{A}\right)^2\varphi(\underline{r},t) = i\hbar\dot{\varphi}(\underline{r},t) \qquad (2.9)$$

Eq.(2.9) says that the unperturbed states are continuum states embedded in the radiation field, and it is the basic new feature of the collision theory in strong fields, which will be presented in this lecture. The choice of V(r) as a transition operator and of continuum states embedded in the field as unperturbed states is considered appropriate for strong field situations. However, this choice is not unique. According to the specific physical situations, with the same right one may choose as transition operator the particle-field interaction W_{pF}, or both operators V(r) and W_{pF}. (This may be particularly convenient in cases when the atom internal structure is weakly modified by the field). Accordingly, the unperturbed states too will be different. Eq.(2.9) is easily solved for any $\underline{A}(t)$ to give

$$\varphi_{\underline{K}}(\underline{r},t) = \exp\left\{i\underline{K}\cdot\underline{r} - \frac{i\hbar}{2m}\int^t\left(K^2 - \frac{2e}{\hbar c}\underline{K}\cdot\underline{A}(\tau)\right)d\tau\right\}\times$$
$$\times\exp\left\{\frac{-i}{\hbar}\int^t\frac{e^2}{2mc^2}A^2(\tau)\,d\tau\right\}, \qquad (2.10)$$

which may be called nonrelativistic Volkov states or field-modulated plane waves(9). For linearly polarized field, as given by (2.5), we have

$$\varphi_{\underline{K}}(\underline{r},t) = \exp\left\{i\underline{K}\cdot\underline{r} - \frac{i}{\hbar}(\mathcal{E}_K + \bar{V}_L)t + i\lambda_K\sin\omega t\right\}\times$$
$$\times\exp\left\{-i\left(\bar{V}_L/2\hbar\omega\right)\sin 2\omega t\right\} \qquad (2.11)$$

where

$$\hbar\underline{K} = \langle \underset{\sim}{p}(t)\rangle_T$$

is the time averaged momentum,

$$\underline{p}(t) = \hbar \underline{K} - \frac{e}{c}\underline{A}(t) = m\,\underline{v}(t)$$

the instantaneous kinetic momentum;

$$\bar{V}_L = e^2 E_0^2 / 4 m \omega^2$$

the time averaged potential energy of oscillation (coinciding with the corresponding classical quantity);

$$\lambda_K = \frac{e}{m\omega^2}\underline{E}_0 \cdot \underline{K} = \underline{\alpha}_0 \cdot \underline{K} \tag{2.12}$$

the basic coupling parameter of the free particle-laser interaction, and

$$\varepsilon_K = \hbar^2 K^2 / 2m$$

φ_K is not a stationary state as in a time-dependent field energy is not conserved, and the quantity $\varepsilon_K^L = \varepsilon_K + \bar{V}_L$ appearing in eq. (2.11) is called 'quasienergy'[10]. It is easily shown, however, that φ_K may be expressed as a superposition of stationary states. With the help of the expansions[11]

$$\exp\{i\lambda_K \sin\omega t\} = \sum_{\ell=-\infty}^{\infty} (-1)^\ell J_\ell(\lambda_K)\, e^{-i\ell\omega t} \tag{2.13}$$

$$\exp\left\{\frac{-i\bar{V}_L}{2\hbar\omega}\sin 2\omega t\right\} = \sum_{\ell'} J_{\ell'}(\bar{V}_L/2\hbar\omega)\, e^{-i\ell' 2\omega t} \tag{2.14}$$

where J_ℓ and $J_{\ell'}$ are Bessel functions of integer index and real argument, φ_K is rewritten as

$$\varphi_K = \sum_{\ell,\ell'} L_{\ell,\ell'} \exp\left\{i\underline{K}\cdot\underline{r} - \frac{i}{\hbar}\left[\varepsilon_K + \bar{V}_L + (\ell+2\ell')\hbar\omega\right]t\right\}, \tag{2.15}$$

with

$$L_{\ell,\ell'} = i^{2\ell} J_\ell(\lambda_K)\, J_{\ell'}(\bar{V}_L/2\hbar\omega) \tag{2.16}$$

The superposition (2.15) says that the charged particle may be found in any of the states with energy $E_K = \varepsilon_K + \bar{V}_L + (\ell+2\ell')\hbar\omega$, the corresponding probability being given by $|L_{\ell,\ell'}|^2$. Actually, these states are virtual, as a free particle cannot exchange energy with an homogeneous plane wave field, filling all the space. This is because the law of conservation of the energy and the momentum cannot be simultaneously satisfied, in the absence of a third body. Hence, the presence of a third body may turn virtual states into real and just at this point comes into play the scattering potential. For not too intense fields, when $\bar{V}_L$ may be neglected $(\bar{V}_L \approx 0)$ φ_K simplify to

$$\varphi_K = \sum_\ell i^{2\ell} J_\ell(\lambda_K) \exp\left\{ i \underline{K}\cdot\underline{r} - \frac{i}{\hbar}(\varepsilon_K + \ell\hbar\omega)t \right\} \tag{2.17}$$

as

$$J_{\ell'}(\bar{V}_L/2\hbar\omega) \approx J_{\ell'}(0) = \delta_{\ell',0} \, .$$

(2.17) amounts to the solution of eq. (2.9) neglecting the $A^2(t)$-term (see also (2.10)). Finally, when no field is present, the Volkov states go, as they must, into the familiar plane waves. In fact

$$E_0 = 0 \; ; \quad \lambda_K = \bar{V}_L = 0 \; ;$$

$$J_{\ell'}(0) = \delta_{\ell',0} \; ; \quad J_\ell(0) = \delta_{\ell,0} \; ; \tag{2.18}$$

$$\varphi_K = \exp\left\{ i \underline{K}\cdot\underline{r} - \frac{i}{\hbar}\varepsilon_K t \right\} .$$

In the theory of laser assisted potential scattering given below the Volkov states enter in the reduced form (2.17), without the $A^2(t)$-term part. However, it must be pointed out that it is so not because of an approximation (the weakness of the field), but simply because the $A^2(t)$-term part cancels exactly from the pertinent matrix elements.

For a right circularly polarized field taken as (2.6), the solution (2.10) becomes

$$\varphi_{\underline{K}}^c(\underline{r},t) = \exp\{ i \underline{K}\cdot\underline{r} \} \exp\left\{ \frac{-i}{\hbar}[\varepsilon_K + 2\bar{V}_L]t \right\} \times$$
$$\times \exp\{ i\lambda_K^x \sin\omega t - i\lambda_K^y \cos\omega t \}$$
$$= \sum_{\ell,\ell'} C_{\ell,\ell'} \exp\left\{ i\underline{K}\cdot\underline{r} - \frac{i}{\hbar}[\varepsilon_K + 2\bar{V}_L + (\ell+\ell')\hbar\omega]t \right\} \tag{2.19}$$

with

$$C_{\ell,\ell'} = i^{2(\ell+\ell')} \, i^{\ell'} \, J_\ell(\lambda_K^x) J_{\ell'}(\lambda_K^y) , \tag{2.20}$$

$$\lambda_K^s = \alpha_0 K_s , \qquad (s = x, y) . \tag{2.21}$$

In (2.19) K_x and K_y are respectively, the X- and Y-components of the particle wavevector. $2\bar{V}_L = \bar{V}_C$ is the average energy of oscillation of an electron in a circularly polarized field, which, thus is twice that of an electron in a linearly polarized field (again, this result coincides with the classical result).

We conclude this Section with few remarks.

(i) To derive the unperturbed states within the adopted (classical) model to represent the field, the latter has been included in the theory exactly, to all orders. Hence, in principle, it opens the possibility of studying in atomic collisions nonlinear effects of radiation-matter interaction with an high degree of accuracy.

(ii) The particle-field coupling is controlled in an essential way by the scalar product $\underline{\alpha}_0 \cdot \underline{K}$, implying the (obvious) fact that some 'geometrical' conditions need to be satisfied to have a strong coupling. In other words, we may have a very strong field but a weak coupling to the particle. On this point, further considerations are delayed until the derivation of the cross sections.

3. SCATTERING THEORY FORMALISM

3.1 S and T matrices

To derive the cross sections for the potential scattering in the presence of a laser field of a charged particle from the initial state $\underline{K}_i$ to $\underline{K}_f$, our starting point is the exact S matrix

$$S = (-i/\hbar)(\varphi_f, V\Phi_i^+) \tag{3.1}$$

where φ_f is the final state Volkov wavefunction, Φ_i^+ is the exact wavefunction satisfying the complete Schrödinger equation (2.8) with causal boundary conditions, and the round brackets indicate both space and time integrations. Formally, we can write

$$\Phi_i^+ = \varphi_i + \bar{G}^+ V \varphi_i , \tag{3.2}$$

where $\bar{G}^+$ is the retarded full Green function (in the presence of both the static potential and the radiation field). $\bar{G}^+$ is now expanded in powers of the scattering potential

$$\bar{G}^+ = \bar{G}_0^+ + \bar{G}_0^+ V \bar{G}_0^+ + \bar{G}_0^+ V \bar{G}_0^+ V \bar{G}_0^+ + \cdots \tag{3.3}$$

with

$$\bar{G}_0^+ = G_0^+(\underline{r},t;\underline{r}'t';E_0)$$
$$= (-i/\hbar)\,\Theta(t-t')\sum_m \varphi_m(\underline{r},t)\,\varphi_m^*(\underline{r}',t') \tag{3.4}$$

the retarded Green function in the presence of the radiation field only. In (3.4) the summation over m is actually an integration over $\underline{K}_m$ according to the prescription

$$\sum_m \longrightarrow (2\pi)^{-3}\int d^3K_m$$

Using (3.2)-(3.4) in (3.1) one has

$$S = (-i/\hbar)\left[(\varphi_f, V\varphi_i) + (\varphi_f, V\bar{G}_0^+ V\varphi_i) + \cdots\right]$$
$$= \sum_{\nu=1}^{\infty} S^{(\nu)} \tag{3.5}$$

where

$$S^{(1)} = (-i/\hbar)(\varphi_f, V\varphi_i)$$
$$= (-i/\hbar)\int_{-\infty}^{\infty} dt \int d^3r\, \varphi_f^*(\underline{r},t)\, V(r)\, \varphi_i(\underline{r},t); \tag{3.6}$$

$$S^{(2)} = (-i/\hbar)^2 \int_{-\infty}^{\infty} dt_1 \int_{-\infty}^{t_1} dt_2 \sum_m \int d^3r_1\, \varphi_f^*(r_1,t_1)V(r_1)\varphi_m(\underline{r}_1,t_1) \times$$
$$\times \int d^3r_2\, \varphi_m^*(\underline{r}_2,t_2)V(r_2)\varphi_i(\underline{r}_2,t_2) \tag{3.7}$$

and so on. All the φ_α's appearing in (3.5)-(3.7) are Volkov wavefunctions, and it is the new feature as compared with the conventional S matrix series of potential scattering. Using (2.17) and identities like (2.13), the time integrations required in (3.5) are easily carried out to give the S matrix expressed now as

$$S = \sum_{\ell=-\infty}^{\infty} \langle f | S_\ell | i \rangle \tag{3.8}$$

with

$$\langle f | S_\ell | i \rangle = (-2\pi i)\, \delta(\varepsilon_f - \varepsilon_i - \ell\hbar\omega) \langle f | T_\ell | i \rangle , \tag{3.9}$$

the brackets $\langle \cdots \rangle$ indicating now space integration only and $|\alpha\rangle = \exp(i\underline{K}\cdot\underline{r})$. $\langle f | T_\ell | i \rangle$ may be considered the generalization of the T matrix to include the presence of the radiation field (however, in general, it is off the energy shell) and it is given by

$$\langle f | T_\ell | i \rangle = \langle f | V(\ell) | i \rangle$$

$$+ \sum_m \sum_{\ell_1} \frac{\langle f | V(\ell-\ell_1) | m \rangle \langle m | V(\ell_1) | i \rangle}{(\varepsilon_i - \varepsilon_m + \ell_1 \hbar\omega + i\eta)}$$

$$+ \sum_{\substack{m \\ n}} \sum_{\ell_1, \ell_2} \frac{\langle f | V(\ell-\ell_1-\ell_2) | n \rangle \langle n | V(\ell_2) | m \rangle}{(\varepsilon_i - \varepsilon_n + (\ell_1+\ell_2)\hbar\omega + i\eta)} \times$$

$$\times \frac{\langle m | V(\ell_1) | i \rangle}{(\varepsilon_i - \varepsilon_m + \ell_1 \hbar\omega + i\eta)} + \cdots \tag{3.10}$$

In the T matrix (3.10) the basic ingredient is the 'compound' matrix element

$$\langle \beta | V(\mu) | \gamma \rangle = J_\mu(\lambda_{\beta\gamma}) \langle \beta | V | \gamma \rangle \tag{3.11}$$

formed: (i) by the Bessel function J_μ with argument

$$\lambda_{\beta\gamma} = \lambda_\beta - \lambda_\gamma = \underline{\alpha}_0 \cdot \underline{Q}_{\beta\gamma} \tag{3.12}$$

$$\hbar \underline{Q}_{\beta\gamma} = \hbar(\underline{K}_\beta - \underline{K}_\gamma) \tag{3.12'}$$

being the transferred momentum in going from state γ to state β ; and (ii) by

$$\langle \beta | V | \gamma \rangle = \int d^3r \exp(-i \underline{Q}_{\beta\gamma} \cdot \underline{r}) V(r), \quad (3.13)$$

the Fourier transform of the scattering potential. Loosely speaking $J_\mu(\lambda_{\beta\gamma})$ may be considered as the matrix element giving the amplitude that μ photons of a given field are exchanged with the particles during the collision in which the momentum $\hbar(\underline{k}_\beta - \underline{k}_\gamma)$ is transferred. A number of remarks are in order concerning the S matrix (3.9).

(i) The presence of the energy conserving delta function allows us to speak of separate and distinct multiphoton processes. It is worth remembering that we started with a classical description of the field; through expansions of exponentials containing time-dependent periodic functions (like (2.13) and (2.14)) we have ended up with a discrete photon picture. However, the assumption of a single-mode, zero bandwidth field is crucial.

(ii) The field-free S matrix is readily recovered if the field is put equal to zero.

(iii) The factored structure of the basic matrix element (3.11) is due to the use of the dipole approximation. Taking the S matrix to the first order only, the factorization will extend to the cross section as well.

(iv) The ubiquitous presence of the Bessel functions with argument $\underline{\alpha}_0 \cdot \underline{Q}$ stems from the use of a model for the radiation field based on the assumption of single-mode, homogeneity and linear polarization. Changing the model for the radiation field, will change to a lesser or a greater extent the mathematical expression accounting in the S matrix for the effect of the field.

(v) The S matrix has a more rich and complicated structure, mostly due to the presence of Bessel functions containing in their arguments the integration variables, and to the sums over the number of photons exchanged during the intermediate scattering events. This last circumstance (i.e. the presence in (3.10) of energy denominators like $\varepsilon_i - \varepsilon_m + l_1\hbar\omega + i\eta$, with l_1 going formally from $-\infty$ to $+\infty$, opens the interesting possibility of 'replicate' a scattering resonance, say, at $\varepsilon_R = \varepsilon_m = \varepsilon_i + l_1\hbar\omega$ with different initial particle energies ε_i and different l_1. This possibility has been proved experimentally in the case of elastic scattering of electrons by Ar and by replicating the two resonances known to exist in the field-free scattering in the energy interval $\varepsilon_i = 11.0 - 11.5$ eV[12] (see Fig.1).

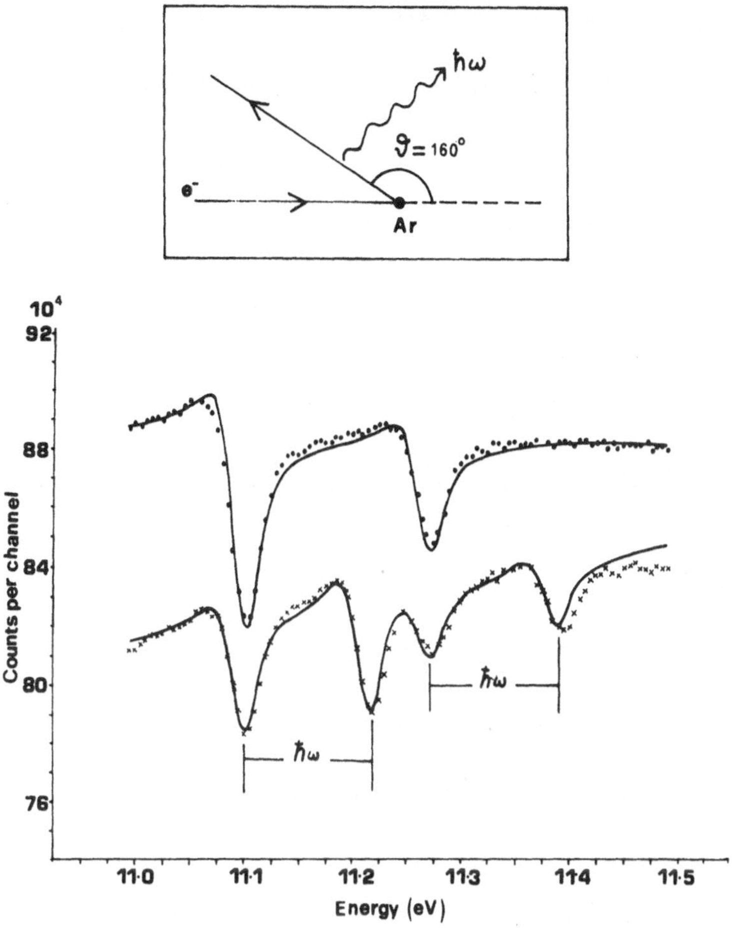

Fig. 1 One photon stimulated emission in elastic e-Ar scattering, showing the doubling of scattering resonances. The incident energy interval is 11.0 – 11.5 eV, around the two known resonances of the Ar doublet $^2P_{3/2,1/2}$. The upper curve corresponds to the field-free elastic cross section, showing the presence of two resonances. The dots are experimental points, the full curve is a calculation plus some adjustements for the instrumental apparatus. The lower curve corresponds to the cross section for one photon emission and shows four resonances. The assisting field is 50 Watt CW CO_2 laser ($\hbar\omega$ = 0.117 eV) focused about 2.5 mm^2, i.e. 2×10^3 Watt/cm^2. The crosses are experimental points, the full curve is calculation similar to that of the upper curve. The scattering angle was ϑ = 160°. To measure the reported field assisted cross sections it took about 100 hours(from Ref. 12).

3.2 Transition Probabilities and Cross Sections

Proceeding from the S matrix (3.8) and following usual procedures it is an easy matter to obtain
(i) the transition probability per unit time

$$\mathcal{P}_{fi} = \frac{2\pi}{\hbar}\sum_{\ell=-\infty}^{\infty} \delta(\varepsilon_f - \varepsilon_i - \ell\hbar\omega)|\langle f|T_\ell|i\rangle|^2 \tag{3.14}$$

and
(ii) the Fermi Golden Rule,

$$P_{fi} = \int \mathcal{P}_{fi}\, \rho(\varepsilon_f)\, d\varepsilon_f$$

$$= \frac{2\pi}{\hbar}\sum_{\ell} |\langle f|T_\ell|i\rangle|^2 \rho[\varepsilon_f(\ell)] \tag{3.15}$$

In (3.15)

$$\rho[\varepsilon_f(\ell)] = m\hbar k_f(\ell)\, d\Omega/(2\pi\hbar)^3 \tag{3.16}$$

is the density of final scattering states;

$$\varepsilon_f(\ell) = \varepsilon_i + \ell\hbar\omega = \hbar^2 k_f^2(\ell)/2m \tag{3.17}$$

is the final energy fixed by the energy conserving delta function; accordingly, the final state wave number is given by

$$k_f(\ell) = \frac{1}{\hbar}(2m(\varepsilon_i + \ell\hbar\omega))^{1/2} \tag{3.18}$$

According to (3.17), $\ell > 0$ corresponds to absorption of ℓ laser photons by the scattered particle, and $\ell < 0$ to emission. We observe that now the final energy ε_f and the momentum $\hbar \underline{k}_f$ depend on the number of transferred laser photons.

To define the differential cross section we first consider the probability current associated with the incident Volkov plane wave $\varphi_i(\underline{r},t)$:

$$\underline{j}(t) = \frac{i\hbar}{2m}(\varphi_i \underline{\nabla}\varphi_i^* - \varphi_i^* \underline{\nabla}\varphi_i) - \frac{e}{mc}\underline{A}(t)\varphi_i\varphi_i^*$$

$$= \hbar \underline{k}_i/m - (e\underline{E}_0/m\omega)\cos\omega t \tag{3.19}$$

The result (3.19) is just the classical instantaneous velocity of a particle in a plane wave field. Defining the incident flux as the time averaged probability of current

$$<\underline{j}> = \frac{\omega}{2\pi}\int_0^{\omega/2\pi} \underline{j}(t)\,dt = \hbar \underline{k}_i/m, \tag{3.20}$$

(iii) the differential cross section (DCS) is obtained by dividing P_{fi} by $<\underline{j}>$,

$$(d\sigma/d\Omega)_F = \sum_\ell (d\sigma/d\Omega)_\ell, \tag{3.21}$$

$$(d\sigma/d\Omega)_\ell = (k_f^{(\ell)}/k_i)\left(\frac{m}{2\pi\hbar^2}\right)^2 |<f|T_\ell|i>|^2 \tag{3.22}$$

(iv) Finally, for the total cross section (TCS) we have

$$\sigma_F = \sum_\ell \sigma_\ell, \tag{3.23}$$

$$\sigma_\ell = (k_f^{(\ell)}/k_i)(m/2\pi\hbar^2)^2 \int d\Omega\, |<f|T_\ell|i>|^2 \tag{3.24}$$

3.3 First Born Approximation

The exact expressions for $\mathcal{P}_{fi}$, P_{fi}, $(d\sigma/d\Omega)_F$ in their general form are too intricate to extract from them further detailed information on the role of the laser field in the collision event. So, for the sake of gaining additional insight into the physics of the field assisted collisions, we consider relatively high energies and weak potential situations, when it is allowed to approximate the T_ℓ operator with the first term of its series: $T_\ell \simeq V^{(\ell)}$ (First Born Approximation, FBA). For $\mathcal{P}_{fi}$, P_{fi} and $(d\sigma/d\Omega)_F$ it gives, respectively, (anything else remaining the same as compared to formulas (3.14)-(3.22)):

$$\mathcal{P}_{fi}^B = \frac{2\pi}{\hbar}\sum_\ell J_\ell^2(\lambda_{fi})|<f|V|i>|^2\,\delta(\varepsilon_f-\varepsilon_i-\ell\hbar\omega); \tag{3.25}$$

$$P_{fi}^B = \frac{2\pi}{\hbar}\sum_\ell J_\ell^2(\lambda_{fi})|<f|V|i>|^2\,\rho[\varepsilon_f^{(\ell)}]; \tag{3.26}$$

$$(d\sigma/d\Omega)_F^B = \sum_\ell (K_f(\ell)/K_i) J_\ell^2(\lambda_{fi}) \left[(m/2\pi\hbar^2)^2 |\langle f|V|i\rangle|^2\right]_\ell \tag{3.27}$$

In (3.27) the expression in squared brackets is the field-free FBA DCS of potential scattering

$$(d\sigma(\ell)/d\Omega)^B = \left[(m/2\pi\hbar^2)^2 |\langle f|V|i\rangle|^2\right]_\ell \tag{3.28}$$

evaluated at shifted energy $\varepsilon_f(\ell)$ and momentum $\hbar \underline{K}_f(\ell)$ given by (3.17) and (3.18), and because of it dependent on ℓ .

We see that in the FBA the DCS of potential scattering with exchange of ℓ photons, $(d\sigma/d\Omega)_\ell^B$, is modified by $J_\ell^2(\lambda_{fi})$ which is an oscillatory function, whose behaviour is controlled by its argument

$$\lambda_{fi} = \underline{\alpha}_0 \cdot (\underline{K}_f - \underline{K}_i) = (e\underline{E}_0/m\omega^2) \cdot [\underline{K}_f(\ell) - \underline{K}_i] \; . \tag{3.29}$$

From λ_{fi} as given by (3.29) it is easy to anticipate that, for a given ℓ , the presence of the field may drastically modify the angular distribution of the scattered particles as compared to the field-free case. Unfortunately, this is only a part of the story. The other part is that J_ℓ^2 never exceeds unity and is much smaller than unity, when the index ℓ differs appreciably from the argument λ_{fi} . It amounts to say that, for a given ℓ , the DCS may be much smaller than the field-free one. This circumstance suggests to look at the full DCS (3.27), i.e. to the cross section summed over all the numbers of exchanged photons. The FBA is justified for relatively high incident energies, so that in this approximation it is reasonable to assume the energy exchanged with the assisting field be a small fraction of the particle energy: $\varepsilon_i \gg \ell\hbar\omega$. This assumption enables us to neglect the ℓ-dependence of the final wavevector $\underline{K}_f(\ell)$, and, accordingly, also of the quantities containing $\underline{K}_f(\ell)$, i.e. the cross section (3.28) and the argument of the Bessel function (3.29). As a result one has

$$(d\sigma/d\Omega)_F^B \approx (d\sigma(0)/d\Omega)^B \sum_\ell J_\ell^2(\lambda_{fi}) \approx (d\sigma/d\Omega)^B, \tag{3.30}$$

where for obtaining the last approximate equality use has been made of the closure property of the squared Bessel function[11]

$$\sum_\ell J_\ell^2(x) = 1 \tag{3.31}$$

The sum rule (3.30) says that the DCS's with given numbers of exchanged photons sum up to give the field-free cross section. The physical content is that, compared to the field-free case, the field does not change the number of particles scattered at a given angle ϑ but changes by $\pm \ell\hbar\omega$ the energy with which they arrive at the detector. Thus the field produces a broadening and a discretization of the particle initial energy distribution (see Fig.2). This result has been proved experimentally within an assessable degree of accuracy[13], but it must not be considered as having a general validity nor as being typical of what happens in collisions when a laser field is present. Actually, the range of validity of the sum rule (3.30) has not yet been assessed in a conclusive way. Nevertheless, it is likely that it is simultaneously connected: (i) to the condition $\varepsilon_i \gg \ell\hbar\omega$; (ii) to the factorization effect of the field on the collision dynamics like in (3.25)-(3.27); and (iii) to the use of the dipole approximation. Provided these conditions are met, (3.30) holds also beyond first-order treatments in the scattering potential[14]. In practice, (3.30) is expected to hold whenever the DCS are not a very sensitive function of particle momentum or of the transferred momentum, or when the external field is not able to modify appreciably these quantities.

4. FIELD DEPENDENCE OF THE CROSS SECTIONS

In applications, it is generally of interest to know how the cross sections depend upon the field parameters such as, for instance, the intensity or the frequency. In the cross sections derived above, the field parameters enter in a nonlinear way (basically, through the arguments of the Bessel functions). So the question how the cross sections depend on the field parameters may be answered exhaustively only by means of numerical calculations. In analytical form, an answer may be given for some limiting case in FBA. Beyond FBA is still possible for the same limiting cases provided the low frequency approximation for the field is additionally made (see below).

4.1 Differential cross sections in FBA

For weak particle-field coupling, we have $\lambda_{fi} \ll 1$, and the

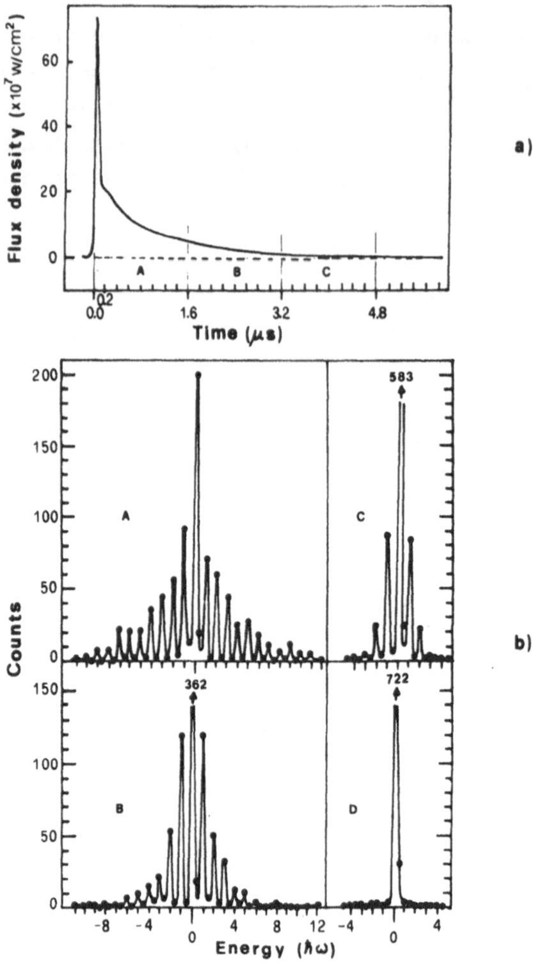

Fig. 2 *Multiphoton free-free transitions in e-Ar elastic scattering in the presence of a CO_2laser ($\hbar\omega$ = 0.117 eV).*
a) Temporal shape of the pulse of the CO_2 laser used in the experiments.
b) Energy gain-loss spectra (counts for 500 1.6 μs pulses plotted against energy in units of laser photons). Incident electron energy $\mathcal{E}_i$ = 9.923 eV; scattering angle ϑ = 155°. Polarization vector $\hat{\epsilon}_L$ parallel to the scattering plane. The spectra A, B and C show the number of scattered electrons (in 500 laser shots) as a function of decreasing flux density corresponding to the time interval A, B and C of the laser pulse shown in the upper part of the figure. Spectrum D shows the energy distribution without the laser. Dots are experimental points; full curves trace out the multiphoton processes approximately (from Ref.13).

squared Bessel functions entering the DCS (3.27) may be approximated as[11]

$$J_\ell^2(\lambda_{fi}) \approx (\lambda_{fi}/2)^{2|\ell|}\,[\Gamma(\ell+1)]^{-2} + \dots$$

Then for the DCS with a given ℓ one finds

$$(d\sigma/d\Omega)_\ell \sim (E_o/\omega^2)^{2|\ell|} \sim (I/\omega^4)^{|\ell|}, \quad (E_o^2 \sim I), \tag{4.1}$$

which is the result expected on the basis of a perturbation-theory treatment.

For strong particle-field coupling, for which we have $\lambda_{fi} \gg 1$, the Bessel functions may be approximated as[11]

$$J_\ell(\lambda_{fi}) \approx (2/\pi\lambda_{fi})^{1/2}\cos(\lambda_{fi} - \ell\tfrac{\pi}{2} - \tfrac{\pi}{4}), \quad (\ell < \lambda_{fi}).$$

Taking the average of the fastly oscillating squared cosine appearing in the approximate squared Bessel function ($\cos^2(\dots) = 1/2$), one has

$$(d\sigma/d\Omega)_\ell \sim \omega^2/E_o \quad . \tag{4.2}$$

(4.2) shows that in the asymptotic regions of very strong coupling: (i) the dependence of the DCS on the number of exchanged photons becomes very weak; and (ii) the DCS decrease as E_o^{-1} with increasing laser electric field.

4.2 Total cross sections in FBA

For weak coupling, the same dependence on the field parameters is found for TCS as for DCS. For strong coupling, the TCS show a nonincreasing behaviour vs E_o only for values of the field approaching the limit at which the dipole approximation breaks down; more exactly, calculations[15] show an oscillatory behaviour. Besides, the TCS continue to depend in a significant way on the number of exchanged photons[15]. The point concerning the different behaviour of differential and total CS vs the field parameters for strong fields is worth stressing. In the FBA, the total CS is given by

$$\sigma_\ell^B = \int_0^{2\pi} d\varphi \int_0^{\pi} d\vartheta \sin\vartheta\, J_\ell^2[\underline{\alpha}_o\cdot(\underline{k}_f-\underline{k}_i)]\left(\frac{k_f}{k_i}\right)\left(d\sigma^{(\ell)}/d\Omega\right)^B \tag{4.3}$$

where, generally, the argument of the Bessel function depends on ϑ and φ . During the integrations over the angles, λ_{fi} varies and in some intervals of integration it may become very small. Thus a single and simple asymptotic expression of $J_\ell^2(\lambda_{fi})$ for $\lambda_{fi} \gg 1$ used in (4.3) may be a rather bad representation The physical content of this point is that the coupling of a particle to a strong field is effective only along a given direction or within some narrow cone. The differential cross sections are quantities related to chosen scattering angles and thus to directions, which may be chosen to maximize the coupling with the field. The total cross sections are quantities integrated over all directions, and thus they include contributions from large regions of space, in which the particle-field interaction is weak in spite of the assisting field being strong. The most spectacular outcome of this circumstance is that in potential scattering in the presence of external laser fields up to very high values of the intensity the total cross section with exchange of only one photon is the largely dominating cross section. Precisely, calculation shows that [15]

$$\sigma_1 > \sigma_2 > \cdots > \sigma_\ell > \cdots \tag{4.4}$$

(see also Fig.3).

5. LASER MODELS IN ATOMIC COLLISIONS

In the remark (iv) after formulas (3.8)-(3.13) it was pointed out that the Bessel functions stemmed from the model used to describe the laser field. On the other hand, the crucial role of $J_\ell^2(\lambda_{fi})$ in the FBA DCS of potential scattering with exchange of ℓ photons says that the modifications of the DCS due to the presence of a radiation field are strongly dependent on the model adopted to represent the field. This suggests to consider other models to describe the assisting field, and see which modifications they predict. This introductory lecture does not allow us to dwell on this interesting aspect of field-assisted atomic collisions, but the possibility of incorporating exactly laser models into the theory of field-assisted atomic collisions is likely to open a new avenue, which may produce unique information also on radiation field properties in highly nonlinear domains [16-27].

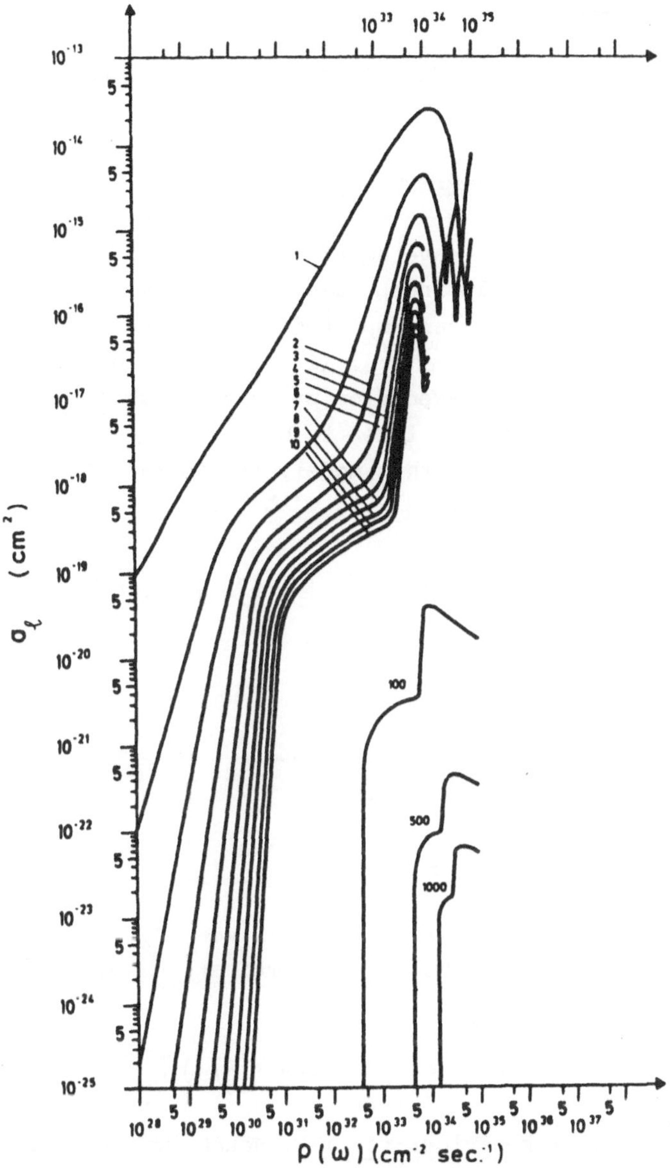

Fig. 3 *Total cross sections* σ_ℓ*, in* cm^2*, for simultaneous absorption of* ℓ *ruby laser photons (* $\hbar\omega$ = 1.8 eV*) by electrons colliding with protons versus photon flux, in* cm^{-2} sec^{-1}*. Electron incident energy* $\mathcal{E}_i$ = 100 eV. *(From Ref.15).*

5.1 Equal Frequency Multimode Field

As an instance, we consider first the following model of a multimode field[17]:

$$\underline{A}_L(t) = \hat{e}_L \sum_{i=1}^{N} (cE_i/\omega)\cos(\omega t + \varphi_i), \tag{5.1}$$

in which: (i) all the modes have equal frequency ω; (ii) the phases φ_i are statistically independent and randomly distributed within $(-\pi, \pi)$; (iii) the amplitudes E_i have a gaussian distribution with dispersion $\alpha^{-1/2}$; (iv) the total power of the field is constrained to be fixed from pulse to pulse $\sum_i E_i^2 = \bar{E}_0^2$; (v) the number of the modes is taken to be very large ($N \to \infty$, 'chaotic field' limit). Using this model, within the FBA, one arrives at formulas like (3.25)-(3.27), with

$$J_\ell^2(\lambda_{fi}) \qquad \text{replaced by} \qquad \exp\{-\bar{\lambda}_{fi}^2/2\} I_\ell(\bar{\lambda}_{fi}^2/2)$$

where

$$\bar{\lambda}_{fi} = \lambda_{fi}(\bar{E}_0/E_0) \tag{5.2}$$

is the same as (3.29) with the variance

$$\bar{E}_0 = (\langle E^2(t)\rangle - \langle E(t)\rangle^2)^{1/2} = (\langle E^2(t)\rangle)^{1/2} \tag{5.3}$$

replacing the constant electric field amplitude E_0, and I_ℓ is a Bessel function of integer index and imaginary argument(16,17,20,21). The functions

$$J_\ell^2(x) \qquad \text{and} \qquad \exp(-x^2/2)\, I_\ell(x)$$

have a quite different behaviour vs x. Hence two different radiation fields will modify the partial DCS $(d\sigma/d\Omega)_\ell^B$ in a quite different way. It opens interesting possibilities. First, useful information may be obtained from scattering data on the laser structure, when the latter is not precisely known, or after long periods of operation. Second, comparing cross sections in different fields, unique information may be obtained on strong radiation field properties in highly nonlinear domains. This possibility is essentially due to the circumstance that the field (and, accordingly, its properties) is treated exactly, to all orders.

As an instance, let us define, for a given ℓ and $\bar{E}_0 = E_0$, the ratio between the DCS (in FBA) in the presence of a chaotic field and the DCS in the presence of a single mode field:

$$\chi_\ell = (d\sigma/d\Omega)^\ell_{CH} \Big/ (d\sigma/d\Omega)^\ell_{SM}$$

$$= \exp\{-\tfrac{1}{2}\lambda^2\} I_\ell(\tfrac{1}{2}\lambda^2) / J_\ell^2(\lambda) \quad . \tag{5.4}$$

(In (5.4) to shorten notations we have used $\lambda_{fi} = \bar{\lambda}_{fi} = \lambda$). The behaviour of χ_ℓ vs λ is rather involved. In fact, to get an idea, it is enough to remember the oscillatory behaviour and the zeros of $J_\ell^2(\lambda)$. Here we will consider only some limiting cases, for which reliable estimates may be obtained in analytical form[17,21].

(i) weak coupling ($\lambda \ll 1$) and $\ell \gg \lambda$

Expanding the Bessel functions of (5.4) in powers of λ, for χ_ℓ we get

$$\chi_\ell \approx \ell! \tag{5.5}$$

(Here and below in this subsection $\ell > 0$). This result says that, for a given number ℓ of exchanged photons, a weak chaotic field of variance $\bar{E}_0$ yields a DCS which is $\ell!$ larger than the DCS in the presence of a single mode field with $E_0 = \bar{E}_0$. This result is well known from other branches of radiation-matter interaction physics[28-31], and often is believed to be characteristic and exhaustive of a chaotic field.

(ii) strong coupling ($\lambda \gg 1$) and $\ell \ll \lambda$

Using asymptotic expansions for the Bessel functions[11], we get

$$\chi_\ell \approx \frac{\sqrt{\pi}}{2} \cos^2\left(\lambda - \ell\frac{\pi}{2} - \frac{\pi}{4}\right) \approx \sqrt{\pi} \tag{5.6}$$

saying that in this domain a chaotic field yield cross sections only slightly larger than those of a single mode field.

(iii) strong coupling ($\lambda \gg 1$) and $\ell \gg \lambda$

Using proper asymptotic expansions for the Bessel functions[11] we get again

$$\chi_\ell \approx \ell!$$

(iv) strong coupling ($\lambda \gg 1$), and $\lambda \gtrsim \ell$, $\lambda \gg 1$, $\ell \gg 1$.

Again using the proper asymptotic expansions for the Bessel functions[11], we get

$$\chi_\ell \approx \sqrt{\pi}\, \exp\{-\ell^2/\lambda^2\} (1 - \ell^2/\lambda^2)^{1/2} \tag{5.7}$$

showing that χ_ℓ is a rapidly decreasing function of (ℓ/λ). (5.7) goes to $\sqrt{\pi}$ when $\ell \ll \lambda$ (formula (5.6)), and approaches zero at $\ell \approx \lambda$. Thus around $\ell \approx \lambda$, χ_ℓ is smaller than unity, meaning that in this domain a chaotic field yields a smaller cross sections compared with the single mode field. We may conclude this point by noting that the information of this subsection and concerning the influence of a chaotic field is largely new, and is confirmed by an exact numerical calculation of (5.4)[21].

5.2 Chaotic field with nonzero bandwidth

The previous laser model may be improved by allowing for a nonzero bandwidth $\Delta\omega$. Assuming as before a large number of modes ($N \to \infty$), in this limit the resulting field may be treated as a complex Gaussian stochastic process, with the first-order correlation function of the electric field values

$$\langle E(\tau') E(\tau) \rangle = \frac{1}{2} E_0^2 \cos[\omega(\tau' - \tau)] \exp(-\Delta\omega|\tau' - \tau|) \tag{5.8}$$

and a Lorentzian energy spectrum with $\Delta\omega$ the full width at half maximum.

Within the First Born Approximation, neglecting the energy spread of the incoming particle beam, and assuming $\Delta\omega \ll \omega$, the following "double" differential cross section is arrived[20] at

$$(d^2\sigma)_{CH} = \sum_\ell (k_f/k_i)(d\sigma/d\Omega)^B \exp(-\tfrac{\lambda^2}{2}) \mathcal{L}_\ell(\omega)\, d\Omega\, d\varepsilon_f \tag{5.9}$$

$$\mathcal{L}_\ell(\omega) = \sum_{K=0}^{\infty} f(L, K, \tfrac{1}{2}\lambda^2) \times$$

$$\times \frac{1}{\pi}\left\{\frac{\hbar(L+2K+\frac{3}{2}\lambda^2)\Delta\omega}{[\hbar(L+2K+\frac{3}{2}\lambda^2)\Delta\omega]^2 + (\varepsilon_f - \varepsilon_i - \ell\hbar\omega)^2}\right\} \quad (5.10)$$

In (5.9) and (5.10)

$$f(L, K, z) = \frac{1}{K!\,(L+K)!}\left(\frac{z}{2}\right)^{L+2K} \quad (5.11)$$

$$L = |\ell| \quad \text{and} \quad \lambda^2 = \bar{\lambda}^2_{fi}$$

Because of the presence of a nonzero bandwidth, the energy conserving delta functions of the type $\delta(\varepsilon_f - \varepsilon_i - \ell\hbar\omega)$ are replaced by $\mathcal{L}_\ell(\omega)$, which are interpreted as the corresponding scattering line shapes. (5.10) shows that the scattering line shape in strong fields may be much broader than $\Delta\omega$, or than $|\ell|\Delta\omega$. This result too is considered to give new information on the broadening effects a strong chaotic field may have on the scattering line shape in nonlinear domains. Actually, we have ignored the energy spread of the incoming particle beam, which usually is considerably larger than $\Delta\omega$. Calculations[20,22] shows that even under these conditions, the broadening due to the field fluctuations may be considerably larger than that originated by the particle energy spread.

For vanishing field bandwidth ($\Delta\omega \to 0$)

$$\mathcal{L}_\ell(\omega) = \delta(\varepsilon_f - \varepsilon_i - \ell\hbar\omega) \sum_{K=0}^{\infty} f(L, K, \tfrac{1}{2}\lambda^2)$$

$$= I_\ell(\tfrac{1}{2}\lambda^2)\, \delta(\varepsilon_f - \varepsilon_i - \ell\hbar\omega), \quad (5.12)$$

and we recover the results of the previous subsection.

Considering, instead, weak coupling ($\lambda \ll 1$)

$$\mathcal{L}_1(\omega) = \frac{1}{L!\,2^L}\left(\frac{\lambda^2}{2}\right)^L \frac{1}{\pi}\left\{\frac{\hbar L\Delta\omega}{(L\hbar\Delta\omega)^2 + (\varepsilon_f - \varepsilon_i - \ell\hbar\omega)^2}\right\} \quad (5.13)$$

which is the result expected on the basis of perturbation treatments.

5.3 Phase Diffusion Field

We now consider briefly the model of a phase diffusion field, corresponding to a well-stabilized laser beam. The phase diffusion field (PDM) has a constant real amplitude but its phase is a Wiener-Levy process[28-32]. Assuming again a Lorentzian energy spectrum for the field and performing as in the previous subsection we obtain[22]

$$(d^2\sigma)_{PDM} = \left(\frac{K_f}{K_i}\right)\left(\frac{d\sigma}{d\Omega}\right)^B \frac{1}{\pi} Re\left\{\frac{i}{(\varepsilon_f-\varepsilon_i) + \xi(q_o+\bar{q}_o)}\right\} d\Omega\, d\varepsilon_f \tag{5.14}$$

with $\xi = \lambda\hbar\omega$ and $\lambda = \lambda_{fi}$, while q_o and $\bar{q}_o$ are quantities which generate continued fractions[22]. (5.14) immediately reproduces the field-free result letting $\xi = 0$. Taking the weak field limit ($\xi \ll 1$) we obtain, instead,

$$(d^2\sigma)_{PDM} = \sum_\ell \left(\frac{K_f}{K_i}\right)\left(\frac{d\sigma}{d\Omega}\right)\left(\frac{\lambda}{2}\right)^{2L}\left(\frac{1}{L!}\right)^2 d\Omega\, d\varepsilon_f \times$$

$$\times \frac{1}{\pi}\left\{\frac{L^2\hbar\Delta\omega}{(L^2\hbar\Delta\omega)^2 + (\varepsilon_f-\varepsilon_i-\ell\hbar\omega)^2}\right\} \tag{5.15}$$

In the same limit of weak coupling, and taking $\bar{E}_o = E_o$, and the same bandwidth the ratio between the cross sections for a given ℓ calculated in the presence of a chaotic and phase diffusion field is obtained as

$$\chi_\ell = \frac{(d^2\sigma)^\ell_{CH}}{(d^2\sigma)^\ell_{PDM}} = \frac{\ell!\,(\varepsilon_f-\varepsilon_i-\ell\hbar\omega)^2 + (\ell^2\hbar\Delta\omega)^2}{\ell\,(\varepsilon_f-\varepsilon_i-\ell\hbar\omega)^2 + (\ell\hbar\Delta\omega)^2} \tag{5.16}$$

when for simplicity we consider $\ell > 0$. When

$$\varepsilon_f - \varepsilon_i - \ell\hbar\omega \approx 0$$

$$\chi_\ell \approx \ell!\,\ell \tag{5.17}$$

At the largest differences of $\varepsilon_f - \varepsilon_i - \ell\hbar\omega$ from zero

$$\chi_\ell \approx (\ell-1)! \tag{5.18}$$

The results (5.16)-(5.18) are typical of perturbation treatments, and are common to other processes[29]. But they generally stop to be true as soon as the strength of the coupling is increased. For instance, formulas (5.13) and (5.15) say that for a given ℓ, the scattering line shape is broadened according to a ℓ^2-law for PDM field, and to ℓ- law for a chaotic field. Thus, a PDM field is more effective in broadening the line shape than a chaotic field. Fig.4 shows[22] that this is no more true for strong fields, and that a strong chaotic field broades a line shape much more effectively than a PDM field.

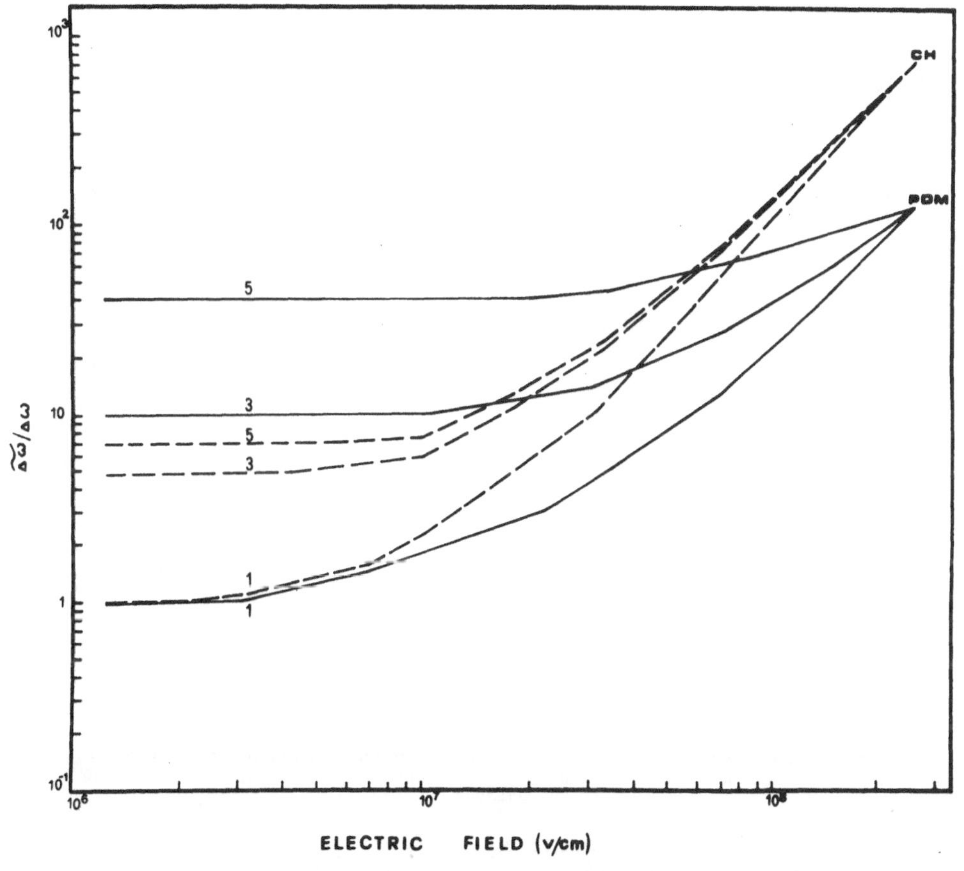

Fig. 4 Width of the scattering line, in units of the laser field bandwidth *<u>vs</u>* *the electric field strength, in V cm^{-1}. The number on the curves refer to the photon multiplicity. Continuous lines - phase diffusion field, PDM. Dashed lines - chaotic field, CH. The incident particle beam width is neglected. (From Ref.22).*

It would be of interest to consider also how the spatial and temporal inhomogeneities of strong fields affect the scattering parameters; and, in turn, which information on the properties of the assisting fields may be extracted from scattering data. These inhomogeneities are generally connected with the focusing and the pulsed regime. For these aspects the reader is referred to the sparse available literature(18,23,25-27). Some results concerning a field with a transverse spatial inhomogeneity of Gaussian type are shown in Fig.s 5, 6, where differential and total CS in an inhomogeneous field are compared with a single mode, homogeneous field with the same peak intensity(26).

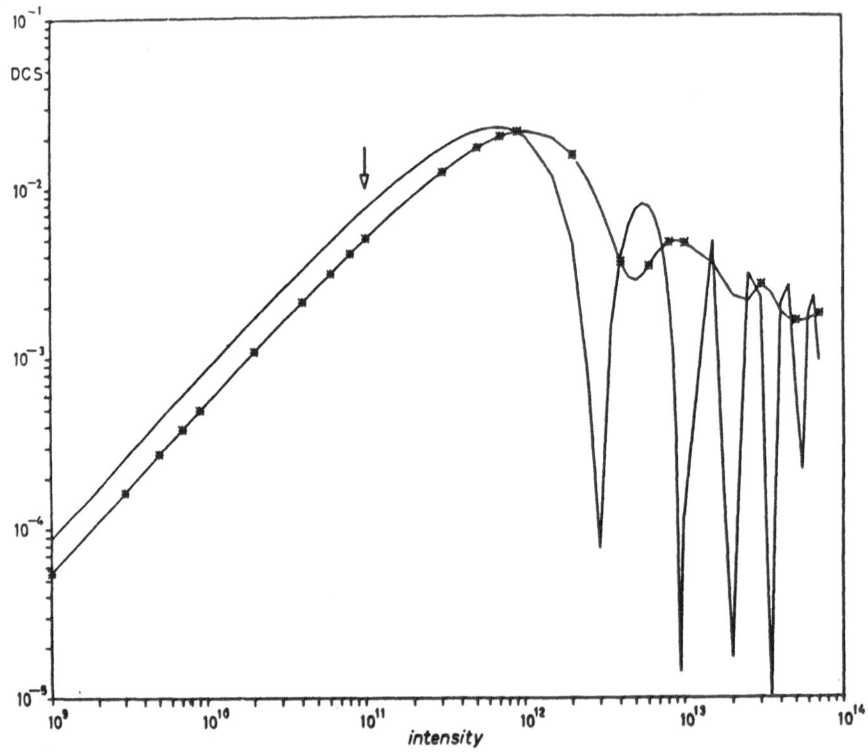

Fig. 5 Differential cross sections (DCS) of potential scattering in units of $\pi a_0^2/ster^{-1}$ as a function of the laser intensity, in Watt cm^{-2}. Number of exchanged photons $\ell = 1$, incoming particle energy $\mathcal{E}_i = 100$ eV; scattering angle $\vartheta = 45°$, photon energy $\hbar\omega = 1.17$ eV.
Curve with asterisk - laser field with a transverse inhomogeneity of Gaussian type. Curve without asterisk - homogeneous field. The two field are taken to have the same peak intensity. (From Ref.26).

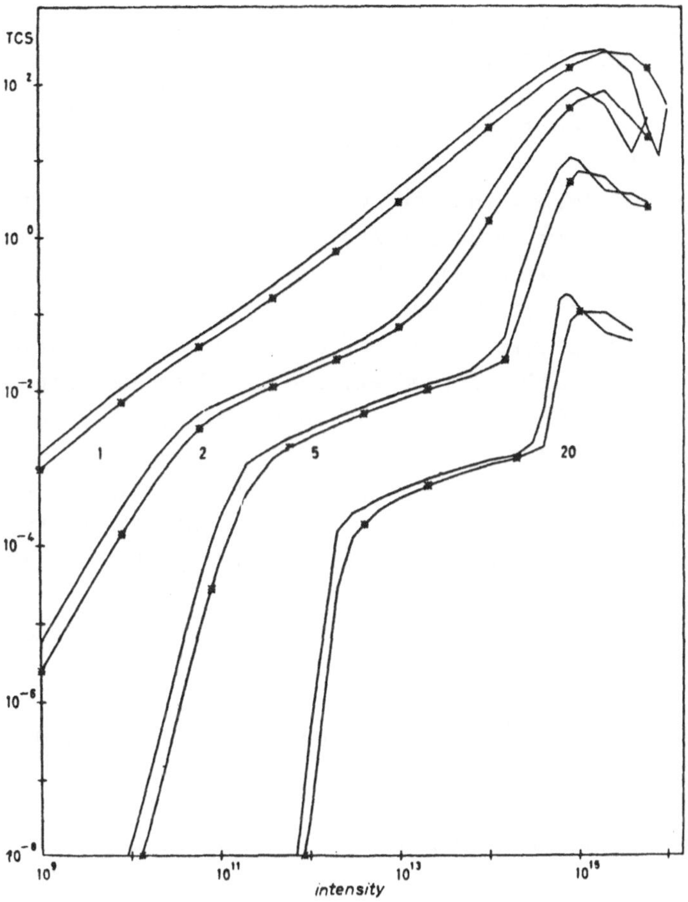

Fig. 6 *Total cross sections (TCS), in πa_0^2, vs the laser intensity, in Watt cm^{-2}. The numbers on the curve refer to the number of exchanged photons. Other explanations as in caption to Fig.5. (From Ref.26).*

6. LOW FREQUENCY APPROXIMATION

Most of the considerations made above and based on the FBA cross sections hold true also for the exact cross sections (especially away from the scattering resonances) under the additional assumption of a 'low frequency (LF)' field. Below we will consider that a low frequency field situation is encountered when the initial particle energy ε_i is much greater than the photon energy $\hbar\omega$ and the particle-field interaction energy $W_0 = (\hbar\omega)\,\underline{\alpha}_0 \cdot \underline{K}$:

$$\varepsilon_i \gg \hbar\omega \; ; \qquad \varepsilon_i \gg (\hbar\omega)\,\underline{\alpha}_0 \cdot \underline{K} \quad . \tag{6.1}$$

In physical terms, the conditions (6.1) amount to saying that the particle-field energy exchanges are a small fraction of the energies typical of the collision event. To some extent, (6.1) reproduces the physical content of the FBA; thus it should be not very surprising that a factorization like that obtained in the first order treatment is eventually arrived at. The low frequency approximation (LFA) to field assisted collisions, in different versions, has attracted until now a lot of attention[33]. It has been due, basically, to two reasons: (i) in many real cases, collision processes and experiments take place under conditions satisfying the inequalities (6.1) (see also Table 1); (ii) the LFA yields exact formal results of a rather general interest.

From a more technical point of view, the LFA outlined below will be based on the following assumptions:

(i) The LF field couples strongly to the particle only in the initial and final states (for which, accordingly, use is made of Volkov states).

(ii) In the intermediate states the field couples weakly to the particle, and their interaction is treated accordingly to pertubration theory.

The physical justification of (i) and (ii) is that the perturbation

$$W(t) = W\cos\omega t = (\hbar\omega)\,\underline{\alpha}_0 \cdot \underline{K}\,\cos\omega t \tag{6.2}$$

has a longer time to act on the initial and final states than on the intermediate states. This time may be estimated to be of order of the field period T, and it eliminates a power of ω in the perturbation (6.2), which then becomes of the order of magnitude of the coupling parameter $\lambda_K = \underline{\alpha}_0 \cdot \underline{K}$.

(iii) Finally, we will account also for the possibility that there is an isolated scattering resonance at energy $E_R = \varepsilon_R - i\Gamma/2$ such that $\varepsilon_R \approx \varepsilon_0 = \varepsilon_i + \ell_0 \hbar\omega$.

In the treatment outlined below, it is also understood that in considering low frequencies, the parameter α_0 which depends on ω^{-2}, is held fixed and finite. This on physical grounds. Thus any decreasing of the frequency implies a corresponding decreasing of the electric field. This fact, together with the truncation of expansion (6.3) to terms linear in ω, makes in practice the contents of this Section a low frequency and low field treatment.

The starting point will be again the exact S matrix (3.1) and the exact wavefunction (3.2). But the difference will be that the full retarded Green function appearing in (3.2) instead of being expanded inpowers of the scattering potential V, will be expanded in powers of the particle-field interaction W :

$$\bar{G}^+ = G^+ + G^+ W G^+ + G^+ W G^+ W G^+ + \cdots \tag{6.3}$$

where now G^+ is the retarded Green function in the absence of the laser

$$G^+(\underline{r}t;\underline{r}'t') = (-i/\hbar)\sum_m \psi_m(\underline{r})\,\psi_m^*(\underline{r}')\exp\left\{\frac{-i}{\hbar}(E_m - i\eta)(t-t')\right\} \tag{6.4}$$
$$\lim \eta \to 0^+$$

In (6.4) $\{\psi_m\}$ is a complete set of normalized wavefunctions of the particle in the presence of the potential, and $\{E_m\}$ their eigenvalues. Using the expansion (6.3), the exact S matrix (3.1) is rewritten as

$$S = (\varphi_f, V\varphi_i) + (\varphi_f, VG^+V\varphi_i) + \sum_{\nu=3}^{\infty}(\varphi_f, V[G^+W]^{\nu-2}G^+V\varphi_i)$$
$$= \sum_{\nu=1}^{\infty} S^{(\nu)}$$

The first two terms of (6.5)

$$S^{(1)} + S^{(2)} = (\varphi_f, V\varphi_i) + (\varphi_f, VG^+V\varphi_i) \tag{6.5}$$

correspond to the complete S matrix, in which only the initial and final states are coupled to the field (through the Volkov wavefunctions φ_i and φ_f). After the time integrations they are readily rewritten as

$$\sum_{\nu=1}^{2} S^{(\nu)} = (-2\pi i)\sum_{\ell}(-1)^{\ell}\,\delta(\varepsilon_f-\varepsilon_i-\ell\hbar\omega)\langle \underline{K}_f|\sum_{\nu=1}^{2} T_\ell^{(\nu)}|\underline{K}_i\rangle, \tag{6.7}$$

$$\langle \underline{K}_f|\sum_{\nu} T_\ell^{(\nu)}|\underline{K}_i\rangle = \sum_{\ell_1} J_{\ell_1-\ell}(\underline{K}_f\cdot\underline{\alpha}_0)\langle \underline{K}_f|T(\varepsilon_i+\ell_1\hbar\omega)|\underline{K}_i\rangle J_{\ell_1}(\underline{K}_i\cdot\underline{\alpha}_0), \tag{6.8}$$

$$\langle \underline{K}_f|T(\varepsilon_x)|\underline{K}_i\rangle = \langle \underline{K}_f|V|\underline{K}_i\rangle + \sum_m \frac{\langle \underline{K}_f|V|\underline{K}_m\rangle\langle \underline{K}_m|V|\underline{K}_i\rangle}{\varepsilon_x-\varepsilon_m+i\eta} + \dots \tag{6.9}$$

.

(6.9) is the conventional field-free T matrix of a particle with incoming energy ε_x. The next and last term, which is considered in the present treatment is

$$S^{(3)} = (\varphi_f, V G^+ W G^+ V \varphi_i) \tag{6.10}$$

This term says that in the intermediate scattering events only one photon is exchanged. After the time integrations are carried out also in (6.10), the first three terms of (6.5) are further elaborated (the details may be found in Ref. 34) according basically to the following two points. (i) It is assumed that there is an isolated resonance at the energy $E_R = \varepsilon_R - i\Gamma/2$ such that $\varepsilon_R \approx \varepsilon_0 = \varepsilon_i + \ell_0\hbar\omega$. It is used to distinguish $\ell_1 = \ell_0$ from all the other terms, and to separate the T matrix into a resonant and a nonresonant part. (ii) It is assumed that $\ell_0 \neq 0$ and $\ell_0 \neq \ell$ (neither the initial nor the final energies are resonant). It is used to expand in power series the nonresonant part of the T matrix around ε_i and ε_f. After a certain amount of algebra, what is left of the exact S matrix yields the following differential cross section

$$\begin{aligned}(d\sigma/d\Omega)^{LF} = {} & \sum_{\ell}(K_f^{(\ell)}/K_i)\, J_\ell^2(\underline{Q}_{fi}\cdot\underline{\alpha}_0)\,(d\sigma^S/d\Omega)_{NR}^{Ex} \\ & + \sum_{\ell}\left(\frac{K_f}{K_i}\right) J_{\ell-\ell_0}^2(\underline{K}_f\cdot\underline{\alpha}_0)\, J_{\ell_0}^2(\underline{K}_i\cdot\underline{\alpha}_0)\,(d\sigma(\varepsilon_0)/d\Omega)_R^{Ex} \\ & + \sum_{\ell}(-1)^{\ell}\left(\frac{K_f}{K_i}\right) J_{\ell-\ell_0}(\underline{K}_f\cdot\underline{\alpha}_0)\, J_{\ell_0}(\underline{K}_i\cdot\underline{\alpha}_0)\, J_\ell(\underline{Q}_{fi}\cdot\underline{\alpha}_0)\,\times \\ & \times (d\sigma/d\Omega)_{IN}^{Ex}\end{aligned} \tag{6.11}$$

In (6.11)

$$(d\sigma^s/d\Omega)^{Ex}_{NR} = \left(\frac{m}{2\pi\hbar^2}\right)^2 |\langle \underline{K}^s_f | T_{NR}(\varepsilon^s_i) | \underline{K}^s_i \rangle|^2 \qquad (6.12)$$

is the exact field-free nonresonant cross section, where the field-free T matrix has momenta and energies shifted by the field according to

$$\begin{aligned} \varepsilon^s_j &= \varepsilon_j - \delta\varepsilon_j \;; \quad \delta\varepsilon_j = (\ell\hbar\omega)\,\underline{K}_j\cdot\underline{\alpha}_0/\underline{Q}_{fi}\cdot\underline{\alpha}_0 \\ \underline{K}^s_j &= \underline{K}_j - \delta\underline{K} \;; \quad \delta\underline{K} = (\ell\hbar\omega)\,m\,\underline{\alpha}_0/\hbar^2\underline{Q}_{fi}\cdot\underline{\alpha}_0 \end{aligned} \qquad (6.13)$$

Thus the first term of (6.11), which is the term originally derived by Kroll and Watson [2], shows to all orders the same factored structure as the FBA cross section (3.27).

$$(d\sigma(\varepsilon_0)/d\Omega)^{Ex}_R = \left(\frac{m}{2\pi\hbar^2}\right)^2 |\langle \underline{K}_f | \overline{T}_R(\varepsilon_0) | \underline{K}_i \rangle|^2 \qquad (6.14)$$

is the exact field-free resonant cross section. $\overline{T}_R(\varepsilon_0) = T_R(\varepsilon_0) - T_{NR}(\varepsilon_i)$ with $T_R(\varepsilon_0)$ the resonant part of the field-free T matrix and $T_{NR}(\varepsilon_i)$ a nonresonant contribution, which tends to cancel $T_R(\varepsilon_0)$ as one moves away from resonance. The second term of (6.11) corresponds to the result originally derived by Kruger and Jung [14].

$$(d\sigma/d\Omega)^{Ex}_{IN} = \left(\frac{m}{2\pi\hbar^2}\right)^2 \left\{ 2\,\mathrm{Re}\,\langle \underline{K}^s_f | T_{NR}(\varepsilon^s_i) | \underline{K}^s_i \rangle \langle \underline{K}_f | \overline{T}_R(\varepsilon_0) | \underline{K}_i \rangle \right\} \qquad (6.15)$$

is the contribution to the cross section originating from the interference of the scattering amplitudes.

The cross section (6.11) may now be simplified exploiting the weak dependence of scattering parameters on ℓ, and some properties of the Bessel functions. In fact, using twice the closure property (3.31) and one addition theorem of the Bessel functions, (6.11) may be rewritten in the form of another sum rule as

$$\sum_\ell \left(\frac{d\sigma}{d\Omega}\right)^{LF}_\ell = \left(\frac{d\sigma}{d\Omega}\right)^{Ex}_{NR} + J^2_{\ell_0}(\underline{K}_i\cdot\underline{\alpha}_0)\left(\frac{d\sigma}{d\Omega}\right)^{Ex}_R + J^2_{\ell_0}(\underline{K}_i\cdot\underline{\alpha}_0)\left(\frac{d\sigma}{d\Omega}\right)^{Ex}_{IN} \qquad (6.16)$$

or, after some rearrangements, as [34]

$$\sum_\ell \left(\frac{d\sigma}{d\Omega}\right)^{LF}_\ell = \left[1 - J^2_{\ell_0}(\underline{K}_i\cdot\underline{\alpha}_0)\right]\left(\frac{d\sigma}{d\Omega}\right)^{Ex}_{NR} + J^2_{\ell_0}(\underline{K}_i\cdot\underline{\alpha}_0)\left(\frac{d\sigma}{d\Omega}\right)^{Ex}_{\alpha_0=0} \qquad (6.17)$$

where

$$\left(\frac{d\sigma}{d\Omega}\right)^{Ex}_{\alpha_0=0} = \left(\frac{m}{2\pi\hbar^2}\right)^2 |\langle \underline{K}_f | T_{NR}(\varepsilon_i) + T_R(\varepsilon_0) | \underline{K}_i \rangle|^2 \tag{6.18}$$

is the field-free cross section, in which the resonant and the non-resonant parts are not separated. The result (6.11) suggests some comments and further elaborations. First, it shows that the presence of a laser produces a cross section in which nonresonant, resonant and interference parts enter separately, contrary to the field-free case. So, in principle, it opens the interesting possibility of a separate observation of only one of them. In fact, choosing the condition $\underline{Q}_{fi} \cdot \underline{\alpha}_0 = 0$ and observing only electrons with given energy $\varepsilon_f = \varepsilon_i + \ell\hbar\omega$, $(\ell \neq 0)$, one would measure only the resonant cross section

$$\left(\frac{d\sigma}{d\Omega}\right)^{Ex}_{\ell, R} = \left(\frac{K_f(\ell)}{K_i}\right) J^2_{\ell-\ell_0}(\underline{K}_f \cdot \underline{\alpha}_0) J^2_{\ell_0}(\underline{K}_i \cdot \underline{\alpha}_0) \left(\frac{d\sigma}{d\Omega}\right)^{Ex}_R \tag{6.19}$$

in which the background is absent. This is the original suggestion by Jung and Taylor[35]. The shortcoming of (6.19) is that the 'resonant' signal is very small (proportional to the product of two squared Bessel functions) and thus difficult to detect. However, considering the two low-frequency and low field assumptions (6.1), it should be possible to fulfill the condition $\underline{Q}_{fi} \cdot \underline{\alpha}_0 \approx 0$ for most of the electrons suffering nonresonant scattering, and not only for electrons with a given $\varepsilon_f = \varepsilon_i + \ell\hbar\omega$. More precisely, choosing $\underline{Q}_{fi} \cdot \underline{\alpha}_0 = 0$ at $\ell = 0$. Then all electrons scattered at a given angle may be divided in two groups. The first group is formed by electrons with energy $\varepsilon_f = \varepsilon_i$, and most of them come from the nonresonant part of the collision process (a fraction comes also from the resonant part taken with $\ell = 0$, and corresponds to electrons which emit (absorb) in the final state the same number of photons absorbed (emitted) in the initial state). The second group is formed by electrons with $\varepsilon_f \neq \varepsilon_i$, and they come mostly from the resonant part of the collision events (a very small fraction may come also from nonresonant collisions). Thus detecting, under the condition $\underline{Q}_{fi} \cdot \underline{\alpha}_0 \approx 0$ all the electrons with $\varepsilon_f \neq \varepsilon_i$ one measures essentially the resonant cross section with the difference that now the signal is more intense (being proportional to only one squared Bessel function). In particular, it may be expressed approximately in the form

$$\left(\frac{d\sigma}{d\Omega}\right)_F = \left(\frac{d\sigma}{d\Omega}\right)^{Ex}_{NR} + (-1)^{\ell_0} J^2_{\ell_0}(\underline{k}_i\cdot\underline{\alpha}_0)\left(\frac{d\sigma}{d\Omega}\right)^{Ex}_{IN} +$$

$$+ J^2_{\ell_0}(\underline{k}_i\cdot\underline{\alpha}_0)\left[1 - J^2_{-\ell_0}(\underline{k}_f\cdot\underline{\alpha}_0)\right]\left(d\sigma/d\Omega\right)^{Ex}_R \quad . \qquad (6.20)$$

The first two terms of (6.20) represent collisions going essentially with $\varepsilon_f = \varepsilon_i$, while the third term (the resonant one) represents collisions with $\varepsilon_f \neq \varepsilon_i$. The differences with (6.17) are worthy to be stressed. In (6.17) all the terms (nonresonant and interference) represent collisions with $\varepsilon_f = \varepsilon_i + \ell\hbar\omega$ and $\ell = 0, \pm 1, \pm 2, \ldots$.

7. PARTICLE-ATOM COLLISIONS IN A LASER FIELD

We now consider particle-atom collisions in the presence of a laser field, when the atom internal structure plays a role. For simplicity, we will consider only the case of one-electron atoms. Besides, we will confine ourselves only to a number of considerations, without developing in detail the full formalism. One of the reasons of this restriction is that the formalism is rather involved and lengthy and several references are available in the literature, which contain a presentation of the formalism with more details than allowed by the limited space of this lecture. Another reason is that, given a limited space, one can learn more about the basic theory by dwelling on basic concepts rather than by developing the formalism in detail.

7.1 Laboratory and center of mass systems

Our starting point will be the following Schrödinger equation (in the laboratory reference frame)

$$\left[\frac{1}{2m_p}\left(\frac{\hbar}{i}\nabla_p - \frac{e_p}{c}\underline{A}\right)^2 + \frac{1}{2m_e}\left(\frac{\hbar}{i}\nabla_e - \frac{e}{c}\underline{A}\right)^2 + \frac{1}{2m_N}\left(\frac{\hbar}{i}\nabla_N - \frac{e_N}{c}\underline{A}\right)^2\right.$$

$$\left. + V(|\underline{r}_e - \underline{r}_N|) + W(\underline{r}_e, \underline{r}_p, \underline{r}_N)\right]\Psi(\underline{r}_e \underline{r}_p \underline{r}_N, t) = i\hbar\,\dot{\Psi}(\underline{r}_e \underline{r}_p \underline{r}_N, t) \qquad (7.1)$$

Eq.(7.1) is meant to describe collisions between a projectile with mass and charge M_p and e_p and a one-electron target with a nucleus having mass and charge m_N and e_N . V is the electron-nucleus interaction, while W stands for the projectile-atom interactions. Fig. 7 illustrates the meaning of the coordinates appearing in (7.1) and in the following equations. As before, the vector potential $\underline{A}(t)$ is taken to be only a function of time (dipole approximation). In this approximation, transforming (7.1) to the cm system with the change

$$(\underline{r}_e, \underline{r}_p, \underline{r}_N) \rightarrow (\underline{R}, \underline{x}_e, \underline{x}_p)$$

where

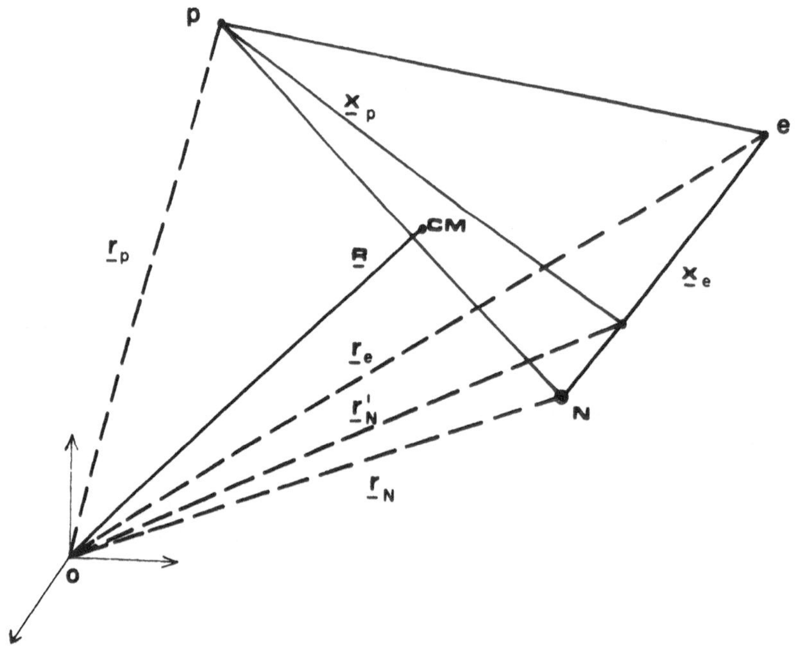

Fig. 7 Laboratory and cm coordinates.

$$\underline{R} = (\underline{r}_e m_e + \underline{r}_p m_p + \underline{r}_N m_N)/M_R \; ; \tag{7.2}$$

$$M_R = m_e + m_p + m_N \; ; \tag{7.2'}$$

$$\underline{x}_p = \underline{r}_p - \underline{r}_N'$$

$$= \underline{r}_p - (\underline{r}_e m_e + \underline{r}_N m_N)/(m_N + m_e) \; ; \tag{7.3}$$

$$\underline{x}_e = \underline{r}_e - \underline{r}_N , \tag{7.4}$$

one obtains

$$\Big[\frac{1}{2M_R} \Big(\frac{\hbar}{i} \underline{\nabla}_R - \frac{Q_R}{c} \underline{A} \Big)^2 +$$

$$+ \frac{1}{2\mu_p} \Big(\frac{\hbar}{i} \underline{\nabla}_p - \frac{Q_p}{c} \underline{A} \Big)^2 + \frac{1}{2\mu_e} \Big(\frac{\hbar}{i} \underline{\nabla}_e - \frac{Q_e}{c} \underline{A} \Big)^2 + W(\underline{x}_e, \underline{x}_p)$$

$$+ V(x_e) \Big] \Psi(\underline{R}\, \underline{x}_e \underline{x}_p, t) = i\hbar \dot{\Psi}(\underline{R}\, \underline{x}_e \underline{x}_p, t) \quad . \tag{7.5}$$

In (7.5)

$$\mu_p = m_p (m_e + m_N)/M_R \; ; \tag{7.6}$$

$$\mu_e = m_e m_N / (m_e + m_N) \; ; \tag{7.7}$$

$$\underline{\nabla}_p \equiv \underline{\nabla}_{x_p} \; ; \quad \underline{\nabla}_e \equiv \underline{\nabla}_{x_e} \quad .$$

To the masses M_R, μ_p and μ_e the following total and reduced charges are, respectively, associated

$$Q_R = e + e_N + e_p \tag{7.8}$$

$$Q_p = [e_p(m_e + m_N) - (e_N + e) m_p]/M_R \tag{7.9}$$

$$Q_e = (e m_N - e_N m_e)/(m_e + m_N) \quad . \tag{7.10}$$

As in the case of potential scattering, considered in Section 2, the change of variables (7.2)-(7.4) has yielded the separation of the cm motion from the internal, relative motion. The considerations presented in the subsection 2.2 and concerning the cm motion of the whole colliding complex in an external field apply to the present case as well, and accordingly, the same simplification is adopted. Here, to neglect the effect of the field on the whole system cm motion, it must be added that this effect be also much smaller than the effect of the field on the target internal structure.

7.2 Gauge considerations

Removing from (7.5) the part accounting for the cm motion, for the internal motion, one is left with the Schrödinger equation

$$\Big[\frac{1}{2\mu_p}\Big(\frac{\hbar}{i}\nabla_p - \frac{Q_p}{c}\underline{A}\Big)^2 + \frac{1}{2\mu_e}\Big(\frac{\hbar}{i}\nabla_e - \frac{Q_e}{c}\underline{A}\Big)^2 + W(\underline{x}_e,\underline{x}_p)$$

$$+ V(x_e)\Big]\psi(\underline{x}_p,\underline{x}_e,t) = i\hbar\dot{\psi}(\underline{x}_p,\underline{x}_e,t) \tag{7.11}$$

Choosing as before the projectile-target interactions $W(\underline{x}_e,\underline{x}_p)$ as the operator responsible for the transitions, the unperturbed states are given by a product of the solutions of the two equations

$$\frac{1}{2\mu_p}\Big(\frac{\hbar}{i}\nabla_p - \frac{Q_p}{c}\underline{A}\Big)^2\varphi(\underline{x}_p,t) = i\hbar\dot{\varphi}(\underline{x}_p,t) \,, \tag{7.12}$$

$$\Big[\frac{1}{2\mu_e}\Big(\frac{\hbar}{i}\nabla_e - \frac{Q_e}{c}\underline{A}\Big)^2 + V(x_e)\Big]\psi(\underline{x}_e,t) = i\hbar\dot{\psi}(\underline{x}_e,t) \,; \tag{7.13}$$

namely

$$\Phi_{\underline{K}n}(\underline{x}_p,\underline{x}_e,t) = \varphi_{\underline{K}}(\underline{x}_p,t)\,\psi_m(\underline{x}_e,t) \tag{7.14}$$

From eq.s(7.12) and (7.13), it is apparent that the unperturbed states, both continuum and discrete, are states embedded in the field.

Within the present formalism, as before, this is the basic new feature of the theory of field assisted particle-atom collisions. $\varphi_K(\underline{x}_p,t)$ is exactly known, being nothing else than the familiar nonrelativistic plane waves (2.10); no exact solutions are instead known for atomic electrons in the presence of a vector potential $\underline{A}(t)$. So one is forced to look for approximate solutions of (7.13). In trying to solve approximately the Schrödinger equation of an atom in an external field is customary to look for choices of the e.m. gauges giving better results or, more exactly, a faster convergence within the adopted approximate treatment.

In particular, as far as bound electrons are concerned, the gauge is more frequently used, in which the field is represented by its electric field $\underline{E}(t)$, (electric field gauge), instead of the gauge in which the field is represented by its vector potential $\underline{A}(t)$, (radiation gauge). 'En passant' we observe that the questions concerned with the choice of the e.m. gauge to correctly describe radiation-matter interactions have received in the last years a lot of attention[36], and this topic is likely to be important also for field-assisted collisions.

From the standpoint of quantum mechanics, a change of gauge amounts to perform a unitary transformation on the corresponding Schrödinger equation. In our case to go from the R-gauge to the E-gauge, the required unitary operator is[37]

$$T(\underline{x}_e,t) = \exp\left\{-i\frac{Q_e}{\hbar c}\underline{A}(t)\cdot\underline{x}_e\right\} \tag{7.15}$$

(the dipole approximation being implied).

Before using (7.15) to transform (7.13), we write (7.13) as

$$H\psi^R = (H_0 + H_i^A)\psi^R = i\hbar\dot{\psi}^R \tag{7.16}$$

with

$$H_0 = \frac{p^2}{2\mu_e} + V(x_e) \tag{7.17}$$

$$H_i^A = -\frac{Q_e}{\mu_e c}\underline{A}\cdot\underline{p} + \frac{Q_e^2}{2\mu_e c^2}A^2 \tag{7.18}$$

The transformed Schrödinger equation is

$$\left[i\hbar \dot{T}T^{\dagger} + THT^{\dagger}\right]\psi^{E}(\underline{x}_{e},t) = i\hbar\,\dot{\psi}^{E}(\underline{x}_{e},t) \tag{7.19}$$

where

$$\psi^{E} = T\,\psi^{R} \tag{7.20}$$

and vice versa

$$\psi^{R} = T^{\dagger}\psi^{E}. \tag{7.20'}$$

As

$$THT^{\dagger} = H_{o} \tag{7.21}$$

$$\begin{aligned} i\hbar\,\dot{T}T^{\dagger} &= \frac{Q_e}{c}\,\underline{\dot{A}}\cdot\underline{x}_e \\ &= -\,Q_e\,\underset{\sim}{E}(t)\cdot\underline{x}_e \\ &= H_i^{E} \end{aligned} \tag{7.22}$$

finally we have

$$\left(-\frac{\hbar^2\nabla_e^2}{2\mu_e} + V(x_e) - Q_e\,\underset{\sim}{E}(t)\cdot\underline{x}_e\right)\psi^{E}(\underline{x}_e,t) = i\hbar\,\dot{\psi}^{E}(\underline{x}_e,t) \tag{7.23}$$

with (7.22) describing the coupling of the electric dipole $Q_e\underline{x}_e$ with the electric field of the laser $\underline{E}(t)$. An order of magnitude estimate shows that

$$H_i^{E} \approx H_i^{A}\,(\omega/\omega_{nm}) \tag{7.24}$$

with ω the frequency of the field, and ω_{nm} a typical transition frequency of the bound electron. Thus, the unitary transformation (7.15) has the appealing feature of transforming the Schrödinger equation to a gauge in which the coupling between the bound electron and the field is reduced provided the frequency of the perturbing field ω is lower than the atomic transition frequencies.

To ensure gauge consistency, the product wavefunction (7.14) must be given in the same gauge. In the E-gauge, for instance, we have

$$\begin{aligned}\phi^{E}(\underline{x}_p, \underline{x}_e, t) &= \varphi^{E}_{\underline{k}}(\underline{x}_p, t)\, \psi^{E}_{n}(\underline{x}_e, t)\\ &= T(\underline{x}_p, t)\, \varphi^{A}_{\underline{k}}(\underline{x}_p, t)\, \psi^{E}_{n}(\underline{x}_e, t)\end{aligned} \qquad (7.25)$$

where

$$\begin{aligned}\varphi^{E}_{\underline{k}} &= T \varphi^{R}_{\underline{k}}\\ &= \exp\left\{ i\left(\underline{k} - \frac{Q_p \underline{A}}{\hbar c}\right)\cdot \underline{x}_e \right\} \times\\ &\quad \times \exp\left\{ -\frac{i}{\hbar} \int^{t} \frac{1}{2\mu_p}\left[\hbar \underline{k} - \frac{Q_p}{c}\, \underline{A}(\tau)\right]^2 d\tau \right\}\end{aligned} \qquad (7.26)$$

is an exact solution in the E-gauge, and $\psi^{E}_{n}(\underline{x}_e, t)$ is now meant to represent an approximate solution in the E-gauge.

To use ψ^{E}_{n} in the R-gauge, one must write

$$\phi^{R}_{\underline{k}n} = \varphi^{R}_{\underline{k}}(\underline{x}_p, t)\, T^{\dagger}(\underline{x}_e, t)\, \psi^{E}_{n}(\underline{x}_e, t) \qquad (7.27)$$

It must be pointed out, however that $T^{\dagger}\psi^{E}_{n}$ is generally not the same as ψ^{R}_{n} , the solution of (7.13) in the R-gauge obtained within the same approximate treatment used to determine ψ^{E}_{n} . $T^{\dagger}\psi^{E}_{n}$ is thus representing in the R-gauge only an approximate result obtained in the E-gauge. In any case, however, the unitary operator needs to be properly included in the product wavefunction. In concluding these considerations, we observe that in collision treatments using states embedded in the field the problems raising from the use of given e.m. gauges are new as compared with those arising in treatments in which the field enters only in the transition operators. In spite of its evident interest, until now these problems have received very little attention.

8. FORMALISM FOR PARTICLE-ATOM COLLISIONS

8.1 The S-Matrix

Here we show briefly how the scattering theory formalism outlined in Section 3 is generalized to include the target internal structure. Let us start again with the exact S-matrix

$$S = (-i/\hbar)(\phi_f , W\Psi_i^+) \tag{8.1}$$

where now ϕ_f is the final state product wavefunction, given by (7.14), determined in a suitable gauge. Ψ_i^+ is the exact wavefunction satisfying the complete Schrödinger equation (7.11) with causal boundary conditions, and the round brackets indicate both space and time integrations. Formally, we have

$$\Psi_i^+ = \phi_i + \bar{G}^+ W \phi_i \tag{8.2}$$

with $\bar{G}^+$ the retarded full Green function (in the presence of both interactions W and the radiation field). $\bar{G}^+$ is now expanded in powers of W :

$$\bar{G}^+ = \bar{G}_0^+ + \bar{G}_0^+ W \bar{G}_0^+ + \bar{G}_0^+ W \bar{G}_0^+ W \bar{G}_0^+ + \cdots \tag{8.3}$$

with

$$\begin{aligned}\bar{G}_0^+ &= (-i/\hbar)\,\Theta(t-t')\sum_m \int \frac{d^3K_m}{(2\pi)^3}\,\phi_m(\underline{x}_p,\underline{x}_e,t)\,\phi_m^*(\underline{x}_p',\underline{x}_e',t') \\ &= (-i/\hbar)\,\Theta(t-t')\,\times \\ &\quad\times \sum_m \int \frac{d^3K_m}{(2\pi)^3}\,\varphi_{\underline{K}}(\underline{x}_p,t)\,\varphi_{\underline{K}}^*(\underline{x}_p',t')\,\psi_m(\underline{x}_e,t)\,\psi_m^*(\underline{x}_e',t)\end{aligned} \tag{8.4}$$

Using (8.2)-(8.4) in (8.1) one has

$$\begin{aligned}S &= (-i/\hbar)\left[(\phi_f , W\phi_i) + (\phi_f , W\bar{G}_0^+ W\phi_i) + \cdots\right] \\ &= \sum_{\mu=1}^{\infty} S^{(\mu)}\end{aligned} \tag{8.5}$$

where

$$S^{(1)} = (-i/\hbar)(\Phi_f, W\Phi_i)$$

$$= (-i/\hbar)\int_{-\infty}^{\infty} dt \int d^3x_p\, d^3x_e\, \varphi^*_{K_f}(\underline{x}_p,t)\psi^*_f(\underline{x}_e,t) \times$$

$$\times W(\underline{x}_p,\underline{x}_e)\, \varphi_{K_i}(\underline{x}_p,t)\psi_i(\underline{x}_e,t) \tag{8.6}$$

$$S^{(2)} = (-i/\hbar)\int_{-\infty}^{\infty} dt \int_{-\infty}^{t} dt' \sum_m \int \frac{d^3K_m}{(2\pi)^3} \times$$

$$\int d^3x_p\, d^3x_e\, \varphi^*_{K_f}(\underline{x}_p,t)\,\psi^*_f(\underline{x}_e,t)\, W(\underline{x}_p,\underline{x}_e)\,\varphi_{K_m}(\underline{x}_p,t)\,\psi_m(\underline{x}_e,t) \times$$

$$\int d^3x'_p\, d^3x'_e\, \varphi^*_{K_m}(\underline{x}'_p,t')\,\psi^*_m(\underline{x}'_e,t')\, W(\underline{x}'_p,\underline{x}'_e)\,\varphi_{K_i}(\underline{x}'_p,t')\,\psi_i(\underline{x}'_e,t'), \tag{8.7}$$

and so on.

If the incoming particle is an electron, the full wavefunction must be antisymmetrized, to take into account the Pauli principle. In this case

$$\Psi^+ = A_{pe}(\Phi_i + \bar{G}^+ W\Phi_i) \tag{8.8}$$

where $A_{pe} = 1 \pm P_{pe}$ and P_{pe} is the operator inter-changing the coordinates $\underline{x}_p$ and $\underline{x}_e$. The upper sign is for singlet states, the lower for triplet. Further

$$S = (-i/\hbar)\left[(\Phi_f, WA_{pe}\Phi_i) + (\Phi_f, WA_{pe}\bar{G}_0^+ W\Phi_i) + \dots\right],$$

and so on.

It should be evident that the series (8.5) will have a structure much more involved than the corresponding series for potential scattering (3.8), and that a wealth of possibilities are offered concerning spacific approximations. First of all, we need a set of approximate wavefunctions $\{\psi_n(\underline{x}_e,t)\}$ for the target embedded in the field. As no exact solution is known, the approximation take advantage of the particular physical conditions considered, and it opens several possibilities, as discussed below. Further, the Green function (8.4) as well may be expanded in powers of the operators giving the coupling of the projectile and of the target with the field. As seen previously, it is particularly true in the cases of low frequency fields.

8.2 Some considerations on laser-atom interactions

In any case, the series (8.5) poses formidable calculational problems; here, however, we will be interested only in the theoretical aspects of it. Namely, we wish to discuss how acceptable approximations for the bound electron wavefunction may be determined. When this is done, it is generally found that the time integrations in (8.5) are easily performed, and explicit expressions for transition probabilities and cross sections may be given.

To discuss problems concerned with the laser modifications of the atomic spectrum, a convenient reference value of electric field strength is the field the electron experiences in the hydrogen atom in the first Bohr orbit,

$$E_c = e/a_0^2 = 5.14 \times 10^9 \text{ V/cm}.$$

It gives also a reference value of intensity

$$I_c = 3.3 \times 10^{16} \text{ W/cm}^2.$$

Concerning specifically the laser-atom interaction energy, an estimate gives

$$eE_0 \langle \underline{x}_e \rangle \simeq eE_0 a_0 = Ry(I/I_c)^{1/2}$$

saying that very large intensities are required to produce interaction energies that are even a small fraction of a rydberg.

Thus, as far as the field intensity is concerned, we can easily envisage two extreme situations. The first: the laser has no effect on the atom internal structure. A significant example is given by the experiments by Weingartshofer and coworkers on elastic electron-Argon collisions, where up to 11 eleven photons have been exchanged(13). In those experiments the field intensity is estimated to be $I \gtrsim 10^8$ Watt/cm^2. These experiments are satisfactorily explained without invoking field modifications of the atomic spectrum. The other extreme situation is encountered when the laser destroys the atom before a collision experiment may be completed. Actually, the above considerations holds true only for external fields with frequencies lower than the atomic transition frequencies. If the laser frequency matches any atomic transition frequency, then the field starts to effect significantly the atomic spectrum at intensities much smaller than those considered above.

8.3 Case of a resonant field

In this case, if in addition the field is not strong, the usual procedure to account for the field effects on the atomic structure is to isolate the two atomic levels resonantly coupled by the field (say, levels 1 and 2), and to treat their interaction with the field within the two-level atom model and (most frequently) the rotating wave approximation[4,38]. The normalized and orthogonal wavefunctions resulting from the field mixing of the states 1 and 2 are

$$\Psi_- = C_1\left[\psi_1^{(0)} - \eta C_2 e^{-i\omega t}\psi_2^{(0)}\right]\exp\left\{-\frac{i}{\hbar}E_- t\right\}, \qquad (8.10)$$

$$\Psi_+ = C_1\left[\psi_2^{(0)} + \eta^* C_2 e^{i\omega t}\psi_1^{(0)}\right]\exp\left\{-\frac{i}{\hbar}E_+ t\right\}, \qquad (8.11)$$

where

$$E_- = \frac{1}{2}(E_1 + E_2 - \hbar\omega) - \hbar\Omega ;$$

$$E_+ = \frac{1}{2}(E_1 + E_2 + \hbar\omega) + \hbar\Omega ;$$

$$C_1 = \left[\frac{1}{2}(1 + \varepsilon/2\Omega)\right]^{1/2} ;$$

$$C_2 = \left[(1 - \varepsilon/2\Omega)(1 + \varepsilon/2\Omega)\right]^{1/2} ;$$

$$\eta = F_{10}/|F_{10}| ;$$

$$F_{10} = e\,\underset{\sim}{E}_0 \cdot \langle \psi_2^{(0)} | \underline{x}_e | \psi_1^{(0)} \rangle / 2i\ ;$$

$$\varepsilon = (E_2 - E_1 - \hbar\omega)/\hbar$$

is the detuning; and

$$\Omega = [\varepsilon^2/4 + |F_{10}|^2/\hbar^2]^{1/2}$$

is the Rabi frequency. E_1 and E_2, the energies of the unperturbed states $\psi_1^{(0)}$ and $\psi_2^{(0)}$, and the field has been taken as $\underset{\sim}{E}(t) = \underset{\sim}{E}_0 \sin\omega t$. ψ_- and ψ_+ are derived assuming an adiabatic switching on of the field. It is easy to verify that for positive detuning ($\varepsilon > 0$), when at $t = -\infty$, $E_0 \to 0$,

$$\psi_- \to \psi_1^{(0)} \exp\{-\tfrac{i}{\hbar} E_1 t\}$$

$$\psi_+ \to \psi_2^{(0)} \exp\{-\tfrac{i}{\hbar} E_2 t\}.$$

With (8.10) and (8.11), as an approximate set of field-modified wavefunctions to be used in (8.5) may serve $\{\psi_-, \psi_+, \psi_n^{(0)}\}$, or $\{\psi_-, \psi_+, \psi_n\}$ ($n \neq 1, 2$), $\psi_n^{(0)}$ standing for the unmodified atom wavefunction corresponding to the state with energy E_n , and ψ_n for a field-modified wavefunction, constructed by the use of conventional methods of perturbation theory.

For not strong resonant fields, the basic features of field-assisted collisions in this case are [39-42]:

(i) the laser-atom interaction dominates; the laser-projectile interaction may be neglected;

(ii) one photon processes dominate (we remind that the wavefunctions (8.10) and (8.11) are determined under assumption $\omega \approx \omega_2 - \omega_1$);

(iii) the main role of the field is to pump the target into state 2;

(iv) from a physical point of view, this pumping opens the possibility of measuring cross sections simultaneously from two different atomic states, or from excited states that are too short lived in absence of a field to be used for a scattering experiment.

For stronger fields, the projectile-laser interaction can not be neglected; and coupling to states other than 1 and 2 and to continuum becomes important. In this case, the situation may be approached when direct laser excitation and ionization begin to interfere significantly with the collision process. When it happens, the theory outlined in these notes needs substantial modifications.

8.4 Case of an off-resonant field

In this case, the projectile-laser interaction is again the dominating interaction, while for relatively low intensities the effect of the field on the atom may be neglected.

In general, the field modifications of the target spectrum may be handled by means of standard perturbation methods. In particular, one can exploit the circumstance that the external field acting on the atom is periodic in time. Then to construct field-modified atomic states in cases of moderately high intensities, one can profitably use the formalism of the quasienergies and steady states (10). Within this formalism, there is a variety of levels of sophistication, depending upon the problem at hand and the desired accuracy (43-46).

Below we consider only two cases.

In the first case we give modified wavefunctions for hydrogenic atoms obtained within a treatment essentially based on a first-order perturbation theory and on periodicity considerations.

In the presence of a linearly polarized field $\underline{E}(t) = \underline{E}_0 \sin \omega t$ with its electric vector lying along the z axis, the component of the orbital angular momentum along the z axis is conserved. As n and m are good quantum numbers, a set of field-modified wavefunctions for hydrogenic atoms $\{ \psi^{(\nu)}_{nm}(\underline{x}_e, t) \}$ may be derived of the following form

$$\psi^{(\nu)}_{n,m} = \sum_{\ell} C^{n,m}_{\ell,\nu} \Big\{ \psi^{(0)}_{n\ell m}(\underline{x}_e) + \sum_{n'\ell'(\neq n,\ell)} \left[a_{nn'\ell'} e^{-i\omega t} + b_{nn'\ell'} e^{i\omega t} \right] \psi^{(0)}_{n'\ell' m}(\underline{x}_e) \Big\} \Lambda^{(\nu)}_{n}(t). \tag{8.12}$$

In (8.12)

$$a_{nn'\ell'} = (i/\hbar)\, e\, \underline{E}_0 \cdot \langle n'\ell' m | \underline{x}_e | n\ell m \rangle / 2(\omega_{nn'} - \omega); \tag{8.13}$$

$$b_{nn'\ell'} = (-i/\hbar)\, e\, \underline{E}_0 \cdot \langle n'\ell' m | \underline{x}_e | n\ell m \rangle / 2(\omega_{nn'} + \omega); \tag{8.14}$$

$$\Lambda^{(\nu)}_{n}(t) = \exp\Big\{ -\frac{i}{\hbar} E_n t - i\, \rho^{(\nu)}_{n} \cos \omega t \Big\} \tag{8.15}$$

$$|n\ell m\rangle \equiv \psi^{(0)}_{n\ell m} \; ; \quad \omega_{n'n} = (E_{n'} - E_n)/\hbar \, .$$

The wavefunction (8.12) implies that the field mixes the substates with the same m belonging to different n', besides the given n. Without the sum over n' and ℓ', eq.(8.12) reduces to the wavefunctions originally derived in Ref. 47, while, without the factor containing $\rho^{(\nu)}_n$, it reduces to generalization, to account for the ℓ-degeneracy, of the first-order perturbed wavefunction as reported by any textbook on quantum mechanics[48] (see also below). The coefficients $\rho^{(\nu)}_m$ have the meaning of first-order AC Stark parameters. $\rho^{(\nu)}_m$ and $C^{nm}_{\ell\nu}$ are obtained by solving the pertinent system of coupled equations and by imposing normalization requirements. ν counts the resulting linear combinations $\psi^{(\nu)}_{n,m}$ for given n and m.

The second relatively simple case is given by one-electron atoms, where the energy levels do not exhibit ℓ-degeneracy (ground state hydrogenic atoms, valence electron in alkali atoms after reduction to one electron atom via pseudopotential or model potential techniques and so on).

The simplification brought about by the absence of ℓ-degeneracy allows to go to a more refined treatment (as compared with that used for (8.12)) giving a set of field-modified wavefunctions $\{\psi_n(\underline{x}_e,t)\}$ of the form

$$\psi_n(\underline{x}_e,t) = \Big[\psi^{(0)}_n + \sum_{n'\neq n} \left(\alpha_{nn'} e^{-i\omega t} + \beta_{nn'} e^{i\omega t}\right) \psi^{(0)}_{n'}\Big] \times \times \exp\{-i\,\gamma_n(t)\} \tag{8.16}$$

where

$$\alpha_{nn'} = (i/\hbar)\, e\underline{E}_0 \cdot \langle n'|\underline{x}_e|n\rangle / 2(\omega_{n'n} - \omega) \, ; \tag{8.17}$$

$$\beta_{nn'} = (-i/\hbar)\, e\underline{E}_0 \cdot \langle n'|\underline{x}_e|n\rangle / 2(\omega_{n'n} + \omega) \, ; \tag{8.18}$$

$$\gamma_n(t) = \hbar^{-1}(E_n + \Delta E_n)t - \xi_n \sin 2\omega t + + i\,\Gamma_n \cos 2\omega t \, ; \tag{8.19}$$

$$\xi_n = (1/4\hbar^2) \sum_{n'\neq n} \frac{|\langle n|-e\underline{E}_0\cdot\underline{x}_e|n'\rangle|^2}{(\omega^2_{nn'} - \omega^2)} \left(\frac{\omega_{nn'}}{\omega}\right) ; \tag{8.20}$$

$$\Gamma_n = (1/4\hbar^2) \sum_{n' \neq n} \frac{|\langle n| - e\underline{E}_0 \cdot \underline{x}_e |n'\rangle|^2}{(\omega_{nn'}^2 - \omega^2)} ; \qquad (8.21)$$

$$\Delta E_n = 2\hbar\omega\xi_m \quad . \qquad (8.22)$$

The part of (8.16) enclosed in square brackets is exactly the first-order perturbed wavefunction usually quoted in standard quantum mechanics textbook[48], and accounts essentially for the spatial modification of the wavefunction. The temporal modification is instead accounted by $\gamma_n(t)$, a quantity derived exploiting periodicity considerations, proper to the quasienergy formalism[10].

9. CONCLUDING REMARKS

(i) Inserting in the series (8.5) one set of approximate wavefunctions describing the atom in the field together with the Volkov wavefunctions describing the relative motion, and performing the required time integrations, it is not difficult to obtain the S and T matrices in detailed forms, ready, in principle, for actual calculations. In practice, however, one soon discoveres that the structure of the T and S matrices is so complicated, to make very difficult calculations beyond first-order approximations, the difficulties being much larger than in the corresponding field-free cases

(ii) Considerable progress is possible instead when the low frequency assumption is legitimate. In that case, following procedures generalizing those used in potential scattering, it is possible to express the field-assisted scattering parameters in terms of the field-free ones times factors accounting for the presence of the field. Moreover, a number of results of rather general interest are obtained, which often are generalizations of previous results obtained in weak field quantum electrodynamics. Low frequency treatments of particle-atom collisions have received a lot of attention, and representative results on this particular subject may be found in Ref.s 5, 6, 49-51.

(iii) In the first stages of the construction of the theory, when experiments are very scarce or even not existing, actual calculations based on first-order treatments may be considered an alternative. They serve to find out which first-order effects, if any, are brought about by the presence of an external field, and to give information on what to look at in a more rigorous way. Some of the results and predictions based on first-order treatments are found in Ref.s 52-54.

(iv) The difficulties encountered in using the S matrix series to compute scattering parameters are, actually, not inherent to field-assisted collision theory. The field-free theory also is faced with huge difficulties; thus no wonder that things become much more involved adding a third agent. This analogy suggests to put effort in working out approximate and manageable scattering methods, similar to those existing in the field-free theory, such as eikonal, close-coupling, optical potential, variational methods and so on. This line of research until now has been not particularly active, and some of the representative contributions may be found in Ref.s 55-62.

(v) The previous remarks implicitly assume that formally the existing theory is adequate, and that only practical, calculational difficulties should be overcome. Actually, as observed in the subsection 8.2, it may be considered true (with a number of improvements) only for weak and relatively strong fields. For intense fields, when the direct excitation and ionization channels are expected to interfere strongly with the collision channels, the theory needs to be modified in a substantial way. Work on this line is only starting[63-65].

(vi) As in any other field of physics, experiments are needed to establish firmly the theory. Because of this, there is need to focus on the theoretical results and predictions in this subject, which are worth studying experimentally. At this preliminary stage, it is impossible to assess the true potential of field-assisted particle-atom collisions. In this context, we would like to finish these considerations by recalling once again the multiphoton free-free transitions measurements, performed at different times by Weingartshofer and coworkers[13]. They appear, in spite of their limitations, as the first clear confirmation of a nonperturbative (off-resonant) theory of radiation-matter interaction[66], and contain a lot of suggestions on the significance of the working and intrinsic properties of a strong radiation field in nonlinear domains.

ACKNOWLEDGEMENTS

The author is much indebted to Dr. Rosalba Daniele for kind

and very helpful assistance in preparing the camera-ready typescript.

This work has been partially supported by the Italian Ministry of Education, the National Group of Structure of Matter and the Sicilian Committee for Nuclear and Structure of Matter Research.

REFERENCES

1. F.V. Bunkin and M.V. Fedorov, Sov. Phys. JETP 22, 844 (1966) and references therein.
2. N.M. Kroll and K.M. Watson, Phys. Rev. A8, 804 (1973).
3. J.F. Seely and E.G. Harris, Phys. Rev. A7, 1064 (1973) and references therein.
4. N.B. Delone and V.P. Krainov, "Atoms in Strong Light Fields", Atomizdat, Moscow (1978), (in Russian).
5. L. Rosenberg, Phys. Rev. A22, 2283 (1981) and references therein.
6. C. Leone, P. Cavaliere and G. Ferrante, J. Phys. B: At. Mol. Phys. 17, 1027 (1984).
7. H.R. Reiss, Phys. Rev. A19, 1140 (1979).
8. F. Ehlotzky, Opt. Comm. 40, 135 (1981).
9. D.M. Volkov, Z. Phys. 94, 250 (1934).
10. H. Sambe, Phys. Rev. A7, 2203 (1973).
11. I.S. Gradshteyn and I.M. Ryzhik, "Tables of Integrals, Series and Products", Academic Press, New York (1973).
12. L. Langhans, J. Phys. B: At. Mol. Phys. 11, 2361 (1978).
13. A. Weingartshofer, J.K. Holmes, J. Sabbagh and S.L. Chin, J. Phys. B: At. Mol. Phys. 16, 1805 (1983);
A. Weingartshofer and C. Jung in: "Multiphoton Ionization of Atoms", Academic Press, Canada (1984).
14. H. Krüger and C. Jung, Phys. Rev. A17, 1706 (1978).
15. H. Breme, Phys. Rev. C3, 837 (1971).
16. P. Zoller, J. Phys. B: At. Mol. Phys. L249 (1980).
17. R. Daniele and G. Ferrante, J. Phys. B: At. Mol. Phys. 14, L635 (1981).
18. R. Daniele, G. Ferrante and S. Bivona, J. Phys. B: At. Mol. Phys. 14, L213 (1981).
19. R. Daniele and G. Ferrante, J. Phys. B: At. Mol. Phys. 15, 2741 (1982).
20. R. Daniele, F.H.M. Faisal and G. Ferrante, J. Phys. B: At. Mol. Phys. 16, 3831 (1983).
21. F. Trombetta, C.J. Joachain and G. Ferrante in: "Collisions and

Half-Collisions with Lasers", eds. N.K. Rahman and C. Guidotti, Harwood, London (1984).

22. F. Trombetta, G. Ferrante, K. Wodkiewicz and P. Zoller, "Free-Free Transitions in a PDM Field" (to be published).
23. C. Jung, Phys. Rev. A21, 408 (1980); A24, 360 (1981).
24. E.L. Belin and B.A. Zon, Kvantobaja Elektronika 9, 1962 (1982), (in Russian).
25. S. Bivona, R. Burlon, R. Zangara and G. Ferrante in: "Contributed Papers", XII Symposium of the Physics of Ionized Gases, ed. M. Popovic, Institute of Physics, Belgrade (1984), pag. 215.
26. R. Daniele, G. Ferrante and R. Zangara, Nuovo Cimento 2D, 1509 (1983).
27. R. Zangara, P. Cavaliere, C. Leone and G. Ferrante, J. Phys. B: At. Mol. Phys. 15, 3881 (1982).
28. B.R. Mollow, Phys. Rev. 175, 1555 (1968).
29. G.S. Agarwal, Phys. Rev. A1, 1445 (1970).
30. V.A. Kowarskii, N.F. Perel'man, I. Sh. Averbukh, S.A. Baranov and S.S. Todirashku, "Non-Adiabatic Transitions in Strong Electromagnetic Fields", Shtiintsa, Kishinev (1980), (in Russian).
31. P. Zoller in: "Laser Physics", eds. D.F. Walls and J.D. Harvey, Academic Press, Sydney (1980), pag.99.
32. K. Wodkiewicz, Z. Phys. B - Condensed Matter 47, 239 (1982).
33. A list of references is found in Ref. 6.
34. M.H. Mittleman, Phys. Rev. A20, 1965 (1979).
35. C. Jung and H.S. Taylor, Phys. Rev. A23, 1115 (1981).
36. The basic references may be traced back through: C.K. An, J. Phys. B: At. Mol. Phys. 16, L563 (1983), and J. Bergou, J. Phys. B: At. Mol. Phys. 16, L647 (1983).
37. M. Goeppert-Mayer, Ann. Phys. Lpz. 9, 273 (1931).
38. L. Allen and J.H. Eberly, "Optical Resonance and Two Level Atoms", Wiley, New York (1975).
39. A.D. Gazagian, J. Phys. B: At. Mol. Phys. 9, 3197 (1976).
40. I.V. Hertel and W. Stoll, Adv. At. Mol. Phys. 13, 113 (1977).
41. M.H. Mittleman, Phys. Rev. A14, 1338 (1976); A16, 1961 (1977).
42. P. Cavaliere, G. Ferrante and C. Leone, J. Phys. B: At. Mol. Phys. 15, 475 (1982).
43. B.A. Zon and E.S. Sholokhov, Sov. Phys. JETP 43, 461 (1976).
44. N.L. Manakov, V.D. Ovsyannikov and L.P. Rapoport, Sov. Phys. JETP 43, 885 (1976).
45. A.G. Fainshtein, N.L. Manakov and L.P. Rapoport, J. Phys. B: At. Mol. Phys. 11, 2561 (1978).
46. J.E. Bayfield, Phys. Rep. 51, 317 (1979).

47. V.A. Kowarskii and N.F. Perel'man, Sov. Phys. JETP 33, 274 (1971); 34, 738 (1972).
48. See, for instance,
D.I. Blokhintsev, "Quantum Mechanics", D. Reidel, Dordrect (1964), chap. XV;
L.D. Landau and E.M. Lifshitz,"Quantum Mechanics", Pergamon Press, London (1965), chap. VI.
49. L. Rosenberg, Phys. Rev. A21, 1939 (1980); A22, 2485 (1980).
50. M.H. Mittleman, Phys. Rev. A21, 79 (1980).
51. J. Banerji and M.H. Mittleman, J. Phys. B: At. Mol. Phys. 14, 3717 (1981).
52. G. Ferrante, in: "Fundamental Processes in Energetic Atomic Collisions", ed.s H.O. Lutz, J.S. Briggs and H. Kleinpoppen, Plenum Press, New York and London (1983).
53. F. Ehlotzy, Can. J. Phys. 59, 1200 (1981), and references therein.
54. P. Cavaliere, C. Leone and G. Ferrante, Nuovo Cimento 4D, 79 (1984).
55. Y.I. Gersten and M.H. Mittleman, Phys. Rev. A12, 1840 (1975); A13, 123 (1976).
56. B.A. Zon, J. Phys. B: At. Mol. Phys. 8, L86 (1975).
57. G. Ferrante, C. Leone and L. LoCascio, J. Phys. B: At. Mol. Phys. 12, 2319 (1979).
58. P. Cavaliere, C. Leone and G. Ferrante, Lett. Nuovo Cimento 26, 321 (1979).
59. L. Rosenberg, Phys. Rev. A26, 132 (1982).
60. M. Zarcone, D.L. Moores and M.R.C. McDowell, J. Phys. B: At. Mol. Phys. 16, L11 (1983).
61. Dz. Belkic, P.S. Krstic and D.B. Milosevic, in: "Contributed Papers", XII Symposium of the Physics of Ionized Gases, ed. M. Popovic, Institute of Physics, Belgrade (1984), pag.223.
62. F.W. Byron, Jr. and C.J. Joachain, J. Phys. B: At. Mol. Phys. (in press).
63. A.D. Gazazian and R.G. Unanyan, J. Exptl. Theoret. Phys. (URSS) 85, 1553 (1983), (in Russian).
64. E. Fiordilino and M.H. Mittleman, J. Phys. B: At. Mol. Phys. 16, 2205 (1983).
65. L. Dimou and F.H.M. Faisal, in: "Collisions and Half-Collisions with Lasers", ed. N.K. Rahman and C. Guidotti, Harwood, London (1984).
66. (The comparison is with the so-called Keldish's tunnelling theory of multiphoton ionization, and some other nonperturbative theories, which until now do not seem to have received any support by the existing experiments).

APPLICATIONS OF ATOMIC COLLISION PROCESSES IN ASTROPHYSICS

David L. Moores

Department of Physics and Astronomy
University College London
Gower Street, London WC1E 6BT, U.K.

INTRODUCTION

The title of these lectures, the application of atomic collision processes in astrophysics, covers a very broad field and consequently it is impossible in the time available to even begin to give a comprehensive review of the subject. The discussion will be limited therefore to those processes which probably have the widest range of application in astrophysical problems; electron impact excitation and ionisation of atoms and ions, and radiative processes (bound-bound, bound-free, free-free transitions,recombination) involving the same systems. Even if the discussion is restricted to just these important processes, the subject is still an enormous one and therefore, a few representative problems have been selected for discussion in some detail. An attempt will be made to highlight those particular collision processes of special importance and to discuss the astrophysical information that can be obtained.

After introducing some general theoretical concepts, the use of the ratios of spectral line intensities to obtain electron temperature and electron density diagnostics will be discussed. Temperature diagnostics will be illustrated by giving a historical account of the calculation of the electron temperatures of gaseous nebulae from the intensity ratios of the forbidden lines in OIII ions. This will be followed by an account of the determination of solar electron densities from the intensity ratios of ultra-violet emission lines of Si III. In both cases, the importance of having accurate atomic data will be stressed.

The third topic to be discussed will be the calculation of ionisation balance curves and the dependence of the temperature of maximum fractional abundance of a given stage of ionisation of an ion on the approximations employed for electron impact ionisation and recombination rates used in the calculation. It will be seen that use of different approximations can lead to differences of factors of two in electron temperatures. A discussion will be given about the information that may be gained from a study of satellite lines in X-ray spectra of high-temperature plasmas such as solar flares.

The fourth topic will concern the determination of stellar opacities, which depend on knowledge of accurate radiative data. It has recently become apparent that conflicting conclusions about, among other things, the masses and periods of Cepheid variable stars arise from the use of different opacity calculations. It is therefore important that the existing opacities, which depend upon old and in some cases dubious atomic data, be re-studied, making full uses of recent advances in atomic theory. An extensive project to do this, involving collaboration between three centres, has recently been set up. This involves the enormous task of re-calculating all radiative data (bound-free, free-free, bound-bound) for all atoms and ions of significant abundance, by modern techniques.

Finally, a discussion will be given of the importance of dielectromic recombination in C N and O ions at nebular temperatures and of its relevance in the interpretation of ultraviolet spectra of planetary nebulae and Nova Cygni 1978, which have been extensively studied with the International Ultraviolet Explorer Satellite.

THEORY

What the processes under discussion have in common is that they all involve the bound- or continuum states of an electron in the field of an atomic system. A theoretical treatment of each process requires solution of a collision problem, or a bound-state problem, which are solved by closely-related techniques. A summary of the relevant theory has been given by Phil Burke in his lectures, so that I will only need to summarize it here. In the partial wave theory of electron-ion collisions, the wave functions corresponding to boundary conditions symbolised by the subscript γ are taken to be eigenfunction expansions of the form

$$\Psi_\gamma = \sum_i \Theta_{i\gamma} + \sum_j C_{j\gamma}\Phi_j \qquad (1)$$

Application of the variational principle then leads to coupled integro-differential equations for the radial wave functions contained in $\Theta_{i\gamma}$ describing the motion of the scattered electron and for the correlation coefficients $C_{j\gamma}$. Solution of these equations gives the S matrix and hence the excitation cross sections. For the other processes, the actual wave functions are required. The partial wave expansion, in the coupled angular momentum representation, of the wave function of an electron scattered by an ion in the state $\alpha L_\gamma M_\gamma S_\gamma M_{S_\gamma}$ may be written

$$\Psi = \Sigma\ 2\pi A(L_\gamma M_\gamma S_\gamma M_{S_\gamma} \ell_\gamma m_\gamma m_{S_\gamma} LM_L SM_S)\ \chi_\gamma^{-\frac{1}{2}}\ Y^*_{\ell_\gamma m_\gamma}(\hat{\chi}_\gamma)\ \Psi_\gamma(\pi LM_L SM_S) \quad (2)$$

where A is an angular coefficient and $\underset{\sim}{\chi}_\gamma$ is the wave vector of the electron. In photoionisation of an ion of charge z and N electrons,

$$A^{+z} + h\nu \rightarrow A^{+(z+1)} + e^-\ (\chi_\gamma^2)$$

the final state wave function is clearly that of an electron of energy χ_γ^2 scattered by the positive ion $A^{+(z+1)}$ and is thus given by an expansion of the form (2). The wave function for A^{+z} may be obtained by solving the same problem that yields the final continuum state, but applying bound-state boundary conditions. (This is referred to as a solution with all channels closed). The photoionisation cross-section depends on the matrix element

$$\langle\Psi_\gamma(\text{out},\ \chi_\gamma^2)|\sum_{i=1}^{N} \underset{\sim}{r}_i|\Psi_a\rangle \quad (3)$$

where Ψ_γ and Ψ_a are respectively continuum and bound solutions of the coupled integro-differential equations for scattering of an electron by the (N-1)-electron ion. To obtain accurate results it is necessary to include a number of channels in the problem, giving complicated resonant behaviour of the cross section. Many good examples are illustrated by Saraph (1984). These examples serve to highlight the fact that to generate accurate radiative data, an accurate solution of the collision problem is required. In the theory of free-free transitions, the function Ψ_a is replaced by another continuum function $\Psi_{\gamma''}$, also obtained from the solution of a collision problem. For bound-bound transitions, the same method may be used or alternatively the wave functions may be obtained by solving the structure problem for the N-electron system, including a number of configurations and diagonalising the complete Hamiltonian. Programs to do this have been published by Hibbert (1975) and by Eissner et al (1974), (SUPERSTRUCTURE). For electron impact ionisation

$$A^{+z} + e^- \rightarrow A^{+(z+1)} + e^- + e^-$$

in a Coulomb-Born or Distorted-Wave approximation, the incident electron is regarded as a fast charged particle which provokes a bound-free transition in the ion. If the incident and scattered electrons are described by Coulomb waves $\phi(z,\ -\underset{\sim}{k},\ \underset{\sim}{x})$, then in this approximation, the ionisation amplitude depends on a matrix element of the form

$$\langle\Psi_\gamma(\text{out},\ \chi^2)|\underset{\sim}{V}_N|\Psi_a\rangle \quad (4)$$

where

$$V_N = \int \phi^*(z, -k_o, x_{N+1}) \sum_{i=1}^{N} \frac{1}{r_{i,N+1}} \phi(z, -k, x_{N+1}) dx_{N+1} \qquad (5)$$

where k_o and k are the wave numbers of the incident and scattered electrons and x denotes the coordinates of the $N+1^{th}$ electron. The theory is then similar to that for photoionisation except that $\sum_{i=1}^{N} r_i$ is replaced by V_N. This however means that many more values of $SL\pi$ have to be included, with a corresponding increase in complexity of the cross section owing to the superposition of many systems of resonances.

General Considerations

Much of the knowledge that we have of a wide range of astronomical objects - stellar atmospheres, gaseous nebulae, the interstellar medium, quasars, galaxies - is derived from observations of their radio, infra-red, optical, ultra-violet and x-ray spectra. From these spectra, information about the conditions existing in the object may be obtained and estimates made of such quantities as electron temperature and electron density, and of the abundances of the elements. This information, together with equations expressing the equilibrium conditions in the object and details of the physical processes believed to be taking place, enables theoretical models to be constructed. These models may be tested by comparing the calculated spectrum with the observed one. In all this work, accurate atomic data are crucial.

Consider a level i in a positive ion, and let the number density (level population) be N_i. The rate of change of N_i is determined by the rates of all the processes which populate and depopulate the level i:

$$\frac{dN_i}{dt} = \sum_{k \neq i} N_k C_{ki} - N_i C_{ik} \qquad (6)$$

where C_{ki} are the rate coefficients for the processes which populate i, which will include excitation by photons and by charged particles, and recombination; and C_{ik} the corresponding coefficients for depopulation, which may be brought about by de-excitation, spontaneous radiative decay, or ionisation. If we assume that changes in population of each ion stage are so slow compared with excited state re-distribution that we need only consider one stage of ionisation of the element at a time and further that the radiation field is sufficiently weak that only spontaneous radiative decay is important, and all that excitation and de-excitation is by electron impact, then (6) may be written

$$\frac{dN_i}{dt} = \sum_{k \neq i} (N_k N_e q_{ki} - N_i N_e q_{ik}) + \sum_{k>i} N_k A_{ki} - \sum_{k<i} N_i A_{ik} \quad (7)$$

where N_e is the electron density, A_{ij} the spontaneous radiative decay rate and q_{ij} is the electron impact excitation rate coefficient. q_{ij} is obtained by averaging the excitation cross section $Q_{ij}(E)$ over the electron velocity distribution $f(E)$; $(E = \frac{1}{2}mv^2)$

$$q_{ij} = \int_{E_0}^{\infty} v\, Q_{ij}(E)\, f(E)dE \quad (8)$$

where E_0 is the threshold energy. If the distribution is Maxwellian,

$$f(E) = 4\pi \left(\frac{m}{2\pi kT_e}\right)^{3/2} e^{-E/kT_e}\, v^2 \quad (9)$$

where T_e is the electron temperature. The temperature dependence of the excitation rate may be emphasised by writing it in the form

$$q_{ij}(T_e) = C_0\, T_e^{1/2}\, e^{-E_0/kT}\, \Gamma_{ij}(T_e) \quad (10)$$

where C_0 is a constant and

$$\Gamma_{ij}(T_e) = \int_0^{\infty} Q_{ij}(E_0 + xkT_e) e^{-x} \left(x + \frac{E_0}{kT_e}\right) dx \quad (11)$$

is a gently-varying function of T_e. In general, the rate will be small except at temperatures for which the cross section has its largest values overlapping with the peak of the Maxwellian, $E \simeq kT_e$. In a steady state, $\frac{dN_i}{dt} = 0$ and we obtain

$$N_i = \frac{\sum_{k \neq i} N_k N_e q_{ki}(T_e) + \sum_{k>i} N_k A_{ki}}{\sum_{k \neq i} N_e q_{ik}(T_e) + \sum_{k<i} A_{ik}} \quad . \quad (12)$$

A corresponding equation exists for each level i, giving an infinite set of simultaneous equations (the equations of statistical equilibrium) for the N_i. In practice, the set may often be restricted to a small number of low-lying states, since the rate coefficients q_{ij} and A_{ij} decrease rapidly as the principal quantum number of state j increases.

In the case in which only collisional excitation from the ground state g and spontaneous radiative decay back to it are significant, we obtain

$$N_i = N_e N_g q_{gi}/A_{ig} \quad . \quad (13)$$

This is known as the coronal excitation equation and applies to low densities. At higher densities where collisions dominate over radiative transitions we obtain

$$N_i \sum_j q_{ij} = \sum_j N_j q_{ji} \quad . \quad (14)$$

But detailed balance requires

$$q_{ji} = q_{ij} \frac{\omega_i}{\omega_j} e^{E_{ij}/kT_e} \quad . \tag{15}$$

The population distribution is then given by that obtained in thermodynamic equilibrium and the relative population statistical. Equations (14) and (15) may typically apply to a subset of states such as the fine structure levels of a multiplet which have high mutual collisional rates. Their population may be treated as statistical.

If the ion has a metastable level k with a long lifetime (small A_{kg}) it can be seen from (13) that the population N_k will be larger than that of a level with an allowed decay. This will affect the other populations N_i since in (12) terms in the numerator with $j=k$ will be non-negligible. The long-lived metastable levels will populate levels other than the ground. The existence of an excess population of metastables may also mean that ionisation and recombination involving them may become important so that the assumptions leading to (12) may no longer be valid and it might be necessary to go back to (7) or even (8).

In an optically thin medium, the intensity of the spectral line due to a transition between levels i and j is

$$\mathcal{I}_{ij} = N_i A_{ij} \quad \text{photons cm}^{-3}\ \text{sec}^{-1} \tag{16}$$

or for an emitting volume V,

$$I_{ij} = \frac{h\nu_{ij}}{4\pi} \int_V N_i A_{ij}\, dV \quad \text{ergs sr}^{-1}\ \text{sec}^{-1} \quad . \tag{17}$$

ELECTRON TEMPERATURE DETERMINATION

Suppose we have two lines, both excited by electron impact from the ground state g and both decaying spontaneously back to it. Suppose also that coronal excitation may be assumed. For upper states i and j

$$\frac{\mathcal{I}_{ig}}{\mathcal{I}_{jg}} = \frac{N_i A_{ig}}{N_j A_{jg}} = \frac{q_{gi}}{q_{gj}} \quad . \tag{18}$$

This is independent of N_e but dependent on T_e. Using (10),

$$\frac{\mathcal{I}_{ig}}{\mathcal{I}_{jg}} = \frac{\Gamma_{ig}}{\Gamma_{jg}} e^{-(E_{gi} - E_{gj})/kT_e} \quad . \tag{19}$$

When $|E_{gi}-E_{gj}| >> kT_e$, this is a sensitive function of T_e. Thus the intensity ratio of two lines for which the upper levels are well separated in energy enables the electron temperatures to be determined. A good example is provided by the 2p-2s and 3p-2s transitions in Li-like ions. A slightly more complicated example has given us a useful tool for determining the temperatures of planetary nebulae. Planetary nebulae have been described as extra-terrestrial laboratories of atomic physics, for in them processes which are very important yet impossible as yet to study in a laboratory on earth lend themselves to observation and analysis. On the other hand the atomic processes give information about the physical conditions in nebulae. Planetaries are formed during a rapid stage of stellar evolution when a star is evolving from a red giant into a white dwarf. In the process, the star sheds its envelope, which becomes the nebula, leaving behind a small, very dense hot star to evolve into the dwarf. The spectra of planetary nebulae are characterised by just a few narrow emission lines. These were first observed by Huggins over 100 years ago. A few weak lines were found to correspond to known solar lines but the stronger ones at 4959Å and 5007Å could not be identified. A suggestion was made that they were due to a new element, nebulium, but this suggestion was refuted by Huggins who considered them to be due to a known element but that the conditions under which they were emitted were totally different from those in the laboratory or in the solar atmosphere. In this Huggins was perfectly correct, and in 1927 Bowen identified them as being due to forbidden transitions in O III, which occur as a result of the very low density of the nebula. The upper states are excited by electron impact, and the levels involved are actually the different terms of the ground configuration $1s^2 2s^2 2p^2$ of the ion. The 5007 and 4959 lines are due to the transition $^1D - {}^3P_{2,1}$ while a weaker line 4363A is due to the transition $^1S - {}^1D$. The ratio

$$R = I(4959,\ 5007)/I(4363) \tag{20}$$

is very sensitive to electron temperatures and insensitive to N_e for $N_e \lesssim 10^4\ \mathrm{cm}^{-3}$. Labelling the ^3P, ^1D and ^1S states by 1, 2 and 3 respectively we have

$$N_2 = \frac{N_e N_1 q_{12}}{A_{21}} \tag{21}$$

$$N_3 = \frac{N_e N_1 q_{13}}{(A_{31} + A_{32})} \tag{22}$$

$$R = \frac{E_{21} N_2 A_{21}}{E_{32} N_3 A_{32}} = \frac{E_{21}}{E_{32}} \frac{q_{12}}{q_{13}} \frac{A_{21}}{A_{31} + A_{32}} \ . \tag{23}$$

The temperature dependence is in q_{12}/q_{13} since

$$\frac{q_{12}}{q_{13}} = \frac{\Upsilon_{21}}{\Upsilon_{31}} \exp(-\ E_{23}/kT_e) \tag{24}$$

where

$$\Upsilon_{ji} = \int_0^\infty \Omega(i,j)\ \exp(-\varepsilon_j/kT_e)\ d(\varepsilon_j/kT_e);\ \Omega(i,j) = \frac{2\varepsilon_j\omega_i}{\pi}\ Q_{ji}(\varepsilon_j)$$

The determination of an accurate value of T_e from the observed intensity ratio and equation (23) clearly depends on having accurate excitation cross sections. The history of this problem illustrates the interplay between atomic physics and astronomical studies characteristic of this work. The first calculations of the cross sections were done by Hebb and Menzel (1940) using a Coulomb-Born-Oppenheimer approximation. This was not really accurate enough but gave reasonable ratios. Seaton (1953) did the first non-perturbative calculation, which was subsequently improved upon by Saraph et al. (1966). The 1953 calculations gave a value of T_e for the Orion nebula (which is not a planetary, but in some aspects similar) of 13000 K. Peimbert and Costero (1969) used the Saraph et al. data to give values of 8050-8600 K, but the difference was principally due to a change of a factor of four in the measured intensity of the 4363 line, rather than changes in the atomic data. In 1963, Burbidge et al., on the basis of thermal balance arguments found temperatures in the 3000-5000 K range. Further observational work on radio surface brightnesses, radio recombination lines, Balmer decrements and Balmer line-continuum ratios also gave low values, appearing to confirm their results. It therefore seemed necessary to re-examine the atomic calculations. If a value of T_e of 5000 K, say, were correct instead of 10000 K, this would require that $\Upsilon_{21}/\Upsilon_{31}$ should be reduced by a factor of about 27. This could be brought about by the existence of a resonance in Υ_{31} not included in the calculation. In the work of Saraph et al. (1) no configuration mixing was allowed for in the target wave functions; (2) the eigenfunction expansion was limited to three states, the three terms of the ground configuration and (3) an exact resonance approximation (for p-waves) or a distorted wave approximation (for other partial waves) was assumed. Eissner and Seaton (1973) carried out improved calculations in which five target configurations $1s^2 2s^2 2p^2$, $1s^2 2p^4$, $1s^2 2s 2p^2 3d$, $1s^2 2s 2p^3$ and $1s^2 2s^2 2p 3d$ were included and all states with dominant configurations $1s^2 2s^2 2p^2$ and $1s^2 2s 2p^3$ retained in the eigenfunction expansion. Full coupled integro-differential equations were then solved. The results did give $1s^2 2s 2p^3 (^3P) 3s\ ^2P$ resonance just above the threshold for excitation of level 3; but also a resonance $1s^2 2s 2p^3 (^3D)\ 3s\,^2D$ just above threshold for excitation of level 2. The resulting values of Υ_{21} and Υ_{31} were in good agreement with the old results of Saraph et al., an agreement which however must be regarded as fortuitous since a great deal more structure went into the newer results. (The value of Υ_{32} was much larger in the new calculations.) As a check on their calculations, Eissner and Seaton compared the energy level separations and f-values for their O III target with experiment and with accurate calculations of Smith and Wiese, good agreement being recorded. In addition to this, the program used to obtain the cross sections was used to calculate the bound states of O II, by a solution with all channels closed, again resulting in good agreement with experiment. Finally, the resonance positions were found to compare well with iso-electronic extrapolations (for the higher members of the O III sequence the resonances become true bound states). Similar results for the iso-electronic ion N I compare well with independent R-matrix calculations by Robb. On the basis of these comparisons, Eissner and Seaton claim that their cross sections should be correct to 10%

The discrepancy in electron temperatures as deduced from the forbidden lines and from certain other methods thus persists. For the nebula NGC 7662 the observed forbidden line ratio gives $T_e = 12{,}500\,K$, compared with the value of $5500 \pm 2000\,K$ obtained from radio observations. However, recent photoionisation models are more consistent with the temperatures deduced from forbidden line ratios.

ELECTRON DENSITY SENSITIVITY

Suppose now that the two upper levels are i and k where k has a long lifetime for radiative decay, so that collisional de-excitation competes with it. The coronal excitation equation is not valid for level k. Instead,

$$N_k = N_e\, N_g\, q_{gk}/(\sum_{m'} A_{km'} + \sum_{m'} N_e\, q_{km'}) \tag{25}$$

If N_i is given by (13) then

$$\frac{I_{ig}}{I_{km}} = \frac{N_i\, A_{ig}}{N_k\, A_{km}} = \frac{q_{gi}}{q_{gk}} \left(1 + \sum_{m' \neq m} \frac{A_{km'}}{A_{km}} + \sum_{m'} \frac{N_e\, q_{km'}}{A_{km}}\right) \quad . \tag{26}$$

If the levels can be chosen such that q_{gi}/q_{gk} and $q_{km'}$ are insensitive to T_e, the line ratio can be sensitive to density, via the third term on the right of (26). In general, a line ratio will depend on both N_e and T_e and it is often necessary to solve the full set of statistical equilibrium equations (12) including as many states as is practical. A good example of density sensitivity occurs in C III in which the ratio I(1176Å)/I(977Å) is sensitive to N_e but insensitive to T_e for $N_e = 2$ to $3 \times 10^9\ cm^{-3}$ and $T_e \sim 7 \times 10^4$ K. These lines are due to the transitions $2p^2\ ^3P^e \rightarrow 2s2p\ ^3P^o$ and $2s2p\ ^1P^o_1 \rightarrow 2s^2\ ^1S_0$ respectively.

Another example is in the Mg-like ion Si III . A rich emission line spectrum at ultra-violet wavelengths, mainly formed in the solar transition region, shows a strong line at 1206Å ($3s^2\ ^1S$-$3s3p\ ^1P$); also observed are lines due to the $3s^2\ ^1S$-$3s3p\ ^3P^o_J$ transitions (around 1892Å) the $3s3p\ ^3P^o_J$ - $3p^2\ ^3P^e_{J'}$ multiplet at 1300Å and the $3s3p\ ^1P$-$3s4s\ ^1S$ transition at 1313Å. Various combinations of these lines give temperature and density diagnostics. Dufton et al. (1983) have considered a 10-level model ion and have solved the equations of statistical equilibrium, using transition probabilities calculated by Baluja and Hibbert, and collision rates calculated from an R-matrix calculation including the lowest 12 states by Baluja, Burke and Kingston (1981). Transitions between fine structure levels were obtained with the aid of the code JAJOM published by Saraph (1978).

Proton excitations among the levels of the $^3P^o$ multiplet were considered but found to be lower than the electron rates by a factor of five at the temperatures and densities of interest, appropriate to the transition region. The following line ratios are density sensitive;

$$R_1 = I(1296)/I(1892)$$
$$R_2 = I(1301)/I(1313)$$
$$R_3 = I(1301)/I(1296)$$
$$R_4 = I(1301)/I(1303)$$

Results obtained were found to differ significantly from earlier calculations by Nicholas (1977) owing to the complex resonant structures, not present in his simple distorted wave calculations, which were found to dominate the calculated cross sections. The theoretical ratios R_1 and R_2 were found to be slightly temperature sensitive, R_3 and R_4 less so.

A lot of data on Si III line intensity ratios was obtained from spectrograms taken with the NRL slit spectrograph on Skylab, for a variety of solar regions including quiet sun, coronal hole, active regions, and flares, in 1973. For the two temperature-insensitive line ratios R_3 and R_4, the derived electron densities are in good agreement with independent measurements while the derived quiet-sun electron pressure (log N_eT_e) is close to the value currently used in solar models. Consistent electron densities can be derived from R_2 if log T_e = 4.7 but for R_1 to give similar values the temperature has to be lowered to log T_e = 4.5

It is interesting that ionisation balance calculations of Jordan (discussed below) give the abundance Si III to be sharply peaked about a value of log T_e = 4.7 whereas Baliunas and Butler (1980) have argued that if the charge exchange processes

$$Si^+ + H^+ \rightarrow Si^{+2} + H$$

$$Si^{+2} + He^+ \rightarrow Si^{+3} + He$$

were important the dominant stage of ionisation could be Si III from log T_e = 4.2 to 4.8.

IONISATION BALANCE PROBLEMS

In a plasma in a steady state which is neither ionising nor recombining, the rate of ionisation by all processes of each ion species is equal to its rate of recombination. In the solar corona, for example, ionisation is by electron impact and recombination radiative

$$A^{+m} + e^- \rightarrow A^{+m+1} + e^- + e^- \tag{27}$$
$$A^{+m+1} + e^- \rightarrow A^{+m} + h\nu \quad . \tag{28}$$

In a steady state the rates of processes (27) and (28) are equal;

$$N_e N(A^{+m})\ q_{ion}(A^{+m}) = N_e N(A^{+m+1})\ \alpha(A^{+m}) \qquad (29)$$

where q_{ion} is the collisional ionisation rate

$$q_{ion}(A^{+m}) = \int_I^\infty v\ q_{ion}(E)\ f(E)dE \qquad (30)$$

and α the radiative recombination rate

$$\alpha = \frac{1}{c^2}\ \left(\frac{2}{\pi}\right)^{1/2} (mkT)^{-3/2}\ \frac{\omega_i}{\omega_+}\ e^{I/kT} \int_I^\infty (h\nu)^2\ a_\nu\ e^{-h\nu/kT}\ d(h\nu) \qquad (31)$$

a_ν being the photoionisation cross section. This gives

$$\frac{N(A^{+m+1})}{N(A^{+m})} = \frac{q_{ion}}{\alpha} = F(T) \quad . \qquad (32)$$

This formula may be used to calculate the fractional abundance of each ionisation stage

$$f_m(A) = \frac{N(A^{+m})}{\sum\limits_m N(A^{+m})} \qquad (33)$$

for each A and m as a function of T. The plots of f_m against T are known as ionisation balance curves. For each stage of ionisation m, f_m has a maximum value at some temperature $T(A^{+m})$ which is called the ionisation temperature, and becomes small when T differs very much from $T(A^{+m})$. If strong lines of an ion A^{+m} are emitted in some region then we may conclude that the temperature of that region cannot differ very much from the temperature of maximum fractional abundance $T(A^{+m})$. We equate $T(A^{+m})$ with the electron temperature since it is the motion of the electrons which determines the ionisation equilibrium.

In order to calculate ionisation curves we need to know the rates of all contributing ionisation and recombination processes. In the case of ionisation the main contribution to the rate comes from electron energies up to about 2-3 times the ionisation energy. Unfortunately it is at these energies, where exchange is important, that ionisation cross sections are least well known. Also, at low energies, excitation-autoionisation

$$A^{+m} + e^- \rightarrow (A^{+m})^{**} \rightarrow A^{+m+1} + e^- + e^- \qquad (34)$$

can also be important, and may even dominate the cross section. After an initial suggestion by Unsöld (1961), it was shown by Burgess (1964) that dielectronic recombination

$$A^{+m+1} + e^- \rightleftarrows (A_d^{+m})^{**} \rightarrow A^{+m} + h\nu \qquad (35)$$

can also be important and its inclusion served to reduce the discrepancy in the temperature of the solar corona deduced from ionisation equilibrium and from the thermal Doppler broadening of spectral lines.

Ionisation balance calculations are complex and require such a large amount of atomic data that it is usually necessary to employ simple formulae for the cross sections which may not be of the highest precision. A number of different sets of ionisation balance calculations which have been published, and the differences between them reflect the use of different cross sections and also the fact that different authors treat different parts of the calculation with different degrees of accuracy. The calculations may be divided roughly into two groups; those by Jordan (1969, 1970), Landini and Monsignori-Fossi (1972), Jacobs et al. (1977-80) and Jain and Narain (1978), all of whom used semi-empirical formulae for the ionisation rates, with a variety of approximations for recombination, which seem to yield temperature curves in reasonable mutual agreement; and those by Summers (1974, 1979), who used the exchange classical impact parameter method (ECIP) for ionisation but concentrated on the recombination rates and in particular on the effects of increasing electron density. Summers' calculations tend to give larger equilibrium temperatures, especially for highly-ionised ions, by up to a factor of two, due mainly to the use of the ECIP ionisation rates which are smaller than the rates calculated from the empirical formulae of Seaton (1964) or of Lotz (1968). The formulae however tend to give results which are in better agreement with approximate ab-initio quantal calculations and with crossed-beam experimental data, although this point is debatable.

The lack of agreement among the calculations has led to confusion and uncertainty in the analysis of solar spectra. Cheng et al. (1979) have found that the line widths of highly-ionised Fe and Si ions in the spectra of different solar atmospheric regions imply temperatures up to a factor of two lower than those predicted by Summers, but this could be explained by the effects of autoionisation to excited levels on dielectronic recombination rates. From analysis of x-ray spectra of solar flares, Feldman et al. (1981) also argue that the Jordan or Jacobs et al. calculations give more plausible temperatures. A lot of solar predictions are based on observations of Satellite Lines, which are spectral lines observed on the long-wavelength side of He-like resonance lines and are due to transitions of the type

$$1s^2 n\ell \rightarrow 1s2pn\ell$$

in Li-like ions. Such lines are observed in the x-ray spectra of solar flares where they are due to highly-ionised Fe and Ca. Comparison of satellite line intensities with those of the He-like ions allows the determination of both electron temperature and the transient ionising state of the plasma, solar flares not generally being in ionisation equilibrium. Gabriel and Paget (1972) have proposed two mechanisms - dielectronic recombination and inner-shell excitation.

(a) Dielectronic recombination. This is the process

$$1s^2\ {}^1S + e^- \rightleftarrows 1s2pn\ell\ {}^{2S+1}L'_{J'} \rightarrow 1s^2 n\ell\ {}^2L_J + h\nu \ . \qquad (36)$$

The lines of the He-like system formed by the transition $1s2p \rightarrow 1s$ are accompanied by dielectronic satellite lines of the Li-like system due to $1s2pn\ell \rightarrow 1s^2n\ell$. Let C_d be the dielectronic capture rate of the state $1s2pn\ell$ and A_a and A_r be its autoionisation and radiative transition probabilities. The intensity I_s of the satellite line is then given by

$$I_s = N_{II} N_e\, C_d \frac{A_r}{A_a + A_r} \qquad (37)$$

where N_{II} is the number density of He-like ions. The intensity of the line $1s2p \rightarrow 1s^2$ is given by

$$I = N_{II} N_e\, q_{12}(T_e) \qquad (38)$$

where q_{12} is the $1s^2 \rightarrow 1s2p$ electron impact excitation rate. Now

$$\frac{C_d}{A_a} = \frac{\text{const.}}{T_e^{3/2}} \frac{\omega_s}{\omega_1} e^{-E_s/kT_e} \qquad (39)$$

where E_s is the energy of $1s2s2p$ relative to $1s^2$.

Hence, from equations (37), (38) and (39) the intensity ratio is

$$\frac{I_s}{I} = F(T_e) \qquad (40)$$

and a function of electron temperature.

(b) Inner-shell excitation.

Under conditions of transient ionisation and especially for more highly-ionised systems, the upper state may be populated by inner-shell excitation of the Li-like system;

$$1s^2 n\ell + e^- \rightarrow 1s2pn\ell + e^- \ . \qquad (41)$$

The satellite line intensity is then given by

$$I'_s = N_{III} N_e q \frac{A_r}{A_a + A_r} \tag{42}$$

where q is the rate for the inner shell process and N_{III} the number density of Li-like ions. We then have

$$\frac{I'_s}{I} = \frac{N_{III}}{N_{II}} \frac{q}{q_{12}} \frac{A_r}{A_a + A_r} \quad . \tag{43}$$

The intensity ratio thus depends on the ratio N_{III}/N_{II}. Ionisation balance calculations give this ratio N_{III}/N_{II} as a function of a temperature parameter T_z which can only be equated to the electron temperature T_e under conditions of ionisation equilibrium. If we write (43)

$$\frac{I'_s}{I} = G(T_z) \tag{44}$$

where $G(T_z)$ is determined by (43) then (44) defines an ionisation temperature T_z. In ionisation equilibrium, $T_z = T_e$; for an ionising plasma, $T_z < T_e$ while for a recombining one, $T_z > T_e$. The function $G(T_z)$ and hence T_z depends on the ionisation balance calculation used to determine N_{III}/N_{II}.

Because of the fine-structure splitting, a number of different satellite lines ${}^{2S+1}L'_{J'} \rightarrow {}^{2}L_J$ will be observed, each corresponding to a different combination of fine-structure levels. Depending on the rate coefficients, some lines will be excited primarily by dielectronic recombination and others by inner-shell excitation. Thus both T_e and T_z may be determined from the same spectrum.

X-ray spectra of Ca XIX and Fe XXV observed in a solar flare on March 25, 1979 have been analysed by the above method. Feldman et al. (1980) whose calculations relied on the Jordan ionisation balance results, and the calculations by Bhalla et al. (1975), obtained vales of T_e and T_z which implied that the plasma was ionising even during the decay phase of the flare. Doyle and Raymond (1981) however, using their own ionisation balance calculations, which employed the ECIP ionisation rates used by Summers, obtained the more plausible results that during the rise phase $T_e > T_z$ in both Fe and Ca, indicating an ionising plasma, while for the decay phase the Ca results indicated it was recombining whereas the Fe results indicated that it was initially recombining but thereafter in ionisation equilibrium. The ionisation balance results of Doyle and Raymond are intermediate between those of Jacobs et al. and those of Summers.

It will be seen from all this that the situation is far from satisfactory and that there is clearly an urgent need here for some further work on the relevant atomic physics. A new semi-empirical formula for ionisation rates suggested by Burgess and Chidichimo (1983) which incorporates autoionisation in a simple fashion and which is in reasonable global agreement with laboratory experiments looks a promising development. It may also be necessary to consider the effects of charge transfer on ionisation balance calculations (Baliunas and Butler 1980).

THE PROBLEM OF THE STELLAR OPACITY

The opacity of a star gives a measure of the amount of impedance to the passage of radiation through the stellar material. This impedance may be produced by scattering of photons by free electrons (Thomson scattering) or by absorption of radiation by atomic systems. In the latter case free-free, bound-free, and bound-bound transitions may all contribute, depending on the conditions. Under extreme conditions of temperature and density other processes such as conduction by degenerate electrons, pair production, photoneutrino processes, and nuclear absorptions may have to be considered, but these, like Thomson scattering, do not involve atomic collisions and will not concern us here. We consider only the radiative opacity. The total absorption coefficient is the sum of contributions from bound-bound, bound-free and free-free absorption;

$$K(\nu) = K_{bb}(\nu) + K_{bf}(\nu) + K_{ff}(\nu) \quad . \qquad (45)$$

The bound-free coefficient is given by

$$\rho\, K_{bf}(\nu) = \sum_A \sum_m \rho(A^{+m}) \sum_j N(A^{+m}, j)\; a_{bf}(A^{+m}, j) \qquad (46)$$

where ρ is the mass density, $N(A^{+m}, j)$ the number density of ions of element A in the m-th stage of ionisation in the level j; and $a_{bf}(A^{+m}, j)$ the photoionisation cross section of A^{+m} from level j.

In thermodynamic equilibrium the number densities are related by the Boltzmann and Saha equations

$$\frac{N(A^{+m}, j)}{N(A^{+m}, g)} = \frac{\omega_i}{\omega_g}\, e^{-X_j/kT} \qquad (47)$$

$$\frac{N(A^{+m+1})N_e}{N(A^{+m})} = \left(\frac{2\pi m_e\, kT}{h^2}\right)^{3/2} \frac{2U_{m+}(T)}{U_m(T)}\, e^{-I(A^{+m})/kT} \qquad (48)$$

where

$$N(A^{+m}) = \sum_j N(A^{+m}, j) \qquad (49)$$

$$U_n(T) = \sum_j \omega_j\, e^{-X_j/kT} \quad . \qquad (50)$$

For free-free absorption

$$\rho\,\kappa_{ff}(\nu) = \sum_A \sum_m \rho(A^{+m})\,N(A^{+m})\,N_e \int_0^\infty f(E)\,a_{ff}(A^{+m},E)\,dE \quad . \tag{51}$$

The bound-bound (or line) contribution to the opacity may also have to be considered at the lower temperatures; this may also require an investigation of those processes responsible for the line profiles.

In stellar interiors, where local thermodynamic equilibrium may be assumed to a good approximation, the relevant quantity is the Rosseland Mean Opacity $\bar{\kappa}$ defined by

$$\frac{1}{\bar{\kappa}} = \frac{\int_0^\infty \frac{1}{\kappa(\nu)(1 - e^{-h\nu/kT})} \frac{dB\nu}{dT}\,d\nu}{\int_0^\infty \frac{dB\nu}{dT}\,d\nu} \tag{52}$$

where

$$B_\nu = \frac{2h\nu^3}{c^2}\,(e^{h\nu/kT} - 1)^{-1} \quad . \tag{53}$$

The radiative energy transport equation then becomes

$$L_r = -4\pi r^2 \frac{4ac}{3} \frac{T^3}{\bar{\kappa}\rho} \frac{dT}{dr} \tag{54}$$

where L_r is the luminosity at radius r or the total energy per sec. flowing through a spherical surface of radius r.

The calculation of the opacity as a function of T_e, N_e and chemical composition is a formidible task which presents considerable data-handling problems, as every process of importance must be included. Not only are a large number of atomic cross sections required but in addition, for each temperature, density, for an assumed chemical composition the absorption coefficients have to be computed using the Boltzmann and Saha equations. Finally the Rosseland Means have to be computed. Two principal sets of calculations have been carried out; by the Los Alamos group (Cox and Stewart 1965); and by Carson and co-workers (1968). (The references just given refer to original papers: much subsequent work has been done by both groups, some of it unpublished). The two sets differ from each other in some regions of temperature and density. As in the case of ionisation balance calculations, in order to make the calculations tractible both sets of authors were obliged to use rather simple approximations for the atomic data, especially for the heavier ions. Recent astrophysical work has brought to light certain discrepancies which could indicate that one or both sets of opacity calculations may be seriously in error, and it has become clear that a new set of calculations is required which makes full use of modern advances in atomic theory and computing capability to incorporate accurate radiative cross sections, particularly for ions of relatively abundant elements such as C, N and O. The discrepancies were first pointed out in the theory of Cepheid variable stars. These are stars for which the observed brightness varies regularly with a period of the order of 1-50 days, and which obey a period-mean luminosity relation which makes them important in establishing a basic distance scale in the universe. A number of independent methods exist for determining the mass of these stars, but they do not give consistent results. For example, the masses obtained from model calculations adjusted to fit the observed position of bumps on the light curves (the "bump masses") give masses 30-40% lower than those estimated from stellar evolution theory. Mass discrepancies may be transformed into period discrepancies: if the bumps are attributed to a resonance between the fundamental mode (period P_0) and the second overtone (period P_2) then calculations give $P_2/P_0 = 0.539$ compared with observed values of between 0.47 and 0.53. In double-mode (beat) Cepheids the ratio of the period of the first overtone, P_1 to that of the fundamental P_0 is observed to lie in the range 0.7 to 0.71 while model calculations give 0.742. The calculations depend sensitively on the opacity, with the Carson opacity models giving slightly smaller discrepancies than their Cox opacity counterparts. Various proposals have been made for removing the discrepancy, such as increasing the He content, or introducing tangled magnetic fields, but all tend to introduce further difficulties. Simon (1982) has produced evidence that the discrepancies would be removed if the contribution to the opacity from ions heavier than H and He were augmented by a factor of 2-3, with the heavy element abundance, however, unchanged. For a double-mode Cepheid model, such a procedure reduces the period ratio from 0.742 to 0.713, much closer to observation, and for a bump Cepheid model reduces P_2/P_0 from 0.539 to 0.498, which lies between the observed limits. On the other hand, the agreement between observed and calculated values (using standard opacity) of P_1/P_0 for a model RR Lyrae variable star is not destroyed. This is because RR Lyrae stars have such a low heavy element abundance anyway that changes in heavy element opacity are not significant. However, increasing the H and He opacities does destroy the agreement for R Lyrae variables, demonstrating that changes in overall opacity are not acceptable. Simon also proposes that the suggested increase in opacity could explain the as yet unsolved problem of the energising mechanism for the β Cephei variables.

The unsatisfactory state of affairs with regard to the currently available opacity tables has stimulated the establishment of a Stellar Opacity Project, the aim of which is the ambitious but perfectly feasible task of creating a completely new set of opacities calculated from atomic data of the highest possible accuracy. This will involve the development of new techniques for handling the enormous amount of data required and for coping with the myriads of resonances occurring in the cross sections, which will be computed by the R-matrix method. The project, which has just been launched, involving co-operation between groups in London, Belfast and JILA will obviously take a number of years to reach completion. Concurrently with the opacity project a programme of work on variable star models is being started in which it is intended to use the new opacities to study the kind of discrepancy discussed above.

DIELECTRONIC RECOMBINATION AT LOW TEMPERATURES

The availability of the International Ultra-Violet Explorer Satellite (IUE) has meant that good ultra-violet spectra of objects such as planetary nebulae can now be obtained with little difficulty. Extensive IUE observations have been made of emission lines of the principal stages of ionisation of elements such as C, N and O. In nebulae the emission lines are excited not just by electron impact but also by recombination, fluorescent excitation or, as recent work has shown, by dielectronic recombination. Previous studies of this process have been principally concerned with high temperature plasmas such as the solar corona where ionisation is dominated by electron impact. In nebulae the temperatures are too low for electron impact ionisation to play a role and ionisation is caused by the intense flux of photons from the central star. The ionisation equilibrium is determined by

$$N(A^{+m}) \int F_\nu \, a_\nu \, d\nu = N(A^{+m+1}) \, N_e \, \alpha_{eff} \tag{55}$$

where F_ν is the radiation flux, a_ν the photoionisation cross section and α_{eff} the recombination rate. Equation (55) then determines

$$N(A^{+m})/N(A^{+m+1}) \; . \tag{56}$$

Observations of the intensities of the UV emission lines provide a useful means for estimating relative abundances of the elements. For example, the emissivity in a collisionally-excited line ij is given by

$$q_{ij}(T_e, A^{+m}) \, N_e \, N(A^{+m}) \, \Delta E_{ij} \; . \tag{57}$$

The emissivity of H_β, excited by recombinaion is

$$N_e \, N(H^+) \, \alpha_{eff}(H_\beta) \, (h\nu)_{H_\beta} \tag{58}$$

and the intensity ratio is

$$\frac{I_{ij}(A^{+m})}{I(H_\beta)} = \frac{q_{ij}(T_e, A^{+m})}{\alpha_{eff}(H_\beta)} \, \frac{\Delta E_{ij}}{(h\nu)_{H_\beta}} \, \frac{N(A^{+m})}{N(H^+)} \; . \tag{59}$$

If T_e can be obtained (e.g. from the O III forbidden line intensity ratios) equation (58) enables $N(A^{+m})/N(H^+)$ to be determined from the observed ratio of the intensity of the line to that of H_β, provided that the rate coefficients are known. For recombination lines

$$\frac{I_{ij}(A^{+m})}{I(H_\beta)} = \frac{\alpha_{eff}(A^{+m})}{\alpha_{eff}(H_\beta)} \frac{\Delta E_{ij}}{(h\nu)_{H_\beta}} \frac{N(A^{+m})}{N(H^+)} \tag{60}$$

If the $N(A^{+m})/N(A^{+m+1})$ are determined from ionisation equilbrium the abundances

$$\frac{N(A)}{N(H)} = \sum_m N(A^{+m})/N(H) \tag{61}$$

relative to H may be obtained.

Abundance determinations of C in planetary nebulae have been made by Torres-Peimbert and Peimbert (1977, TPP) Pottasch, Wesselius and van Duinen (1978, PWD) and Harrington et al. (1980, HLSS). TPP deduced values of $N(C^+)/N(H^+)$ from the C II 4267 line which they assumed to be due to recombination; PWD deduced values of $N(C^{3+})/N(H^+)$ from the collisionally-excited C IV 1548-51 doublet. For the nebula NGC 7009 , TPP found a value of N(C)/N(H) which was 250 times larger than that of PWD. These values were both deduced from an assumed value of the ratio $N(C^{2+})/N(C^{3+}) = 0.4$ based on existing ionisation equilibrium models. A larger value of this ratio would remove the discrepancy. From their IUE observations of the collisionally-excited lines C III 1908Å, C II 2326Å and O II 2470Å in the nebula IC 418 , HLSS deduced a value of $N(C^{2+})/N(H^+)$ which was four times smaller than that obtained by TPP. They also obtained a value for $N(C^+)/N(C^{2+})$ from the intensities of the 1908Å and 2326Å lines 5 times larger than that obtained from ionisation equilibrium models. All the HLSS calculations were characterised by the use of up-to-date, elaborate, accurate collision rates. HLSS suggested that there must be some additional $C^{2+} \rightarrow C^+$ recombination process not included in the models which might also enhance 4267Å and remove the other discrepancy.

Although estimates of dielectronic recombination, based on the general formula of Burgess (1964) had led to the conclusion that the process was negligible it was pointed out that the general formula is not expected to be valid at nebular temperatures. Harrington, Lutz and Seaton (1981) then identified the 2297Å line due to the transition $2p^2\ {}^1D \rightarrow 2s2p\ {}^1P$ in C III in the IUE spectrum of the planetary nebula NGC 7009 . At the temperatures deduced for this nebula from the N II and O III forbidden lines, the collisional excitation rate of this transition is much too small to account for the line intensity.

The first new nova to be observed by the IUE satellite, Nova Cygni 1978, when in its nebular phase, showed in its spectrum strong collisionally-excited lines of C, N and O together with other much weaker lines, among them C III 2297Å and C II 1335Å. The pair 2297Å and 1909Å in C III were assumed to be both collisionally excited and their intensity ratio was used to determine the temperature of formation of the lines, giving values in the range 4 to $5 \cdot 10^4$K, much higher than the temperatures of planetary nebulae. The C II pair 1335, 2326Å gave similar results. The conditions however in the nova shell were certainly very different from those in planetary nebulae, the CNO abundances being enhanced by a factor of 10, the density higher and the source of ionising photons hotter and more luminous in the nova (Stickland et al. 1981).

Storey (1981) investigated the effects of dielectronic recombination at nebular temperatures. The rate α_{ab} for the processes

$$A^{+m} + e^- \rightarrow A_a^{+m-1} \rightarrow A_b^{m-1} + h\nu \qquad (62)$$

is given by

$$\alpha_{ab} = \frac{\omega_a}{2\omega_+} \left(\frac{h^2}{2\pi m k T_e}\right)^{3/2} e^{-E_a/KT} \frac{\Gamma_a^A \Gamma_{ab}^R}{\Gamma_a^A + \sum_{b'} \Gamma_{ab'}^R} \qquad (63)$$

and the total α_d is obtained on summing over a and b.

Under the conditions of validity of the Burgess formula, the free electrons have energies comparable with the ionisation potentials of the abundant ions. Burgess considered resonances (the states a) for which the parent term is connected to the ground state of the recombining ion by an optically allowed transition and assumed that stabilisation takes place via a core transition, the highly-excited external electron acting as a spectator. In the sum over a, Γ_a^R is independent of a and the sum converges since $\Gamma_a^R \rightarrow 0$ for large $n_a \ell_a$. At the temperatures of interest to Burgess, $n_a \sim 100$, $l_a \leq 10$. In the Burgess formula the energies E_a were replaced by a constant term close to the parent term energy difference ΔE and α_d then behaves as $e^{-\Delta E/kT}$ as $T \rightarrow 0$. This leads to a negligible α_d for $T \sim 10^4$K. At nebular temperatures however the exponential factor in (63) picks out the low-lying autoionising states, which have larger values of Γ_a^A and decay radiatively by outer electron transitions. Optically forbidden parent ion transitions are also important (Storey 1983). In $C^{3+} + e^-$, the range of contributing states includes 2p4l, l > 0. Seaton and Storey (1976) have shown that $\Gamma^A = c/\nu^3$ in atomic units where ν is the effective quantum number and c is of the order unity and depends on l. Γ^A is independent of the charge z. The radiative rates, on the other hand, are given in the same units by $\Gamma^R = \alpha^3 R z^p$ for core decays and $\Gamma^R = \alpha^3 R' z^p/\nu^3$ for outer electron decays where α is the fine-structure constant R and R' are of order unity and p = 1 or p = 4 according to whether there is or is not a change of principal quantum number in the core decay. Hence when z and ν are small, $\Gamma^A >> \Gamma^R$ because of the factor α. Hence in (63) the factor Γ^A cancels out and only knowledge of Γ^R is required.

Making use of this fact, Storey (1981) calculated α_d assuming the resonance states to be purely bound with no interaction with the continuum by computing all significant radiative transition probabilities with the program SUPERSTRUCTURE., for C, N and O ions. The resulting total dielectronic recombination rates were found to exceed the radiative recombination rates calculated without resonances by factors of 2 to 4.

The results for $C^{+3} + e^-$ showed that at nebular temperatures dielectronic recombination provides an important contribution to the ionisation balance of C^{2+}/C^{3+} and leads to a change in population of C III $2p^2\ ^1D$ sufficient to explain the observed intensity of C III 2297A in planetary nebulae. Harrington, Lutz and Seaton have estimated C abundances in NGC 7009 and NGC 7662 and find results larger than those of PWD and smaller than those of TPP. This is attributed to an underestimate of $N(C^{3+})/N(H^+)$ by PWD due to neglect of dust absorption of C IV 1549 radiation; an over-estimate of the observed flux of 4267 by TPP; and the use of the incorrect ratio $N(C^{3+})/N(C^{2+})$ due to neglect of dielectronic recombination in the ionisation equilibrium models. A model of NGC 7662 has been constructed which gives satisfactory agreement with observed UV and other spectra except for poor agreement with the observed intensities of C IV 1549 and C II 4267. The ionisation equilibrium was calculated using Storey's results and the C abundance adjusted to give the observed fluxes in 2297, 1908 and 2326. The calculated flux in 1549 is then 2.8 times larger than that observed with IUE. This is attributed to reduction of 1549Å flux by dust absorption. In the nova Nova Cygni 1978 it was found that only if dielectronic recombination was invoked as the mechanism for populating the upper states of the emission lines C II 1335Å, N IV 1718Å and O V 1371Å was it possible to explain the observed intensities of these lines and obtain a consistent model of the nebular shell of the nova. In the model the temperatures are comparable with those of planetary nebulae and consistent with those obtained from the intensities of the forbideen lines of N II and O III.

The total CNO abundance was found to be enhanced relative to the solar value by factors of 20 and the N abundance by a factor of 200. This provides evidence in support of the thermo-nuclear runaway mechanism (Starrfield et al. 1978) for production of the nova.

References

Baliunas, S.L., and Butler, S.E., 1980, Ap. J., 235:L45.
Baluja, K.L., and Hibbert, A., 1980, J. Phys. B., 13:L327.
Baluja, K.L., Burke, P.G. and Kingston, A.E., 1980, J. Phys. B., 13;829.
Berrington, K.A., Burke, P.G., Dufton, P.F., and Kingston, A.E., 1977, J. Phys. B., 10:1465.
Bhalla, C.P., Gabriel, A.H., and Presnyakov, L.P., 1975, Mon. Not. R. astr. Soc., 172:359.
Bowen, I.S., 1927a. Phys. Rev., 29:231; 1927b, Publ. Astr. Soc. Pacific, 39:295; 1928, Ap. J., 67:1.
Burbidge, G.R., Gould, R.J., and Pottasch, S.R., 1963, Ap. J., 138: 945.
Burgess, A., 1964, Ap. J., 139:776.
Burgess, A., and Chidichimo, M., 1983, Mon. Not. R. astr. Soc., 203:1269.
Carson, T.R., Mayers, D.F., and Stibbs, D.W.N., 1968, Mon. Not. R. astr. Soc., 140:483.
Cheng, C.C., Feldman, U., and Doschek, G.A., 1979, Ap. J., 233:736.
Cox, A.N., and Stewart, J.N., 1965, Ap. J. Supp., 11:22.
Doyle, J.G., and Raymond, J.C., 1981, Mon. Not. R. astr. Soc., 196:907.
Dufton, P.L., Hibbert, A., Kingston, A.E., and Doschek, G.A., 1983, Ap. J., 274:420.
Eissner, W., Jones, M., and Nussbaumer, H., 1974, Comput. Phys. Commun., 8:270.
Eissner, W., and Seaton, M.J., 1973, Mem. Soc. R. Liège, 5:203.
Feldman, U., Doschek, G.A., and Kreplin, R.W., 1980, Ap. J., 238:365.
Feldman, U., Doschek, G.A., and Cowan, R.D., 1981, Mon. Not. R. astr. Soc., 196:517.
Gabriel , A.H., and Paget, T.M., 1972, J. Phys. B., 5:673.
Harrington, J.P., Lutz, J.H., Seaton, M.J., and Stickland, D.J., 1980, Mon. Not. R. astr. Soc., 191:13.
Harrington, J.P., Lutz, J.H., and Seaton, M.J., 1981, Mon. Not. R. astr. Soc., 195:21P.
Hebb, M.H. and Menzel, D.H., 1940, Ap. J., 92:408.
Hibbert, A.. 1975, Comput. Phys. Commun., 9:141.
Jacobs, V.L., Davis, J., Kepple, P.C. and Blaha, M.,1977, Ap. J., 211:605; Ap. J., 215:106,
Jacobs, V.L., Davis, J., Rogerson, J.E., and Blaha, M., 1978, Ap. J., 230:627.
Jacobs, V.L., Davis, J., Rogerson, J.E., Blaha, M., Cain, J., and Davis, M., 1980, Ap. J., 239:1119.
Jain, N.K., and Narain, U., 1978, Astron. Astrophys., Suppl., 31:1.
Jordan, C., 1969, Mon. Not. R. astr. Soc., 142:499; 1970, Mon. Not. R. astr. Soc., 148:17.
Landini, M., and Monsignori-Fossi, B.C., 1972, Astron. Astrophys., Suppl., 7:291.
Lotz, W. 1968, Z Phys. 216, 241.

Nicholas, K.R., 1977, Ph. D. Thesis, University of Maryland.
Nussbaumer, H., and Storey, P.J., 1978, Astron. Astrophys., 64:139.
Peimbert, M., and Costero, R., 1969, Bol. Obs., Tonantzintla Tacubaya, 31:3.
Pottasch, S.R., Wesselius, P.R., and van Duinen, R.J., 1978, Astron. Astrophys., 70:629.
Saraph, H.E., 1978, Comput. Phys. Commun., 15:247.
Saraph, H.E., 1984, Physica Scripta, T8:134.
Saraph, H.E. and Seaton, M.J., and Shemming, J., 1966, Proc. Phys. Soc., 89:27.
Seaton, M.J., 1983, Proc. Roy. Soc., A218:400.
Seaton, M.J. and Storey, P.J., 1976 in "Atomic Processes and Applications" ed. P.G. Burke and B.L. Moiseiwitsch. North-Holland Publishing Co.,135.
Seaton, M.J. 1964, Planet. Space Sc. 12, 55.
Simon, N.R., 1982, Ap. J., 260:L87.
Smith, M.W., and Wiese, W.L., 1971, Ap. J., 23:103.
Starrfield, S., Truran, J.W., and Sparks, W.M., 1978, Ap. J., 226:186.
Stickland, D.J., Penn, C.J., Seaton, M.J., Snijders, M.A.J., and Storey, P.J., 1981, Mon. Not. R. astr. Soc., 197:107.
Storey, P.J.,1981, Mon. Not. R. astr. Soc., 198:27P.
Storey, P.J., 1983 in "Planetary Nebulae" ed. D.R. Flower. D. Reidel Publishing Co., 195.
Summer, H., 1974, Mon. Not. R. astr. Soc., 169:633; 1979, Appleton Lab. Report, AL-R-5.
Torres-Peimbert, S., and Peimbert, M., 1977, Rev. Mex. Ast. Astrofis., 2:181.
Unsöld, A. 1961, Prive letter to Dr. M.J. Seaton.

THE EINSTEIN-PODOLSKY-ROSEN PARADOX FIFTY YEARS LATER

Franco Selleri

Istituto Nazionale di Fisica Nucleare
Dipartmento di Fisica dell'Università
via Amendola, 173 - 70126 BARI

ABSTRACT

The purpose of this paper is to provide an introduction to the Einstein-Podolsky-Rosen paradox which, in recent times, has developed into a theorem of incompatibility between the empirical predictions of quantum theory and the assumption that our universe can be analyzed in terms of separable real objects. The theorem is expressed by Bell's inequality for ideal detectors, but exists also for real detectors and no additional assumptions are needed.

INTRODUCTION

Although the debate about the nature of quantum properties of atomic systems has never ceased, there are two periods in which it has been particularly intense: the years in which quantum mechanics was founded and the present years. In 1954 Max Born recalled the depth of the splitting which divided the famous quantum theorists in two camps: "When I say that physicists had accepted the way of thinking developed by us at that time, I am not quite correct. There are a few most noteworthy exceptions - namely among those very workers who have contributed most to the building up of quantum theory. Planck himself belonged to the sceptics until his death. Einstein, de Broglie and Schrödinger have not ceased to emphasize the unsatisfactory features of quantum mechanics ..."[1]

The existence of such important critical voices could not stop the general acceptance of quantum mechanics: nobody was ready to accept that the theory had to be modified in times in which its triumphal successes were recent, the fields to which it could be applied were numerous and the difficulties and the ambiguities were yet to be discovered. In fact still today it can be said that the successes of quantum mechanics are so numerous and so accurate that never has a scientific theory been even remotely comparable with it from this point of view.

The critical researches of Einstein, Schrödinger and de Broglie were directed much more against the interpretation than against the validity of the theory. What they refused was more than anything else the "philosophical flavour" of the Copenhagen-Göttingen paradigm, while they were willing to accept the theory as correct in its quantitative predictions.

The situation has been changing somewhat in recent times because of experimental and theoretical developments. Researches into neutron interferometry, ultracold neutrons, laser physics, superfluorescence, low intensity beams, and so on, have sharpened our concrete understanding of the atomic world and rendered more acute some fundamental problems: typical from this point of view is the situation of neutron interferometry concerning the nature of wave-particle dualism.

On the theoretical front there has been a dramatic development of the so-called Einstein-Podolsky-Rosen paradox, which has shown that some simple and elementary physical ideas lead to testable predictions in gross disagreement with the predictions of quantum mechanics. This disagreement is expressed, for instance, by Bell's inequality.

These experimental and theoretical developments have brought quantum physics to a new and highly interesting situation. The old debate which saw Einstein and Bohr on opposite sides has been developed to a point where experiments can decide who is right.

THE PROBLEM OF CAUSALITY

Those physicists who did not accept the final formulation of quantum mechanics disagreed with the opinions of the Copenhagen and Göttingen schools in particular on one point: they thought that it was possible and useful to complete the theory in such a way as to make it causal.

In order to understand as clearly as possible the need for causality, let us consider the following example: neutrons are unstable particles and each of them lives its individual lifetime at the end of which it disintegrates into proton - electron - antineutrino. The average of the lifetimes of many neutrons is about 1000 seconds, in the same sense that the average lifetime of the Europeans is about seventy years: some people live longer, some other ones live less but the average over millions of people is seventy years. Similarly, individual neutrons can live much less (say, 100 sec.) or much more (say, 3000 sec.) than the average lifetime of a thousand seconds. The problem which presents itself very naturally is to understand the reasons for these differences, or , in other words, to understand the causes which determine different individual lifetimes of different unstable systems.

Today's physics does not provide any understanding of these causes and accepts an acausal philosophy: every decay is a spontaneous process and does not admit a causal explanation. The question about the different individual lifetimes of different unstable systems, like neutrons, should, according to this line of thought, remain forever without answer and should be considered a "non-scientific" question.

Einstein and the other critics took instead a different point of view and believed that our present knowledge is not complete and that, in becoming wider, it will naturally lead to an understanding of the individual lifetimes and their causes. The causes of such different individual behaviour are usually called hidden variables. One finds sometimes in the literature statements according to which these hidden variables would not be measurable in principle. We think that such a point of view does not have any justification: if they exist they will also be measurable, when some insight has been gained into their physical nature. Therefore when we speak about "hidden variables" we mean only that they are hidden now, in the same sense that atoms were hidden until about 1910 and that pulsars were hidden until a few years ago. It could be that quantum mechanics will turn out to be an incomplete theory of the same type as classical thermodynamics.

The latter is successful in predicting the equilibrium properties of matter, but is unable to describe such phenomena as thermal fluctuations, brownian motion and so on, which can be explained only by taking into account the atomic structure of matter.

The danger deriving from the idea of hidden variables, by declaring quantum mechanics incomplete, and thus proposing a reversal of the philosophical basis of theoretical physics, was immediately understood by the physicists of the Copenhagen-Göttingen schools. A defense against this causal philosophy was provided by J von Neumann who proposed in 1932 his famous theorem stating the impossibility of a deterministic completion of quantum mechanics given few very general assumptions about the mathematical structure of physical theories.[2] Bohr, Born, Pauli, Heisenberg and Jordan forcefully stressed the importance of the theorem. The great authority of these physicists together with the mathematical complexity of von Neumann's theorem had the practical effect of outlawing the idea of hidden variables. Bohr expressed his opinion on this matter for instance in 1938 when the conference on "New Theories in Physics" was held in Warsaw.[3] In a book published in 1948 Born discussed the axioms of von Neumann's theorem and concluded: "The result is that the formalism of quantum mechanics is uniquely determined by these axioms; in particular, no concealed parameters can be introduced with the help of which the indeterministic description could be transformed into a deterministic one."[4] The unconditional acceptance of von Neumann's theorem had the result of eliminating all scientific dialectics: the philosophy of Copenhagen and Göttingen triumphed almost everywhere and all opposition was virtually eliminated. Einstein, who firmly kept his position of rejecting acausality found himself rather isolated in Princeton. From 1935 to about 1970 the physicists who worked on the problem of causality were therefore very few. Nevertheless important results were obtained by Bohm[5] and by de Broglie[6] who were able to do what von Neumann had declared impossible to do: find hidden-variable models that do not contradict quantum mechanics in their statistical predictions and which, at the same time, provide causal formulations for the individual behaviour of quantum systems. It took some time before these results were fully appreciated, that is until it was understood that von Neumann's theorem, although mathematically correct, could in no way forbid deterministic generalizations of quantum mechanics simply because one of its axioms was in general not physically reasonable.

VON NEUMANN'S THEOREM

Consider a general observable R corresponding to the linear hermitean operator $\tilde{R}$. If we perform a series of N measurements of R on similarly prepared particles (that is, particles all having the same initial wave-function) we will obtain the results

$$R = r_1, r_2, \dots r_N \quad .$$

According to quantum mechanics, whose correctness is not questioned here in any way, these results will in every case coincide with one of the eigenvalues of $\tilde{R}$.

It is useful to consider also the observable R^2 which is by definition that observable which can be measured by measuring R and taking the square of the obtained result. The previous N measurements for R therefore constitute N measurements of R^2:

$$R^2 = r_1^2, r_2^2, \dots r_N^2 \quad .$$

Assume now that the results $r_1, r_2 \dots r_N$ are actually determined by a hidden variable λ, more precisely by the values which λ had immediately before the measurements. Thus we can write $r_1 = r(\lambda_1)$, $r_2 = r(\lambda_2), \dots, r_N = r(\lambda_N)$, where $\lambda_1, \lambda_2, \dots, \lambda_N$ are the values of λ before each individual measurement process. If we prepare the particles in such a way that λ is always the same before the measurements (say $\lambda = \lambda_0$) we will obviously always obtain the same result:

$$R = r_1 = r_2 = \ldots = r_N = r(\lambda_0)$$

It follows for such an ensemble with fixed λ

$$\langle R \rangle = \frac{1}{N} \Sigma_i r_i = r(\lambda_0); \quad \langle R^2 \rangle = \frac{1}{N} \Sigma_i r_i^2 = r^2(\lambda_0)$$

Therefore the root mean squared deviation of the observable R vanishes:

$$\Delta R = [\langle R^2 \rangle - \langle R \rangle^2]^{\frac{1}{2}} = 0$$

An equivalent but shorter name for ΔR is dispersion. We can thus say that there is no dispersion for our observable R once the hidden variable has a fixed value.

More generally we can consider a situation in which there are several different observables (R, S, T, ...) pertaining to the considered physical systems. In the spirit of hidden variables we can say that the results obtained by measuring R, or S, or T,... are fixed by the values of some hidden variables collected in the set $(\lambda, \mu, \nu, \ldots)$. Naturally, the number of hidden variables is in general not related to the number of observables. However considering a set of particles in which all the hidden variables have fixed values $(\lambda = \lambda_0, \mu = \mu_0, \nu = \nu_0, \ldots)$ all the observables must be found, when measured, to have a well defined value

$$R = r(\lambda_0, \mu_0, \nu_0, \ldots) \equiv r_0$$

$$S = s(\lambda_0, \mu_0, \nu_0, \ldots) \equiv s_0$$

$$T = t(\lambda_0, \mu_0, \nu_0, \ldots) \equiv t_0$$

$$\cdots\cdots\cdots\cdots\cdots$$

Repeating the reasoning made above for R alone, we conclude now that all the observables R, S, T,... have zero dispersion. Since these are all the observables of the particles forming the considered statistical ensemble, we say that this ensemble is dispersion-free. We need not worry about the fact that different observables may be represented by noncummuting operators. This merely means that it is not possible to build instruments capable of a simultaneous measurement of them. If we have two such observables R and S represented by the two noncommuting operators $\tilde{R}$ and $\tilde{S}$ and just one hidden variable λ with the fixed value λ_0 we will perform on some of the systems measurements of R and on the remaining systems measurements of S. In the first case we will always obtain r_0, in the second always s_0. Therefore we will have $\Delta R = 0 = \Delta S$, but this implies no contradiction with the noncommutativity of $\tilde{R}$ and $\tilde{S}$, because the measurements of R and S were performed on different systems. If we believe in hidden variables we can always think of a given ensemble as a sum of dispersion-free ensembles. The familiar results of Quantum Mechanics (e.g. $\Delta R \, \Delta S \gtrsim \hbar$) will hold for the full ensemble but not for its dispersion-free parts.

We can conclude quite generally that the existence of hidden variables implies necessarily the existence of dispersion-free ensembles. It was precisely against the existence of such ensembles that von Neumann's theorem was directed.

The axioms on which von Neumann's theorem is based are the following three:

N1 - There is a one to one correspondence between observables and hermitean operators.

N2 - If the observable R corresponds to the operator $\tilde{R}$, then the observable f(R) corresponds to $f(\tilde{R})$.

N3 - If R and S are arbitrary observables and a and b real numbers the following relation (expressing the linearity of the averages) is true

$$\langle aR + bS\rangle = a\langle R\rangle + b\langle S\rangle \ .$$

From these axioms von Neumann deduces that dispersion-free ensembles cannot exist. The proof is very simple for spin - $\frac{1}{2}$ and is given in the closing part of the present section.

Before coming to it, it is better to gain a better understanding of the axioms. The following four definitions should clarify their meaning.

D1 - Definition of function of an observable R. Given a function f(x) of the real variable x one defines f(R) in the following operative way: in order to measure f(R) one measures R; if one finds R = r one computes f(r) and says that the latter is the measured value of f(R).

D2 - Definition of function of an operator $\tilde{R}$. The function $f(\tilde{R})$ is defined by means of a power development of f(x) The n-th power of $\tilde{R}$ can easily be obtained from standard definitions of elementary Hilbert-space theory.

D3 - Definition of sum of two observables R and S. If the two observables R and S correspond, respectively, to the hermitean operators $\tilde{R}$ and $\tilde{S}$ one assumes that R + S is the observable corresponding to the operator $\tilde{R} + \tilde{S}$.

D4 - Definition of linear combination of two observables. The definition of the product of a real constant a with observable R is obtainable as a particular case of D1. The product of a with the operator $\tilde{R}$ is well known. Having so defined aR and bS (and $a\tilde{R}$ and $b\tilde{S}$) one uses D3 to define aR + bS.

Notice that N1 is intended to mean also that the measured values of an observable R cannot be different from the eigenvalues of its associated operator $\tilde{R}$. As far as N2 is concerned we wrote it for historical completeness, but it will not be used in our simplified proof of the theorem.

We come now to the proof of von Neumann's theorem, that is we show that hidden variables cannot exist for the spin - ½ system if N1 and N3 are true.

Let us suppose that a hidden variable λ exists which determines the results of the measurements of all conceivable spin observables. Consider an ensemble in which λ has a fixed value λ_0, that is to say a dispersion-free ensemble. Consider any three observables R, S and T and their associated 2 x 2 matrices $\tilde{R}$, $\tilde{S}$ and $\tilde{T}$. Measurements of these observables in the dispersion-free ensemble lead necessarily to the conclusion that

R has a fixed value $r(\lambda_0)$ which is an eigenvalue of $\tilde{R}$
S " " " " $s(\lambda_0)$ " " " " " $\tilde{S}$
T " " " " $t(\lambda_0)$ " " " " " $\tilde{T}$

Two facts should be kept in mind when discussing spin observables. Firstly, the most general hermitean 2 x 2 matrix is

$$\tilde{X} = \alpha I + \vec{\beta} \cdot \vec{\sigma} \tag{1}$$

where α, β_1, β_2, β_3 are arbitrary real numbers, I is the 2 x 2 unit matrix and σ_1, σ_2, σ_3 are the Pauli matrices. Secondly, the eigenvalues of such an X are

$$\alpha \pm |\vec{\beta}| \, .$$

Specializing this general conclusion assume that

$$\tilde{R} = \sigma_1; \quad \tilde{S} = \sigma_2; \quad \tilde{T} = \vec{\sigma} \cdot \vec{n} \tag{2}$$

where $\vec{n} = (1,1,0)$, whence $|\vec{n}| = \sqrt{2}$. It follows from (2) that $\tilde{T} = \tilde{R} + \tilde{S}$. Therefore one concludes, using D3, that

$$T = R + S \tag{3}$$

Now in our dispersion-free ensemble with $\lambda = \lambda_0$, every time one measures R one finds $r(\lambda_0)$. This means that the ensemble average of R is also given by $r(\lambda_0)$: $\langle R\rangle = r(\lambda_0)$. Similarly, one must have $\langle S\rangle = s(\lambda_0)$ and $\langle T\rangle = t(\lambda_0)$. Applying now N3 one deduces from (3) $\langle T\rangle = \langle R\rangle + \langle S\rangle$ which implies that $t(\lambda_0) = r(\lambda_0) + s(\lambda_0)$. The latter relation cannot be satisfied, however, since the eigenvalues of $\tilde{T} = \vec{\sigma} \cdot \vec{n}$ are $\pm\sqrt{2}$, and no choice of the signs can satisfy the relation

$$\pm\sqrt{2} = \pm 1 \pm 1$$

Starting from the axioms of von Neumann's theorem and from the existence of dispersion-free ensembles we have reached an absurd conclusion. If we believe the axioms to be correct we must therefore conclude that dispersion-free ensembles cannot exist.

THE THEOREM IS NOT GENERAL ENOUGH

If the axioms of von Neumann's theorem were so general that no conceivable physical situation could violate them we should obviously give up all hope of finding a deterministic completion of quantum mechanics. Fortunately, as we have already anticipated several times, this is not the case. If we consider superficially N1 and N3 nothing seems to be possibly wrong with them.

N1 establishes a very simple correspondence between spin observables and 2 x 2 hermitean matrices: if this axiom were wrong we would be forced to build a theory different from quantum mechanics which makes the correspondence [observables $\leftrightarrow$ linear hermitean operators] the very basis of its theoretical formalism. N3 postulates average linearity properties which are true within quantum mechanics and which seem to be valid also in classical physics.

Still it is precisely N3 which fails when one considers certain types of classical measurements.

In order to see this as clearly as possible we consider a concrete physical model of spin - ½ particles and of spin-measurements[7]. This model reproduces all the quantum mechanical predictions for spin measurements (results, probabilities) and does therefore exactly what von Neumann's theorem tried to forbid, that is provides a causal completion of quantum mechanics. Of course, the model itself should not be taken too seriously because, as we will see, it is for many reasons very different from the reality of sub-atomic entities: e.g. it does not take into account wave-particle dualism. It is however totally accurate in reproducing the quantum mechanical predictions for spin measurements and constitutes therefore a real counter-example to von Neumann's theorem since this theorem, as we saw, can be formulated also for spin alone.

The properties of our model are contained in the following three hypotheses.

M1 - Spin - ½ particles can be simulated with spinning spheres which are assumed to propagate along the x-axis. The intrinsic angular momentum $\vec{\lambda}$ of each sphere (spin) is assumed to lie in all cases in the (yz)-plane perpendicular to the direction of motion.

M2 - In order to reproduce the probabilistic properties of the $\binom{1}{0}$ state of quantum mechanics we assume that a statistical ensemble of such spheres described as $\binom{1}{0}$ has a statistical distribution of vectors $\vec{\lambda}$ such that the density $\rho(\theta)$ of the angle θ that $\vec{\lambda}$ forms with the z-direction is given by

$$\begin{aligned} \rho(\theta) &= \frac{1}{2}\cos\theta \quad , \text{ if } \; -\frac{\pi}{2} \leq \theta \leq \frac{\pi}{2} \; ; \\ \rho(\theta) &= 0 \qquad\qquad , \text{ otherwise.} \end{aligned} \tag{4}$$

M3 - An instrument built for measuring the quantum mechanical observable X corresponding to the operator $\tilde{X}$ given in (1) works in the following way: it measures <u>the sign</u> of the projection to $\vec{\lambda}$ onto $\vec{\beta}$, multiplies the obtained result by $|\vec{\beta}|$ and adds α to the result.

The model so formulated leads to interesting consequences. Firstly, the procedure of "measurement" defined in M3 implies that the results obtained will not be different from $\alpha \pm |\vec{\beta}|$, thus guaranteeing that no disagreement with quantum mechanics can be obtained at this level. Secondly, as a particular case of the previous result and as a consequence of (4) if one measures σ_3 [whence $\alpha = 0$, $\vec{\beta} = \hat{k}$ (unit vector along z)] the angle between the $\vec{\lambda}$'s and $\vec{\beta}$ is in all cases between $-\pi/2$ and $\pi/2$ implying that the projection of $\vec{\lambda}$ over $\vec{\beta}$ is never negative. It follows that all measurements will give $\sigma_3 = +1$, a result consistent with the fact that the state $\binom{1}{0}$ is an eigenstate of σ_3 with eigenvalue +1.

Finally we can calculate the probabilities p_1 and p_2 of finding, respectively, the results $\alpha + |\vec{\beta}|$ and $\alpha - |\vec{\beta}|$ when the general observable X is measured. Notice that M3 consists of two parts: a true measurement when the sign of $\vec{\lambda}\cdot\vec{\beta}$ is determined and a mechanical process which consists of the multiplication of this sign by $|\vec{\beta}|$ and of the addition of α to the result. Therefore p_1 and p_2 coincide respectively with the a-priori probabilities of finding sign $(\vec{\lambda}\cdot\vec{\beta}) = \pm 1$, resp. In order to calculate p_1 one needs only to calculate the integral of $\rho(\theta)$ over the hatched area σ_0 because by M2 the $\vec{\lambda}$'s can lie only in the upper half-circle. (see Figure 1) Obviously then

$$p_1 = \int_{\sigma_0} d\,\theta\, \frac{1}{2}\cos\theta = \frac{1+\cos\,\theta_\beta}{2}$$

Remembering that $p_1 + p_2 = 1$ one obtains immediately also

$$p_2 = \frac{1 - \cos\,\theta_\beta}{2} \quad .$$

These results for p_1 and p_2 coincide exactly with the quantum mechanical probabilities.

This model gives therefore results of measurements and pro-

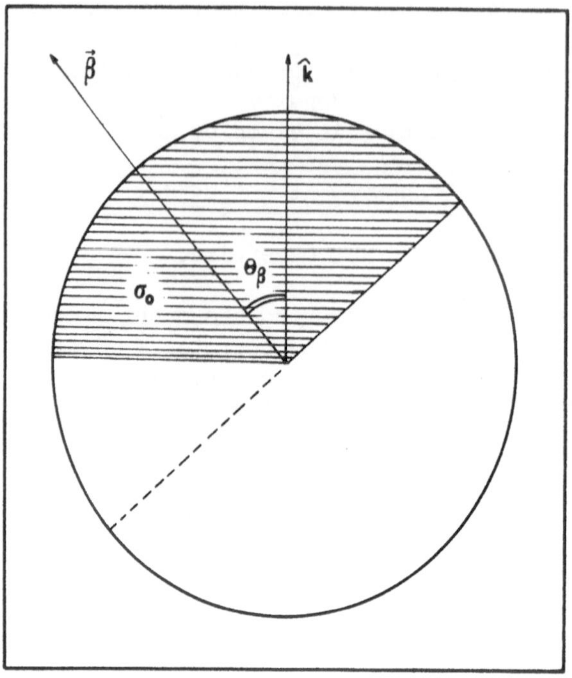

Fig. 1. The integral of the density function $\rho(\theta)$ [see eq.(4)] over the dashed area gives the probability of the first eigenvalue.

babilities for the different results in perfect agreement with quantum mechanics. Still, the outcome of an act of measurement of $\alpha I + \vec{\sigma}\cdot\vec{\beta}$ on a given sphere is completely fixed by the value of $\vec{\lambda}$ for that sphere. Therefore measurements are deterministic processes in this model and we have obtained a causal generalization of the quantum mechanical theory of spin - $\frac{1}{2}$.

The model does therefore exactly what von Neumann's theorem forbids, and this requires that we discuss the validity of von Neumann's axioms within the model. The hidden variable is here the vector $\vec{\lambda}$. Consider then a dispersion-free ensemble of particles all having $\vec{\lambda}$ pointing along the bisectrix of the (yz)-quadrant and measure on part of this ensemble the observable σ_2, on a second part the observable σ_3 and on a third part of the observable $\vec{\sigma}\cdot\vec{m}$, with $\vec{m} = (0,1,1)$. Notice that these observables have been chosen in such a way that their associated operators satisfy

$$\sigma_2 + \sigma_3 = \vec{\sigma}\cdot\vec{m} \ .$$

Since the $\vec{\lambda}$'s of our dispersion-free ensemble form acute angles with the y- and the z-axis, and naturally also with the $\vec{m}$ direction, the result "$\alpha+|\vec{\beta}|$" will be obtained in all these three types of measurements. Therefore

Meas. of σ_2 gives always +1 $\rightarrow \langle\sigma_2\rangle = +1$

Meas. of σ_3 gives always +1 $\rightarrow \langle\sigma_3\rangle = +1$

Meas. of $\vec{\sigma}\cdot\vec{m}$ gives always $+\sqrt{2} \rightarrow \langle\vec{\sigma}\cdot\vec{m}\rangle = +\sqrt{2}$

The conclusion is obviously that $\langle\vec{\sigma}\cdot\vec{m}\rangle \neq \langle\sigma_2\rangle + \langle\sigma_3\rangle$ and this means that N3 is not satisfied for our dispersion-free ensemble.

THE ORIGINAL FORMULATION OF THE EPR-THEOREM

"If, without in any way disturbing a system, we can predict with certainty (i,e., with probability equal to unity) the value of a physical quantity, then there exists an element of physical reality corresponding to this physical quantity."

This is the famous "criterion of physical reality" proposed by Einstein, Podolsky and Rosen in 1935[8]. It is a very weak criterion of reality, very carefully formulated and extremely general. It appeared to them (and still appears today) "far from exhausting all possible ways of recognizing a physical reality". This criterion can have many applications to the macroscopic domain where it appears as trivially true: for instance we can say that if we can predict with certainty that the length of this table is two meters (up to some error of, say, half a centimeter) then there exists an element of physical reality corresponding to the length of the table. Notice that EPR would not say that the length itself is real. For one thing the unit of length is conventional and contains therefore subjective aspects. Furthermore what appears to us immediately as extension in space might result from properties, actually different from our perception, giving rise to a final sensation which is only a topological distorsion of the true physical property. Even in this complicated case the EPR criterion would apply, since it assumes only the existence of an element of physical reality without specifying its nature and its relationship to the measured value of the chosen physical quantity. Similar considerations could easily be made in many other macroscopic situations.

Einstein, Podolsky and Rosen considered their criterion in agreement with classical as well as with quantum-mechanical ideas of reality.

In quantum physics the existence of the quantum of action h implies a finite interaction between the atomic object and the measuring apparatus. This puts limits in the precision with which the position (x) and the momentum (p) of the object can be measured simultaneously; such limits can be expressed by the relation

$$\Delta x \, \Delta p \geq \hbar \quad .$$

The validity of Heisenberg's relations is compatible with the existence in quantum theory of two particular types of wave-functions:

i) those describing a particle with fixed momentum p_0 but with position completely unknown

$$\psi = u_{p_0}(x) \quad ; \tag{5}$$

ii) those describing a particle with position x_0 fixed but with momentum completely unknown

$$\psi = v_{x_0}(x) \quad . \tag{6}$$

It can be shown that x disappears from the squared modulus of (5), so that all positions are equally probable, while (6) vanishes everywhere except for $x = x_0$ so that the position x_0 is certain.

The EPR criterion of reality can be applied to (5) since such a wave function predicts that a measurement of the momentum will give p_0 with certainty. It also can be applied to (6) since this wave function predicts that a measurement of the position will give x_0 with certainty. Consequently we can say that

1) An element of physical reality corresponds to the value p_0 of the momentum contained in (5);

2) Another element of physical reality corresponds to the value x_0 of the position contained in (6).

We can obviously conclude that if some further elements of reality exist in the physical systems described by (5) and (6), different from momentum (in the case of (5)) and from position (in the case of (6)), then the quantum mechanical theory must necessarily be considered incomplete. We call "complete" a theory satisfying the following criterion: every element of physical reality must have a counterpart in the physical theory.

Consider next a system M (molecule, atom, particle), capable of decaying into two new systems S_1 and S_2. Hundreds of examples exist in different branches of physics of such decay processes: the molecule NO can decay from an excited state to a state of two free atoms (N and O) which propagate in opposite directions, the Λ-hyperon decays into a proton and a negatively charged π-meson, and so on. The wave-function will in general be of the type

$$\psi = \psi(x_1, x_2)$$

where x_1 and x_2 are co-ordinates of S_1 and S_2 respectively. The physical interpretation of ψ is the usual one: $|\psi(x_1,x_2)|^2$ provides the probability density for finding S_1 at x_1 and S_2 at x_2.

There exist wave-functions for which the sum of the momenta of S_1 and S_2 has a fixed value p and the difference of the positions of S_1 and S_2 has simultaneously a fixed value x. We write one of these wave-functions as

$$\psi = \psi(x,p;\ x_1,x_2) \tag{7}$$

Notice that (7) predicts with certainty the observables x and p which can therefore be considered as two simultaneous elements of reality. No fixed value of the momentum p_1 of S_1 and of p_2 of S_2 is instead contained in ψ, only their <u>sum</u> being given as fixed and equal to p. Similarly, no fixed value of the positions x_1 of S_1 and x_2 of S_2 is contained in ψ, only their <u>difference</u> being given as fixed and equal to x.

We can now prove that individual positions and momenta of S_1 and S_2 do possess a physical reality, so reaching the conclusion that the quantum-mechanical description of reality is not complete. The reasoning is the following:

Suppose we are given a very large set E of similar decays $M \to S_1 + S_2$ to all of which the wave-function ψ given in (7) applies. Quantum mechanics predicts therefore that measurements of positions of S_1 and S_2 performed on individual decay processes will give results x_1 and x_2 satisfying <u>always</u> the relation $x_2 - x_1 = x$. Similarly, measurements of momenta of S_1 and S_2 performed on (other) individual decay processes will give results p_1 and p_2 which satisfy <u>always</u> the relation $p_1 + p_2 = p$.

Consider now a subset E_1 of E on which no previous measurement has been done and perform position measurements on the system S_1 of each individual decay. Let x_1', x_1'',... be the obtained results. We can then predict <u>with certainty</u> that subsequent measurements of the position of S_2 will give $x + x_1'$ for the first pair, $x + x_1''$ for the second pair, and so on.

Therefore, we can apply the criterion of physical reality and conclude that the position of S_2 is an element of physical reality.

It is natural to conclude that this element of physical reality of S_2 exists independently of the fact that a measurement of S_1 has been made. Therefore the position of S_2 is an element of reality <u>for all the decays of the ensemble E.</u>

A parallel argument can be made for the momenta and it can also be concluded that the momentum of S_2 is an element of reality for all the decays of the ensemble E.

It has been proved that to positions and momenta of S_1 and S_2 correspond elements of reality. We must then conclude, with EPR, that the quantum-mechanical description of reality given by wave functions is not complete.

BOHR'S ANSWER

Bohr's answer to the original EPR paper[9] does not question the correctness of the EPR reasoning once all its implicit and explicit premises are accepted. Rather Bohr implies that the quantum mechanical formalism cannot be adapted to a philosophical point of view like that of Einstein. Bohr argues in favour of "a final renunciation of the classical ideal of causality and a radical revision of our attitude towards the problem of physical reality", and stresses that his notion of complementarity is a "new feature of natural philosophy" which implies a "radical revision of our attitude as regards physical reality".

His line of reasoning is the following.
Firstly, one has to notice that all measurements performed on atomic and subatomic systems must necessarily be prepared, carried out and, when results are obtained, expressed in classical terms. This is so because the physicist lives in a macroscopic world where classical laws and, even more important, classical conceptions (space, time, causality,...) hold and have become unavoidable means of expression of all human experience. The experimental physicist will therefore naturally try to express his experimental results in classical terms, that is he will try to express the regularities of the microscopic world as causal processes in space and time. This is however impossible to do and the physicist finds himself suddenly exposed to an irrational element because the existence of the quantum of action h implies a finite mutual interaction between the measured object and the measuring instrument. This gives rise to effects on the observed system which cannot even be eliminated logically. In other words the existence of h implies a perturbation of the measured object which is completely unpredictable and thus, in a way, irrational. These facts not only set a limit to the extent of the information obtainable by measurements, but they also set a limit to the meaning which can be attributed to such information. Bohr's opinion is that an independent reality cannot be attributed to atomic phenomena: the word phenomenon should exclusively be used to refer to the observations obtained under specified circumstances, including an account of the whole experimental apparatus.

From this point of view physics must only deal exclusively with acts of observation, all reference to an unobserved elementary reality being banished from a truly scientific reasoning. Obviously, then, no paradox exists when one considers two correlated systems described by the wave-function(7).

From Bohr's point of view the EPR assumption about the elements of reality appears immediately useless: one can make it, but it means only that there is an element of reality associated with an act of measurement concretely performed, for there is no other reality which can be considered. In particular, the EPR conclusion that position and momentum correspond to two simultaneously existing elements of reality appears totally unjustified (Bohr writes that it contains "an essential ambiguity") because one can never perform simultaneous measurements of position and momentum; if there is no concrete measurement there is also nothing real to which an element of reality can be attributed.

INEQUALITIES FROM EINSTEIN LOCALITY

If Bohr's solution of the problem raised by Einstein, Podolsky, Rosen appears as perfectly rational from a narrow logical point of view, it must be admitted that it is after all not "natural" from a more general point of view. The idea of an objective reality which science studies with its techniques is the driving force of all branches of research and should not be given up easily.

It might appear that the problem is solved by the proved obsolescence of von Neumann's theorem: EPR tried to show that quantum mechanics was not complete, Bohr defended the completeness by giving up the usual idea of reality. Since today we have learnt that von Neumann's theorem is not applicable, perhaps the problem is solved in favour of Einstein in a very simple and painless way. The situation is however much more serious since the incompatibilty between the idea of a separable reality and the empirical predictions of quantum mechanics exists also if quantum theory is assumed incomplete. The first discovery of this important theorem was made by J S Bell in 1965.

Here we will give a general proof[10] of this fact which will lead to infinitely many inequalities violated by the predictions of quantum theory and of which Bell's inequality is a particular case.

Consider two measurements in the spacetime regions R_1 and R_2 with a spacelike separation. In R_1 the first observer measures the dichotomic observable A(a) dependent on the instrumental parameter a, while in R_2 the second observer measures the similar observable B(b). These measurements are performed on correlated systems, for instance on two photons produced in the same atomic cascade. The only possible values of the dichotomic observables are assumed to be ± 1. In a deterministic approach we assume that the results of the two measurements are completely fixed once the "hidden variable" λ is given. Thus the equations

$$\begin{cases} A(a\lambda) = \pm 1 \\ B(b\lambda) = \pm 1 \end{cases} \tag{8}$$

mean that A and B have well defined values once their arguments are specified (and that these values can be either +1 or -1).

If $\rho(\lambda) \geq 0$ is the probability density of the variable λ, the correlation function $P(a,b)$, defined as the average product of A and B, is given by

$$P(a,b) = \int d\lambda\, \rho(\lambda)\, A(a\lambda)\, B(b\lambda) \tag{9}$$

It should be noticed that λ in the previous equation is just a general symbol, which could cover several "hidden variables". In fact a given local realistic theory of the type (8) - (9) above is specified by the following

i) Number and nature of the additional parameters $\lambda_1, \lambda_2, \ldots$

ii) Functional dependence of $A(a, \lambda_1, \lambda_2, \ldots)$ and $B(b, \lambda_1, \lambda_2, \ldots)$ on these parameters.

iii) Probability density $\rho(\lambda_1, \lambda_2, \ldots)$.

Two theories are different if they differ in any one of the three previous specifications. Denoting again with a single symbol λ the parameters $\lambda_1, \lambda_2, \ldots$, we can more simply say that two local realistic theories are different if they are based on different functions $A(a,\lambda)$, $B(b,\lambda)$ and/or different probability densities $\rho(\lambda)$.

The only interesting inequalities deduced from Einstein locality are those which hold true for all conceivable local realistic theories of the type (8) - (9). Obviously, it is not possible today to say which one (if any) of the infinitely many theories based on Einstein locality is the correct one. Therefore inequalities deduced from a particular theory (or from a particular set of theories) are not interesting.

Lemma: Given a real number M, the inequality

$$\sum_{ij} c_{ij}\, P(a_i, b_j) \leq M \tag{10}$$

can be true for all the conceivable local realistic theories if and only if the inequality

$$\sum_{ij} c_{ij}\, A(a_i, \lambda) B(b_j, \lambda) \leq M \tag{11}$$

is true for arbitrary values of λ and for arbitrary dependences of A and B on their arguments.

In fact (11) is a necessary consequence of (10) since among all the conceivable local realistic theories there are those in which the density function ρ is a delta-function. Viceversa, if (11) is true for arbitrary λ and arbitrary dependences of $A(a_i,\lambda)$ and $B(b_j,\lambda)$ on their arguments it is enough to multiply it by $\rho(\lambda)$ and integrate in order to obtain (10) as true for an arbitrary local realistic theory. The proof is thus completed.

Of course, the previous lemma does not specify the value of M. Obviously if it holds for a given M it holds as well for a different M'>M.

We are interested in the smallest possible value of M which is obviously given by

$$M_0 = \mathrm{Max}\{\sum_{ij} c_{ij}\xi_i\eta_j\} \tag{12}$$

where $\xi_i = \pm 1$ and $\eta_j = \pm 1$ are sign factors which must be chosen in such a way to maximize the r.h.s. of (12). Of course there cannot exist an $M_0'<M_0$ for which (10) and (11) are true since the value (12) is actually obtained in a particular local realistic theory for the linear combination of correlations appearing in the left-hand side of (10). The latter theory would violate the inequality of the type (10)+(11) with M_0' in place of M_0, which is absurd since only inequalities valid for <u>all</u> local realistic theories deduced from Einstein locality are given by

$$\sum_{ij} c_{ij}P(a_i,b_j) \leq M_0$$

if M_0 is given by (12). Actually, one can easily show that the stronger inequality

$$\left|\sum_{ij} c_{ij}P(a_i,b_j)\right| \leq M_0 \tag{13}$$

is also a consequence of Einstein locality.

It can be shown that the inequalities deduced from (12)-(13) are trivial if the coefficients c_{ij} have factorable signs, because in such cases one has

$$M_0 = \sum_{ij} |c_{ij}|$$

and the resulting inequalities are always satisfied by all conceivable values of $P(a_i,b_j)$, since these correlation functions are by definition not larger than 1, in modulus.

If the signs of c_{ij} are not factorable, the inequalities (12)-(13) are in general violated by quantum mechanical predictions. If, for instance, the indexes i, j run over the values 1, 2, one gets from (13)

$$\left|\sum_{i=1}^{2} \sum_{j=1}^{2} c_{ij} P(a_i, b_j)\right| \leq M_0 \tag{14}$$

where

$$M_0 = \underset{\xi,\eta}{\mathrm{Max}} \left\{ \sum_{i=j}^{2} \sum_{j=1}^{2} c_{ij} \xi_i \eta_j \right\} =$$

$$= \underset{\eta}{\mathrm{Max}} \left\{ \sum_{i=1}^{2} \left| \sum_{j=1}^{2} c_{ij} \eta_i \right| \right\}$$

$$= \sum_{ij} |c_{ij}| - 2 \underset{\ell m}{\mathrm{Min}} |c_{\ell m}| \tag{15}$$

In the previous particular case in which

$$c_{11} = -c_{12} = c_{21} = c_{22} = 1$$

(notice that signs are not factorable) one obtains Bell's inequality:

$$|P(a_1 b_1) - P(a_1 b_2) + P(a_2 b_1) + P(a_2 b_2)| \leq 2 \tag{16}$$

which is well known to be violated by the quantum mechanical predictions, for instance in the case of the singlet state of two spin-½ particles.

EXPERIMENTS ON SEPARABILITY

Actual experiments on the validity of Bell's inequality have always been carried out with photons. The quantum mechanical treatment of photon polarization is similar to that of spin-½ in one important respect: both observables can assume only two values. This property of photon polarization is actually due to the absence of a photon mass, a fact which has the practical effect of eliminating from the theoretical scheme the photons with longitudinal polarization. Thus Bell's inequality can be formulated also for correlated photons.

Consider a spin - 0 system that decays into two photons: the quantum state of the two photons is in certain cases

$$\frac{1}{\sqrt{2}} (x_1 x_2 + y_1 y_2) \tag{17}$$

where x_1, y_1(x_2, y_2) are linear polarization states of the first (second) photon along the x and y axis, respectively. The previous state predicts that the probabilities of finding the photons with parallel and perpendicular polarizations are one and zero respectively.

Experimentally feasable tests of Bell's inequality were proposed in 1969 by Clauser, Horne, Shimony and Holt[11] who considered a J=0 → J=1 → J=0 electric dipole cascade in calcium and showed that

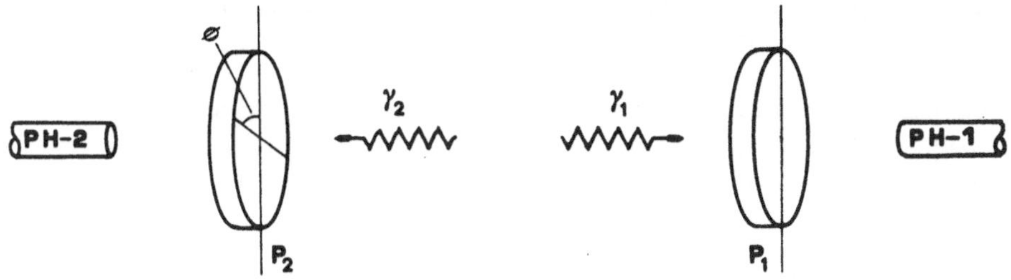

Fig. 2. Set-up of an EPR-type experiment for photon coincidences. PH-1, PH-2 are photomultipliers and P_1, P_2 are polarizers.

for photons emitted near opposite directions the predicted quantum-mechanical polarization correlation was given by

$$\frac{R(\phi)}{R_0} = 0.25 + 0.22 \cos 2\phi \ . \tag{18}$$

The photons are first transmitted (or absorbed) through two linear polarizers whose axes form the relative angle ϕ (see Figure 2.). Beyond these polarizers two photomultipliers are placed which reveal the impinging photons with certain efficiencies, typically of the order of 20%. $R(\phi)$ is the coincidence rate at the photomultiplier outputs when the relative angle of the polarizer axes is ϕ; R_0 is the coincidence rate with both polarizers removed; the numerical factors in (18) are determined mainly by the efficiencies of the polarizers.

Experiments of this type have been performed by Freedman and Clauser,[12] Holt and Pipkin,[13] Clauser,[14] Fry and Thomson[15] and Aspect, Grangier and Roger.[16] With the exception of the experiment by Holt and Pipkin good agreement with quantum mechanics has always been found; Bell's inequality in the case of the function (18) can be shown to lead to the prediction

$$\delta \equiv \left|R(22°.5) - R(67°.5)\right|/R_0 - \frac{1}{4} \leq 0 \ . \tag{19}$$

The best experimental result obtained before 1982 was

$$\delta_{exp} = (5.72 \pm 0.43) \times 10^{-2}$$

violating equality (19) by more than 13 standard deviations and in perfect agreement with quantum-mechanical predictions.

Two comments should be added to the published experiments. Firstly the majority of them agrees with quantum mechanics, but no satisfactory explanation has been found of the reported disagreement of the experiment by Holt and Pipkin.

Secondly, all the performed experiments have not been direct experimental checks of the conflict between Einstein locality and quantum mechanics, but have always needed additional assumptions in order to compare theory and experiment.[17]

For instance the CHSH assumption is, given that a pair of photons emerges from two polarizers, that the probability of their joint detection from two photomultipliers is independent of the polarizer orientations.

As Clauser and Shimony noted,[18] it is noteworthy that there exists an important hidden-variable theory - the semiclassical radiation theory - which correctly predicts a large body of atomic physics data, but which violates the CHSH assumption. We can add that such an assumption is contrary to the spirit of all hidden-variables theories: emergence from two polarizers whould in all cases imply a selection of the two-photon hidden variables, but these variables could well be those that determine the photomultiplier discharge.

An alternative assumption has been formulated by Clauser and Horne[19] and consists of the idea that for every pair of particles, the probability of a count with the polarizer in place is less than or equal to the corresponding probability with the polarizer removed.

Again it can be objected that the polarizer in place implies a selection of the hidden variables and that the probability of a count may be larger with the selected rather than with the "normally distributed" hidden variables.

Besides atomic cascades two further sources of separability - violating quantum mechanical correlations have been used experimentally:

A) Positronium annihilation experiments

A test of Bell's inequality using the high-energy photons produced by positronium annihilation ($e^+e^-\to\gamma\gamma$) is possible if one studies correlated Compton scatterings of the two γ-rays. The first experiments of this type were performed by Kasday, Ullman and Wu[20] followed by Faraci, Gutkowski, Notarrigo and Pennisi,[21] by Wilson, Lowe and Butt,[22] by Bruno, d'Agostino and Maroni[23] and by Bertolini.[24]

With the exception of the Catania experiment good agreement has been found with the prediction of quantum mechanics.

B) Proton - proton scattering experiments

Following the suggestion originally made by Fox[25] an experiment designed to test the validity of Bell's inequality for spin correlations of two protons has been carried out by Lamehi-Rachti and Mittig.[26] Once more, agreement with quantum mechanics was obtained within the limited statistics of this experiment.

The evidence obtainable from these experiments in favour or against separability is of limited interest because local hidden variable models are known to exist which are able to reproduce the quantum mechanical predictions in these cases. In fact nothing equivalent to a polarizer exists for 0.5 Mev γ-rays or for protons. The experiments of the type A) and B) above are therefore carried out by letting the final particles propagating in opposite directions scatter on two targets. Therefore what is directly measured is the direction of propagation of the scattered particle, that is essentially a momentum. But momentum components commute and a theorem exists [27] saying that it is in principle impossible with these experiments to prove or disprove Bell's inequality. The same point is shown by the existence of local hidden variable models which can reproduce the observed distributions of correlated scatterings. One such model was due to Kasday.[28]

The situation is somewhat better with the atomic cascade experiments discussed in the beginning, even though also there conceptual difficulties arise from the low efficiency of the photomultipliers.

There are three experiments which can eliminate at least part of these difficulties. The first one is Aspect's experiment which puts on the path of each photon an optical commutator.[29] This device consists of a solid in which an acoustic standing wave is generated: the interaction of the photon beam with the standing wave gives rise to a transmitted and to a diffracted beam which are made to impinge on polarizers with suitable orientation. In this way for every pair of photons the choice of the polarizers with which to interact is made during a very short time interval, so short that there is practically no possibility for the information to be transmitted from the first to the second act of measurement. Very good agreements with quantum predictions have been found.

The second experiment is being made by Rapisarda's group[30] and is based on the idea that a calcite cube can be used as a polarizer. This cube acts as a beamsplitter and is built in such a way that pairs of right angle prisms are cemented together, hypotenuse-face-to-hypotenuse-face, with a special multilayer dielectric film in between. Monochromatic unpolarized light which is orthogonally incident upon the external faces of the resulting cube (internally) incident at 45° upon the multilayer film) is separated into two polarized beams which emerge from the cube through adjacent faces and in directions which are exactly 90° apart. The linear polarizations of the emerging beams are accurately known and mutually orthogonal. The great advantage of this set-up is that now four different coincidence rates can be measured simultaneously.

The third experiment is being performed by the Stirling group[31] and has been discussed at this conference by Duncan.

PHOTON RESCATTERING IN ATOMIC-CASCADE EXPERIMENTS

Resonant photon rescattering in the beam is important in the published atomic-cascade experiments of Bell-type inequalities.[32] The ratio γ between the geometrical half-width of the atomic beam, L/2, and the mean free path l of the "second" photon emitted in the two-photon cascade can easily be shown to have the value

$$\gamma = \frac{1}{2} L n_0 \sigma, \tag{20}$$

where σ is the photon-atom cross-section and n_0 the atomic density of the beam (referred to atoms in the lowest level of the cascade). The value of n_0 is high for calcium, since for this element the lowest level of the cascade coincides with the ground state, but negligibly small for mercury, where no such a coincidence exists. The value of γ gives the average number of rescatterings of the "second" photon in the beam. In the case of the Orsay experiment one has L = 1mm. Aspect's thesis[33] indicates that the most probable value of n_0 is $8 \cdot 10^{10}$ atoms/cm³.

If one takes for σ the resonant value σ_r ($\sigma_r \simeq \lambda^2$, where λ = 4277 Å is the photon wave-length), one obtains $\gamma \simeq 7$, a very large value. The conclusion seems, therefore, to be that photon rescattering in the beam is important and that the good agreement between the experimental results and quantum theory is somewhat fortuitous, since the theoretical predictions used hold for the no-rescattering case. One must, however, consider the Doppler effect which is not negligible in the atomic beam: its contribution reduces σ in (20), since only a fraction of the atoms can actually give rise to resonant scattering if the photon Doppler shift is taken into account. The velocity distribution of the calcium atoms has been measured[34] in conditions similar to those of the Orsay experiment. From the results of ref.[34] we deduce that a Maxwellian distribution

$$F(v) = \frac{2v^3}{\alpha^4} \exp[-v^2/\alpha^2] \tag{21}$$

is able to provide a reasonable fit to the velocity distribution of the beam of the Orsay experiment if α is chosen to satisfy the relation

$$\langle v \rangle = \int_0^\infty dv\, v\, F(v) = \frac{3\sqrt{\pi}}{4} \alpha \tag{22}$$

For $n_0 = 8 \cdot 10^{10}$ atoms/cm³ Aspect shows that one should take $\langle v \rangle = 0.70\ 10^5$cm/s.
The value of α can then be deduced from (22).

The average relative velocity η of two atoms in the beam is

$$\eta = \int_0^\infty dv \int_0^\infty dv' \, |v'-v| F(v) F(v') \ . \tag{23}$$

A direct calculation using (21) gives $\eta = 0.54\alpha$, whence using (22) and (23)

$$\eta = 2.84 \cdot 10^4 \text{cm/s} \ . \tag{24}$$

A photon emitted by atom A is, therefore, expected to be seen Doppler-shifted according to the formula

$$\nu' = \nu(1 - \frac{\eta}{c}\cos\theta) \tag{25}$$

by an atom B moving in the beam with velocity η relative to atom A, if ν is the unshifted frequency and θ is the relative angle between the photon momentum and the B-atom momentum.

The cross-section for photon-atom scattering can be represented, near resonance, by the Breit-Wigner formula

$$\sigma(\nu) \simeq \frac{(c/\nu_0)^2(\frac{1}{2}\Delta\nu)^2}{(\nu-\nu_0)^2+(\frac{1}{2}\Delta\nu)^2} \ ,$$

whence one deduces the $\sigma(\nu)$ values of the following table, where also the corresponding $\gamma(\nu)$'s, calculated from (20), and $l(\nu)$'s, calculated from $[n_0\sigma(\nu)]^{-1}$ are shown.

Table 1.

	$\sigma(\nu)$	$\gamma(\nu)$	$l(\nu)$ (mm)
$\nu = \nu_0$	σ_r	7.15	0.07
$\nu = \nu_0 \pm \Delta\nu$	$\sigma_r/5$	1.43	0.35
$\nu = \nu_0 \pm 2\Delta\nu$	$\sigma_r/15$	0.42	1.19

Remembering that γ gives the average number of rescatterings in the beam, we see that rescattering is very important in the $\nu_0-\Delta\nu \leq \nu \leq \nu_0+\Delta\nu$ range (where $\gamma \cong 4.3$, on the average) and important in the ranges $\nu_0-2\Delta\nu \leq \nu \leq \nu_0-\Delta\nu$ and $\nu_0+\Delta\nu \leq \nu \leq \nu_0+2\Delta\nu$ (where $\gamma \simeq 0.9$, on the average). Limiting the considered frequency interval to $\nu_0 \pm 2\Delta\nu$ one has

$$|\cos\theta| \leq 2\frac{c}{\eta} \simeq 0.095 \ . \tag{26}$$

Therefore, only photons propagating within the angles

$$84°.5 \leq \theta \leq 95°.5 \tag{27}$$

can undergo resonant rescattering, in this approximation.

The third column in Table 1 gives the mean free path in a medium with density $n_0 = 8 \cdot 10^{10}$ atoms/cm^3 for photons of different frequencies near resonance. The values of $l(\nu)$ should be compared with the size of the atomic beam which is 1 mm thick in the Orsay experiment.

Next we estimate the fraction of the 4227 Å photons with emerge outside the angular interval (27) and are observed by the counter ("clean" photons).

The source of photon pairs is a cylinder of length 1 mm and thickness 60 μm just in the middle of the beam, where the laser light generates atomic excitation followed by practically immediate de-excitation. The photons emitted by the source hitting the focusing lens move practically within a cone of half-opening angle 30°. From the full sample of such photons, we must subtract those which satisfy (27), that is photons emitted in the region included between the planes passing through the source's symmetry axis (we neglect the small thickness of the source) and forming angles of 84°.5 and 95°.5 with the beam's direction. The fraction Γ of "clean" photons can be shown to be given by:

$$\Gamma = \frac{\pi tg30° - 4tg5°.5}{\pi tg30°} \simeq 0.79 \ . \tag{28}$$

We conclude that in the case of Orsay's experiment about 21% of the photons with wave-length 4227 Å emitted by the source and travelling towards the lens undergo rescattering in the beam.

This conclusion has been strengthened from the rigorous calculation performed by S Pascazio[35] who found that 33% of these photons undergo rescattering. It is of course very important to stress that rescattering is for various reasons totally negligible in the case of Stirling's experiment.

PHOTONS WITH VARIABLE DETECTION PROBABILITY

As we saw in a previous section all the experiments performed with atomic cascades in order to test Bell's inequality have been analyzed with the help of some additional hypothesis.

The first assumption of this type was proposed by Clauser, Horne, Shimony and Holt (CHSH) in 1969.[11] It consisted of the following:

If a pair of photons emerges from two polarizers, the probability of their joint detection is independent of the orientation of the polarizers' axes.

A slightly different assumption was proposed by Clauser and Horne (CH) in 1974[19] and was formulated in the following way:

For every atomic emission, the probability of a count with a polarizer in place is not larger than the probability with the polarizer removed.

There is no logical reason for believing that these hypotheses are true in nature. Once one accepts the idea that the present-day quantum theory predicts only averages and that a more detailed theoretical description should be possible by taking into account individual physical properties, one can describe the *detection* process as being different for different objects.[36]

This is enough to violate in an natural way both the CHSH and the CH assumption.

In order to give a more detailed physical description of photons we start by assuming that a linear polarization perpendicular to the direction of motion can always be attributed to any given photon. This assumption brings us beyond quantum theory in two important respects:

i) Elliptically polarized photons in quantum theory have, strictly speaking, no linear polarization at all. Rather their state-vector results from a linear combination of the two possible linear polarization states. This is not a real difficulty if quantum theory provides only a description valid for statistical ensembles since models describing individual photons with (eventually rotating) linear polarization vectors do not need to contradict the quantum theoretical *empirical predictions* in any way.

ii) The two photons composing an EPR-pair do not possess an *individual* polarization state of any type. Again, *a priori* this is not necessarily a real difficulty. In the case of EPR-pairs, we will however see that some residual descrepancy at the empirical level is necessarily left, but for experiments which have not yet been performed.

Let $D(\ell)$ be the photon detection probability by some given detector (e.g. a photomultiplier) if the photon has polarization ℓ at the time in which it enters in the detector. [We denote polarizations with scalar symbols which are intended to represent the angle that the polarization *vector* forms with some fixed direction.] If the photon had no further physical attribute besides ℓ, we should assume that $D(\ell)$ is not dependent on ℓ, since the detection process of polarized photon beams is experimentally known to take place with a polarization-independent detection probability. We should so lose all possibilities of violating the CHSH and CH assumptions.

The only way out is therefore to assume that photons have at least two physical properties, the linear polarization ℓ and a new variable λ and that the photon detection probability can be written

$$D(\ell,\lambda).$$ [37]

Consistency with the available data on the detection of polarized photon beams requires that the λ-average of D be ℓ-independent. We write

$$\langle D(\ell,\lambda)\rangle_\lambda = \eta \tag{29}$$

where η is the well known ℓ-independent quantum efficiency of the counter. Obviously we are also assuming, as it is natural to do, that beams of photons present a random distribution of λ-values.

A second physical restriction can be put on λ. When a polarized photon beam crosses a polarizer before entering the detector the detection probability is observed to be the product of the quantum efficiency η with the transmission probability

$$T(\ell,a) = \cos^2(\ell-a) \tag{30}$$

where a is the polarizer's axis direction. The latter equation expresses Malus' law. No correlation between transmission and detection, of the type one would expect to exist if T depended on λ, has ever been seen. Therefore we assume that T does not depend on the new variable λ, as it happens in (30). In this way the whole single photon physics can be reproduced.[37] The "additional assumptions" with which the present section started can now be expressed by means of the probabilities T and D: if we refer to an incoming photon with linear polarization ℓ we should write the CH condition as

$$T(\ell,a)D(a\lambda) \leq D(\ell\lambda) \tag{31}$$

since the polarizer changes the photon polarization from ℓ to a. This is a non-obvious condition and it is easy to find examples of D-functions not satisfying it!

Similarly, the CHSH condition can be written

$$D_1(a\lambda) \cdot D_2(b\lambda') \quad : \text{independent of } a,b, \tag{32}$$

if $D_1(a\lambda)$ $[D_2(b\lambda')]$ is the detection probability of the first (second) photon having detection parameter $\lambda(\lambda')$ and having crossed a polarizer with axis a (a polarizer with axis b). But (32) can be true if D_1 does not depend on a and D_2 on b, which is again a far from obvious condition!

We will next show that our general assumptions give rise to theoretical models which are necessarily divergent from quantum theory at the empirical level even though the differences need not to be in the set of the experiments performed up to the present time.

The starting point is the following inequality

$$-1 \leq xy - xy' + x'y + x'y' - x' - y \leq 0 \qquad (33)$$

which holds without exception if the four real numbers x, x', y, y' satisfy

$$0 \leq x, x', y, y' \leq 1 \qquad (34)$$

The proof is very simple, since the quantity

$$F \equiv xy - xy' + x'y + x'y' - x' - y$$

is linear in the four variables x,x',y,y': Its extreme values can therefore be found in the vertices of the 4-dimensional cube defined by (34). A direct substitution of the values assumed by x, x', y, y' in these 16 vertices shows that (33) holds.

We apply the inequality (33) to the case in which

$$\begin{cases} x = H_1(aa_1)\ T_1(a_1\ell) \\ x' = H_1(aa_2)\ T_1(a_2\ell) \\ y = T_2(\ell' b_1)\ H_2(b_1 b) \\ y' = T_2(\ell' b_2)\ H_2(b_2 b) \end{cases} \qquad (35)$$

where the index 1 (2) attached to T and H refers to the first (second) photon of an EPR pair arriving with linear polarization $\ell(\ell')$ on a polarizer with axis orientation a_1 or a_2 (b_1 or b_2) and being transmitted with probability $T_1(T_2)$.

After this, the first (second) photon is assumed to impinge on a half-wave plate oriented in such a way that the photon linear polarization is in all cases rotated from a_1 or a_2 (b_1 or b_2) to a (to b). The probability that the first (second) photon crosses the half wave plate has been denoted by $H_1(H_2)$. The variables (35) are therefore probabilities for a double crossing (polarizer + half-wave plate): They must satisfy (34) and (33) becomes

$$\begin{aligned}
-1 \leq\; & H_1(aa_1)\; T_1(a_1\ell)\; T_2(\ell' b_1)\; H_2(b_1 b) \\
& - H_1(aa_1)\; T_1(a_1\ell)\; T_2(\ell' b_2)\; H_2(b_2 b) \\
& + H_1(aa_2)\; T_1(a_2\ell)\; T_2(\ell' b_1)\; H_2(b_1 b) \\
& + H_1(aa_2)\; T_1(a_2\ell)\; T_2(\ell' b_2)\; H_2(b_2 b) \\
& - H_1(aa_2)\; T_1(a_2\ell)\; - T_2(\ell' b_1)\; H_2(b_1 b) \leq 0 \qquad (36)
\end{aligned}$$

Notice that all symbols in (36) appear exactly in order which is relevant to the experimental set-up shown in Figure 3.

The half-wave plates have been inserted in order to eliminate all variations in the double-photon detection probabilities. Four experiments are performed by combining the two orientations (a_1 and a_2) of the first polarizer with the two orientations (b_1 and b_2) of the second one: in all cases, however, the action of the $\lambda/2$-plates is such that the photons enter in the detectors with a fixed polarization (a for the first photon, b for the second one).

We perform now on (36) the following operations.

1. Multiply all terms by $D_1(a\lambda) \cdot D_2(b\lambda')$.

2. Multiply all terms by $\rho(\ell\ell'\lambda\lambda')$, the density function of the four parameters $\ell\ell'\lambda\lambda'$ which describes the singlet state.

3. Integrate over $\ell, \ell', \lambda, \lambda'$.

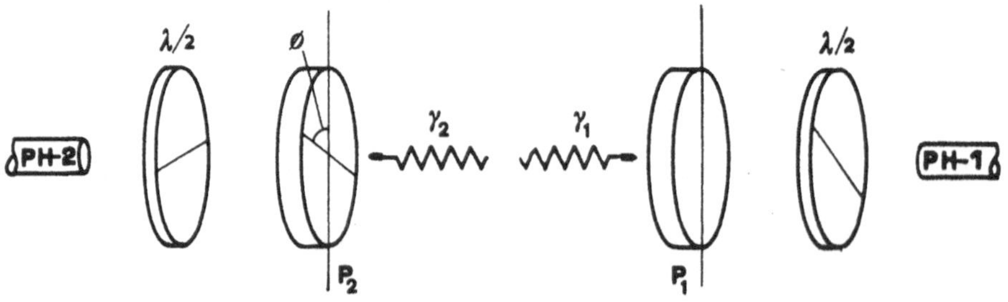

Fig. 3. Set-up of the proposed experiment on variable detection probabilities with polarizers, half-wave plates and detectors.

We obtain

$$\omega(aa_1|b_1b) - \omega(aa_1|b_2b) + \omega(aa_2|b_1b)$$
$$+\omega(aa_2|b_2b) \leq \omega(aa_2|\infty b) + \omega(a\infty|b_1b) \quad (37)$$

where

$$\omega(aa_1|b_1b) = \int d\rho \; D_1(a\lambda)\cdot H_1(aa_1)\cdot T_1(a_1\lambda)\cdot T_2(\ell' b_1)\cdot H_2(b_1b)\cdot D_2(b\lambda) \quad (38)$$

is the coincident photon detection probability when the polarizers are set at a_1 and b_1 and the $\lambda/2$-plates rotate a_1 to a and b_1 to b. We have adopted the simplified notation

$$d\rho = \rho(\ell\ell'\lambda\lambda')\, d\ell\, d\ell'\, d\lambda\, d\lambda'.$$

Definitions strictly similar to (38) hold for the second, third and fourth term on the l.h.s. of (37).
Similarly we have

$$\omega(aa_2|\infty b) = \int d\rho \; D_1(a\lambda)\cdot H_1(aa_2)\cdot T_1(a_2\ell)\cdot D_2(b\lambda), \quad (39)$$

$$\omega(a\infty|b_1b) = \int d\rho \; D_1(a\lambda)\cdot T_2(b_1\ell')\cdot H_2(b_1b)\cdot D_2(b\lambda). \quad (40)$$

The physical meaning of (39) is the following: a photon pair is emitted in the singlet state, as it is shown by $d\rho$. The first photon crosses a polarizer with probability $T_1(a_2\ell)$ and a $\lambda/2$-plate with probability $H_1(aa_2)$ and is detected from a photomultiplier with probability $D_1(a\lambda)$. The second photon has its polarization vector changed, in flight and with efficiency 1, from the (variable) value ℓ' to b and is then detected from a photomultiplier with probability $D_2(b\lambda)$. The meaning of (40) is similar. The discussion of real experiments in which the previous situations can be closely approximated is discussed elsewhere.[38]

Quantum mechanics gives unambiguous predictions for all well definded physical experiments. As far as the six coincidence probabilities appearing in (37) are concerned their quantum theoretical expressions are

$$\omega(aa_i|b_jb) = \frac{\eta_1\eta_2}{4}\,[\alpha + \beta\cos 2(a_i - b_j)]H^2 \quad (i,j = 1,2) \quad (41)$$

$$\omega(aa_2|\infty b) = (a\infty|b_1b) = \frac{\eta_1\eta_2}{2}\cdot H \quad (42)$$

where η_1 and η_2 are the quantum efficiencies of the two detectors, H is the flux reduction factor due to the $\lambda/2$-plate (assumed the same for the two plates for simplicity) and α and β are well known products of parameters describing the polarizers and geometrical factors. Typical values are

$$\alpha \simeq 1; \quad \beta \simeq .9; \quad H \simeq .99 \tag{43}$$

The factor $\eta_1\eta_2$ can be eliminated when (41) and (42) are inserted in (37). For a suitable choice of the four angles (a_i-b_j) the inequality (37) would become

$$\frac{\alpha+\beta\sqrt{2}}{2} \cdot H \leq 1$$

which is violated, since the values (43) give to the l.h.s. of the previous inequality the value 1.125.

We see thus that the incompatibility between quantum theory and Einstein locality exists at the empirical level also if the usual additional assumptions are not made.

An experiment designed to test this incompatibility should measure the six coincidence rates appearing in (37) and should be carried out by inserting $\lambda/2$-plates between polarizers and detectors on the trajectories of the two photons.

REFERENCES

1. M Born "Physics in my Generation", Springer, New York (1969), p. 173.
2. J von Neumann, "Die Matematische Grundlagen der Quantenmechanik", Springer, Berlin (1932).
3. N Bohr in "New Theories in Physics, Warsaw 1938", Int. Inst. of Intellectual Cooperation, Paris (1939).
4. M Born, "Natural Philosophy of Cause and Chance", Dover, New York (1964).
5. D Bohm, Phys Rev 85:166 (1952) and 85:180 (1952).
6. L de Broglie, Journ Phys Rad 20:963 (1959).
7. F Selleri, "Die Debatte um die Quantentheorie", Vieweg, Braunschweig (1983), pages 50-52.
   J S Bell, Rev Mod Phys 38:447 (1966).
8. A Einstein, B Podolsky, N Rosen, Phys Rev 47:777 (1935).
9. N Bohr, Phys Rev 48:696 (1935).
10. A Garuccio and F Selleri, Found Phys 10:209 (1980).
11. J F Clauser, M A Horne, A Shimony and R A Holt, Phys Rev Lett 23:880 (1969).
12. S J Freedman and J F Clauser, Phys Rev Lett 28:938 (1972).
13. R A Holt and F M Pipkin, Harvard University preprint (1973).
14. J F Clauser, Phys Rev Lett 36:1223 (1976).
15. E S Fry and R C Thomson, Phys Rev Lett 37:465 (1976).
16. A Aspect, P Grangier and F Roger, Phys Rev Lett 47:460 (1981).
17. These assumptions will be discussed in detail in the last section of the present paper.
18. J F Clauser and A Shimony, Rep Progr Phys 41:1881 (1978).

19. J F Clauser and M A Horne, Phys Rev D10:526 (1974).
20. L R Kasday, J D Ullman and C S Wu, Nuovo Cim 25B:633 (1975).
21. G Faraci, D Gutkowski, S Notarrigo and A R Pennisi, Nuovo Cim Lett 9:607 (1974).
22. A R Wilson, J Lowe and D K Butt, Jour Phys G2:613 (1976).
23. M Bruno, M D'Agostino and C Maroni, Nuovo Cim 40B:142 (1977).
24. G Bertolini, E Diana and A Scotti, Nuovo Cim 63B:651 (1981).
25. R Fox, Nuovo Cim Lett 2:565 (1971).
26. M Lamehi-Rachti and W Mittig, Phys Rev 14:2543 (1976).
27. V Capasso, D Fortunato and F Selleri, Int Jour Theor Phys 7:319 (1973).
28. L R Kasday, in Rendiconti della scuola internazionale di fisica "Enrico Fermi", IL Corso, "Fondamenti di Meccanica Quantistica", Academic Press, New York (1971), p195.
29. A Aspect, P Grangier and G Roger, Phys Rev Lett 49:91 (1982).
A Aspect, J Dalibard and G Roger, Phys Rev Lett 49:1804 (1982).
30. F Falciglia, A Garuccio and L Pappalardo, Lett Nuovo Cim 34:1 (1982).
31. A J Duncan, Polarization Correlation in Simultaneous Two-Photon Decay Process and Tests of Bell's Inequality, paper presented at Symposium on New Trends in Atomic Collision Physics (20-21 September 1984, Santa Flavia, Sicily).
32. T Marshall, E Santos and F Selleri, Lett Nuovo Cim 35:417 (1983).
F Selleri, Lett Nuovo Cim 39:252 (1984).
33. A Aspect, Thése, Université de Paris-Sud, Centre d'Orsay (1983).
34. G Giusfredi, P Minguzzi, F Strumia and M Toselli, Z Phys A274:279 (1975).
35. S Pascazio, A careful estimate of photon rescattering in atomic cascade experimental tests of Bell's inequality, Nuovo Cim, in print.
36. T W Marshall, E Santos and F Selleri, Phys Lett 98A5 (1983).
T W Marshall, Phys Lett 99A:163 (1983).
37. A Garuccio and F Selleri, Phys Lett 103A:99 (1984).
38. F Selleri, Photons with Variable Detection Probability, Phys. Lett., in print (1985).

NEW APPLICATIONS OF ELECTRON-ELECTRON COINCIDENCES

Wolfgang Sandner

Fakultät für Physik
Universität Freiburg
D-7800 Freiburg West Germany

I INTRODUCTION

Although the coincidence technique is known as a well established tool in the field of electron atom collisions, it is still true that its application has mostly been limited to impact processes in the outermost shells of atoms. As a consequence, two distinctly different types of coincidence experiments have emerged in the past: the (e,2e) experiment in the case of valence shell *ionization*, and electron-photon coincidences in the case of *excitation*, respectively. In the (e,2e) experiment both the scattered and the ejected electron are detected in coincidence, leading to the determination of triply differential ionization cross sections. These, in turn, yield information on the scattering process or on the initial valence shell wave function, depending on the kinematics chosen in the experiment.[1-3] The residual ion, being typically in its ground state, remains unobserved. Electron-photon coincidences, on the other hand, focus on the measurement of scattering amplitudes by determining the complete quantum mechanical states of both the scattered electron and the excited atom, the latter being deferred from the subsequent deexcitation process.[4] In this way quantum mechanically "complete" information on the scattering process can be obtained, if one accepts the restriction to those excited states whose geometrical properties can be completely deduced from their single photon decay (more sophisticated probing methods usually lead to unacceptably low coincidence rates). Speaking in terms of statistical tensors or state multipoles[5] the restriction is, in general, equivalent to a maximum rank $k \leq 2$ of the statistical tensor describing the anisotropy of the excited atom.

The situation changes in several ways if we consider coincidence experiments involving inner atomic shells. First we note that now both excitation *and* ionization lead to excited states of the target system, which may be analyzed by their subsequent decay products. Furthermore it is known from the fluorescence yields[6] that by far the preferred decay mode is the nonradiative autoionization or Auger decay. The only exceptions of limited practical importance are the innermost shells of the heaviest atoms and few excited states which are stable against Coulomb decay. In practice, this means for the case of impact *excitation* that electron-photon coincidences will frequently be replaced by electron-electron coincidences involving autoionization electrons. Electrons, in contrast to photons, can carry away more than one unit of angular momentum, thus making collisionally excited systems with state multipoles of rank $k>2$ accessible for investigation in coincidence experiments. This important advantage is, however, balanced by the fact that the polarization measurement of electrons, necessary for the determination of any odd k state multipole component, is still so inefficient that there is little hope for a "complete" inner shell electron-electron coincidence experiment in the near future.

Electron impact *ionization* experiments, on the other hand, may substantially increase their information content by moving from outer to inner shells. While the common (e,2e) technique remains applicable,[7-9] the additional Auger emission now bears information on the anisotropy of the initially created ion, as well as on continuum correlation effects involving Auger electrons. Apart from straightforward alignment calculations performed within the framework of 1. Born approximation,[10,11] coincident angular correlations with Auger electrons have not been considered theoretically in the past. Some recent experiments will be presented in chapter III.

Up to date the total number of reported inner shell coincidence experiments following electron impact remains rather limited. Apart from experiments which did not focus on the angular dependence in both channels[12-17] it was only in the last years that several coincident angular correlation experiments involving inner shells have been reported.[18-24] One of the reasons for this late development may lie in the low coincidence rates, which are typically orders of magnitudes smaller than in outer shells. Under these conditions the efficiency of the experimental setup plays a particularly important role, and it is not surprising that some recent advances in the technique of electron-electron coincidence experiments have been reported in connection with inner shell experiments, as will be outlined in the following chapter.

II RECENT ADVANCES IN THE ELECTRON-ELECTRON COINCIDENCE TECHNIQUE

Let us consider a "classical" coincidence experiment designed to measure correlations between pairs of electrons, which are emitted from some physical process not specified here. Both electrons are state selected with respect to their energies E_i and emission directions $\underline{\hat{k}}_i$ (i=1,2), and with respect to their emission times t_i from the target region. The time difference $t=t_1-t_2$ is evaluated to identify correlated pairs of particles, characterized for instance by equal emission times, t=0. Experimentally, for reasons discussed below, they will be spread over a finite time window Δt of typically few nsec around t=0. As a consequence, a certain (and sometimes substantial) background of uncorrelated electron pairs will add to the signal, with a rate proportional to Δt. The quantity of interest is the rate of correlated pairs or "true coincidences", which must be extracted from the total coincidence rate by subtraction of the background. Its dependence on the state parameters is usually explored in a series of experiments covering some portion of the multi dimensional parameter space $(\underline{\hat{k}}_1,\underline{\hat{k}}_2,E_1,E_2)$.

Starting from this concept several successful attempts have been reported in recent times to increase the overall efficiency by substantial amounts, usually one or more orders of magnitude. The key ideas are background reduction by improving the time resolution Δt, and/or the use of multi parameter coincidence techniques.

Völkel and Sandner[25] have investigated the contributions leading to a finite time resolution Δt in an assumed coincidence experiment with electrostatic electron energy analyzers. They found that, out of a total time uncertainty Δt = 6.8 nsec, as much as 74% (5 nsec) were caused by different trajectories inside the electrostatic analyzers. While the absolute numbers may vary substantially in different experiments it appears to be true that the trajectory geometry is generally the major source for the finite time resolution Δt. We note that this geometrical time spread Δt_{geom} is proportional to both the acceptance solid angle $\Delta\Omega$ and the overall size of an analyzer, and inversely proportional to the square root of its transmission energy E. Therefore, any straightforward approach of reducing Δt_{geom} conflicts with the requirements for a high coincidence rate, or for a good energy resolution, respectively. This dilemma has been solved by the introduction of time spread compensation methods which do not affect other properties of the analyzer.

One of these compensation methods is sketched on fig. 1, using the 30° parallel plate electrostatic analyzer as an example. It is easily verified that electrons entering the entrance slit S1 under the extreme angles $\theta_0 + 1/2\Delta\theta$ and $\theta_0 - 1/2\Delta\theta$, respectively, will have experienced a relative time delay Δt_{geom} at the exit slit S2 which is given by

$$|\Delta t_{geom}| = (W_0/v_z)\ \tan\theta_0\Delta\theta. \qquad (1)$$

Here $v_z = v\cos\theta_0$ is the mean electron velocity component parallel to the plates, and W_0 is the lateral distance between the slits S1 and S2. It has been shown that this time delay can almost totally be eliminated by utilizing the existence of "isochrones" (surfaces of equal flight times) behind the exit slit: time spread compensation is simply achieved by tilting an extended detector (microchannel plate) such that its surface approximately conforms with one of the isochrones. A comprehensive tabulation of all necessary design parameters (inclination angle and curvature of the isochrones) has been reported for the most common types of electrostatic analyzers.[25]

For some rotationally symmetric analyzers with large acceptance solid angles the isochrones are so strongly curved in space that they cannot be approximated by flat microchannel plate surfaces. For these cases an electronic time spread compensation has been developed. It utilizes separate detection of different trajectories by means of a channel plate multianode, together with suitably chosen electronic delay lines for pulses from the individual anodes. In a first application of both methods an improvement of more than a factor of 6 in the *overall* time resolution Δt of the experiment has been reported.[25] It can be shown that this is equivalent to an efficiency gain of up to a factor of 6, depending somewhat on the actual ratio of true to random coincidences. A typical resolution is now 800 psec without any loss in the acceptance solid angle.[22]

In the same experiment[26] the multi parameter coincidence technique has been employed, where, in particular, correlations between more than one pair of emission directions ($\hat{\underline{k}}_1, \hat{\underline{k}}_2$) are recorded simultaneously. For this purpose, one branch of the experiment, the Auger electron branch, has been equipped with 12 independent detectors, while one single detector detecting scattered electrons was kept in the other branch. In the actual setup the 12 Auger detectors are not physically independent but consist of one cylindrical mirror analyzer (CMA), whose azimuth has been divided into 12 sectors which are separately imaged onto corresponding sectors of a channel plate multianode (fig. 2). The idea has been

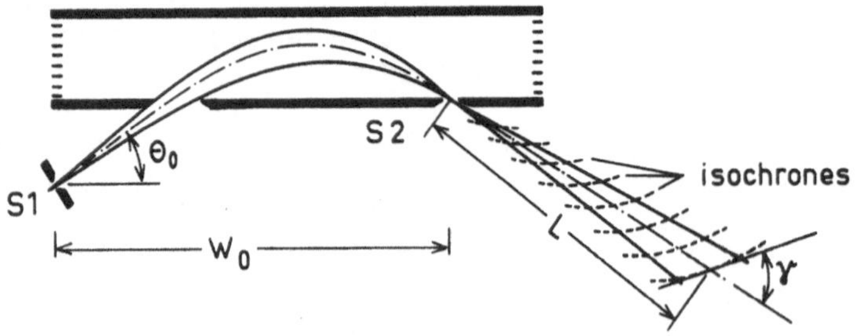

Fig. 1. Illustration of the time spread compensation using isochrones (surfaces of equal flight time). For optimum timing performance of the analyzer, a channel plate, mounted in a distance L behind the exit slit, has to be tilted by an angle γ to conform to the isochrone.

adopted from similar designs reported earlier.[27,28] This way the energy analysis is common to all 12 detectors, which differ only by their mean emission directions for the Auger electrons. In the coordinate frame of the CMA the directions $(\tilde{\theta}_i,\tilde{\phi}_i)$ are characterized by $\tilde{\theta}_i=\tilde{\theta}_0=39.2^0$ (with $\Delta\theta = \pm 2.5^0$), and $\tilde{\phi}_i = i\cdot 30^0$ (with $\Delta\phi_i = \pm 15^0$). In the coordinate frame of the incident beam, which is more convenient for data analysis, the following range of emission directions (θ_i,Φ_i) is covered

$$50.8^0 < \theta_i < 129.2^0 \tag{2a}$$

with

$$\cos \Phi_i = \cos(39.2^0)/\sin \theta_i \tag{2b}$$

where $\theta = 0$ is the incident beam direction and $\Phi = 0$ defines the scattering plane.

The use of the multi parameter coincidence technique does increase the efficiency of the apparatus by another factor of 12, allowing the simultaneous measurement of a complete set of 12 angular correlations. However, it is worth mentioning that, even though the 12 Auger detectors are not confined to the scattering plane, the *cylindrical symmetry* of the whole arrangement sets certain limits to the determination of unknown angular distributions. For instance, a superposition of low order spherical harmonics (as in experiments involving Auger or autoionization electrons) cannot be completely analyzed from one single measurement with the present setup. This limitation can be removed by incorporating multi parameter detection in the scattered electron branch as well, where the scattered electrons have the same scattering angle θ_s but different azimuths Φ_s, thus defining additional scattering planes and, hence, additional sets of 12 Auger emission directions (θ'_i,Φ'_i).

A second type of multi parameter coincidence has recently been reported, where, instead of a range of emission directions, a certain energy band has been recorded simultaneously in each branch of the experiment.[29] The key idea for maintaining a good energy resolution was the application of a position sensitive multichannel detection system, yielding a resolution of 2.9 eV FWHM over a 15.2 eV wide energy band (center energy was 600 eV). The efficiency gain of this technique is substantial and is maximum in those experiments where the signal in both detectors is a smooth function of the energy (as in (e,2e) experiments), whereas it may be somewhat reduced if one branch of the experiment is kept fixed on the peak of a narrow Auger or autoionization line. Unfortunately, this technique does not allow for a complete compensation of the geometrical time spread, even though the time resolution may eventually be improved over the value of 30 nsec reported in the first application.

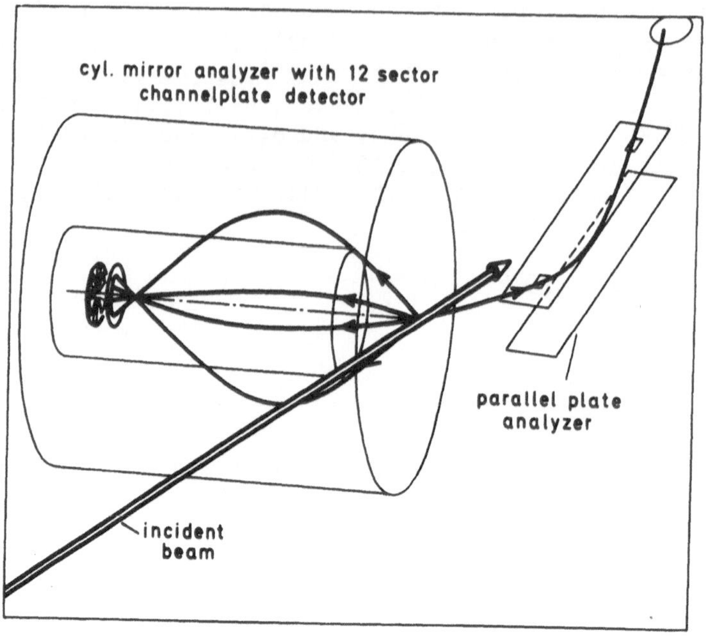

Fig. 2. Schematic drawing of a multi parameter coincidence apparatus using one parallel plate analyzer and one cylindrical mirror analyzer with 12 independent azimuthal sectors (from ref. 26).

Both multi parameter coincidence techniques described use only one channel plate detector and one dispersive energy analyzer in each branch of the experiment. Since there is no easy way of combining both techniques the next step towards even higher efficiencies will require either the use of expensive multi detector systems, or some new concept for state selection of the electrons. We note in passing that a different concept with comparable efficiency is already in use for inner shell (e,2e) experiments at relativistic electron energies.[9] There, the energy analysis is achieved by two surface barrier detectors whose acceptance solid angles have been maximized by the aid of triply focusing non dispersive magnets. This setup does allow for the simultaneous recording of a very wide energy band (37 to 430 keV), thereby maintaining an energy resolution of 7.5 keV, which is sufficient to separate (e,2e) contributions from different inner shells in high Z elements.

We may conclude that the latest generation of electron-electron coincidence experiments is characterized by an efficiency which is at least one, often more than one order of magnitude higher than that of the "conventional" machines used in the pioneering days of this technique. Technological advances of this kind are essential in a field which is frequently plagued by run times in the order of weeks or months for one single data point.

III COINCIDENT ANGULAR CORRELATIONS BETWEEN AUGER ELECTRONS AND SCATTERED ELECTRONS

As already mentioned above, coincident angular correlation experiments in inner shells following electron impact are rather scarce. Under these circumstances it must be considered fortuitous that three of these experiments have focused on the same system, namely angular correlations between scattered electrons and L_3-Auger electrons in argon. The kinematical conditions in all three cases were similar and chosen such as to allow a direct comparison between them, as well as comparison with existing theories.

Theory

The process under investigation is

$$\begin{aligned} e_o(T) + Ar \rightarrow Ar^+(2p_{3/2})^{-1} &+ e_o(T-E) + e_{ej}(E-E_B) \\ &\hookrightarrow Ar^{++}(^1S_o) + e_{Auger}(201\ \mathrm{eV}) \end{aligned} \qquad (3)$$

Coincidences are measured between scattered electrons of kinetic energy (T-E), and L_3-$M_{2,3}M_{2,3}(^1S_o)$ - Auger electrons of 201 eV kinetic energy, respectively. The problem has been treated theoretically already some time ago,[10,11] as a continuation of previous work on the noncoincident distribution of Ar Auger electrons. In view of this development it is quite natural that several basic assumptions,which have already proven to be successful in the non-coincident case, have been adopted in the theoretical treatment of the coincidence experiment. These basic assumptions can be identified as follows:

1. First, and most importantly, the ionization and subsequent Auger decay are treated as a two step process. This is generally considered to be true if a) the direct excitation to the final state is negligible, and b) the lifetime of the intermediate ionic state is long compared with any characteristic time of the collision.[32] We note, that the direct excitation to the final state corresponds to a *double ionization* of the atom, where the Ar^{++} ion must be left in a final state identical to that of the specific Auger transition under consideration. We further note that the requirement concerning the lifetime excludes any continuum correlation involving the Auger electron, in particular the post collision interaction (PCI).

The immediate consequence of this first assumption is that the Auger decay is governed by internal energy and angular momentum conservation rules only. The theoretical analysis of the *geometry* of such a decay process is most conveniently performed within the framework of the general angular correlation theory,[30,31] where both the initial and final state are expanded in terms of irreducible statistical tensor operators. The most transparent case is, of course, the Auger transition to an isotropic final state, in which case no interference between different Auger partial waves occurs and the statistical tensor of the emitted Auger electrons is an exact image of the statistical tensor of the emitting intermediate state. However, as already mentioned above, electron-electron experiments are far from being "complete", and it can be shown[10] that for the present experiment only three independent components $\mathcal{A}_{20}, \mathcal{A}_{21}$ and $\mathcal{A}_{22}$ of the alignment tensor of rank 2 can be measured with today's techniques.

2. Further assumptions have been made in the interpretation of the relevant tensor components. First, electron exchange effects have been neglected, in which case the scattered and the ejected electrons can be treated independently.[10,11] This approximation should hold better the higher one chooses the energy difference between these two electrons. Experimentally, this means that by detection of the fast scattered electron one obtains information on the energy E and the momentum $\underline{K}$ being transferred during the collision.

3. Finally, at this point the 1. Born approximation can be applied. This approximation is not only the traditional starting point for any numerical calculation, but it bears an inherent symmetry which may be used by itself for comparison with experimental results. It is the axial symmetry around the momentum transfer direction vector $\underline{K}$, which follows from the transition operator $\exp(i\underline{K}\cdot\underline{r})$ and, as a consequence, makes any statistical tensor component $\mathcal{A}^K_{k\kappa}$ with $\kappa \neq 0$ vanish in a system with $\underline{K}$ as quantization axis. In the present experiment, assuming that the above chain of approximations holds, this would leave only one independent tensor component $\mathcal{A}^K_{20}$ to be determined, together with one angle θ_K defining the quantization axis $\underline{K}$.

Even though theoretical predictions[10] were available for the alignment tensor component $\mathcal{A}^K_{20}$, it was certainly the possibility of testing the 1. Born approximation very directly by its symmetry requirements which attracted attention to the first angular correlation experiments between Auger electrons and scattered electrons.

Experiment

Angular correlation experiments for the system under consideration have been carried out by three different groups: The Belfast[19,20] and Freiburg[21,22] group both used 1 keV incident electron energy and scattering angles of 15° [20,22] and 21°,[20] whereas the Paris/Rome collaboration[23] used 8.255 keV incident energy and scattering angles of 1.5° and 5.5°, respectively. In all cases the energy loss was 253 eV, only 5 eV above the minimum energy transfer necessary for

L_3-shell ionization. Angular correlations have been measured both in in-plane[20,23] and out-of-scattering-plane geometry.[22]

The very first measurements of this kind[19,21] appeared to exhibit a rather spectacular failure of the 1. Born approximation: almost complete alignment was found where zero alignment was predicted, and the symmetry of the alignment tensor $A_{k\kappa}$ deviated from the momentum transfer direction by as much as 43°. Soon it was realized,[21,24] however, that these results were possibly influenced by an unexpected (e,2e) background, and measures were taken to account for this problem: the Belfast group measured two independent angular correlations for the Auger- and background intensities, respectively, which were then subtracted,[20,24] whereas the Freiburg group investigated details of the Auger line shape and background in a simultaneous five-energy-point coincident angular correlation experiment.[22] The Paris/Rome experiment, carried out at considerably higher impact energies, was reportedly free from (e,2e) background processes.

The current status of the experiments is summarized on figs. 3 through 5. In fig. 3 the angular correlations recorded *in the scattering plane* are shown. The four different experimental results are arranged such that the value of the momentum transfer increases from the bottom to the top. In general, one expects 1. Born approximation to work better the higher the incident energy and the smaller the value of K is. From these arguments, the Paris/Rome experiment at 8.2 keV incident energy and 1.5° scattering angle (K=0.5 a.u.[23]) should be in best agreement with theory (fig. 3d). In fact, the observed Auger angular distribution is symmetric with respect to the momentum transfer direction θ_K=64°. For a quantitative comparison with calculated alignment values we shall express the coincident Auger angular distribution $W(\psi)$ in 1. Born approximation as

$$W(\psi) = \frac{W_0}{4\pi} \left[1 + \alpha_2 A^K_{20} P_2(\cos\psi) \right] \qquad (4)$$

Here ψ is the angle between the Auger emission direction (θ,Φ) and the momentum transfer direction (in particular, $\psi = \theta-\theta_K$ in the scattering plane), $P_2(\cos\psi)$ is the second Legendre polynomial, and α_2 is a factor depending on the Auger transition under consideration. For the L_3-$M_{2,3}M_{2,3}(^1S_0)$ transition of Ar one obtains[33]

$$\alpha_2 = -1 \qquad (5)$$

Equation (4) is a special case of the general angular correlation formula (eq. 2 of ref. 10), and can easily be derived using the rotation transformation of the alignment tensor $A_{k\kappa}$, together with the addition theorem for spherical harmonics. We note that

$W(\psi)$ displays a minimum at $\psi=0$ (in direction of $\underline{K}$) only for positive values of A^K_{20}, whereas a maximum is found for negative A^K_{20}. Thus, the minimum around $\theta = \theta_K = 64°$, found by the Paris/Rome collaboration at K=0.5 a.u.[23], does not agree with the negative A^K_{20} calculated by Berezhko et al.[10] for K < 2 a.u. (a fact which has previously[23] been obscured by a double valued squareroot in eq. (7) of ref. 10). The experimental accuracy is high enough to make the difference between theory and experiment appear to be significant.

A rather interesting aspect of this discrepancy arises from the observation that $(Ka_o)^2$ is in the order of 0.1 for the process under consideration, and thus approaches the region of the dipole approximation,[34] characterized by $(Ka_o)^2 \ll 1$. Apart from the case of photon impact the dipole approximation also governs the *non-coincident* electron impact alignment at high incident energies, as discussed in detail by Sandner and Schmitt.[35] The decisive quantity there is the dipole calculated tensor component $A^K_{20}(E)$, integrated over all allowed values of the energy transfer E. The main contribution to this integral comes from small values of E, which is exactly where the coincidence experiment under consideration has been carried out. Interestingly enough, the noncoincident alignment found at high energies (around 50 keV) was also of opposite sign from what was predicted in dipole approximation.[35] The discrepancy has then been qualitatively attributed to relativistic kinematics. however, the new results would alternatively suggest a reinvestigation of the nonrelativistic dipole cross sections for the Ar 2p magnetic substates. Efforts are presently under way to finally clarify the situation.

Turning back to the coincidence experiments displayed on fig. 3, we observe a more serious discrepancy between theory and experiment at higher values of the momentum transfer, K=2.38 a.u. and K=3.12 a.u., respectively. The experimental angular correlations do not only deviate strongly from the numerical calculations, but even exhibit considerable violations of the symmetry required by the 1. Born approximation. Equally important, Born approximation predicts the alignment to be independent of the incident energy T, and to depend on K and E only. Therefore, the anisotropies found by the Belfast group at T = 1 keV, $\theta_S = 15°$, and by the Paris/Rome collaboration at T = 8.25 keV, $\theta_S = 5.5°$ should be of equal amplitude due to the identical value of K, and should be shifted by about 25° with respect to each other, following the difference in the momentum transfer directions. Again, this behaviour is notclearly reproduced by the experiments. In general, the coincident angular correlations at higher momentum transfers are characterized by rather large error bars, reflecting the experimental difficulties arising from the low cross sections under these kinematical conditions.

Apart from problems associated with experimental statistics, more basic questions concerning the interpretation of the measured

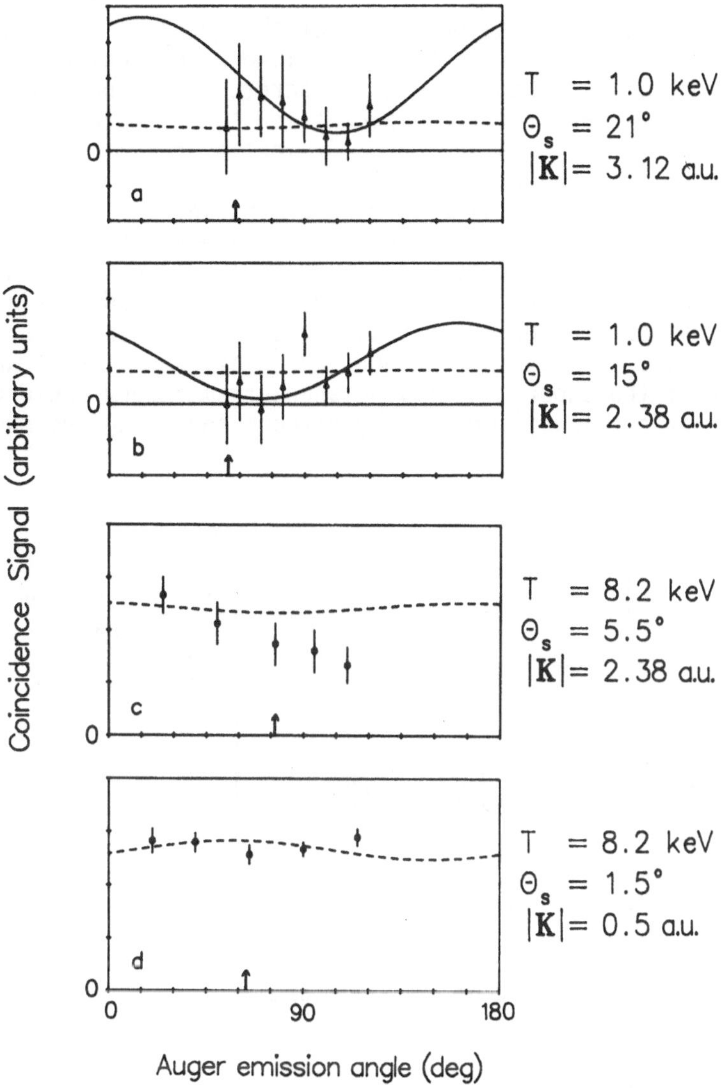

Fig. 3. Coincident Auger angular distributions measured *in the scattering plane*. The incident energy T, scattering angle θ_S, and momentum transfer value K are indicated to the right of each diagram, direction of K is indicated by a vertical arrow, energy transfer E was 253 eV in all cases. Exp.: • Paris/Rome collab.,[23] ▴ Belfast,[20] solid line: fit through the data (from ref. 20). Theory: dashed line 1. Born approximation[10].

Auger angular correlations have recently been raised in connection with an experiment which focused on the angular dependence of the coincident Auger line shapes[22]. This experiment, carried out by the Freiburg group, was again performed at T = 1 keV, θ_S = 15°, leading to a momentum transfer of 2.38 a.u. In contrast to the other experiments the transmission energy of the Auger detector was not held fixed on the center of the Auger line under consideration, but was periodically scanned over five discrete energy points covering the whole line shape and the adjacent low energy background region. A full scanning cycle lasted only 10 minutes, resulting in a quasi simultaneous measurement of the Auger energy distribution, combined with a truely simultaneous recording of the angular distribution as described in chapter II. Great efforts have been made to obtain an accurate theoretical expression for the Auger line shape, which was then fitted to the data. This technique was expected to yield rather reliable values for the angular distribution of both the Auger line area and the coincident background. Moreover, any significant deviation of the experimental line shape from theory would give evidence for continuum correlation effects involving the Auger electrons.

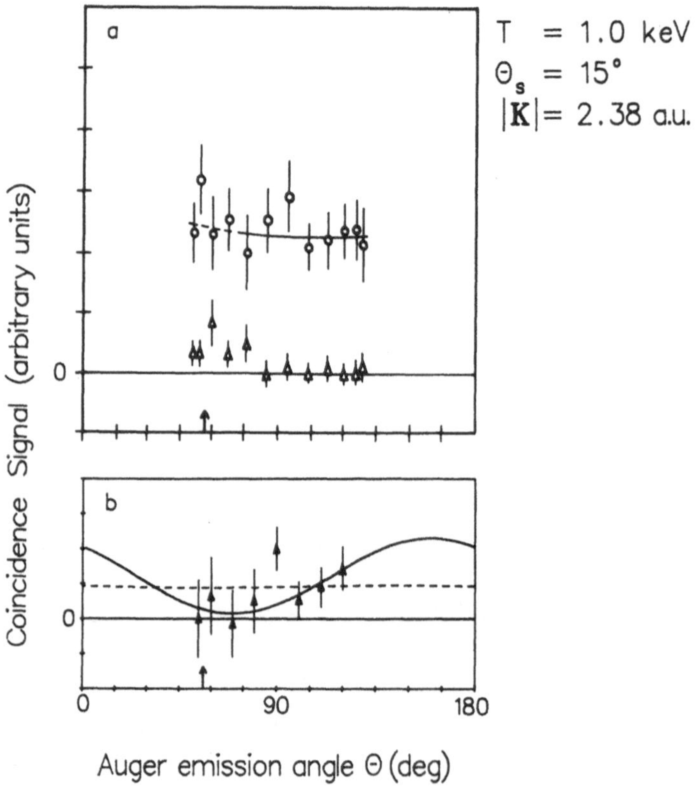

Fig. 4a) Coincident Auger (○) and background (Δ) angular correlation measured out of the scattering plane, the azimuth Φ of the Auger direction is to be calculated from eq. (2) (from ref. (22)). Solid line: fit through the data.

4b) same as fig. 3b, shown for comparison.

As a result, the angular distribution of the coincident Auger line areas (○) and of the coincident background (Δ) is shown on fig. 4a. Note that this background denotes *correlated* electron pairs and is not to be confused with the background of accidential coincidences referred to in chapter II. Several processes have been identified to contribute to the background,[22] and previous experimental results of the Belfast group[24] were in qualitative agreement with fig. 4a: significant coincident background only occurs in the vicinity of the momentum transfer direction. Disagreement, however, was found in the coincident Auger intensity, especially around $\theta = \theta_K = 54°$: Here an almost vanishing Auger intensity was left after background subtraction in the Belfast experiment (fig. 4b). We note that direct comparison between the angular distributions of fig. 4a and 4b is possible for $\theta = 50.8°$ and $\theta = 129.2°$, respectively, which is where the Freiburg experiment also detects Auger electrons in the scattering plane (cf. eq. (2)).

For a comprehensive discussion of these angular distributions the observed line shapes must be taken into account. Fig. 3 shows coincident Auger energy spectra obtained at two representative emission directions, which are chosen such as to lie within (fig. 5a) or outside (fig. 5b) the region where coincident background was found. First one notes that in either case the coincident line shape deviates considerably from the noncoincident spectrum recorded under identical conditions (dashed lines in fig. 5). This already gives evidence for continuum correlation effects occurring in the coincidence experiment. In fact, an isotropic post collision interaction line shape[36] yields very good agreement over the whole angular region $80° \leqq \theta \leqq 130°$, the zero background region, as exemplified by the solid line on fig. 5b. Two conclusions may be drawn from this observation: a) there exists post collision interaction in the coincident Auger transition under consideration (which, in fact, is to be expected from the existence of a 5 eV ejected electron in process (3), and has meanwhile been confirmed by independent experiments[37,38]), and b) the PCI-influenced Auger line *shape* is the same for all emission angles within the experimental accuracy. This does not necessarily imply that PCI has no effect on the Auger line *intensities*, even though the flat angular distribution of fig. 4a is in good agreement with 1. Born approximation. This may be fortuitous, especially, since considerable PCI-induced anisotropies have already been found in other systems[39] and the existence of post collision interaction by itself violates the basic two step assumption made in the theoretical treatment of the process under consideration.

The situation appears to be even worse in those regions where coincident background has been observed. Here, in general, the same PCI line shape, together with a constant background contribution, was not able to produce fits of similar quality as in the region without background. This has been interpreted as evidence for a *coherent* superposition between the coincident Auger line and the coincident background, much like the well known coherence between autoionization excitation and direct ionization. In fact, the tentative fit of a Beutler-Fano profile to the data yielded acceptable agreement in some emission directions, as exemplified on fig. 5a (solid line). The assumption of at least partial coherence between the Auger line and the background seems to be justified since each of the background processes considered so far[22] ultimately leads to the same final state as process (3), with the only difference that the Ar^{++} ion may be in any of the states (^1D), (^1S), or (^3P), out of which only the (^1S) state can interfere with the Auger process (3). Unfortunately, in the presence of interferences the time consuming technique of measuring coincident Auger line shapes is inevitable for a reliable Auger intensity determination.

In conclusion, the investigation of the coincident Auger line shapes yielded continuum correlation and coherence effects involving the Auger electrons: the post collision interaction and, possibly, interference effects with direct excitation of the final state. Both effects constitute a violation of a very basic assumption made in the interpretation of the experiments, which was the two step assumption for ionization and Auger decay in the primary collision. As a

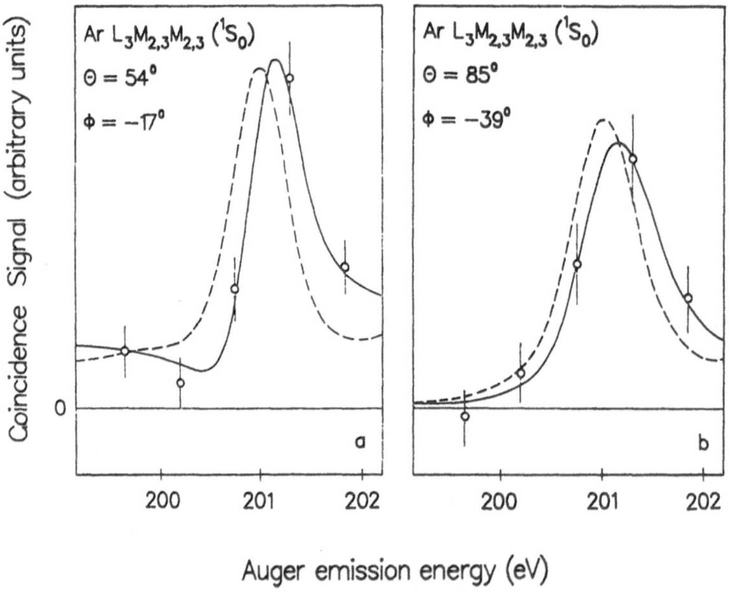

Fig. 5. Coincident Auger line shapes (from ref. 22). Auger emission direction is indicated on each diagram, collision kinematics is the same as in fig. 4a. Dashed line: noncoincident line shape, solid line fig. 5a: Beutler-Fano profile, solid line fig. 5b: PCI line shape.

consequence, the close relationship between Auger angular distribution and inner shell alignment, which was derived using the two step approximation, can not a priori be assumed in any of the experiments discussed above. In particular, conclusions about the validity of 1. Born approximation may need to be postponed until more experimental data on the continuum correlation effects are available.

ACKNOWLEDGEMENT

The author acknowledges stimulating discussions and correspondence with Drs. A. Crowe, G. Stefani, and A. Lahmam-Bennani, as well as their submission of experimental results prior to publication. He especially thanks M. Völkel for critical discussions and his assistance during the preparation of this manuscript.

REFERENCES

1 I.E. McCarthy and E. Weigold, Phys.Rep. 27C 277 (1976)

2 A. Giardini, R. Fantoni, R. Camilloni and G. Stefani, Comments At.Mol.Phys. 10 107 (1981)

3 H. Erhardt, M. Fischer and K. Jung, Zeit.Phys. A304 119 (1982)

4 H.J. Beyer and H. Kleinpoppen, in *Fundamental Processes in Energetic Atomic Collisions* Eds. H.O. Lutz, J.S. Briggs, H. Kleinpoppen (New York: Plenum Press) p. 531 (1983)

5 J. Macek in *Fundamental Processes in Energetic Atomic Collisions* Eds. H.O. Lutz, J.S. Briggs, H. Kleinpoppen (New York: Plenum Press) p. 39 (1983)

6 W. Bambynek, B. Craseman, R.W. Fink, H.U. Freund, H. Mark, C.D. Swift, R.E. Price and P.V. Rao, Rev.Mod.Phys. 44 716 (1972)

7 R. Camilloni, A. Giardini-Guidoni, R. Tribelli and G. Stefani, Phys.Rev.Lett. 29 618 (1972)

8 A. Lahmam-Bennani, H.F. Wellenstein, A. Duguet, A. Daoud, XIII ICPEAC Berlin Book of abstracts p. 153 (1983)

9 E. Schüle and W. Nakel, J.Phys. B 15 L639 (1982)

10 E.G. Berezhko, N.M. Kabachnik and V.V. Sizov, J.Phys. B 11 1819 (1978)

11 E.G. Berezhko and N.M. Kabachnik, J.Phys. B 12 2993 (1979)

12 C.A. Quarles and J.D. Faulk, Phys.Rev.Lett. 31 859 (1973)

13 W. Hink, G. Ulm, K. Brunner and T. Ebding, Abstracts of International X-XUV Conference, Sendai Japan (J.Appl.Phys. 17 Suppl. 17-2,341) (1978)

14 W. Sandner and C.E. Theodosiou, IX ICPEAC Kyoto Book of abstracts p. 263 (1979)

15 M.J.v.d. Wiel and G. Wiebes, Physica 53 225 (1971)

16 M. Komma and W. Nakel, J.Phys. B 15 1433 (1982)

17 L. Ungier and T.D. Thomas, Phys.Rev.Lett. 53 435 (1984)
18 N.L.S. Martin, T.W. Ottley and K.J. Ross, J.Phys. B 13 1867 (1980)
N.L.S. Martin and K.J. Ross, J.Phys. B 17 4033 (1984)
19 E.C. Sewell and A. Crowe, J.Phys. B 15 L357 (1982)
20 E.C. Sewell and A. Crowe, J.Phys. B 17 2913 (1984)
21 M. Völkel and W. Sandner, XIII ICPEAC Berlin Book of abstracts p. 142 (1983)
22 W. Sandner and M. Völkel, J.Phys. B 17 L597 (1984)
23 A. Lahmam-Bennani, G. Stefani and A. Duguet
Lecture Notes in Chemistry 35
Wavefunctions and Mechanisms from Electron Scattering Processes
Eds. F.A. Gainturco and G. Stefani (Berlin: Springer) p. 191 (1984)
24 A. Crowe in *Electronic and Atomic Collisions*, eds. J. Eichler, I. Hertel and N. Stolterfoht (North Holland) p. 97 (1984)
25 M. Völkel and W. Sandner, J.Phys. E: Sci.Instr. 16 456 (1983)
26 M. Völkel and W. Sandner, XIII ICPEAC Berlin Book of abstracts p. 703 (1983)
27 J.H. Moore, M.A. Coplan and E.D. Brooks III, Rev.Sci.Instr. 49 463 (1978)
28 H.A. van Hoof and M.J. van der Wiel, J.Phys. E: Sci.Instr. 13 409 (1980)
29 J.P.D. Cook, I.E. McCarthy, A.T. Stelbovics and E. Weigold, J.Phys. B 17 2339 (1984)
30 U. Fano, Phys.Rev. 90 577 (1953)
31 A.J. Ferguson, *Angular Correlation Methods in Gamma Ray Spectroscopy* (Amsterdam: North-Holland, New York: Wiley) (1965)
32 T. Åberg in *Inner Shell and X-Ray Physics of Atoms and Solids* eds. D.J. Fabian, H. Kleinpoppen and L.M. Watson (New York: Plenum Press) p. 251 (1981)
33 E.G. Berezhko and N.M. Kabachnik, J.Phys. B 10 2467 (1977)
34 M. Inokuti, Rev.Mod.Phys. 43 297 (1971)
35 W. Sandner and W. Schmitt, J.Phys. B 11 1833 (1978)
36 K. Helenelund, S. Hedman, L. Asplund, U. Gelius and K. Siegbahn, Phys.Scr. 27 245 (1983)
37 E.C. Sewell and A. Crowe, J.Phys. B 17 L 547 (1984)
38 G. Stefani and A. Lahmam-Bennani, priv. communication (1984)
39 P.J.M. van der Burgt, J. van Eck and H.G.M. Heideman, J.Phys. B (1984, to be published)

COLLISIONAL SPECTROSCOPY OF LASER EXCITED ATOMS

Maria Allegrini

Istituto di Fisica Atomica e Molecolare del C.N.R.
Via del Giardino, 7 - 56100 Pisa, Italy

INTRODUCTION

This lecture will be devoted to the study of collisions between two atoms, both in an excited state. For their simple structure (only one external electron) and strong interaction with the light, we will consider here only experiments involving alkali atoms. Recently a great deal of work has been done also on more complex atoms such as calcium, strontium and barium[1]. Collisions between two excited atoms were first observed using powerful discharge lamps as excitation sources, (for a review see for example ref./2/), but it is with the laser that high concentrations of excited reactants have been normally obtained. The study of collisions in a laser field is now a rapidly growing field of research[3-5]. The simplest case of collisions between laser excited alkali atoms regards two atoms both in the first P state. A near resonance, of the order of few kT, between the colliding atoms and the final products of the collision is required. Indeed the observed effects have been:

i) excitation transfer to higher atomic levels nL, with energy close to the sum energy of the two nP atoms;
ii) formation of the molecular ion which is energetically achievable from two atoms in the lowest P state in all the alkalis, but lithium.

Thus we are considering the binary inelastic collisions

$$A(nP) + A(nP) \longrightarrow A(nL) + A(nS) \quad (1)$$

$$A(nP) + A(nP) \longrightarrow A_2^+ + e \quad (2)$$

Many other phenomena are also produced in a dense vapor excited by resonant laser radiation. For their understanding it is important to identify in the experiment the parameters determining the dominant effects. These parameters are the laser power density W, the atom density N and the temporal characteristics of the laser. We are interested in experiments where the laser provides a sufficient concentration of excited atoms to enable the observation of their collisions but it is not powerful enough to give direct multiphoton ionization or other processes induced by the strong electromagnetic field associated with the laser. With a laser power density $\lesssim 10^3 W/cm^2$ this aim is certainly achieved and an atom density of $\lesssim 10^{13} cm^{-3}$ usually assures that secondary collisional processes, either with other atoms or ions and electrons produced in the experiments, are negligible. Attention has to be paid also to the concentration of alkali dimer which may obscure the results of the atomic collisions we wish to study[6-8]. For cw laser irradiation there is a continuous production of excited atoms with a steady state density dependent upon the effective atomic lifetime; for pulsed excitation the pulse duration has to be considered in the time dependence of the process[6].

There are many reasons for interest in studying processes (1) and (2). From a theoretical point of view we are in presence of a rare mechanism where two excitations in the atomic system are combined to give one single higher excitation. In classical physics processes leading to a more uniform distribution of the energy or towards the exchange of excitation from one atom (or molecule) to another, (see for example the sensitized fluorescence phenomenon[9]), are more common. Contrary to the multiple excitations obtained with high power lasers, the higher excitation achieved in (1) and (2) is produced internally in the system through the collisions. In this case the laser is merely a tool to create simultaneously the excited partners for the collision. A more specific theoretical interest arises because these collisions depend upon long range interactions between the two atoms. The atomic energy levels for the two atoms at infinite distance are of course well known. Also the potential curves for the two atoms at short interatomic distances, when they are bound to form the neutral molecular dimer, are reasonably well known. In the intermediate range where the asymptotic approximations fail these collisions may provide an experimental test for the calculations of potential curves and of their crossing points. Another important result is that the determination of reliable cross sections for processes (1) and (2) has indirectly given the incentive to study the intriguing phenomena of radiation diffusion and saturation in optically thick vapors excited by laser radiation[10].

From an experimental point of view the interest is even more immediate. The processes (1) and (2), followed by photoionization or collisional ionization have been proved[11] to provide, together with the multistep ionization of the dimers, the seed electrons for complete ionization of dense vapors with moderate laser powers[12,13]. Efficient breakdown and plasma formation in gases or vapors is an actual goal of applied physics; any process which may improve the ion yield without increase of laser intensity supplied from outside is of interest. Infrared laser emission has been observed[14] upon excitation of the 3P level in sodium vapor, suggesting that 3P/3P collisions may also be important in this phenomenon. Application of these collisional processes to laser isotope separation is obvious since the laser selects the excited state of the reactants and the collision yields a selected product. A(nP)/A(nP) collisions are also a promising way to populate triplet electronic excited states of the neutral molecule A_2 resulting in an excimer configuration with possible application to new tunable laser sources.

We will restrict ourselves in the following to few basic experiments and ideas, while updated results on this subject are easily found in the open literature or references /3-5/.

ENERGY POOLING COLLISIONS

The first experiment involving collisions between laser excited atoms were performed on sodium vapor, but all the alkalis have since been widely studied[3-5]. Lucatorto and McIlrath[12] observed nearly complete ionization in a 10-cm column of Na at $\simeq 10^{16}cm^{-3}$ atom density, irradiated with a pulsed laser of $\simeq 1MW/cm^2$, tuned to the 3S-3P resonance line. Allegrini et al.[15], in a cell with a sodium density of $\simeq 10^{12}cm^{-3}$, observed fluorescence from atomic states lying higher than the 3P level, optically excited with a cw laser of less than $10W/cm^2$. Na energy levels are shown in figure 1. As can be seen from figure 1, direct photoionization requires three yellow photons. Rigorous calculations[16,17] show that the rate of ion production by this process is negligible in both the above experiments, even if the laser power density available in the pulsed experiment is five or six orders of magnitude bigger than in the cw experiment. Also the difference between pulsed and cw excitation is not the key parameter in this case, because the pulse duration of the flash lamp pumped dye laser used by Lucatorto and McIlrath was long ($\sim 1\mu sec$) compared to the lifetime of the 3P excited state, giving a pool of excited reactans as in the cw excitation experiment. The dominant effect in the experiment of ref./12/ was the associative ionization

$$Na(3P) + Na(3P) \longrightarrow Na_2^+ + e \qquad (3)$$

When energetically possible, two excited atoms associate during a collision to form a molecular ion plus one electron. The outcoming electron is slow because it carries an energy of few kT, namely the energy defect between the colliding 3P atoms and the state in which the molecular ion is formed. However, in presence of a high density of excited atoms, the electron, before recombination, undergoes one or more superelastic collisions.

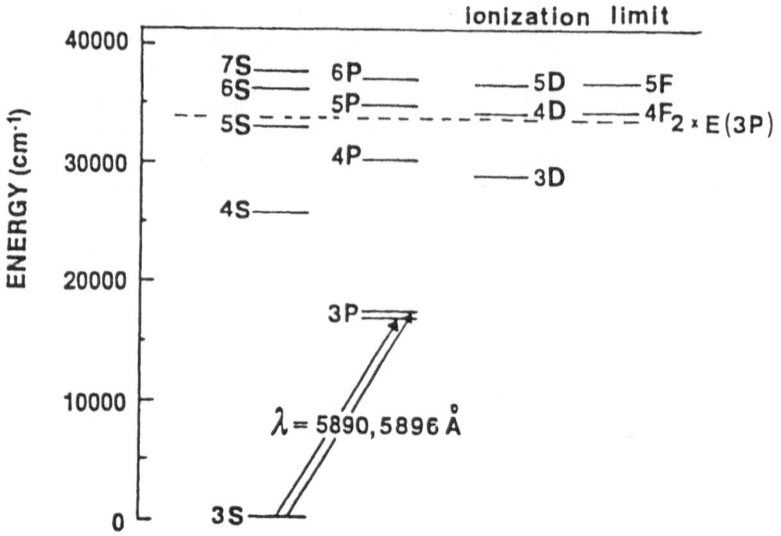

Fig. 1. Na energy levels showing the laser excitation, the ionization limit and the energy of two 3P atoms. The fine structure of the 3P level is not in scale.

$$e\ (\text{slow}) + Na(3P) \longrightarrow e\ (\text{fast}) + Na(3S) \qquad (4)$$

Fast electrons then produce, by avalanche mechanisms, the atomic ions. In the experiment of Allegrini et al.[15] the following process was observed

$$Na(3P) + Na(3P) \longrightarrow Na(nL) + Na(3S) \qquad (5)$$

where nL indicates a highly excited state with energy close to twice the 3P energy. The energy defect (positive or negative) is supplied by conversion of relative kinetic energy of the two interacting Na (3P) atoms or carried away by the final Na(nL) and Na(3S) atoms. Thus this is a peculiar process involving pooling of internal and relative kinetic energy of the atoms. The nL levels were identified simply by their characteristic radiative emission lines. A quadratic dependence of the fluorescence intensity upon the laser power and upon the atom density plus comparison with the resonance fluorescence 3P-3S versus the laser wavelength through the atomic resonance, has clearly established that two 3P excited atoms are involved in the population mechanism of the highly excited states. In process (5) there is a transfer of electronic energy as in the sensitized fluorescence phenomenon

$$A^* + B \longrightarrow A + B^* \qquad (6)$$

where during a collision between one excited atom A^* and one in the ground state B the excitation is transferred from one atom to the other. We will see later that in analogy with process (6), the rate constant K or cross section σ of process (5) has been derived by solving the rate equations at the steady state. However, the main characteristic of process (5) is the combination in the collision of two excitations to give a single higher excited product as in the associative ionization. The name "energy pooling collisions" has been used first by Leventhal[18,19] to describe both processes (3) and (5). In a few words this expression contains the physical meaning of the observed phenomena: the highly excited nL states and the molecular ions are two different products of the same collisional reaction in which the energy of the reactants peakes in one product instead of being distributed.

DETECTION METHODS

The energy pooling collisions produce atoms in excited states, which radiatively decay to the ground state, ions and electrons. Detection of photons, ions or electrons requires different techniques and apparatus and each experiment has its own particular geometry and original technical solutions that will be too long to report. We will try to give few illustrative examples of the three detection techniques.

Fluorescence

In order to look at the fluorescence spectrum of a laser excited alkali vapor the collision cell containing the metal may be an ordinary pyrex cell sealed under vacuum and heated in a temperature controlled oven. This system reaches easily a thermodynamic equilibrium, the vapor is saturated and the atom density can be determined with reasonable accuracy from the temperature, using the Nesmeyanov tables [20]. The fluorescence is collected at right angles to the laser beam, dispersed with a monochromator mounted with the proper grating and detected with a photomultiplier in the visible range or with a suitable detector for the infrared. It is common to control the laser output with a photocell for power variation checking and to attenuate with a filter the resonance fluorescence, i.e. the radiation emitted directly from the laser excited first P level, because it is orders of magnitude more intense than the fluorescence from the levels populated through process (1). The fluorescence signal coming from the nL levels is very weak and usually its detection needs amplification by techniques that differ substantially for excitation by pulsed or cw lasers. A schematic diagram of the apparatus is shown in figure 2 and two examples of such spectra are reported in figures 3 and 4. Figure 3 is taken from ref. /7/ and shows the fluorescence spectrum of sodium vapor excited to the $3P_{1/2}$ level with a broadband cw dye

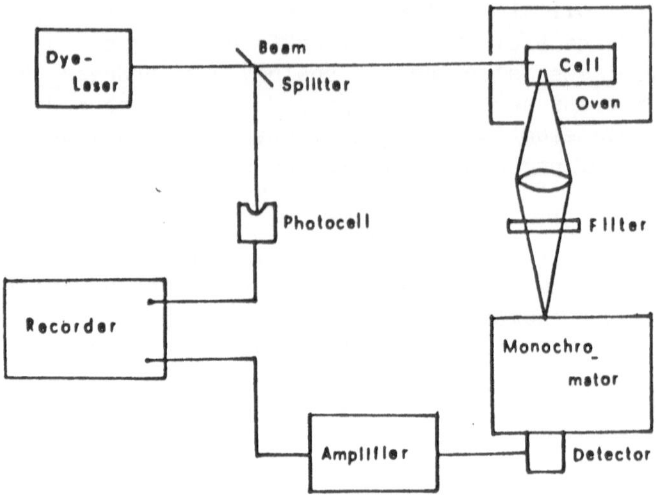

Fig. 2. General scheme for detection of the fluorescence emitted by the highly excited nL levels, populated through the energy pooling collisions.

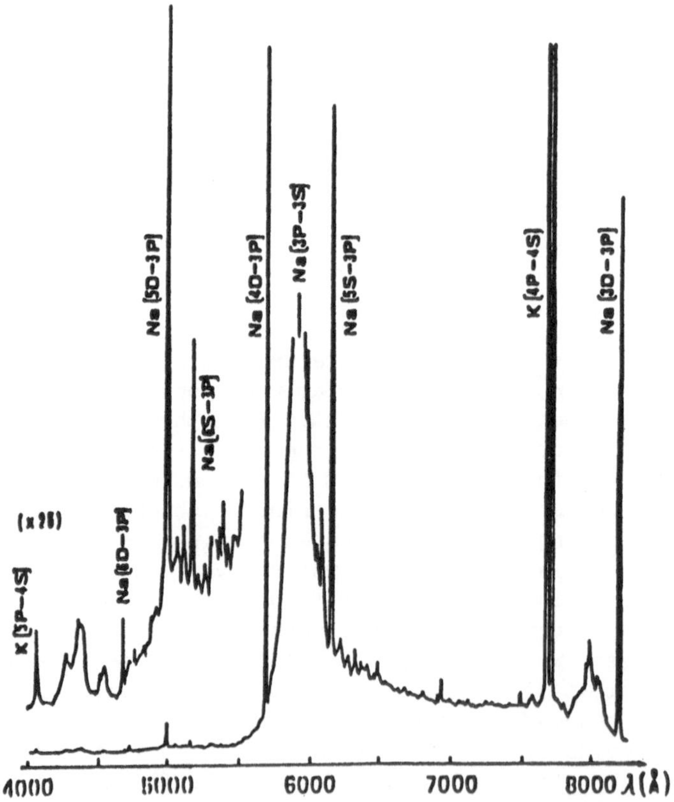

Fig. 3. Fluorescence spectrum of sodium vapor excited by a cw dye laser tuned to the $3S_{1/2} - 3P_{1/2}$ transition[7].

laser of $\sim$40mW; since this spectrum was recorded at relatively high atom density ($N \sim 10^{15} cm^{-3}$), at which secondary collision process occur, beside the fluorescence lines emitted by the nL levels populated directly by the energy pooling collisions, it contains also lines from higher sodium levels, molecular sodium bands and potassium atomic lines, although potassium was present only as an impurity. The resonance fluorescence line, corresponding to the 3P-3S transition, was not recorded to avoid saturation and possible damage to the detection apparatus; its position is pointed out in the diagram. A similar fluorescence spectrum[21] of potassium vapor excited to the $4P_{3/2}$ level with a broadband cw dye laser of $\sim$100mW is shown in figure 4. The resonance fluorescence, not shown in the diagram, is at ~7700Å and and it was, after correction for the spectral response of the detection apparatus, ~10^4 times more intense than the strongest line in the spectrum (6S-4P transition at 6911-6939Å).

Ions

A simple way to detect the ions formed in the laser excited vapor is to insert in the collision cell a current collecting electrode and read the total ion yield with an electrometer. A small transverse electric field may be necessary to assure complete collection. For the identification of the different species of ions mass resolved spectra are taken, for example with a quadrupole mass filter. Another common mass analysis is carried out with a time of flight drift tube which delivers current pulses whose height is proportional to the collected ion number. The collision cells used by various groups are quite different and many experiments have also been performed on ion crossing beams[3-5]. As an example we report here the experimental set up used by Leventhal and coworkers[6,18,19] on sodium vapor. A cylindrical cell is mounted on an oven consisting of a resistance heater imbedded in a Cu block and located beneath the cell. Sodium vapor effuses into the cell through a slot in the bottom of the cell. Apertures in the cell permit the laser beam to enter and leave along the axis of the cylinder. Ions formed in the cell are collected through a side hole of ~5mm and a window on the opposite side permits also observation of photons for fluorescence spectrum recording. Because of the apertures, the cell works in non equilibrium conditions withconsequent nonuniformities in the atom density. The density can thus not be determined simply from a temperature reading as in the sealed pyrex cell of Figure 2. The ions are extracted from the cell with a set of electrostatic lenses, mass analyzed with a quadrupole mass filter and detected with a CuBe particle detector. A scheme of the apparatus is shown in figure 5. Different modes of operation are possible with this apparatus. In one mode the laser is fixed at the atomic transition wavelength and mass scans are taken; alternatively the quadrupole mass spectrometer is fixed at a specific ion mass and the laser wavelength is scanned. Figure 6 shows a spectrum of sodium vapor taken with the apparatus of figure 5 set in the

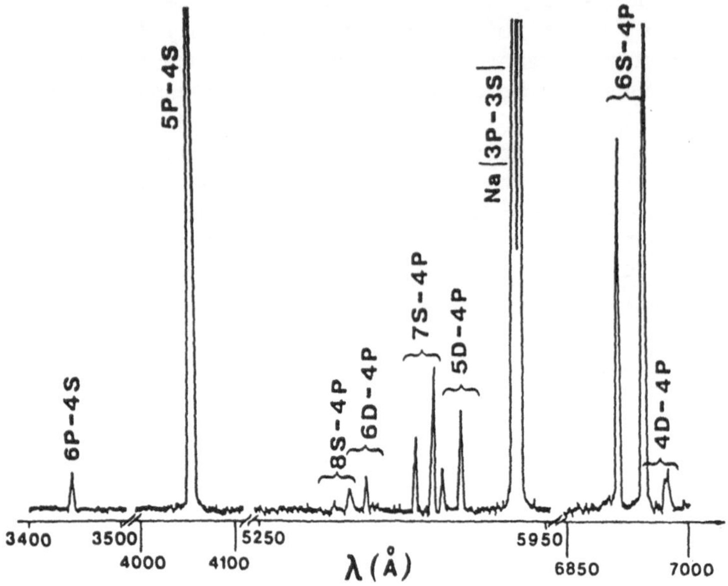

Fig. 4. Fluorescence spectrum of potassium vapor excited by a cw dye laser tuned to the $4S_{1/2}$ - $4P_{3/2}$ transition[21].

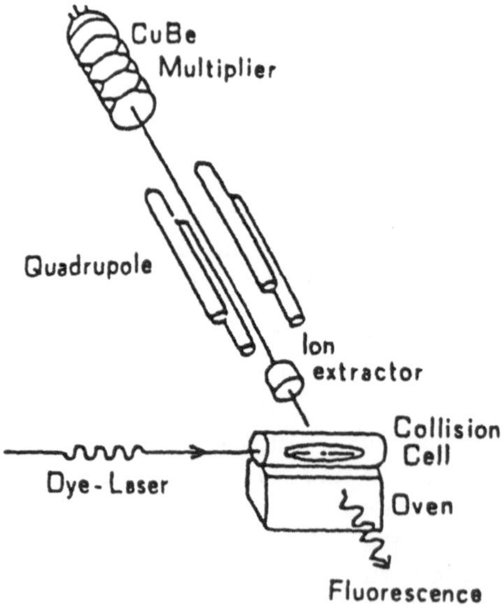

Fig. 5. Schematic diagram of the apparatus for the detection of ions formed in collisions between laser excited sodium atoms[6].

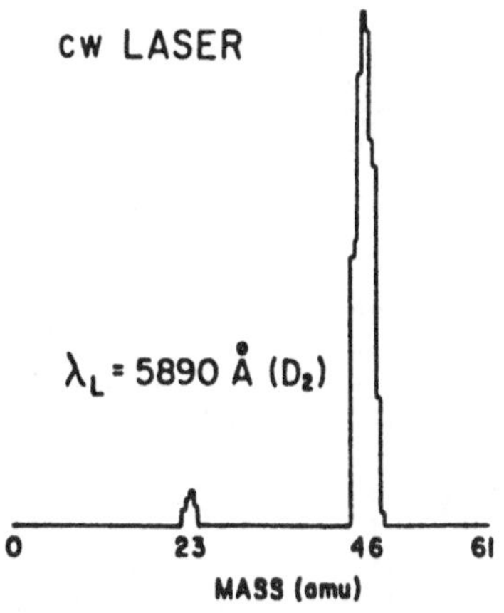

Fig. 6. Mass spectrum taken with a multimode cw laser tuned to the D_2 sodium line[6]. The spectrum is corrected for relative transmission as a function of mass. Ordinate scale is in arbitrary units.

first mode of operation. Na_2^+ ions were formed by associative ionization (process (3)) only when the laser was tuned to D_1 or D_2 sodium lines. Na_2^+ were the dominant ionic product, however a small yield of Na^+ ions, resulting from the photodissociation of nascent Na_2^+ ions, was detected[22].

Electrons

The third method to study the energy pooling collisions is the detection of the electrons created in the vapor. A simple measurement of the total current does not provide relevant information on the collisional processes and the results may be masked by the photoelectric effect. An original way to use electron detection, although it provides only indirect and qualitative information, was applied to an experiment on potassium[21]. The atom density in this experiment was between 10^{13} and $10^{14} cm^{-3}$, an intermediate value at which secondary collisional processes become important. It was our intention to distinguish the levels populated directly by the primary 4P/4P collisions, which should have a clear quadratic dependence upon the 4P density, from the levels populated through other mechanisms. The cylindrical pyrex cell was supplied from the entrance window with two internal electrodes, parallel to the cylinder axis. Optogalvanic signals, taken with a standard apparatus, as shown in figure 7, were recorded simultaneously with the fluorescence signals. Optogalvanic signals[23] arise in an electrical discharge at low pressure whenever light tuned across some characteristic transition frequency of the atoms in the cell is absorbed.

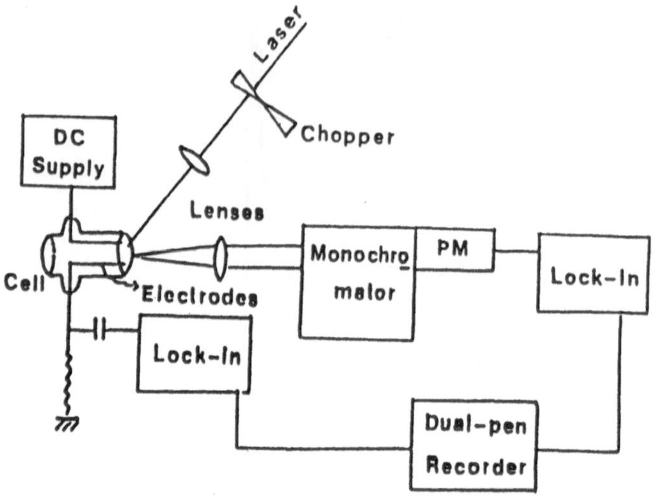

Fig. 7. Schematic diagram of the apparatus for simultaneous detection of optogalvanic and fluorescence signals.

We have compared the linewidth of the optogalvanic and fluorescence signals as a function of the laser wavelength, tuned across the resonance atomic transition 4S-4P. The ratio between the halfwidth of the resonance fluorescence signal $\Delta\nu(RF)$ and that of the optogalvanic signal $\Delta\nu(OG)$ was $\Delta\nu(RF)/\Delta\nu(OG)=\sqrt{2}$, which demonstrated that, as a consequence of the irradiation of the vapor by the laser resonant with the 4P level the electrons were produced by a process involving two 4P excited atoms, as the associative ionization

$$K(4P) + K(4P) \longrightarrow K_2^+ + e \qquad (7)$$

The direct photoionization process requires three photons (see figure 8) and was completely negligible because the laser power density was $\lesssim 10^3 W/cm^2$. The ratio between $\Delta\nu(OG)$ and the halfwidth of the fluorescence signal from one level close in energy to the 4P+4P energy (for example the 6S level) was $\Delta\nu(OG)/\Delta\nu(6S)=1$. Since the optogalvanic signal had a quadratic dependence upon the 4P atom density this demonstrated that also the population of the 6S level depends upon $N^2(4P)$, as in the process

$$K(4P) + K(4P) \longrightarrow K(6S) + K(4S) \qquad (8)$$

Finally looking at the fluorescence from a level far from the energy

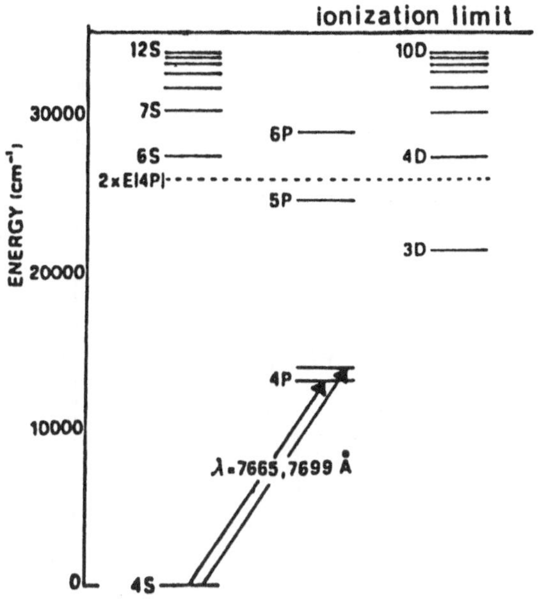

Fig. 8. Partial energy level diagram for K. The fine structure of the 4P level is not in scale.

sum of two 4P atoms (for example the 7S level) it resulted $\Delta\nu(OG)/\Delta\nu(7S)\neq 1$, which showed that higher levels were not populated directly by the energy pooling collisions. These results[24] are shown in figure 9.

In the above experiment the detection of the variation of the electron density produced by the collisions has been indirectly used to interpret the collisional processes. A direct evidence of these processes was obtained in an important experiment[25] where an electron spectrometer was introduced for the first time to resolve the energy spectrum of the electrons created in the vapor. The experimental set up consisted in a weakly collimated effusive sodium beam excited by an orthogonal cw dye laser locked to an hyperfine component of the D_2 resonance line. The electrons emerging from the interaction zone were detected by a cylindrical mirror analyzer. The original apparatus[25] did not transmit electrons of low energy, but it has been recently modified[26] to detect also electrons of near zero energy. The absolute energy scale was accurately established and direct observationof superelastic collisions (4) was obtained. The resolved peaks in the electron energy spectrum have unambiguously proved that the primary electrons were created by purely collisional processes involving excited atoms; the first mechanism is the associative ion-

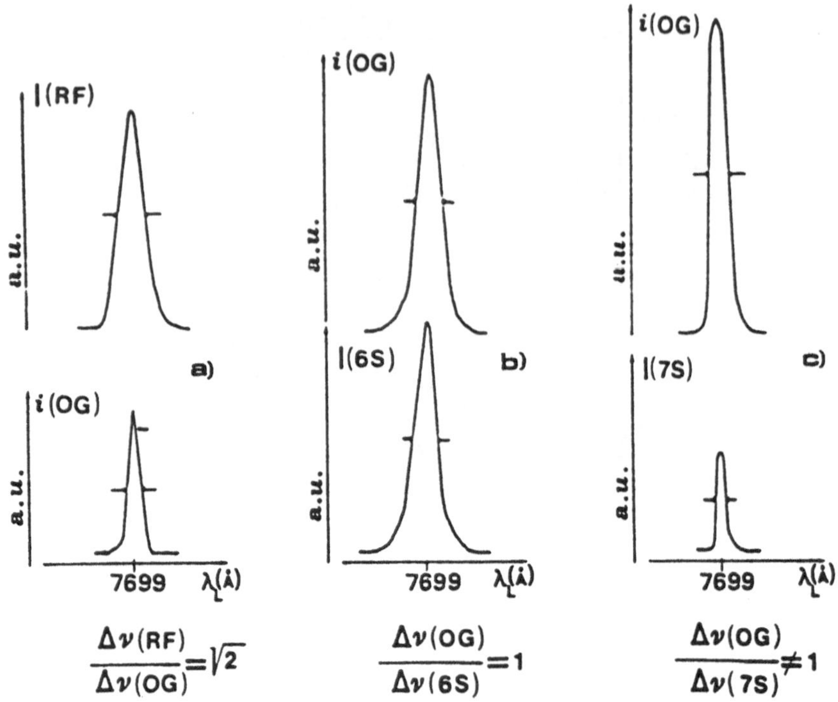

Fig. 9. Traces of the optogalvanic and resonance fluorescence ($4P_{1/2}$-$4S_{1/2}$) signals (a), of the optogalvanic and 6S-4P fluorescence signals (b) and of the optogalvanic and 7S-4P fluorescence signals (c) versus the laser wavelength, tuned across the $4S_{1/2}$-$4P_{3/2}$ atomic transition at 7699Å. The ratios between the relative widths at half maximum are also reported[24]. The shape of the signals is not uniform because the laser was scanned by hand; however this fact is insignificant for these results.

ization (3) and the second mechanism is the Penning ionization

$$Na(3P) + Na(nL) \longrightarrow Na^+ + Na(3S) + e(2.11eV - E(nL)) \quad (9)$$

where Na(nL) are provided by the energy pooling collisions (5) and 2.11eV is the energy of a 3P atom. A typical energy spectrum of electrons ejected from the laser irradiated sodium vapor is shown in figure 10.

CROSS SECTION DETERMINATION

In the introduction we have mentioned some reasons for which it is interesting to investigate the energy pooling collisions. However, for any serious application or comparison with the theory,

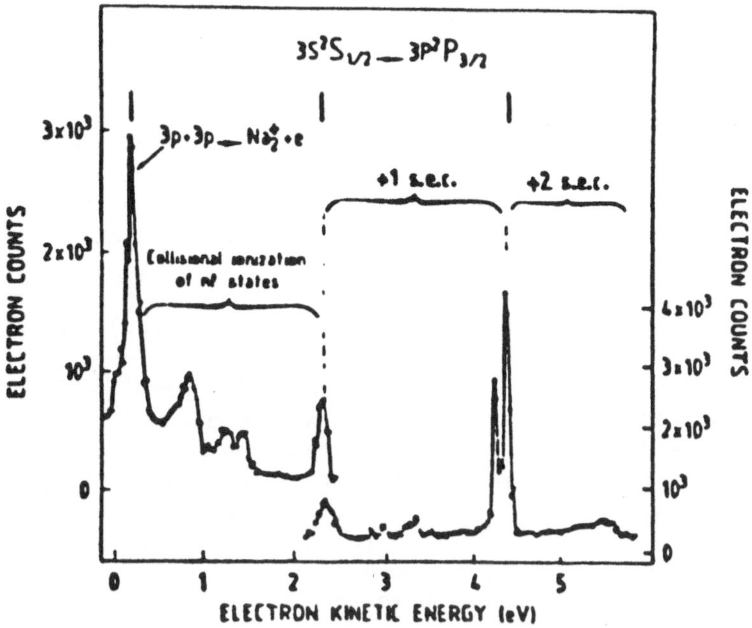

Fig. 10. Energy spectra of the electrons emitted from a cw laser excited Na vapor. Electrons from associative ionization and Penning ionization of nL states are also observed after 1 or 2 superelastic collisions with 3P state[27].

the rate coefficient K or cross section σ for the reactions (1) and (2) have to be measured with great accuracy. Since we are dealing with atoms having thermal energies we assume that K and σ are related simply by $K=\langle\sigma v\rangle=\sigma\bar{v}$, where $\bar{v}=\sqrt{8kT/\pi\mu}$ is the relative mean interatomic velocity. A common method used for the determination of the rate constants is based on the solution of the rate equations, in analogy with the familiar treatment of the sensitized fluorescence process (6)[28]. The rate equations for process (6) are

$$\dot{N}(A^*) = Z - N(A^*)/\tau(A^*) - KN(A^*)N(B) \tag{10}$$

$$\dot{N}(B^*) = KN(B)N(A^*) - N(B^*)/\tau(B^*) \tag{11}$$

where Z is the number of atoms excited from A to A^* per unit time and τ is the radiative lifetime of an excited level. Solution of equation (11) at the steady state gives

$$K = (1/N(B)N(A^*))(N(B^*)/\tau(B^*)) \tag{12}$$

Usually the quantity measured in the experiments is the intensity of the fluorescence lines. Generally speaking the intensity per unit solid angle of a fluorescence line from a level i to a level j is related to the population of the level i by[29]

$$I(i,j) = \hbar\omega(i,j)A(i,j)\ N(i)\varepsilon(i,j)V/4\pi \tag{13}$$

where $\omega(i,j)$ is the transition frequency, $A(i,j)$ is the spontaneous transition probability which is equal to $1/\tau$, $\varepsilon(i,j)$ is a factor which takes into account the instrumental response of the detecting apparatus and V is the fluorescence volume. Using relation (13) in equation (12) the rate coefficient K may be determined from the intensity ratio of the fluorescence from the excited states A^* and B^* to the ground state as

$$K = (1/N(B)\tau(A^*))(I(B^*)/I(A^*))(\omega(A^*)/\omega(B^*))(\varepsilon(A^*)/\varepsilon(B^*)) \tag{14}$$

Therefore the measurement of K depends upon the density of ground state atoms B and the lifetime of the excited atoms A^*. While ground state atom density can be measured with great accuracy, the measurement of the lifetime of an excited state may present serious problems because of the radiation self-trapping phenomenon[30,31], which changes the natural lifetime to an effective lifetime. The error in the sensitized fluorescence cross sections is mainly due to the fact that it is difficult to measure the effective τ with great accuracy.

This same procedure can be applied to the collisions between excited atoms, however many more difficulties complicate a measurement of the rate coefficients K(nL) for process (1) and K(AI) for process (2). For sake of semplicity let us consider the experiments where all the processes other than the energy pooling can be neglected; then the following simplified rate equations can be written

$$\dot{N}(nL) = K(nL)N^2(nP) - N(nL)\sum_{n'L'}A(nL,n'L') \tag{15}$$

$$\dot{N}(A_2^+) = K(AI)N^2(nP) \tag{16}$$

Solution of equation (15) at the steady state gives

$$K(nL) = \sigma(nL)\bar{v} = (N(nL)/N^2(nP))\sum_{n'L'}A(nL,n'L') \tag{17}$$

and integration of equation (16) over the time of A_2^+ production yields

$$K(AI) = N(A_2^+)/(N^2(nP)\ \Delta t) \tag{18}$$

The population of the nL level can be determined by measuring the fluorescence intensity I(nL) and knowing the transition probabilities A(nL,n'L'), as shown in equation (13), since radiation trapping for the nL levels populated through the collisions (1) can be neglected in most of the experiments. The integrated Na_2^+ ion signal can be precisely measured in any experiment, provided the collection efficiency and the instrumental response of the overall apparatus is known. The crucial point for both K(nL) and K(AI) is the quadratic dependence upon the excited atom density N(nP), which is a quantity difficult to measure with high accuracy. Incorrect assumptions about the excited atom density and spatial distribution are the primary cause of orders of magnitude differences in the value of the cross sections measured in different experiments. In the experiments where fluorescence spectra are detected it is an advantage to take the ratio of the intensities I(nL)/I(nP) so that the absolute efficiency of the detection system and the volume do not enter directly in the calculations. However the laser-vapor interaction volume must be well definied for the determination of the excited atom density N(nP) and the fluorescence emitted from the levels nL and nP must be collected from the same volume. In a vapor the optical absorption length varies considerably with changes in the density of the absorbing atoms or in the laser intensity; as a consequence also the volume of laser-vapor interaction varies. This is one problem to be solved in the experiment. The second problem arises because of the radiation self-trapping and diffusion of the laser excited atoms. Owing to these phenomena the excited P atoms show an effective lifetime longer than the natural lifetime, as given by the transition probability, and are not confined to the laser beam cylinder. Then the radiation from the nL levels to the first P state is attenuated and the effective volume for nP/nP collisions extends outside the laser beam. A certain number of different experimental methods have been introduced to resolve all these problems and we will summarize here some of them.

In the experiments where ions can be detected a common method for the determination of the nP excited atom density relies on the measurement of the atomic ions A^+ produced by direct photoionization of the nP level with a second laser. This approach has been used for example by Weiner and coworkers[32] for the determination of the absolute rate constants for process (3). Their apparatus consisted in two crossed effusive atomic sodium beams excited by a tunable flash-lamp dye laser to the $3P_{3/2}$ state from which the associative ionization (3) takes place. A second Nd:YAG pumped dye laser provided the uv source to directly photoionize the 3P level by the process

$$Na(3P) + \hbar\omega \longrightarrow Na^+ + e \qquad (19)$$

The Na^+ ion signal was linear with the 3P atom density and the volume of production coincided with the volume swept out by the photoionizing laser. The Na_2^+ ions, before saturation, showed a quadratic dependence upon the 3P atom density and were produced in an enlarged effective volume because of the diffusion of the trapped 3P atoms outside the laser beam. The actual volume of excited atoms was determined by comparing the Na^+ signal when only the photoionizing laser was on with the Na^+ signal when both lasers were on. The correct intensity of Na_2^+ ion signal, necessary to determine the associative ionization cross section, was then given by the total Na_2^+

yield divided by this volume. The effect of radiation trapping on the lifetime of the excited atoms was determined with the same apparatus by monitoring the decrease of the Na^+ ion signal versus the time delay between the laser tuned to the $3S_{1/2}$-$3P_{3/2}$ transition and the uv laser which photoionized the 3P atoms. A similar approach, based on the direct photoionization given by a second laser, has been extensively used in the studies of Penning and associative ionizations of highly excited rubidium atoms[33] and energy pooling collisions between 5P rubidium atoms[34].

Two original methods for the excited atom density determination have been introduced by Huennekens and Gallagher[35] and by Allegrini et al.[36] during the measurement of the energy pooling cross section for the 5S and 4D states in sodium

$$Na(3P) + Na(3P) \longrightarrow Na(5S,4D) + Na(3S) \qquad (20)$$

The first approach takes advantage of radiation trapping to measure the spatial and temporal distribution of the 3P atoms, the second uses the saturation of the resonant 3P-3S fluorescence for an absolute calibration of the apparatus at low temperature, when radiation trapping is avoided. In both experiments a special collision cell was built to overcome the difficulties related to the determination of the laser-vapor interaction volume. The cell of Huennekens and Gallagher was a cross shaped 5-cm stainless steel block with vacuum sealed sapphire windows which do not react with the sodium vapor. Two sapphire rods were also placed in the opposite arms of the cross (see figure 11) to create the correct geometry for radiation trapping calculations. Pulsed excitation from a dye laser, (peak power $\simeq$ $60\mu J$, pulse duration $\sim$5nsec, linewidth $\sim 0.5cm^{-1}$), tuned to the D_2 resonance line was used to produce the Na(3P) atoms. The experiment was performed in radiation trapping conditions which were well described by the Holstein theory[31]. Thus the spatial distribution of the 3P atoms was obtained by measuring the resonance fluorescence intensity and using the Holstein theory for the diffusion of the trapped atoms. The time dependence of the 3P atom density was obtained by looking at the change in the transmission of a cw laser, following the pulsed excitation. The cw laser was a single mode dye laser, highly attenuated ($\simeq 100\mu W$) and tuned to the far wings of the D_2 resonance transition. Once the absolute value for spatial and time dependences of the 3P atom density was measured the cross sections for 5S and 4D formation by 3P/3P collisions were obtained by measuring the fluorescence ratios of 4D-3P and 5S-3P to 3P-3S and solving the rate equations (15), specifically written for process (20).

The cell of Allegrini et al.[36] was a pyrex capillary cylindrical tube of internal radius $\sim$0.9mm with two polished slices of glass rods as windows. Sodium atoms were excited to the $3P_{3/2}$ state with a broadband ($\Delta\lambda$ could be varied from 0.3Å to 0.03Å)

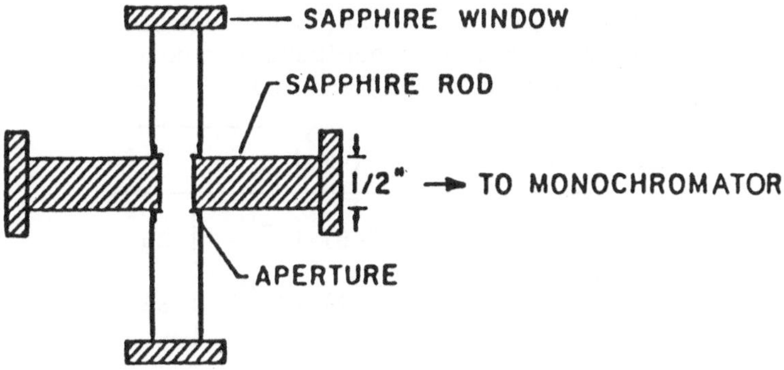

Fig. 11. Top view of the collision cell used by Huennekens and Gallagher (taken from ref. /37/).

cw dye laser of moderate power $\leq 10 W/cm^2$. The laser beam illuminated the whole cross section of the cell uniformly and the effective volume from which the fluorescence was detected was further defined by a narrow slit $\simeq 150 \mu m$ placed near the entrance window. Excited atom density was determined absolutely at a low temperature T_o, in absence of radiation trapping, by observing the saturation of the resonance fluorescence signal. In fact at saturation the $P_{3/2}$ population density is related to the ground $S_{1/2}$ atom density simply by the level degeneracies. The 3P density at temperatures T', high enough to allow observation of fluorescence from the collision populated 5S and 4D states, was determined by comparing the intensity of the resonance fluorescence signals at T_o and T'. The radiation trapping present at T' was considered by correcting the natural lifetime of the 3P level with the trapped lifetime calculated from the Milne[30] theory, valid in the conditions of that experiment. Once the excited state density was known the cross sections for process (20) were obtained with the same procedure of rate equation solution and fluorescence intensity ratio measurements. The two above experiments have given results which agree within the experimental errors (see table 1), although the methods used for the excited atom density determination were very different and excitation was pulsed in one case[35] and cw in the other[36]. Previous measurements were different as much as five orders of magnitude; this fact evidences the difficulties in measuring the cross sections for energy pooling collisions, but it also proves that the solution to the many problems is not unique.

CONCLUSIONS

In these notes the background aspects of the energy pooling collisions rather than a complete review of the full work done in this field, have been presented. This presentation has been inten-

Table 1. Comparison of the Na(5S,4D) energy pooling cross sections experimentally determined in references /35/ and /36/.

Level	T(K)	σ (cm^2)	Ref.
5S	597	$(1.6 \pm 0.6)x10^{-15}$	35
	483	$(2.0 \pm 0.7)x10^{-15}$	36
4D	597	$(2.3 \pm 0.8)x10^{-15}$	35
	483	$(3.2 \pm 1.1)x10^{-15}$	36

tionally oversimplified by treating the energy pooling processes as singled out from multiphoton ionization, laser induced and other collisional processes. Seldom this is the case in the actual experiments; however the aim of these notes was to give the basic idea of the energy pooling processes and the general detection methods. While the qualitative interpretation of the phenomena involving collisions between excited atoms was immediate, many problems were met to get quantitative reliable values of the cross sections. A special emphasis has been placed here on these problems and on some approaches used for their solution. A table with updates results for the energy pooling cross sections for the alkalis can be found in ref.38.

Theoretical work has been very limited; the cross section has been calculated only for few levels populated through the collisions between excited atoms (for a complete list see ref. /38/). Although the order of magnitude of these cross sections agrees with the experimental determinations, the theoretical method used is not fully valid at the interatomic distances relevant to energy pooling collisions. Complete calculations, based on the Multi-Configuration-Self-Consisted-Field approach, are in progress[39] and they should provide a useful map of the interatomic potential energy curves for any internuclear distance.

ACKNOWLEDGMENTS

This work has been partially supported by a U.S.-Italy Cooperative Science Program, C.N.R. Grant no. 47423.

I am particularly grateful to Professor Ugo Fano for his generous contributions to this lecture. I am also indebted to Dr. Silvia Gozzini for assistance in preparing the typescript and to Dr. Mike Kelley for a critical reading of the manuscript.

REFERENCES

1. L. Jahreiss and M.C.E. Huber, Phys. Rev. A28, 3382 (1983) and references therein
2. A. Kopystynska and L. Moi, Phys. Rep. 92, 135 (1982)
3. "Photon-Assisted Collisions and Related Topics", N.K. Rahman and C. Guidotti eds., Harwood Academic, Chur (1982)
4. "Collisions and Half-Collisions with Lasers", N.K. Rahman and C. Guidotti eds., Harwood Academic, Chur (1984)
5. "Atomic and Molecular Collisions in a Laser Field", J.L. Picqué, G. Spiess and F. Wuilleumier eds., Journal de Physique, Serie des Colloques (1984)
6. M. Allegrini, W.P. Garver, V.S. Kushawaha and J.J. Leventhal, Phys. Rev. A28, 199 (1983)
7. M. Allegrini, G. Alzetta, A. Kopystynska, L. Moi and G. Orriols, Opt. Comm. 22, 329 (1977)
8. J. Keller and J. Weiner, Phys. Rev. A30, 2134 (1984)
9. A.C.G. Mitchell and M.W. Zemansky, "Resonance Radiation and Excited Atoms", Cambridge University Press (1971)
10. J. Huennekens and A. Gallagher, Phys. Rev. A28, 238 (1983)
11. R.M. Measures and P.G. Cardinal, Phys. Rev. A23, 804 (1981)
12. T.B. Lucatorto and T.J. McIlrath, Phys. Rev. Lett. 37, 428 (1976)
13. T.J. McIlrath and T.B. Lucatorto, Phys. Rev. Lett. 38, 1390 (1977)
14. W. Müller and I.V. Hertel, Appl. Phys. 24, 33 (1981)
15. M. Allegrini, G. Alzetta, A. Kopystynska, L. Moi and G. Orriols Opt. Comm. 19, 96 (1976)
16. C. Laughlin, J. Phys. B: At. Mol. Phys. 11, 1399 (1978)
17. M.H. Nayfeh, private communication
18. G.H. Bearman and J.J. Leventhal, Phys. Rev. Lett. 41, 1227 (1980)
19. V.S. Kushawaha and J.J. Leventhal, Phys. Rev. A25, 346 (1982)
20. A.N. Nesmeyanov, "Vapor Pressure of the Chemical Elements", Elsevier, New York (1963)
21. M. Allegrini, S. Gozzini, I. Longo, P. Savino and P. Bicchi, Nuovo Cimento D1, 49 (1982)
22. V.S. Kushawaha and J.J. Leventhal, J. Chem. Phys. 75, 5966 (1981)
23. R.B. Green, R.A. Keller, G.G. Luther, P.K. Schenck and J.C. Travis, Appl. Phys. Lett. 29, 727 (1976)
24. M. Allegrini and P. Bicchi, in ref. /3/ pag. 227
25. J.L. Le Goüet, J.L. Picqué, F. Wuilleumier, J.M. Bizeau, P. Dhez, P. Koch and D.L. Ederer, Phys. Rev. Lett. 48, 600 (1980)
26. J.M. Bizeau, B. Carré, P. Dhez, D.L. Ederer, P. Gerard, J.C. Keller, P.M. Koch, J.L. Le Goüet, J.L. Picqué, F. Roussel, G. Spiess and F. Wuilleumier, in ref. /5/
27. J.M. Bizeau, B. Carré, P. Dhez, D.L. Ederer, J.C. Keller, P. Koch, J.L. Le Goüet, J.L. Picqué, G. Spiess and F. Wuilleumier, Proc. XIII Int. Conf. Physics Electronic and Atomic Collisions, Berlin 1983
28. M. Elbel, in "Progress in Atomic Spectroscopy", W. Hanle and H. Kleinpoppen eds., Plenum Press, New York (1979), pag. 1299
29. W.L. Wiese, in "Progress in Atomic Spectroscopy", W. Hanle and

H. Kleinpoppen eds., Plenum Press, New York (1979), pag.1101
30. E.A. Milne, J. London Math. Soc. 1, 40 (1926)
31. T. Holstein, Phys. Rev. 72, 1212 (1947); 83, 1159 (1951)
32. R. Bonanno, J. Boulmer and J. Weiner, Phys. Rev. A28, 604 (1983)
33. M. Cheret, L. Barbier, W. Lindinger and R. Deloche, J. Phys. B: At. Mol. Phys. 15, 3463 (1982)
34. L. Barbier and M. Cheret, J. Phys. B: At. Mol. Phys. 16, 3213 (1983)
35. J. Huennekens and A. Gallagher, Phys. Rev. A27, 771 (1983)
36. M. Allegrini, P. Bicchi and L. Moi, Phys. Rev. A28, 1338 (1983)
37. J.P. Huennekens, Ph.D. thesis, University of Colorado, 1982
38. M. Allegrini, C. Gabbanini and L. Moi, in ref. /5/
39. R. Colle, private communication.

ANISOTROPY IN PHOTOFRAGMENTATION

Chris H. Greene

Department of Physics and Astronomy
Louisiana State University
Baton Rouge, Louisiana 70803

INTRODUCTION

The past decade has seen intensive study of photoionization and photodissociation processes in atoms and molecules. The degree of experimental difficulty and the degree of theoretical sophistication needed to characterize a photofragmentation process both increase rapidly with the total number of correlated directions observed simultaneously. From this point of view the simplest processes involve only one direction, the incident photon polarization axis $\hat{\varepsilon}$. For this class of processes the quantity of interest is the total cross section σ (or perhaps the isotropic partial cross sections σ_i in alternative fragmentation channels i). The next step up in complexity involves two directions, the incident polarization axis and one other direction. For photofragment angular distributions the second axis is of course $\hat{k}$, the escape axis of the separating fragments. Another class of experiments in which just two directions are relevant is the class of alignment and orientation experiments. In these a photon, emitted after the photoeffect has produced an excited fragment state, is observed along some axis $\hat{k}'$, thereby providing the second direction. The polarization axis $\hat{\varepsilon}'$ of the emitted photon may or may not be detected, but this apparent introduction of a third direction does not complicate the theory since the photon polarization aspects are purely geometrical and so do not reflect any additional information about the fragmentation dynamics.

This review is concerned with the theoretical description of these latter two processes involving two directions: one in the initial state correlated with one in the final state. The next class of higher complexity has three directions correlated, such as measurements of the spin polarization and angular distribution of photoelectrons, but despite much current interest this subject will be excluded from the present discussion.

The fundamentals of atomic mechanics were used long ago to describe photofragmentation in the present context of single photon electric dipole processes. This description, reviewed in depth by Starace,[1] is based on reduced dipole matrix elements $\langle \Psi_f || r || \Psi_o \rangle$ connecting the initial bound state Ψ_o to a final continuum state Ψ_f satisfying the incoming wave boundary condition and normalized per unit energy.[1] Since the formal theory of the photoeffect is well in hand to the extent that Ψ_f and Ψ_o can be determined, the main questions of interest have been of a more qualitative nature. That is, to what extent can the angular distribution or the alignment of photofragments be understood in advance of the very detailed and often difficult procedures needed to solve the Schroedinger equation? Moreover, it is important to sort out the range of possible phenomena, in order to pinpoint as readily as possible the dynamical implications of any particular experimental outcome.

The angular momentum transfer formulation of Dill and Fano[2,3] provides a convenient tool for simplifying as far as possible the dynamical description of anisotropies. Somewhat different versions of the approach are needed for angular distribution studies[2,3] than for alignment studies[4,5], but these are linked by a conceptual unity which I stress here.

PHOTOFRAGMENT ANGULAR DISTRIBUTIONS

The differential cross section for electric dipole photofragmentation of a randomly oriented atom or molecule into two separating fragments has a simple dependence on direction:

$$\frac{d\sigma}{d\Omega} = \frac{\sigma}{4\pi} [1 + \beta P_2(\cos\theta)] \quad . \tag{1}$$

Here θ is the angle between the escape axis and the electric vector of the incident linearly polarized photons. The photofragmentation anisotropy is wholly represented by an asymmetry parameter β, which multiplies a second Legendre polynomial. Experimental and theoretical studies of this process aim chiefly at determining β, notably its dependence on the incident photon energy, and on the initial state or the final fragment states.

The second Legendre moment is the highest present in (1) because the only dynamical element in the initial state which singles out a nonrandom direction is the electric dipole photon, whose corresponding transition operator is an irreducible tensor of unit rank. The final state probability density, quadratic in the transition operator, has therefore second rank at most. [This argument does not rule out another term in (1) proportional to $\cos\theta$, but parity conservation does not permit such a term if the initial state has definite parity.]

For definiteness, consider the following specific process

$$h\nu(j_\gamma=1) + AB(J_o) \rightarrow [A(j) + B(s)](\hat{k}) \quad , \tag{2}$$

in which the initial state of the atom or molecule AB has angular momentum J_o, the incident electric dipole photon has $j_\gamma = 1$, and the final state consists of two fragments with respective internal angular momenta j and s, with A and B separating along an axis $\hat{k}$. For example, in photoionization experiments s is the photoelectron spin and j is the angular momentum of the residual atomic or molecular fragment A in the bound state corresponding to the observed channel. Quantum mechanical calculations as usually performed obtain a final state Ψ_f having a definite orbital angular momentum ℓ of the separating fragments instead of a definite escape axis $\hat{k}$. The final state usually has a definite total angular momentum J of all particles as well, since its squared operator commutes with the field free Hamiltonian. Accordingly the final state $|\Psi_f\rangle$ is denoted in a descriptive angular momentum coupling notation by $|[j,(s\ell)j_s]J\rangle$. This denotes a jj-coupled wavefunction in which a particular coupling scheme has been chosen with j_s the vectorial sum of s and ℓ, and with J the vectorial sum of j and j_s. The reduced dipole matrix elements connecting this wavefunction to the initial state via the dipole operator will be represented by $D(j,s,\ell,j_s,J) \equiv D(\ell,j_s,J)$ for brevity.

Several theoretical formulations[6,7] have been presented which derive expressions for the asymmetry parameter β in terms of these reduced dipole matrix elements. For example Lee[6] obtains a result of the form

$$\beta = N^{-2}\sum_J \sum_{J'} \sum_\ell \sum_{\ell'} \sum_{j_s} \sum_{j_s'} D(\ell',j_s',J')^* D(\ell,j_s,J)\ \omega(\ell\ell';j_s j_s';JJ') \tag{3}$$

where the normalization factor is

$$N^2 = \sum_\ell \sum_{j_s} \sum_J |D(\ell,j_s,J)|^2 \ . \tag{4}$$

Here all dynamical information about the fragmentation dynamics is contained in the reduced dipole matrix elements D. Instead the factor $\omega(\ell\ell';j_s j_s';JJ')$ is a standard but complicated geometrical function, involving 6j-coefficients and Clebsch-Gordan coefficients; it is given explicitly in the formulas of Ref.6.

While equation (3) correctly expresses β in terms of the calculated matrix elements $D(\ell,j_s,J)$, it is not particularly transparent owing to the multiple summations over angular momentum quantum numbers. It would be a challenging task to discern even the well-known range of β $(-1 \leq \beta \leq 2)$ starting from Eq.(3). The presence of double summations, for example over ℓ and ℓ', implies that each partial wave ℓ contributes coherently to the asymmetry parameter. That is, each element $D(\ell, j_s, J)$ represents a quantum mechanical amplitude for photofragmentation via a specific intermediate state:

$$|h\nu + AB(J_o)\rangle \rightarrow |[j,(s\ell)j_s]J\rangle \rightarrow |[A(j)+B(s)](\theta)\rangle \ . \qquad (5)$$

Viewed in this fashion, it is clear that the summation over all intermediate paths must be coherent in ℓ, j_s, and J since the corresponding operators do not commute with the observed quantity θ:

$$[\vec{\ell}^{\,2},\theta] \neq 0; \quad [\vec{j}_s^{\,2},\theta] \neq 0; \quad [\vec{J}^{\,2},\theta] \neq 0 \ . \qquad (6)$$

This reasoning is no different from the elementary interpretation applied to the wave-mechanical two-slit experiment.

The angular momentum transfer formulation amounts to a systematic procedure to characterize as much as possible the photofragmentation amplitudes by alternative quantum numbers whose operators commute with the observable quantity of interest. In particular the angular momentum transfer j_t is defined as the difference between all unobserved angular momenta in the initial and final states. Because all of these unobserved angular momenta are independent of the observed quantity (in this case θ), the squared angular momentum transfer operator clearly commutes with the observable. Consequently alternative angular momentum transfer channels, by design, contribute incoherently to the observed quantity.

For photofragment angular distributions the only initial state angular momentum not observed is J_o, while the unobserved final state angular momenta are j and s, with ℓ the only "observed" angular momentum. Thus the prescription of Fano and Dill[2,3] leads to the choice

$$\vec{j}_t = \vec{j} + \vec{s} - \vec{J}_o = \vec{j}_\gamma - \vec{\ell} \ . \qquad (7)$$

Note that the second equality in (7) follows from the angular momentum conservation equation

$$\vec{j}_\gamma + \vec{J}_o = \vec{j} + \vec{s} + \vec{\ell} = \vec{J} \ , \qquad (8)$$

and that j_t is a quantum number of the transition, not of the separate initial or final state. Reference 2 derives in detail the angular distribution in terms of transfer probabilities $|S(j_t)|^2$

characterized by a definite value of j_t. As anticipated above, the asymmetry parameter takes a far simpler and more transparent form:

$$\beta = \sum_{j_t} |S(j_t)|^2 \beta(j_t) \; / \sum_{j_t} |S(j_t)|^2 \quad . \tag{9}$$

For any chosen value of j_t, only three orbital angular momenta can contribute, $\ell = j_t \pm 1$ or $\ell = j_t$.

Let the parity of the initial state be π_o and the parities of the separating fragments A and B be π_A and π_B respectively. Parity conservation in an electric dipole transition requires that the only contributing partial waves ℓ are those satisfying

$$(-1)^{\ell} \pi_A \pi_B = -\pi_o \quad , \tag{10}$$

so that for each j_t only even ℓ or odd ℓ contribute. The two possibilities have been termed parity-favored or parity-unfavored depending on whether the "parity-favoredness" quantum number

$$\pi_f = (-1)^{j_t - j_{obs} + j_\gamma} = (-1)^{j_t - j_{obs} + 1} \tag{11}$$

is +1 or -1, respectively, with $j_{obs} \equiv \ell$ representing the observed angular momentum here. A remarkable result obtained by Dill and Fano[3] is that for parity-unfavored transitions ($\ell = j_t$), $\beta_{unf}(j_t) = -1$ in (9), independently of any dynamical considerations. On the other hand for parity-favored transitions ($\ell = j_t \pm 1$) the asymmetry parameter depends on dynamics through the relative amplitude and phase of the two contributing partial waves. Explicitly, Ref. 3 gives

$$\beta_{fav}(j_t) = \frac{(j_t+2)|S_+|^2 + (j_t-1)|S_-|^2 - 6[j_t(j_t+1)]^{1/2} \operatorname{Re}(S_+ S_-^*)}{(2j_t+1)[|S_+|^2 + |S_-|^2]} \quad . \tag{12}$$

In Eq.(12) $S_\pm$ are a shorthand notation for $S_\ell(j_t)$, whose actual definition in terms of reduced dipole matrix elements can be found in Ref.8. It involves of course a coherent summation over J, as expected from (3), but the coherence in ℓ is immediately apparent in expression (12) for $\beta_{fav}(j_t)$.

The qualitative properties of this treatment of β are simple. Parity-unfavored transitions always have $\beta = -1$, implying that Eq.(1) reduces to a $\sin^2\theta$ distribution. Considering that in photoionization the force exerted on the electrons is along the polarization axis $\hat{\varepsilon}$, it is most interesting that the ejected electron tends to emerge at right angles to $\hat{\varepsilon}$. Indeed this is not the most common situation, though it is by no means rare. Most measured angular distributions have in practice yielded β values in the positive range, closer to the value 2 which implies ejection along $\hat{\varepsilon}$ in a $\cos^2\theta$ distribution. (This is of course the value obtained by photoionizing hydrogen atoms in their ground state.) Positive values of the asymmetry parameter β clearly indicate that parity-favored transitions dominate. Strictly speaking the converse is not true. Depending on the relative amplitude and phase of $S_{\pm}$, $\beta_{fav}(j_t)$ can lie anywhere in its full range $-1 \lesssim \beta \lesssim 2$. Still, parity-favored contributions have tended to be positive. This might be traced to the fact that photoionization often preferentially excites one partial wave ℓ of the photoelectron, and inspection of (12) shows that $\beta_{fav} \geq 0$ in that event.

In recent years, however, an example has become apparent in which this simple expectation appears to be violated in the extreme.[9,10] The example concerns one-electron photoionization of the two-electron systems H^- and He, accompanied by simultaneous excitation of very high states H(n) or $He^+(n)$ of the residual one-electron atom. Detection of the lowest energy photoelectrons just above a threshold with high n should, in accordance with arguments in Refs.9 and 10, produce an asymmetry parameter $\beta \simeq -1$. This is the predicted value for $^1P^o$ final states even though all j_t contributing to β are parity-favored. (Instead the opposite value $\beta \simeq 2$ is expected for $^3P^o$ final states.) At finite $n<\infty$ the threshold value $\beta(n)$ differs from the limiting value by an amount predicted in Ref.9. Some numerical values are given in Table I:

Table I

n	$\beta(n)$ He	$\beta(n)$ H^-
1	2.000	2.000
2	0.412	-0.455
3	-0.037	-0.614
4	-0.271	-0.704
5	-0.413	-0.761
6	-0.509	-0.799
7	-0.578	-0.827
8	-0.630	-0.848
9	-0.671	-0.865
10	-0.704	-0.878

This process has been the subject of extensive experimental investigation only for n=1 and n=2 of helium. For n=1 the value $\beta=2$ is obtained as required by rather trivial considerations. Figure 2 shows several experimental[11-15] and theoretical[16,13] results for n=2. The theoretical curves of Jacobs and Burke[16] and of Chang[13] are the results of lengthy calculations involving numerical solution of coupled differential equations. At energies above 75 eV the calculations agree reasonably well with each other and with the experiments. In the near threshold region near 65 eV, however, electron correlations are increasingly important and only the Jacobs-Burke theoretical result agrees with experiment. The much simpler considerations of Ref.9 yield the value $\beta=0.412$ from Table I, though this result is expected to be reasonable only near the n=2 threshold. This value is outside of the experimental error bars in Figure 1. The discrepancy probably originates in the small (10%) coherent contribution from the "-" dipole channel neglected in Ref.9 (see Ojha[17], and the related study of Berrington et al.[18]). At higher n the "-" channel amplitudes decrease rapidly in accordance with the Wannier theory[19,20], and the predictions of Ref. 9 implying $\beta \rightarrow -1$ should become more realistic in this limit. Indeed Refs.11 and 12 have also measured $\beta(n=3) = -0.2\pm0.2$ at a photon energy of E = 80 eV, which agrees with the predicted value shown in Table I, $\beta(n=3) = -0.037$, to within experimental error. Aside from the quantitative agreement or disagreement with Ref.9, the surprising tendency for a parity-favored β to be small and even negative appears to be borne out by experiment. This is only possible because many partial waves ℓ contribute almost

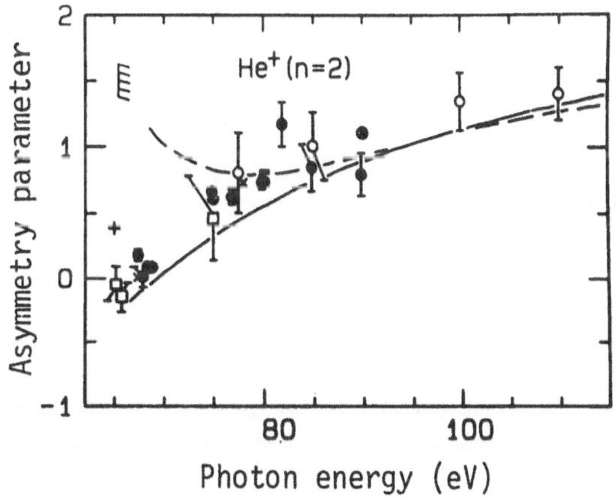

Fig. 1. Asymmetry parameter of the $He^+(n=2)$ channel as a function of the incident photon energy in a helium photoionization experiment. Experimental results: solid circles - Lindle[11] and Shirley et al.[12]; open circles - Bizau et al.[13]; X - Schmidt et al.[14] ; open squares - Morin et al.[15] Theoretical results: solid curve - Jacobs and Burke[16]; dashed curve - Chang from Bizau et al.[13]; cross - threshold value predicted by Greene[9].

equally to the near-threshold photoionization, as stressed a decade ago by Fano.[21]

PHOTOFRAGMENT ALIGNMENT AND ORIENTATION

The analysis of photofragment fluorescence polarization is subject to many of the same complications as encountered for angular distributions. The Fano-Macek[22] description of fluorescence polarization is convenient for this problem. Two dynamical parameters describe the angular distribution and polarization of fluorescence from a cylindrically symmetric ensemble of photofragments A. These are the orientation parameter

$$O_o(j) = [j(j+1)]^{-1/2} \sum_m m\sigma(jm) \;/\; \sum_m \sigma(jm) \tag{13}$$

and the alignment parameter

$$A_o(j) = [j(j+1)]^{-1} \sum_m [3m^2-j(j+1)]\sigma(jm) \;/\; \sum_m \sigma(jm) \quad , \tag{14}$$

with $\sigma(jm)$ the partial cross section for exciting fragment A in the substate with angular momentum j and z-component m. The distribution of radiation is expressed in terms of these anisotropy parameters in Ref.22 or Ref.5. Note that the normalization of O_o in (13) differs from that of Fano and Macek[22], as discussed in Ref.5.

Equations (13) and (14) show that the observed quantity here is j_z. The "observed angular momentum" in this problem must therefore be $j_{obs} = j$, the angular momentum of the excited fragment A, instead of ℓ as in the preceding example. The cross section $\sigma(jm)$ needed to evaluate Eqs.(13) and (14) is again coherent in J (see Ref. 23, for instance), since

$$[\vec{J}^{\,2}, j_z] \neq 0 \quad . \tag{15}$$

Nevertheless $\sigma(jm)$ is clearly incoherent in ℓ and j_s since these quantum numbers are compatible with j_z. The angular momentum transfer in this problem differs then from the choice in Eq.(7) above:

$$\vec{t} = \vec{s} + \vec{\ell} - \vec{J}_o = \vec{j}_\gamma - \vec{j} \quad . \tag{16}$$

Again the second equality follows from the angular momentum conservation equation (8).

After a straightforward application of Racah-Wigner angular momentum recoupling algebra, Reference 5 obtains the alignment and orientation parameters as an incoherent average over three values t = j, j±1:

$$A_o(j) = \sum_t |S(t)|^2 A_o(j;t) \Big/ \sum_t |S(t)|^2 , \qquad (17)$$

$$O_o(j) = \sum_t |S(t)|^2 O_o(j;t) \Big/ \sum_t |S(t)|^2 . \qquad (18)$$

Here the S(t) are photoionization amplitudes into each angular momentum transfer channel given explicitly in Ref.5. Equations (17) and (18) introduce two purely geometrical quantities, the universal alignment and orientation functions given explicitly for left-circularly polarized incident photons (q=1) by

$$A_0(j;t) = \begin{cases} 1/5 - 3/10(j+1) , & t = j + 1, \ \pi_f = +1 \\ 1/5 + 3/10j , & t = j - 1, \ \pi_f = +1 \\ -2/5 + 3/10j(j+1), & t = j, \ \pi_f = -1 \end{cases} , \qquad (19)$$

$$O_o(j;t) = \begin{cases} -[j/4(j+1)]^{1/2} , & t = j + 1, \ \pi_f = +1 \\ [(j+1)/4j]^{1/2} , & t = j - 1, \ \pi_f = +1 \\ [4j(j+1)]^{-1/2} , & t = j, \ \pi_f = -1 \end{cases} . \qquad (20)$$

[For right-circularly polarized incident light, multiply the values of O_o only by -1; for unpolarized incident light O_o vanishes. For linearly polarized incident light O_o still vanishes and A_o must be multiplied by -2.] These formulas (17)-(20) constitute the main result of this approach, showing how dynamics and geometry have been explicitly factored into $|S(t)|^2$ and $\{A_o(j;t), O_o(j;t)\}$ respectively. Just two independent ratios of the three $|S(t)|^2$ determine A_o and O_o in general, and these two ratios are uniquely specified by a measurement of both A_o and O_o. This is a simpler situation than in Eq.(9) for β since it is not possible there to obtain all of the angular momentum transfer channel probabilities from an angular distribution measurement.

The main reason that this alignment and orientation formulation turns out to be simpler than the angular distribution formulation is that j_{obs} has only one value in the former, while several values of j_{obs} contribute coherently to β in the latter. Still, the two treatments show some important similarities. For example, the second rank alignment parameter A_o has opposite sign for parity-favored and unfavored contributions, just as for β when one ℓ dominates. Reference 5 discusses the qualitative implications of this formulation in somewhat greater detail. Suffice it to say here that the parity-favored processes tend to be associated with direct fragmentation processes, while the parity-unfavored processes reflect additional torques in the many-particle dynamics. These additional torques are neglected in the simpler, less general

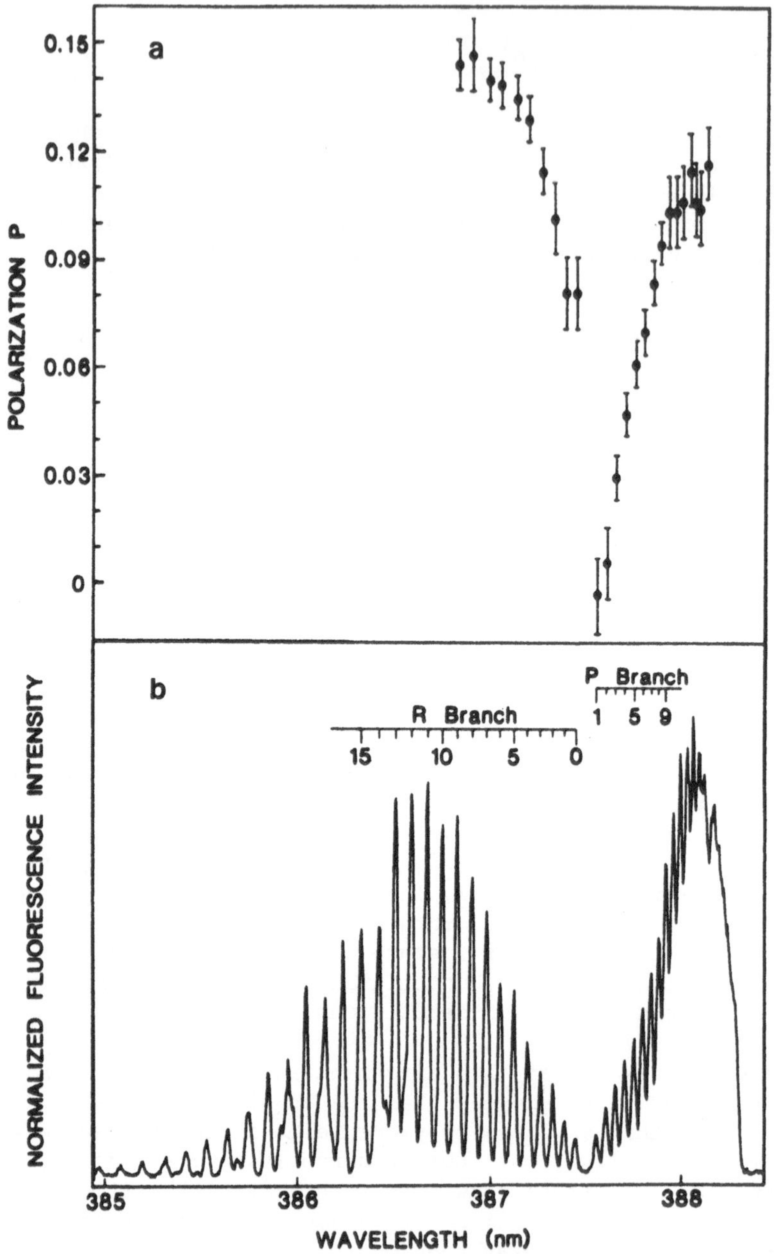

Fig. 2. (a) CN B-X photofragment fluorescence polarization measured for different rotational transitions in the (0,0) band at 0.03 nm resolution. (b) The normalized CN B-X photofragment emission spectrum following ClCN dissociation at 157.6 nm, measured at the magic angle to eliminate alignment dependences. (From Guest et al.[26])

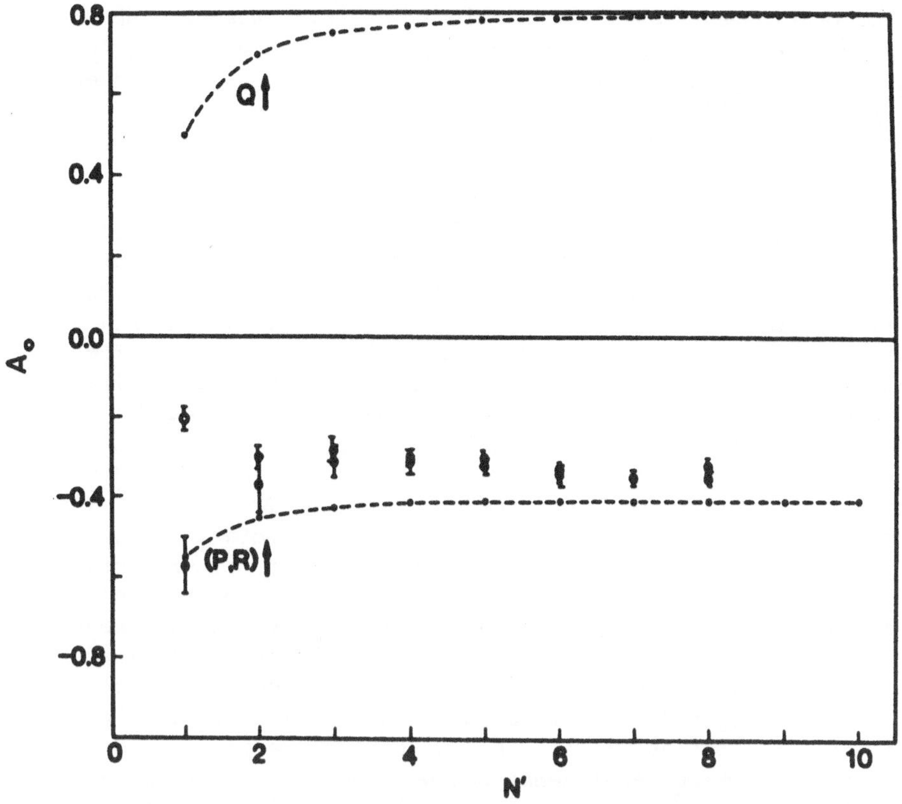

Fig. 3. CN B $^2\Sigma^+(v'=0)$ rotational alignments determined in Hund's case $b_{\beta J}$ (closed circles with error bars) compared to the average of parity-favored angular momentum transfer results (dashed curve). The rotational alignment for N'=1 in case $b_{\beta S}$ is also shown (open circle). (From Guest et al.[26])

formulations of photofragmentation anisotropies presented by Cooper and Zare[24] and by Caldwell and Zare[25].

The utility of the angular momentum transfer treatment for interpreting fragment alignment has been demonstrated by a recent experiment of Guest, O'Halloran, and Zare.[26] This study considered the photodissociation of the triatomic molecule ClCN, using monochromatic linearly polarized photons of wavelength 157.6 nm produced by an F_2 excimer laser. The resulting CN fluorescence was dispersed and its polarization and intensity were analyzed as a function of the excited CN rotational state. These are shown for one band in Figure 2. Note the complicated dependence of the fluorescence polarization on the CN rotational quantum number N'. In particular Figure 2(a) shows a dramatic decrease of the polarization to essentially zero for the lowest CN rotational levels.

At first appearance this might seem to signify some striking dependence of the fragmentation dynamics on N'. Before reaching such a conclusion, however, it is important to consider the depolarizing effect of the nitrogen nuclear spin and also the analogous depolarizing effect of the lone unpaired electron spin in the molecule. These spins are presumably not aligned by the initial photoabsorption and so they cause the initially aligned angular momentum vector $\vec{N}$ to precess about a resultant $\vec{F}$; consequently this reduces the initial rotational alignment to a smaller average value. The physical considerations needed to analyze this effect are described by Refs. 5, 22, and 26. The analysis of Guest et al.[26] extracts a purely "rotational" alignment $A_o(N')$ which is shown in Fig. 3. Two features stand out. First, the alignment $A_o(N')$ is now seen to be nearly independent of N', a much simpler result than would have been expected from Fig. 2(a) alone. Second, the values $A_o(N')$ are close to the average of the two parity-favored alignment functions in Eq.(19), multiplied by -2 to account for the linear polarization of the incident photons in Ref. 26. No hint of parity-unfavoredness is apparent, which makes sense because the photodissociation of ClCN is expected[27] to belong to the class of direct fragmentation processes.

REFERENCES

1. A. F. Starace, "Theory of Atomic Photoionization," in Handbuch der Physik, Vol. 31, W. Mehlhorn, Ed. (Springer, Berlin, 1982). pp 1-121.
2. U. Fano and D. Dill, Phys. Rev. A 6, 185 (1972).
3. D. Dill and U. Fano, Phys. Rev. Lett. 29, 1203 (1972).
4. D. Dill, unpublished correspondence with R. N. Zare (1977).
5. C. H. Greene and R. N. Zare, Ann. Rev. Phys. Chem. 33, 119 (1982).
6. C. M. Lee, Phys. Rev. A 10, 1598 (1974).
7. V. L. Jacobs, J. Phys. B 5, 2257 (1972).
8. D. Dill, Phys. Rev. A 7, 1976 (1973).
9. C. H. Greene, Phys. Rev. Lett. 44, 869 (1980).
10. U. Fano and C. H. Greene, Phys. Rev. A 22, 1760 (1980).
11. D. W. Lindle, "Inner-Shell Photoemission from Atoms and Molecules using Synchrotron Radiation",Doctoral Dissertation, Lawrence Berkeley Laboratory, University of California (1983).
12. D. A. Shirley, P. H. Kobrin, C. M. Truesdale, D. W. Lindle, T. A. Ferrett, P. A. Heimann, U. Becker, H. G. Kerkhoff, and

S. H. Southworth, "Gas-Phase Photoemission with Soft X-Rays: Cross Sections and Angular Distributions", presented at the Brookhaven Conference, Advances in Soft X-Ray Science and Technology, Brookhaven National Laboratory, Upton, New York (1983).
13. J. M. Bizau, F. Wuilleumier, P. Dhez, D. L. Ederer, E. N. Chang, S. Krummacher, and V. Schmidt, Phys. Rev. Lett. 48, 588 (1982).
14. V. Schmidt, H. Derenbach, and R. Malutzki, J. Phys. B 15, L523 (1982).
15. P. Morin, M. Y. Adam, I. Nenner, J. Delwiche, M. J. Hubin-Franskin, and P. Lablanquie, Nucl. Instrum. Meth. 208, 761 (1983).
16. V. L. Jacobs and P. G. Burke, J. Phys. B 5, L67 (1972).
17. P. C. Ojha, J. Phys. B 17, 1807 (1984).
18. K. A. Berrington, P. G. Burke, W. C. Fon, and K. T. Taylor, J. Phys. B 15, L603 (1982).
19. H. Klar and W. Schlecht, J. Phys. B 9, 1699 (1976).
20. C. H. Greene and A. R. P. Rau, Phys. Rev. Lett. 48, 533 (1982), and J. Phys. B 16, 99 (1983).
21. U. Fano, J. Phys. B 7, L401 (1974).
22. U. Fano and J. H. Macek, Rev. Mod. Phys. 45, 553 (1973).
23. J. Vigué, J. A. Beswick, and M. Broyer, J. Phys. (Paris) 44, 1225 (1983).
24. J. Cooper and R. N. Zare, "Photoelectron Angular Distributions", in Lectures in Theoretical Physics: Atomic Collision Processes, Vol. XI-C, S. Geltman, K. T. Mahanthappa, and W. E. Brittin, Eds. (Gordon and Breach, 1969), pp 317-337.
25. C. D. Caldwell and R. N. Zare, Phys. Rev. A 16, 225 (1977).
26. J. A. Guest, M. A. O'Halloran, and R. N. Zare, Chem. Phys. Lett. 103, 261 (1984).
27. W. S. Felps, S. P. McGlynn, and G. L. Findley, J. Mol. Spectry. 86, 71 (1981).

ACKNOWLEDGMENT

I thank D. W. Lindle for providing results in advance of their publication. This work was supported in part by the National Science Foundation and in part by an Alfred P. Sloan Foundation Fellowship.

LOW ENERGY ATOM COLLISIONS

Manfred Faubel

Max-Planck-Institut für Strömungsforschung
3400 Göttingen, West Germany

INTRODUCTION

Low energy atom scattering experiments provide a straightforward and sensitive means for investigating intermolecular potentials and the collision dynamics of thermal velocity molecular collisions. Such studies are of basic interest for the theory of the bulk properties of gases, liquids and solids, and, for a detailed understanding of the mechanisms of chemical reactions.

At collision energies between 20 meV and 100 meV, except for the elastic scattering, the rotational excitation of a molecule is the most probable inelastic process in an atom-molecule collision. For e.g. the j=0 to j=2 rotational transition of N_2 molecules the energy gap is $\Delta E_{0\to 2}=1.5$ meV corresponding to a period of internal motion $\tau=h/\Delta E \approx 2.8 \cdot 10^{-12}$ s . This is longer than the collision time $\tau_c \lesssim 10^{-12}$ sec for a thermal velocity collision and the *Massey* criterion[1]) predicts strong excitation. The energy gaps between vibrational levels are in the order of several hundred meV. Thus, a significant amount of vibrational excitation is only to be expected for collision energies of several eV. As a particular case of vibrational states of large, highly symmetrical molecules one may consider the phonons of a crystal lattice. Their typical excitation energies are in order of a few meV and, thus, the excitation of phonons in atom surface collisions can be studied with similar high resolution molecular beam techniques as the rotational excitation of diatomic molecules. Further, more complicated inelastic scattering processes in atom-molecule collisions can involve electronic transitions, ionization and charge exchange or can lead to chemical reactions or to the dissociation of the molecule. Often, however, these processes have energy barriers of several eV and can be neglected in low energy collisions of collision partners which are not free radical atoms or molecules and are in their electronic ground state.

Both experimental techniques[2,3] and theoretical methods[4] for dealing with low energy atom molecule collisions have been considerably improved during the past decade and a number of inelastic and reactive molecular scattering cross sections could be studied in great detail. In the following, three different inelastic scattering experiments from our Göttingen laboratory will be discussed in order to illustrate some typical molecular beam scattering techniques and results. These are experiments on the (i) vibrationally and (ii) rotationally inelastic scattering of small molecules and (iii) on the excitation of surface phonons in a lithium fluoride crystal by the impact of helium atoms. All three experiments have in common that they use the time of flight (tof) method for identifying individual inelastic scattering channels. Further methods, in particular, laser state selective schemes are briefly discussed in reference[3].

VIBRATIONAL EXCITATION OF SMALL POLYATOMIC MOLECULES

For the study of vibrationally inelastic collisions ion scattering experiments are particularly well suited because ion beams can be readily accelerated to the required collision energies of several eV. A schematic view of an ion molecule scattering apparatus is shown in Fig. 1. The ions, in this case protons, are produced in the ion source and pass a small magnetic mass filter in the lower left corner of Fig. 1. Similarly as in electron scattering experiments an electrostatic energy selector is employed for reducing the ion beam energy spread to 20 - 50 meV. For the tof-analysis of inelastic scattering events the ion beam is then chopped by a pair of electric deflection plates into short, 100 ns duration pulses before it is focused onto the neutral molecule target beam. Scattered ions are detected with nearly 100% efficiency by an open secondary electron multiplier in a distance of one to several meters from the scattering center. The high vacuum of 10^{-6} mbar in the scattering chamber and of 10^{-8} in the tof-drift tube is sustained by four differential pumping stages not shown here.

In the earliest vibrationally inelastic scattering experiments with this and similar machines[2] the primary experimental goal was to investigate very simple ion-diatomic molecule collision systems as H^+-H_2 and Li^+-H_2 in order to test 'ab initio' ion molecule interaction potentials. Notably, in these studies of diatomics it was often found that favorable scattering angles exist where almost pure vibrational excitation with little rotational broadening is observed. Exact calculations are not feasible for larger molecules and experiments were done here more recently with the intention to gain some qualitative insights into their complex collision dynamics.

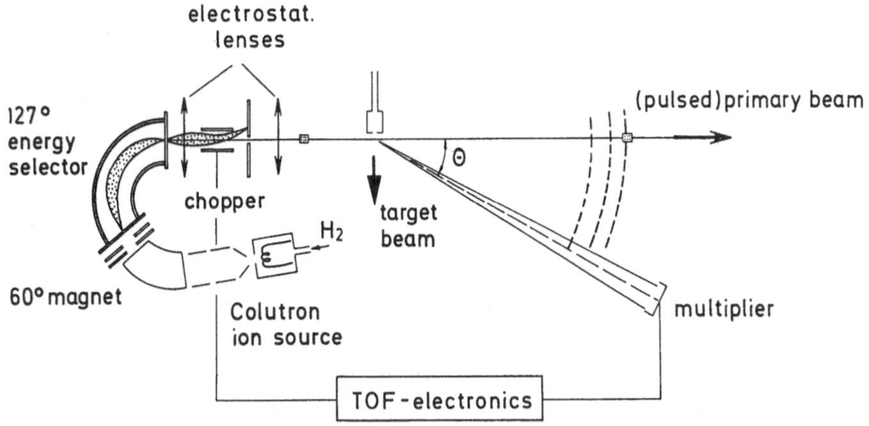

Fig. 1. Scattering apparatus for time of flight studies of inelastic ion-molecule collisions.

Polyatomic molecules have, in contrast to diatomics, several different vibrational modes. For example, the still quite simple, tetrahedral CF_4 molecule has four different vibrational modes (ν_1,ν_2,ν_3,ν_4) with a widest energy level spacing of 0.159 eV for the ν_3 mode vibrations. Except for the quantum states of the isolated modes also combination mode vibrations can be excited. Thus, with increasing internal energy the density of vibrational states becomes very high and at 0.5 eV internal energy the CF_4 has already a density of 25 vibrational states per meV.

In view of the accessible vibrational phase space it was, thus, quite unexpected to observe for some polyatomic molecules well separated vibrational structures as illustrated in Fig. 2 with a tof -spectrum[5] for the scattering of 18.5 eV protons on CF_4 at a scattering angle of $\Theta=5°$. The spectrum shows five strong and many more weaker peaks with a maximum observable energy loss of up to 2.5 eV . From their respective energy loss all peaks can be assigned to the excitation of the subsequent quantum states n=0 up to n=14 of the ν_3 mode of the CF_4. A barely noticeable, ten times weaker excitation of a second mode in the deep valleys between the first five maxima of the tof-spectrum Fig. 2 has been identified[6] as the excitation of the ν_3-ν_4 combination mode (0,0,n,1). The mode selective vibrational excitation is seen strongest for small scattering angles. It is persistent for larger scattering angles up to about the rainbow angle. Beyond the rainbow angle the sharp structures start to be scrambled by other modes or by rotational transitions.

The observation that the excitation probabilities for the individual states of the ν_3-mode follow closely a *Poisson* distribution led to a forced oscillator model interpretation of the mode select-

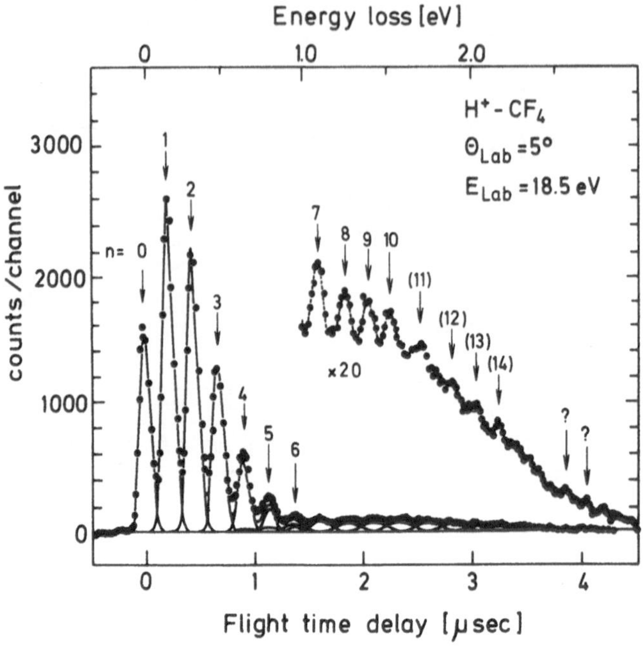

Fig. 2. H^+-CF_4 tof-spectrum showing mode selective vibrational excitation of the ν_3-mode of CF_4. A straightforward and quantitative explanation of this mode selectivity is provided by a forced oscillator model.

ivity phenomenon. This model[7)] assumes that the force acting during a collision onto the ν-th normal coordinate q_ν of the molecule can be approximated as the interaction of the ion point charge with the electric dipole moment $\vec{\mu}$ of the molecule. For a straight trajectory with impact parameter b classical mechanics then yields for the energy transfer ε onto a harmonic oscillator the expression

$$\varepsilon_\nu \simeq \frac{1}{2} \left|\frac{\partial\vec{\mu}}{\partial q_\nu}\right|^2 \frac{e^2}{b^4} \tau_{coll}{}^2 \qquad (1)$$

when the half width duration τ_{coll} of the electric pulse produced by the passing ion is short in comparison to a vibrational period. For the CF_4 molecule the dipole moment derivative $|\partial\vec{\mu}/\partial q_\nu|$ is largest for the infrared active ν_3-mode explaining, thus, the strong preference of the ν_3-mode excitation for this molecule. In addition, for a forced quantum oscillator the average energy transfer is identical to the classical value ε while the excitation probabilities for different vibrational states of one mode is a *Poisson* distribution in ε : $P_n = 1/(n!)\ \varepsilon^n\ e^{-\varepsilon}$, as experimentally observed. Working out this model in greater detail with numerical trajectories for a realistic spherical potential a perfect agreement with experiments was achieved over a wide range of scattering angles and collision energies.[7] Because in the rainbow scattering region three different trajectories contribute to one scattering angle, the model is even capable to account for slight deviations of the experimental vibrational transition probabilities from an ideal single *Poisson* distribution.[5]

Another, quite different, mechanism of mode selective excitation is observed in e.g. H^+-O_2 or H^+-C_2H_4 collisions. For these molecules a near resonant charge exchange with the proton is possible. The C_2H_4 has twelve different vibrational modes. Fig. 3 shows that for a proton energy of 23.7 eV and a scattering angle $\Theta_{lab}=6°$ ten to twelve vibrational quanta of the ν_{12}-mode of C_2H_4 are selectively excited in the H^+-tof-spectrum.[8] The transition probabilities do not follow a *Poisson* distribution but, resemble more closely a *Franck-Condon* distribution. Very recently, it was noticed

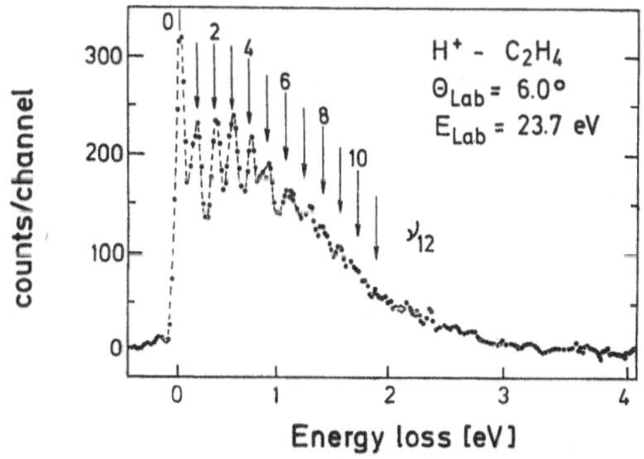

Fig. 3. H^+-C_2H_4 energy loss spectrum showing preferential excitation of the ν_{12}-mode. Here, a more complicated mode selective excitation mechanism involves charge exchange processes.

that the secondary electron multiplier detector of the tof-apparatus, Fig. 1, responds with an efficiency of 1% to the 20eV-neutralized hydrogen atoms of the $H^+-C_2H_4 \rightarrow H^0-C_2H_4^+$ channel. The H^+-ions can be blocked off by an electric field and, thus, it became possible to observe the vibrational excitation structure of the $C_2H_4^+$, of O_2^+ and, of several other ions in the H^0-tof-spectra.[8] These charge exchange tof-spectra show a structure resembling the vacuum ultraviolet photo-electron spectra of the respective molecules. In the future, charge exchange energy loss spectroscopy could, therefore, well become a more flexible and, possibly, less expensive competitor for the conventional photo-electron spectroscopy of ions.

ROTATIONAL EXCITATION OF SMALL MOLECULES

The well separated vibrational excitation structures discussed thus far are rather an exception than the rule in low energy atom -molecule scattering. For example, strong rotational excitation is observed in the scattering of 16 eV Li^+-ions on N_2 and on CO [9] shown in Fig. 4a,b for the scattering angle $\Theta=60°$. Quite surprisingly, in view of the close similarity of the molecular properties of N_2 and CO, the Li^+-N_2 spectrum has one broad inelastic peak while the inelastic part of the Li^+-CO spectrum shows two maxima. These structures could either be interpreted as unresolved vibrational excitation of up to 10 vibrational quanta,[10] or as rotational excitation of as many as 100 rotational quanta.

Classical trajectory calculations show the inelastic processes here are almost pure rotational excitation[9] with the experimentally observed maxima being 'rotational rainbows'.[11,12] The rotational rainbow structure is arising because for a given potential deformation the rotational excitation at a fixed scattering angle is limited by the maximum available angular momentum transfer. This leads to a singularity in the classical rotational energy loss cross section at the maximum rotational state j_R. When the anisotropic potential for the atom-diatomic molecule interaction is simplified to a hard wall ellipsoid with major and minor semiaxes A and B the very simple expression

$$j_R = 2k[A - B \pm \delta[A/(A + B)]^{1/2}] \sin(\vartheta/2) \qquad (2)$$

is obtained for the position of the rotational rainbow. k is here the collision wave number. δ represents the distance between the center of mass and the center of symmetry of the ellipsoid. Thus, for a homonuclear molecule like N_2 one rotational rainbow maximum appears in the energy loss tof-spectrum and, as experimentally observed, two maxima have to be expected in the case of the heteronuclear CO molecule. For a rough quantitative estimate of the ro-

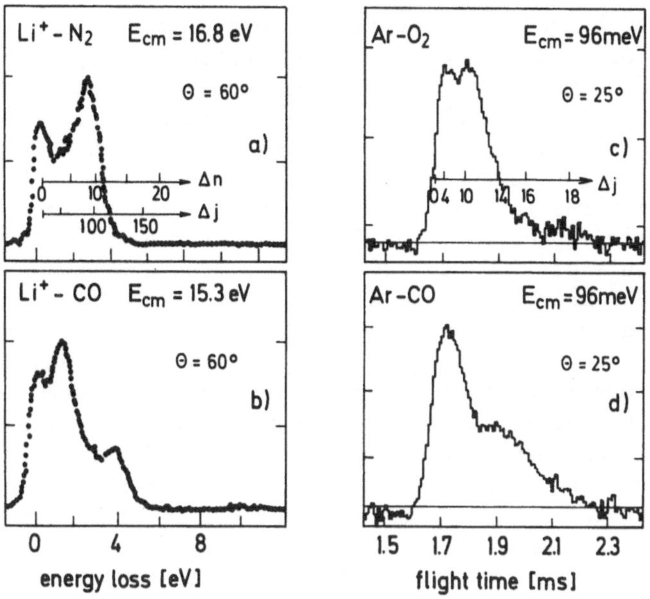

Fig. 4. 'Rotational Rainbow' structures in (a) Li^+-N_2 and (b) Li^+-CO energy loss spectra at ca. 16 eV collision energy, and, in (c) Ar-O_2 and (d) Ar-CO tof -spectra at 0.096 eV .

tational rainbow position in the Li^+-N_2 spectrum, Fig. 4a, half of the molecular bond length $r_e/2 \approx 0.5$ Å of the N_2 molecule may be used for the deformation factor (A-B) in eq.(2). With the Li^+-N_2 collision number $k \approx 207$ Å^{-1} the rotational rainbow is then quite well predicted at $\Delta j_R \approx 103$.

The validity of this rotational rainbow picture extends over a wide range of energies. For example, very similar energy loss structures were observed in K-N_2,CO collisions at $E \approx 1$ eV [12] and led to the discovery of the hard ellipsoid model. Recently we observed, [13] in the apparatus to be described next, the rotational rainbow structure at a still lower energy E=96 meV in Ar-O_2 and Ar-CO tof-spectra shown in Fig. 4c,d for a laboratory scattering angle of 25° . The collision number k is here some ten times smaller than in the Li^+-N_2,CO experiment. Therefore, the rotational rainbows appear at, approximately, also ten times smaller Δj values near $\Delta j_R \approx 10$ while the general shape of the spectra is essentially unchanged.

For resolving individual rotational transitions of molecules other than H_2 the energy resolution of a tof-experiment would have to be improved by one order of magnitude over the present space charge limit of $\Delta E \gtrsim 10$ meV of ion scattering experiments. A resolution of 1 meV can, however, be achieved in neutral-neutral crossed molecular beam machines as the apparatus[3)] shown schematically in Fig. 5a, and, in technical detail in Fig. 5b. In this experiment a high intensity, nearly monochromatic beam of He-atoms with a velocity resolution of $\Delta v/v \lesssim 0.5\%$ is produced in a high pressure nozzle expansion. The 30 μm diameter nozzle operates at a backing pressure of 100 to 200 atmospheres and expands into a vacuum of 10^{-3} mbar which is sustained by a 12000 ℓ/sec diffusion pump. In such an expansion almost all energy of the stagnation gas is converted into translational flow velocity of the gas stream while the internal gas temperature is cooled to below 20 °mK. For gases other than helium a nozzle beam starts to show dimerization when internal gas temperatures below 1 to 5 °K are reached in the expansion. The target beam nozzle is therefore 100 μm in diameter and operates at a maximum pressure of 6 atmospheres for the target gases N_2 , O_2 , CO and CH_4 . An additional benefit of nozzle beam sources is that the internal state temperature is also cooled to a few Kelvin and the beam molecules are prepared in their rotational ground states. The primary and the target beam are collimated by several differential pumping stages to angular spreads $\lesssim 2°$ before they intersect at right angles in the center of the scattering apparatus. For tof-measurements the primary beam is chopped with a 4 slit, 10 cm diameter mechanical tof-wheel into pulses of 10 μs duration. The neutral particle detector, shown at the right hand side of Fig. 5b, is a high performance, ultrahigh vacuum mass spectrometer in a distance of 165 cm from the scattering center.

In Fig. 6 four representative rotational state resolved tof -spectra are shown for the scattering of He on the diatomics N_2 , CO , O_2 and on the spherical top molecule CH_4 . The He-N_2 spectrum, Fig. 6a, shows for E=27.7 meV and Θ=39.5° a large elastically scattered peak at a flight time of 1.82 ms . This peak is followed by two smaller maxima at longer flight times which originate from the excitation of the $0 \rightarrow 2$ and of the $1 \rightarrow 3$ rotational transitions of N_2 with the respective energy gaps of 1.5 and of 2.5 meV . For He-CO at the same angle and energy in Fig. 6b the excitation of both odd and even Δj transitions with the energy level spacing of 0.5 meV are observed, but, can just not be fully resolved. The scattering of He on O_2 , Fig. 6c, yields a very well resolved tof-spectrum showing the elastic $1 \rightarrow 1$ and the $1 \rightarrow 3$ rotational transition without too much broadening from the O_2 fine structure transitions ("F.S.T." in top of this spectrum). Full rotational state resolution is also feasible for He-CH_4 collisions. At E=34.8 meV and Θ=19.6° in Fig. 6d the $1 \rightarrow 2$ and the $1 \rightarrow 3$ rotational transitions of the T-symmetry modification of methane and the $0 \rightarrow 3$ of the A-modification are observed.

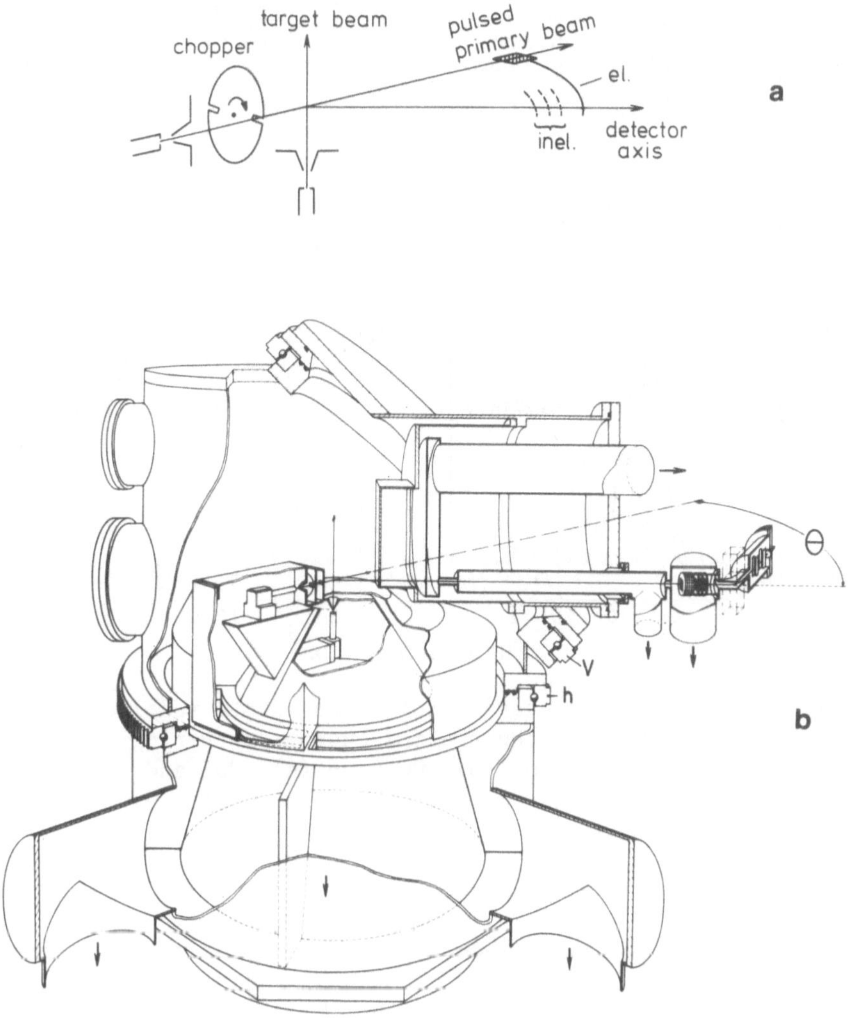

Fig. 5. Crossed molecular beams machine for 1 meV resolution tof-studies of rotationally inelastic collisions. Fig. 5a, in top, shows the principle, Fig. 5b an approximately to scale technical view of this apparatus.

The He-N_2 rotational state to state differential cross sections for the $0 \rightarrow 2$ (o), $1 \rightarrow 3$ (■) and $0 \rightarrow 4$ (Δ) transition in Fig. 7 are derived from a large number of tof-measurements at scattering angles between 5° and 58°. The attractive potential well for He-N_2 is, with a depth of 2 meV, very shallow. Thus, the total differential cross section (•) which is also shown in Fig. 7 is dominated by the characteristic diffraction oscillations for the scat-

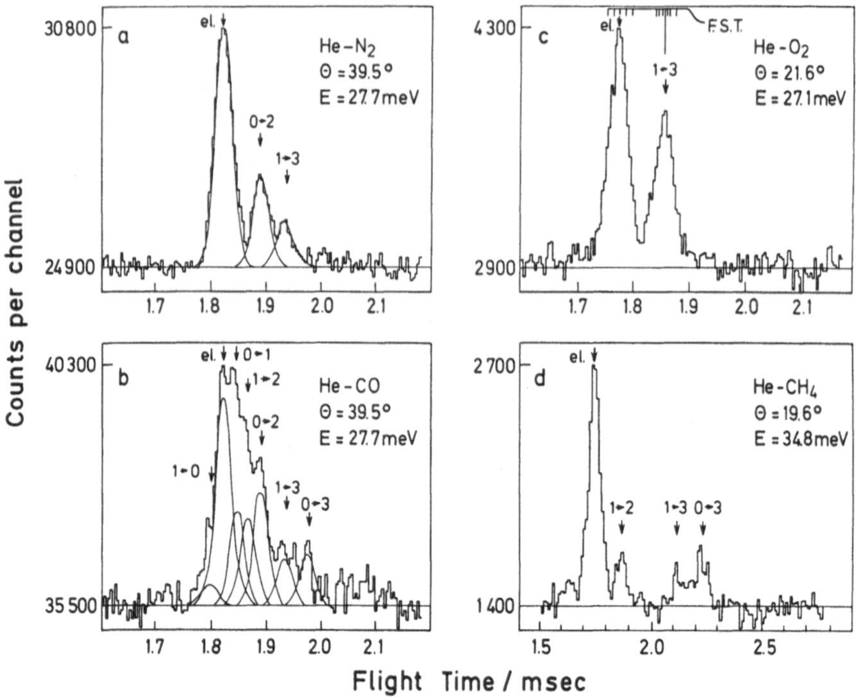

Fig. 6. Four representative rotational state resolved tof -spectra for the thermal energy scattering of He on (a) N_2 , on (b) CO , on (c) O_2 and, on (d) CH_4 . The (fwhm) energy resolution is 0.75 meV .

tering of a wave on a repulsive potential. These oscillations had been treated first for a quantum system by *Massey* and *Mohr* in 1933.[14] Later on, for the purposes of α-particle scattering on medium size nuclei, this *Fraunhofer* model has been generalized for the inelastic scattering on a deformed sphere,[15] and, was recently applied to the present cases of low energy molecular scattering.[16] For a deformed sphere of the shape $R(\alpha,\beta)=R_0+\delta_2P_2(\cos\alpha)$ the model yields the analytical cross section expressions:

$$d\sigma/d\omega\ (\text{elastic}) \simeq (kR_0{}^2)^2 \cdot 2/(\pi x^3) \cdot \cos^2(x+\pi/4) \quad \text{and,}$$

$$d\sigma/d\omega\ (0 \rightarrow 2) \simeq (kR_0)^2 \cdot \delta_2{}^2/(2\pi^2 x) \cdot \sin^2(x+\pi/4)\ , \qquad (3)$$

where x is the reduced scattering angle: $x=k\cdot R_0\cdot\vartheta$. Thus, this model provides an explanation for the, at first, quite surprising observation of a phase shift between the oscillations of the $\Delta j=2$ rotationally inelastic and the predominantly elastic total cross section for He-N_2 in Fig. 7. When fitted to the experimental He-N_2 cross sections this model gives reasonable values of $R_0=3.5$ Å for the average radius and, of $\delta_2=0.65$ Å for the deformation of the interaction potential.[16]

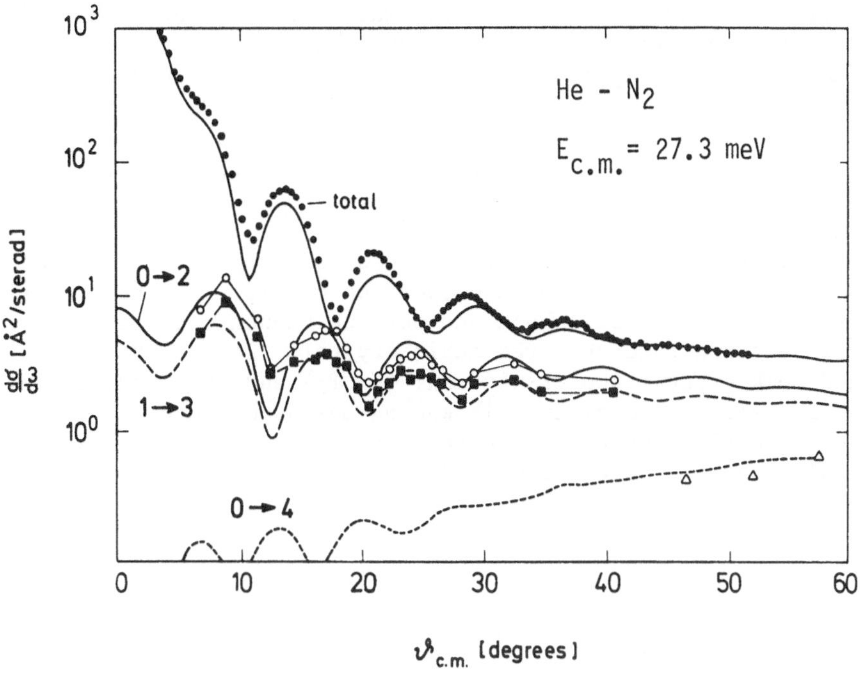

Fig. 7. Experimental, rotational state-to-state differential cross sections for He-N_2 collisions at 27.3 meV collision energy. The smooth curve theoretical results (see text) are in reasonable agreement with the experimental data points.

For the rotational excitation cross sections at the low collision energy of the present experiment, also, exact quantum cross section calculations are feasible. For a semi-'ab initio' He-N_2 potential, based on a SCF calculation of the repulsive wall and on *Hartree-Fock* results for the long range dispersion part, the exact quantum cross sections[17)] are shown as the smooth curves in Fig. 7. The comparison shows a good general agreement between these 'first principle' results and the experiment. As discussed elsewhere[3,17)] the removal of the maximum discrepancies of 80% in the cross section amplitudes requires only very minor changes in the mean radius, the deformation and the average well depth of this potential.

PHONON EXCITATION IN THE SCATTERING OF ATOMS ON CRYSTAL SURFACES

A quite different and, very interesting new application of low energy atom scattering are investigations of inelastic collision processes on well defined single crystal surfaces.[18)] Such studies can be undertaken with very similar experimental techniques as used in the rotationally inelastic scattering experiments when the target beam of Fig. 5 is replaced by a crystal surface which is kept in an ultrahigh vacuum for preventing a rapid coverage of the surface. A surface scattering apparatus is shown in Fig. 8. The differentially pumped, 0.3° wide, chopped He-beam enters the ultrahigh vacuum crystal chamber from the left. The crystal is mounted on a manipulator for aligning the crystal surface orientation with respect to the primary beam and the tof-detector axis.

As in the bulk crystal the phonons on a crystal surface are characterized by their phonon dispersion curves. A schematic illustration of the acoustic bulk and surface phonon dispersion is given in Fig. 9 for a face-centered cubic crystal. In the infinitely extended bulk crystal two transversal (T_1,T_2) and one longitudinal mode L exist for a given bulk phonon momentum vector q . The three dispersion curves for these, respective, bulk modes are schematically shown at the bottom left corner of Fig. 9 for the irreducible portion of the *Brillouin* zone. On a crystal surface the

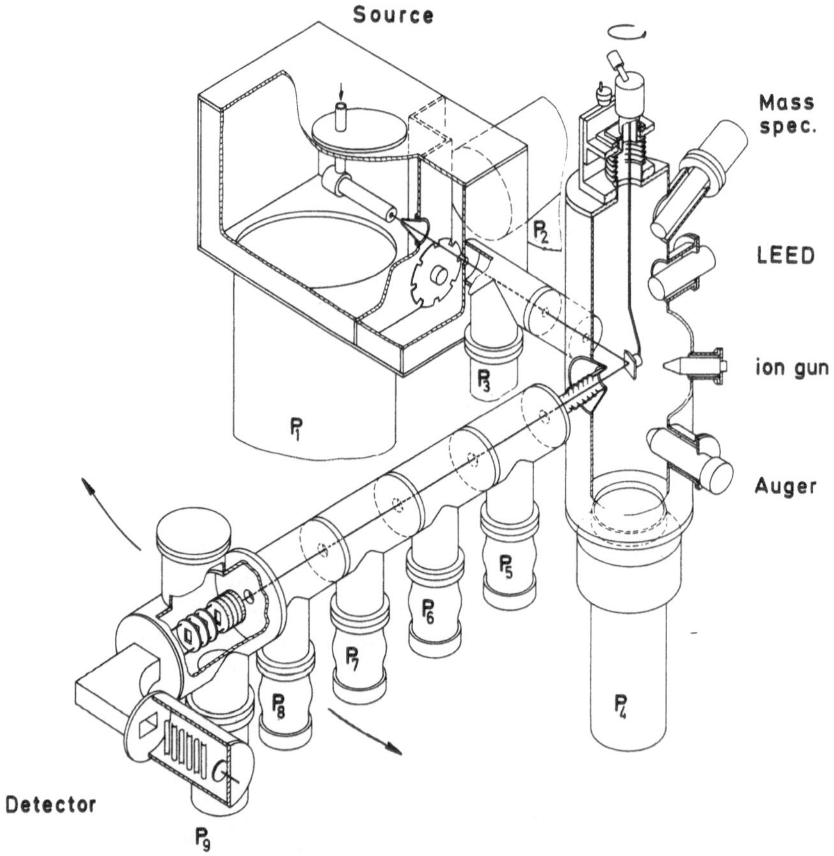

Fig. 8. High resolution He atom-surface scattering apparatus.

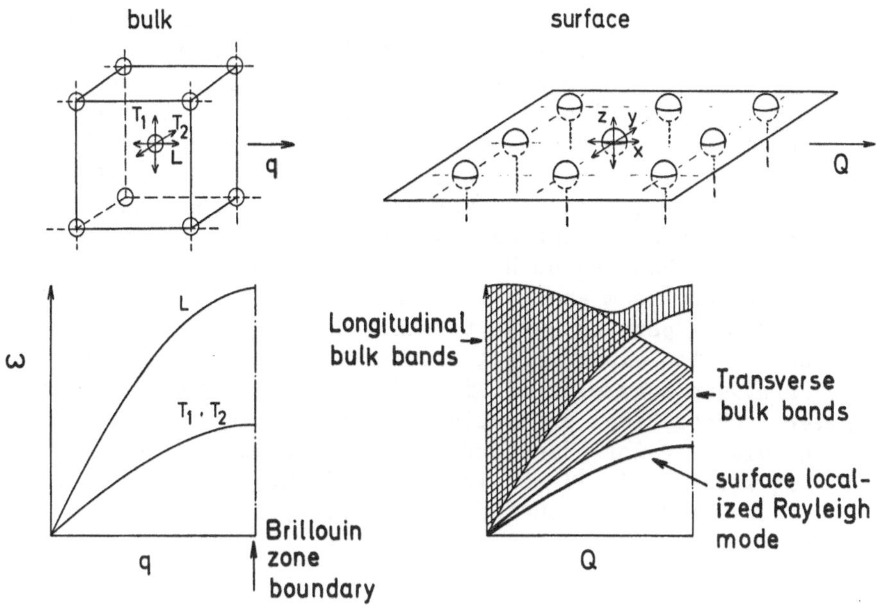

Fig. 9. Bulk and surface phonon dispersion curves for a face-centered cubic crystal.

symmetry of the bulk crystal is broken and the surface phonon dispersion curves with a momentum vector Q parallel to the surface split into the broad band structures shown on the right hand side of Fig. 9. In lattice dynamics calculations[19] it is found that isolated, surface-localized modes exist within, or, separated from the broad bulk phonon bands as, for example, the acoustic surface -localized *Rayleigh* mode shown in Fig. 9. These penetrate only a few Ångstroms deep into the crystal and, thus, can be expected to interact preferentially with a projectile hitting a crystal surface.

Experimentally the energy $\hbar\omega$ and the momentum $\vec{Q}$ of a collision induced surface phonon is derived from the energy loss of the scattered atom and from the change of its momentum component parallel to the crystal surface.[18] Fig. 10a represents one of the first experimental helium tof-spectra showing creation and annihilation of individual *Rayleigh* surface mode phonons on the surface of a LiF single crystal. Here the wave number of the incident He-beam is k_i=6.06 Å^{-1} . The crystallographic <100> orientation of the (001) crystal surface is aligned parallel to the scattering plane. The spectrum is taken at an (azimuthal) angle of incidence Θ_i=64.2° while the angle between the primary beam and the detector is $\Theta_i+\Theta_f$=90° . This angle is slightly off from the first order $(\bar{1},\bar{1})$ elastic surface diffraction maximum. In comparison to the intensity in this nearby $(\bar{1},\bar{1})$ *Bragg* diffraction peak the intensity of the elastically scattered helium has dropped here already by four orders of magnitude. At the right hand side from the elastic peak in Fig. 10a three distinct phonon excitation peaks are observed. A phonon annihilation peak occurs toward shorter flight times. An additional 'quasi elastic' peak has been identified as an experimental artifact resulting from the very weak intensity in the tails of the primary He-beam velocity distribution. Except for the intense *Rayleigh* phonon peaks also some weaker excitation of the broad bulk phonon bands is observed between the creation peaks at -5 meV and at -10 meV. By changing both the angle of beam incidence Θ_i and/or the primary beam energy in different tof-measurements the complete <100> dispersion curve for the surface *Rayleigh* mode was mapped out.[18] The experimental results (o) are shown in the irreducible *Brillouin* zone plot of Fig. 10b. They are compared here with several theoretical predictions (smooth curves) of lattice dynamics calculations for this surface-localized mode. The experiment agrees best with the (-·-) result of Ref. 20 where the relaxation of the lattice bond force on the surface has properly been taken into account.

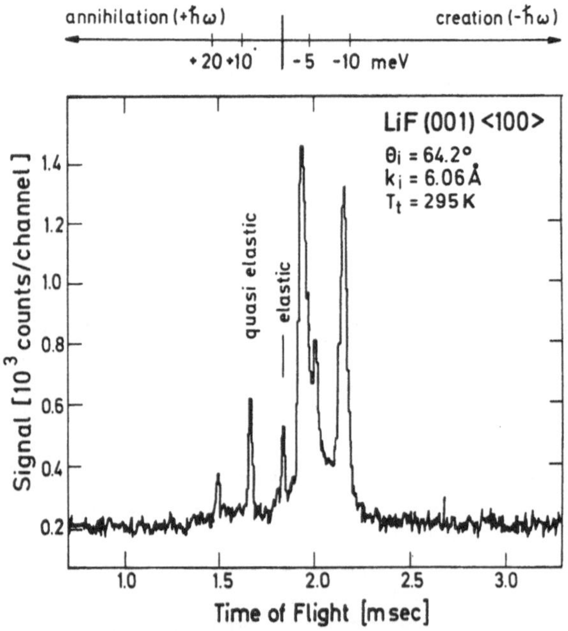

Fig. 10a. tof-spectrum of scattered He showing both excitation and annihilation of surface phonons on an oriented LiF crystal surface.

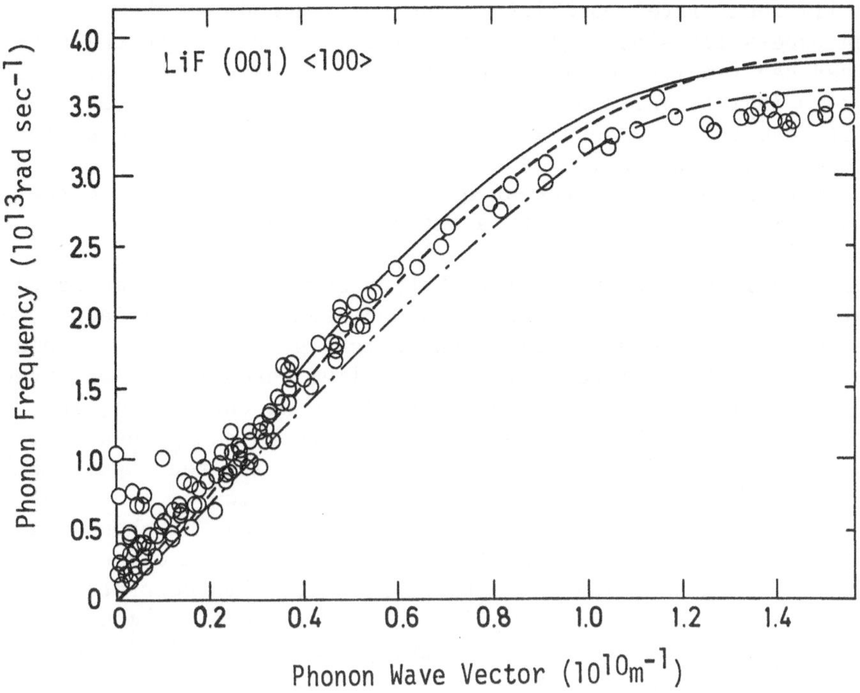

Fig. 10b. Irreducible *Brillouin* zone plot of the experimental *Rayleigh* surface phonon dispersion curve (o) for the <100> direction of a LiF(001) single crystal surface. Smooth lines are theoretical predictions (see text).

While this evaluation of the peak positions in the tof-spectra provides only - hitherto unknown - spectroscopic information on the surface phonon dispersion of LiF and many other crystals, also, a large amount of dynamical information is revealed by such scattering experiments. One process which is of particular interest as a first step in surface catalytic reactions is the phonon-assisted adsorption of a gas atom on a surface. The interaction potential for a helium atom with the (001) surface of LiF has an average well depth of 8 meV and allows for four He-surface bound states with the vibrational energy levels $\varepsilon_0=-5.90$, $\varepsilon_1=-2.46$, $\varepsilon_2=-0.78$ and $\varepsilon_3=-0.21$ meV . Elastic, selective adsorption in one of these surface bound states occurs when the direction of one of the *Bragg* diffraction maxima coincides with the surface plane. For an incident energy E_i and an angle Θ_i the resonance condition for elastic, selective adsorption is $E_i+|\varepsilon_v| = \hbar^2/(2m)\cdot(\vec{K}_i+\vec{G}_{mn})^2$ with $K_i=(2m/\hbar^2 E_i)^{1/2}\cdot\sin\Theta_i$ and with $\vec{G}_{mn}$ being a reciprocal lattice vector. Thus, by choosing the correct incidence angle one specific bound state v can be selectively populated with He-atoms. The atoms will then travel parallel to the surface and, as they still have an excess translational energy component parallel to the surface they can finally be diffracted back into the free state. This process is indistinguishable from a direct scattering process. However, it results in phase delays which lead to constructive or destructive interferences in the elastic *Bragg*-diffraction spots. A second exit channel, of interest here, is the reemission into the continuum by the collision with a surface phonon.

An experimental angular distribution for this phonon-assisted desorption process for He-atoms which had been selectively adsorbed in the $v=0$ vibrational state is shown in Fig. 11a as a polar diagram of the scattered intensity. The maximum intensity is observed at a final angle $\Theta_f=74°$. The tof-spectrum at this angle, in Fig. 11b, shows that here surface phonons with an energy near 10 meV have been annihilated. The two minima in the scattered angular distribution, Fig. 11a, originate from flux competition of off-plane scattering channels. Similarly, the reverse process of phonon-assisted, selective adsorption could be investigated. A closer inspection of the kinematical conditions for selective desorption, with the aid of a *Celli* diagram,[22] shows that isolated "magic" final angles Θ_f exist. At these angles elastically diffracted flux from only one specific surface bound state is observed over a wide range of He final translational energies. Keeping the detector fixed at this magic angle the dependence of the phonon-assisted adsorption on the angle of incidence can be investigated. The experimental result for the phonon-assisted adsorption into the $v=0$ state is shown in Fig. 11c with the polar diagram now representing the adsorption probability as a function of the beam incidence angle Θ_i. The sample tof-spectrum for $\Theta_i=74.5°$ in Fig. 11d shows that the selective adsorption at this angle is assisted by the creation of a phonon with $\hbar\omega \approx 6$ meV. Further details of this experiment are discussed in Ref. 21. It was even possible to identify a scattering process where the He-atom is selectively adsorbed into the $v=2$ surface bound state, collides with one of two possible surface phonons and, jumps by this collision into the $v=0$ state before it finally leaves the surface by elastic diffraction into the magic angle. Thus, one has here a model case for studying surface physisorption processes in great detail.

Except for investigations on several other ionic crystals inelastic atom-surface scattering experiments have recently also been extended to metal surfaces[23] and to semiconductors. Their full theoretical interpretation will still require a considerable amount of computational work.

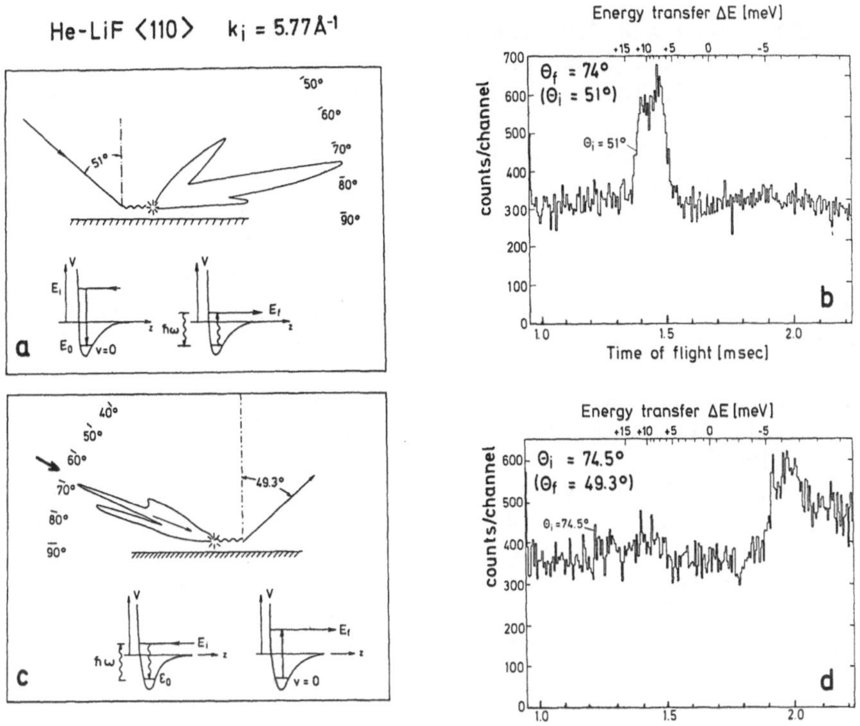

Fig. 11. Experimental angular distributions for the phonon -assisted desorption (a) and adsorption (c) of He atoms from and into surface resonant bound states on a LiF surface. The sample tof-spectra confirm the assistance of surface phonons in the desorption (b) and adsorption (d) processes.

ACKNOWLEDGEMENT

The author is very grateful to his colleagues D. Eichenauer, G. Kraft, M. Noll, J.P. Toennies, and M. Wilde for many stimulating discussions on these topics.

REFERENCES

1. H.S.W. Massey and E.H.S. Burhop, "Electronic and Ionic Impact Phenomena", p. 441, Oxford Univ. Press (Clarendon), London and New York (1952).
2. M. Faubel and J.P. Toennies, Adv.At.Mol.Phys. 13, 229 (1978).
3. M. Faubel, Adv.At.Mol.Phys. 19, 345 (1983).
4. R.B. Bernstein, "Atom-Molecule Collision Theory", Plenum, New York (1979).
5. M. Noll and J.P. Toennies, Chem.Phys.Lett. 108, 297 (1984).
6. U. Gierz, M. Noll, and J.P. Toennies, submitted to J.Chem.Phys..

7. T. Ellenbroek and J.P. Toennies, Chem.Phys. 71, 309 (1982).
8. M. Noll and J.P. Toennies, to be published.
9. U. Gierz, J.P. Toennies, and M. Wilde, Chem.Phys.Lett. 110, 115 (1984).
10. R. Böttner, U. Ross, and J.P. Toennies, J.Chem.Phys. 65, 733 (1976).
11. J.M. Bowman and R. Schinke, Top.Curr.Phys. 33, 61 (1982).
12. D. Beck, in: "Physics of Electronic and Atomic Collisions", S. Datz, ed., North Holland Publ., Amsterdam (1982).
13. M. Faubel, G. Kraft, and J.P. Toennies, to be published.
14. H.S.W. Massey and C.B.O. Mohr, Proc.R.Soc. London, Ser. A141, 434 (1933).
15. J.S. Blair, in: "Nuclear Structure Physics", P.D. Kunz, ed., Vol. VIII C, P. 343, Univ. of Colorado Press, Boulder (1966).
16. M. Faubel, J.Chem.Phys., scheduled for Nov. 15 issue (1984).
17. M. Faubel, K.H. Kohl, J.P. Toennies, K.T. Tang, and Y.Y. Yung, Faraday Discuss.Chem.Soc. 73, 205 (1982).
18. G. Brusdeylins, R.B. Doak, and J.P. Toennies, Phys.Rev. B27, 3662 (1983).
19. R.E. Allen, G.P. Alldrege, and F.W. de Wette, Phys.Rev. B4, 1648 (1971).
20. G. Benedek, Phys.Status Solidi, B58, 661 (1973).
21. G. Lilienkamp and J.P. Toennies, J.Chem.Phys. 78, 5210 (1983).
22. V. Celli, in: "Dynamics of Gas Surface Interaction", G. Benedek and U. Valbusa, ed., p. 2, Springer Verl., Berlin (1982).
23. R.B. Doak, U. Harten, and J.P. Toennies, Phys.Rev.Lett. 51, 578 (1983).

ELECTRON CORRELATION EFFECTS IN

ELECTRON-ATOM COLLISIONS

H.G.M. Heideman

Fysisch Laboratorium
Rijksuniversiteit Utrecht
3584 CC Utrecht, The Netherlands

INTRODUCTION

What do we mean when we are talking about electron correlation effects and when do they occur? Speaking of electron-atom collisions electron correlation effects may be defined as effects, which result from the detailed interaction between the projectile electron and one or more of the individual atomic electrons. So they are in fact due to departures from the independent-electron model in which each electron is considered to move in the combined field of the atomic nucleus and the average charge distribution of the other electrons. The importance of correlation effects is mainly determined by the collision energy. If the velocity of the incident electron is large (i.e. much larger than the velocity of the outer-shell atomic electrons), the target atom experiences a rapidly varying electric field and may still be in the process of evolving to an excited state when the scattered electron is already far away. So there is no time for correlations to be developed between the projectile and one or more of the atomic electrons.

The situation changes appreciably when the velocity of the incident electron becomes of the same order of magnitude as the velocity of the outer-shell atomic electrons. As long as the incident electron is at finite distance (of the order of a few diameters of the target atom) the target is polarised and, in turn, distorts the incoming plane wave. At this point the interaction between target and projectile may still be represented by some static multipole field. This is no longer the case when the electron comes very close to the atom or even penetrates into the atomic electron cloud. If now an appreciable part of the projectile's kinetic energy is transferred to one of the target electrons, a negative-ion complex is formed with two slowly moving excited electrons (either in discrete or in continuum states) outside a positive ionic core. In this situation the detailed interaction between these two excited electrons becomes of great importance. Contrary to the case where all atomic electrons are in closed shells and pinned down to their positions by the Pauli principle, the two excited electrons may now easily adjust their motions to slight changes in the acting forces; i.e. their mutual repulsion and the net attraction by the ionic core, which are comparable in size. This is so because there is a manifold of (nearly) degenerate orbitals available for the two excited electrons. As a result they may become strongly correlated both in angle (due to the availability of (nearly) degenerate orbitals of different orbital angular momentum) and in distance (due to the availability of closely spaced orbitals of (slightly) different excitation energy. - It is important to note that this stage of the electron-atom collision process cannot be distinguished from the situation where the projectile electron had been initially combined with the target atom and the excitation energy had been contributed by a photon or some other external source. There is therefore a great similarity between low energy inelastic electron-atom collisions and two-electron excitations induced by photon absorption. In fact no separate theoretical study of these two processes is required (Fano 1983). - In the final stage of the collision process the doubly excited complex may break up and evolve to the final state, which is detected in our apparatus; for instance to a singly excited atom plus a scattered electron or to a singly charged ion plus two free electrons. It is now to be expected that the (detected) final state will reflect the correlations developed during the existence of the doubly excited complex. So the study of electron correlation effects in electron-atom collisions in fact boils down to the study of the above described two-electron excitations (three-electron excitations if three electrons are involved, etc.).

For an extensive review on the theoretical aspects of such studies we refer to a recent article by Fano (1983). In this paper we will mainly concentrate on the experimental evidence for the occurrence of electron correlation effects.

THE ROLE OF ELECTRON CORRELATIONS IN NEAR THRESHOLD IONISATION AND EXCITATION

More than 30 years ago Wannier (1953) emphasized that the near threshold behaviour of the ionisation of atoms by electrons is largely determined by the occurrence and decay of correlations in the motion of the two escaping electrons. These correlations result from the long-range Coulomb forces between the two slowly receding particles, which may interact during a fairly long time. The important point emphasized by Wannier is that near threshold actual ionisation can only occur if during the escape process the distances of the two electrons to the residual ion remain approximately equal. This is called the radial correlation and this radial correlation must persist up to a distance at least equal to the distance where the potential energy of the two electrons has dropped below their kinetic energy (the "Wannier radius"). The radial correlation is highly unstable as can readily be seen in terms of what Rau (1971) called a dynamical screening effect. If one of the electrons lags behind, the attractive field of the residual ion is less effectively screened for the slower electron than for the faster one, resulting in a further deceleration of this slower electron. So the initial unbalance is enhanced with time and the end of the story may be that the slower electron looses so much energy that it gets captured by the residual ion, so that excitation has taken place rather than ionisation. The effect is that the instability of the radial correlation suppresses the ionisation probability near threshold, the more so when the excess energy is reduced. In the theories of Wannier (1953), Rau (1971) and Peterkop (1971) it is assumed that it is precisely this effect that determines the energy behaviour of the ionisation cross section near threshold and that this behaviour is independent of the initial preparation of the doubly excited complex. On the basis of this assumption Wannier derived the following dependence for the ionisation cross section σ on the excess energy E above the ionisation threshold: $\sigma \sim E^{1.127}$. A theory, in which the electrons are assumed to move uncorrelated, would give a linear threshold law.

The region in configuration space, where the distances r_1 and r_2 of the two electrons with respect to the residual ion are approximately equal, is called the Wannier region. In the Wannier region the potential energy of the two electrons has a maximum as a function of the ratio r_1/r_2 with deep valleys at either side. This maximum is called the Wannier ridge and plays a crucial role in the theoretical description of the correlation effects under discussion.

Another manifestation of the electron-electron correlation results from the repulsive Coulomb interaction between the two electrons, which forces them to move in nearly opposite directions. Contrary to the radial correlation this angular correlation is stable. It has been argued by Fano (1974) that as a result of the angular correlation the two electrons may acquire very large orbital angular momenta, even though the total arbital angular momentum is small. This may be understood by considering the θ_{12}-dependence of the two-electron wave function (θ_{12} = angle between the two electron directions) to be expanded in spherical harmonics Y_ℓ^m. High-order harmonics (which correspond to high ℓ-values) are required to represent a sharply peaked function.

Apart from its effect on the near-threshold ionisation the above described correlation mechanism may also influence the near threshold excitation of discrete atomic states (Fano 1974). If the radial correlation between the escaping electron pair breaks down within a critical radius (Wannier radius) the slower electron will eventually fall back into a bound state and excitation has taken place instead of ionisation. This possibility of a seemingly ionising collision to end up in the excitation channel may be viewed as an indirect excitation mechanism for bound states, that may cause structures on the excitation curves of these states. Fano (1980) has considered the occurrence of quasi-standing wave patterns on the Wannier ridge formed when the wave-packet, representing the two-electron system, propagates up the ridge and becomes reflected at the classical turning point (assuming the collision energy is below the ionisation energy). Due to the instability of the radial correlation these standing waves would then readily diverge into the excitation channel giving rise to resonances there at energies near to and below the ionisation threshold.

For the same reason as actual ionisation just above threshold can only occur if the radial correlation persists up to at least the Wannier radius, high-lying Rydberg states can only be excited near their thresholds if the radial correlation persists up to distances of the order of their radii. If the radial correlation breaks down in an early stage, when the atomic electron is still close to the residual core, it will, as a result of the dynamic screening effect, be pushed back into a low-lying state. In order to reach a high-lying state, far away from the core, the two electrons, atomic and projectile, must keep themselves mutually in balance by equally screening for each other the attractive field of the ionic core until a distance has been reached about equal to the radius of the highly-excited state concerned. So the excitation of states lying near the maximum available energy is suppressed when this energy approaches the ionisation threshold from below in much the same way as the ionisation probability is reduced when the energy approaches the threshold from above. Moreover, as argued by Fano (1974), it is to be expected that, due to the stability of the angular correlation, there will be a large probability for the excitation of states with large orbital angular momenta at collision energies near the ionisation threshold.

HYPERSPHERICAL COORDINATES

The dynamics of the correlated motion of the two escaping electrons is most conveniently described when the cartesian coordinates $\underline{r}_1$ and $\underline{r}_2$ are replaced by hyperspherical coordinates R, α, and θ_{12}.

$$R = (r_1^2 + r_2^2)^{\frac{1}{2}}$$

$$\alpha = \tan^{-1}(r_2/r_1)$$

$$\theta_{12} = \cos^{-1}(\hat{r}_1 . \hat{r}_2)$$

The hyperradius R measures the "size" of the two-electron state, while the angle α measures the radial correlation of the two electrons. Note that when $\alpha = \pi/4$, $r_1 = r_2$; when $\alpha = 0$ or $\pi/2$, one of the electrons is at much larger distance from the nucleus than the other.

In terms of these coordinates double escape of the two electrons near the ionisation threshold does only occur for values of α restricted to a narrow cone around $\alpha = \pi/4$, this cone being more and more expanded as the excess energy above threshold increases.

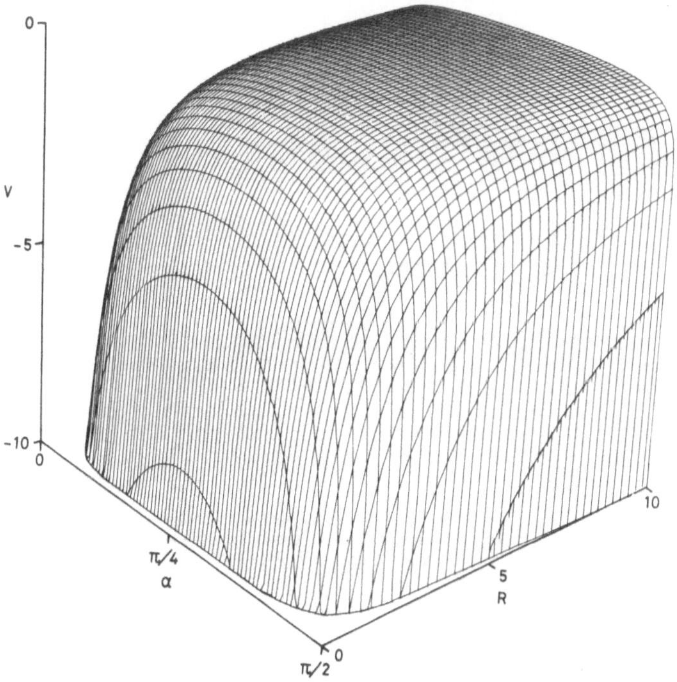

Fig. 1. Three-dimensional view of the potential surface $V(R, \alpha, \theta_{12})$ as a function of the hyperspherical coordinates R and α. The mutual angle θ_{12} is fixed at $\theta_{12} = \pi$.

In cartesian coordinates the Coulomb potential of the two-electron system in the field of the ionic core is (using au):

$$V(\underline{r}_1,\underline{r}_2) = \frac{1}{r_1} + \frac{1}{r_2} - \frac{1}{\underline{r}_1 - \underline{r}_2}$$

In hyperspherical coordinates:

$$V(R,\alpha,\theta_{12}) = \frac{1}{R}\left[\frac{1}{\cos\alpha} + \frac{1}{\sin\alpha} - \frac{1}{(1 - \sin 2\alpha\ \cos\theta_{12})^{\frac{1}{2}}}\right]$$

A 3-dimensional plot of this potential surface provides a nice illustration of the discussed correlation effects (see also Lin 1974). In fig. 1 the potential is shown as a function R and α for $\theta_{12} = \pi$. The unstable radial correlation is reflected by the ridge at $\alpha = \pi/4$ with deep valleys at either side. At zero residual energy a two-electron orbit, which diverges away from the ridge, inevitably disappears in one of the valleys, which means that one of the electrons is captured by the residual ion. For finite energies the

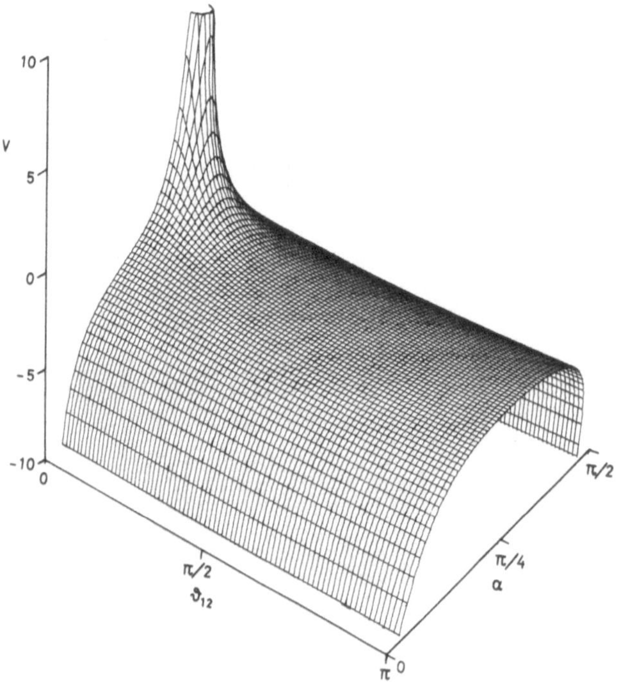

Fig. 2. Three-dimensional view of the potential surface $V(R, \alpha, \theta_{12})$ as a function of the hyperspherical coordinates α and θ_{12}. The radius R is fixed at R = 1 au.

range of orbits leading to double escape is expanded: slightly diverging orbits may, due to their finite energy, still reach infinity.

In fig. 2 the potential is shown as a function of α and θ_{12} for a fixed value of R (1 au). One again sees the instability in the coordinate α, whereas the dependence on θ_{12} has a minimum for $\theta_{12} = \pi$ with high peaks at $\theta_{12} = 0$ and 2π (where the two electrons are close to each other). This minimum reflects the stable angular correlation. So on its way to infinity an electron orbit is more and more confined to a narrow cone around $\theta_{12} = \pi$ anabling the two electrons to attain large individual orbital angular momenta, even when the total orbital angular momentum of the two-electron system is zero (see also Fano (1974)).

EXPERIMENTAL RESULTS

The assumption by Wannier that the threshold behaviour of the ionisation cross section is determined by the instability of the radial correlation between the two escaping electrons was for the first time verified by Cvejanović and Read (1974). These authors directed an electron beam of adjustable velocity through helium gas and detected the yield of very low-energy (< 50 meV) scattered electrons when sweeping the incident energy across the ionisation threshold. The result is shown in fig. 3. The discrete peaks below the ionisation energy (unresolved for $n > 9$) are due to the excitation of bound states of helium very close to their thresholds. The smoothly varying yield above the ionisation threshold represents the ionisation continuum. The interesting feature is the cusp-like structure centered around the ionisation threshold. This structure very nicely reflects the unstable radial correlation. In section 2 we have seen that, in order to reach ionisation or high-Rydberg excitation near threshold, the motion of the projectile and excited electron must remain tightly correlated up to large distance from the ionic core. However, due to the instability of the radial correlation the probability for that becomes smaller and smaller as the energy approaches the ionisation threshold from above or from below, respectively. This is precisely what is observed in figure 3.

Using a time-of-flight coincidence technique Cvejanović and Read also showed that the energy distribution of the two escaping electrons is uniform (as predicted by the Wannier theory) in the energy range from 0.2 to 0.8 eV above threshold (the two limits being determined by the experimental set-up) and that up to 3 eV above the threshold the width of the distribution of the angle θ_{12} between the two outgoing electrons varies with the excess energy E in a way, which is consistent with a $E^{\frac{1}{4}}$ dependence (also predicted by the Wannier theory).

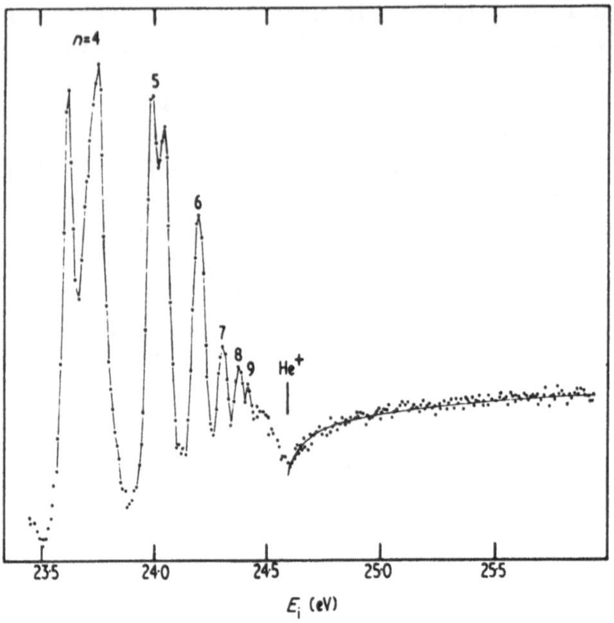

Fig. 3. The measured yield of very low energy (≤ 50 meV) electrons as a function of the incident energy resulting from electron-helium collisions. The curve drawn through the points above the ionisation threshold is proportional to $E^{0.127}$, where E is the energy excess above the ionisation energy (Cvejanović and Read 1974).

If the energy distribution of the two escaping electrons is uniform near threshold the yield of the low-energy electrons above the ionisation threshold in fig. 3 must be proportional to the first derivative of the ionisation cross section with respect to the excess energy. Thus by fitting a power function to the measured data in fig. 3 the exponent n in the threshold law $\sigma \sim E^n$ for the ionisation cross section can be determined. Cvejanović and Read found $n = 1.131 \pm 0.019$, a convincing verification of the value $n = 1.127$ predicted by Wannier.

Measurements on energy and angular distributions of ionisation electrons have later also been performed by Pichou et al. (1978) and by Walker et al. (1982). These authors also found that these distributions were fairly uniform in restricted energy ranges above the ionisation threshold.

As was pointed out in section 2 the correlation mechanism may also affect the near-threshold excitation of bound states. The decay of the radial correlation at rather large distance or the occurrence of standing waves on the Wannier ridge diverging into the excitation channel may be viewed as indirect excitation mechanisms, which may cause structures on the excitation curves concerned. Because of the stable angular correlation, which relates to the occurrence of large orbital angular momenta, such a mechanism is expected to influence in particular states with high orbital angular momenta. More specifically it has been argued by Fano (1974) that the above mechanism is the predominant mechanism for the near-threshold excitation of high-L Rydberg states.

The first experimental evidence for the occurrence of such

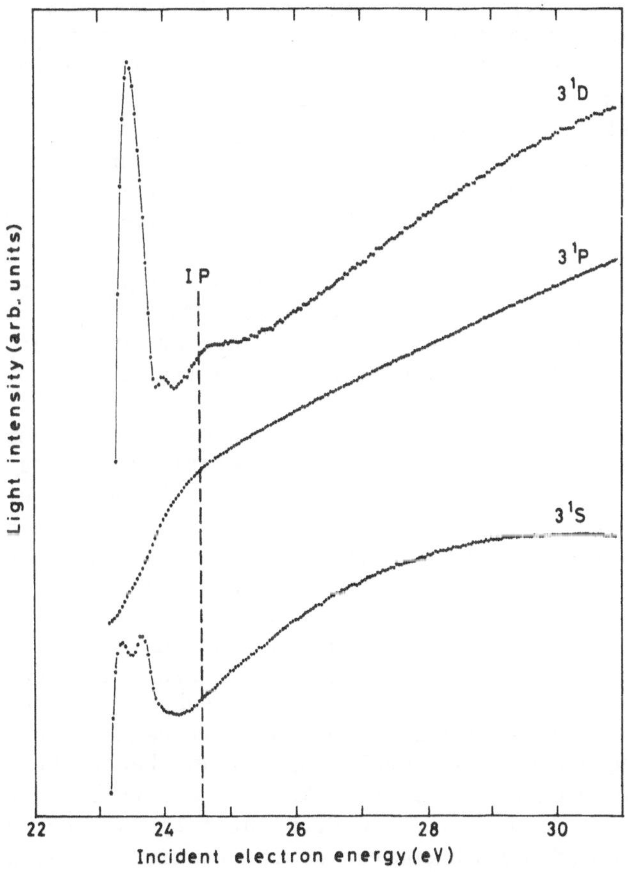

Fig. 4. Optical excitation functions of the 3^1S, 3^1P and 3^1D states of helium from threshold to about 8 eV above. The first data point of each curve lies at about 0.2 eV above the excitation threshold. The dashed vertical line indicates the ionisation threshold.

effects was presented by our laboratory (Heideman et al. 1976). Fig. 4 shows our measured optical excitation functions of the 3^1S, 3^1P and 3^1D states of helium at incident-electron energies between 0.2 eV above threshold and 31 eV. The prominent structures just above threshold in the excitation of the 3^1S and 3^1D states are caused by negative-ion resonances, whose configurations are probably of the type ($1s4\ell4\ell'$). More interesting for the present study, however, is the direct vicinity of the ionisation threshold (24.59 eV). The 3^1S and 3^1P curves appear to vary rather smoothly with energy there (although the 3^1P curve exhibits an apparent change of slope). In the 3^1D curve, however, a relatively broad structure can be observed, which starts below the ionisation threshold and extends up to about 1 eV above. It is very likely that this structure has exactly the same origin as the cusp-like dip structure observed in the Cvejanović and Read experiment (see fig. 3): namely, the instability of the radial correlation between the projectile and atomic electron. This instability causes a great deal of orbits, which originally were propagating closely along the Wannier ridge and thus were heading for ionisation or high-Rydberg excitation, to diverge away from the ridge and to end up in an excitation channel where one electron is finally captured in a lower excited state and the other reaches infinity. So to the loss of flux from the ionisation and high-Rydberg channels there corresponds an increase of flux in the lower bound state channels. This also explains the reversed shape of the structure around 24.6 eV in the 3^1D curve as compared to the corresponding structure in fig. 3.

Detailed evidence for the occurrence of electron correlation effects in the excitation channel has recently been obtained by the Manchester group (Buckman et al. 1983). Fig. 5 shows their measured yield of metastables (2^3S and 2^1S) produced in electron-helium collisions in the energy range from 22 to 25 eV. As can be seen from the figure the measured curve is dominated by narrow resonances. Around the ionisation threshold a global cusp-like structure is observed, which is very similar in shape as the structure we observed at the same energy in the optical excitation curve of the 3^1D state (see fig. 3). There is no doubt that this structure is caused by the same mechanism as the one that caused the 3^1D structure. The narrow resonances near to, and below, the ionisation threshold have been analysed by Buckman et al. (1983) with help of a modified Rydberg formula and are attributed by them to the formation of quasi-standing waves on the Wannier ridge, as predicted by Fano (1980) (See one but last paragraph of section 2). That these structures are not seen in the curves shown in fig. 3 is probably due to insufficient energy resolution of the electron beam.

So the results of figs. 4 and 5 tend to indicate that the occurrence and decay of correlations between the projectile and excited electron may have two different effects on the near-

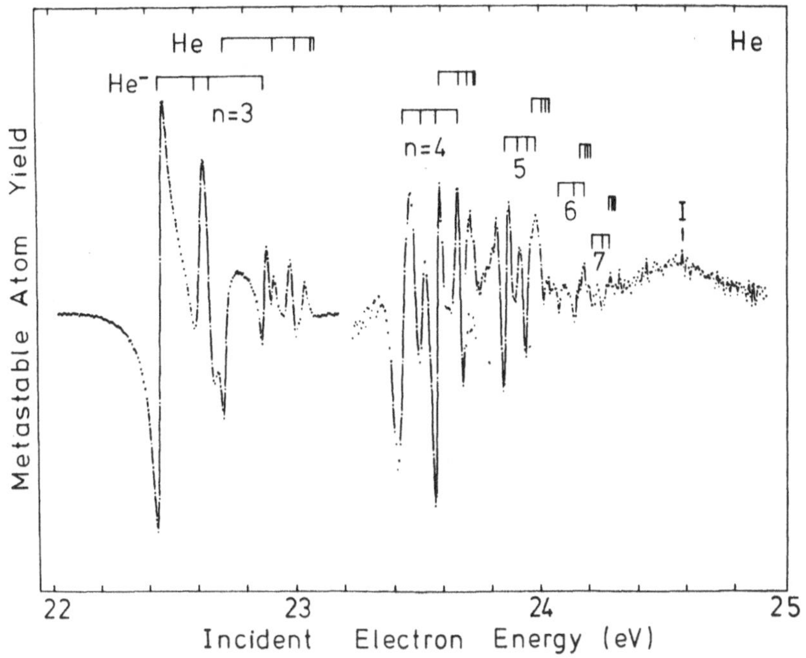

Fig. 5. Yield of metastable helium atoms resulting from electron impact taken over four restricted energy regions. The normalisation of the four regions is arbitrary (Buckman et al. 1983).

threshold excitation of bound states; firstly, the leakage of flux from the ionisation and high-Rydberg excitation channel into the lower excitation channels (due to the instability of the radial correlation) may give rise to relatively broad cusp-like structures centered around the ionisation threshold, and secondly, the formation of quasi-standing waves on the Wannier ridge, which readily diverge into the valleys (see fig. 1) and thus into the excitation channel, may cause narrow resonances just below the ionisation threshold. Fano's suggestion that the first mechanism may be the dominant mechanism for the excitation of high-L Rydberg states is supported by experiments of Tarr et al. (1980) and Hammond et al. (1981) who directly demonstrated the enhanced threshold excitation probability of high-L states in molecules.

A few years ago van de Water and Heideman (1982) performed near-threshold excitation measurements on krypton. Of particular

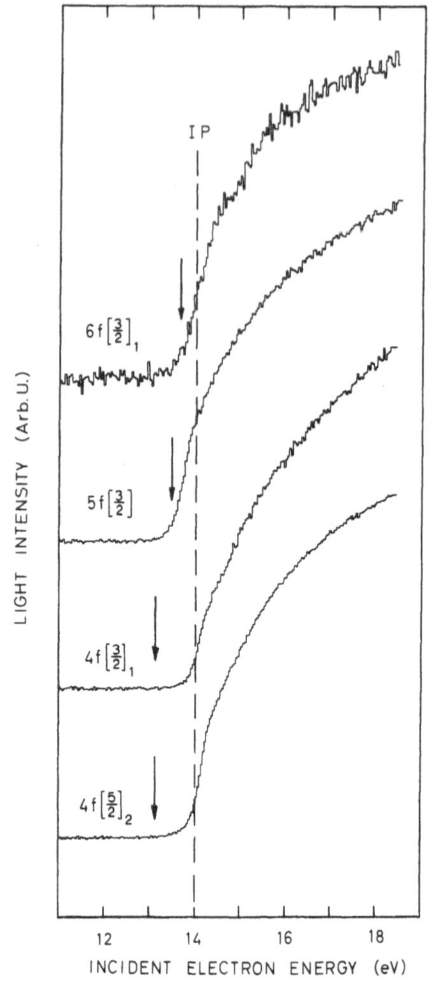

Fig. 6. Measured optical excitation curves of 4, 5 and 6f states in krypton in the energy range from threshold to 18.5 eV. The 5f[3/2] curve represents a mixture of the excitation curves of the $5f[3/2]_1$ and $5f[3/2]_2$ states. The broken line indicates the ionisation threshold.

interest are their measured optical excitation functions of the f-states, which are shown in fig. 6. The arrows in the figure indicate the spectroscopically known threshold energies of the states. It looks like the excitation functions of the 4f states have their onsets at about 13.8 eV rather than at 13.14 eV, which is the spectroscopically known threshold energy of the 4f states. The 5f and 6f curves do not show this anomalous threshold behaviour. Within the uncertainty in the incident-electron energy the onsets of these curves coincide with the spectroscopically known threshold energies of these states. The 4f anomaly can not be explained by energy calibration errors since all four curves in fig. 5 were separately calibrated against the onset of the 437.6 nm transition ($6p[1/2]_0 \rightarrow 5s[3/2]_1$). Apparently the excitation cross sections for the 4f states under consideration are very small, or even zero, in the energy range from threshold to about 0.7 eV above.

A possible explanation for the anomalous threshold behaviour of the 4f states in krypton is the following. Let us assume (as suggested by Fano (1974)) that higher-L states can only be excited near their thresholds via a correlated motion of the projectile and one of the atomic electrons, which enables them to exchange the required amount of orbital angular momentum. The electronic configuration of the ground state of krypton is $(1s^2\ 2s^2\ \ldots.\ 4s^2\ 4p^6)^1S$, sothat the 4f excitation remains within the ground state n = 4 shell. On the other hand the excitation of a 4f state in krypton requires an orbital angular momentum exchange between the projectile and excited electron of 2 au. However, within the n = 4 shell this is virtually impossible because no (nearly) degenerate orbitals of different ℓ are available. The 4s, 4p, 4d and 4f orbitals are widely separated in energy, mainly because of the Pauli-exclusion principle. Thus excitation of the 4f states via the correlation mechanism is impossible. As to the excitation of higher f states such as 5f, 6f, etc., closely spaced 5s, 5p, 5d and 5g (or 6s, 6p, ... 6h) orbitals are available, sothat the excited electron can effectively exchange orbital angular momentum with the projectile, facilitating 5f and 6f excitation via the correlation mechanism.

Finally we will discuss the possible influence of electron-electron correlations on the polarisation or angular distribution of the impact radiation. Because the long-range electron correlations cause effective angular momentum exchanges it is to be expected that they also affect the relative population of the different magnetic substates and this should be reflected in the polarisation of the emitted impact radiation. As the behaviour of the polarisation near threshold has presented an unsolved problem for more than half a century (Skinner and Appleyard 1927) we will discuss it in some detail.

Theory provides clear predictions for the polarisation in two limiting cases: at high energy where the born approximation is valid and at the threshold where the polarisation is fixed by conservation of angular momentum. Interpolation between the results, valid in the two limits, would seem plausible, but leads to predictions, which are in most cases in disagreement with the experimental results. In the cases of helium and mercury the polarisation of all spectral lines studied exhibits a large dip, which extends from threshold to

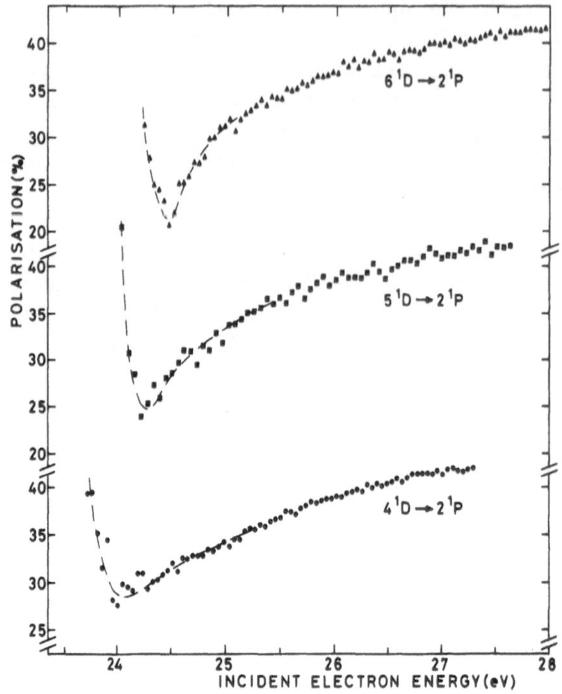

Fig. 7. Measured polarisation near threshold of three $n^1D \rightarrow 2^1P$ transitions in helium as a function of the incident electron energy. The dip in the polarisation curve becomes deeper with increase of the principal quantum number n.

a few eV above. Fig. 7 shows our measured polarisation curves for the $n^1D \rightarrow 2^1P$ transitions in helium with n = 4, 5 and 6. In all three curves the threshold dip is clearly present. Within the last few tenths of an eV the polarisation rises very steeply when the energy is lowered towards the excitation threshold. The theoretical threshold value of 60% is not reached due to the finite energy resolution of the electron beam.

The anomalous threshold dip is not observed in the cases of the resonance lines of lithium (Hafner and Kleinpoppen 1967), sodium (Enemark and Gallagher 1972) and calcium (Ehlers and Gallagher 1973). For these lines the polarisation rises monotonically to the expected threshold value when the energy is lowered towards the excitation threshold. Fig. 8 shows the results of Gould (1970) and of Enemark and Gallagher (1972) for the $3p \rightarrow 3s$ resonance transition in sodium. Also for the $(4s4p)^1P_1 \rightarrow (4s^2)^1S_0$ transition in calcium the measured polarisation rises very clearly to the theoretical value at threshold, although in this case a narrow structure is

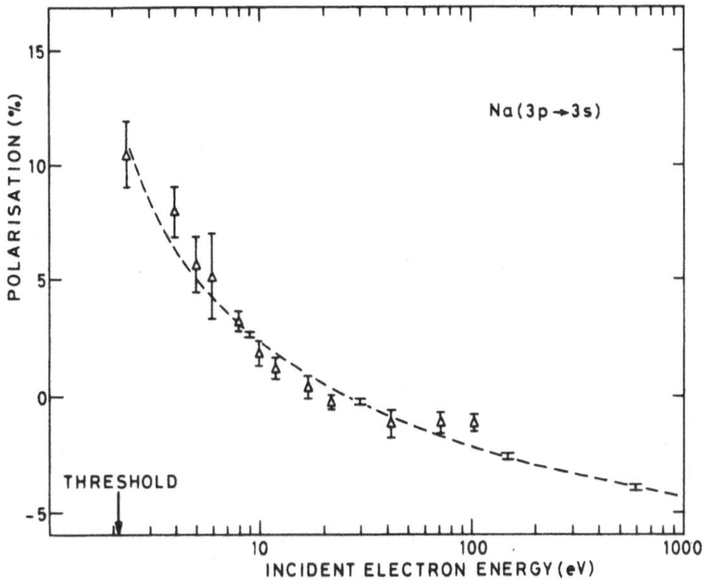

Fig. 8. Measured polarisation of the 3p → 3s transition in sodium as a function of the incident electron energy. Δ : Gould (1970); - -I- -: Enemark and Gallagher (1972).

observed near 4.5 eV (Ehlers and Gallagher 1973). However, contrary to the broad dip in the case of helium and mercury, this structure is only a few tenths of an eV wide and can be explained in a straightforward manner by resonance and cascade effects.

Previous attempts to explain the threshold anomaly in the polarisation of helium and mercury lines in terms of negative-ion resonances or secondary processes such as cascade or collisional transfer of excitation energy were not satisfactory in that they were unable to account for the whole effect, in particular for that part of the dip extending beyond the ionisation threshold where, in the case of helium, no discrete negative-ion resonances occur.

We have suggested (Heideman et al. 1980) to explain the polarisation anomaly in terms of the electron correlation effects which we are presently discussing and which are effective in a limited energy range around the ionisation threshold, precisely where the anomaly occurs. As already mentioned, at the excitation threshold the polarisation can be calculated exactly as the scattered electron then has zero orbital angular momentum and consequently (in the case of helium) only m = 0 substates can be excited. However, a little above threshold the stable angular correlation between the projectile and excited electron may cause the scattered electron to acquire a significant orbital angular momentum. For n = 4, 5 and 6 several closely spaced orbitals of different ℓ are available to facilitate effective orbital angular momentum exchanges. Now if the scattered electron despite of its low velocity can carry away an orbital angular momentum unequal to zero, the excited atom is not necessarily left in a m = 0 magnetic substate. So even though the impact energy is only slightly above the excitation threshold the probability for excitation of $m \neq 0$ substates may become quite appreciable resulting in a decrease of the polarisation of the emitted radiation. It is to be expected that the effect becomes more and more important for the excitation of states with larger principal quantum numbers (larger radii). Namely, the excitation of these states in the threshold region requires that the radial correlation persists up to larger distances, so that the angular correlation can be more effective. This is precisely what is observed in fig. 6. The anomalous dip in the polarisation curves seems to be come deeper with increase of the principal quantum number of the upper state of the transition concerned. Additional evidence for such a trend is provided by the polarisation measurements on the $3^1D \rightarrow 2^1P$ transition by Heddle et al. (1977).

Now the question remains to be answered why the polarisation anomaly is not observed in the polarisation curves of the first alkali and alkaline-earth resonance lines. The answer is similar as in the case of the anomalous behaviour of the 4f states in krypton discussed above. The excitation of the lowest resonance levels of alkali and alkaline-earth atoms remains within the ground state shell (n = 2, 3 and 4 for Li, Na and Ca, respectively). Thus effective orbital angular momentum exchanges between projectile and excited electron are hindered because no closely spaced orbitals with different ℓ are available. In other words the mechanism as discussed above for n^1D excitation in helium does not work here. But what about the higher lying states of the alkali atoms? Polarisation measurements by Hafner and Kleinpoppen (1967) on the $3^2D \rightarrow 2^2P$ transition in lithium show that the polarisation anomaly is present there again, as expected.

ACKNOWLEDGEMENT

I gratefully mention the fruitful discussions with Professor U. Fano and Dr. S. Watanabe.

REFERENCES

Buckman, S.J., Hammond, P., Read, F.H. and King, G.C., 1983, J. Phys. B: At. Mol. Phys. 16:4039.

Cvejanovic, S. and Read, F.H., 1974, J. Phys. B: at. Mol. Phys. 7:1841.

Ehlers, V.J. and Gallagher, A., 1973, Phys. Rev. A 7:1573.
Enemark, E.A. and Gallagher, A., 1972, Phys. Rev. A 6:192.
Fano, U., 1974, J. Phys. B: At. Mol. Phys. 7:L401.
Fano, U., 1980a, Phys. Rev. A 22:2660.
Fano, U., 1980b, J. Phys. B: At. Mol. Phys. 13:L519.
Fano, U., 1983, Rep. Prog. Phys. 46:97.
Hafner, H. and Kleinpoppen, H., 1967, Z. Phys. 198:315.
Hammond, P., King, G.C. and Read, F.H., 1981, in "Proc. 12th Int. Conf. on Physics of Electronic and Atomic Collisions", Gatlinburg, Abstracts, p.142.
Heddle, D.W.O., Keesing, R.G.W. and Parking, A., 1977, Proc. R. Soc. A 352:419.
Heideman, H.G.M., van de Water, W., Nienhuis, G. and Peeters, P.H., 1976, J. Phys. B: At. Mol. Phys. 9:L523.
Heideman, H.G.M., van de Water, W. and van Moergestel, L.J.M., 1980, J. Phys. B: At. Mol. Phys. 13:2801.
Lin, C.D., 1974, Phys. Rev. A 10:1986.
Peterkop, R., 1971, J. Phys. B: At. Mol. Phys. 4:513.
Pichou, R., Huetz, A., Joyez, G. and Landau, M., 1978, J. Phys. B: At. Mol. Phys. 11:3683.
Rau, A.R.P., 1971, Phys. Rev. A 4:207.
Skinner, H.W.B. and Appleyard, E.T.S., 1927, Proc. R. Soc. A 117:224.
Tarr, S.M., Schiavone, J.A. and Freund, R.S., 1980, Phys. Rev. Lett. 44:1660.
Wannier, G.H., 1953, Phys. Rev. 90:817.
van de Water, W. and Heideman, H.G.M., 1982, J. Phys. B: At. Mol. Phys. 15:2285.

THE ROLE OF ELECTRON SPIN IN ATOMIC COLLISIONS

Wolfgang Jitschin

Fakultät für Physik, Universität Bielefeld

D-4800 Bielefeld 1

1. INTRODUCTION

Experimental investigations of ion-atom collisions have evolved from "simple" total cross section measurements to studies of increasing complexity and sophistication. The detailed information obtained by advanced experiments provides a basis for an improved understanding of the various interaction mechanisms in a collision. One ultimate aim of these investigations might be a so-called perfect experiment in which the states of all collision partners are completely known prior to and after the collision (Bederson 1969, 1970). Experiments which employ an analysis of states and an identification of particle trajectories have been performed (see, e.g., Andersen and Nielsen 1982). For a perfect experiment it is necessary to specify all intrinsic properties of the collision partners, including the spin states. Investigations which explicitely study the conservation or the change of spin during the collision may reveal novel and otherwise unaccessible information about the collisional interaction. Corresponding experiments may be called spin-experiments.

In the field of electron-atom collisions a variety of spin-experiments have been performed during the past years; these experiments have become feasible due to significant progress in the development of sources and detectors for polarized electrons (Kessler 1984). The preparation and analysis of electronic and atomic spins have proved to be the experimental tool for separating various reaction channels, i.e. direct, spin-flip, and exchange channel. In the field of ion-atom collisions the collisional interaction is in general rather different from the interaction in electron-atom collisions; nevertheless, basic ideas of the spin-experiments in the study of electron-atom collisions can be transferred to the field of ion-atom collisions.

In the present paper, first some objectives of spin-experiments are pointed out, then the behaviour of spins during a collision is discussed. In the remaining part of the paper a few selected experiments are presented.

2. OBJECTIVES OF SPIN-EXPERIMENTS

Consider the collision of two atoms (ions) which are in the states A and B before the collision; after the collision the atomic states may have changed to A' and B' (Fig. 1). If the state preparation and analysis include spin properties it may become possible to derive information about the spin as well as about coherences between spin and orbital angular momentum. The spin is characterized by its magnitude and its projection on a quantizing axis (polarization). One can now raise the following question: What physical information can be obtained from an observed conservation or change of the spins? Two aspects of this basic question will now be discussed in more detail with suggestions for corresponding experiments.

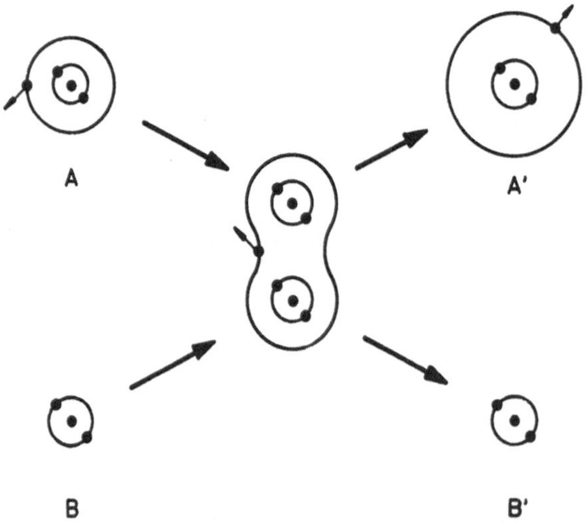

Fig. 1: Scheme of a collision; small arrows indicate spin direction of the valence electron.

2.1 Total spin information

The total spin is defined as vectorial sum of the spins of the two atoms. It is conserved in a collision if any coupling between spin-dependent and orbital angular momenta during the collision can be neglected

$$\vec{S}(A) + \vec{S}(B) = \vec{S}(A') + \vec{S}(B') \quad (1)$$

This spin-conservation rule (Wigner 1927) is frequently cited for the conservation of the magnitude of the total spin. As an example, consider a charge exchange experiment performed with polarized collision partners:

$$Ne^{+}(2p^{5}\uparrow) + Na(3s\uparrow) \rightarrow Ne(2p^{6}) + Na^{+} \quad (2)$$

An arrow indicates the polarization of a particle; experimental techniques for polarizing both collision partners are established (Hils et al 1981, Ganz et al 1982). The final state of the collision system is a singlet, and if Wigner's rule is fulfilled the initial state also has to be a singlet. Thus in case of impinging atoms having parallel spin direction, i.e. a triplet state, Wigner's rule forbids the reaction. In experiments as the one described the spin may provide a possibility to block reaction channels.

2.2 Information from the spins of individual electrons

The spin orientation (polarization) of an individual electron may be employed as a label for a specific electron if there are no forces affecting the spin during a collision. By giving a specific spin-orientation to an electron it becomes possible to trace individual reaction channels.

Consider for example the collision of a quasi-one-electron atom with a rare gas atom A in which the former one is excited:

$$Li(2s\uparrow) + A \rightarrow Li^{*}(2p\uparrow) + A \quad (3)$$

At a first glance one would expect that the excitation occurs by "direct" excitation of the Li 2s electron to the excited 2p state. However, experimental data obtained for different rare gases suggest that, depending on the rare gas species a "molecular" excitation mechanism may also be important in which the excited Li 2p electron originates from the rare gas atom (Menner et al 1981). In order to clarify the excitation process one may employ spin-properties: experimentally one can prepare an initially polarized Li 2s electron and perform a spin-analysis of the Li 2p electron after the collision. For the direct excitation mechanism one expects spin conservation leading to a polarized 2p state, whereas for excitation by electron transfer from the rare gas core one would expect no polarization of the 2p state. Thus the spin-polarization of the 2p state allows to distinguish between the two excitation mechanisms as has been proposed (for the target atom A = Na) by Bartschat et al (1983).

3. SPIN COUPLING DURING A COLLISION, SINGLE ELECTRON CASE

Now we turn to the behaviour of spin during a collision; for simplicity we concentrate on (quasi-)one-electron systems and neglect external magnetic fields. The electron spin is inherently connected with its magnetic moment. Thus magnetic fields, e.g. produced by orbital angular momentum, may change the spin orientation. The question is what happens with the spin during a collision. Let us first consider a colliding system at different internuclear separations.

At <u>large separations</u> the colliding partners behave as free atoms. If the orbital angular momentum of a (quasi-)one-electron atom is nonvanishing, the electron spin is affected by the magnetic field associated with the orbital momentum. This <u>spin-orbit interaction</u> has an energy $\hbar\omega$ and leads to the formation of eigenstates of the total angular momentum J. Since the vector J is space-fixed the spin-orbit interaction causes a precession of L and S in space.

At <u>smaller internuclear separations</u> the other collision partner perturbs the spherical symmetry of the atom. The <u>electrostatic interaction</u> ΔV may be strong enough to fix the orbital motion L of the electron to the internuclear axis (sticking). The spin S then precesses around L, i.e. the internuclear axis.

At <u>rather small internuclear distances</u> the internuclear axis may rotate in space faster than the classical electron orbital motion. If the <u>rotational</u> energy $h\dot{\phi}$ is sufficiently large the electron orbital motion will (at least partially) decouple from the internuclear axis.

As follows from the above discussion, different interactions are dominant at different internuclear distances. The angular momentum coupling for the various interaction cases can be conveniently described by the appropriate case of the well known Hund's coupling schemes (Masnou-Seeuws and McCarroll 1974). The coupling schemes are merely a basis for describing the states; a transformation between different bases can be performed by angular momentum algebra (Aquilanti and Grossi 1980).

The dynamics of a collision process determines the behaviour of the system when it moves through the different coupling cases. For small collision velocities the spin will always be locked to the magnetic field caused by the electron orbital motion and thus the spin will change adiabatically during the collision. In the regime of intermediate velocities the spin behaviour has to be calculated by a quantum-mechanical treatment (Nikitin 1965, Masnou-Seeuws and McCarroll 1974). At higher collision velocities the collision process may become sudden: the comparatively weak forces acting on the spin are not sufficient to cause a significant spin change during the short collision time. Thus the spin properties immediately before and after the collision are the same.

Guided by the spin behaviour at high collision velocities one may put the hypothesis that the electron spin is decoupled in a collision and thus its orientation (polarization) is conserved. This hypothesis was applied to electron-atom collisions by Percival and Seaton (1958), where it was confirmed in numerous experiments. For ion-atom collisions the spin-forces are of the same order of magnitude as for electron-atom collisions. However, for ion impact experiments at considerably lower velocities than for electron impact are feasible due to the larger ion mass. Apparently the validity of the sudden approximation for the field of ion-atom collisions has to be investigated in corresponding studies. Some examples will now be discussed.

4. EXPERIMENTS WITH UNPOLARIZED COLLISION PARTNERS

The angular momentum coupling in a collision and the complications arising in the interpretation of experimental data are illustrated at two examples. Tolk et al (1976) studied the excitation of the rare gas Ne by Na^+ ions:

$$Na^+ + Ne \rightarrow Na^+ + Ne^*(2p^5 3p) \qquad (4)$$

Initially both collision partners are in a singlet state. The final Na^+ ions are not detected but are expected to remain in the singlet ground state. The final excited Ne^* $2p^5 3p$ state consists of 10 multiplet levels corresponding to different coupling cases of orbital angular momentum and spin of the 2p hole and the 3p electron. Experimentally the excitation cross sections of all 10 levels were measured. In order to explain the data the authors suggested the following excitation mechanism (Fig. 2): At small internuclear distances two 2p electrons of Ne are strongly promoted via the 4fσ molecular state. At about 1.6 Å the 4fσ orbital crosses the excited $^1\Pi(\Omega=1)$ and $^3\Pi(\Omega=2)$ states which are eigenstates of the total spin and the projection of orbital angular momentum on the internuclear axis. At the crossings one electron may be transferred to these excited states. On the outgoing part of the trajectory at rather large internuclear distances the electrostatic interaction vanishes and the molecular states are subject to a recoupling to eigenstates of the free atom (unfortunately the 10 Ne $2p^5 3p$ multiplet states may not be described in a simple angular momentum coupling scheme but only in an intermediate coupling scheme (Schectman et al 1973)). The transition from molecular to atomic states was assumed to be instanteous. This proposed simple excitation mechanism yields a reasonable description of the experimental data. The finite excitation cross section for triplet states (although the initial state is singlet) has been attributed to a violation of Wigner's rule. However, a more rigorous analysis of the Na^++Ne collision system revealed that the excitation of Ne triplet states is presumably due to a simultaneous excitation of Na^+ ions to non-singlet states. Accordingly the observed excitation of the Ne triplet states is no indication for a violation of Wigner's rule (Olsen et al 1979).

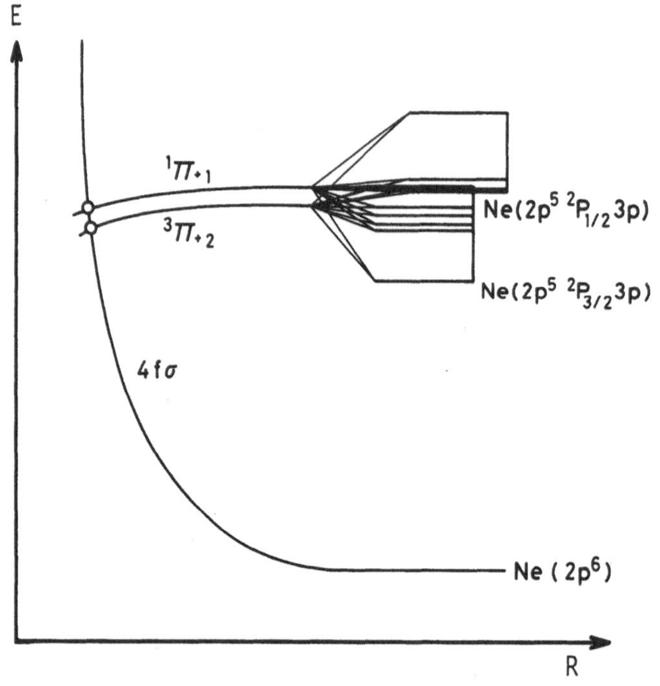

Fig. 2: Simplified correlation diagram showing the energies of some Ne states in a Na+Ne collision.

(Quasi-)one-electron systems are popular candidates for studies of collision dynamics (confer, e.g., Andersen and Nielsen 1982). For the collision system Na+Ne, comprehensive experimental information as well as detailed <u>ab initio</u> calculations are available (see, e.g., Courbin-Gaussorgues et al 1979 and references given therein). An interesting feature has been observed experimentally by Mecklenbrauck et al (1977): these authors studied the total excitation cross sections of the Na $3P_{3/2}$ and $3P_{1/2}$ levels which are separated by as less as 2 meV. They found that the ratio of the two cross sections at about 40 eV E_{CM} impact energy differs from the ratio for a statistical level population (Fig. 3). A non-statistical population is incompatible with a complete spin-decoupling during the collision, i.e. the Percival and Seaton hypothesis. This result demonstrates that the transition from molecular states to atomic states is not sudden at small collision velocities (Fig. 4).

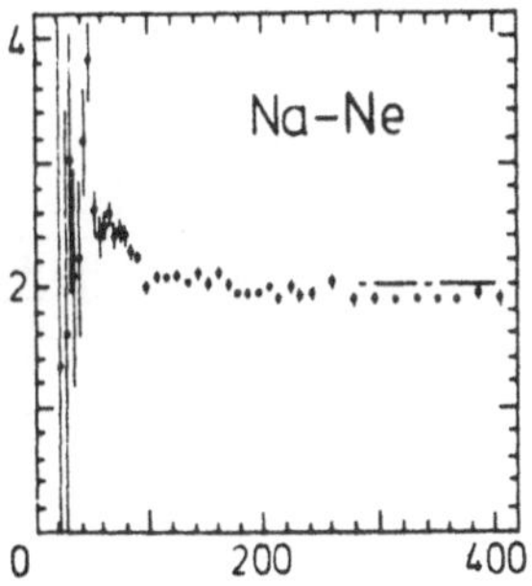

Fig. 3: Cross section ratio for emission of the Na $3\ ^2P_{3/2} \rightarrow 3\ ^2S_{1/2}$ and $3\ ^2P_{1/2} \rightarrow 3\ ^2S_{1/2}$ resonance lines in Na+Ne collisions vs. CM collision energy (in eV) (from Mecklenbrauck et al 1977).

5. EXPERIMENTS WITH POLARIZED COLLISION PARTNERS

In the field of energetic ion-atom collisions experiments employing polarized collision partners are just beginning. Recently Jitschin et al (1984) and Osimitsch et al (1984) studied the 3p excitation of spin-polarized Na atoms either by electrons or by Ne^+ ions.

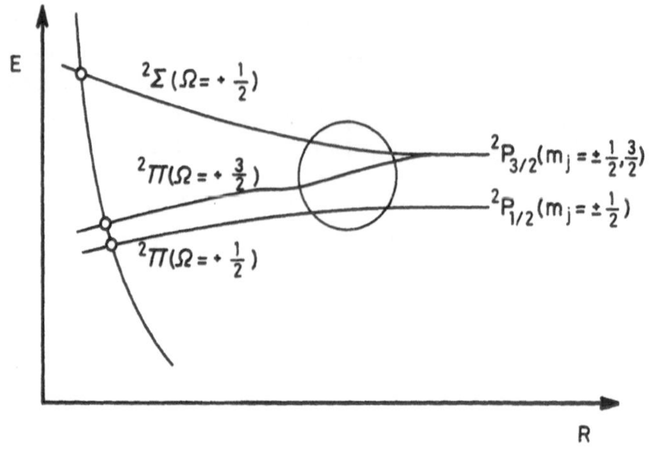

Fig. 4: Schematic diagram of the adiabatic potential curves with indicated diabatic transitions for an alkali atom-rare gas collision system (based on a diagram given by Nikitin 1965).

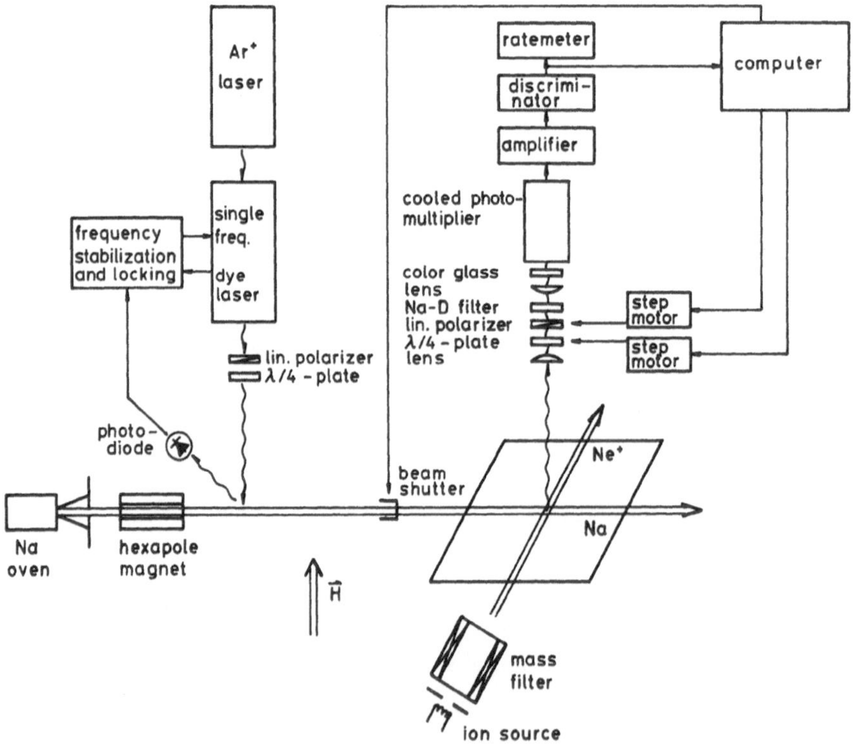

Fig. 5: Experimental setup for studying the excitation of polarized Na atoms by electron and Ne^+ ion impact.

$$Ne^+ + Na(3s\uparrow) \rightarrow Ne^+ + Na(3p\uparrow) \qquad (5)$$
$$\downarrow$$
$$Na(3s) + h\omega$$

Fig. 5 shows the experimental setup. The Na atoms are spin-polarized by state selection in a hexapole magnet resulting in a spin-polarization of about 20% for the Na 3s electron. The spin-polarization of the excited Na 3p state can be derived from the polarization properties of the induced fluorescence light. In experiments employing polarized collision partners and performing a polarization analysis of the fluorescence light it is convenient to use the following geometrical arrangement (Fig. 6):

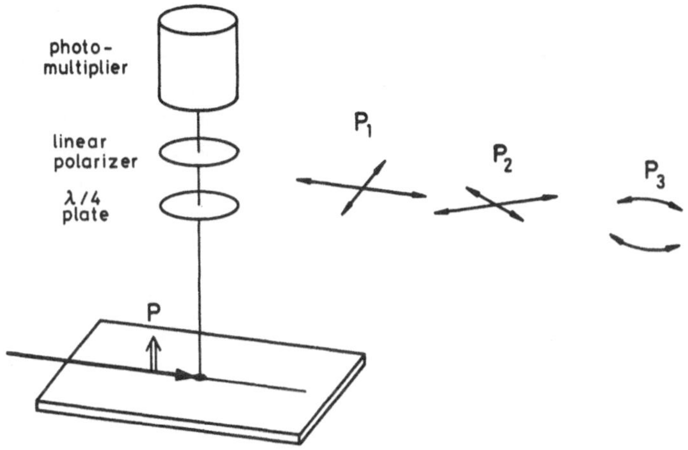

Fig. 6: Collision and light detection geometry for experiments employing transversally polarized beams.

- the spin polarization P is transversal,
- the fluorescence light is detected in the direction of the spin.

For this arrangement the different polarization properties of the induced fluorescence light can be simply related to physical quantities: The Stokes polarization parameters P (McMaster 1954) or, in a different notation, η (Blum and Kleinpoppen 1979) deliver the following information in a non-coincident experiment (Bartschat and Blum 1982):

- The linear polarization $P_1 = \eta_3 = [I(0^o)-I(90^o)]/I$ with respect to an axis parallel to the beam axis is the standard linear polarization which reflects a non-statistical population of m_l magnetic substates of the excited state, i.e. a collisionally induced anisotropy (alignment).

- The linear polarization $P_2 = \eta_1 = [I(45^\circ)-I(135^\circ)]/I$ with respect to an axis at 45° to the beam axis must vanish in a non-coincidence experiment for symmetry reasons except if the scattering symmetry is broken by the spin. A finite P_2 value directly reflects a "left-right" scattering asymmetry due to interaction of spin and projectile orbital angular momentum <u>during</u> the collision.

- The circular polarization $P_3 = \eta_2 = [I(RHC)-I(LHC)]/I$ reflects an orientation of the excited state. In case of a spin-orbit coupled state the orientation results from the spin orientation (polarization) and thus P_3 is a measure for the spin-polarization of the excited electron.

Let us now first discuss experimental results for <u>electron impact</u> (Fig. 7). P_1 reflects the collisional induced alignment and is similar to the results one obtains for unpolarized collisional partners. Measured P_2 values are compatible with zero giving no evidence for a spin-orbit interaction during the collision as expected for light atoms (Bransden and McDowell 1978). P_3 reflects the spin-polarization of the 3p state. The relationship between measured light polarization P_3 and spin-polarization of the 3p electron is rather complex due to relaxation into hyperfine levels after the collision (Jitschin et al 1984). We note that for high impact velocities excitation theories predict excitation by direct interaction only, in which case the excited 3p electron has the same polarization as the initial 3s electron. For impact energies close to threshold the excitation cross sections for direct interaction and for exchange interaction between incoming electron and atomic electron are almost equal resulting in zero polarization of the excited 3p electron. The predictions for the circular polarization P_3 at high impact velocity and at threshold are in reasonable agreement with the experimental data (Fig. 7).

The same apparatus has been used to investigate the Na 3p excitation by <u>Ne^+ ion impact</u> (Fig. 8, Jitschin et al 1985). In the investigated energy range E_{lab} = 0.2 to 6 keV the measured absolute cross section increases monotonically from 1.5 to 15 $Å^2$. The rather large cross section shows that excitation occurs at comparatively large internuclear distance, and accordingly is mainly a one-electron process involving the outer Na 3s electron. We suppose that this exitation occurs by a two-step process (Figs. 9,10): at rather large internuclear distance the Na 3s electron is transferred from the 3sσ state to the 4pσ state which essentially is an excited Ne $2p^5 3s$ state (quasi-resonant Demkov-Meyerhof charge transfer; Demkov 1964, Meyerhof 1973); at smaller distances rotational coupling to the 3pπ state occurs which correlates to the excited Na 3p final state. The light polarization P_1 observed in the experiment indicates a preferential population of states with magnetic quantum number $m_l = \pm 1$ in accordance with the proposed excitation mechanism. The observation of vanishing P_2 values

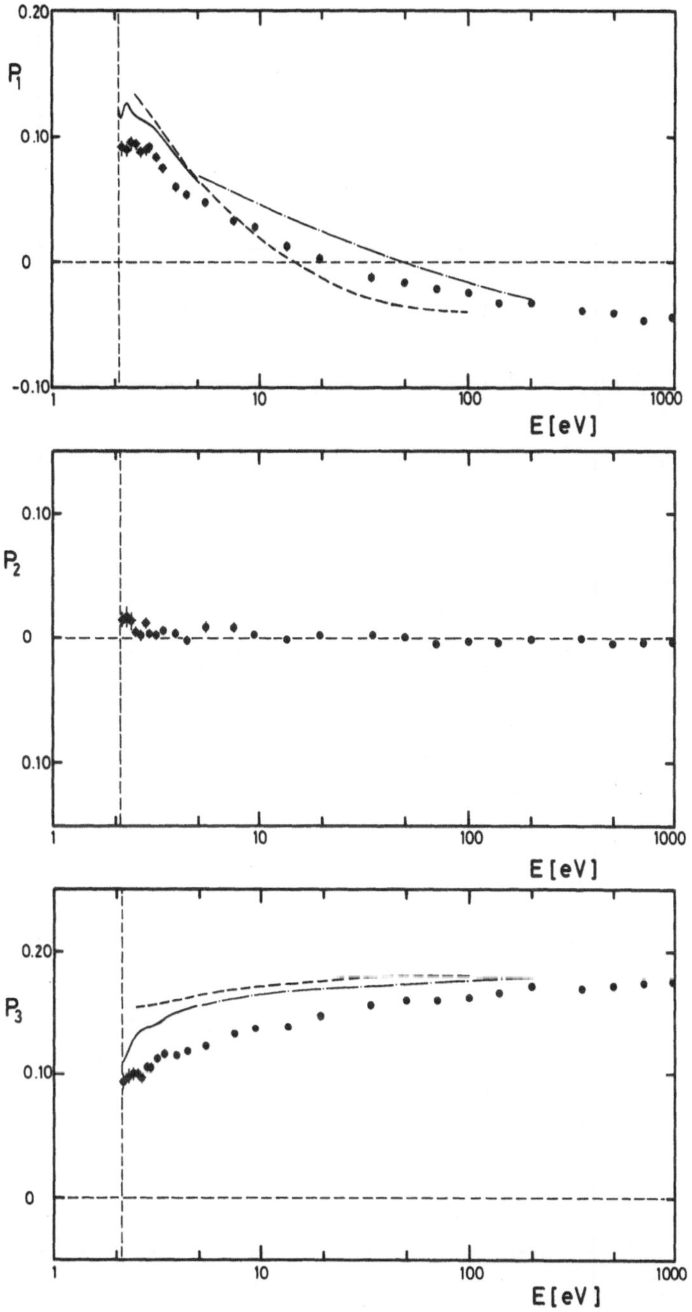

Fig. 7: Measured Stokes polarization parameters of the D line of polarized Na atoms (P≈20%) induced by electron impact. Error bars correspond to counting statistics and do not take possible instrumental effects into account (from Jitschin et al 1984).

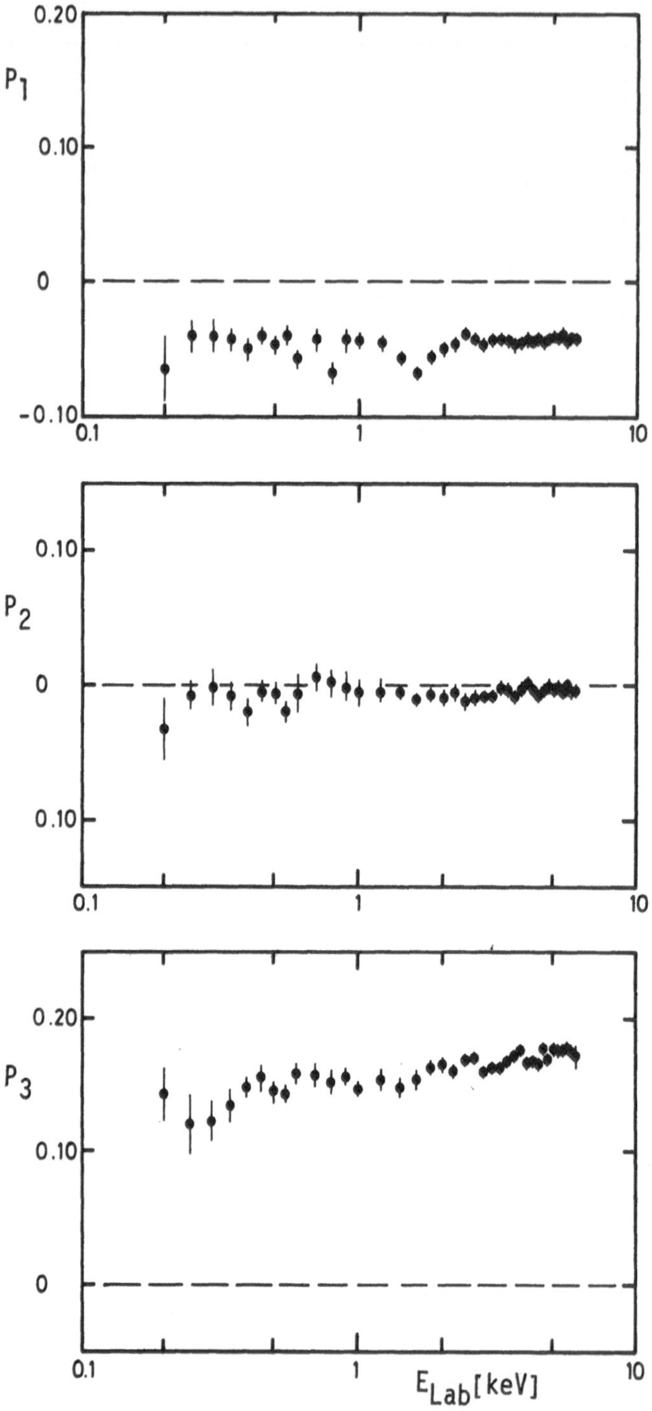

Fig. 8: As Fig. 7, but for Ne^+ ion impact (from Osimitsch et al 1984).

corresponds to a negligible left-right scattering asymmetry. This result is not surprising since for ion-atom collisions one expects only small interaction between target spin and projectile orbit as compared to the dominant electrostatic interaction and thus no scattering asymmetry. Interestingly enough P_3 shows a similar energy dependence as for electron impact: At higher impact energies the observed P_3 is compatible with spin conservation in the collision. The decrease of P_3 at smaller impact energies, however, indicates a non-conservation of spin in the collision.

In order to explain the observed spin non-conservation we suggest the following mechanism. During the collision the outer electron is for some time in the 4pσ state which evolves to the $2p^5$ 3s configuration of the free Ne atom (Fig. 9). In this configuration the 3s electron is subjected to two spin-affecting interactions with the $2p^5$ electrons (or, equivalently, with the $2p^{-1}$ hole): firstly, the electrostatic interaction causes an exchange of electrons, and, secondly, the magnetic interaction of the 3s spin with the magnetic moment of the $2p^{-1}$ hole causes a spin precession. Both interactions are estimated to be of comparable magnitude, say ΔE. A measure for the strength of the resulting depolarization is the ratio of interaction time to the collision time:

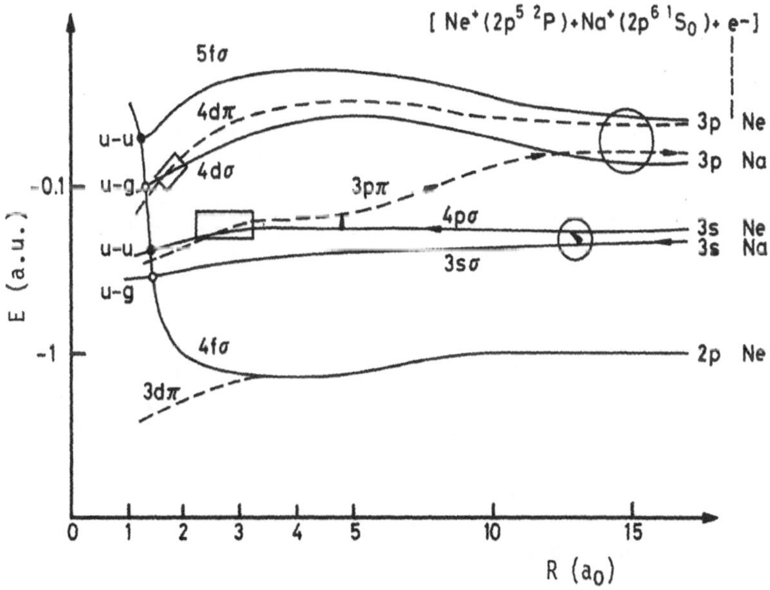

Fig. 9: Schematic correlation diagram for the $[NaNe]^+$ system. Circles indicate radial couplings, and rectangles rotational couplings (based on the diagrams given by Olsen et al 1979).

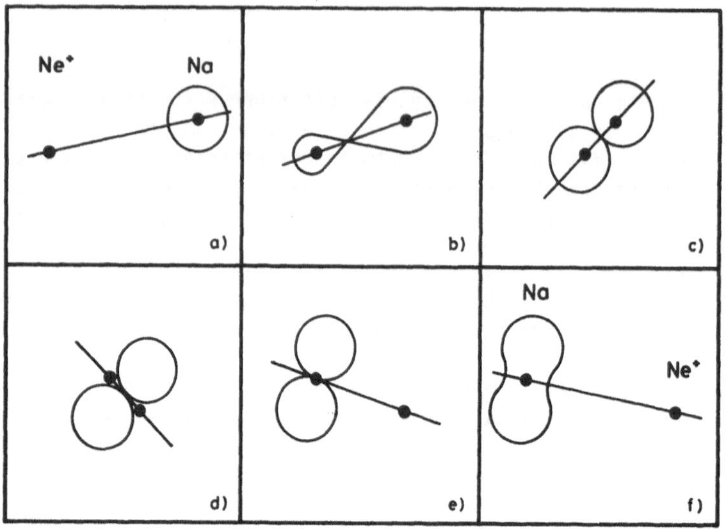

Fig. 10: Time evolution of the valence electron state in a collision. Pictures a) to f) correspond to various times from before to after the collision; in f) fine-structure coupling after the collision is indicated.

$$t_{int} / t_{coll} = (2\pi\hbar/\Delta E) / (s/v) \qquad (6)$$

For a rough estimate of the depolarization effect one may take the interaction energies from the level splitting of the $2p^5\,3s$ configuration of a free Ne atom. Furthermore, assume $v = 6*10^{-2}$ a.u. (corresponding to 200 eV_{Lab} = 100 eV_{CM} collision energy), and s = 10 a.u. (compare Fig. 9) one finds for t_{int}/t_{coll} a value of 3. Although this value has been obtained by making crude assumptions, it confirms the effectiveness of the proposed mechanism for spin depolarization.

The process of non-negligible spin precession within the collision time has recently been suggested to explain observed incoherences in superelastic Na^++Na scattering (Bähring et al 1983). For a detailed understanding of the spin behaviour in collisions it seems necessary to perform comprehensive theoretical studies on the excitation mechanisms which include a treatment of the spin. It is expected that experimental results will stimulate corresponding studies.

ACKNOWLEDGEMENTS

The author is indebted to Prof. H. Kleinpoppen, Prof. H.O. Lutz, Prof. U. Fano, Dr. D. W. Mueller, and Dr. K. Blum for numerous clarifying and stimulating discussions. The work has been supported by the Deutsche Forschungsgemeinschaft (Sonderforschungsbereich 216).

REFERENCES

Andersen, N., and Nielsen, S. E., 1982, Adv. At. Mol. Phys. 18, 265-308.

Aquilanti, V., and Grossi, G., 1980 J. Chem. Phys. 73, 1165-1172.

Bähring, A., Hertel, I. V., Meyer, E., and Schmidt, H., 1983, Z. Phys. A 312, 293-304.

Bartschat, K., and Blum, K., 1982 ,Z. Phys. A 304, 85-88.

Bartschat, K., Andrä, H. J., and Blum, K., 1983, Z. Phys. 314, 257-266.

Bederson, B., 1969, Comments At. Mol. Phys. 1, 41-44.

Bederson, B., 1970, Comments At. Mol. Phys. 2, 7-9.

Blum, K., and Kleinpoppen, H., 1979, Phys. Rep. 52, 203-261.

Bransden, B. H., and McDowell, M. R. C., 1978, Phys. Rep. 46, 249-394.

Courbin-Gaussorgues, C., Wahnon, P., and Barat, M., 1979, J. Phys. B 12, 3047-3062.

Demkov, Yu. N., 1964, Soviet Physics JETP 18, 138-142.

Ganz, J., Lewandowski, B., Siegel, A., Bussert, W., Waibel, H., Ruf, M.-W., and Hotop, H., 1982, J. Phys. B 15, L485-489.

Hils, D., Jitschin, W., and Kleinpoppen, H., 1981, Appl. Phys. 25, 39-47.

Jitschin, W., Osimitsch, S., Reihl, H., Kleinpoppen, H., and Lutz, H. O., 1984, J. Phys. B 17, 1899-1912.

Jitschin, W., Osimitsch, S., Reihl, H., Kleinpoppen, H., and Lutz, H. O., 1985, to be published.

Kessler, J., 1984, Comments At. Mol. Phys. 14, 275-284.

Masnou-Seeuws, F., and McCarroll, R., 1974, J.Phys. B 7, 2230-2243.

McMaster, W.H., 1954, Am. J. Phys. 22, 351-362.

Mecklenbrauck, W., Schön, J., Speller, E., and Kempter, V., 1977, J. Phys. B 10, 3271-3281.

Menner, B., Hall, Th., Zehnle, L., and Kempter, V., 1981, J. Phys. B 14, 3693-3706.

Meyerhof, W.E., 1973, Phys. Rev. Lett. 31, 1341.

Nikitin, E. E., 1965, J. Chem. Phys. 43, 744-750.

Olsen, J. O., Andersen, T., Barat, M., Courbin-Gaussorgues, C., Sidis, V., Pommier, J., Agusti, J., Andersen, N., and Russek, A., 1979, Phys. Rev. A 19, 1457-1484.
Osimitsch, S., Jitschin, W., Reihl, H., Kleinpoppen, H., and Lutz, H. O., 1984, poster presented at the Ninth International Conference on Atomic Physics, Seattle WA.
Percival, I. C., and Seaton, M. J., 1958, Phil. Trans. R. Soc. A 251, 113-138.
Schectman, R. M., Shoffstall, D. R., Ellis, D. G., and Chojnacki, D. A., 1973, J. Opt. Soc. Am. 63, 80-84.
Tolk, N. H., Tully, J. C., White, C. W., Kraus, J., Monge, A. A., Simms, D. L., Robbins, M. F., Neff, S. H., and Lichten, W., 1976, Phys. Rev. A 13, 969-984.
Wigner, E., 1927, Nachr. Ges. Wiss. Göttingen 374-381.

POLARIZATION CORRELATION IN SIMULTANEOUS TWO-PHOTON DECAY PROCESSES AND TESTS OF BELL'S INEQUALITY

A.J. Duncan, W. Perrie, H.J. Beyer and H. Kleinpoppen

Atomic Physics Laboratory
University of Stirling
Stirling, FK9 4LA, Scotland

INTRODUCTION

This lecture considers the correlation in the polarization of two photons emitted simultaneously from a common source. The only two processes of this kind on which experiments have been carried out, are the annihilation of para-positronium, in which an electron and positron in a singlet state are converted into two identical photons each with energy 0.511 MeV, and the decay of metastable atomic hydrogen. The former process has been studied experimentally quite extensively and the results are well documented[1,2] so we shall concentrate our attention on the decay of metastable atomic hydrogen, for which the polarization correlation has only recently been observed at Stirling[3], to illustrate the main features of this type of process and its application to tests of so called "hidden variable" theories and Bell's inequality. We shall not consider the extensive work[4] which has been carried out using atomic cascades in which the two photons are not emitted simultaneously, although many of the conclusions we shall reach also apply to these cascades.

Interest in the polarization correlation of photons goes back to the measurement of Wu and Shaknov[5] in 1950 who following the suggestion of Wheeler[6] measured the linear polarization correlation of the two photons produced by the annihilation of para-positronium. As pointed out by Yang[7] measurements of this type are capable of giving information on the parity properties of nuclear particles. In addition, the polarization correlation properties of two photons have been used to illustrate fundamental ideas in atomic and nuclear physics. In particular, Fano[8] uses such properties in his discussion of pairs of two level systems and Feynman[9] does likewise to show the power of symmetry arguments in physics.

However, undoubtedly the main stimulus to measurement of the polarization correlation in two-photon processes came first from the gedanken experiment of Bohm[10] and the paper in 1957 by Bohm and Aharanov[11] in which the so called paradox of Einstein, Podolsky and Rosen[12] was couched in terms of the polarization of photons and subsequently, from the paper by Bell[13] in 1964 and its interpretation in experimental terms by Clauser, Horne, Shimony and Holt[14] in 1969 which allowed actual experimental tests to be carried out to distinguish between the predictions of quantum mechanics and hidden variable theories.

We shall now proceed to consider the theory and main characteristics of the two-photon decay process, then describe some recent measurements on hydrogen and finally discuss the implications of these measurements for the debate on hidden variable theories.

THEORY OF THE TWO-PHOTON DECAY PROCESS

For metastable atomic hydorgen [H(2S)] electric dipole and electric quadrupole transitions from the $2S_{\frac{1}{2}}$ to $1S_{\frac{1}{2}}$ state are forbidden. Consequently H(2S) has a long lifetime of about 1/8 second and decays primarily by the simultaneous emission of two photons, as indicated in Figure 1.

Contributions to the decay of the $2S_{\frac{1}{2}}$ state are also possible by

(1) single photon magnetic dipole transitions which only become significant, however, when relativistic effects are important, for example in high Z hydrogenic ions. For hydrogen itself the lifetime of this type of transition is about 4×10^5 seconds[15] and can be neglected for most purposes.

(2) a cascade, involving the sequential emission of two photons through the $2P_{\frac{1}{2}}$ state which, because of the Lamb shift, lies slightly below the $2S_{\frac{1}{2}}$ state in energy. For hydrogen, the associated lifetime is about 5×10^9 seconds[16] and hence this process can also be effectively neglected.

To describe the two-photon decay process itself we expand the scattering operator S to second order in terms of the interaction Hamiltonian H_i between the atomic electron and the radiation field and calculate the matrix element of the second order term between the initial and final state of the atom plus radiation field. The relevant part of the interaction Hamiltonian is given by $H_i = -(e/m)\underline{p}.\underline{A}$ where -e, m and $\underline{p}$ are the electron charge, mass and canonical momentum respectively and $\underline{A}$ is the vector potential of the radiation field. Using this expression for H_i we find that the matrix element describing the two-photon decay process is of

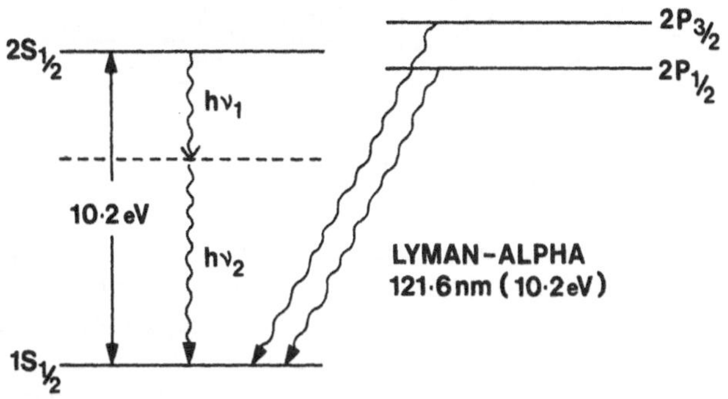

Fig. 1. Energy level diagram for atomic hydrogen neglecting hyperfine structure (not to scale).

the form

$$\frac{e^2}{m^2} \sum_j \frac{\langle\Phi_f|p.A|\Phi_j\rangle\langle\Phi_j|p.A|\Phi_i\rangle}{E_i - E_j} \tag{1}$$

where $|\Phi_i\rangle$, $|\Phi_f\rangle$ represent the initial and final states respectively of the atom and radiation field together and E_i is the energy of the initial state. The summation is taken over all allowed virtual states $|\Phi_j\rangle$ of energy E_j which in our case contain one more photon than the initial state $|\Phi_i\rangle$. If $|\Psi_i\rangle$, $|\Psi_f\rangle$ and $|\Psi_j\rangle$ represent the electronic states of energies W_i, W_f and W_j respectively, in the case that photons of frequencies $h\nu_1$ and $h\nu_2$ are emitted, we can write

$$|\Phi_i\rangle = |\Psi_i\rangle|0\rangle, \quad |\Phi_f\rangle = |\Psi_f\rangle|h\nu_1\rangle|h\nu_2\rangle$$

$$|\Phi_j\rangle = |\Psi_j\rangle|h\nu_1\rangle \text{ or } |\Psi_j\rangle|h\nu_2\rangle \tag{2}$$

where $|0\rangle$ represents the initial state of the radiation field, $|h\nu_1\rangle$ its state when a photon of frequency ν_1 has been added etc.

In terms of Feynman diagrams the process of two-photon emission can be represented as shown in Figure 2, which demonstrates that the two photons can be emitted in "either order" leading to two terms in the final expression for the probability per second $A(\nu_1)d\nu_1$ for the spontaneous, simultaneous emission of two photons with one photon in the range $d\nu_1$ in the neighbourhood of frequency ν_1. Making the dipole approximation and changing from the dipole velocity to the dipole length form of the matrix element we obtain

$$\overline{\overline{A(\nu_1)}}d\nu_1 = \frac{1024\pi^6 e^4 \nu_1^3 \nu_2^3}{c^6} \Big(\Big|\sum_j \frac{\langle\Psi_f|\underline{r}.\hat{e}_1|\Psi_j\rangle\langle\Psi_j|\underline{r}.\hat{e}_2|\Psi_i\rangle}{W_i-W_j-h\nu_2}$$

$$+ \sum_j \frac{\langle\Psi_f|\underline{r}.\hat{e}_2|\Psi_j\rangle\langle\Psi_j|\underline{r}.\hat{e}_1|\Psi_i\rangle}{W_i-W_j-h\nu_1}\Big|^2\Big)_{av.} d\nu_1 \qquad (3)$$

where the average is taken over the directions of propagation and over the polarization directions as indicated by the double bar over $A(\nu_1)$. Note that the quantities $\hat{e}_1$ and $\hat{e}_2$ represent unit vectors in the direction of the polarization of the two photons and that the intermediate states $|\Psi_j\rangle$ are P states of atomic hydrogen.

The above expression for $\overline{\overline{A(\nu_1)}}$ was originally derived of course by Maria Göppert Mayer[17] in 1931 in her paper which pioneered the field of multi-photon transitions.

Before proceeding further, it is important to realise that the idea, suggested by the theoretical expression for the transition probability and the Feynman diagrams, that the photons are emitted one after the other is purely a convenient picture representing the mathematical calculation. The system never really assumes the virtual intermediate states and the two photons are, in fact, emitted simultaneously.

It should also be noted that other types of two-photon transition are possible. For example, the $2S_{\frac{1}{2}}$ state could decay by emitting two quadrupole photons. However, the effect of such processes is negligible for our purposes. For instance, Au[18] has shown that taking the possibility of quadrupole transitions into account, the angular correlation would be altered from the dipole prediction only by terms quadratic in the fine structure constant.

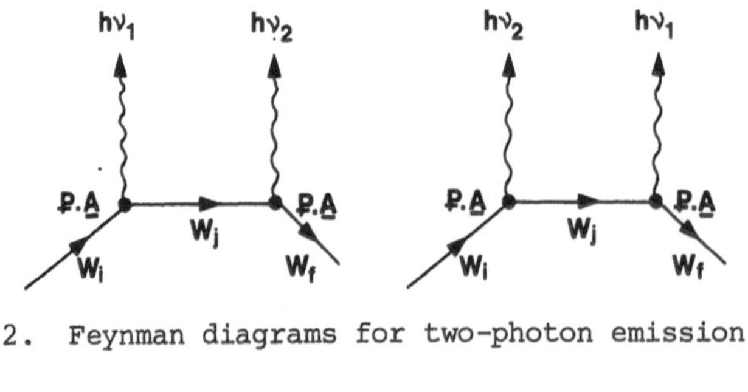

Fig. 2. Feynman diagrams for two-photon emission.

As a consequence of the above theoretical considerations the two-photon decay process has the following characteristics.

(1) The energies $h\nu_1$ and $h\nu_2$ of the two photons satisfy

$$h\nu_1 + h\nu_2 = 10.2eV$$

(2) Each photon can have any energy from zero up to 10.2eV with the spectral distribution shown in Figure 3. Experimentally the fact that the spectral distribution has its maximum at a wavelength of 243 nm is important, since it allows measurements of photon polarization to be made in air using simple lenses and pile-of-plates polarizers. The hydrogen decay is unique in this respect, all other hydrogenic two-photon decay processes and positronium annihilation producing photons of much shorter wavelength in regions of the spectrum where it is difficult to measure polarization directly.

(3) Since the electric dipole operator $e\underline{r}$ is diagonal in the nuclear and electronic spin, the effects of hyperfine and fine structure may be neglected.

(4) If $\hat{e}_1$ and $\hat{e}_2$ are unit linear polarization vectors for photons of energies $h\nu_1$ and $h\nu_2$ emitted at angle α relative to each

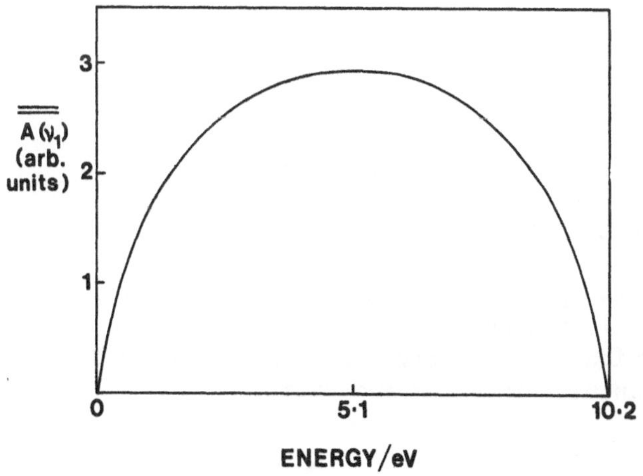

Fig. 3. Predicted two-photon spectrum for hydrogen.

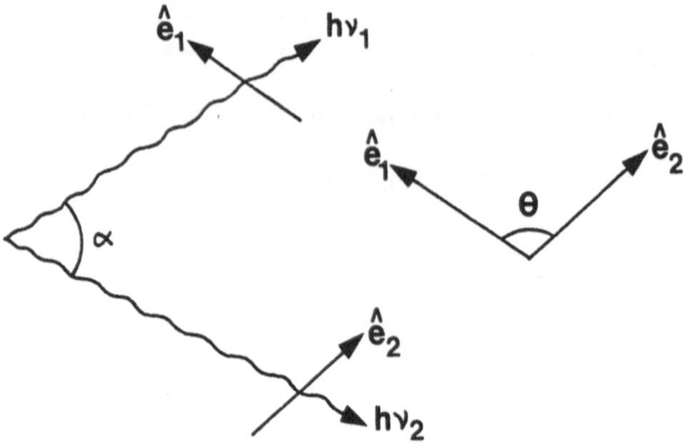

Fig. 4. Diagram to illustrate the angles involved in two-photon emission.

other as shown in Figure 4, then it follows that the transition probability has a dependence on the polarizations of the photons given by

$$A(\nu_1) \propto |\hat{e}_1.\hat{e}_2|^2 = \cos^2 \theta \tag{4}$$

where θ is the angle between $\hat{e}_1$ and $\hat{e}_2$, while averaged over polarization we find

$$\overline{A(\nu_1)} \propto \langle|\hat{e}_1.\hat{e}_2|^2\rangle_{av} \propto (1+\cos^2 \alpha) \tag{5}$$

In contrast, for the two γ-rays emitted in positronium annihilation the dependence of the transition probability A on the angle between the polarization vectors is given by

$$A \propto |\hat{e}_1 \times \hat{e}_2| = \sin^2\theta \tag{6}$$

and the angular correlation by

$$\overline{A} \propto \delta\ (\pi-\alpha) \tag{7}$$

where δ represents the Dirac delta function.

TWO-PHOTON STATE VECTOR FOR $\alpha = \pi$

When the two photons are emitted in diametrically opposite directions ($\alpha=\pi$), the photon wavefunction has a particularly interesting form which can be derived from simple consideration of conservation of angular momentum and parity.

In order to conserve angular momentum when two photons are emitted in opposite directions from a source whose consituents are isotropic before and after emission, it is necessary for the photons to have equal helicity. In terms of right handed ($|R\rangle$) and left handed ($|L\rangle$) helicity states we thus expect to find photon pairs in states represented by $|R\rangle_1|R\rangle_2$, $|L\rangle_1|L\rangle_2$ or a superposition of these where $|R\rangle_1$ denotes a photon of right handed helicity propagating to the right, $|R\rangle_2$ a photon of right handed helicity propagating to the left etc.

In addition, since, in the processes we are considering, the initial and final states of the constituents of the source have definite parity, the two-photon state must also have definite parity. If P is the parity operator, $P|R\rangle_1 = |L\rangle_2$, $P|L\rangle_1 = |R\rangle_2$ etc. Hence, to ensure definite parity we conclude that the photon wavefunction must be in one of the forms:

$$|\Psi\rangle_\pm = \frac{1}{\sqrt{2}}[\,|R\rangle_1|R\rangle_2 \pm |L\rangle_1|L\rangle_2\,] \tag{8}$$

since then $P|\Psi\rangle_+ = |\Psi\rangle_+$ (even parity) and $P|\Psi\rangle_- = -|\Psi\rangle_-$ (odd parity). In the case of the decay of metastable atomic hydrogen the initial and final atomic states are both of even parity so that the photon wavefunction is given by $|\Psi\rangle_+$ whereas for para-positronium the initial state is odd and the final vacuum state even so that the photon wavefunction is given by $|\Psi\rangle_-$.

Assuming the photons propagate in the +z and -z directions, it is also possible to describe the photon wavefunction in terms of the linear polarization states $|x\rangle$, $|y\rangle$ by using the relations

$$|R\rangle_1 = \frac{1}{\sqrt{2}}(|x\rangle_1 + i|y\rangle_1),\quad |R\rangle_2 = \frac{1}{\sqrt{2}}(|x\rangle_2 - i|y\rangle_2)$$

$$|L\rangle_1 = \frac{1}{\sqrt{2}}(|x\rangle_1 - i|y\rangle_1),\quad |L\rangle_2 = \frac{1}{\sqrt{2}}(|x\rangle_2 + i|y\rangle_2 \tag{9}$$

Substituting these relations into the previous expression for $1\Psi\rangle_\pm$ gives:

$$|\Psi\rangle_+ = \frac{1}{\sqrt{2}}(|x\rangle_1|x\rangle_2 + |y\rangle_1|y\rangle_2)$$

and $$|\Psi\rangle_- = \frac{1}{\sqrt{2}}(|x\rangle_1|y\rangle_2 - |y\rangle_1|x\rangle_2) \tag{10}$$

We shall return to consider the implications of these wavefunctions in more detail later but first the measurement of polarization correlation will be illustrated by reference to the experiment on the two-photon decay of metastable atomic hydrogen.

TWO-PHOTON APPARATUS

A schematic diagram of the apparatus is shown in Figure 5. A 1keV beam of metastable atomic deuterium[D(2S)] is produced by charge exchange in cesium vapour of deuterons extracted from a radiofrequency ion source. The charge exchange process

$$D^+ + Cs \rightarrow Cs^+ + D(2S)$$

is nearly resonant and has a maximum cross-section for an incident deuteron energy of about 700eV with approximately 25% of the deuterons being neuteralized into the 2S state. However, larger extraction voltages result in a more intense beam which more than compensates for the decrease in charge transfer cross-section into the 2S state, and the metastable beam density, in fact, increases with energy up to about $10^5 cm^{-3}$ at 5keV. However, higher beam energies result in a proportionately greater increase in noise generated by collisions with the background gas in the vacuum system and it is found empirically that, in a given time, the best statistical accuracy in the measurements can be obtained at an energy of 1keV.

Deuterium is used instead of hydrogen (protium) for two reasons. Firstly, the beam extracted when using deuterium has a higher deuteron fraction (∿90%) and a lower fraction of the molecular ions D_2^+ and D_3^+ than is the case for protium where as much as 30% of the beam may consist of the molecular ions H_2^+ and H_3^+. Secondly, for a given metastable density and hence two-photon signal, the noise generated by interaction of the beam with the background gas in the vacuum system is less in the case of deuterium, partly because of its slower speed, partly because of its higher purity.

Once the D(2S) beam is formed it passes through electric field pre-quench plates which by Stark mixing of the 2S and 2P states allow the metastable component of the beam to be switched on and off. After traversing the detection region the beam is quenched at the end of the apparatus by an electric field and the resulting Lyman-α signal is used to normalize the two-photon coincidence rate. The Lyman-α signal is detected using a photomultiplier in front of which is placed an oxygen filter which has a transmission window at the appropriate wavelength.

The detection optics and electronics are shown schematically in Figure 6. The two photon signal is collected and collimated by two 50 mm diameter lenses of focal length 50mm at 500nm. Each lens is placed approximately 50mm from the beam and subtends a half angle $\delta \cong 25°$ at the source. For the linear polarization measurements, two high transmission (∿95%) ultraviolet polarizers, each consisting of twelve plates set at Brewster's angle are placed as shown on either side of the source. The lenses and plates are made from high

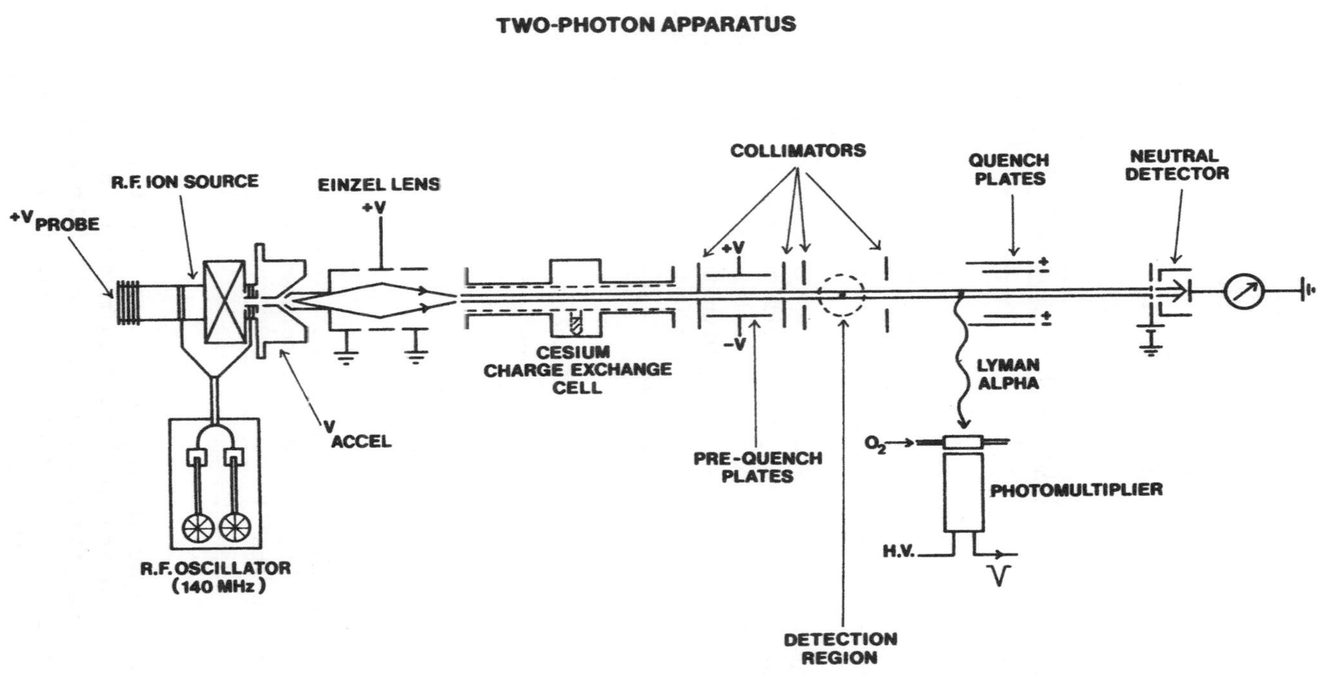

Fig. 5. Schematic diagram of the apparatus (not to scale).

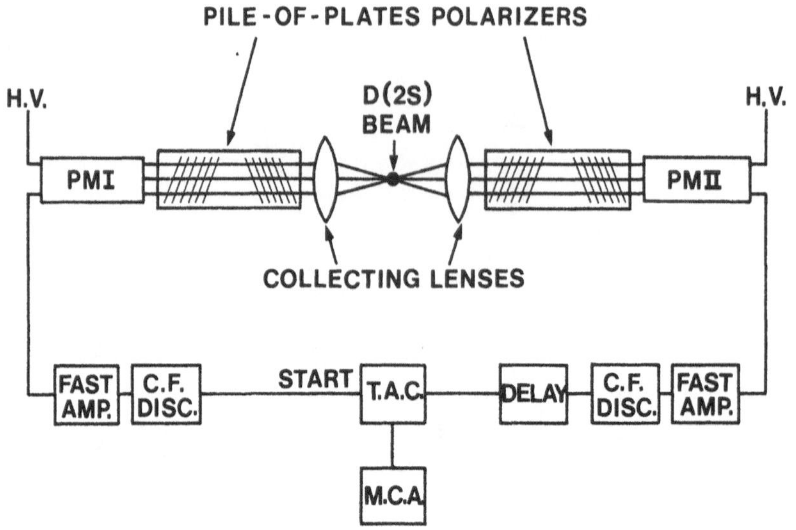

Fig. 6. Detection optics and electronics (not to scale).

quality fused quartz with a short wavelength cut-off at 160nm, but in practice, the short wavelength cut-off occurs at about 180nm due to obsorption in oxygen. For the circular polarization measurements the linear polarizers are used along with two achromatic quarter-wave plates which operate in the range 180-320nm but reduce the detection solid angle and the overall transmission efficiency of the system considerably.

On either side, the signal is detected by fast rise-time photomultipliers and the time correlation spectra obtained with the metastable beam switched on and off are stored in separate segments of the multichannel analyser memory and subtracted at the end of a run. A typical spectrum obtained in this way is shown in Figure 7. Note that as we should expect for a simultaneous emission process, the coincidence peak is symmetrical and shows no lifetime effects. Note also that the background signal is due mainly to radiation excited by interaction of the atomic beam with the background gas and only slightly to uncorrelated photons from the two-photon decay process itself.

MEASUREMENTS AND RESULTS

The circular polarization correlation is obtained by first setting one detection arm to respond to right hand circularly polarized light then varying the relative angle ϕ between the optical axis of the quarter-wave plate and the transmission axis of the pile-of-plates polarizer in the other arm from -45° to +45°, detecting right hand circularly polarized light at one extreme, left hand circularly polarized light at the other. The results obtained are shown in Figure 8, along with a least squares fit of the form

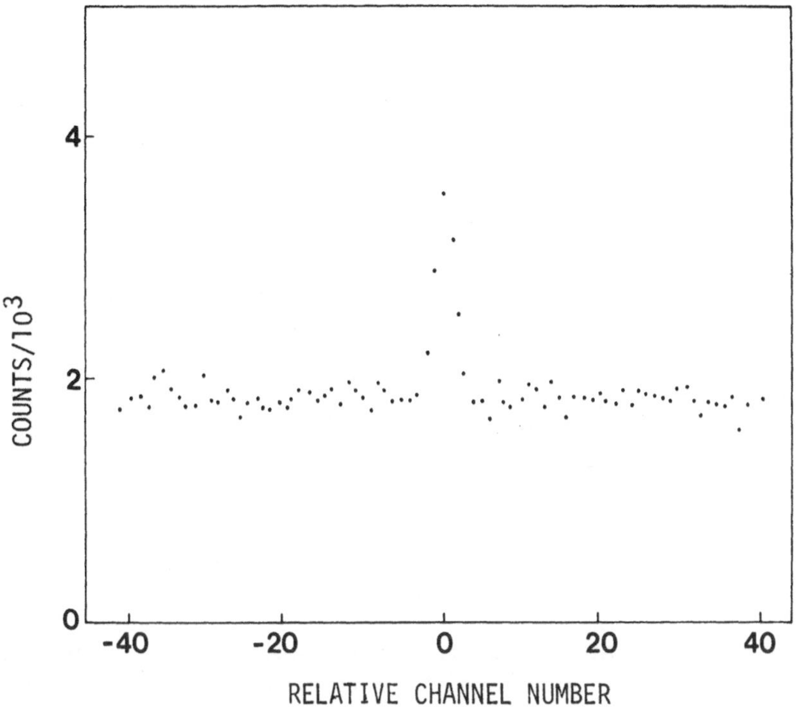

Fig. 7. A typical time correlation spectrum. Time delay per channel: 0.8 ns. FWHM: <3.0 ns. Integral under peak: 5274 ± 174. Time to acquire spectrum: 21.5 hrs. Coincidence rate: 490 ± 16 hr^{-1}.

$A + B\cos^2(\pi/4+\phi)$. They are consistent with conservation of angular momentum of the photons along their common axis and the collapse of the state vector from

$$\frac{1}{\sqrt{2}}\left(|R\rangle_1|R\rangle_2 + |L\rangle_1|L\rangle_2\right) \text{ to } |R\rangle_1|R\rangle_2 \tag{11}$$

which, in the ideal case, leads us to expect a $\cos^2\phi$ variation in the intensity. From the fitted curve $A = 3.4 \pm 0.8$ hr^{-1}, $B = 21.7 \pm 1.2$ hr^{-1} but, without knowledge of the coincidence signal with the linear polarizer plates removed, it is not possible to quote expected values for A and B.

The measured linear polarization correlation is shown in Figure 9, along with a least squares fit of the form $A + B\cos^2\theta$, where θ is the relative angle between the transmission axes of the polarizers. On the basis of the collapse of the state vector from

$$\frac{1}{\sqrt{2}}\left(|x\rangle_1|x\rangle_2 + |y\rangle_1|y\rangle_2\right) \text{ to } |x\rangle_1|x\rangle_2 \tag{12}$$

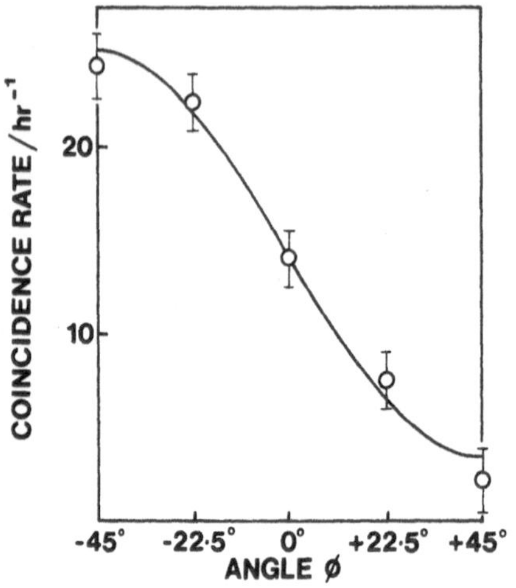

Fig. 8. Circular polarization correlation. The coincidence rate is normalized to a typical Lyman-α count rate.

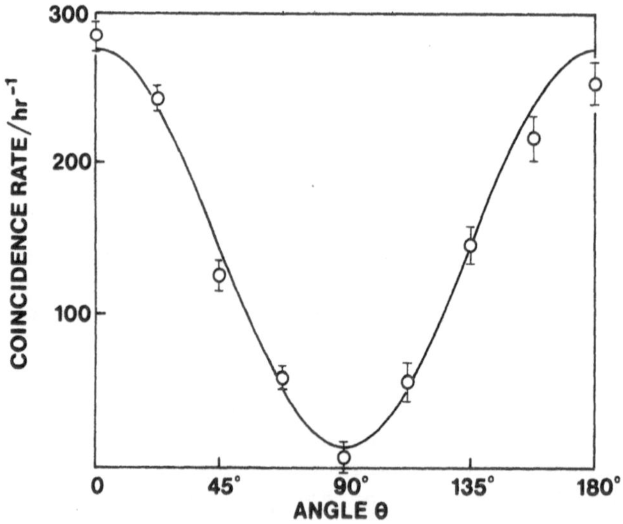

Fig. 9. Linear polarization correlation. The coincidence rate is normalized to a typical Lyman-α count rate.

we expect, in the ideal case, a $\cos^2\theta$ variation of the signal intensity $R(\theta)$. In practice, taking into account the transmission efficiencies ε_M, ε_m of the polarizers for light polarized parallel and perpedicular to the transmission axis respectively and the half-angle δ subtended by the lenses at the source, quantum mechanics predicts[14]

$$\frac{R(\theta)}{R_o} = \frac{1}{4}(\varepsilon_M + \varepsilon_m)^2 + \frac{1}{4}(\varepsilon_M - \varepsilon_m)^2 F(\delta)\cos 2\theta \tag{13}$$

where R_o is the coincidence rate with both polarizers removed and $F(\delta)$ is a function of δ. In our case $\varepsilon_M = 0.91 \pm 0.02$, $\varepsilon_m = 0.031 \pm 0.002$, $\delta \cong 25°$, $F(\delta) = 0.998$ and $R_o = (670 \pm 13)\ \text{hr}^{-1}$.

The fitted and predicted values for A and B, shown in Table 1, demonstrate agreement between the quantum mechanical theory and experiment and confirm the correctness of the above form for the state vector.

Table 1. Fitted and predicted values for the constants A and B in the case of the linear polarization correlation measurements.

	A	B
Fitted	13.3 ± 7.0	262.9 ± 10.7
Predicted	19.2 ± 1.2	258.3 ± 11.6

Additional confirmation of the correctness of the quantum mechanical analysis and the form of the state vector was obtained in an experiment where a linear polarizer was placed in one arm, a circular polarizer in the other. The signal in the circular polarizer arm was found, within the limits of experimental error, to be of equal intensity for left and right circular polarizations.

IMPLICATIONS OF THE POLARIZATION CORRELATION MEASUREMENTS

We have seen theoretically that in the cases we have considered we expect the two-photon state to be represented by

$$|\Psi\rangle_\pm = \frac{1}{\sqrt{2}}\left(|R\rangle_1|R\rangle_2 \pm |L\rangle_1|L\rangle_2\right)$$

or $$|\Psi\rangle_+ = \frac{1}{\sqrt{2}}\left(|x\rangle_1|x\rangle_2 + |y\rangle_1|y\rangle_2\right),\quad |\Psi\rangle_- = \frac{1}{\sqrt{2}}\left(|x\rangle_1|y\rangle_2 - |y\rangle_1|x\rangle_2\right) \tag{14}$$

Experimentally, the measurements using positronium and atomic hydrogen have produced strong confirmation of the correctness of these state vectors. On examining their form, first of all we note that they represent pure quantum mechanical states, not mixtures of say $|R\rangle_1|R\rangle_2$ and $|L\rangle_1|L\rangle_2$ states. We can not, for instance, describe the situation by saying that atoms are emitting equal numbers of right-right and left-left circularly polarized photons. Secondly, as we have discussed, according to the usual quantum mechanical approach, on making a measurement, the state vector collapses instanteously. For example, for the even parity state in the linear polarization representation

$$|\Psi\rangle_+ \to |x\rangle_1|x\rangle_2 \text{ or } |y\rangle_1|y\rangle_2 \tag{15}$$

each possibility occurring with a probability of one half. However, the collapse in this case implies that the result of a measurement by one detector determines instantaneously the result for the other detector irrespective of the separation of the detectors when the first measurement is made. In addition, since the choice of directions x and y is arbitrary, it is clear that the result of the second measurement also depends on which direction of polarization we choose for the first measurement i.e. on which property we choose to measure in the first detector. This effect highlights the intrinsic non-locality built in to the quantum mechanical formalism which may seem strange but is quite consistent with the polarization correlation being a property not of single photons but of the photon pair and the measuring apparatus together. Before a measurement is made it is incorrect to think of individual photons with a separate existence and properties which do not depend on any measurements that may be made on the system.

Although the above interpretation is well supported by experiment the non-locality and lack of reality in the philosophical sense that it implies has led some physicists to question the range of validity of quantum mechanics. For example, it has been suggested by Schrödinger[19] and Furry[20] that over a "large enough distance" the state vector $|\Psi\rangle_+$ might collapse spontanteously, to a mixture of states of the form $|x\rangle_1|x\rangle_2$ with x being any direction perpendicular to the z -axis. Such a collapse would cause a change in the observed polarization correlation but the experimental evidence is strongly against such an effect despite the use of detector-source separations of up to 6m in positronium and cascade experiments.

Attempts have also been made to find theories capable of explaining the experimental results, which are both local and realistic (often referred to as hidden variable theories). A local theory is one in which the result of a measurement at some point A in space-time is not influenced by a measurement made at another point B spatially separated from A in a relativisitc sense. A realistic theory is one in which it is assumed that the world is made up of objects with properties which exist independently of any observations which may be made on them. Quantum mechanics is neither local nor realistic but everyday life is organised on the basis that the macroscopic world is both local and realistic. The following treatment of the two-photon polarization correlation due originally to Holt[21] is a specific example of a local realistic theory.

Assume we have an isotropic source emitting pairs of photons in the +z and -z directions each with the same polarization vector at an angle ϕ to the x-axis, as shown in Figure 10. The angle ϕ takes on all values from 0 to π with equal probability. If there are R_o decays per second in the source and if the polarizers and detectors D_A, D_B are 100% efficient we expect a coincidence signal rate

$$dS = R_o \cos^2(\theta_A - \phi) \cos^2(\theta_B - \phi) d\phi/\pi \tag{16}$$

for photons with polarization angle between ϕ and $\phi + d\phi$.

Integrating over all ϕ,

$$\frac{S}{R_o} = \frac{1}{\pi} \int_0^{\pi} \cos^2(\theta_A - \phi) \cos^2(\theta_B - \phi) d\phi = \frac{1}{4} [1 + \tfrac{1}{2} \cos 2(\theta_A - \theta_B)] \tag{17}$$

In contrast, in the ideal case, the corresponding result using quantum mechanics is

$$\frac{S}{R_o} = \frac{1}{4} [1 + \cos 2(\theta_A - \theta_B] \tag{18}$$

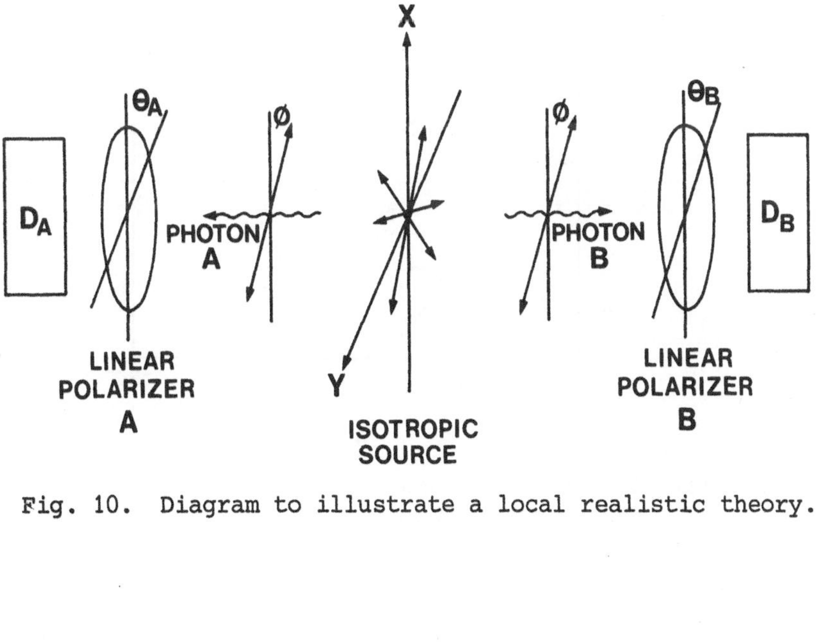

Fig. 10. Diagram to illustrate a local realistic theory.

The local realistic and quantum theory predictions are compared in Figure 11 as a function of the angle $\theta_A - \theta_B$ from which we see immediately that this specific local realistic theory does not give rise to the strong polarization correlation predicted by quantum theory.

With regard to the construction of local realistic theories in general, it was shown by Bell[13] in the form of his now famous inequality, that all so called local deterministic hidden variable theories (which are a sub-class of local realistic theories) predict a weaker correlation between the polarization of photons than that given by quantum mechanics. The work of Bell together with that of Clauser, Horne, Shimony and Holt[14] who demonstrated, by making some plausible additional assumptions, how Bell's inequality could be treated in a real experimental situation, allowed experiments to be carried out to distinguish between the predictions of quantum mechanics and local deterministic hidden variable theories and, as it turned out, also for a much wider class of local realistic theories.

The most convenient form of Bell's inequality for experimental purposes is that due to Freedman[22] who showed that, for local realistic theories in general, the quantity

$$\eta = \left|\frac{R(22.5°) - (R(67.5°)}{R_o}\right| \leqslant 0.250 \tag{19}$$

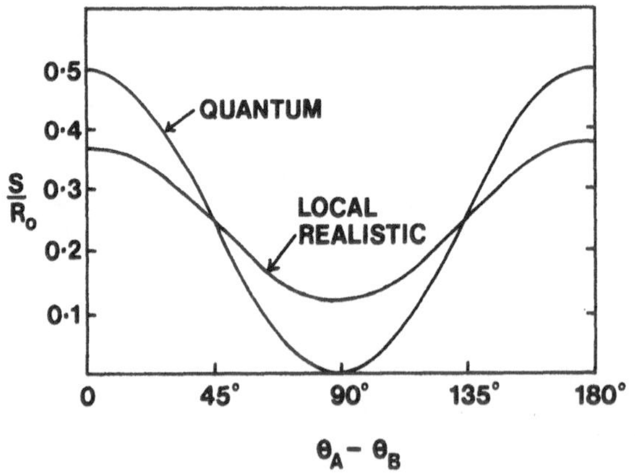

Fig. 11. Comparison of local realistic and quantum theory predictions.

where $R(22.5°)$ = coincidence rate for $\theta_A-\theta_B = 22.5°$
$R(67.5°)$ = coincidence rate for $\theta_A-\theta_B = 67.5°$
R_o = coincidence rate with polarizer plates removed.

In contrast, in the ideal case quantum mechanics predicts a $\cos^2(\theta_A-\theta_B)$ variation in the coincidence rate and results in

$$\eta = 0.354 \tag{20}$$

In the non-ideal case, however, taking into account the solid angle of detection and the efficiency of each polarizer we expect quantum mechanically from equation 13 that

$$\eta = 0.273 \pm 0.011 \tag{21}$$

Experimentally, on the other hand, from the results shown in Figure 9 at angles 22.5° and 67.5° and using $R_o = (670 \pm 13)\ hr^{-1}$ we find

$$\eta = 0.275 \pm 0.016\ (1\sigma) \tag{22}$$

in agreement with the quantum mechanical results (equation 21) and in disagreement with the prediction (equation 19) of any local realistic theory by more than one standard deviation.

The vast majority of experiments using positronium annihilation and atomic cascades has also produced results in agreement with quantum mechanics. However, positronium annihilation experiments are open to the criticism that the validity of quantum mechanics itself must be assumed in order to analyse the experimental results. In addition, some doubts have been expressed[23,24] about the correctness of the results from cascade experiments where rescattering effects may be important. The metastable deuterium experiment, cannot be criticised on either of the above grounds but in general the low efficiency of the photon detectors leaves open the possibility that the results could be explained in terms of local realistic theories if the assumptions of Clauser, Horne, Shimony and Holt[14] are brought into question. Consequently, no experiment to date can be considered to have provided grounds for a completely unequivocal rejection of such theories.

In any case, whatever the final outcome of the debate between quantum theory and local realism it is as well to bear in mind that the questions raised have troubled many of the greatest physicists of our time. The doubts of de Broglie, Schrödinger and Einstein are well known, those of Dirac and Feynman less so. However, Dirac[25] in referring to the apparent non-locality inherent in quantum mechanics stated, "It is against the spirit of relativity but it is the best we can do ...... and, of course one is not satisified with such a theory", while Feynman[26] recently wrote, "We have always had a great deal of difficulty in understanding the world view that quantum mechanics represents ...... It has not yet become obvious to me that there's no real problem".

REFERENCES

1. J F Clauser and A Shimony, Rep.Prog.Phys. 41, 1881 (1978).
2. F M Pipkin, Atomic Physics Tests of Basic Concepts in Quantum Mechanics, in: Advances in Atomic and Molecular Physics, D R Bates and B Bederson eds., Academic Press, New York (1978).
3. W Perrie, A J Duncan, H J Beyer and H Kleinpoppen, Ninth International Conference on Atomic Physics, Seattle, Washington (1984).
4. See, for example, A Aspect, J Dalibard and G Roger, Phys. Rev. Lett. 49 , 1804 (1982).
5. C S Wu and I Shaknov, Phys. Rev. 77, 136 (1950).
6. J A Wheeler, Ann. N.Y. Acad. Sci. 48, 219 (1946).
7. C N Yang, Phys. Rev. 77, 242 (1949).
8. U Fano, Rev. Mod. Phys. 55, 855 (1983).
9. R P Feynman, R S Leighton and M Sands, The Feynman Lectures in Physics Vol III, Addison-Wesley, Reading, Mass. (1965).
10. D Bohm, Quantum Theory, Prentice-Hall, Engelwood Cliffs, N.J (1951).
11. D Bohm and Y Aharonov, Phys. Rev. 108, 1070 (1957).
12. A Einstein, B Podolsky and N Rosen, Phys. Rev. 47, 777 (1935).
13. J S Bell, Physics 1, 195 (1964).
14. J F Clauser, M A Horne, A Shimony and R A Holt, Phys. Rev. Lett. 23, 880 (1969).
15. F A Parpia and W R Johnson, Phys. Rev. A26, 1142 (1982).
16. J Shapiro and G Breit, Phys. Rev. 113, 179 (1959).
17. M Göppert-Mayer, Ann. Phys. 9, 273 (1931).
18. C K Au, Phys. Rev. A14, 531 (1976).
19. E Schrödinger, Proc. Camb. Phil. Soc. 31, 555 (1935).
20. W H Furry, Phys. Rev. 49, 393 (1936).
21. R A Holt, PhD. Thesis, Harvard University (1973).
22. S J Freedman, PhD. Thesis, University of California, Berkeley (1972).
23. T W Marshall, E Santos and F Selleri, Lett. Nuovo Cimento 38, 417 (1983).
24. F Selleri, Lett. Nuovo Cimento 39, 258 (1984).
25. P A M Dirac, Chap. 1 in: "The Physicst's Conception of Nature", J. Mehra ed., Reidel, Dordrecht (1973).
26. R P Feynman, Int. J. Theor. Phys. 21, 467 (1982).

STOKES PARAMETERS $\eta_{1,2,3}$ AND SCATTERING PARAMETERS λ,χ FOR POSITIVE AND NEGATIVE SCATTERING ANGLES IN ELECTRON-PHOTON COINCIDENCE EXPERIMENTS

H-J Beyer[1], K Blum[2] and M C Standage[3]

[1]Atomic Physics Laboratory
University of Stirling, Stirling, Scotland
[2]Institut für Theoretische Physik I
Universität Münster, D-4400 Münster, West Germany
[3]School of Science, Griffith University
Nathan, Queensland, Australia, 4111

Introduction

Inelastic electron-atom scattering processes usually transfer alignment and orientation to the atoms. If alignment and orientation are measured in coincidence with the scattered electron which excited the atom, the scattering amplitudes can be determined completely (apart from an overall phase factor). To determine the sign of the orientation a measurement of the circular polarisation of the photon emitted by the excited atom is required, but the alignment and the absolute value of the orientation can be derived either from linear polarisation correlation measurements or from angular correlation measurements, always provided that a sufficient number of independent measurements can be made on the system to extract all scattering parameters. For excitation from a S ground state, the orientation is a measure of the angular momentum $\langle\vec{L}\rangle$ transferred to the atom in the scattering process. The sign of $\langle\vec{L}\rangle$ has attracted renewed interest recently in connection with scattering models by Steph and Golden[1] and by Kohmoto and Fano[2] linking small angle scattering with an attractive and large angle scattering with a repulsive potential. The two potentials result in opposite signs of $\langle\vec{L}\rangle$, so that there should be an intermediate scattering angle with $\langle\vec{L}\rangle = 0$. It has also been shown[3] that non-zero $\langle\vec{L}\rangle$ should lead to a left-right scattering asymmetry of the electrons.

The first coincidence measurement of the complete scattering parameters including the sign of $\langle\vec{L}\rangle$ was carried out by Standage and Kleinpoppen[4] on the 3^1P state of He using 80 eV electrons and scattering angles between 15° and 27.5°. It is also the only polarisation correlation measurement so far covering both positive and negative scattering angles. In view of the recent interest, a new series of measurements has been carried out on the same state. This study, which is nearly completed, considerably extends the range of electron energies and scattering angles[5] and provides more data for negative scattering angles at 80 eV including the left-right scattering asymmetry[6]. These measurements depart in one point from the results of Standage and Kleinpoppen[4] and have lead to the realisation that the equations used to link the polarisation measurements to the scattering parameters require modification for negative scattering angles. These general relations are discussed in this report.

General Stokes parameter equations for λ and χ

Figure 1 shows the scattering and detector co-ordinates commonly used to describe polarisation correlation measurements.[4,7] The photons are detected in +y-direction ($\theta_\gamma=\phi_\gamma=\pi/2$) at right angles to the (x,z) - scattering plane so that in this case the scattering and detector co-ordinate systems are identical. A positive scattering angle (scattering to the left) is shown in Figure 1. This is

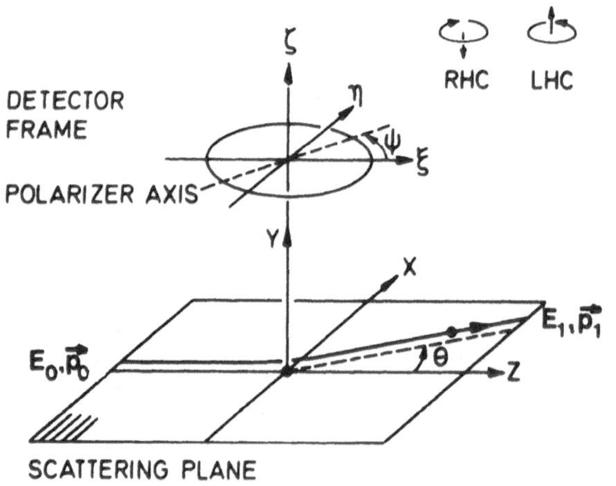

Fig. 1 (from Ref. 10) Scattering and detector co-ordinate systems for polarisation correlation studies. The spectroscopic definition of left and right hand circular polarisation is used.

described by $0 \leq \theta_e \leq \pi$, $\phi_e = 0$ while a negative scattering angle (scattering to the right) is described by $0 \leq \theta_e \leq \pi$, $\phi_e = \pi$. The linear polarisation analyser angle ψ is measured in the same way as θ_e.

For n^1P excitation of helium the 3 scattering amplitudes $f_{M_L}(E_1, \theta_e, \phi_e, E_o)$ for $M_L = 0, \pm 1$ and the phases between f_{M_L} can be completely described by 3 independent parameters:[7-9]

$$\sigma = |f_o|^2 + 2\,|f_1|^2 \tag{1}$$

and $$\lambda = |f_o|^2/\sigma \tag{2}$$

are absolute and relative cross sections while χ is the phase angle between f_o (set real and positive) and f_1:

$$f_1 = |f_1| e^{i\chi}. \tag{3}$$

The polarisation of the light measured in the coincidence experiment is conveniently described by the Stokes parameters and these are related to the scattering parameters:[4,7,9]

$$\eta_1 = \frac{I_{45} - I_{135}}{I_{45} + I_{135}} = -\,2\sqrt{\lambda(1-\lambda)}\,\cos\chi \tag{4a}$$

$$\eta_2 = \frac{I_R - I_L}{I_R + I_L} = 2\sqrt{\lambda(1-\lambda)}\,\sin\chi \tag{4b}$$

$$\eta_3 = \frac{I_0 - I_{90}}{I_0 + I_{90}} = 2\lambda - 1 \tag{4c}$$

where I_0, I_{45}, I_{90} and I_{135} relate to the linear polariser angle ψ, as shown in Figure 1 and I_R, I_L correspond to right hand and left hand circularly polarised light. The spectroscopic definition of circularly polarised light is used (as shown in Figure 1 and used by Standage and Kleinpoppen[4]). The helicity definition used by Blum and Kleinpoppen[7] and Blum[9] results in the opposite sign of equation (4b).

Equations (4a-c) are correct for positive scattering angles since the definition of χ (equation 3) implicitly assumes $\phi_e=0$. It remains to be checked whether equation (4a-c) also apply for negative scattering angles (θ_e, $\phi_e=\pi$). The general situation with $\phi_e \neq 0$ can be related to the situation with $\phi_e=0$ by a rotation around z by the angle $-\phi_e$. In terms of the electron momenta $\vec{p}_0$ before and $\vec{p}_1$ after the collision, the scattering amplitudes can be written as matrix elements of the interaction T which is invariant under rotations:

$$R(-\phi_e)_z^+ \; T \; R(-\phi_e)_z = T \tag{5}$$

Thus:

$$f_{M_L}(E_1, \theta_e, \phi_e, E_0) = f_{M_L}(\vec{p}_1, \vec{p}_0)$$

$$= \langle M_L\, \vec{p}_1 | T | \vec{p}_0 \rangle$$

$$= \langle M_L \vec{p}_1 | R(-\phi_e)_z^+ \; T \; R(-\phi_e)_z | \vec{p}_0 \rangle$$

$$= e^{iM_L\phi_e} \langle M_L \vec{p}_1 (\phi_e{=}0) \; | T | \vec{p}_0 \rangle$$

$$= e^{iM_L\phi_e} f_{M_L}(E_1, \theta_e, \phi_e{=}0, E_0)$$

$$= |f_{M_L}(E_1, \theta_e, \phi_e{=}0, E_0)| e^{i(\chi + M_L\phi_e)} \quad (6)$$

Thus:

$$\chi(\phi_e) = \chi + M_L\phi_e \quad (7)$$

In the case of n^1P states of He, M_L is 1 and $\phi_e = 0$ (π) for positive (negative) scattering angles. Since the photon detection co-ordinates are not affected by the transition from positive to negative scattering angles, the Stokes parameter equations (4a-c) would in fact return $\chi(\phi_e{=}\pi)$ for negative scattering angles, i.e. an apparent shift of χ by π. In order to obtain the scattering parameters in line with the original definition of χ(eq. 3), the general $\chi(\phi_e)$ has to replace χ in the Stokes parameter equations 4a-b; eq. 4c is not affected. Thus the general equations are:

$$\eta_1 = -2\sqrt{\lambda(1-\lambda)}\cos(\chi+\phi_e) \quad (8a)$$

$$\eta_2 = 2\sqrt{\lambda(1-\lambda)}\sin(\chi+\phi_e) \quad (8b)$$

$$\eta_3 = 2\lambda - 1 \quad (8c)$$

with $\phi_e{=}0$ (π) for positive (negative) scattering angles.

Discussion

Using the general Stokes parameter equations (8a-c) a sign change is expected for both, η_1 and η_2 when moving from positive to negative scattering angles. The sign change of η_2 was observed by Standage and Kleinpoppen,[4] but the sign of η_1 remained unchanged. In contrast, the new measurements by Silim et al[6] show sign changes for both η_1 and η_2. For positive scattering angles the new and old results agree in all respects.

The values of $|\chi|$ obtained by Standage and Kleinpoppen[4] remain valid in spite of the points discussed above, and so does the sign of χ for positive scattering angles where it was found to be negative. However, had the sign of χ been evaluated for negative scattering angles, it would have become positive. Correcting the sign of η_1 in line with the new measurements[6] and using eq. (8a-c), χ remains unaffected by the change from positive to negative scattering angles and negative throughout the range of angles observed.

Acknowledgments

We would like to thank Professor H Kleinpoppen and Mr H Silim for valuable discussions. The support of the British Science and Engineering Research Council is gratefully acknowledged.

References

1. N C Steph and D E Golden, Phys. Rev. A 21, 1848 (1980)
2. M Kohmoto and U Fano, J. Phys. B 14, L447 (1981)
3. H-J Beyer, H Kleinpoppen, I McGregor and L C McIntyre Jr, J. Phys. B 15, L545 (1982)
4. M C Standage and H Kleinpoppen, Phys. Rev. Lett. 36, 577 (1976)
5. K S Ibraheim, H-J Beyer and H Kleinpoppen, to be published.
6. H A Silim, H-J Beyer and H Kleinpoppen, to be published.
7. K Blum and H Kleinpoppen, Phys. Reports 52, 203 (1979)
8. M Eminyan, K B MacAdam, J Slevin and H Kleinpoppen, J. Phys. B 7, 1519 (1974)
9. K Blum, "Density Matrix Theory and Applications", 1981, Plenum Press, New York.
10. K Blum and H Kleinpoppen, Phys. Reports, 96, 251 (1983)

A TEST OF THE INFLUENCE OF THE NUCLEAR SPIN IN

ELECTRON IMPACT EXCITATION PROCESSES

K. Bartschat[+], K. Blum[+], P. G. Burke[†] and N. S. Scott[†]

[+] Institut für Theoretische Physik I
Universität Münster
Domagkstr. 71, D-4400 Münster
West Germany

[†] Department of Applied Mathematics & Theoretical Physics
The Queen's University of Belfast
Belfast BT7 1NN, Northern Ireland

INTRODUCTION

The Percival-Seaton hypothesis (Percival and Seaton 1958) on the role of the nuclear spin $\underline{I}$ in electron impact excitation of atoms can be summarised as follows: during the excitation and decay processes $\underline{I}$ is assumed to be "frozen" and only the total electronic angular momentum $\underline{J}$ of the atom is changed. Between excitation and decay, however, $\underline{I}$ and $\underline{J}$ couple to give the total angular momentum $\underline{F}$; this coupling then allows a transfer of orientation and alignment from the electronic to the nuclear spin system and vice versa. These effects can be described by "perturbation coefficients" (Frauenfelder and Steffen 1968, Steffen and Alder 1975, Blum 1981) and for collisionally oriented electronic angular momenta and unpolarised nuclear spins in general lead to a depolarisation of the emitted radiation.

Recently doubt has been thrown on the validity of the Percival-Seaton hypothesis by McLucas et al (1982a,b). They observed the $6^1D_2 \rightarrow 6^3P_1$ decay in Hg where the initial state was produced by first exciting the $6\ ^1P_1$ state by electron impact followed by laser excitation of the $6\ ^1D_2$ state from the $6\ ^1P_1$ state. It was then pointed out by Wolcke et al (1984) that electron-photon coincidence experiments can also be used to test the dynamical influence of the nuclear spin on collision processes. The purpose of this paper is to extend the work of Wolcke et al and to show how the combination of general theoretical considerations, explicit numerical calculations and experimental data can be used for a direct test of the validity of the Percival-Seaton hypothesis.

STOKES' PARAMETERS FOR ELECTRON-PHOTON COINCIDENCE EXPERIMENTS

The following experiment has recently been performed in Münster: Hg-atoms are excited from the $(6s^2)\,^1S_0$-ground state to the $(6s6p)\,^3P_1$-state by transversally polarised electrons with polarisation vector $\underline{P} = P_y \hat{e}_y$ (see fig. 1). The electrons scattered in the forward direction ($\theta_e = 0^o$) and the light emitted in the y-direction (i.e. parallel to the polarisation vector of the incident electrons) are detected in coincidence.

The Stokes' parameters of the light emitted in the subsequent decay of the excited atoms are defined as:

i) the degree of linear polarisation with respect to the z and x axes

$$\eta_3 = \frac{I(0) - I(90)}{I(0) + I(90)} \tag{1a}$$

where $I(\beta)$ denotes the intensity transmitted by a Nicol prism orientated at an angle β with respect to the z axis;

ii) the degree of linear polarisation with respect to axes at 45° and 135° to the z axis

$$\eta_1 = \frac{I(45) - I(135)}{I(45) + I(135)} \tag{1b}$$

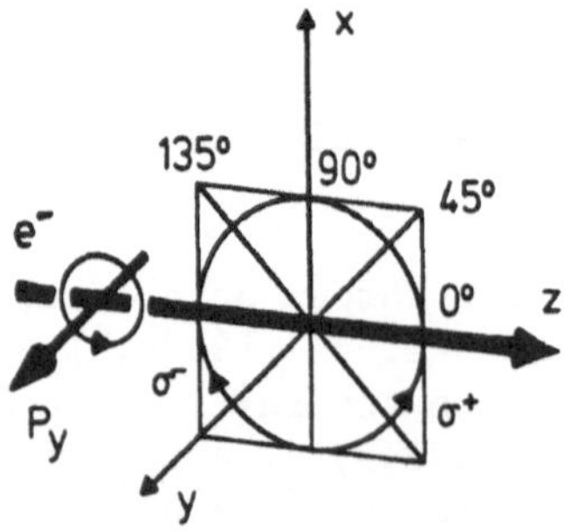

Figure 1: Geometrical arrangement of the electron-photon coincidence experiment performed in Münster (Wolcke et al. 1984).

iii) the degree of circular polarisation

$$\eta_2 = \frac{I_+ - I_-}{I_+ + I_-} \tag{1c}$$

where I_+ (I_-) is the intensity of light transmitted by polarisation filters which only admit photons with positive (negative) helicity.

For unpolarized electrons η_1 and η_2 are zero, while the general description of the excitation process in terms of scattering amplitudes and state multipoles has been given elsewhere (Bartschat et al 1981). For the case of interest here the special geometry of the experiment leads to considerable simplification in the general formulae, as shown below.

Assuming the validity of the Percival-Seaton hypothesis we define the scattering amplitude $f(M,m_1,m_o)$, where M is the z-component of the total electronic angular momentum of the $(6s6p)\,^3P_1$-state, m_o and m_1 are the initial and final z-components of the spin of the continuum electron and all other fixed quantum numbers are omitted. From the conservation of the total electronic angular momentum of the combined system consisting of target electrons plus continuum electron we find, assuming that the nuclear spins are frozen, the selection rule

$$m_o + m_{\ell_o} = M + m_1 + m_{\ell_1} \tag{2a}$$

where m_{ℓ_o} and m_{ℓ_1} are the z-components of the orbital angular momentum of the continuum electron. The special geometry of the experiment under consideration (electron scattering angle $\Theta_e = 0^o$) yields $m_{\ell_o} = m_{\ell_1} = 0$ with the z-axis as the quantisation axis. Hence

$$m_o = M + m_1 \tag{2b}$$

From eq. (2b) it follows that only the scattering amplitudes $f(1,-\frac{1}{2},\frac{1}{2})$, $f(0,\frac{1}{2},\frac{1}{2})$, $f(-1,\frac{1}{2},-\frac{1}{2})$ and $f(0,-\frac{1}{2},-\frac{1}{2})$ can be different from zero. Further reflection invariance in the scattering plane gives $f(-1,\frac{1}{2},-\frac{1}{2}) = f(1,-\frac{1}{2},\frac{1}{2})$ and $f(0,\frac{1}{2},\frac{1}{2}) = f(0,-\frac{1}{2},-\frac{1}{2})$ (see Bartschat et al 1981). Hence, for the geometry under discussion the scattering process is characterised by two independent amplitudes $f(1,-\frac{1}{2},\frac{1}{2})$ and $f(0,\frac{1}{2},\frac{1}{2})$ which correspond to f_2 and f_5 of Bartschat et al (1981). Finally using eqs. (13) and (20) of their paper we obtain the following results for the Stokes' parameters:

$$\eta_1/P_y = \frac{-6.69\,\mathrm{Im}\{f_2 f_5^*\}}{3.21|f_2|^2 + 5.58|f_5|^2} \tag{3a}$$

$$\eta_2/P_y = \frac{7.53\ \mathrm{Re}\{f_2 f_5^*\}}{3.21|f_2|^2 + 5.58\ |f_5|^2} \qquad (3b)$$

$$\eta_3 = \frac{-2.37(|f_2|^2 - 2|f_5|^2)}{3.21|f_2|^2 + 5.58|f_5|^2} \qquad (3c)$$

where Re{x} and Im{x} denote the real and imaginary parts of the quantity x, and the star denotes the complex conjugate.

The symmetry properties of the experiment as well as the general structure of eqs. (3) has been discussed elsewhere (Bartschat et al 1984, Nagy et al 1984). It should be noted, however, that eqs. (3) are only valid, if i) the scattering amplitudes are independent of the nuclear spin and if ii) the description of the depolarisation of the emitted radiation by perturbation coefficients is correct. The calculation of these perturbation coefficients for the natural isotope mixture of Hg (70% $I = 0$, 17% $I = \frac{1}{2}$, 13% $I = \frac{3}{2}$) is described in the appendix of Wolcke et al (1983).

From eqs. (3) it follows that for our special geometry the three Stokes' parameters are not independent but can be related by the following expression:

$$(\eta_2/P_y)^2 = \frac{1.573x}{0.286x^2 + 0.995x + 0.865} - 1.263\ (\eta_1/P_y)^2 \qquad (4a)$$

with

$$x = \frac{0.789 - 0.930\eta_3}{0.395 + 0.535\eta_3} \qquad (4b)$$

The important point is that eqs. (4) hold only if the excitation and decay processes are not influenced by the nuclear spin. Our main result is therefore that an experimental test of eqs. (4) allows conclusions to be drawn on the importance of the nuclear spin in the excitation and decay processes. Any deviation from eqs. (4) outside the experimental errors indicates that the Percival-Seaton hypothesis is violated.

RESULTS AND DISCUSSION

In fig. 2 (a-c) the Stokes' parameters η_1, η_2 and η_3, discussed in the previous section, are plotted as a function of the incident electron energy; η_1 and η_2 are normalised to an electron polarisation of 100%. The arrows mark the thresholds for the Hg-states $(6s6p)\,^3P_1$ (4.89 eV), $(6s6p)\,^3P_2$ (5.46 eV) and

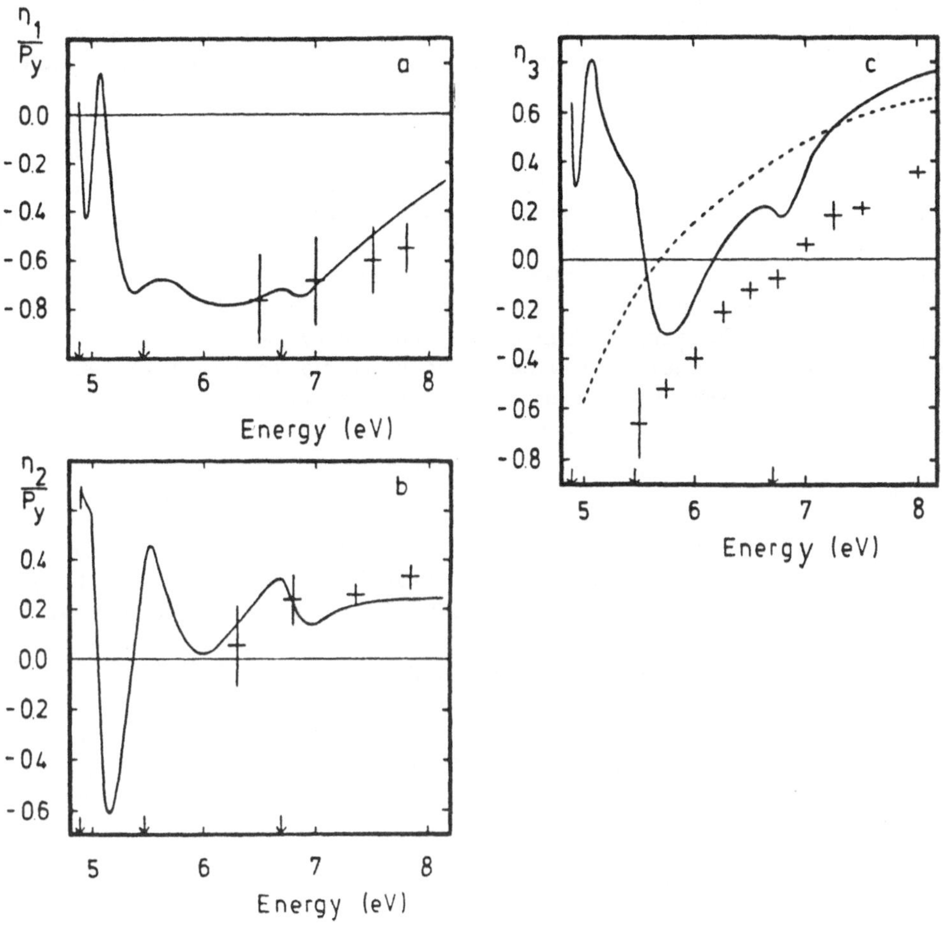

Figure 2: Stokes' parameters for electron-photon coincidence experiments with transversally polarised electrons. Excitation of the $(6s6p)^3P_1$-state of Hg, electron scattering angle $\Theta_e = 0^o$, observation of the emitted light in y-direction.

a) linear light polarisation η_1 normalised to P_y = 100%;

b) circular light polarisation η_2 normalised to P_y = 100%;

c) linear light polarisation η_3.

——— R-matrix calculation;

------- calculation of Bonham (1982);

+ experimental values of Wolcke et al. (1984) and Vollmer (1984).

The arrows mark the excitation thresholds (see text).

$(6s6p)\,^1P_1$ (6.70 eV). The experimental data were taken from Wolcke et al (1984) and Vollmer (1984) and the theoretical curves were calculated using the relativistic R-matrix method of Scott and Burke (1980) as described in Bartschat et al (1984).

It can be seen from fig 2. that the overall agreement between experiment and theory is fairly good. The interesting point with regard to the Percival-Seaton-hypothesis is the following: the theoretical values for the three Stokes' parameters satisfy of course eqs. (4). The experimental data for η_1 and η_2 are in almost complete agreement with the theoretical results within the experimental error bars. Therefore, the experimental data for η_3, which lie significantly below both the R-matrix calculation and the results obtained by Bonham (1982) in the Ochkur-approximation, are not consistent with eqs. (4).

CONCLUSIONS

It has been shown that the electron-photon coincidence experiment in the geometry of Wolcke et al (1984) can be used to test the validity of the Percival-Seaton hypothesis in the $(6s^2)\,^1S_0 \rightarrow (6s6p)\,^3P_1$-electron impact excitation of Hg. The present experimental data do not satisfy eqs. (4). However, before definitive conclusions on the role of the nuclear spin can be drawn, a careful check of possible systematic errors in the experiment is necessary. In particular, the influence of the finite detector resolutions (energy and angles) has to be examined. Furthermore, measurements using an isotope cell (Zaidi et al 1978, 1980), which absorbs the radiation of isotopes with nuclear spin $I \neq 0$, might be useful. These checks are currently in progress in Münster.

ACKNOWLEDGEMENTS

This work was stimulated by discussions with J. Goeke, G. F. Hanne, J. Kessler, W. Vollmer and A. Wolcke. Financial support from the British Council and the Deutsche Forschungsgemeinschaft in Sonderforschungsbereich 216, "Polarisation and Correlation in Atomic Collision Complexes", is gratefully acknowledged.

REFERENCES

Bartschat, K., Blum, K., Hanne, G. F. and Kessler, J, 1981, J.Phys.B. 14:3761.

Bartschat, K., Scott, N. S., Blum, K. and Burke, P. G., 1984, J. Phys. B. 17:269.

Blum, K., 1981, Density Matrix Theory and Applications, Plenum, New York.

Bonham, R. A., 1982, J. Phys. B. 15:L361.

Frauenfelder, H. and Steffen, R. M. in "Alpha-, Beta- and Gamma-Ray Spectroscopy" (ed. K. Siegbahn), North Holland, Amsterdam.

Nagy, O., Bartschat, K., Blum, K., Burke, P. G. and Scott, N. S., 1984, J. Phys. B. 17:L527.

McLucas, C. W., MacGillivray, W. R. and Standage, M. C., 1982a, Phys. Rev. Lett. 48:88.

McLucas, C. W., Wehr, H.J.E., MacGillivray, W. R. and Standage, M. C., 1982b, J. Phys. B. 15:1883.

Percival, I. C. and Seaton, M. J., 1958, Phil. Trans. Roy. Soc. A251:113.

Scott, N. S. and Burke, P. G., 1980, J. Phys. B. 13:4299.

Steffen, R. M. and Alder, K. in "The Electromagnetic Interaction in Nuclear Spectroscopy" (ed W. D. Hamilton), North Holland, Amsterdam.

Vollmer, W., 1984, diploma thesis, University of Münster, unpublished.

Wolcke, A., Bartschat, K., Blum, K., Borgmann, H. Hanne, G. F. and Kessler, J., 1983, J. Phys. B. 16:639.

Wolcke, A., Goeke, J., Hanne, G. F., Kessler, J., Vollmer, W., Bartschat, K. and Blum, K., 1984, Phys. Rev. Lett. 52:1108.

Zaidi, A. A., McGregor, I and Kleinpoppen, H., 1978, J. Phys. B. 11:L151.

Zaidi, A. A., McGregor, I. and Kleinpoppen, H., 1980, Phys. Rev. Lett., 45:1168.

RELATIVISTIC THIRD BORN APPROXIMATION FOR ELECTRON CAPTURE

W.J. Humphries and B.L. Moiseiwitsch

Department of Applied Mathematics and Theoretical Physics
The Queen's University of Belfast
Belfast BT7 1NN, Northern Ireland

INTRODUCTION

In previous papers we have investigated the 1s state to 1s state capture of an electron from a hydrogenic ion by an incident fully stripped positive ion using relativistic generalisations of the Oppenheimer-Brinkman-Kramers (OBK) approximation (Moiseiwitsch and Stockman 1980) and the second Born approximation (Humphries and Moiseiwitsch 1984). In the present paper we analyse the same problem to a greater level of accuracy by using a relativistic form of the third Born approximation, expanding in powers of αZ_A, αZ_B and retaining leading terms, where Z_A, Z_B are the target and projectile charges respectively and α is the fine structure constant. In the limit of non-relativistic but high impact energies our formulae for the scattering amplitude and the total capture cross section without spin flip are in agreement with the results obtained by Shakeshaft (1978) and, as far as the total cross section is concerned, also with the work of Drisco (1955) and Dettman (1971) using the non-relativistic third Born approximation.

PRELIMINARY ANALYSIS

We are concerned with the capture of an electron from a target hydrogenic ion with atomic number Z_A by an incident completely stripped positive ion B with atomic number Z_B which is moving with a very high velocity $\underline{v}$ relative to the nucleus A. The position vectors of the electron referred to the inertial frames S_A and S_B' of the target and the incident ion with origins at A and B are denoted by $\underline{r}_A$ and $\underline{r}_B'$ respectively. $\psi_A(\underline{r}_A,t)$ denotes the initial electronic state vector in frame S_A, $\psi_B(\underline{r}_B',t')$ denotes the final electronic state vector in frame S_B' and $S\psi_B(\underline{r}_B',t')$ denotes the final electronic state in frame S_A where

$$\{i\hbar \frac{\partial}{\partial t}_{\underline{r}_A} + ic\hbar\, \underline{\alpha}.\nabla_{\underline{r}_A(t)} - \beta mc^2 - V_A(\underline{r}_A)\}\psi_A(\underline{r}_A,t) = 0 \tag{1}$$

$$\{i\hbar \frac{\partial}{\partial t'}_{\underline{r}'_B} + ic\hbar\, \underline{\alpha}.\nabla_{\underline{r}'_B(t')} - \beta mc^2 - V_B(\underline{r}'_B)\}\psi_B(\underline{r}'_B,t') = 0 \tag{2}$$

$$\{i\hbar \frac{\partial}{\partial t}_{\underline{r}_A} + ic\hbar\, \underline{\alpha}.\nabla_{\underline{r}_A(t)} - \beta mc^2 - LV_B(\underline{r}'_B)\}S\psi(\underline{r}'_B,t') = 0 \tag{3}$$

Here

$$L = \gamma - (\gamma^2 - 1)^{\frac{1}{2}}\underline{\alpha}.\hat{\underline{v}} \tag{4}$$

transforms the potential $V_B(\underline{r}'_B)$ from S'_B into S_A,

$$S = (\frac{\gamma+1}{2})^{\frac{1}{2}} + (\frac{\gamma-1}{2})^{\frac{1}{2}}\ \underline{\alpha}.\hat{\underline{v}} \tag{5}$$

transforms the wavefunction $\psi_B(\underline{r}'_B,t')$ from S'_B into S_A and

$$V_A(\underline{r}_A) = -\frac{e^2Z_A}{r_A} \tag{6}$$

$$V_B(\underline{r}'_B) = -\frac{e^2Z_B}{r'_B} \tag{7}$$

Finally if $\psi(\underline{r}_A,t)$ denotes the exact electronic wavefunction in frame S_A, then

$$\{i\hbar \frac{\partial}{\partial t}_{\underline{r}_A} + ic\hbar\, \underline{\alpha}.\nabla_{\underline{r}_A(t)} - \beta mc^2 - V_A(\underline{r}_A) - LV_B(\underline{r}'_B)\}\psi(\underline{r}_A,t) = 0 \tag{8}$$

RELATIVISTIC THIRD BORN APPROXIMATION FOR ELECTRON CAPTURE

The transition amplitude a is given by

$$a = \frac{1}{i\hbar}\int dt\ \underline{dr}_A\ [S\psi_B(\underline{r}'_B,t')]^{\dagger}V_A(\underline{r}_A)\psi\ (\underline{r}_A,t) \tag{9}$$

and the third Born approximation to a is

$$a=\frac{1}{i\hbar}\int dt\; d\underline{r}_A [S\psi_B(\underline{r}'_B,t')]^{\dagger} V_A(\underline{r}_A)\psi_A(\underline{r}_A,t)$$

$$+\left(\frac{1}{i\hbar}\right)^2 \int dt\; dt_1\; d\underline{r}_A\; d\underline{r}_{A_1} [S\psi_B(\underline{r}'_B,t')]^{\dagger} V_A(\underline{r}_A) K(\underline{r}_A t,\underline{r}_{A_1} t_1) LV_B(\underline{r}'_{B_1})$$

$$\psi_A(\underline{r}_{A_1},t_1)$$

$$+\left(\frac{1}{i\hbar}\right)^3 \int dt\; dt_1\; dt_2\; d\underline{r}_A\; d\underline{r}_{A_1}\; d\underline{r}_{A_2}$$

$$[S\psi_B(\underline{r}'_B,t')]^{\dagger} V_A(\underline{r}_A) K(\underline{r}_A t,\underline{r}_{A_2} t_2)[V_A(\underline{r}_{A_2}) + LV_B(\underline{r}'_{B_2})]$$

$$K(\underline{r}_{A_2} t_2,\underline{r}_{A_1} t_1) LV_B(\underline{r}'_{B_1})\psi_A(\underline{r}_{A_1},t_1) \qquad (10)$$

where $K(\underline{r}_A t,\underline{r}_{A_1} t_1)$ is the relativistic free particle propagator (Feynman 1949) satisfying

$$\{i\hbar \frac{\partial}{\partial t_{\underline{r}_A}} + ic\hbar\, \underline{\alpha}.\nabla_{\underline{r}_A(t)} - \beta mc^2\} K(\underline{r}_A t,\underline{r}_{A_1} t_1)=i\hbar\delta(t-t_1)\delta^3(\underline{r}_A-\underline{r}_{A_1})$$

(11)

It is interesting to note from (10) that the presence of positron plane wave states in the expansion of the free particle propagator allows the electron to scatter backwards in time during capture:

Second Born Approximation

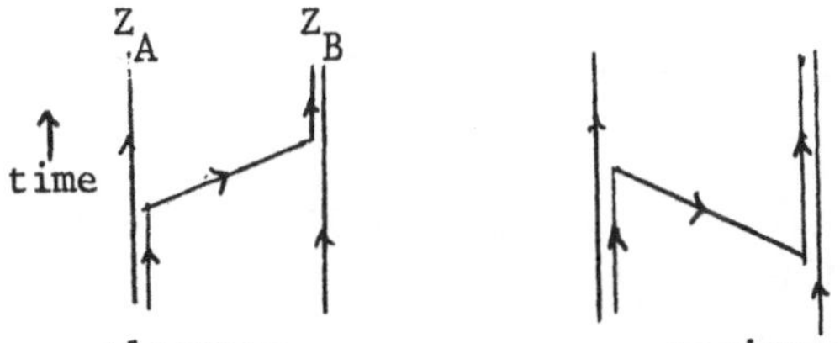

electron intermediate state — a

positron intermediate state — b

(12)

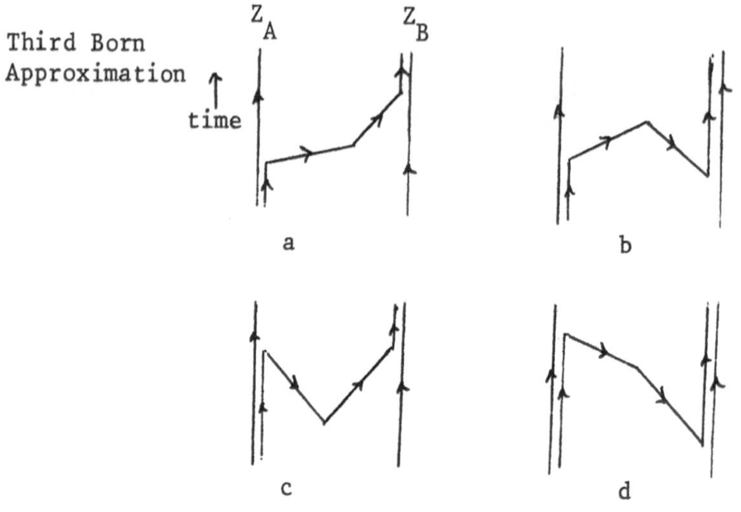

(13)

Process a is dominant in all cases.

TOTAL CROSS-SECTIONS

The evaluation of the transition amplitude and the total cross-section from (10) is lengthy. The integrations are performed most easily using the Fourier transformation method to convert into an equivalent time independent formalism, and details may be found elsewhere (Moiseiwitsch 1966, Humphries and Moiseiwitsch 1984). We find that the total third Born cross-section σ may be written

$$\sigma = \sigma_{2Born} + \sigma_{cor} \tag{14}$$

where σ_{2Born} is the second Born cross-section obtained by retaining only the first two terms in (10), and

$$\sigma_{cor} = \pi^3 a_0^2 \alpha^2 (\alpha^2 Z_A Z_B)^5 \frac{\gamma(2\gamma + \frac{\gamma-1}{\gamma+1})^2}{(\gamma+1)(\gamma-1)^6} \tag{16}$$

is the correction which arises from the inclusion of the third term in (19). In the non-relativistic limit $\gamma \to 1 + \frac{1}{2}(\frac{v}{c})^2$,

$$\sigma_{cor} \to \frac{2^7 \pi^3 a_o^2 \alpha^2 (\alpha^2 Z_A Z_B)^5}{(\frac{v}{c})^{12}} \tag{16}$$

which is in agreement with the result found by Drisco for high but non-relativistic velocities, whilst in the ultra high relativistic limit

$$\sigma_{cor} \to \frac{4\pi^3 a_o^2 \alpha^2 (\alpha^2 Z_A Z_B)^5}{\gamma^4} \tag{17}$$

which is a faster decay than σ_{2Born} which falls of as $\frac{1}{\gamma}$.

CONCLUSIONS

The principal interest in the third Born approximation is that it provides a good indication as to whether or not the second Born approximation gives the correct behaviour of the total cross-section at high energies. This has been confirmed in the non-relativistic case where it is found that the inclusion of the third order term in the Born series does not affect the $E^{-11/2}$ decay of the total cross-section in the asymptotic high energy limit $E \to \infty$. We find a similar result in the relativistic case provided the charges Z_A, Z_B are not too large, that is the inclusion of the third order term in the relativistic Born series has only a minor effect and does not alter the E^{-1} decay of the total cross-section, thus indicating that the second Born approximation is sufficient to describe capture between light ions at relativistic energies.

REFERENCES

Dettman, K., 1971 Springer Tracts in Modern Physics, ed. G. Hohler, Vol. 58, 119.

Drisco, R.M., 1955, Ph.D. Thesis, Carnegie Institute of Technology, Pittsburgh.

Feynman, L.P., 1949, Phys. Rev. 76, 749-

Humphries, W.J. and Moiseiwitsch, B.L., 1984, J. Phys. B: At. Mol. Phys. 17, 2655-69.

Moiseiwitsch, B.L., 1966, Proc. Phys. Soc. 87, 885.

Moiseiwitsch, B.L. and Stockman, S.G., 1980, J. Phys. B: At. Mol. Phys. 13, 2975-81.

Shakeshaft, R., 1978, Phys. Rev. A17, 1011-7.

INNER-SHELL IONIZATION BY TRANSITION RADIATION*

H. Genz

Institut für Kernphysik
Technische Hochschule Darmstadt
6100 Darmstadt, Germany

INTRODUCTION

Recently it has been suggested[1] that photons of the transition radiation might cause K-shell ionization to such an amount that the number of vacancies are not any longer negligible compared to the number caused by ultra relativistic electrons. These assumptions are based on the observation of the X-ray yield following ionization by $e^{\pm}$, $\pi^{\pm}$ and p beams of 5 GeV/c and the calculation of the particle and the transition-radiation photon induced cross section.

Since this possibility for ionization is of interest for the understanding of the interaction between ultra relativistic electrons and the target material - especially, in connection with the often raised question on the absence of the density effect[2] - we have calculated the number of K- and L-shell vacancies caused by transition radiation inside the targets that we have used in our previous experiment. In the present work we want to demonstrate how this effect is reduced by the inclusion of the interference which results from the matching of the generated electromagnetic field at the two boundaries of the target surfaces.

THEORETICAL EXPRESSIONS

Transition radiation is the electromagnetic radiation which is emitted when a charged particle traverses the boundary between two media of different dielectric or magnetic properties. In case the charged particle traverses a slab of material in form of a target,

* Supported by Deutsche Forschungsgemeinschaft

the influence of two boundaries described by the two surfaces of the target has to be taken into account. The phase relation between the field amplitudes originating at the individual interfaces forms a linear superposition of the single interface expression. This phase factor has to be included since it describes properly that the particle traverses different interfaces at different times. Following the thorough investigations by Artru et al. (Ref.3) we obtain under these conditions the transition-radiation photon yield from the expression (Eq.3.25 of Ref.3)

$$\frac{dW}{d\omega} = \frac{4\alpha}{\pi}\int_a^\infty (y-a)\left(\frac{1}{y} - \frac{1}{y+V}\right)^2 \cdot \sin^2 \frac{y+V}{2}\, dy, \tag{1}$$

where $y = (\gamma^{-2}+\Theta^2+\xi^2)\cdot\omega\ell_1/2$, $a = (\omega/\omega_1)/(\gamma/\gamma_1)^2$ and $V = \omega_1/\omega$, with Θ being the angle between the particle trajectory and the direction of observation, $\xi = \omega_p/\omega$, ω_p being the plasma frequency, ℓ_1 is the target thickness, $\gamma_1 = \ell_1\omega_p/2$ and $\omega_1 = \gamma_1\omega_p$. (We are using units with $\hbar = c = 1$).

As an example we have calculated the photon yield according to Eq.(1) caused by electrons of 900 MeV traversing a $_{29}$Cu target with a thickness of 379µg/cm^2 placed inside a vacuum. The result is displayed in Fig.1. One recognizes clearly the oscillating behaviour at very low photon energies and the rapid decrease of the yield over several magnitudes with increasing photon energy. Only transition-radiation photons with energies in excess of the binding energy I can ionize the K-shell (i.e. I>9 keV).

The K-shell ionization cross section due to these real photons can be obtained now from the photon spectrum defined as

$$\frac{dI^{TR}}{d\omega} = \frac{1}{\omega}\cdot\frac{dW}{d\omega} \tag{2}$$

by folding the spectrum with the photoionization cross section[4] according to

$$\sigma_{TR} = \int_I^{E_o} \sigma_{photo} \frac{dI^{TR}}{d\omega}\, d\omega \quad , \tag{3}$$

where E_o denotes the electron impact energy. Since the photoionization cross section drops off exponentially with increasing energy, the main contribution to the total cross section, i.e. about 99%, comes from the photon energy region $I < \hbar\omega < 10I$, which is indicated in Fig.1. Applying Eqs.(1)-(3) in the relevant region of integration we deduce for the example of the Cu target, as displayed in Fig.1, for the K shell σ_{TR} = 11 barn. We have also performed the calculation of σ_{TR} for transition radiation originating from electrons passing only one interface[5], in a very similar approach as the one of the authors of Ref.1, and obtain for photons from 900 MeV electrons on Cu σ_{TR}= 53 barn. Our previous experimental value[2] for the direct K-shell ionization of Cu by ultra relativistic electrons of 900 MeV was σ_K= 461 ± 30 barn. From this example we learn two things: first, the proper inclusion of the interference effect reduces σ_{TR} to less than

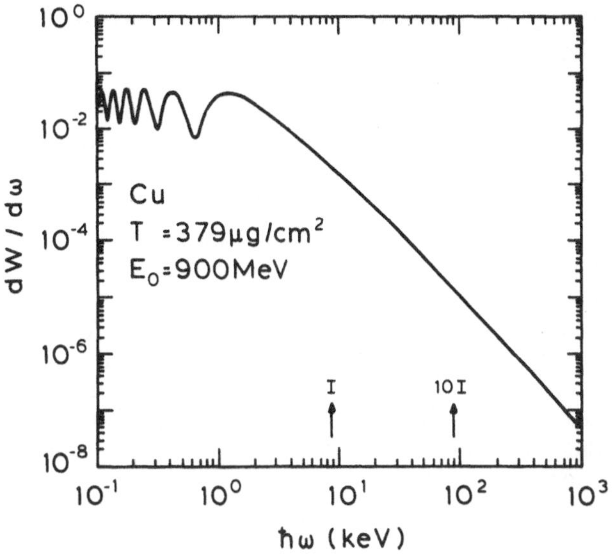

Fig. 1. Transition-radiation photon yield calculated for electrons of 900 MeV passing through a Cu target of 379 μg/cm^2 thickness placed in a vacuum. The K-shell binding energy is denoted by I.

20% of its value; second, the number of vacancies produced by transition radiation amounts only to 2.4% of the electron induced number.

RESULTS AND DISCUSSION

For the low Z targets($_{16}$S, $_{20}$Ca, $_{25}$Mn, $_{26}$Ni, $_{29}$Cu, $_{32}$Ge, $_{47}$Ag) applied in our previous work[2] we have calculated the transition-radiation induced K-shell ionization cross section according to Eqs (1) - (3). These data together with the calculations for transition radiation originating from electrons passing only through one interface are presented in detail elsewhere[5]. Here we will quote only the main consequences that follow from these investigations: 1) The proper inclusion of the phase relation between the field amplitudes originating at the two interfaces of the target - as has been applied to the probability of inner-shell ionization in the present paper for the first time - results in a destructive interference which yields a much smaller value of σ_{TR} for photons created in a single foil as compared to those of a single interface[1]; 2) The fraction of K-shell vacancies caused by transition radiation amounts at most to 3% of the number originating from direct electron impact in the energy region $0.9<E<2.0$ GeV and for our targets; 3) In our previous work we defined the quantity R as the ratio of the electron induced cross section taken at 2.0 GeV and at 0.9 GeV. Correcting our electron induced data by the small amount of vacancies caused by transition radiation we now deduce on the average $\bar{R} = 1.07 \pm 0.02$ in accordance with our former result.

From these findings we learn that the electron induced cross section increases by 7% in the energy region $0.9 \leq E_0 \leq 2.0$ GeV. Furtheron, the initially quoted assumption[1] that within the number of detected K-X rays in our previous experiment a non negligible quantity results from photons of the transition radiation does not hold. Therefore our former conclusion on the behaviour of the electron induced K-shell cross section appear still to be correct. The cross section is increasing in the investigated energy region in contrast to theoretical predictions according to which a saturation of the cross section should occur due to the polarization of the target medium caused by the electromagnetic field of the incident electrons. This phenomenon is also called density effect.

CONCLUSION

From the calculations performed in the present work we deduce for the targets investigated in our previous work that the fraction of inner-shell vacancies caused by transition radiation amounts at most to 3% by which our former electron induced data have to be reduced. However, this is only a small correction which is still within the quoted error. Our conclusion is therefore different from the one of Ref.1 because of the proper treatment of the interference effects in the present paper. At photon energies which give the main contribution to σ_{TR}, the thin targets used in our previous experiment lead to $\exp[-z/\lambda_a(\omega)] \approx 1$ in Eq.1 of Ref.1 while the factor $dI^{TR}/d\omega$ is negligibly small when interference is taken into account. The discrepancy between theory and experiment as quoted in our article[2] for the electron induced cross section can therefore still not be explained by the effect of transition radiation, contrary to the arguments of Bak et.al.[1]

ACKNOWLEDGMENT

Many stimulating discussions with P. Eschwey, A. Richter and A.H. Sørensen are highly acknowledged. For computational work I am thankful to S. Reusch.

REFERENCES

1. J.F. Bak, F.E. Meyer, J.B.B. Petersen, E. Uggerhøj, K. Østergaard, S.P. Møller, and A.H. Sørensen, Phys.Rev.Lett. 51:1163 (1983).
2. H. Genz, C. Brendel, P. Eschwey, U. Kuhn, W. Löw, A. Richter, P. Seserko, and R. Sauerwein, Z.Phys. A305:9 (1982).
3. X. Artru, G.B. Yodh, and G. Mennessier, Phys. Rev. D12:1289 (1975).
4. J.H. Scofield, Lawrence Livermore LAB REPORT, UCRL, 51326 (1973).
5. H. Genz, P. Eschwey, S. Reusch, and A. Richter, Z. Phys., to be published.

APPLICATIONS OF STEPWISE ELECTRON AND LASER EXCITATION OF ATOMS TO ATOMIC COLLISION STUDIES

M.C. Standage

School of Science
Griffith University
Nathan, Queensland, Australia, 4111

1.0 INTRODUCTION

Over the past decade, the application of tunable dye lasers to the field of electron-atom collision physics has produced a range of new techniques for investigating collision processes. One of these techniques utilizes stepwise excitation of target atoms by a combination of electron and laser excitation. Information about the collision processes involved in the electron excitation step is obtained from intensity and polarization measurements made on fluorescence emitted from the stepwise excited atoms. Such experiments fall into two categories determined by whether laser excitation is used in the first or second excitation steps. In the former case, electron excitation from the laser excited state to higher lying states provides information on excited state collision parameters. In this paper, we consider only the latter case which provides new techniques for studying the electron impact excitation of ground state atoms (1-7) and allows several aspects of such collision processes to be studied in new detail. The narrow bandwidth of laser radiation permits the fine and hyperfine structure of many atomic transitions to be resolved in the laser excited step providing a method for studying the effects of such structure on atomic collisions, as in the case of tests of the Percival-Seaton hypothesis, or for eliminating such effects in the measurement of atomic collision parameters. Stepwise excitation techniques of this type provide new methods for investigating the electron impact excitation of vacuum-ultraviolet (VUV) transitions and of metastable states. A brief review of the theory associated with stepwise excitation techniques is given in § 2.0-4.0 and experimental results obtained for electron-mercury atom collisions are presented in § 5.0-6.0.

2.0 THEORY

We consider the stepwise excitation scheme shown in figure 1(a) in which target atoms are excited from the ground state $|g\rangle$ to the first excited state $|e\rangle$ by electron impact, while the second excitation step to state $|i\rangle$ is carried out by laser radiation tuned to resonance. The stepwise excited atoms decay radiatively to either state $|e\rangle$ or $|f\rangle$. For reasons of experimental convenience the fluorescent intensity and polarisation of the $|i\rangle \to |f\rangle$ transition is usually monitored. Figure 1(b) depicts the excitation and emission geometry used in this paper in which the electron and laser beams propagate collinearly along the z axis and radiation is detected along the y axis. It is assumed that the laser radiation may be either circularly or linearly polarised; in the latter case the polarisation vector $\underline{e}$ makes an angle α with the x axis. In the case of linear analysis of the scattered radiation the polarisation vector $\underline{f}$ makes an angle β with the z axis.

The fluorescent intensity emitted by a stepwise excited atom at some time t after the initial electron impact is given by

$$I(t) = \sum_{\mu\mu'} \rho_{\mu\mu'}(t)\, F_{\mu'\mu} \tag{1}$$

where $\rho_{\mu\mu'}(t)$ is the density matrix element taken between substates $|\mu\rangle$ and $|\mu'\rangle$ of the upper excited state $|i\rangle$ and F is an emission operator which describes the fluorescent emission from the atom. The matrix elements of F are given by

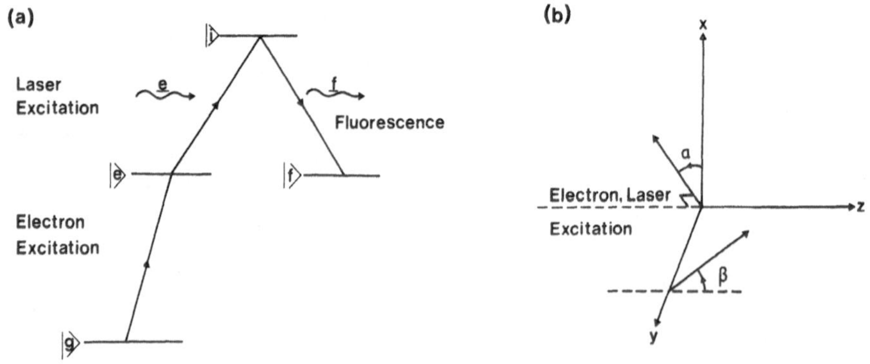

Fig. 1. (a) General stepwise excitation scheme.
(b) Excitation and emission geometry.

$$F_{\mu'\mu} = \sum_n \langle \mu' | \underline{f}.\underline{P} | n \rangle \langle n | \underline{f}^*\underline{P} | \mu \rangle \qquad (2)$$

where $\underline{f}$ is the polarisation vector of the analyser, P is the electric dipole moment operator and $|n\rangle$ are substates of the final state $|f\rangle$.

Our aim is to evaluate equation (1) in terms of the density matrix elements which describe the electron excitation step. The approach used here is similar to that of Hertel and Stoll (1977) (8), for example, in that a quantisation axis is selected such that the fluorescent emission operator is diagonal. For a linear analyser, F is diagonalised by choosing a quantisation axis parallel to the polarisation vector of the polariser. For a circular analyser, the quantisation axis is chosen along the propagation direction of the fluorescence. For such a choice of quantisation axis, equation (1) becomes

$$I(t) = \sum_{\mu} \rho'_{\mu\mu}(t)\, F_{\mu\mu} \qquad (3)$$

where $\rho'_{\mu\mu}(t)$ is determined in the 'emission' frame of reference. This expression may be evaluated in the 'excitation' frame of reference for which the quantisation axis is chosen along the z axis of figure 1(b). Equation (1) may be rewritten to give, (6)

$$I(t) = \sum_{\mu k Q} (-1)^{J'-\mu} \langle J'\ J'\ \mu{-}\mu | ko \rangle\, D^k_{oQ}(\alpha\beta\gamma)\, \langle T^{\dagger}_{kQ} \rangle\, F_{\mu\mu} \qquad (4)$$

where $0 \leqslant k \leqslant 2J'$, $-k \leqslant Q \leqslant k$, $\langle J'J'\mu{-}\mu | ko \rangle$ is a Clebsch-Gordan coefficient and $D^k_{oQ}(\alpha,\beta,\gamma)$ are matrix elements of the rotation operator. The Euler angles (α,β,γ) give the angular position of the z-axis of the 'emission' frame relative to the 'excitation' frame. $\langle T^{\dagger}_{kQ} \rangle$ are state multipoles which may be expressed in terms of density matrix elements by, (9)

$$\langle T^{\dagger}_{kQ} \rangle = \sum_{\mu\mu'} (-1)^{J'+\mu'} \langle J'J'\ \mu{-}\mu' | kQ \rangle\, \rho_{\mu\mu'} \qquad (5)$$

We consider the electron impact excitation of atoms initially in their ground state $|J_g M_g\rangle$ at time t = 0. The electrons have momentum P_i and spin component m_i and the quantisation axis is taken parallel to P_i. Since electrons and atoms are assumed to be uncorrelated prior to the collision, the initial density matrix operator ρ_i can be represented as

$$\rho_i = \frac{1}{2(2J_g+1)} \sum_{M_g m_i} |J_g M_g\rangle\langle J_g M_g|\, |m_i\rangle\langle m_i| \qquad (6)$$

The electron scattering process may be represented by a scattering operator T such that the density matrix before and after scattering is related by the expression $\rho = T\rho_i T^\dagger$. The matrix elements of ρ are given by, (6)

$$\rho_{MM'} = \frac{1}{2(2J_g+1)} \sum_{\substack{M_g m_i \\ m_f}} \langle Mm_f|T|M_g m_i\rangle\langle M_g m_i|T^\dagger|M' m_f\rangle \tag{7}$$

where the state $|Mm_f\rangle \equiv |JM\rangle|m_f\rangle$. The state $|JM\rangle$ is the excited atomic state produced during the collision, while $|m_f\rangle$ is the final electron state.

If the electrons are undetected, the appropriate reduced matrix element is obtained by integrating over the electron scattering angles to give

$$\langle M|\rho|M'\rangle = \sum_{m_f} \int \langle Mm_f|\rho|M'm_f\rangle \, d\Omega_e \tag{8}$$

The symmetry properties of the reduced electron density matrix elements have been discussed by a number of authors. Briefly summarised, the Hermitian character of p implies that $\rho_{MM'} = \rho^*_{-M-M'}$ and symmetry with respect to reversal of the z axis implies that $\rho_{MM} = \rho_{-M-M}$. For axial symmetry conditions, p is diagonal in M, while in the case of planar symmetry electron scattering, $\rho_{MM'} = (-1)^{M+M'}\rho_{-M-M'}$.

Following excitation by electron impact, the atom either decays back to the ground state or is further excited by laser radiation to the state $|i\rangle$. If the laser excitation is sufficiently weak such that no optical Rabi nutations or optical pumping can occur, then an exact expression may be derived which relates the upper excited state density matrix elements $\rho_{\mu\mu'}$ to the electron impact excited density matrix elements $p_{MM'}$. Webb et al 1984a (6) have shown that the connection between the upper and lower excited state density matrix elements $\rho_{\mu\mu'}$ and $\rho_{MM'}$ is given by

$$\rho_{\mu\mu'} = \frac{1}{\Gamma_i\Gamma_e} \sum_{MM'} \langle\mu|\underline{e}.\underline{P}|M\rangle\langle M'|\underline{e}^*\underline{P}|\mu'\rangle \, \rho_{MM'} \tag{9}$$

where Γ_i and Γ_e are decay rates associated with states $|i\rangle$ and $|e\rangle$.

Webb et al (1984b) (7) have considered the case where the laser interaction with the atoms is strong and an adequate description requires the representation of the optical excitation process in terms of optical density matrix equations.

3.0 STEPWISE EXCITATION/EMISSION SCHEMES

In this section, we consider excitation schemes in which electron impact excitation from a ground state $|J_g = 0\rangle$ to a first excited state $|J_e = 1\rangle$ is followed by laser excitation to the state $|J_i = 2\rangle$. The intensity and polarisation of fluorescence emitted in the decay to the state $|J_f = 1\rangle$ is calculated. It is assumed that optical excitation and emission occurs via optically allowed transitions. We first consider those cases for which the electron excitation step has axial symmetry. Hyperfine-structure effects and the Percival-Seaton hypothesis are then considered.

A discussion of theoretical expressions appropriate to planar symmetry experiments is to be found in Webb et al (1984 a,b) (6,7).

The excitation and emission geometry is shown in figure 1(b). For circularly polarised laser excitation and linear analysis of the stepwise excited fluorescence emitted in the $(J_i = 2)-(J_f = 1)$ transition, equation (4) can be shown to yield (6)

$$I_\beta = C[6\rho_{00} + 14\rho_{11} + \tfrac{1}{2}(3\cos^2\beta-1)(3\rho_{00}-10\rho_{11})] \tag{10}$$

where ρ_{00} and ρ_{11} are given by equation (8) and correspond to the partial total cross sections for the M = 0 and M = ±1 sublevels of the electron impact excited $(J_e = 1)$ state. The term C includes such factors as the reduced dipole moments for the optical transitions, decay constants, the atomic density and geometrical factors. The expression for stepwise fluorescent intensity, given by equation (10), is independent of whether the direction of propagation of the laser beam is collinear or anti-collinear to the electron beam direction.

The total excitation cross section is defined to be proportional to the total intensity I, where

$$I = I_0 + 2I_{90} = 6C(3_{00} + 7_{11}) \tag{11}$$

The line polarisation is given by

$$P = (I_0 - I_{90})/(I_0 + I_{90}) = (9\rho_{00}/\rho_{11}-30)/(27\rho_{00}/\rho_{11}+46) \tag{12}$$

Measurements of the total intensity and polarisation of fluorescence emitted in the $(J_i = 2)-(J_f = 1)$ transition allow the relative excitation and the line polarisation of the electron impact excited $(J_g = 0)-(J_e = 1)$ transition to be determined with the aid of equations (11) and (12). The total excitation cross section for the $(J_g = 0)-(J_e = 1)$ excitation step is proportional to I_e, where

$$I_e \sim \rho_{00} + 2\rho_{11} \tag{13}$$

and the line polarisation of the $(J_g = 0)-(J_e = 1)$ transition is given by

$$P_e \sim \frac{\rho_{00}/\rho_{11} - 1}{\rho_{00}/\rho_{11} + 1} \tag{14}$$

The relative excitation cross section Q_e for the $(J_g = 0)-(J_e = 1)$ transition may be expressed in terms of the directly measured intensity I for the $(J_i = 2)-(J_f = 1)$ transition such that

$$Q_e \sim \frac{\rho_{00}/\rho_{11} + 2}{3\rho_{00}/\rho_{11} + 7} \times I \tag{15}$$

Values of the ratio ρ_{00}/ρ_{11} are obtained from line polarisation measurements made on the fluorescence emitted in the $(J_i = 2)-(J_f = 1)$ transition and applied to equation (12).

The high spectral resolution which is possible in the laser-excited transition of a stepwise excitation process allows the role of hyperfine structure in collisions to be studied. McLucas et al (1982 a,b) (4,5) compared line polarisation measurements for stepwise excited transitions in the I = 0 and I = ½ isotopes of Hg. Such measurements can in suitable cases provide a test of the Percival-Seaton hypothesis. This hypothesis assumes that during a collision between an atom possessing nuclear spin and an electron, the coupling between the nuclear-spin angular momentum and the electronic angular momentum is broken. The angular momenta are assumed to recouple after the collision with no change in the nuclear-spin quantum numbers. Hyperfine structure and the Percival-Seaton hypothesis may be included in the theory of the electron excitation process in the following way. The initial density matrix may be represented as a direct product of the decoupled ground-state electronic and nuclear-spin wavefunctions together with the wavefunctions for the colliding electrons, such that

$$\rho_i = \frac{1}{2(2J_g+1)(2I+1)} \sum_{\substack{M_g M_i \\ m_i}} |J_g M_g\rangle\langle J_g M_g| \, |IM_I\rangle\langle IM_I| \, |m_i\rangle\langle m_i| \tag{16}$$

Noting that $\rho = T\rho_i T^\dagger$, it is assumed under the Percival-Seaton hypothesis that the T operator does not act on the nuclear-spin states and it can be shown that (6)

$$\rho_{M_F M_F'} = \sum_{\substack{MM' \\ M_I}} \langle IJM_I M|FM_F\rangle\langle IJM_I M'|FM_F'\rangle \, \rho_{MM'} \tag{17}$$

where $\rho_{MM'}$ is given by equation (7).

The role of hyperfine structure in stepwise excitation expressions may be illustrated by considering the hyperfine stepwise excitation scheme in ^{199}Hg studied by McLucas et al (1982 a,b) (4,5). The stepwise transitions were electron excitation of the $6^1S_0(F = \frac{1}{2})$ $-6^1P_1(F = \frac{3}{2})$ transition followed by laser excitation of the 6^1P_1 $(F = \frac{3}{2})-6^1D_2(F = \frac{5}{2})$ transition with fluorescence from the $6^1D_2(F = \frac{5}{2})$ $-6^3P_1(F = \frac{3}{2})$ transition being monitored for line polarisation. The breakdown of LS coupling in Hg makes the latter transition optically allowed. The excitation and emission geometry is shown in figure 1(b). We consider only the axial symmetry case with circularly polarised laser excitation and linear analysis of the scattered fluorescence. For weak laser excitation, it can be shown that the intensity is given by

$$I_\beta = \frac{c'}{4}\left[\frac{55}{3}\rho_{\frac{3}{2}\frac{3}{2}} + 15\rho_{\frac{1}{2}\frac{1}{2}} - (3\cos^2\beta-1)\left(\frac{23}{3}\rho_{\frac{3}{2}\frac{3}{2}} - 3\rho_{\frac{1}{2}\frac{1}{2}}\right)\right] \qquad (18)$$

where the density matrix elements $\rho_{\frac{1}{2}\frac{1}{2}}$, $\rho_{\frac{3}{2}\frac{3}{2}}$ are associated with the electron-excited $6^1P_1(F = \frac{3}{2})$ state. Under the Percival-Seaton hypothesis, the density matrix elements may be written in terms of the I = 0 density matrix elements using equation (17) to give,

$$\rho_{\frac{3}{2}\frac{3}{2}} = \rho_{11} \qquad \rho_{\frac{1}{2}\frac{1}{2}} = \frac{2}{3}\rho_{00} + \frac{1}{3}\rho_{11} \qquad (19)$$

where ρ_{00} and ρ_{11} are J = 1 density matrix elements. The line polarisation is given by

$$P = \frac{3\rho_{00}/\rho_{11} - 10}{11\rho_{00}/\rho_{11} + 20} \qquad (20)$$

This expression is valid only if equation (17) holds. A direct test of the Percival-Seaton hypothesis can be made by comparing ratios of the partial total cross sections ρ_{00}/ρ_{11} obtained from I = 0 and I = $\frac{1}{2}$ polarisation measurements.

4.0 STEPWISE EXCITATION OF METASTABLE STATES

Stepwise excitation techniques provide a new means of studying the electron impact excitation of metastable states as the work of Phillips et al (1981) (1-3) on Ne has demonstrated. Webb et al (1985) (10) have reported on the study of electron impact excitation of the 6^3P_2 metastable state of mercury using a stepwise excitation scheme in which the $6^1S_0-6^3P_2$ transition is excited by electron impact followed by single-mode laser excitation of the $6^3P_2-7^3S_1$ (546.1nm) transition. Stepwise-excited fluorescence is observed in the $7^3S_1-6^3P_0$ (404.7nm) transition. Here, this stepwise excitation scheme is analysed for the excitation and emission geometry of figure 1(b) with linearly polarised laser excitation and linear analysis of the stepwise excited fluorescence. Only axial symmetry is considered for the electron excitation step. For the laser-excited step, the 6^3P_2 state is metastable and Γ_e corresponds to the transit time of the atom through the laser beam. Webb et al 1984a (6) show that for this case equation (9) becomes

$$\rho_{\mu\mu'} = \frac{1}{\Gamma_i} \sum_{MM'} \langle \mu | \underline{e}.\underline{P} | M \rangle \langle M' | \underline{e}^*.\underline{P} | \mu' \rangle \, \rho_{MM'} \tag{21}$$

For linearly polarised excitation and linear analysis it can be shown that (6)

$$I_\beta \sim 2\rho_{22}+\rho_{11}+\frac{1}{3}\rho_{00}+\frac{1}{2}\rho_{00}\cos 2\alpha \, \mathrm{Sin}^2\beta - (3\cos^2\beta-1)(\rho_{22}-\rho_{11}+\frac{1}{6}\rho_{00}) \tag{22}$$

where ρ_{00}, ρ_{11}, ρ_{22} are metastable-state density matrix elements.

The excitation cross section is proportional to I where

$$I = C''[6\rho_{22} + 3\rho_{11} + (1+\cos 2\alpha)\rho_{00}] \tag{23}$$

The line polarisation is given by

$$P = \frac{6\rho_{11}/\rho_{00} - 6\rho_{22}/\rho_{00} - (1+\cos 2\alpha)}{6\rho_{11}/\rho_{00} + 6\rho_{22}/\rho_{00} + (1+\cos 2\alpha)} \tag{24}$$

The ratios of the partial total cross sections ρ_{11}/ρ_{00} and ρ_{22}/ρ_{00} may be deduced from line polarisation measurements made for selected values of the input polariser angle α. Thus, stepwise excitation techniques offer not only a new method for obtaining metastable excitation cross sections, but also provide new information about metastable partial total cross sections.

5.0 EXPERIMENTAL RESULTS

In this section, stepwise excitation experiments on atomic mercury are reviewed (4,5,10). A schematic diagram of the experiment is shown in figure 2. A turntable system mounted off an end flange of the vacuum system carried the electron gun, Faraday cups and the atomic beam source. Most components including the vacuum chamber were fabricated out of non-magnetic stainless steel. A set of Helmholtz coils placed around the vacuum chamber cancelled residual magnetic fields in the central region of the chamber. The laser beam was sent through an optical window and intersected the electron and atomic beam. Fluorescence scattered from the interaction region was collected by a quartz lens and focused through another optical window onto a photon detection system mounted on the end flange of the vacuum system opposite to the turntable system. The photon count rate was recorded as a function of the polariser angle using standard photon-counting equipment. An electron gun of the electrostatic aperture lens type was used. The electron source was a heated

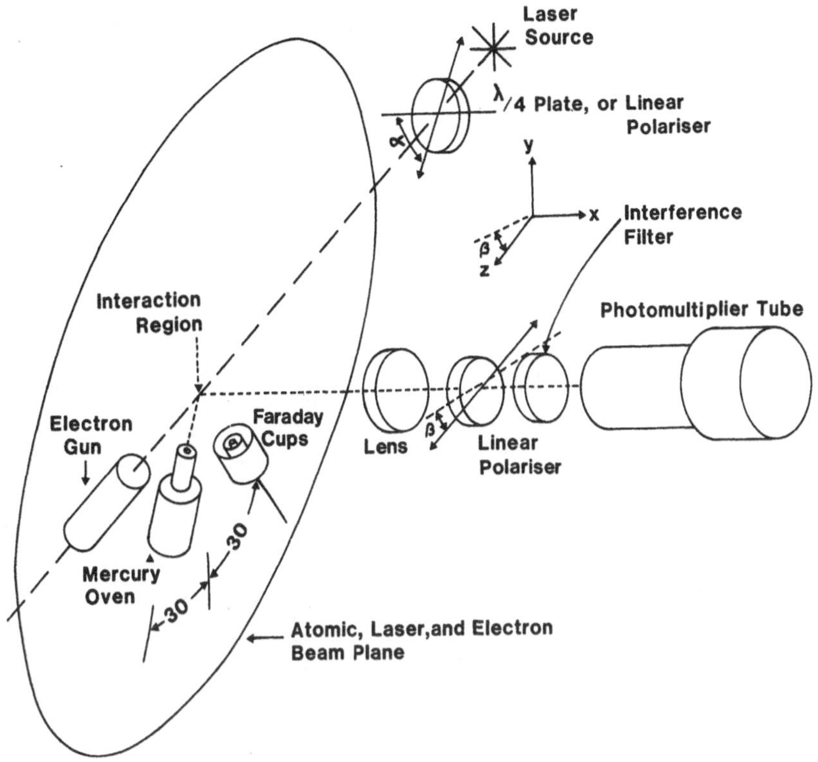

Fig. 2. The experimental geometry.

tungsten hair pin filament. Electrons were accelerated and focused by a two-stage system consisting of a two aperture lens followed by a three aperture 'zoom' lens. The gun was designed to operate in the 5-20eV energy range with currents of 1-5μA.

Laser radiation was provided by a Spectra-Physics 380A single-mode ring dye laser for the initial experiments and by a 380D actively stabilized dye laser for later experiments. The first experiments were performed on the stepwise excitation scheme consisting of electron excitation of the 6^1S_0-6^1P_1 transition of Hg followed by laser excitation of the 6^1P_1-6^1D_2 (579nm) transition. Stepwise fluorescence was detected from the 6^1D_2-6^3P_1 (313nm) transition. Figure 3 shows the hyperfine structure of the 6^1P_1-6^1D_2 transition obtained by recording the stepwise fluorescent intensity as a function of laser frequency (14). Experiments were performed for the I = 0 and I = ½ isotopes by tuning the laser to either peak C corresponding to the I = ½ isotope hyperfine transition $6^1P_1(F = \frac{3}{2})$-$6^1D_2(F = \frac{5}{2})$.

The total fluorescence recorded by the photon detector consists in general of a signal due to direct electron excitation of the $^1D_2$, $^3D_1$ and $^3D_2$ levels as well as the stepwise excited signal. The relative strength of the stepwise to direct signal was typically 50%

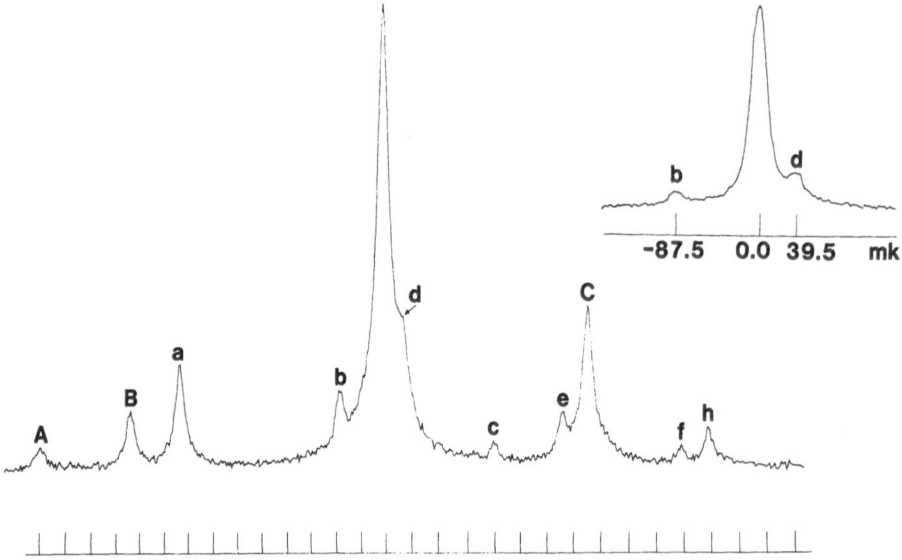

Fig. 3. Stepwise excitation spectral scan of the 6^1D_2-6^1P_1 (579.1nm) line of mercury. Insert shows a higher resolution scan of the central feature. Frequency marker interval is 1.4537 GHz.

at 9eV and increased at lower energies as the direct signal diminished. Measurements at a particular energy were performed by taking repeated runs of fixed duration as a function of polariser angle with the laser tuned and detuned to eliminate the direct electron excitation signal. Four measurements, at polariser angles separated by 90° to eliminate small instrumental asymmetries, were used for each polarisation measurement. Total cross section measurements were normalised to the electron beam current measured at the Faraday cup. Experimental conditions were varied to establish sources of systematic errors. Three sources were investigated; pressure depolarisation due to radiation trapping, magnetically-induced Hanle effects and optical pumping and operating conditions were chosen which minimized these effects. A full discussion is given in ref. 4, 5.

To reduce experimental uncertainties, up to twenty individual measurements of the 313nm line polarisations were taken at a given incident electron energy. The mean and standard deviation of these measurements were determined and used in equation (12) to determine the mean and standard deviation of the ratio of partial total cross sections ρ_{00}/ρ_{11} for the I = 0 isotopes under circularly polarised laser excitation. These values were then used in equation (14) to determine the line polarisation and its standard deviation for the 185nm line. The polarisation obtained was corrected for pressure depolarisation and the added uncertainty was incorporated into the experimental uncertainty.

The total excitation cross section for the I = 0 isotopes was obtained by using the expression given by equation (15).

Figure 4 shows theoretical and experimental polarisation values of the 185nm line for the I = 0 isotope. The theoretical predictions of McConnell and Moiseiwitsch (1968) (11) used the Coulomb approximation with intermediate coupling wavefunctions to represent the mercury atom. The Ochkur approximation was used to determine the polarisation. This theory is not expected to be good for low incident electron energies. The I = 0 experimental data exhibit a decrease in polarisation towards threshold in contrast to the theoretical predictions. The experimental results confirm the behaviour near threshold observed by Skinner and Appleyard (1927) (12) and Ottley et al (1974) (13), using direct electron excitation of the naturally occurring isotopic mixture. Comparison with polarisation data obtained by Ottley et al (1974) for the direct excitation of the naturally occurring isotope mixture, which is dominated by the I = 0 isotopes, is in reasonable agreement with the I = 0 data except within about 2eV of threshold. Some evidence of structure is apparent in the vicinity of 8eV incident electron energy.

A direct test of the Percival-Seaton hypothesis can be made by comparing the ratios of partial total cross sections ρ_{00}/ρ_{11} obtained from the I = 0 and I = $\frac{1}{2}$ polarisation data using equations (12) and (20) respectively. As mentioned in § 3.0, equation (20) depends explicitly on the Percival-Seaton hypothesis. Table 1 shows a comparison of ρ_{00}/ρ_{11} data. At higher energies the agreement is good, but nearer threshold a marked discrepancy exists which lies well outside 90% confidence limits for the data indicating a breakdown in the Percival-Seaton hypothesis at these energies.

Incident Electron Energy (eV)	ρ_{00}/ρ_{11} (I = 0)	ρ_{00}/ρ_{11} (I = $\frac{1}{2}$)
8.0	$5.60 \pm ^{0.62}_{0.52}$	$3.05 \pm ^{0.42}_{0.36}$
8.5	$3.90 \pm ^{0.70}_{0.54}$	$2.43 \pm ^{0.15}_{0.14}$
9.0	$3.64 \pm ^{0.26}_{0.23}$	$3.36 \pm ^{0.23}_{0.21}$
9.5	$3.38 \pm ^{0.34}_{0.30}$	$3.54 \pm ^{0.58}_{0.45}$
10.0	$3.51 \pm ^{0.67}_{0.51}$	$3.52 \pm ^{0.66}_{0.51}$
10.5	$4.01 \pm ^{0.44}_{0.37}$	$3.42 \pm ^{0.50}_{0.41}$

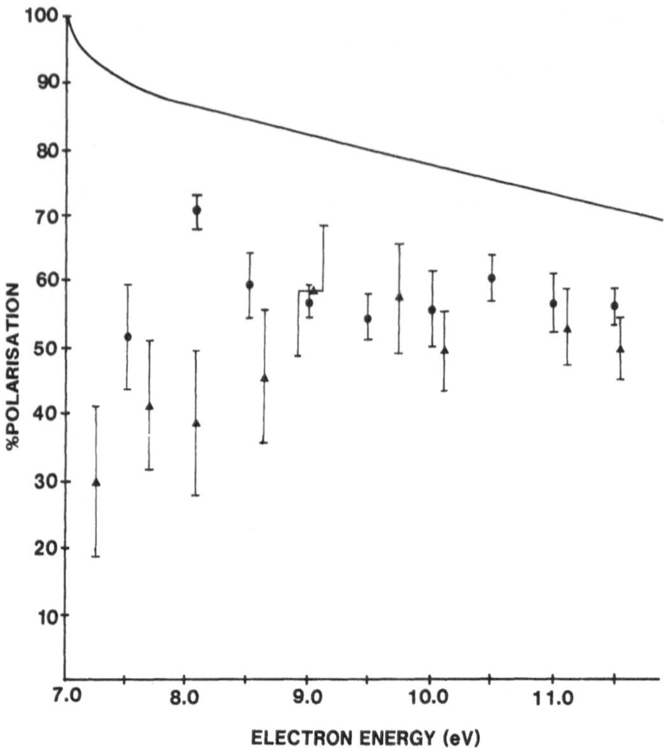

Fig. 4. Experimental and theoretical polarisation values for the 185nm line. Experimental uncertainties are 90% confidence limits. 0, I = 0 isotope data. Δ, data for natural isotope mixture using direct electron excitation (13). -, theoretical values for the I = 0 isotope (11).

Several authors have speculated on the conditions under which a breakdown of the Percival-Seaton hypothesis might occur. Hertel and Stoll (1977) (8) concluded that appropriate conditions could arise if the collision time between the electron and atom was of the same order of magnitude as the inverse of the hyperfine splittings of the transition under study and pointed out that long range interactions could prolong the duration of the collision time. The presence of negative-ion resonances could also prolong the collision time. The observation of a breakdown in the Percival-Seaton hypothesis in mercury is therefore not entirely unexpected because of the strong spin-orbit interactions which give rise to a large hyperfine structure and the presence of very strong negative-ion resonances for electron excitation of many states in mercury. The 6^1P_1 splittings are of the order of 10 GHz and although no high resolution electron excitation studies of the 6^1P_1 level have been carried out, our polarisation data show evidence of structure at 8eV which might well be due to resonances. It is also possible that electron-electron correlations play a role in the breakdown of the Percival-Seaton hypothesis. The anomalous behaviour near threshold further suggests the influence of such effects.

6.0 EXPERIMENTS ON THE 6^3P_2 METASTABLE STATE OF Hg

The stepwise excitation scheme used in these experiments (10) consisted of electron impact excitation of the 6^1S_0-6^3P_2 transition of Hg followed by laser excitation of the 6^3P_2-7^3S_1 (546.1nm) transition. Stepwise excited fluorescence was observed on the 7^3S_1-6^3P_0 (404.7nm) transition. Experiments were carried out on the (I = 0) ^{200}Hg isotope. Because the 6^3P_2 state is metastable, the effects of radiation trapping, optical pumping and magnetic fields on the stepwise excited fluorescence signal are quite different than for the experiments discussed in § 5.0. The absence of an optically allowed transition in the electron excitation step and the small fraction of metastable to ground state atoms present in the interaction region minimize the effect of radiation trapping. However, because the effective lifetime of the 6^3P_2 state is determined by the transit time of the mercury atoms through the laser beam (~ 5μs), the laser excited transition saturates at low laser intensities and optical pumping effects can readily occur. Experimental studies of such effects established that they could be neglected below a laser intensity of ~ 1 mW mm^{-2}. Another consequence of the long effective lifetime of the 6^3P_2 state is that line polarisation measurements are particularly sensitive to magnetic fields. Stray magnetic fields were cancelled by Helmholtz coils tuned with the aid of a magnetic probe and by measurements of the depolarisation of the stepwise excited fluorescence induced by small variations of the applied magnetic fields.

The theoretical expression for the line polarisation is given by equation (24). It is apparent that line polarisation measurements made for selected values of the input polariser angle α (see figure 1) can yield the ratios of the partial total cross-sections ρ_{11}/ρ_{00} and ρ_{22}/ρ_{00}. Measurements were made using two values of α ($\alpha = 0°$ and $\alpha = 90°$) and figure (5) shows the experimental data obtained together with results from a recent close-coupling calculation (15). The agreement between theory and experiment seems quite reasonable with the energy spread of the electron beam precluding observation of the sharp rise in cross-section ratios near threshold shown in the theoretical curves.

7.0 CONCLUSION

In summary, particular applications of the stepwise excitation techniques described in this paper are in the study of electron-impact excited VUV transitions, metastable states and where the spectroscopic structure of transitions is a consideration. The problems associated with linear and circular polarisation measurements at VUV wavelengths are well-known and, subject to the availability of suitable excitation schemes, stepwise excitation techniques provide a new method for investigating electron-impact

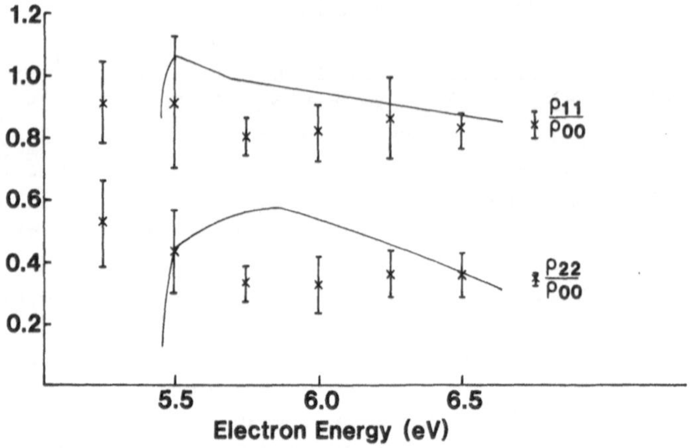

Fig. 5. Experimental and theoretical values for partial total cross-section ratios ρ_{11}/ρ_{00} and ρ_{22}/ρ_{00} of the 6^3P_2 metastable state of mercury. Experimental uncertainties are 90% confidence limits. - Theoretical values from a close-coupling calculation (15).

excited VUV transitions. Stepwise excitation techniques provide an entirely new method of studying the electron-impact excitation of metastable states which gives not only total cross-sections, but also previously inaccessible information about partial total cross-sections. The very high spectral resolution possible in the laser excitation step provides a new method of investigating the role of spectroscopic structure in electron-atom collision physics.

ACKNOWLEDGEMENTS

The work reported in this paper has been very much a team effort and was performed in collaboration with my colleague, Dr W.R. MacGillivray and the following former and present research students, Dr C.W. McLucas, Mr C.J. Webb and Mr H.J.E. Wehr. The support of the Australian Research Grants Scheme is also gratefully acknowledged.

REFERENCES

1. Phillips, M.H., Anderson, L.W. and Lin, C.C., 1981, Phys. Rev. A23, 2751.

2. Miers, R.W., Gastineau, J.E., Phillips, M.H., Anderson, L.W. and Lin, C.C., 1982, Phys. Rev. A25, 1185.
3. Phelps, J.O., Phillips, M.H., Anderson, L.W. and Lin, C.C., 1983, J. Phys. B: At. Mol. Phys. 16, 3825.
4. McLucas, C.W., MacGillivray, W.R. and Standage, M.C., 1982a, Phys. Rev. Lett. 48, 88.
5. McLucas, C.W., Wehr, H.J.E., MacGillivray, W.R. and Standage, M.C., 1982b, J. Phys. B: At. Mol. Phys. 15, 1983.
6. Webb, C.J., MacGillivray, W.R. and Standage, M.C., 1984a, J. Phys. B: At. Mol. Phys. 17, 1675.
7. Webb, C.J., MacGillivray, W.R. and Standage, M.C., 1984b, J. Phys. B: At. Mol. Phys. 17, 2377.
8. Hertel, I.V. and Stoll, W., 1977, Adv. At. Mol. Phys. B, 113.
9. Blum, K., 1981, Density Matrix Theory and Applications (N.Y. Plenum).
10. Webb, C.J., MacGillivray, W.R. and Standage, M.C., 1985, submitted to J. Phys. B: At. Mol. Phys.
11. McConnell, J.C. and Moiseiwitsch, B.L., 1968, J. Phys. B: At. Mol. Phys. 81, 406.
12. Skinner, H.W.B. and Appleyard, E.T.S., 1927, Proc. Roy. Soc. (London), A117, 224.
13. Ottley, T.W., Denne, D.R. and Kleinpoppen, H., 1974, J. Phys. B: At. Mol. Phys. 7, L179.
14. Webb, C.J., MacGillivray, W.R. and Standage, M.C., 1985, submitted to J. Phys. B: At. Mol. Phys.
15. Bartschat, K., 1983, Private communication.

THE IDENTIFICATION OF COLLISIONALLY EXCITED AUTOIONIZING STATES IN ALUMINIUM

M. Wilson

Physics Department
Chelsea College
London S.W.6

and

G. K. James, L. F. Forrest and K. J. Ross

Physics Department
The University
Southampton

INTRODUCTION

The study of autoionization processes in metal vapour atoms has direct relevance to astrophysical and fusion data. In addition, many new laser systems are dependant on a detailed knowledge of autoionizing levels of such atoms.

By using electron impact to excite inner-shell levels we have been able to measure the ejected electron spectrum resulting from autoionizing transitions in aluminium vapour atoms for the incident electron kinetic energies 20 and 500 eV. In order to identify features observed in the spectrum we have performed single-configuration and few-configuration Hartree-Fock based calculations.

METHOD AND RESULTS

The measurements were performed using the crossed-beam ejected-electron spectrometer which has been described earlier[1], comprising an electron gun and a hemispherical electrostatic electron velocity analyser together with a radio- frequency-heated oven to produce the atomic beam. For the present study the oven was fabricated from a titanium diboride/boron nitride alloy, the high reactivity of aluminium with more conventional oven materials precluding their use in this work.

The spectrum obtained in the ejected-electron energy range 0 to 4.2 eV, following excitation by an electron beam of 20 eV kinetic energy, is shown in Fig. 1. The overall energy resolution is 30 meV.

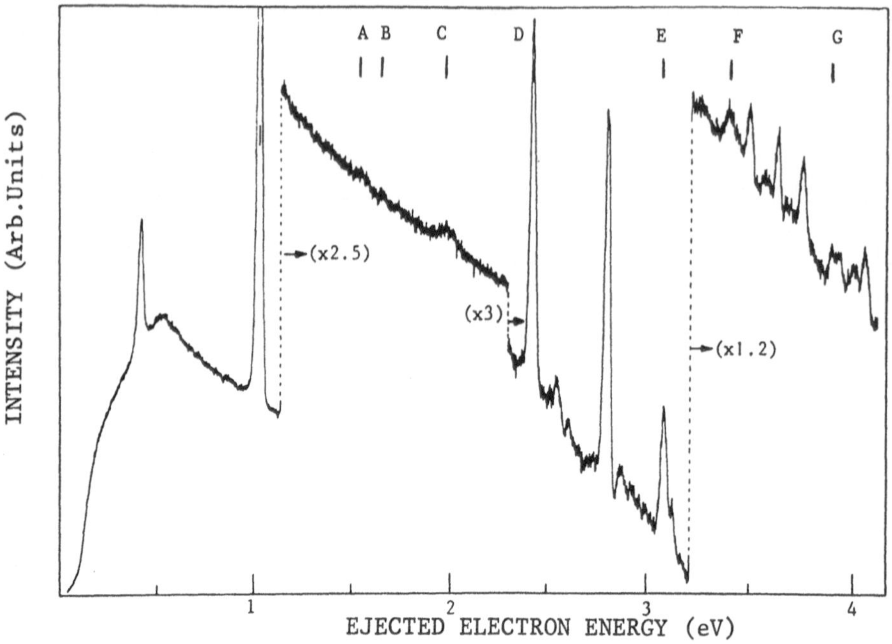

Fig. 1 Ejected-electron spectrum of aluminium vapour atoms in the range 0 to 4.2 eV, observed at 75° with respect to an incident electron beam of 20 eV kinetic energy.

DISCUSSION

The most likely process leading to the production of ejected-electrons in Al is 3s excitation from the ground state $3s^23p\ {}^2P_{1/2,3/2}$ to configurations of the type $3s3pn\ell$, followed by de-excitation to the $3s^2\ {}^1S_0$ ground state of the Al II ion, or, where energetically possible, to an excited state of the ion such as $3s3p\ {}^3P$. In addition, two-electron transitions, such as the simultaneous excitation of both 3s electrons to states with the configuration $3p^3$, can also decay through the autoionization process.

The even parity states of the configurations $3s3p^2$ and 3s3pnp (n>3) have been fully investigated by ultraviolet absorption spectroscopy[2]. On the other hand, odd parity states arising from 3s excitation from the ground state to the 3s3pns (n>3), 3s3pnd (n>2) and (possibly) $3p^3$ autoionizing states cannot be observed by photon absorption; electron spectroscopy provides a means of studying them.

For the 3s3p4s configuration the $^4P_{1/2,3/2,5/2}$ levels have been assigned by Paschen[3] based on arc spectrum data. We assigned the remaining $^2P_{1/2,3/2}$ levels based on the 3s3p(^{3}P) and 3s3p(^{1}P) core states using the results of a three-configuration Hartree-Fock based CI calculation. The intensity of these lines was observed to increase at low incident beam energy, supporting their assignment to levels resulting from dipole-forbidden transitions. Once the 3s3p4s configuration was completely assigned quantum defect calculations were used to predict the positions of higher series members, 3s3pns (n>4), based on the (3s3p)^{3}P and ^{1}P core limits. In this way the $^2P_{1/2,3/2}$ levels of the 3s3p(^{3}P)5s and 3s3p(^{3}P)6s configurations have been assigned. Unfortunately the 3s3p5s $^4P_{5/2}$ level has not been identified in our ejected electron spectrum.

The level structure of the 3s3p3d configuration was first investigated in arc spectrum studies by Paschen[3] and Eriksson and Isberg[4] and eleven levels assigned. In an attempt to identify the remaining twelve levels of this configuration in our spectrum we used the results of the three-configuration CI calculation in which the structure of 3s3p3d is expected to be strongly perturbed by interaction with $3p^3$. These calculations, however, indicate that the 3s3p(^{3}P)3d $^2P_{1/2,3/2}$ levels lie close to the previously assigned (^{3}P)$^2F_{5/2,7/2}$ levels (line E in Fig. 1 and Table 1) and cannot be resolved in our spectrum.

An interesting point arose when we considered the intensity of line D (3s3p(^{3}P)3d ^{2}D) which appears to be anomalously high when compared to that of line E (3s3p(^{3}P)3d ^{2}F); 3s3p(^{3}P)3d ^{2}F can be expected to autoionize strongly into the $3s^2\varepsilon f$ ^{2}F continuum, whereas there is no ^{2}D continuum available for autoionization of the 3s3p(^{3}P)3d $^2D_{3/2,5/2}$ levels. This situation is represented in Fig. 2 which shows the results of autoionization probability calculations performed using the method in §13 of Cowan[5]. However, the three-configuration CI calculation which includes (3s3p4s x 3s3p3d x $3p^3$) shows that 3s3p(^{3}P)3d ^{2}D contains a small admixture, approximately 5%, of $3p^3$ ^{2}D. Since this state decays strongly into the 3s3p(^{3}P)εd continuum its mixing with the lowest ^{2}D of 3s3p3d probably accounts for the suprisingly high intensity of the latter.

In order to obtain an estimate of the position and structure of $3p^3$ we have made a series of ab initio CI calculations using Hartree-Fock Slater integrals (scaling the electrostatic interactions by 0.75). Since the levels of $3p^3$ are already known in Si II[6] and P III[7] we made the same CI calculations for these ions to see how well the results of such a procedure agree with observations. We found that the estimated position and structure of $3p^3$ in Si II and P III is fairly well reproduced by such calculations. Since there is no reason to suppose that the Al I calculations should be in any way worse than those for Si II and P III they suggest that $3p^3$ lies in

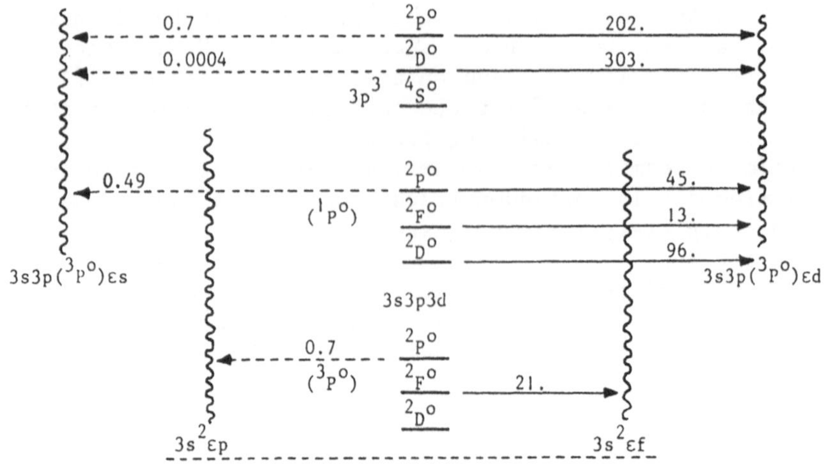

Fig. 2 Summary of autoionization probability calculations in Al I (units of 10^{-13} s^{-1}). Level spacings are schematic.

the approximate energy range 4.9 eV to 6.8 eV above the Al II $3s^2$ state. However, we did not observe any structure in our spectrum over this range. This is perhaps not too suprising since there exists no suitable continuum state based on Al II $3s^2$ for $3p^3$ to autoionize into. The possibility of the levels of $3p^3$ autoionizing into continua based on Al II $3s3p(^3P)$ i.e. $3s3p(^3P)\varepsilon s,\varepsilon d$ must, however, be considered. As shown in Fig. 2 the ^2D and ^2P terms of $3p^3$ can autoionize strongly into the $3s3p(^3P)\varepsilon d$ continuum and thus might be expected to be a strongly broadened source of ejected electrons.

Using the three-configuration CI calculation as a guide to the positions and separations of these ^2D and ^3P terms, lines A and B have been assigned to $3p^3$ $^2D_{3/2,5/2}$ and $^2P_{1/2,3/2}$ respectively decaying to the 3s3p ^3P excited state of Al II. The $3p^3$ ^4S term has not been observed in the ejected-electron spectrum which is no real suprise since it is most likely to be metastable against autoionization.

ACKNOWLEDGEMENTS

This work is supported by a grant from the Science and Engineering Research Council.

Table 1 List of levels of the 3p3pns(n>3) $3p^3$ and 3s3p3d configurations observed in the present work.

Line	Ejected Electron Energy (eV)	Assignments	
A	1.56	$3p^3\ ^2D_{3/2,5/2} \rightarrow 3s3p(^3P)$	
B	1.66	$3s3p(^3P)4s\ ^4P_{1/2,3/2,5/2}$	(Ref.3)
C	1.99	$3p^3\ ^2P_{1/2,3/2} \rightarrow 3s3p(^3P)$; $3s3p(^3P)4s\ ^2P_{1/2,3/2}$	
D	2.40	$3s3p(^3P)3d\ ^2D_{3/2,5/2}$	(Ref.3)
E	3.06	$3s3p(^3P)3d\ ^2F_{5/2,7/2}$ (Ref.4), $^2P_{1/2,3/2}$	
F	3.37	$3s3p(^3P)5s\ ^2P_{1/2,3/2}$	
G	3.89	$3s3p(^3P)6s\ ^2P_{1/2,3/2}$	
H	4.48	$3s3p(^1P)3d\ ^2D_{3/2,5/2}$	

REFERENCES

1. G.K. James, K.J. Ross and M. Wilson, J.Phys.B. 16, 4237 (1983)
2. W.C. Martin and R. Zalubas, J.Phys.Chem.Ref.Data 8, 817 (1979)
3. F. Paschen, Ann.Phys. (Leipzig) 12, 509 (1932)
4. K.B.S. Eriksson and H.B.S. Isberg, Arkiv.Fys. 23, 527 (1963)
5. R.D. Cowan, Theory of Atomic Structure and Spectra, University of California Press, Berkeley, U.S.A. (1981)
6. W.C. Martin and R. Zalubas, J.Phys.Chem.Ref.Data 12, 323(1983)
7. C.E. Moore, Atomic Energy Levels N.B.S. Circular No. 467, Vol. 1, Washington D.C. : Govt. Printing Office (1949)

ORBITAL ANGULAR MOMENTUM EXCHANGE

IN POST-COLLISION INTERACTION

P.J.M. van der Burgt. J. van Eck and H.G.M. Heideman

Fysisch Laboratorium
Rijksuniversiteit Utrecht
3584 CC Utrecht, The Netherlands

INTRODUCTION

When an autoionising state is excited near its threshold by electron impact the scattered electron is still close to the atom when the latter decays by ejecting an electron. After the autoionisation the post-collision interaction (PCI) causes the ejected and the scattered electron to exchange energy and angular momentum in the Coulomb field of the residual ion. The energy exchange may even be large enough for the scattered electron to be captured into a singly excited state of the neutral atom. This post-collision interaction plays a role in several other atomic collision processes (see Heideman 1980 and references therein).

Several studies have been published in which quantum mechanical models for post-collision interaction have been fitted to measured spectra of ejected electrons. Much attention is paid to the effects of post-collision interaction resulting from the energy exchange between the scattered and the ejected electron. However, little is known about the exchange of orbital angular momentum between both electrons. The shake-down model (Read 1977) is based on the idea that the scattered electron experiences an instantaneous change of charge of the autoionising atom. According to this model the transition probability is proportional to an overlap integral of the initial and final state wavefunctions of the scattered electron. Therefore angular momentum exchanges are excluded in the shake-down model. The semiclassical model formulated by Morgenstern et al. (1977) is equivalent to this model. An optical potential description of post-collision interaction leading to the excitation of singly excited helium states showed that angular momentum exchanges may be significant in some cases (van de Water et al. 1981).

In order to find evidence for angular momentum exchange during the post-collision interaction we have studied the angular distribution of ejected electrons from the $He^{**}(2s^2)^1S$ state after its excitation via the $He^-(2s2p^2)^2D$ resonance. Since the $He^{**}(2s^2)^1S$ state decays to the ground state $He^+(1s)$ of the helium ion, the initial angular momentum of the ejected electron is fixed at $\ell_{ej} = 0$. Angular momentum exchange during the post-collision interaction results in $\ell_{ej} \neq 0$ and may therefore lead to a deviation from isotropy of the angular distribution of ejected electrons. However, a measurement of an angular distribution is greatly complicated because the ejected electrons may interfere with electrons from the direct ionisation of helium. In the following sections we discuss this interference, we present measurements of the angular distribution of electrons ejected after resonant excitation and we give a brief analysis taking into account interference with electrons from the direct ionisation of helium. A more detailed analysis is given by van der Burgt et al. (1984).

EXCITATION AND DECAY OF THE $He^{**}(2s^2)^1S$ AUTOIONISING STATE

As the PCI-structures in ejected electron spectra are interference structures, the differential cross section for the production of ejected electrons may be written as:

$$\frac{d\sigma}{d\Omega} = |B|^2 + |A + aqe^{i\chi}|^2 \qquad (1)$$

where a is the amplitude for excitation of the autoionising state and $qe^{i\chi}$ refers to the decay of the autoionising state and subsequent PCI, with the ejected electron eventually moving in a solid angle element $d\Omega$. χ is the relative phase between q and the interfering part A of the direct ionisation. B is the non-interfering part of the direct ionisation. The variations in the amplitude $qe^{i\chi}$ are slow compared to the variations in a when the incident electron energy is swept across the narrow resonance. This is clearly seen in the measurements of Baxter et al. (1979). The effective excitation functions of the $He^{**}(2s2p)^3P$ state measured by them show significant variations over an incident energy range of the order of 1 eV. The $He^-(2s2p^2)^2D$ resonance, however, has a width of 0.05 eV. So the variations caused by the resonance in the amplitude a for excitation of the $He^{**}(2s^2)^1S$ state will not be disturbed by the variations with the incident energy in the PCI amplitude $qe^{i\chi}$. Therefore we are able to measure the resonance profile by detecting ejected electrons as a function of the incident electron energy.

Now consider the interference between autoionisation and direct ionisation. When measuring ejected electron spectra we do not detect the scattered electron so we must integrate over all angular variables of the scattered electron. This integration causes the interference term to vanish unless the slow electrons in the autoionisation channel and the direct ionisation channel have the same angular momentum. This is clear if the dependences of A and $aqe^{i\chi}$ on the angle of the scattered electron are expanded in spherical harmonics and orthogonality of the spherical harmonics is used. After excitation of the $He^{**}(2s^2)^1S$ state via the $He^-(2s2p^2)^2D$ resonance the scattered electron recedes with an angular momentum $\ell_{sc} = 2$. As the slow electron in the direct ionisation channel has a very low energy it most probably has an angular momentum $\ell_1 = 0$ *. Suppose there would be no angular momentum exchange during PCI. In that case the resonant scattered electron will still have an angular momentum $\ell_{sc} = 2$ after PCI and since $\ell_{sc} \neq \ell_1$ the term describing interference with the direct ionisation channel will vanish. As the angular momenta of the autoionising state and the He^+ ion are both zero, the ejected electron has $\ell_{ej} = 0$ and the angular distribution of the resonance profile in ejected electron spectra will be isotropic. So any deviation from isotropy of the angular distribution of the resonance will provide strong evidence for the occurrence of angular momentum exchange during PCI.

EXPERIMENT

The electron spectrometer used for the present work employs two hemispherical energy selectors, a monochromator and an analyser, which are operated at an energy resolution of approximately 90 and 65 meV, respectively. Interactions between the incident electrons and the helium atoms take place in the interaction chamber filled with helium gas at a pressure of 2.6 Pa. The analyser part of the spectrometer is rotatable within a range of 0 to 60 degrees with respect to the incoming electron beam.

Ejected electron spectra of the $He^{**}(2s^2)^1S$ autoionising state may be measured by detecting the yield of ejected electrons as a function of the incident electron energy at a fixed ejected electron energy. In the absence of post-collision interaction ejected electrons from the $He^{**}(2s^2)^1S$ state have an energy of 33.23 eV but the electrons may acquire a higher energy during the post-collision interaction. Above 33.71 eV also ejected electrons from the

---

* Based on calculations of tripple differential ionisation cross sections Tweed (1980, 1984) estimates by linear extrapolation to an electron energy of 0.5 eV, that the $\ell_1 = 2$ partial wave at all angles contributes only to the fourth significant figure of the tripple differential cross section.

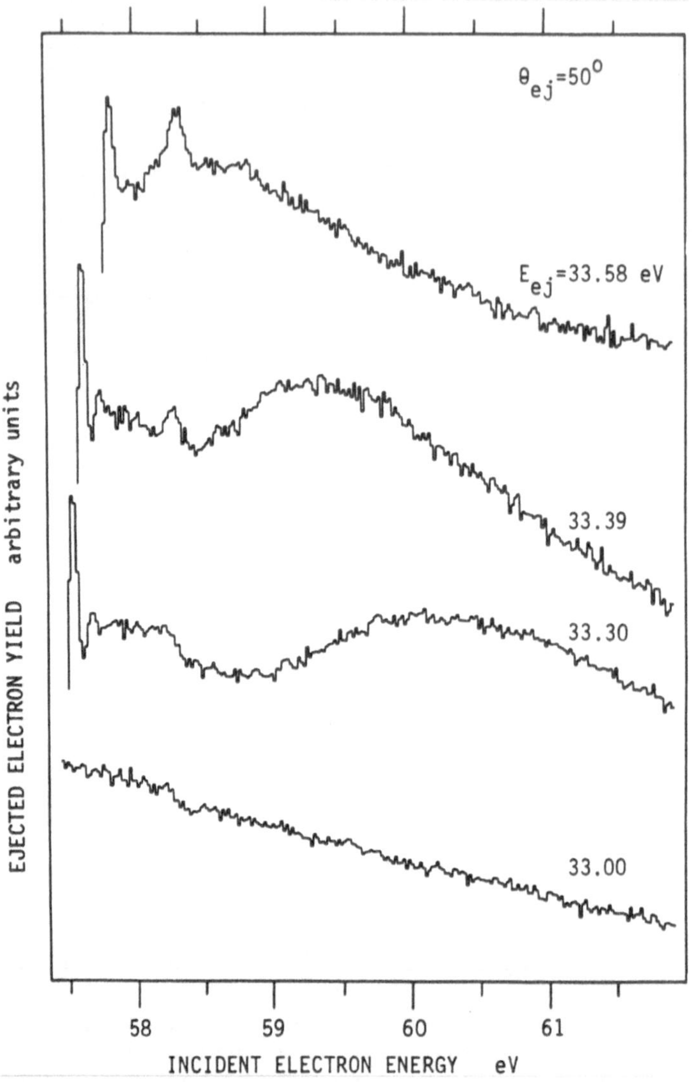

Fig. 1. Ejected electron spectra measured with constant ejected electron energy at $\theta_{ej} = 50°$. The upper three spectra show electrons ejected by the $He^{**}(2s^2)\,^1S$ autoionising state, in the lowest spectrum only electrons due to the direct ionisation process are observed. All spectra were normalised on the direct ionisation cross section so that the vertical scales are comparable. The narrow peaks at the left show electrons having excited the n = 6 helium bound state.

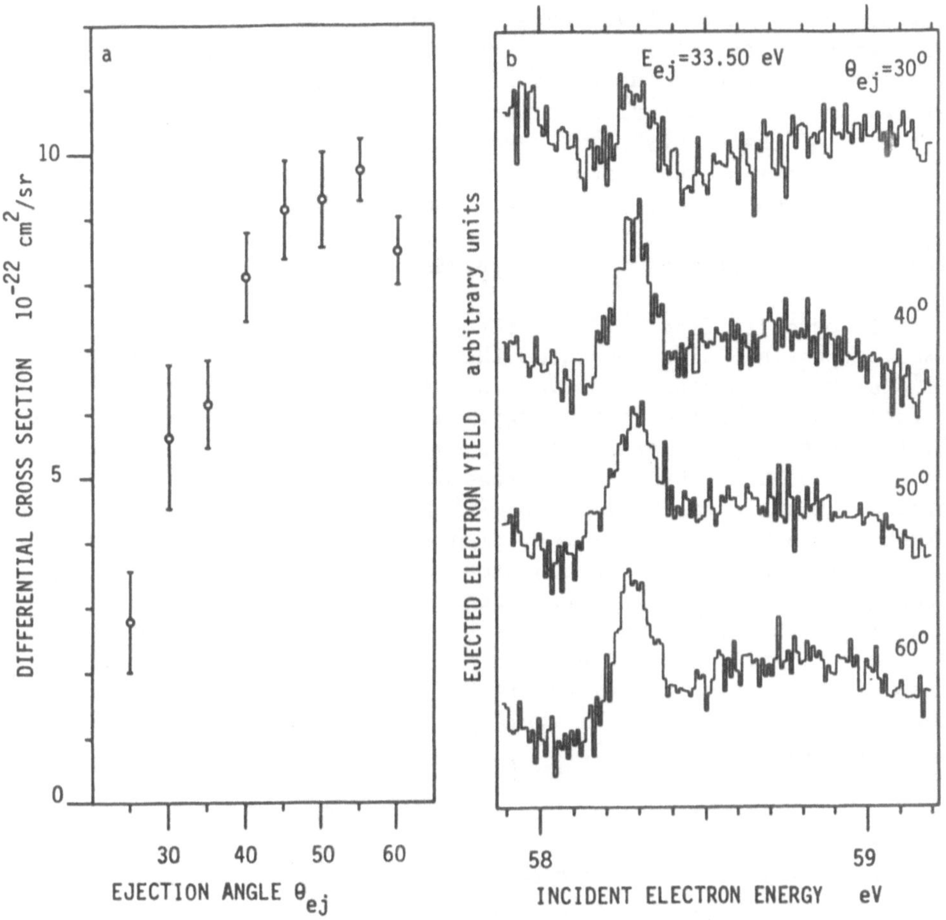

Fig. 2. a) Differential cross sections for resonant excitation and decay of the $He^{**}(2s^2)^1S$ autoionising state. The cross sections are determined for ejected electrons having acquired an extra energy of 0.27 eV during PCI.
b) Spectra measured with constant ejected electron energy at several angles. All spectra are normalised on the differential cross section for excitation of the helium 2^1P singly excited state. A linear sloping background was subtracted from the spectrum at θ_{ej} = 30°.

$He^{**}(2s2p)^3P$ state are present, so when measuring ejected electron spectra of the $He^{**}(2s^2)^1S$ state we may choose the fixed ejected electron energy in a small range of 0.48 eV upward from 33.23 eV. Ejected electron spectra are presented in figure 1. In the lowest spectrum, measured at E_{ej} = 33.00 eV, only electrons from the direct ionisation of helium are present. No sloping background was subtracted from these spectra. The broad oscillations are due to the variation of the relative phase χ in eq. (1) with incoming energy. Similar oscillations were observed by Baxter et al. (1979) in the effective excitation functions of the $He^{**}(2s2p)^3P$ autoionising state. The rapid variations around E_i = 58.30 eV are caused by the presence of the $He^-(2s2p^2)^2D$ resonance. Note that structure due to decay of the resonance into the direct ionisation channel is very small or absent in the lowest spectrum so that the variations around E_i = 58.30 eV in the upper three spectra are probably solely due to resonant excitation of the autoionising state.

In order to find evidence for possible angular momentum exchange during PCI we measured the angular distribution of the $He^-(2s2p^2)^2D$ resonance observed in the ejected electron yield as a function of the incident energy. The results are presented in fig. 2. The measurements were done at a constant ejected electron energy of E_{ej} = 33.50 eV, because the resonance is largest there. The scattering angle was varied from 25° to 60°. Measurements at smaller angles are hampered by the very high direct ionisation background, whereas the mechanical construction of our electron spectrometer excludes measurements at larger angles. After normalizing all spectra on the differential cross section for excitation of the helium 2^1P singly excited state a differential cross section for resonant excitation and decay of the $He^{**}(2s^2)^1S$ state (at a PCI energy exchange of 0.27 eV) could be obtained (for more details see van der Burgt et al. 1984).

Fig. 2a clearly shows that the angular distribution of the $He^-(2s2p^2)^2D$ resonance in the $He^{**}(2s^2)^1S$ ejected electron spectra is anisotropic. According to the discussion in the previous section this angular distribution shows that the electron ejected by the $He^{**}(2s^2)^1S$ autoionising state after resonant excitation must have exchanged orbital angular momentum with the scattered electron. The anisotropic angular distribution is the consequence of both angular momentum exchange during PCI and interference with electrons from the direct ionisation process. As a result the size of the resonance is expected to depend strongly on the energy the ejected electron acquires during the post-collision interaction. This is indeed seen in figure 1. Only if the post-collision interaction is sufficiently strong there is an angular momentum transfer between the scattered electron and the ejected electron so that the interference between resonant excitation followed by autoionisation and direct ionisation becomes visible in the spectra.

CONCLUSION

We have measured the angular distribution of electrons ejected by the $He^{**}(2s^2)^2S$ autoionising state after its excitation via the

$He^-(2s2p^2)^2D$ resonance. Taking into account interference with electrons from the direct ionisation of helium, we were able to show that the measured anisotropic angular distribution is the result of an orbital angular momentum exchange during the post-collision interaction. This result casts doubt on theoretical models for post-collision interaction, which exclude angular momentum exchange at incident electron energies closer than 0.5 eV from threshold.

ACKNOWLEDGEMENT

This work was performed as a part of the research programme of the "Stichting voor fundamenteel Onderzoek der Materie" (F.O.M.) with financial support from the "Nederlandse Organisatie voor Zuiver Wetenschappelijk Onderzoek" (Z.W.O.).

REFERENCES

Baxter, J.A., Comer, J., Hicks, P.J. and McConkey, J.W., 1979, J. Phys. B: At. Mol. Phys. 12:2031.
Heideman, H.G.M., 1980, in: "Coherence and Correlation in Atomic Collisions", H. Kleinpoppen and J.F. Williams, eds., Plenum, New York:493.
Morgenstern, R., Niehaus, A. and Thielmann, U., 1977, J. Phys. B: At. Mol. Phys. 10:1039.
Read, F.H., 1977, J. Phys. B: At. Mol. Phys. 10:L207.
Tweed, R.J., 1980, J. Phys. B: At. Mol. Phys. 13: 4467.
Tweed, R.J., 1984, private communication.
van der Burgt, P.J.M., van Eck, J. and Heideman, H.G.M., 1984, J. Phys. B: At. Mol. Phys., to be published.
van de Water, W., Heideman, H.G.M. and Nienhuis, G., 1981, J. Phys. B: At. Mol. Phys. 14:2935.

INTERMEDIATE ENERGY ELECTRON IMPACT SPECTROSCOPY AT RIO DE JANEIRO

A.C. De A. E Souza and G.G.B. De Souza

Instituto De Quimica - UFRJ - Ilha Do Fundão
Rio De Janeiro - 21910 - Brasil

INTRODUCTION

A new electron impact spectrometer has been built at the Federal University of Rio de Janeiro with the following characteristics :

- incident energy = 0,5 - 3,0 KeV
- angular range = 0 - 60^0
- energy resolution = 0,5 - 3,0 eV
- energy loss = 500 eV

The spectrometer, a crossed beam apparatus features a Möllenstedt type electron velocity analyzer and has been used on the measurement of electron energy-loss and differential cross sections for the excitation of some gaseous compounds (argon , nitrogen, ethane, ethylene, n-pentane, tetra-metylsilane).

A versatile data acquisition system based on a microcomputer (Motorola Mek 6802 D5E) has been incorporated into the spectrometer.

APPARATUS

The spectrometer is shown in figure 1. It consists basically of an electron source, a gas inlet system, a scattering chamber an electron velocity analyzer, an electron detector, a vacuum pumping system and a data acquisition system.

The electron source is a commercial electron gun (SE 3 K/SU)

(Massey et al. 1969) which has been modified in order to allow the use of replaceable tungsten hairpin filament. It is shielded from electrostatic fields by a grounded aluminum cylinder. The electron beam has a diameter of typically $4x10^{-4}$m., when measured at the gas nozzle. The maximum beam current is of the order of 50 μA at 1 KeV. The sample is introduced into the chamber as an atomic or molecular beam formed by the expansion of the selected substance through a hypodermic needle.

For a given scattering angle the electrons pass through a system of double aperture (200 and 300 μ) and are velocity analyzed by a Möllenstedt (Metherell 1971) electron analyzer. This analyzer which has been described earlier (Souza 1979) is an electrostatic, dispersive device, simple to built and operate. Its energy resolution is limited effectively by the intrinsic energy spread of the thermo-ionically produced electron beam namely 0,5 eV.

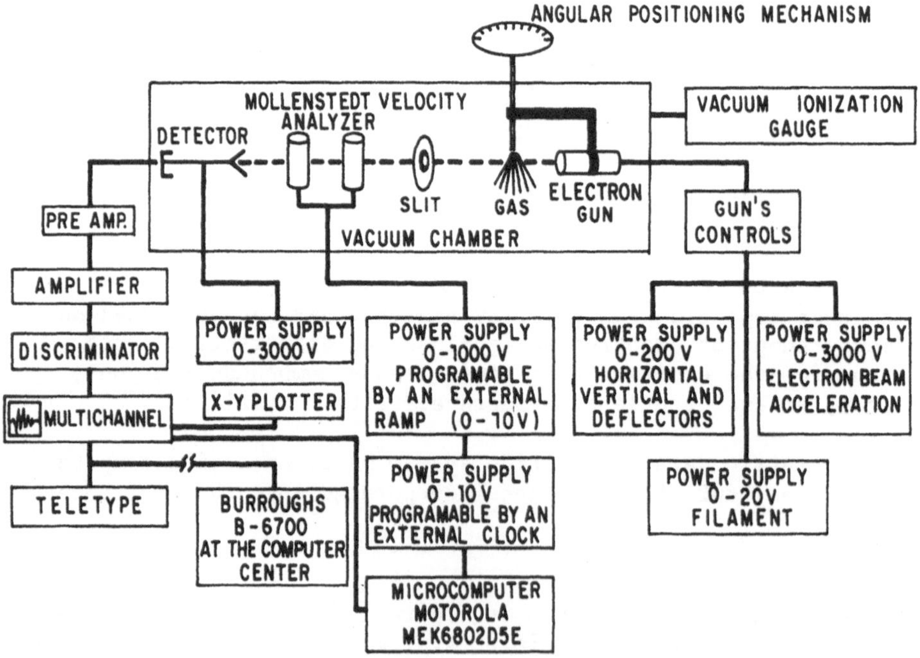

Fig. 1. Scheme of the Spectrometer.

The energy-selected electrons are detected by a spiraltron electron multiplier (model SEM 4219, Galileo Electro-Optics). The pulses coming out from the detector are fed into conventional pulse counting electronics consisting of preamplifier, amplifier and discriminator and are then storaged in the memory of a multichannel analyzer (TN 1705), or using the computerized system described below.

In order to obtain an energy-loss spectrum an inexpensive microcomputer (Motorola MEK6802D5E) is used to scan the analyzer power supply and the time base of the multichannel analyzer operating in the multiscaling mode.

The electron scattering angle Θ, can be changed in the range (-60^0, $+60^0$) by rotating the gun with a ($1/60^0$) precision.

The electron gun, gas nozzle, velocity analyzer and detector are housed in a 19 ℓ brass scattering chamber. All metal surfaces inside the chamber are coated with coloidal graphite in order to minimize contributions from secondary electrons. The pumping system consists of a 6" diffusion pump (VARIAN - VHS - 6) backed by a rotatory pump (Welch Model 1397).

A liquid nitrogen trap attached to the diffusion pump strongly reduces oil backstreaming to the chamber. The effective pumping speed is 1000 ℓ/s which results in the attainment of a base pressure of approximately $2x10^{-6}$ torr.

Control of the residual gases and also a check on the sample purity is routinely done with a quadrupole residual gas analyzer (QMG 311 - Balzers) attached to a side flange.

Three pairs of Helmholtz coils (1,80 m side lenght) surround the spectrometer in order to reduce the ambient magnetic field to a value of less than 5 mG in the scattering chamber to less than 1 mG at the scattering center.

PERFORMANCE

The angular behaviour of the energy-loss spectra of the nitrogen molecule ($X^1 \Sigma_g^+$) has been studied at 1 KeV incident energy in the angular range $2\text{-}14^0$. At least seven peaks are well-defined. For each one the differential cross section has been obtained. As an example, figure 2 shows the energy-loss spectra obtained at 2^0 for N_2. Figure 3 shows the elastic differential cross section; notice the extremelly good agreement with the absolute data published by Jansen (1976)

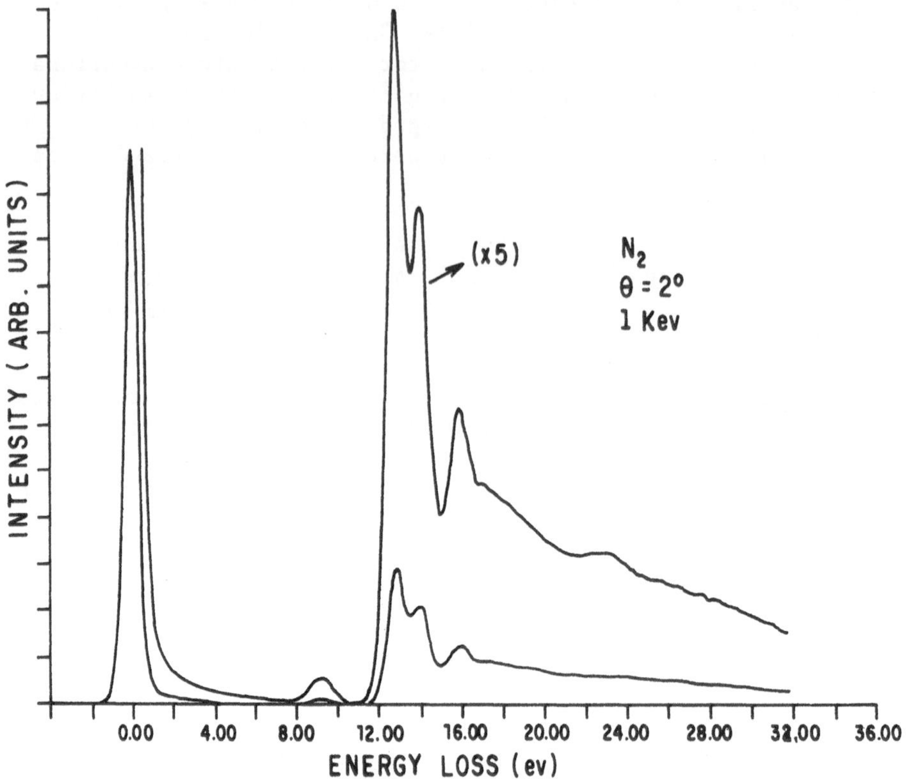

Fig. 2. Electron energy-loss of N_2.

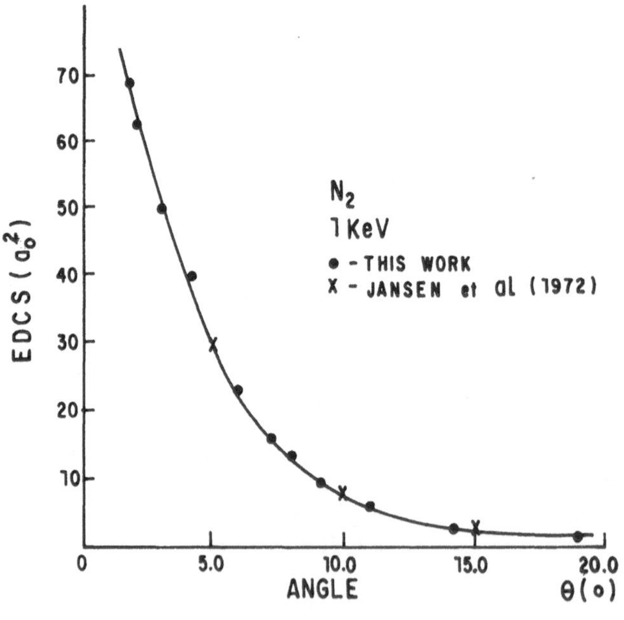

Fig. 3. Elastic differential cross section of N_2 .

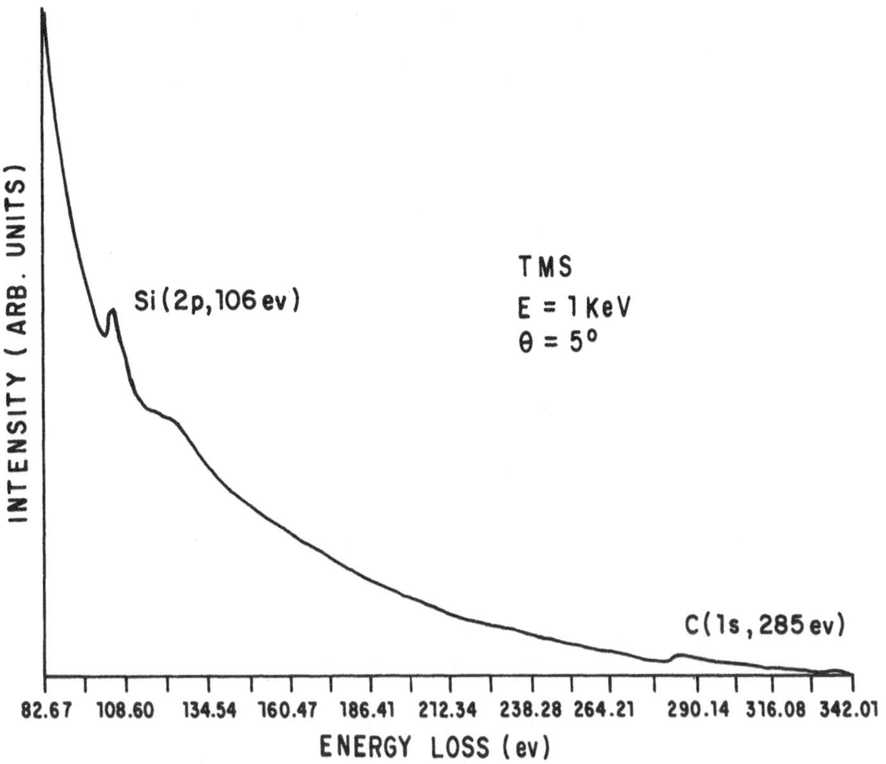

Fig. 4. Electron energy-loss of TMS (inner-shell).

Finally, figure 4 shows the electron energy-loss spectrum of $Si(CH_3)_4$, tetramethyl-silane, measured in the energy-range of 80 to 350 eV, at an incident energy of 1 KeV.

The inner-shell excitation of the silicon (Si, 2p, 106 eV) and carbon (C, 1s, 285 eV) atoms can be clearly seen in this spectrum, showing a good agreement with the results of Sodhi et al. (1983)

We have been working also with ethane, ethylene, acethylene and n-pentane which will be published later.

CONCLUSIONS

A new, variable angle, low resolution electron energy-loss spectrometer for the intermediate energy (0,5 - 3,0 KeV) has been described and some energy-loss spectra presented.

The new machine has demonstrated its capability for the measurement of cross sections and generalized oscillator strengths for the electronic excitation of molecules and atoms at both, valence and inner-shells.

ACKNOWLEDGMENTS

I would like to acknowledge the following agencies which gave me the necessary support for presenting this paper : FINEP, CNPq, FUJB, NATO.

REFERENCES

Jansen, R.H.J., de Heer F.J., Lyken H.J., Van Wingerder, B. and Blaauw, H.J., J.Phy.B, 9 (1976).

Massey, H.S.W., Burhop, E.H.S., Gilbody H.B. "Electronic and Ionic Impact Phenomena", 2nd Ed., vol.I and II, Oxford at the Clarendon Press (1969)

Metherell, A.J.F., Advances in Opt. and El.Microscopy, 4 (1971), 263.

Sodhi, R., Daviel S., Brion,C.E., Souza G.G.B. de. J.El.Spec.& Related Phenomena, 1983,(in press).

Souza, G.G.B. de, Santos M.C.A., Peixoto E.M.A. OPTIK, 53 (1979) vol.5, 405.

Souza, G.G.B. de, Souza A.C. de A.e. J. Phy.E (accepted).

e-e CORRELATION IN (e-2e) REACTION:

A SEMICLASSICAL APPROACH

L. Avaldi, R. Camilloni, Yu V. Popov[+] and
G. Stefani

IMAI CNR, Area della Ricerca di Roma
CP 10, 00016 Monterotondo Scalo (Roma) ITALY
[+]Permanent address:
Lab of Education Systems (box ΦΠΚ)
Moscow State University, Moscow 119899 USSR

INTRODUCTION

The (e,2e) experiments constitute the most detailed probe to investigate electron induced ionization processes. At low and intermediate incident energies correlation effects between the final particles are crucial features to be taken into account in order to achieve the correct description of these phenomena.

In recent years several experimental and theoretical works have been devoted to study this subject. While a comprehensive review can be found in several papers[1], this work reports on an approach to the triple differential cross section calculation which maintains the exact factorization of the Plane Wave Impulsive Approximation and accounts for the long range Coulomb interactions among the three final charged particles in a semiclassical way.

The ideas for this approach stem from the assumption that, in impulsive regime, the ionization takes place within a small domain of space whose radius r_0 is determined by the wave function of the atomic orbital involved in the reaction. This region is the crucial one for the dynamics of the process. Outside, the two electrons can be treated as semiclassical particles interacting with the residual ion and with each other. These interactions result in a "deflection" of their paths and in a dumping factor affecting the absolute value of the cross section.

As a first check of this theory, its predictions are compared with a large body of experimental (e,2e) data on Helium.

THEORY

The present approach is based on previous works by Presniakov[2] and Popov[3]. According to them the ionizing interaction in the inner region of radius r_o can be treated as short-range one and the (e,2e) scattering amplitude is given by the product of the half-off shell scattering matrix element times a structure factor. When the distance of the final electrons from the ion is larger than r_o semi-classical wavefunctions are used to account for the correlation effects due to the mutual interaction of two electrons moving in the central field of the ion[4]. The mathematical frame of this approach is found in Merkurieve's paper[5].

In this model the triple differential cross section is given by:

$$\frac{d^3\sigma}{d\Omega_1\, d\Omega_2\, dE} = \frac{1}{C_1 C_2}\left[\frac{d^3\sigma\,(\tilde{\theta}_1,\tilde{\theta}_2)}{d\Omega_1\, d\Omega_2\, dE}\right]_{PWIA} \qquad \begin{aligned}\tilde{\theta}_1 &= \theta_1 - \Delta\theta_1\\ \tilde{\theta}_2 &= \theta_2 - \Delta\theta_2\end{aligned} \qquad (1)$$

where

$$\left(\frac{d^3\sigma}{d\Omega_1 d\Omega_2\, dE}\right)_{PWIA} = \frac{4\, p_1 p_2}{p_o}\; f_t\; \rho(\bar{q}) \qquad (2)$$

f_t is the e-e factor calculated on the basis of the t-matrix[6]. q is the recoil momentum defined by the relationship

$$\bar{p}_o + \bar{q} = \bar{p}_1 + \bar{p}_2$$

$\rho(\bar{q})$ is the square of the Fourier transform of the overlap integral between the initial atomic state and the final ionic one.

The angles θ_1 and θ_2 are the angles between the momenta $\bar{P}1$, $\bar{P}2$ and the projection of $\bar{P}_o$ in the plane defined by the momenta of the outgoing particles. The analytical expression of the angular deflection[4] is

$$\Delta\theta_i = \frac{2\, m\, e^2}{E_i\, r_o}\; \frac{p_i \cos(\chi_0/2)\,\{|\bar{p}_1-\bar{p}_2| + 2\, p_j \sin(\chi_0/2)\}}{4\,|\bar{p}_1-\bar{p}_2|\,\sin(\chi_0/2)\,\{\,|\bar{p}_1-\bar{p}_2| + (p_1+p_2)\sin(\chi_0/2)\}}$$

$$\chi_0 = \theta_1 + \theta_2$$

while the dumping factor C_i is given[7] by

$$C_i = \left\{1 + \frac{2\, m\, e^2}{E_i\, r_o}\left[Z(r_o) - \zeta_i\right]\right\}^{1/2}$$

$$\zeta_i = \frac{p_i}{2}\; \frac{\{\,|\bar{p}_1-\bar{p}_2|\sin(\chi_0/2) + (\,p_i - p_j\cos\chi_0)\,\}}{|\bar{p}_1-\bar{p}_2|\sin(\chi_0/2)\,\{\,|\bar{p}_1-\bar{p}_2| + (p_1+p_2)\sin(\chi_0/2)\,\}}$$

This factor accounts for the momentum change of a particle that moves in a Coulomb field from r_o to infinity.

RESULTS AND DISCUSSION

The radius r_o is the only free parameter to be allowed for by the present approach to the impulsive (e,2e) cross section. To compare its predictions with the large body of experimental results already existing for He a suitable choice of r_o has to be done. For this purpose it has been used as a fitting parameter to reproduce the experimental cross-section as measured at 400 eV incident energy and fully symmetric coplanar condition (i.e. $E_1=E_2$; $\theta_1=\theta_2$)[8]. A least χ^2 criterion has shown r_o=0.70 a.u. to be the best choice for this parameter in reproducing both absolute value and angular shape of the experimental cross-section. This value is very close to the

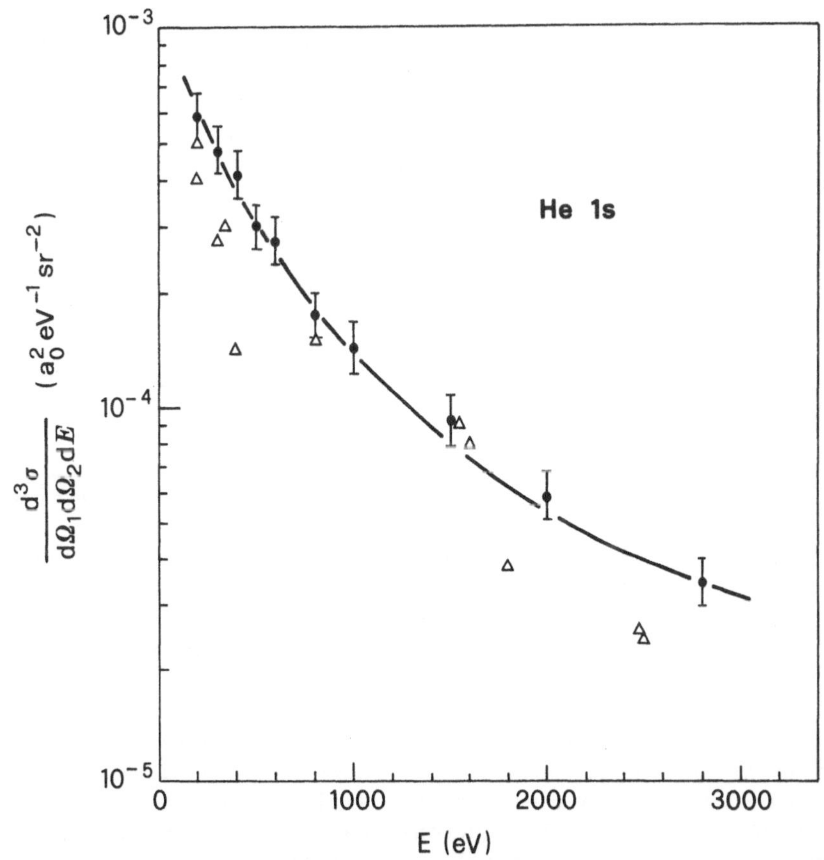

Fig. 1 : (e,2e) Absolute triple differential cross-section at 45° scattering angles. The full line is the present model prediction, dots (ɸ) are experimental data by ref. (10) and triangles (Δ) by ref. (11).

mean radius of the He 1s orbital as calculated from the Clementi and Roetti H F wave function[9]. This fact suggests to assume, in general, in evaluating the cross-section (1) a value r_o not smaller than the mean radius of the orbital involved in the reaction ($r_o > r_n$). In the following comparisons with the He data the value r_o=0.70 a.u. has been used to predict the cross-sections for all the other kinematical conditions investigated.

In Figure 1 comparison with the absolute experimental data of van Wingerden et al[10] and Stefani et al[11] is shown. The data are relative to symmetric coplanar reactions of 45° scattering angle for incident energies varying from 200 to 2800 eV. The agreement is noticeable all over the energy range investigated. Limited to the region of recoil momenta q $\cong$ 0 explored by these data, the Eikonal Distorted Wave Impulse Approximation (EWIA)[12] gives predictions which are indistiguishable from the present model.

For what concerns the angular distributions in coplanar symmetric geometry, where the inematical conditions are widely changed, the model is still satisfactory. In Figure 2, experimental data relative to 200 eV incident energy are reported and compared with the present calculations and with PWIA and EWIA)distorting potential V=20 eV) and fully Distorted Wave Impulse Approximation (DWIA)[13]. The experimental data are reported on absolute scale by normalising them to the corresponding absolute value at 45°. The correct prediction of the cross-section at energies as low as 200 eV and scattering angles lower than 40° is the most relevant result of the present calculation. None of the previously used models correctly accounts for the results in this region, where correlation effects between final electrons are mostely relevant. Not even DWIA by Fuss et al[13] achieves a better agreement, as it under-estimates the absolute angles (>50°) and by a factor 2.8 at the smaller angles ($\simeq$30°).

An overall comparison between theory and experiments performed at several different energies in symmetric coplanar conditions is summarized in Figure 3. Ratios between measured and calculated cross-section are reported as a function of the scattering angle. Disregarding the extreme regions, where uncertainties on the experimental data are large, the theoretical predictions always agree with the experiments within 15%.

A check even more sensitive is represented by those (e,2e) experiments in which the recoil momentum q is kept constant[6]. This condition can be achieved in coplanar asymmetric reactions at equal energies E_1=E_2 for the final electrons, when the scattering angles are suitably varied. These peculiar kinematics, keeping constant in (2) the structure factor, provide us with experiments which are mainly sensitive to the correct description of the collisional term of the (e,2e) amplitude. The data in Figures 4 and 5 are relative to a set of such experiments performed at different incident energies, where

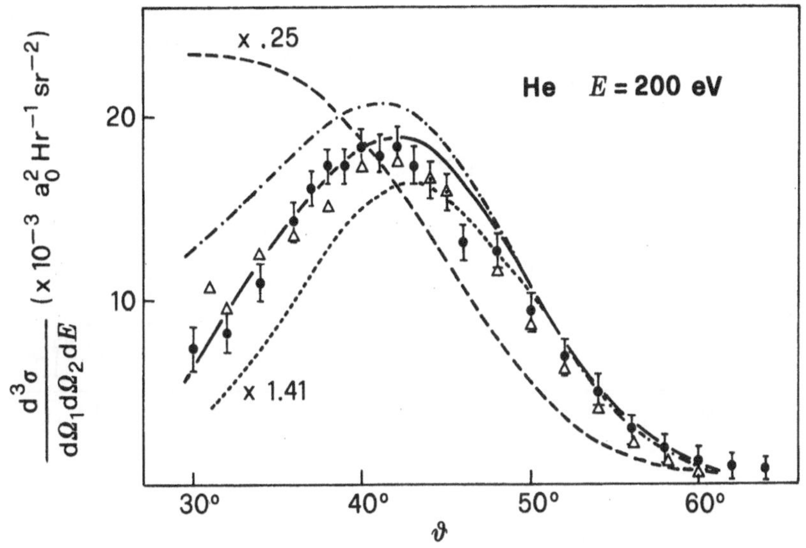

Fig. 2 : Angular correlation measured at 200 eV incident energy. Coincidence rate[8] is compared with theoretical predictions of PWIA (- - -), EWIA (-•-•-), DWIA (•••••) and present model (——)

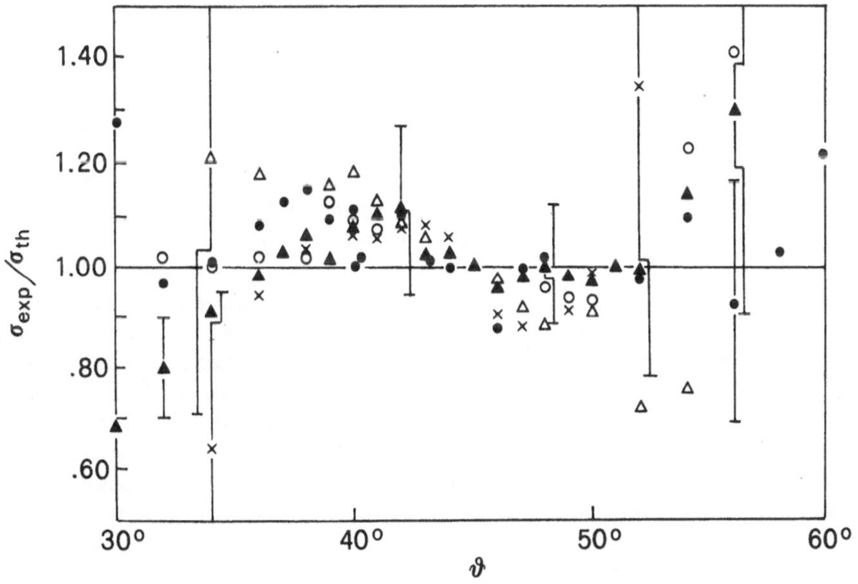

Fig. 3 : The ratio σ_{exp}/σ_{th} is reported versus the scattering angle ϑ for symmetric coplanar reactions at 200 (●), 400 (▲), 800 (○), 1600 (△) and 2500 (×) eV incident energies. Experimental data are taken by ref. (8).

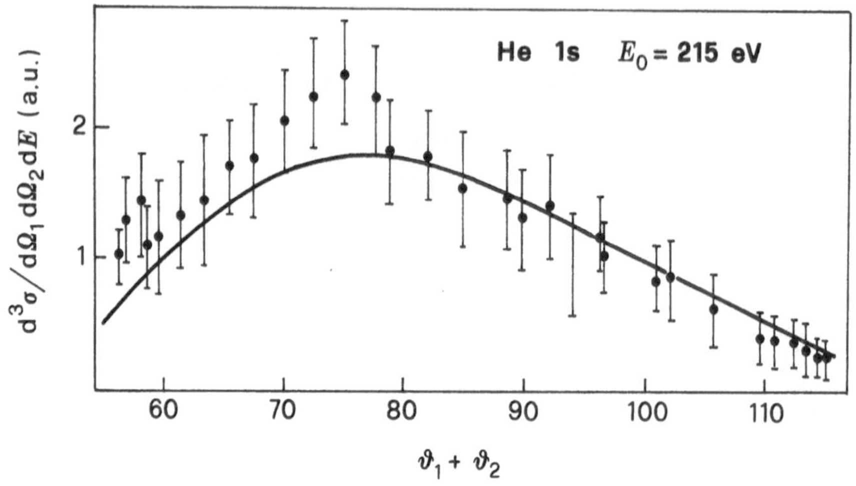

Fig. 4. (e,2e) triple differential cross-sections versus the relative scattering angle $\theta_1 + \theta_2$ for 215 eV[14] incident energy. The full curve represents the theoretical predictions of the present model.

angles have been varied in such a way that the recoil momentum q always retains a value close to 1 a.u. For incoming electrons of 215[14], 409 and 790 eV, the relative angle $\theta_1 + \theta_2$ between the final electrons varies from 56° and 110°. As already done for the coplanar symmetric reactions, we separately present the comparison between theoretical predictions and experimental results at the lowest incident energy (Figure 4) and the global result in form of the ratio between experimental data and theoretical predictions versus the relative scattering angle $\theta_1 + \theta_2$ (Figure 5). In this case, due to the absence of sufficiently accurate cross-section the experimental results, in both the figures, are normalized to the theoretical points at the largest relative scattering angle.

From Figure 4 it is clear that the e-e factor as calculated by this model is fairly good in accounting for the experimental data, even at incident energy as low as 200 eV and scattering angle $\theta_1 + \theta_2 < 70°$. This fact represents a further improvement with respect to the EWIA, which already fails in fitting data at 400 eV incident energy and scattering angles smaller than 70°[6]. This conclusion is confirmed by the results reported in Figure 5, where most of the experimental data agree with the theoretical predictions with 20%.

CONCLUSION

The pure PWIA model has been improved through a simple correction to include the long range tails of the Coulomb interaction between the final electrons. This model has given satisfactory predictions

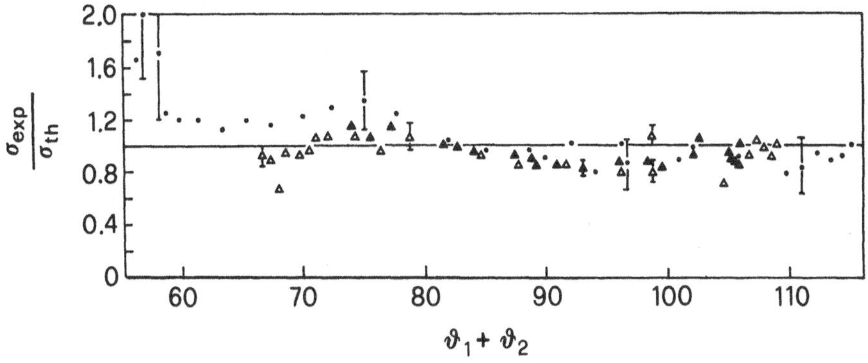

Fig. 5. The ratio $\sigma_{exp}/\sigma th$ versus the relative scattering angle $\theta_1 + \theta_2$ for asymmetric coplanar reactions at 215[14] (•), 406 (△) and 790 (▲) eV incident energies.

for a large part of the existing body of (e,2e) experimental data on He, both for symmetric and not extremely asymmetric reactions and for incident energy as low as 200 eV.

However the model is expected to fail at low kinetic energies of the final electrons and in very asymmetric reactions, where the distortion effect, due to the residual interactions with the ion, become more and more relevant with respect to correlation effects between the two final electrons.

References

1. A Giardini-Guidoni, R Fantoni, R Camilloni and G Stefani Comm. Atom. Mol. Phys. 10 (1981=, 107 and references therein quoted. C J Joachain Lecture Notes in Chemistry 35 (1984), 239
2. L P Presniakov Proc. Lebedev Inst. (FIAN. USSR) 51 (1970), 20
3. Yu V. Popov J Phys. B 14 (1981), 2449
4. Yu V Popov and J J Benayoun J Phys. B 14 (1981), 4673
5. S P Merkuriev Report FUB/NEP 2/80 Inst. Theor. Phys. Berlin (West Germany)
6. R Camilloni, A Giardini-Guidoni, I E McCarthy and G Stefani Phys. Rev. A 17 (1978), 1634
7. L Avaldi, R Camilloni, Yu V Popov and G Stefani to be published
8. R Camilloni, A Giardini-Guidoni, G Missoni, G Stefani, G Tribelli and D Vinciguerra in Momentum Wave Functions 1976, D W Devins Editor AIP conf. Proc. 36 (1977), 205
9. E Clementi and C Roetti At. Data and Nucl. Data Tables 14 (1981), 177
10. B van Wingerden, J T Kimman, M Van Tilburg and F J de Heer J Phys. B 14 (1981), 2475
11. G Stefani, R Camilloni and A Giardini-Guidoni Phys. Lett. 64A (1978), 364

12. I E McCarthy and E Weigold Phys. Rep. C 27 (1976)
13. I Fuss, I E McCarthy, C J Noble and E Weigold Phys. Rev. A 17 (1978), 604
14. R Camilloni, G Stefani and A Giardini-Guidoni private communication 1984.

Work partically supported by an Italy-France scientific co-operative programme.

STIMULATED FREE-FREE TRANSITIONS

IN A COULOMB FIELD

V. Véniard and A. Maquet

Laboratoire de Chimie Physique*
Université P.et M. Curie
11 Rue P.et M. Curie
F 75231 Paris Cedex 05 FRANCE

and

M. Gavrila

FOM-Institute for Atomic and Molecular physics
Kruislaan 407
1098 SJ Amsterdam
The Netherlands

I. INTRODUCTION

The simple model of potential scattering of electrons in the presence of a laser field provides a good account of the main features of multiphoton free-free transitions processes in a wide range of laser frequencies and projectile energies.[1,2] The exchange of photons between the electron and the field occuring during the scattering process can be accounted for on using either a perturbative or a non-perturbative approach, depending on the laser radiation field intensity and frequency.

Up to now, owing to obvious computational difficulties, perturbative calculations were almost exclusively limited to one-photon transitions. In this respect one should note that, since the celebrated works of Kramers, [3]and Sommerfeld,[4]the calculations of the bremsstrahlung spectrum in a Coulomb field still play a key role as a reference in the discussion of more sophisticated models.On the other hand, non-perturbative results, intended to hold in the limit of very intense laser fields, have been established under two restrictive assumptions: the soft-photon approximation and short range potentials.[5] Although recent advances have been obtained in the theory towards higher frequencies,[6]the removal of the second restriction, i.e. a consistent treatment

* Laboratoire "Matière et Rayonnement", Associé au CNRS.

of the infinite range Coulomb potential, which brings additional difficulties, is still under active consideration. [7-9]

In this paper we present a progress report on exact perturbative calculations of one- and two-photon free-free transition amplitudes in a Coulomb field. The corresponding cross-sections are associated to the experimental situation schematized in the figure 1: while scattered in the Coulomb field of a nucleus, incoming electrons absorb (or emit) one or two photons of an external monomode laser field.

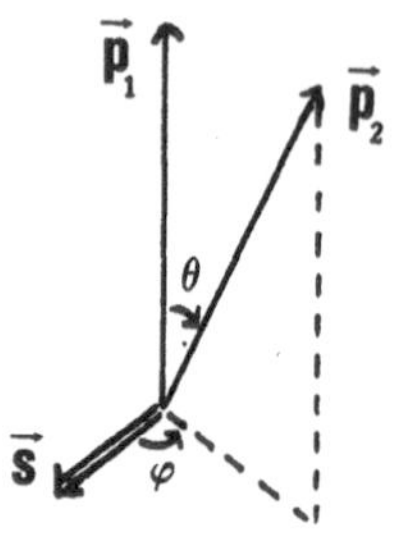

Figure 1

For the sake of comparison with previous works,[5] we have adopted the following definition of the cross-section:

$$\frac{d\sigma_\nu}{d\Omega(\widehat{p_2})} = \frac{\text{Prob of scatt. in } d\Omega(\widehat{p_2}) \text{ with absorption of } \nu \text{ photons}}{\text{density of incoming electronic current}} \tag{1}$$
$$= \frac{p_2}{p_1}\left|f_\nu(\omega)\right|^2$$

where $\vec{p_1}$ (resp. $\vec{p_2}$) is the momentum of the incoming (resp. outgoing) electron; $\widehat{p_2} = \vec{p_2}/p_2$; $p_2^2/2m = p_1^2/2m + \nu\hbar\omega$; $|\nu| = 1, 2, ...$; ω is the laser frequency and $f_\nu(\omega)$ is the scattering amplitude.
In the above figure $\widehat{s}$ represents the polarization of the linearly polarized laser and the incoming electron momentum is chosen such that $\vec{p_1}. \perp \widehat{s}$. The choice of this particular geometry leads to a notable simplification of the angular algebra, without altering the generality of the discussion. Note also that the validity of both the nonrelativistic and dipole approximations are assumed.

Within this framework our calculation is exact, which means more precisely that we have used exact Coulomb wave functions for describing the incoming and outgoing electron states. In addition the infinite summation running over the complete hydrogen spectrum, which appears into the second order amplitude, is implicitly performed by using Schwinger's representation of the Coulomb Green's function.[10] In this latter instance, our main motivation for performing such an exact computation, was primarily to carry out the analytical calculation as far as possible in order to get a closed form expression of the second order amplitude in a physically well defined test case. In this perspective our results represent in some sense the two-photon generalization of Sommerfeld's celebrated formulas for the bremsstrahlung in a Coulomb field. Before going further, let us recall the main features and present the basic formalism we have used in the simpler one-photon calculation.

II. ONE-PHOTON FREE-FREE TRANSITIONS

If $\nu = 1$, i.e. one photon of the laser is absorbed, the transition amplitude reads explicitly:

$$f_1(\omega) = \frac{m}{2\pi\hbar^2}\left(\frac{ea}{2mc}\right) < \vec{p_2}^{(-)}|\hat{s}.\vec{p}|\vec{p_1}^{(+)} > \tag{2}$$

Here a is the amplitude of the vector potential associated to the laser field:

$$\vec{A} = a\hat{s}\cos\omega t \tag{3}$$

The wave functions $|\vec{p_i}^{(\pm)}>$ are the Coulomb wave functions for asymptotic momentum $\vec{p_i}$, with the appropriate asymptotic ingoing or outgoing waves properties; their momentum space representation is explicitly:

$$< \vec{p}'|\vec{p_i}^{(+)} >= N_i \lim_{\epsilon\to 0} \oint \left(\frac{\xi}{\xi - 1}\right)^{-i/p_i} \frac{d\xi}{\left((\vec{p'} - \vec{p_i}\xi)^2 + (\epsilon - ip_i(1-\xi))^2\right)^2}$$

$$N_i = \frac{2p_i}{\pi\sqrt{p_i(1 - exp(-2\pi/p_i))}} \tag{4}$$

where the contour encircles the points (0,1). One has also $< \vec{p}'|\vec{p_i}^{(+)}>= < \vec{p_i}^{(-)}|\vec{p}' >$.The amplitude is then expressed in terms of a triple integral of the general form:

$$f_1(\omega) \sim N_1 N_2 \oint d\xi_1 ... \oint d\xi_2 ... \int d\vec{p}' ... \tag{5}$$

which, after some algebra, reduces to a Gauss hypergeometric function. When inserting the resulting expression of $f_1(\omega)$ in Eq.(1), one gets the one-photon cross section, which is proportional to:

$$\frac{d\sigma_1}{d\Omega(\hat{p_2})} \sim \left(\frac{I}{I_0}\right)\frac{1}{K^4}a_0^2 \left|{}_2F_1(1 - \frac{i}{p_1}, 2 - \frac{i}{p_2}; 2; z)\right|^2 (\hat{s}.\hat{p_2})^2 \tag{6}$$

which corresponds to Sommerfeld's result when specialized to the geometry chosen here ($\hat{s} \perp \vec{p_1}$). Here $I = \omega^2 a^2/8\pi c$ is the time averaged intensity of the laser field, $I_0 = ce^2/8\pi a_0^4 \sim 3.5\,10^{16} W/cm^2$ is the time averaged atomic unit of field strength intensity, a_0 is the Bohr radius, $K = \hbar\omega/1Ry.$ and $z = -4p_1p_2 sin^2(\theta/2)/(p_1 - p_2)^2$ and all other quantities are expressed in atomic units.

In the context of laser stimulated free-free transitions, the range of validity of this standard perturbative result depends on several parameters, which are the electron momentum and the laser intensity and frequency. As already mentioned, both the nonrelativistic ($p_i << mc = 137 a.u.$) and the dipole approximation ($K < 1$) are assumed throughout. On the other hand, the perturbative approach is valid as long as the expectation value of the interaction hamiltonian $< H_{int} >$ remains small with respect to a characteristic energy of the unperturbed system. In our context $< H_{int} >$, which is intensity-dependent, reads (in Rydberg) :

$$< H_{int} >= \frac{2}{K}\sqrt{\frac{I}{I_0}}\hat{s}.(\vec{p_1} - \vec{p_2})$$

and the characteristic energy is the laser photon energy: $\hbar\omega$ The relevant condition ensuring the validity of a perturbative approach is thus:[11]

$$\frac{< H_{int} >}{\hbar\omega} = \frac{4}{K^2}\sqrt{\frac{I}{I_0}}\vec{s}.(\vec{p_1} - \vec{p_2}) \quad << 1 \tag{7}$$

If this condition is not fulfilled, i.e. at higher intensities $I > I_0$,or (and) at lower frequencies, one has to resort to non perturbative approaches; see Sec IV below.

In order to better understand the transition between the perturbative and non-perturbative regimes, it is interesting to study the limits $(p_1^2, p_2^2) >> K$. In this approximation the ${}_2F_1$ function entering the expression becomes:

$${}_2F_1(...)\sim{}_2F_1(1,2;2;z) = (1-z)^{-1} = \frac{(p_1 - p_2)^2}{(\vec{p_1} - \vec{p_2})^2}$$

and the cross-section itself becomes accordingly:

$$\frac{d\sigma_1}{d\Omega(p_2)}\sim 4\frac{I}{I_0}\frac{1}{K^4}\frac{p_2}{p_1}(\hat{s}.\vec{p_2})^2\frac{4a_0^2}{(\vec{p_1} - \vec{p_2})^4} \tag{8}$$

This result displays two interesting features connected to the soft-photon limit of the cross-section:

i)for large values of the incoming p_1 and the outgoing p_2 electron momenta, the Rutherford scattering cross-section,for momentum transfer $\vec{Q} = \vec{p_1} - \vec{p_2}$, factors out:

$$\left(\frac{d\sigma}{d\Omega(\hat{p_2})}\right)_{el} = \frac{4a_0^2}{(\vec{p_1} - \vec{p_2})^4} \tag{9}$$

ii)In the limit of small photon energies $K \to 0$, the cross-section diverges.One recovers in that way the so-called Low theorem related to the divergence of the bremsstrahlung cross-section.[12]

III. TWO-PHOTON STIMULATED FREE-FREE TRANSITIONS

If $\nu = 2$, i.e. two photons of the laser field are absorbed,the transition amplitude reads:

$$f_2(\omega) = \frac{m}{2\pi\hbar^2}\left(\frac{ea}{2mc}\right)^2 < \vec{p_2}^{(-)}|\hat{s}.\vec{p}G^+(\vec{p},\vec{p}';\Omega)\hat{s}.\vec{p}|\vec{p_1}^{(+)} > \tag{10}$$

where $\Omega = p_1^2/2m + 2\hbar\omega + i\epsilon$ and $G^+(\vec{p},\vec{p}';\Omega)$ is the momentum space representation of the Coulomb Green's function.We have found convenient to use Schwinger's representation:[10]

$$G^+(\vec{p},\vec{p}';E) = F\oint_1^{(0^+)} d\rho\rho^{-i/p_0}\frac{d}{d\rho}\left(\frac{\rho^{-1}(1-\rho^2)}{(p_0^2(\vec{p}-\vec{p}')^2+(p_0^2+p'^2)(p_0^2+p^2)(1-\rho^2)/4\rho)^2}\right)$$

where $p_0 = \sqrt{-E}$, $F = p_0^3 exp(-\pi/p_0)/4\pi^2 \sin(i\pi/p_0)$, and the limit $\epsilon \to 0_+$ is understood. The second-order amplitude is then expressed as a multiple integral of the general form:

$$f_2(\omega) = N_1 N_2 F \oint_1^{(0^+)} d\rho ... \oint d\xi_1 ... \oint d\xi_2 ... \int d\vec{p_1}' ... \int d\vec{p_2}' ... \tag{12}$$

The double integral over $\vec{p_1}'$ and $\vec{p_2}'$ has been evaluated in a similar context,[13] and results in an algebraic function of the variables ξ_1,ξ_2 and ρ. Again, the special geometry chosen here ($\hat{p}_1 \perp \hat{s}$) simplifies the angular dependence of the amplitude which becomes:

$$f_2(\omega) = A(\omega) + (\hat{s}.\hat{p_2})^2 B(\omega) \tag{13}$$

where $A(\omega)$ and $B(\omega)$ contain triple integrals of the general form:

$$\begin{pmatrix} A(\omega) \\ B(\omega) \end{pmatrix} = \oint_1^{(0_+)} d\rho\rho^{-i/p_0} \oint d\xi_1 \oint d\xi_2 \begin{pmatrix} a(\omega) \\ b(\omega) \end{pmatrix} \tag{14}$$

After some algebra, the two contour integrals over ξ_1 and ξ_2 can be expressed in terms of Gauss hypergeometric functions ${}_2F_1$,times algebraic factors and the reduced amplitudes $A(\omega)$ and $B(\omega)$ contain in turn simpler integrals of the form:

$$J = \int_0^1 d\rho\rho^{q-i/p_0}(1-x_0\rho)^\alpha(1-x_1\rho)^\beta(1-x_2\rho)^\gamma\, {}_2F_1(a,b;c;z(\rho)) \tag{15}$$

Integrals of this type can be conveniently computed numerically with the help of standard complex variables Gauss quadrature routines. In addition, they lend themselves easily to an analytical study of their limiting behaviour in the physically interesting limit $p_i >> 1$, $K/p_i^2 << 1$.As a matter of fact, in this limit, one easily shows that:

$$\frac{d\sigma_2}{d\Omega(\hat{p_2})} \sim 4\frac{p_2^5}{p_1}\frac{1}{K^8}\left(\frac{I}{I_0}\right)^2(\hat{s}.\hat{p_2})^4\frac{4a_0^2}{(\vec{p_1}-\vec{p_2})^4} \tag{16}$$

Again, one observes that Rutherford's elastic scattering cross- section can be factorized and that in the soft-photon limit ($K \to 0$) the cross-section diverges. This result represents a generalization of the Low theorem to the case of the two-photon bremsstrahlung.Still in this limit, $p_2 \sim p_1$, one notices that the cross-section becomes:

$$\frac{d\sigma_2}{d\Omega(\hat{p_2})} \sim \frac{16}{K^8}\left(\frac{I}{I_0}\right)^2(\hat{s}.\hat{p_2})^4\frac{a_0^2}{(\vec{p_1}-\vec{p_2})^4} \tag{17}$$

result which is independant on the magnitude of $\vec{p_1}$.

Such expressions are useful for establishing the connection between respectively the low-frequency limit of the perturbative result and the low frequency, low intensity limit of the non-perturbative approach.

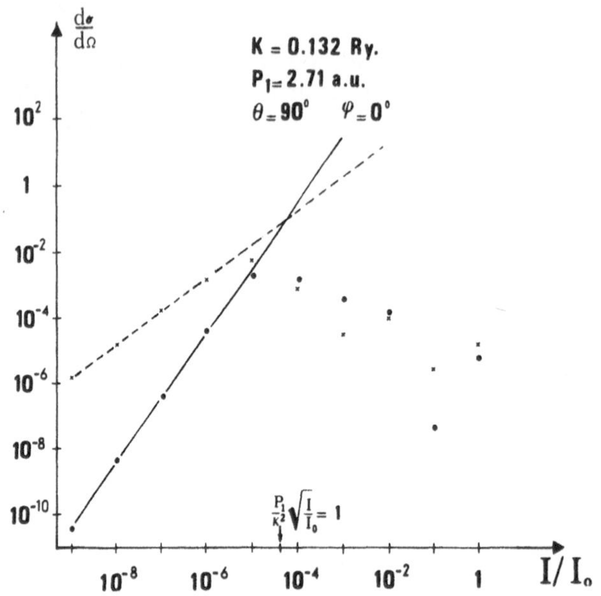

Figure 2

Variations of the one - and two photon cross-sections in terms of the ratio (I/I_0). For the sake of illustration, we have chosen the values of the laser frequency and incident electrons energy as the same as in Brehme's paper.[14] The crosses × and dots • correspond respectively to the one - and two-photon non-perturbative results obtained from Eq. (18) The broken and solid lines correspond respectively to the one - and two-photon perturbative results obtained from Eqs. (6) and (13).

IV. NON PERTURBATIVE APPROACHES

If the criterion Eq.(7) is not met, the perturbative approach is no longer valid and, particularly in the limit of very strong laser fields, one has to take into account the "dressing" of the electron by the field in the initial and final states. The cross-section are accordingly modified and if both the soft-photon approximation and the short range potential model for describing the target are valid, one gets the following result:[5]

$$\frac{d\sigma_\nu}{d\Omega(\hat{p_2})} = \frac{p_2}{p_1}\left|J_\nu(\vec{\alpha}.\vec{Q})\right|^2\left(\frac{d\sigma}{d\Omega(\hat{p_2})}\right)_{el} \tag{18}$$

where $\vec{\alpha} = \frac{4}{K^2}\sqrt{\frac{I}{I_0}}\hat{s}$ and $\vec{Q} = \vec{p_1} - \vec{p_2}$. Again, it is an easy matter to study the low intensity limit of this expression:the argument of the Bessel function J_ν becomes small,($\vec{\alpha}.\vec{Q} << 1$), and one can retain only the first term of its power series expansion. Interestingly enough, though the above expression Eq.18, has been obtained under the assumption of a short range potential, this limiting procedure enables us to recover our perturbative results Eqs.(8) and (17) for the Coulomb field (see the figure (2)).Note also that, besides providing an independent check of our calculation, the comparison of the numerical results obtained by the two methods permits us to assess their respective domains of validity. Detailed results obtained along these lines will be published elsewhere.

V. CONCLUSION

We have presented a progress report on an exact perturbative calculation of one- and two-photon free-free amplitudes and cross-sections in a Coulomb field.Our main results concern the evaluation of second order perturbative amplitudes in the Coulomb continuous spectrum and are intended to provide reference marks for the future discussions of the dynamics of the electron-ion laser-assisted collisions. In particular, the precise comparison in a well defined test case, between the perturbative and non-perturbative approaches will permit to study the transition from one regime to another.

REFERENCES

1. M. Gavrila and M. Van der Wiel, Comments At. Mol. Phys. 8, 1 (1978).
2. L. Rosenberg, in "Advances in Atomic and Molecular Physics", edited by D. Bates and B. Bederson, Academic Press, New York, Vol 18, 1 (1982).
3. H. A. Kramers, Phil. Mag. 46, 836 (1923).
4. A. Sommerfeld, "Atombau und Spektrallinien", Vieweg, Braunschweig,Vol. II, 495 (1939).
5. N. M. Kroll and K. M. Watson, Phys. Rev. A 8, 804 (1973).
6. M. Gavrila and J. Z. Kaminski, Phys. Rev. Lett. 51, 613 (1984).
7. S. Geltman, J. Phys. B 8, L374 (1975); J. Research, N.B.S. 82, 173 (1977).

8. L. Rosenberg, Phys. Rev. A 20, 457 (1979); Phys. Rev. A 26, 132 (1982); Phys. Rev. A 27, 1879 (1983).
9. J. Banerji and M. H. Mittelman, Phys. Rev. A 26, 3706 (1982); E. Fiordilino and M. H. Mittelman, Phys. Rev. A 28, 229 (1983); L. F. Saez and M. H. Mittelman, Phys. Rev. A 29, 2228 (1984).
10. J. Schwinger, J. Math. Phys. 5, 1606 (1964).
11. L. Rosenberg, Phys. Rev. A 23, 2283 (1981); See also R. Shakeshaft, Phys. Rev. A 29, 383 (1984). The ratio appearing in our Eq.(7) is denoted δ_1/δ_2 in these references.
12. F. E. Low, Phys. Rev. 110, 974 (1958).
13. M. Gavrila, Phys. Rev. A 8, 804 (1973).
14.H. Brehme, Phys. Rev. C 3, 837 (1971).

MULTIPHOTONIONIZATION OF XENON

H. J. Humpert, R. Hippler, H. Schwier, and H. O. Lutz

Fakultät für Physik, Universität Bielefeld

D-4800 Bielefeld 1, F. R. Germany

INTRODUCTION

With the advent of high-intensity lasers multiphoton ionization (MPI) processes have attracted considerable interest experimentally as well as theoretically. The study of the angular distribution of ejected photoelectrons yields information about the target atomic or molecular structure, as well as the various ionization channels involved.[1] As has been found experimentally, measurements of the dependence of the MPI signal on the laser intensity deviate considerably from perturbation theory for laser intensities greater than 10^{13} W/cm^2.[2]

In the following, we report on the investigation of photoionization processes of xenon atoms using radiation from a Nd:YAG laser. Photoelectron spectra have been measured at two different laser wavelengths (λ = 1064 nm and λ = 532 nm). The kinetic energy E of a photoelectron is

$$E = N \cdot h\nu - E_I$$

where N is the number of absorbed photons, ν the frequency of the laser light and E_I the ionization energy of the atom. The ionization energies of xenon are 13.44 eV for the $^2P_{1/2}$ and 12.13 eV for the $^2P_{3/2}$ state. For the fundamental line (λ = 1064 nm, hν = 1.165 eV) the minimum number of photons necessary to ionize a xenon atom is eleven and twelve for ionization from the $^2P_{3/2}$ and $^2P_{1/2}$ state, respectively. Six photons are needed in case of the frequency-doubled line (λ = 523 nm, hν = 2.33eV) for both core states (Fig. 1). We have also observed ionization processes with an additional absorption of up to three (λ = 532 nm) or ten photons

(λ = 1064 nm). These processes have been dubbed continuum-continuum transition (C-C) or above threshold ionization (ATI).[3] Using the frequency-doubled line, the angular distributions of the photoelectrons have been obtained while the intensity dependence of the MPI signal has been investigated employing the fundamental line of the laser.

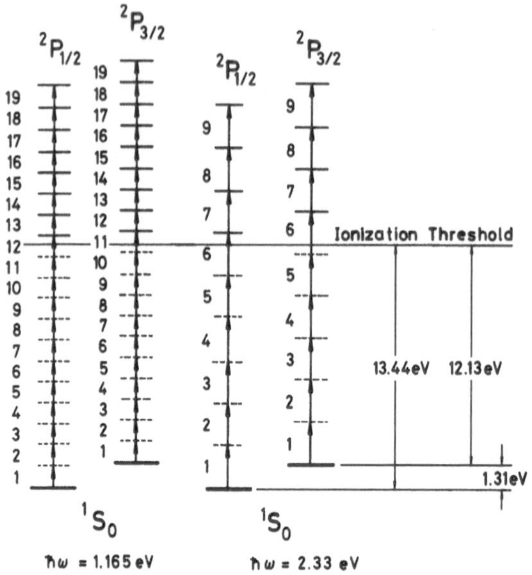

Fig. 1: Multiphoton ionization processes in xenon (schematic)

EXPERIMENTAL METHOD

The laser light has been generated by a Nd-YAG laser consisting of an oscillator with unstable resonator configuration and two amplifiers. The line width was smaller than 1 cm^{-1} and the laser oscillated on about 150 longitudinal modes. A Q-switch provided a pulse width of about 8 nsec. For some experiments the laser light (fundamental line λ = 1064 nm) has been frequency-doubled in a KDP crystal (λ = 532 nm). A half-wave plate was used to rotate the linear polarization direction of the laser light. An additional Glan linear polarizer has been employed to vary the laser intensity. With a 100 mm lens the light was focused onto a xenon gas target effusing from a nozzle or, alternatively, from a pulsed beam source. The gas pressure in the target region was varied between $1 \cdot 10^{-7}$ mbar and $5 \cdot 10^{-4}$ mbar. The angular distributions of the photoelectrons were measured with a 45^{0} parallel plate spectrometer which had been arranged perpendicular to the laser beam direction. In addition, the laser light intensity dependence of photoelectron ejection has been investigated by use of a double stage cylindrical mirror analyzer. The laser power was measured with a calorimeter.

RESULTS AND DISCUSSION

a) Angular Distribution of Photoelectrons

Fig. 2a,b show photoelectron spectra obtained with the frequency-doubled line and the fundamental line, respectively. In Fig. 2a the four lines correspond to the absorption of six and seven photons (λ = 532 nm) with the doublet character of the lines being caused by the fine-structure splitting of the ion core states (1.31 eV).

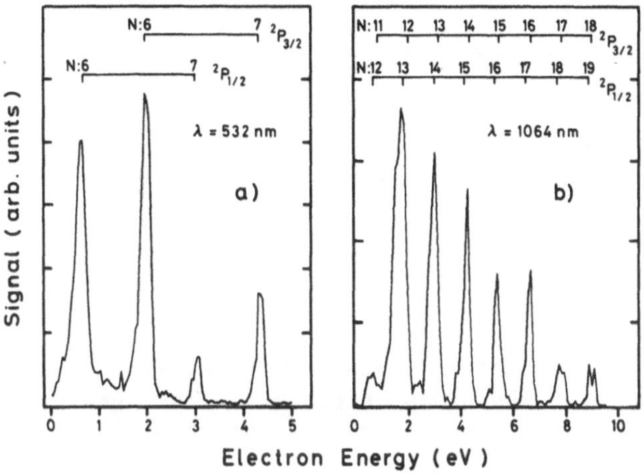

Fig. 2: Electron spectra generated by multiphoton absorption of xenon (a: λ = 532 nm, b: λ = 1064 nm)

The intensities of four transitions (N = 6,7) in Fig. 2a as a function of the polarization direction have been integrated, and were fitted by a sum of Legendre polynomials of order 2k with $0 \leq k \leq L$; L is the maximum angular momentum of the system, i.e. the number of absorbed photons.

$$P(\theta) = \Sigma_k \ \beta_{2k} \ P_{2k}(\cos\theta)$$

The coefficients β_{2k} as a function of the order k are plotted in Fig. 3 . The two N = 6 transitions differ considerably in their β_{2k} values, whereas in the case of the N = 7 transitions the two parameter sets are almost identical, at least for $k \leq 5$. At present, no theory exists for comparison with these data.

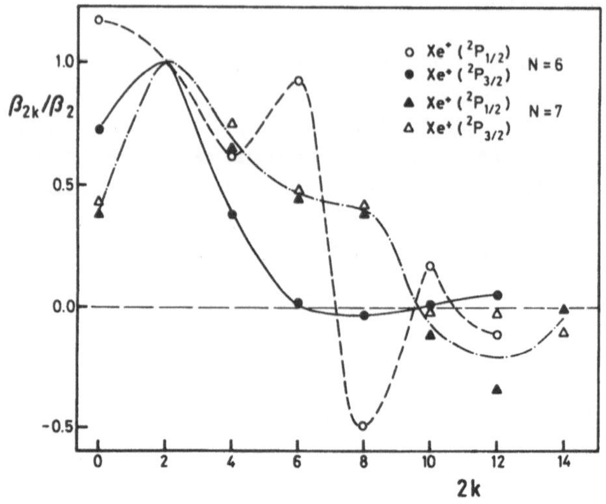

Fig. 3: Amplitudes β_{2k} of the Legendre-Polynomials fitted to the experimental data (λ = 532 nm).

b) Dependence of Photoelectron Ejection on the Laser Light Intensity

An electron spectrum obtained with the fundamental laser line (λ = 1064 nm) is shown in Fig. 2b . The energy difference of electrons resulting from ionization of the $^2P_{3/2}$ state (after absorption of N photons) and from the $^2P_{1/2}$ state (after absorption of N+1 photons) is 0.15 eV. This is of the order of the spectrometer resolution and has thus not been resolved. The data show that with increasing order of the ionization process the heights of the peaks decrease as is expected from perturbation theory. We note, however, that the lowest order peak should appear at 0.54 eV ($^2P_{1/2}$) and 0.69 eV ($^2P_{3/2}$) with an expected intensity higher than the first C-C peak. In contrast, the lowest order peak in Fig. 2b has nearly vanished. This phenomenon has been investigated under different experimental conditions. Fig. 4 displays four photoelectron spectra measured at different laser pulse energies. With increasing laser intensity (at about 10^{13} W/cm^2) the first peak in the spectrum dissapears. This occurs already at intensities just above the onset of the MPI process; therefore, it is difficult to observe the first peak at all. In our experiment we used a pulsed beam source to achieve higher signal counting rates in the low intensity measurements.

The vanishing of the lowest-energy peak was explained by an increase of the ionization threshold in the intense laser field[5]. The vector potential term A^2 of the electro-magnetic field

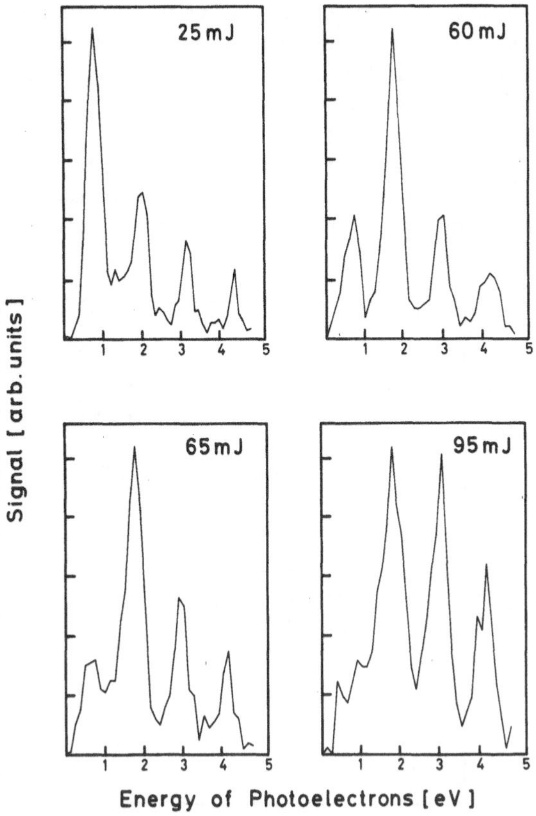

Fig. 4: Electron spetra (λ = 1064 nm) at different laser pulse energies

contributes to the atomic potential and the xenon atoms can no longer be ionized by the absorption of eleven ($^2P_{3/2}$) and twelve photons ($^2P_{1/2}$). One might expect, that the higher ionization energy should not only result in a vanishing of the first peak but also in a shift of the position of the other electron peaks. However, the electrons which have been ionized, gain energy by ponderomotive force acceleration when leaving the laser focus ($F=-[e/2m\omega^2]\nabla[E^2]$). As has been pointed out by Muller et al., this energy gain cancels the increase of the ionization potential. Therefore, the higher order peaks remain relatively unaffected.

In Fig. 5 the integrated intensities S of several peaks are plotted as a function of the laser (λ = 1064 nm) pulse energy. Perturbation theory predicts the dependence of S on the laser intensity to be

$$S \propto I^N$$

where N is the number of absorbed photons. Thus one would expect a slope of about eleven for the intensity of the first peak. The slope of the other lines should increase with increasing N, i.e. the number of absorbed photons. In contrast as one can see in Fig. 5, the slopes of the different peak intensities are nearly identical except for the first one. The order of nonlinearity of the ionization signal was found to be approximately six for the first peak and eight to nine for the others, whereas Kruit et al.[2] measured eight and ten to eleven.

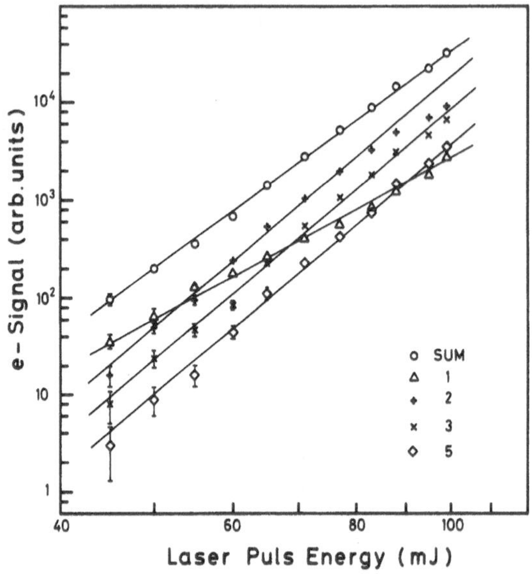

Fig. 5: Intensity dependence of some selected peaks of the MPI (λ = 1064 nm)

The interpretation for the disappearance of the first peak given above is corroborated by the behavior of the second peak at high laser pulse energies. As can be seen in Figs. 4 and 5, its intensity also decreases if compared, for example, to the third one. This phenomenon can clearly be seen in Fig. 6 which was taken at a pulse energy of 300 mJ.

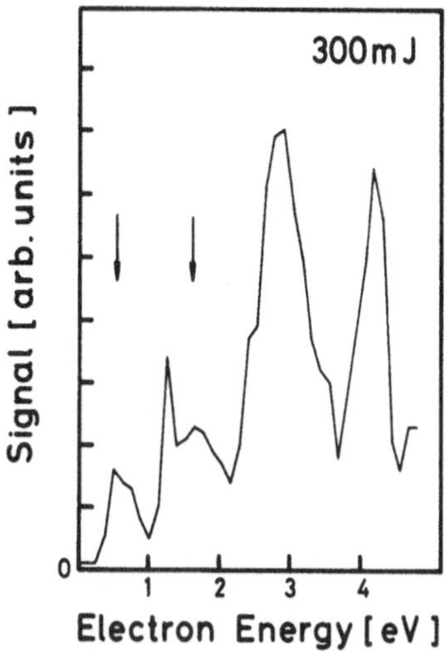

Fig. 6: Electron spectrum (λ = 1064 nm) at 300 mJ pulse energy

REFERENCES

1. P. Lambropoulos, Adv. At. Mol. Phys. 12, 87 (1976)
2. P. Kruit, J. Kimman, H.G. Muller, M.J. van der Wiel; Phys. Rev. A 28, 248 (1983)
3. P. Agostini, M. Clement, F. Fabre, G. Petite; J. Phys. B 14, L491 (1981)
4. R. Hippler, H.J. Humpert, H. Schwier, S. Jetzke, H.O. Lutz; J. Phys. B 16, L713 (1983)
5. H.G. Muller, A. Tip, M.J. van der Wiel; J. Phys. B 16, L679 (1983)

THEORETICAL STUDY OF THE PHOTOELECTRON SPECTRUM OF CESIUM

P. Decleva and A. Lisini

Istituto di Chimica
Università di Trieste
I-34127 Trieste, Italy

Correlation effects can be usefully divided into two groups: those deriving from the instantaneous interactions among the electrons and those deriving from the presence of quasi degenerate configurations[1]. The first are difficult to be treated accurately but are expected to change smoothly from state to state and therefore largely cancel when considering energy differences. The second cause strong deviations from the Hartree-Fock predictions but can be treated exactly within a limited CI.

Correlation effects of this kind are clearly seen in the valence photoelectron spectrum of the alkali atoms[2,3]. Starting from K, because of the availability of the empty 3d orbitals, they become more and more evident going along the series. Although unexpected, such effects cannot be ruled out also in the core ionization. In fact their occurrence has been suggested in the 3d ionization of Cs[4]. We have examined the possibility of treating these correlation effects in Cs within a CI limited to the quasi degenerate configurations, i.e., for the valence region, $5p^5 6s$ and $5p^5 5d$ together with $5p^5 7s$ which is expected to derive substantial intensity from relaxation. Additionally the states originating from the $5p^5 6p$ configuration have been considered as they are observed in the spectrum. 5d, 6p and 7s orbitals were generated in the frozen core of the primary ionic configuration. The numerical approach of Froese Fischer[5] and Grant et al.[6,7] was employed.

The relativistic results are reported in the figures 1,2, together with the experimental values[2,3]. The first peak (calculated energy 16.37 eV) has been shifted in order to coincide with the experimental one. Intensities have been calculated within the sudden approximation, apart from those for the $5p^5 6p$ states, forbidden in this approximation, which have been taken from the experiment.

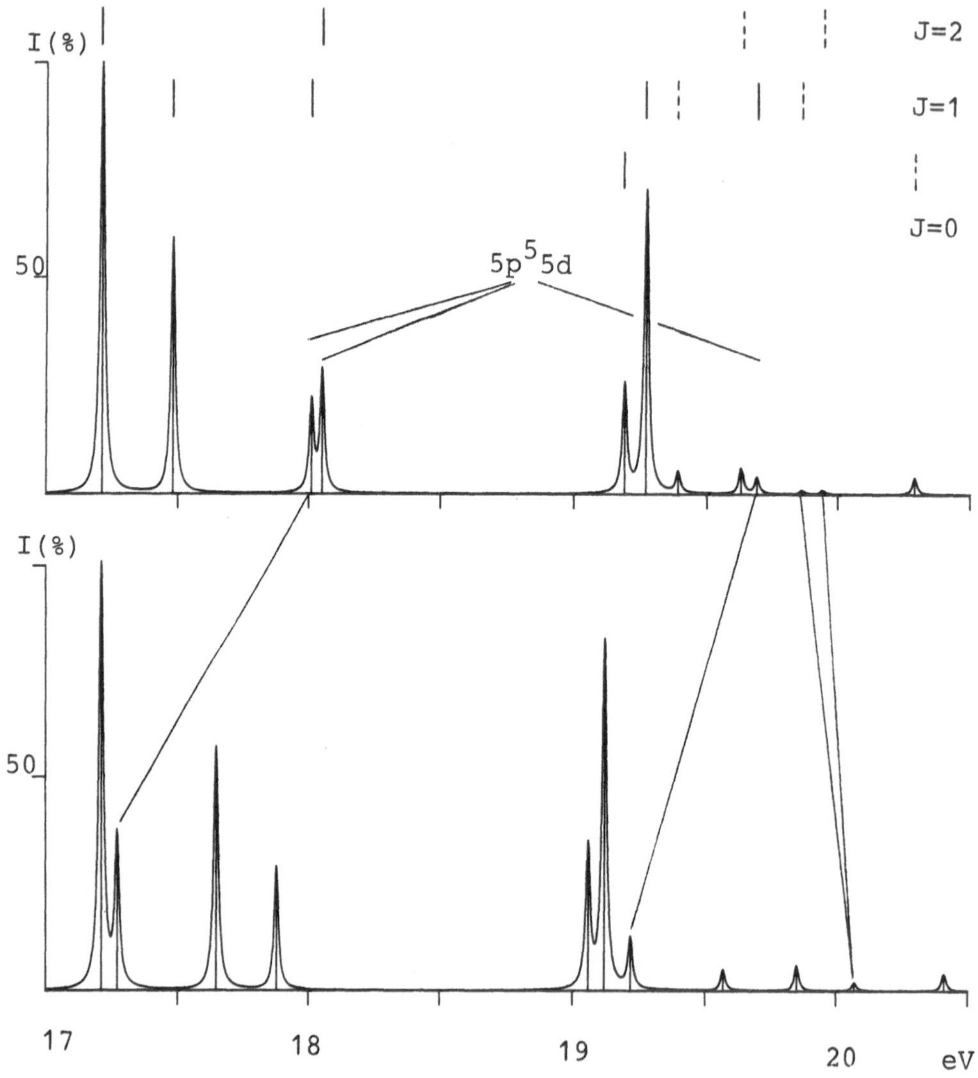

Fig.1. Theoretical (top) and experimental (bottom) spectrum relative to the 5p ionization in Cesium. Full line: odd parity, broken line: even parity. Experimental: up to 19.5 eV Ref.2, the rest from Ref.3.

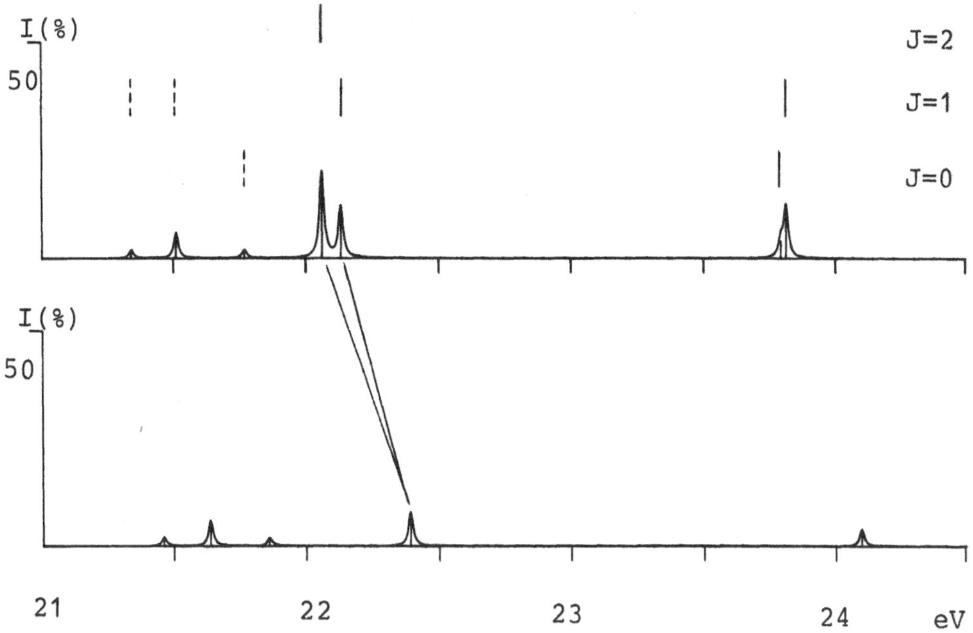

Fig.2. See Fig.1 for explanation.

A general overall agreement is achieved both for the energies and the intensities. The most notable exception is presented by the states $5p^5 5d$ J=1, for which the intensities are calculated too low and the energy too high. This causes two inversions with the $5p^5 6s$ and $5p^5 6p$ J=1 states and indicates the presence of an additional interaction not included in the limited CI employed. It may be noted that a slight worsening is apparent also in the results for these states obtained by least squares fitting within the same model space[8]. The inclusion of higher states of the $5p^5 nd$ Rydberg series in the CI did not give any improvement, so other mechanisms are responsible for this discrepancy. A smaller disagreement is apparent also for the energies of the $5p^5 7s$ states, being due to the loss in correlation energy in going to higher Rydberg states. On the contrary the intensities are calculated higher by a factor of four. As these intensities derive only through relaxation, their amount should not depend substantially on the position of the hole and in fact a very similar intensity was calculated also in the case of the 3d hole[9]. As in the latter case the agreement with the experiment is good, we believe that the discrepancy is due to the experimental inaccuracy in the higher energy region.

A notable intensity is observed in the spectrum[3] also for the $5p^56p$ states which have different parity and therefore cannot mix with the previous ones. A similar feature is seen also in the valence spectrum of Na[10]. In both cases this has been attributed to the presence of strong configuration interaction, particularly in the initial state. In principle, there are five mechanisms giving intensity to these states:

1) Initial state CI (ISCI), e.g.:

$$5p^66s \longleftrightarrow 5p^55d6p$$

2) Final state CI (FSCI), e.g.:

$$5s5p^66s \longleftrightarrow 5s^25p^56p$$

3) Shake-up relative to the 6s ionization:

$$5p^66s \longrightarrow 5p^56p \qquad \begin{array}{l} 6s \longrightarrow \varepsilon p \\ 5p \longrightarrow 6p \end{array}$$

4) Conjugate shake-up:

$$5p^66s \longrightarrow 5p^56p \qquad \begin{array}{l} 5p \longrightarrow \varepsilon p \\ 6s \longrightarrow 6p \end{array}$$

5) CI in the continuum (CSCI), e.g.:

$$5p^56s\varepsilon s \longleftrightarrow 5p^56p\varepsilon p$$

As concerns ISCI, an MCSCF treatment of the ground state shows the absence of strong configuration mixing. In fact the coefficient of the next important configuration, i.e. $5p^55d6p$, is only 0.11 and therefore this mechanism seems to be of little importance in this case. A very small mixing was also obtained for the FSCI involving the 5s ionization. Furthermore this mechanism cannot lead to J=2 final states, apparently observed in the spectrum.

As pointed out for Ne[11], monopole excitations from the shells deeper than the ionized one are of negligible intensity, because the relaxation of the former is very small. Therefore the most important contribution is probably given by the two last mechanisms, which are, however, outside the scope of the sudden approximation. Considerations on the behaviour of the primary peaks tend to exclude the CSCI in Na[10], leaving the conjugate shake-up as the dominant process.

Table 1. 6s ⟶ 7s shake-up energy (eV) relative to the 3d ionization.

HF	MCSCF	DF	MCSCF+DF	EXP
4.55	4.89	4.75	5.09	5.3(.2)

A similar CI for the 3d hole[12] gave no evidence for strong mixing between quasi degenerate configurations. To confirm the absence of such effects, we attempted a treatment of the dynamical correlation (core polarization) with an MCSCF calculation along the line of Froese Fischer[13].

By adding the correlation change so obtained and the relativistic effects at the HF level (Table 1), good agreement with the experiment is obtained for the 6s ⟶ 7s shake-up. Therefore the relativistic and correlation effects appear to be additive in this case and no evidence of additional correlations, specific to the ion, is found. These results tend to exclude the presence of a second shake-up.

REFERENCES

1. D. R. Beck and C. A. Nicolaides, in: "Excited States in Quantum Chemistry", C. A. Nicolaides and D. R. Beck, eds., D. Reidel, Dordrecht, Holland (1978) p. 105.
2. S. Süzer, B. Breuckmann, W. Menzel, C. E. Theodosiou and W. Mehlhorn, J. Pys. B 13 (1980) 2061.
3. E. P. F. Lee and A. W. Potts, Chem. Phys. Letters 66 (1979) 553.
4. R. D. Mathews, A. R. Slaughter, R. J. Key and M. S. Banna, J. Chem. Phys. 78 (1983) 62.
5. C. Froese Fischer, Comp. Phys. Commun. 14 (1978) 145.
6. I. P. Grant, B. J. McKenzie, P. H. Norrington, D. F. Mayers and N. C. Pyper, Comp. Phys. Commun. 21 (1980) 207.
7. B. J. McKenzie, I. P. Grant and P. H. Norrington, Comp. Phys. Commun. 21 (1980) 233.
8. J. Reader, Phys. Rev. A 13 (1976) 507.
9. G. De Alti, P. Decleva and A. Lisini, Chem. Phys. 80 (1983) 229.
10. S. Krummacher, V. Schmidt, J. M. Bizau, D. L. Ederer, P. Dhez and F. Wuilleumier, J. Phys. B 15 (1982) 4363.

11. U. Gelius, J. Electron Spectrosc. Relat. Phenom. 5 (1974) 985.
12. G. De Alti, P. Decleva and A. Lisini, J. Chem. Phys. in press.
13. C.Froese Fischer, Can. J. Phys. 54 (1976) 1465.

PHOTOIONIZATION OF POSITIVE ATOMIC IONS

Ian C. Lyon

Department of Atomic Physics
University of Newcastle upon Tyne
Newcastle upon Tyne, NE1 7RU, England

INTRODUCTION

There have been many experiments in the last eighty years on the photoionization of atoms, molecules and negative ions. They were possible because the low ionization energies correspond to visible or near ultraviolet radiation so that conventional light sources and lasers could be used. But, measurements of the photoionization of positive atomic ions are extremely rare because the lowest threshold of a positive ion (Ba^+) is 10eV. This corresponds to VUV radiation and until the recent advent of dedicated synchrotrons no intense, tunable sources of radiation existed in this spectral region.

A few positive atomic ion photoionization experiments were performed by photographing the absorption spectrum of the ion[1,2] using a weak continuum source such as a discharge lamp. A high ion density ($\sim 5 \times 10^{15}$ ions cm^{-3}) was needed to show significant absorption and this was achieved by photoionizing a dense vapour of neutral atoms with an intense laser pulse and photographing the absorption spectrum before the plasma recombined. This approach suffered many defects, the main one being the population of excited and metastable states by the high electron density of the plasma so that many extraneous lines were seen.

We will describe a much 'cleaner' technique using crossed beam methods. To study the reaction $Ba^+ + h\nu \rightarrow Ba^{2+} + e^-$, a well defined beam of Ba^+ ions was produced with energies of 3keV and this was merged with a monochromatic ($\Delta\lambda \sim 3Å$) VUV beam.

Space charge limits the ion density in the beam to $\sim 10^5 - 10^6$ ions cm^{-3} so absorption of the VUV beam cannot be measured. However, Ba^{2+} ions formed by photoionization can be separated from their parent beam and be detected by a single particle counter. This allowed absolute photoionization cross sections to be measured. The technique is still in its infancy and absolute accuracies obtained in preliminary experiments is of order ±50%, but the method is potentially capable of ±10% or better.

The experiment used a VUV beam from the Daresbury Synchrotron Radation Source (SRS) and yielded the first measurements of photoionization cross-sections for positive atomic ions to use crossed beams.

APPARATUS

The apparatus is illustrated by figure 1.

An unexcited beam of Ba^+ ions produced by a surface ionization source (S), was mass selected by an electromagnet (M1) and merged with a beam of monochromatic VUV radiation from the SRS, the intensity of which was monitored by measuring the current it produced in the photodiode (PD). The Ba^+ ions were collected in a Faraday cup (C) whilst Ba^{2+} ions were deflected by a second electromagnet (M2) and electrostatic analyser (E) and counted by the particle multiplier (D). (Johnston Laboratories MM-1SG). The photoionization cross-section could then be deduced from measurements of the intensities of the ion and VUV beams, the ion velocity, the path length over which the two beams interacted and a 'form factor' which accounted for the spatial overlap of the two beams. The form factor was determined by scanning two slits through the beams, one horizontally and the other vertically. The beam currents were measured whilst the vertical slit was fixed and the horizontal slit was scanned across their widths. This process was then repeated for many positions of the vertical shutter to build up a map of the overlap and profiles of the beams at a fixed position on the beam axis. The construction of the shutters then allowed the measurement to be made at two more points along the beam axis. The slits were mounted on two linear drives and moved by stepping motors under computer control.

Background counts arose from charge stripping reactions, $Ba^+ + R \rightarrow Ba^{2+} + R + e^-$, with residual gas molecules (R). Even with vacua of 10^{-10} torr, the signal to background ratio (SBR) was generally very poor ($10^{-2} - 10^{-3}$) and so a beam modulation technique was developed to separate the signal from the background. Background counts arose from the ion beam alone, so it was necessary only to modulate the VUV beam with a rotating light chopper (L) and arrange for two scalar counters to record

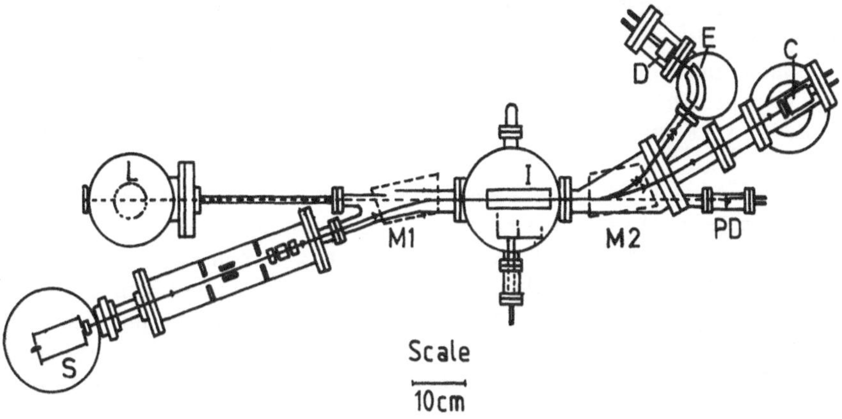

Figure 1 Plan View of the Apparatus.

the countrate, one with the light on and the other with it off. Subtraction of the two rates yielded the signal.

The points at which the beams merged and separated in the two magnetic fields were ill defined so an uncertainty existed in the measurement of the interaction length. To overcome this problem, the beams were made to travel coaxially within a cylinder (I) of known length (12cm) which was biased to potentials up to ±600V. This changed the energy of Ba^{2+} ions formed within the cylinder so that careful tuning of the magnetic field of M2 and voltage of the electrostatic analyser E allowed discrimination against all Ba^{2+} ions formed outside the interaction region.

RESULTS

The apparatus was first tested using a helium resonance lamp (λ = 584Å) and a photoionization cross section of around 7 x $10^{-16}cm^2$ (±50%) was measured at this wavelength but the result was in some doubt because spurious signals arising from modulations of gas from the lamp may have existed. However, it transpired that the magnitude of the cross section was confirmed by results obtained with synchrotron radiation.

The first measurements used monochromatic radiation selected by a 5M McPherson monochromator. Problems with the beam line optics however limited the radiaion flux to $\sim$5 x 10^8 photons

s^{-1} in 3Å bandwidth and this severely restricted the statistical accuracy of the results. Six wavelengths between 375Å and 830Å were chosen and the cross sections ranged between 5 - 13 x $10^{-16}cm^2$ (±50%) except at 830Å where no signal was observed. The results obtained are illustrated in figure 2.

A second set of measurements were made using the monochromator grating simply as a mirror. The photon beam was now broad band ($350 \leq \lambda \leq 1200$Å) and was sufficiently intense for diagnostic checks to be performed. The most important of these was to verify the expected linear dependence of signal against photon flux; all other experimental parameters being constant, and the result is illustrated by figure 3. A cross section of 9 x $10^{-17}cm^2$ averaged over the bandwidth was calculated from these results.

These cross sections are three orders of magnitude greater than predicted by calculations of direct outershell photoionization[4,5] and the difference is attributed to autoionization.

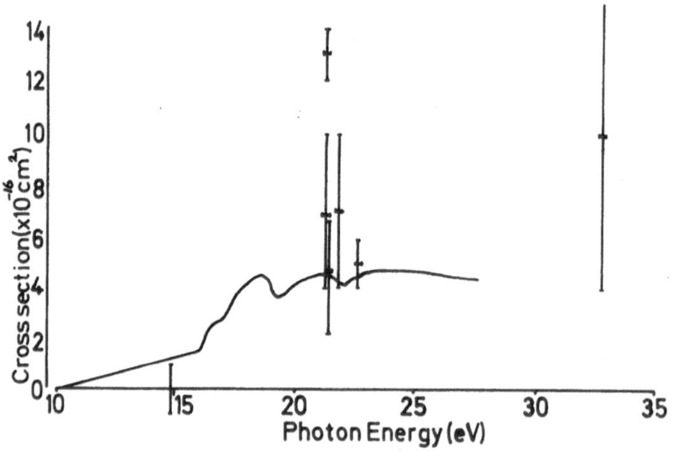

Figure 2 Measured photoionization cross sections of Ba^+. For comparison the electron impact ionization cross sections measured by Peart et al[6] are represented by the full curve.

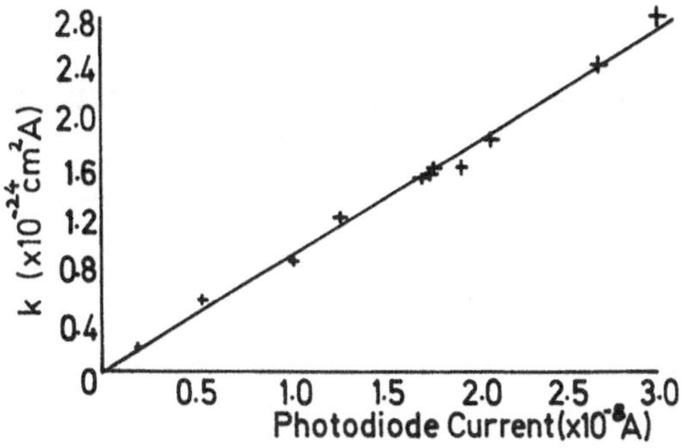

Figure 3 Results of an experimental check to verify the linear dependence between 'k' (signal per unit form factor per unit ion current) against photon flux.

Autoionization was observed by Peart et al[6] in the similar process of electron impact ionization of Ba^+. Large jumps in the ionization cross section were observed at energies of 15.8, 16.8, 19.2 and 21.9eV which was interpreted by Hansen[7] to correspond to different configurations of the autoionization process $Ba^+(5p^6 6s) + e^- \rightarrow Ba^+(5p^5 5d6s) + e^- \rightarrow Ba^{2+}(5p^6) + 2e^-$, although 'capture ionization'[8] may also be relevant.

A similar autoionizing process was expected to occur during photoionization although many differences would be apparent. During photoionization, only configurations allowed by dipole selection rules can be populated, and excitation will only occur when the photon energy equals that of the transition. So a series of resonance peaks are predicted, the width of each dependent only on the lifetime of the autoionizing state. By contrast, exchange excitation during electron impact autoionization can populate any configuration, and above each autoionizing threshold the cross section will fall only slowly because the 'impacting' electron can carry away excess energy.

Some preliminary calculations of the position and strength of the photoionized autoionizing resonances have been made[9] and these are reproduced in table 1.

Table 1

Excitation energies and oscillator strengths for the ground state of Ba^+

Excited State	Excitation energy (eV)	Oscillator strength
$(4d)^{10}(5p)^5(6s)^2\ {}^2P$	19.5	0.26
$(4d)^{10}(5p)^5(5d)({}^3P)6s\ {}^2P$	16.7	0.001
$(4d)^{10}(5p)^5(5d)({}^1P)6s\ {}^2P$	25.0	10.0

A large resonance is predicted at an energy of 25eV with a very short lifetime ($\sim 10^{-16}$ secs) and a consequent large width. If the width is represented by $\Delta\varepsilon$, then the resonance corresponds to a cross section of $\frac{1.1 \times 10^{-15}\text{cm}^2}{\Delta\varepsilon}$ ($\Delta\varepsilon$ in eV). If the 'broad band' measurement is assumed to be solely due to this transition and that no autoionization occurred below 16eV, we obtain an experimental estimate of $\frac{1.2 \times 10^{-15}\text{cm}^2}{\Delta\varepsilon}$. The close agreement is probably fortuitous in view of the wide margins of error in both theory and experiment and the presence of other transitions.

CONCLUSIONS

The photoionization cross sections of Ba^+ have been measured at six wavelengths, five of which yielded results of order 10^3 times greater than values predicted from calculations of direct outershell photoionization. The difference is ascribed to autoionization, which when included in calculations suggest cross sections of a similar order of magnitude to the experimental results. The cross section at 830Å was too small to measure ($\leq 10^{-16}$ cm ^{2}) but since this energy lies below the autoionizing threshold, this result is compatible with the autoionization model.

The experiment and theory have as yet only yielded preliminary results but new data and calculations should soon be forthcoming to explore the process more fully. Measurements will also be made on the astrophysically interesting Ca^+ ion where similar effects are expected.

ACKNOWLEDGEMENTS

The author is pleased to recognise the huge contribution to this project made by Professor K. Dolder and Dr. B. Peart and to thank them for much assistance and advice.

The calculations were performed by Professor A.E. Kingston and I am grateful for many helpful discussions. I also wish to thank Dr. J.B. West and the staff at the Daresbury Synchrotron Radiation Source for much assistance.

A research grant from the Science and Engineering Research Council is also gratefully acknowledged.

REFERENCES

1. Roig, R.A. J. Opt. Soc. Am. (1976) 66, 1400.
2. Lucatorto, T.B., McIlrath, T.J., Sugar, J. and Younger, S.M. Phys. Rev. Lett. (1981) 47, 1124.
3. Lyon, I.C. and Peart, B. J. Phys. E. (1984), 17, 920.
4. Black, J.H., Weisheit, J.C. and Laviana, E. Astrophys. J. (1972) 177, 567.
5. Shevelko, V.P. Opt. Spec. (1974) 36, 7.
6. Peart, B., Stevenson, J.G. and Dolder, K. J. Phys. B. (1973) 6, 146.
7. Hansen, J.E. J. Phys. B. (1974) 7, 1902.
8. Burke, P.G., Kingston, A.E. and Thompson, A. J. Phys.. B. (1983) 16, L385.
9. Lyon, I.C., Peart, B., West, J.B., Kingston, A.E. and Dolder, K. J. Phys. B (1984) 17, L345.

CALCULATION OF FREE-FREE RADIAL DIPOLE TRANSITION AMPLITUDES: AN L^2 BASIS APPROACH

John T. Broad and Juergen Hinze

Fakultät für Chemie
Universität
D 4800 Bielefeld, BRD

As part of a larger project to develop efficient procedures for computing strong-field multiphoton ionization of atoms, we have discovered a way of calculating radial free-free transition amplitudes using a Slater-type basis of much the same kind as is used in existing atomic structure codes. This will allow the method to be extended from the hydrogen atom test results presented here to many-electron atoms. Since finite L^2 basis approximations of free-free amplitudes cannot be expected to converge, an acceleration scheme based on the epsilon algorithm was introduced and a few important lessons were learned about how best to extrapolate from a ten-to-fifteen-basis-function expansion to completeness. In particular, to investigate how to project out channel amplitudes from complex-coordinate calculations, we allowed the Slater exponent in the basis to take on complex values.

The reduced, radial dipole transiton amplitudes,

$$T_{L,L-1} = \langle\Psi_L^-(E)||x||\Psi_{L-1}^+(E')\rangle$$

$$= -i\langle\Psi_L^-(E)||p||\Psi_{L-1}^+(E')\rangle/(E-E') \quad (1)$$

between two hydrogen-atom continuum wave functions can be expressed analytically as a sum of two hypergeometric functions,[1] which are difficult to calculate at low energies. We chose, however, to expand the Coulomb scattering wave functions in the basis,

$$\varphi_n^L(r;\lambda) = y^{L+1}\exp(-y/2)\ L_n^{2L+1}(y),\ n = 0,1,2\ \ldots \quad (2)$$

where $y = \lambda r$. For real λ, such a choice is closely related to the Slater basis typical of atomic structure calculations, and has been used with complex λ in dilatation analytic descriptions of resonances.[2,3] The coefficients can be determined analytically:

$$\Psi_L^+(r;E) = c_L(E,\lambda) \sum_{n=0} \varphi_n^L(r;\lambda) \frac{p_n^L(E;\lambda)n!}{(n+2L+1)!} \tag{3}$$

where,

$$p_n^L = \binom{n+2L+1}{n} (-\xi)^n {}_2F_1(-n, L+1-i/k; 2L+2; 1-\xi^2), \tag{4}$$

is a Pollaczek polynomial[1] with $\xi = (\lambda+2ik)/(\lambda-2ik)$, and

$$c_L = \exp(\pi/2k)\Gamma(L+1-i/k)\xi^{i/k}[(\xi-1/\xi)/i]^{L+1}/\sqrt{2\pi k}, \tag{5}$$

contains the Coulomb phase and spectral density.

Since the radial momentum becomes a matrix of only two bands in the basis set, the transition amplitude reduces to a single infinite sum,

$$T_{L,L-1} = \frac{\sqrt{L}\, c_L c_{L-1}}{(E-E')} \sum_{n=0} \frac{[p_n^L(E;\lambda)-p_{n-2}^L(E;\lambda)]p_n^{L-1}(E';\lambda)n!}{(n+2L-1)!} \tag{6}$$

As might be expected from the purely oscillatory form of the integrand in Eq. (1), however, this infinite sum shows no sign of converging as the number of basis functions increases. For real λ, the sequence of partial sums oscillates sinusoidally, whereas for complex λ it blows up.

Faced with this failure to converge and yet convinced that the first ten to fifteen terms in the expansion must contain high-quality information about the transition amplitude, it is natural to try summing the series by a Padé-approximant approach. The epsilon[4] algorithm provided a convenient way to achieve this goal, but did not work well on the sequence of partial sums as they stand in Eq.(6). Rather, it was first necessary to split the terms in the sum into geometric-series-like components, where a rational fraction fit can be expected to work well. Using an analytic continuation of the hypergeometric series, the Pollaczek polynomials can be expressed as

$$p_n^L = (q_n^{+L} - q_n^{-L})/2\pi i\rho_L \tag{7}$$

where

$$\pi\rho_L = [(\xi-1/\xi)/i]^{2L+1}\xi^{2i/k}\exp(\pi/k)|\Gamma(L+1-i/k)|^2 \tag{8}$$

and

$$q_n^{\pm L} = -2\frac{(n+2L+1)!(-\xi^{\pm 1})^{n+1}\ (L+1\mp i/k)}{\Gamma(n+L+2\mp i/k)}\ {}_2F_1(-L\mp i/k, n+1; n+L+2\mp i/k; \xi^{\pm 2}), \tag{9}$$

whose n dependence is dominated by the $(-\xi^{\pm 1})^{n+1}$ terms even at moderately large n. The four sums resulting from inserting the splitting of Eq.(7) into the single sum in Eq.(6) are summed extremely efficiently by the epsilon algorithm, as can be seen in Table 1 below for real λ. In contrast, note the failure to approximate the single sum directly in the first column. From the definition of ξ below Eq.(4), it is evident that for real λ, ξ is of modulus one, and hence that p_n behaves asymptotically as $\sin(n \arg\xi)$. While the epsilon algorithm can apparently generate the correct $E + i\varepsilon$ limit to sum a geometric-like series in ξ on its radius of convergence, it cannot find the right branch for the single sum.

The same calculation performed with complex λ is more enlightening. For $\mathrm{Im}(\lambda) < 0$, $|\xi| < 1$, so that the sum containing both q_n^- branches diverges, but the epsilon algorithm still suffices. Moreover, because that one sum now dominates over the other three, the epsilon algorithm begins to work moderately well on the single p_n sum directly, as can be seen in Tables 2 and 3.

The epsilon algorithm is thus able to extrapolate an L^2 basis set calculation for free-free dipole transition amplitudes to completeness. For accurate determination of amplitudes for many-electron atoms, it will be necessary to split the expansion of the physical continuum wave functions into two Jost-like components before convergence acceleration. The theory for their calculation has been developed,[3] including an explicit representation of the radial Coulomb resolvent in the basis of Eq.(2).

The authors acknowledge helpful discussions with Prof. Farhad Faisal and the support of the Universität Bielefeld and of the Deutsche Forschungsgemeinschaft, Sonderforschungsbereich 216, "Polarisation and Correlation in Atomic and Molecular Collision Complexes".

Table 1. Padé Approximants to Basis Set Sum Using Real Exponent

Iterate	Single Sum	
0	0.61217	
1	1.37074	gives $T_{L,L-1}$ = 0.1079 -0.0791i
2	0.59057	

Iterate	$\Sigma q^+ q^+$	$\Sigma q^+ q^-$
0	3.90769 -6.25328i	5.71567 -3.25286i
1	1.42264 -5.55171i	9.54821 -7.18798i
2	1.42355 -5.54345i	9.53941 -7.15095i
3	1.42362 -5.54342i	9.53985 -7.15161i
4	1.42362 -5.54342i	9.53981 -7.15162i

Iterate	$\Sigma q^- q^+$	$\Sigma q^- q^-$
0	5.71567 +3.25286i	3.90769 +6.25328i
1	9.54821 +7.18798i	1.42264 +5.55171i
2	9.53941 +7.15095i	1.42355 +5.54345i
3	9.53985 +7.15161i	1.42362 +5.54342i
4	9.53981 +7.15162i	1.42632 +5.54342i

gives $T_{L,L-1}$ = 0.501977 -0.368125i

Comparison of epsilon algorithm convergence acceleration of basis-set sum for radial dipole elements from (E,L) to (E',L-1) for E = 1.0, E' = 0.1 and L = 1. Computed with up to 15 basis functions with real basis set exponent λ = 2.0. The zeroth iteration corresponds to the partial sum with fifteen terms, while the further iterations display successively better Padé approximants.

Table 2. Padé Approximants to Basis Set Sum Using Complex Exponent

Iterate	Single Sum
0	-3.9145 -3.0922i
1	1.2622 +0.3346i
2	-0.0742 -0.1086i
3	2.6885 +0.3805i

gives $T_{L,L-1}$ = 0.4740 -0.4185i

Iterate	$\Sigma q^{+}q^{+}$	$\Sigma q^{+}q^{-}$
0	0.80464 -4.85655i	5.01212 -2.37258i
1	0.67239 -4.74113i	7.63696 -8.42237i
2	0.67273 -4.74056i	7.64312 -8.37849i
3	0.67274 -4.74056i	7.64331 -8.37937i
4	0.67274 -4.74056i	7.64326 -8.37936i

Iterate	$\Sigma q^{-}q^{+}$	$\Sigma q^{-}q^{-}$
0	7.32678 +3.36258i	42.0648 +11.1947i
1	11.45215 +5.83659i	2.39714 +6.62573i
2	11.43569 +5.80693i	2.39083 +6.51732i
3	11.43624 +5.80739i	2.39108 +6.51689i
4	11.43621 +5.80741i	2.39109 +6.51690i

gives $T_{L,L-1}$ = 0.501977 - 0.368125i

As in Table 1, except with complex basis-set exponent, λ = 2.0 -0.2i

Table 3. Padé Approximants to Basis Set Sum with Complex Exponent

Iterate	Single Sum
0	32714.7 +83218.2i
1	-80.0143 +88.1024i
2	0.36414 +0.56619i
3	1.15724 +1.46013i
4	0.85455 +1.19736i

gives $T_{L,L-1} = 0.5014 - 0.3480i$

Iterate	$\Sigma q^+ q^+$	$\Sigma q^+ q^-$
0	-0.97014 -2.49176i	42.6611 -13.3449i
1	-0.97014 -2.49178i	-1.37406 -12.0823i
2	-0.97014 -2.49178i	-1.19507 -12.2220i
3	-0.97014 -2.49178i	-1.19486 -12.2205i
4	-0.97014 -2.49178i	-1.19832 -12.2204i

Iterate	$\Sigma q^- q^+$	$\Sigma q^- q^-$
0	18.4981 -1.1381i	-1005300. +1231291.i
1	20.4459 -0.24693i	-2039.02 -459.018i
2	20.4352 -0.25776i	5.64486 +7.42002i
3	20.4355 -0.25766i	9.50338 +14.3635i
4	20.4355 -0.25766i	9.49471 +14.4354i
5	20.4355 -0.25766i	9.49385 +14.4367i

gives $T_{L,L-1} = 0.501978 - 0.368126i$

As in Table 1, except computed with complex basis exponent, $\lambda = 2.0 - 1.0i$.

REFERENCES

1. Higher Transcendental Functions, A. Erdélyi ed. (McGraw-Hill, New York, 1953).
2. H.A. Yamani and W.P. Reinhardt, Phys. Rev. A11, 1144 (1975).
3. J.T. Broad, submitted for publication to Phys. Rev. A.
4. G.A. Baker and P.R. Graves-Morris, Padé Approximants Basic Theory, Part I, Vol. 13 of Encyclopedia of Mathematics and its Applications, edited by Gian-Carlo Rota (Addison-Wesley, New York 1981)

BUFFER GAS INFLUENCE ON ION COLLECTION IN CESIUM TWO PHOTON IONIZATION AND ON UF_6 PHOTODISSOCIATION

M. I. Schisano

Istituto di Fisica Sperimentale dell' Università
Pad. 20 Mostra d'Oltremare
80125 Napoli, Italy and
E.N.E.A. Energy Research Center
Casaccia P.O. Box 2400
Roma A.D., Italy

INTRODUCTION

Multiphoton ionization (MPI) and resonant ionization spectroscopy (RIS) are largely used as probes of molecular dissociation.[1] In both techniques, ions and electrons are produced by ionizing atoms or molecules through a laser-induced absorption process and then collected by an electric field. The multiphoton excitation may or may not be resonant, depending on whether or not real intermediate levels are involved.

RIS experiments are performed in either the absence or the presence of collisions. In the latter case some specific features appear:

i) collisional depopulation of the intermediate levels occurs,
ii) a reduction of the number of collected electrons and ions is obtained as a result of attachment, diffusion and (or) recombination processes.

The analysis of these two collisional processes provides information about the ionization and (or) dissociation yields.

BUFFER GAS INFLUENCE ON ION COLLECTION IN THE Cs IONIZATION

At the University of Naples we are investigating the effect of a buffer gas on the two photon ionization of Cs atoms via the resonant excitation of the 7P level.[2]

We have observed that in the presence of Ar gas the ionization signals are drastically reduced owing to the attachment and recombination of ions and electrons. The temporal evolution of these processes is analysed by collecting ions and electrons at times subsequent to the ionization.

Experimental procedure

The experimental set-up is composed of a cylindrical cell containing the alkali vapour. A buffer gas is added to the Cs atoms. Two parallel plane electrodes, 1 cm apart, are inserted into the cell. A N_2-pumped dye laser is directed along the axis of the cell, perpendicular to the applied electric field. The laser beam has an energy of 0.3 mJ with a pulse width of 6 nsec and a repetition rate up to 40 Hz. The beam is focused on the region between the electrodes by a lens with a focal length of 15 cm. The power density of the focused beam is of the order of 10^7 W/cm^2.

The dye laser is tuned to the 4593 Å or the 4555 Å wavelength so as to excite resonantly the Cs atoms from the ground state $6^2S_{1/2}$ to either the $7^2P_{1/2}$ or the $7^2P_{3/2}$ states. The same atoms, once excited, can be ionized by a second photon from the same laser pulse.[3] The Cs vapour density is approximately 10^{10}at/cm^3. At this density and at the largest available laser intensity, 10^7 Cs ions and electrons are produced in the excitation volume. The ions so produced are collected on the electrodes by applying an electric field of several tens of volts per cm. This field can be time delayed with respect to the laser pulse in the range 0-100 μsec.

The ionization signals are directly observed on an oscilloscope as a function of the buffer gas pressure.

Analysis

The electron and ion densities produced by the laser excitation at the time $t = 0$ evolve in an electric field-free regime until the field between the electrodes is applied at $t = \tau$. In presence of this field ions and electrons are rapidly collected and the measured photocurrent gives the number of remaining ions and electrons in the volume inside the electrodes.

The loss of charges during the time τ is determined by diffusion, attachment and (or) recombination processes.

The electron and ion densities will be denoted by n_e and n_i, respectively. Local neutrality imposes $n_e(t) = n_i(t) = n(t)$.

The ion diffusion is governed by the following equation

$$\frac{\partial n}{\partial t} = D\nabla^2 n \qquad (1)$$

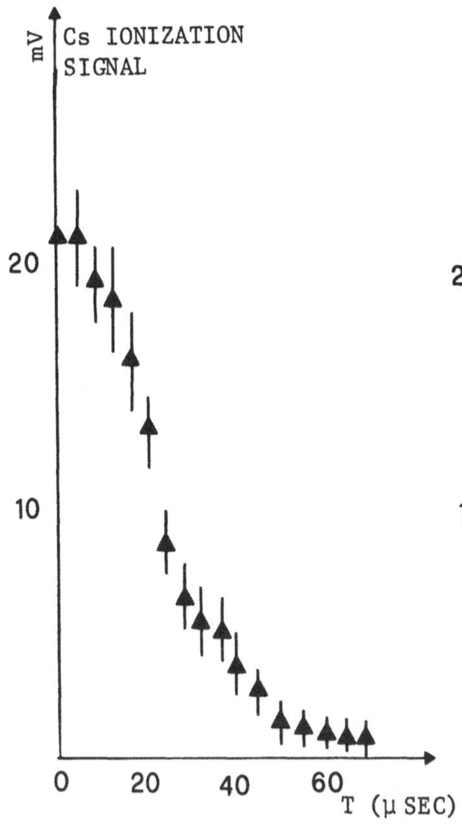

Fig. 1a.

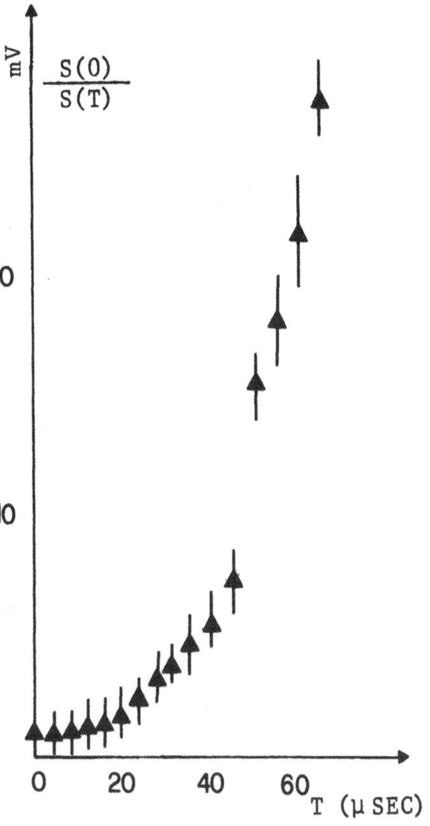

Fig. 1b.

where D is the diffusion coefficient. The resulting time evolution of the ion density in a cylindrical volume of radius R is proportional to $\exp(-t/\tau')$, where $\tau' = R^2/(\pi^2 D)$.

The attachment process is described by the following equation:

$$\frac{\partial n}{\partial t} = -\beta n \tag{2}$$

where the attachment coefficient β is a function of the pressure of the partners involved in the attachment.

Finally the electron-ion recombination coefficient is defined through the recombination equation

$$\frac{\partial n}{\partial t} = -\alpha n^2 \tag{3}$$

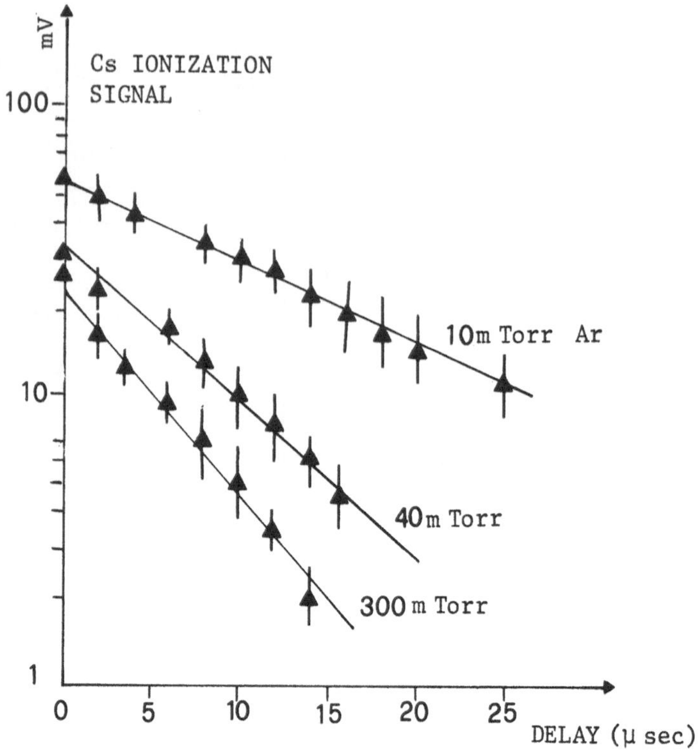

Fig. 2.

The solution of this equation is

$$n(t) = n(0)(1+\alpha n(0)t)^{-1} \quad (4)$$

where n(0) is the initial electron density. The linear dependence of n(0)/n(t) on time should be noted.

Result

The time behaviour of the photoionization signal s(t) and of the quantity s(0)/s(t), which is proportional to n(0)/n(t), is shown in Fig. 1a and Fig. 1b. The non-linear behavior of s(0)/s(t) shows that the recombination process is negligible.

The reduced amplitude of the signal is therefore to be attributed either to diffusion or to attachment. In either case the number of collected charges versus the delay time is a straight line on a semi-log scale. However its slope decreases or increases with the buffer gas pressure accordingly, that either diffusion or attachment is the relevant process.

The results of the measurements carried out at different Ar pressures are shown in Fig. 2. The figure shows that the attachment process is the dominant one.

Our results suggest that a three body reaction may occur through:

$$Cs^+ + Cs + Ar \rightarrow Cs_2^+ + Ar + \Delta E \qquad (5)$$

In such a case the attachment coefficient would depend on the product of the Cs and Ar pressure. This reaction should be followed by a fast recombination process of Cs_2^+ through the following mechanism:[4]

$$Cs_2^+ + e^- \rightarrow Cs + Cs + \Delta E \qquad (6)$$

As a result the reverse of the process (5) is not detected in our experiment.

BUFFER GAS INFLUENCE ON THE UF_6 DISSOCIATION

In an experiment made at E.N.E.A. Laboratories in Rome, the MPI was used to detect UF_6 dissociation in both collisional and collisionless regime with an ionization chamber.[5]

A similar technique was already used in a molecular beam-time of flight spectrometer apparatus.[6]

In order to obtain reliable results we carried out an analysis of the MPI features over a wide range of wavelengths and studied the flux dependence of the ionization rate so as to select the correct wavelength and flux.

Experimental procedure

The experimental set-up used to detect the UF_6 photodissociation is shown in Fig. 3.

A Xe-Cl laser pumps two dye lasers in the visible region. The output of the first dye laser (few μJoules around 260 nm) was frequency doubled and sent into the gaseous sample to obtain the photodissociation of UF_6 in UF_5 + F. The amplified output of the second dye laser was time delayed and sent collinear with the first one into the sample. If appropriate intensity and focusing conditions are met, multiphoton excitation of the newly formed UF_5 brings these molecules above the ionization threshold (11.3 eV). If the U.V. beam is turned off the signal disappears. However a further increase of the visible beam intensity gives rise to a new signal, corresponding to MPI of UF_6, whose ionization potential is 14 eV.

Application of an electric field between the two parallel plates of an ionization chamber allows us to collect the UF_5 ions. The

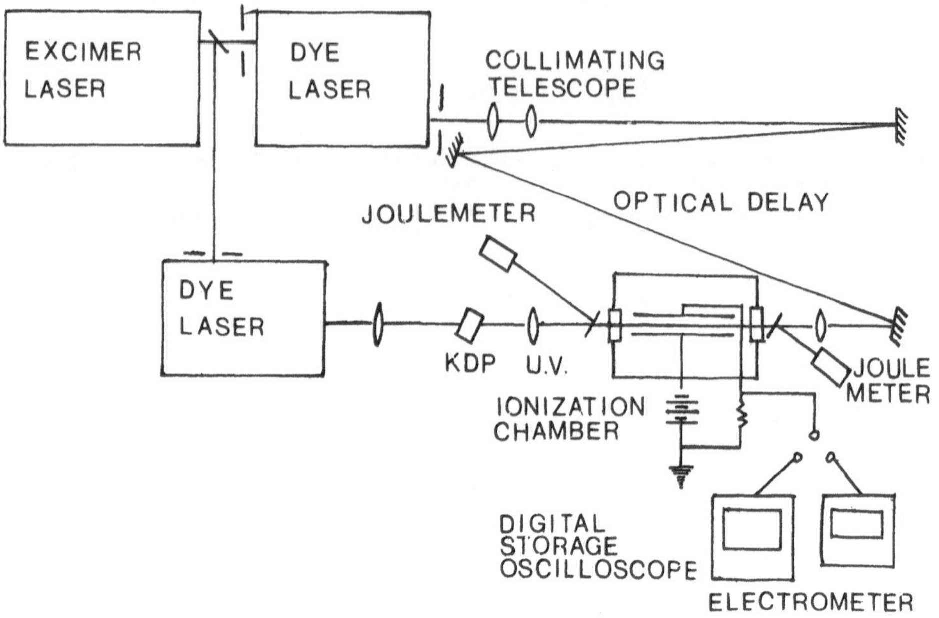

Fig. 3.

charges so collected were measured by an electrometer. Their typical values are 10^{-13} - 10^{-14} Coulombs per shot. The time evolution of UF_5 ionization signal could be observed on an oscilloscope. With a reasonable compromise between the input impedence of the charge amplifier and the velocity of the signal it was possible to observe simultaneously the electronic and ionic signal, as shown in the insert of Fig. 4. The intensity of the collection field was chosen in the range of 30 - 50 V/cm, well below the charge amplification threshold. We checked that below this threshold the signal does not depend on the electric field between the electrodes.

Result and discussion

In order to find out the wavelength which gives the best efficiency in the UF_5 ionization we studied the dependence of UF_5 ionization signal on the wavelength in the range 532 - 440 nm, as well as its dependence on the laser flux. For almost all values of the wavelength the ionization threshold (11.3 eV) is reached with about five photons. Only for the maximum value λ = 440 nm the ionization is achieved with just four photons. The flux dependence is linear at those wavelengths corresponding to higher ionization rates, but strongly non-linear at larger wavelengths. Resonance of the radiation with intermediate states is well known to enhance the ionization rate and to cause linearity of the flux dependence curves

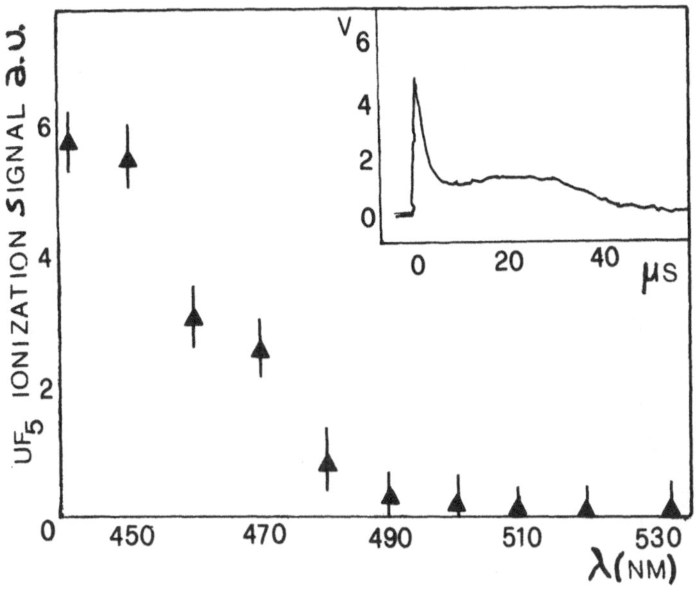

Fig. 4.

due to easy saturation of the intermediate states. Therefore the increase of the ionization rate around 440 nm, shown in Fig. 4, seems to indicate the occurrence of multiple resonances in which the energy redistribution processes practically do not deplete the intermediate states. The decrease of the MPI rate with increasing wavelength may imply either a lowering of the density of the intermediate states or a greater efficiency of the dissipative processes.

A comparison of our experimental data with theoretical predictions on UF_5 electronic structure is not entirely meaningful. A self-consistent field Xα scattered wave calculation predicts a transition at 439 nm.[7] The same calculation however predicts an ionization potential which is much lower than the one experimentally found. The broad structure that we observed is likely to correspond to the first charge transfer transitions from the fluorines to the uranium 5f orbital, that are believed to set in at about 3 eV.[8]

We performed our measurements with a fixed pressure of 10^{-1} Torr of UF_6 and a variable pressure of Ar. In Fig. 5 the plot of the UF_5^+ ion signal versus Ar pressure up to 10 Torr shows a decreasing signal with increasing pressure. This behaviour can arise from three different steps of the process, namely UF_6 dissociation, UF_5 ionization and ion collection. In order to check that the observed trend is not due to a decrease in the collection efficiency we turned the U.V. beam off and increased the visible intensity to scanning the UF_6 photoionization versus Ar pressure. As shown in Fig. 4 this scanning does not depend on pressure. One could still surmise that collisions may influence the UF_5 ionization process but not that of UF_6. However measurements of the dissociation of

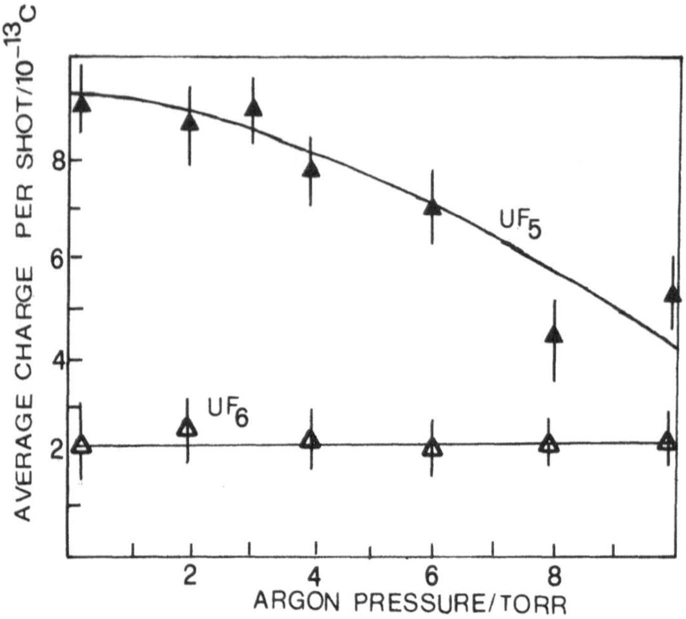

Fig. 5.

UF_6 irradiated with a KrF laser, performed in the same range of pressure, by detecting $HF^*_{v=1}$ 2.8 μm fluorescence, supply much the same results.[9] This gives us confidence in our technique.

REFERENCES

1. R. B. Bernstein, Systematic of multiphoton ionization-fragmentation of poliatomic molecules, J. Phys. Chem. 86:1178 (1982), and references cited therein.
2. A. Sasso, M. I. Schisano, B. M. Tescione and E. Arimondo, Collisional M_J-mixing in the resonance ionization of cesium atoms, to be published.
3. J. S. Hurst, M. G. Payne, S. D. Kramer and J. P. Young, Resonance ionization spectroscopy and one-atom detection, Rev. Mod. Phys. 51:767 (1979).
4. J. M. Hammer and B. B. Aubrey, Ion beam measurements of Cesium recombination cross section, Phys. Rev. 141:146 (1966).
5. P. Dore, M. I. Schisano, M. Menghini, P. Morales, Multiphoton ionization of UF_5, submitted to Chem. Phys. Lett.
6. J. S. Chou, D. Sumida, M. Stuke and C. Wittig, "Single shot" laser multiphoton ionization detection of UF_5 following the 266 nm photodissociation of UF_6, Laser Chem. 1:2 (1982).
7. H. Maylotte, R. L. Peters, R. L. Messmer, Theoretical calculations of the electronic structure and optical transitions of UF_6 and UF_5, Chem. Phys. Lett. 38:181 (1976).
8. W. R. Wadt and P. J. Hay, Ab initio studies of the electronic structure and geometry of UF_5 using relativistic core potentials, J. Am. Chem. Soc. 101:5198 (1979).
9. N. R. Greiner, J. H. Lyman and G. A. Laguna, J. Chem. Phys. (to be published).

THEORETICAL TREATMENT OF ATOMIC COLLISION PROCESSES

R.M. Dreizler, A. Henne, H.J. Lüdde,
W. Stich and A. Toepfer

Institut für Theoretische Physik der Universität
Frankfurt/M. Federal Republic of Germany

INTRODUCTION

The standard approach to the discussion of ion atom scattering problems in the intermediate energy range is the semiclassical approximation. After specification of a classical trajectory $\underline{R}(t)$ for the internuclear motion one is, nevertheless, faced with the solution of a time dependent Schrödinger equation in order to describe the development of the quantum mechanical many electron system

$$i\partial_t \,\Psi(\underline{x}_1 \ldots \underline{x}_N, R(t), t) = H_{el}(R(t), t)\Psi(\underline{x}_1 \ldots \underline{x}_N, R(t), t). \qquad (1)$$

This equation can be integrated directly only for the case of one electron systems[1]. The treatment of systems with more than one electron requires further approximations.

In view of the success of the stationary Hartree-Fock method for atomic systems, the application of the time dependent Hartree-Fock (TDHF) method[2] could be considered as a realistic avenue for this purpose. In this approximation one assumes that the system under consideration can be described at all times by a single Slater determinant.

$$\Psi(\underline{x}_1 \ldots \underline{x}_N, R(t), t) = \hat{A}\{\psi_1(\underline{x}_1, R(t), t) \ldots \psi_N(\underline{x}_N, R(t), t\} \ . \qquad (2)$$

In consequence the solution of the many particle Schrödinger equation can be replaced by the solution of a set of (nonlocal) effective one particle problems

$$i\partial_t \psi_j(\underline{x},R(t),t) = h^{HF}(\hat{\rho}(t),t)\psi_j(\underline{x},R(t),t). \quad (3)$$

First attempts to treat two electron collision systems in terms of this approximation are indicated in section 1. We note that the Pauli principle is fully taken into account in this method, if the final determinantal wavefunction is used for the extraction of cross sections. Implications of the Pauli principle for the discussion of inclusive cross sections are detailed in section 2.

Even this approximation is, in view of the time consuming numerical aspects, at present not sufficient to tackle systems with a larger number of electrons. As a further approximation we suggest the use of effective (local) single particle potentials, i.e. the replacement

$$h^{HF} \rightarrow h^{eff} . \quad (4)$$

We then have to solve a set of linear and local effective one electron problems

$$i\partial_t \psi_j(t) = \{ -\frac{1}{2}\Delta + v_{eff}(R(t))\}\psi_j(t) . \quad (5)$$

The question naturally arises, how one should construct such effective potentials, which are supposed to describe the time variation of screening effects in a sensible fashion. In section 3 we show results for the collision system $Ne^+ + Ne$, which are obtained with a potential of the form $v_{eff}(R(t))$ evaluated with the aid of density functional methods[3]. We note that, in the final analysis, the Pauli principle has again to be implemented by consideration of a determinantal wavefunction constructed from the time developed orbitals.

First attempts to obtain access to the calculation of explicitly time dependent effective potentials $v_{eff}(t,b,E)$ on the basis of a time dependent Thomas Fermi model are outlined in section 4.

1. TDHF Calculations for Two Electron Systems.

We have solved the TDHF equations for impact energies in the range of 5-40 keV/amu by basis expansion techniques for the systems

$$H^+ + He(1s^2) \rightarrow H(\Sigma nl) + He^+(\Sigma nl)$$

$$H^+ + He(\Sigma nln'l')$$

$$He^{2+} + He(1s^2) \rightarrow He(\Sigma nln'l') + He^{2+}$$

$$He^{+}(\Sigma nl) \quad + He^{+}(\Sigma nl)$$

$$He^{2+} + He(\Sigma nln'l')$$

$$Li^{3+} + He(1s^2) \rightarrow Li^{+}(\Sigma nln'l') + He^{2+}$$

$$Li^{2+}(\Sigma nl) \quad + He^{+}(\Sigma nl)$$

$$Li^{3+} + He(\Sigma nln'l').$$

Technical details can be found in Ref.4. The results for global one electron capture in the system $H^+ + He$, for global one and two electron capture in the system $He^{2+} + He$, as well as global one and two electron capture in the system $Li^{3+} + He$ are summarised and compared with experiment and other theoretical results in Table 1 and Figs 1 and 2, respectively. One notes very reasonable agreement between theory and experiment except for the one electron capture channel in $He^{2+} + He$ below 8 keV/amu. Suitable techniques for the extraction of global capture probabilities from the determinantal wavefunction are found in Ref. 5.

In order to illuminate the role of the exchange term of the TDHF Hamiltonian, we have also investigated the (experimentally not accessible) system $He^+(2s) + He^+(2s)$ for the spin singlet and triplet channels. Fig. 3 shows the distribution of the two electrons over the scattering plane at an internuclear separation of 1 a_o (for the collision parameters E = 10keV/amu, b = 0.8 a_o). This snapshot of the time developing system clearly stresses the role of the exchange potential, which leads to a marked antibonding structure during the collision in the triplet case.

2. Evaluation of Inclusive Probabilities.[6]

As a starting point we assume that the final wavefunction of a N electron scattering system has been obtained in an effective single particle model (e.g. TDHF) in the form of a single Slater determinant. It is a simple matter to show that the transition probability to a specific final configuration $|f_1 \ldots f_N\rangle$ can be calculated as

$$P_{f_1 \ldots f_N} = \begin{vmatrix} \langle f_1|\hat{\gamma}|f_1\rangle & \ldots & \langle f_1|\hat{\gamma}|f_N\rangle \\ \vdots & & \vdots \\ \langle f_N|\hat{\gamma}|f_1\rangle & \ldots & \langle f_N|\hat{\gamma}|f_N\rangle \end{vmatrix} \qquad (6)$$

where $\hat{\gamma}$ represents the one particle density operator constructed from the final wavefunction, which has developed from a given initial configuration

$$\hat{\gamma} = \sum_{j=1}^{N} |\psi_j(t\to\infty)><\psi_j(t\to\infty)| \ . \tag{7}$$

On the basis of equation (6) a number of inclusive probabilities (and hence cross sections) can be calculated directly (see also Ref.7). Among these we quote the standard inclusive probability, i.e. the probability of finding q of the N electrons in a specific subconfiguration $|f_1 \ldots f_q>$, while the remaining N-q electrons occupy arbitrary orbitals

$$P_{f_1 \ldots f_q} = \sum_{f_{q+1}<\ldots<f_N} P_{f_1 \ldots f_N} = \begin{vmatrix} <f_1|\hat{\gamma}|f_1>\ldots<f_1|\hat{\gamma}|f_q> \\ \vdots \qquad\qquad \vdots \\ <f_q|\hat{\gamma}|f_1>\ldots<f_q|\hat{\gamma}|f_q> \end{vmatrix}, \tag{8}$$

as well as the extended inclusive probability, i.e. the probability of finding q electrons in a specific subconfiguration, while h vacancies are created in the subconfiguration $|\overline{f}_1 \ldots \overline{f}_h>$.

$$P_{f_1 \ldots f_q}^{\overline{f}_1 \ldots \overline{f}_h} = P_{f_1 \ldots f_q} + \sum_{l=1}^{h} P_{f_1 \ldots f_q \overline{f}_l} - \sum_{l_1<l_2}^{h} P_{f_1 \ldots f_q \overline{f}_{l_1} \overline{f}_{l_2}} \pm \ldots (-)^h P_{f_1 \ldots f_q \overline{f}_1 \ldots \overline{f}_h} \ . \tag{9}$$

3. Calculation of Total Inclusive Cross Sections for the 2p-2s Vacancy Transition in the Ne^+ + Ne System.

We calculate the time development of the orbitals of both atoms with eq. (5). The effective potential is obtained by determining the density of the two centre collision system as a function of the internuclear separation with the aid of the Thomas-Fermi-Dirac-Weizsäcker model[8] and constructing an effective HF Slater type potential from this predetermined density. The solution of the effective time dependent one electron problems is carried through with the following steps

1) diagonalisation of the effective Hamiltonian h^{eff} as a function of the internuclear separation, in order to generate a quasi-molecular basis[9]

2) subsequent solution of the time dependent problem by expansion in terms of this MO basis[10] .

Total cross sections are then evaluated with the method described in Section 3.

In Figs 4 and 5 we indicate the results obtained. Fig. 4 shows different theoretical cross sections in the energy range from 5 to 500 keV. The basis size is, except otherwise indicated, 18 molecular orbitals, from which in the separated atom limit the atomic orbitals

1s, 2s, 2p, 3s, 3p

of both target and projectile can be constructed. The initial state is specified by the configuration

$$i \rightarrow (1s^2\ 2s^2\ 2p^5)_P\ (1s^2\ 2s^2\ 2p^6)_T\ ,$$

where the initial 2p-hole is distributed statistically in the magnetic quantum numbers.

The theoretical cross sections given, correspond to the following situations.

$\sigma_{channel}$: We assume that the full final configuration of the 19 electrons can be identified as

$$f_1 \rightarrow (1s^2\ 2s^1\ 2p^6)_P\ (1s^2\ 2s^2\ 2p^6)_T$$

$$f_2 \rightarrow (1s^2\ 2s^2\ 2p^6)_P\ (1s^2\ 2s^1\ 2p^6)_T$$

and we calculate:

$$\sigma_{channel} = \sigma(i \rightarrow f_1) + \sigma(i \rightarrow f_2)\ .$$

σ_{inc}: We assume that the full final configuration is not identified, but rather only the subconfigurations $(2s^1 2p^6)$ in both target and projectile. We thus use eq. (9) of section 2. σ_{2s}: We assume that only the subconfiguration $(2s^1)$ is identified, i.e. the only information available is the probability for finding any 2s hole in target or projectile. For comparison we also give the cross section calculated previously in a two molecular state approximation[11] .

As the experimental technique for the identification of the 2p-2s vacancy transfer is observation of the 2p-2s spectral lines, we identify σ_{inc} with the experimental cross section. Comparison with experiment is indicated in Fig. 5. As one finds different values for absolute experimental cross sections in the literature, we have normalised all experimental results to the values obtained in Ref. 12 at 200 keV. Our calculated cross section σ_{inc} (reduced by a factor of 0.33 in order to fit the experiment at 200 keV) reproduces the variation with energy quite reasonably.

4. Fluid Dynamical Concepts in Atomic Scattering Problems.

In order to incorporate dynamical effects into the effective screening potential more fully, we have initiated the following programme.

1) Derivation of time dependent density functional models
2) Solution of the corresponding variational equations (of hydrodynamical structure)
3) Calculation of effective potentials with the density obtained in 2) as $v_{eff}(t,b,E)$.

The status of the foundation of the theory can be gleaned from Ref. 13. Some first results for the scattering of protons from atoms indicating the numerical aspects is found in Ref. 14. In Fig. 6 we show, as an indication of the more complete range of results spanning different systems and the velocity range of 1.6 a.u. $\leq v \leq$ 3.3a.u.[13], an illustrative example for the time development of the density distribution in the system $Ne^+ + Ne$. The results are obtained with a two fluid model, encompassing the fluid dynamical variational equations

$$\frac{\partial}{\partial t}\rho_k = -\nabla(\rho_k \nabla\chi_k)$$

$$k = 1,2 \quad , \qquad (10)$$

$$\frac{d}{dt}\chi_k = -\frac{1}{2}(\nabla\chi_k)^2 - \frac{\delta}{\delta\rho_k}E[\rho]$$

where the intrinsic energy is represented by the Thomas Fermi model

$$\frac{\delta}{\delta\rho_k} E[\rho] = \frac{5}{3} c_1 \rho^{2/3} - V$$

$$\Delta V = 4\pi\rho \qquad (11)$$

So far the calculations are carried through with the assumption of rotational symmetry with respect to the internuclear axis. The figure shows the time development of the difference density in the scattering plane (the reference is an adiabatic Thomas-Fermi-model) for the given energy and impact parameter value. One clearly observes the polarisation of the atom by the impinging ion and in the later stages of the process a fairly complex distribution of the electronic charge. Further work is, however, required until step 3) of the programme envisaged can be completed.

Table 1: Total cross sections for one electron capture in $H^+ + He(1s^2)$.

Energy(KeV)	de Heer et al[20] (1966)	Stier and Barnett[21] (1956)	present results
10	1.15	.95	.95
15	1.77	1.7	
20	2.21	1.8	1.8
25	1.92	1.9	
30	1.75	1.7	1.52
35	1.65	1.6	
40	1.35	1.4	1.4

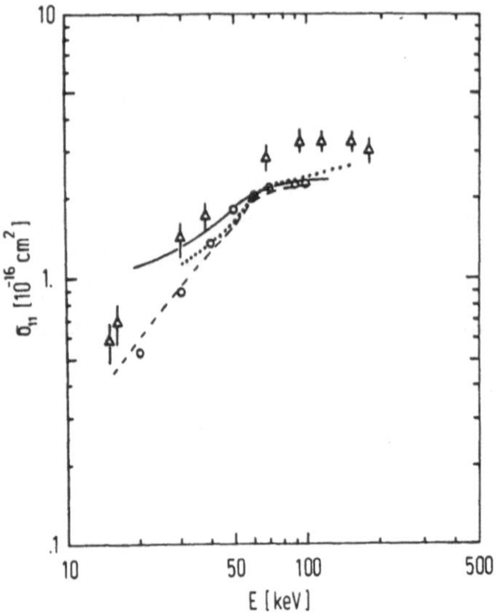

Fig. 1a

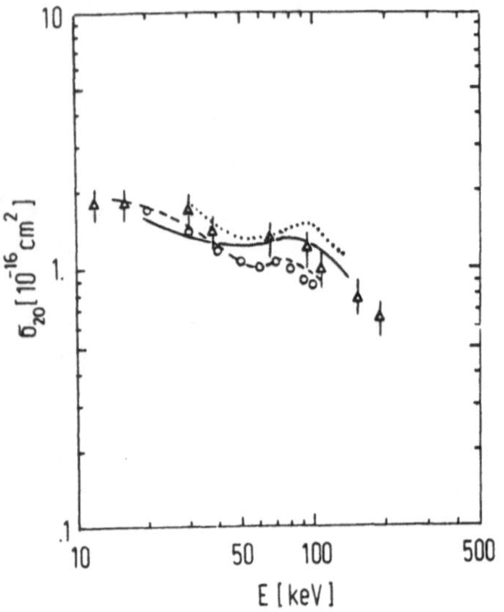

Fig. 1b

He^{2+} + He($1s^2$) total cross sections for one (1a) and two(1b) electron capture.—— present results,..... Sandhya Devi and Garcia[15], --- Harel and Salin[16], o Afrosimov et al[17], Δ Berkner et al[18].

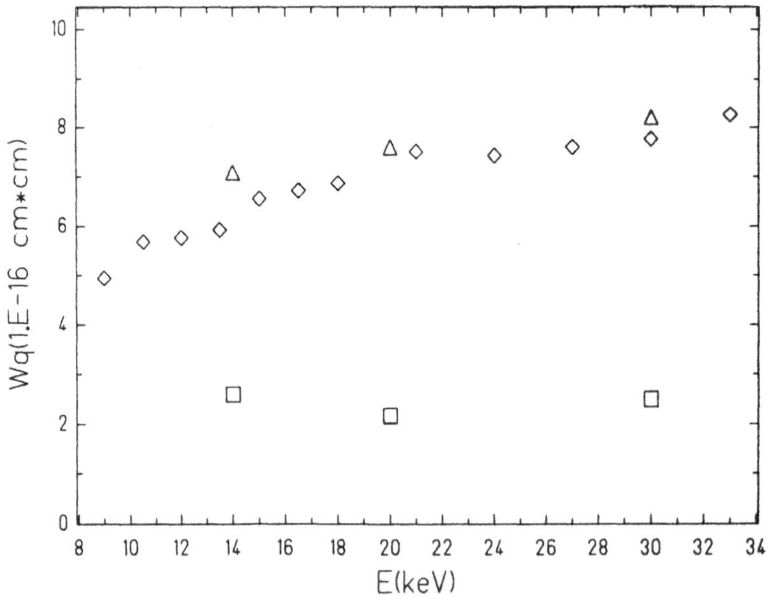

Fig. 2: Li^{3+} + He($1s^2$) total cross sections for one electron capture Δ present results,◇ Wirkner-Bott et al[19],□ present results for two electron capture.

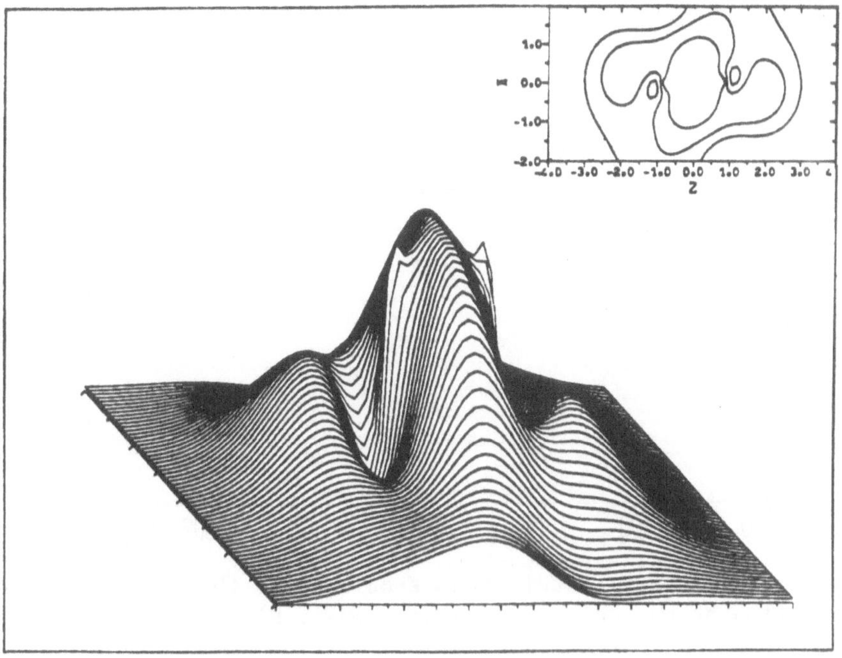

Fig. 3a: $He^{+}(2s) + He^{+}(2s)$ singlet case, electronic density.

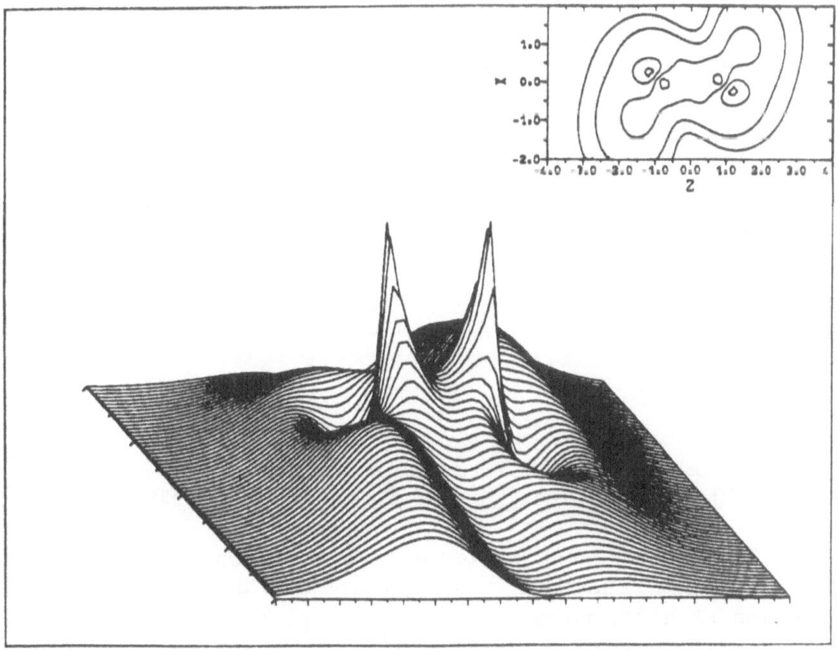

Fig. 3b: $He^{+}(2s) + He^{+}(2s)$ triplet case, electronic density.

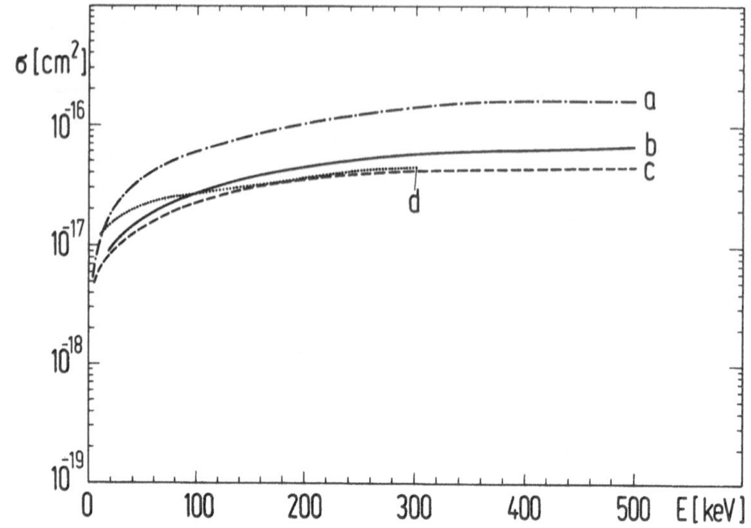

Fig. 4: Theoretical total cross sections for the 2p-2s vacancy transition of the Ne^+ + Ne system, a)σ_{2s}, b)$\sigma_{inc}(2s^1 2p^6)$, c)$\sigma_{channel}$, d) 2 state calculation.

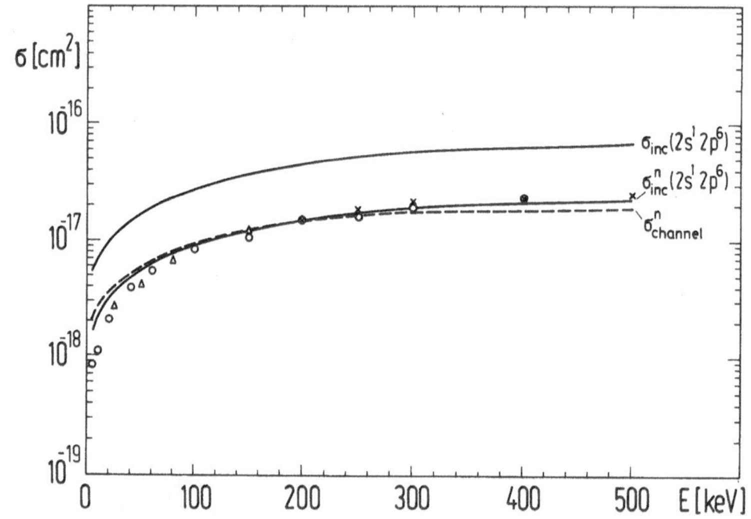

Fig. 5:Calculated cross sections in comparison with experiment x: Hippler and Schartner[12], o: Andersen et al[22], normalised Δ:Bloemen et al[23], normalised, σ_{inc}: Fig. 4, curve b; σ^n_{inc}: Fig. 4, curve b, normalised, $\sigma^n_{channel}$: Fig. 4, curve c, normalised.

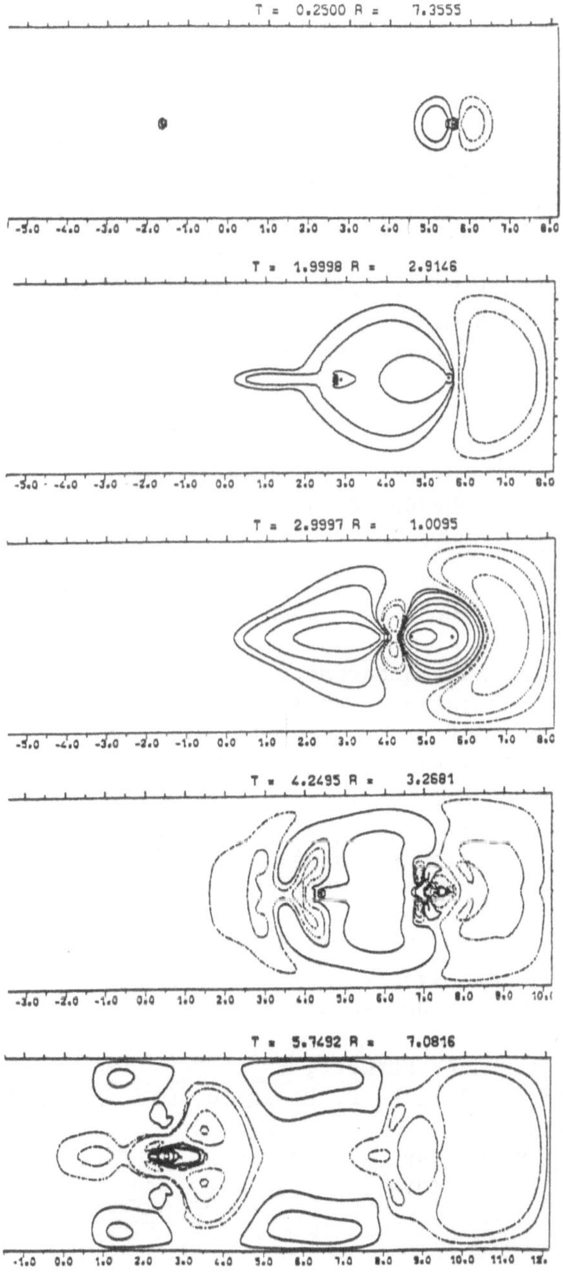

Fig. 6: Example: Ne^{+} + Ne, v_{rel} = 2.6 a.u., b=1.a.u., density contour plots for difference density $\rho_d = \rho - \rho_A^{TF} - \rho_B^{TF}$, — $\rho_d > o$, --- $\rho_d < o$, position of nuclei.

REFERENCES

1. e.g. W. Fritsch, C.D. Lin and L.N. Tunnell, IEEE Trans. Nucl. Sci. NS 28:1146 (1981).
   H.J. Lüdde and R.M. Dreizler, J. Phys. B15:2713 (1982).
2. K.C. Kulander, K.R. Sandhya Devi and S.E. Koonin, Phys. Rev. A 25:2968 (1982).
3. E.K.U. Gross and R.M. Dreizler, in "Proceedings of the NATO Advanced Study Institute on Density Functional Methods in Physics", Alcabideche (Portugal) (1983).
4. W. Stich, H.J. Lüdde and R.M. Dreizler, Phys. Lett.99A:5 (1983).
   W. Stich, H.J. Lüdde and R.M. Dreizler, to be published in J. Phys. B.
5. H.J. Lüdde and R.M. Dreizler, J. Phys. B16:3973 (1983).
6. H.J. Lüdde and R.M. Dreizler, to be published in J. Phys. B.
7. R.L. Becker, A.L. Ford and J.F. Reading, Phys. Rev.A29:3111 (1984).
8. E.K.U. Gross and R.M. Dreizler, Phys. Rev. A20:1798 (1979).
9. A. Toepfer, E.K.U. Gross and R.M. Dreizler, Phys. Rev. A20: 1808 (1979).
10. A. Toepfer, H.J. Lüdde, B. Jacob and R.M. Dreizler, to be published in J. Phys. B.
11. W. Fritsch and U. Wille, J. Phys. B11:4019 (1978).
    A. Toepfer, B. Jacob, H.J. Lüdde and R.M. Dreizler, Phys. Lett. 93A:18 (1982).
12. R. Hippler and K.H. Schartner, J. Phys. B8:2528 (1975).
13. A. Henne, Dissertation Frankfurt 1984 unpublished.
14. M. Horbatsch and R.M. Dreizler, Z. Phys. A308:329 (1982).
15. K.R. Sandhya Devi and J.D. Garcia, J. Phys. B16:2837 (1983).
16. C. Harel and A. Salin, J. Phys. B13:785 (1980).
17. V.V. Afrosimov, A.A. Bansaleau, G.A. Leiko and M.N. Panov, Sov. Phys. JETP 47:837 (1978).
18. H. Berkner, R.V. Pyle, J. Warren Stearns and J.C. Warren, Phys. Rev. 166:44 (1968).
19. I. Wirkner-Bott, W. Seim, A. Müller, P. Kester and E. Salzborn, J. Phys.B14:3987 (1981).
20. F.J. de Heer, J. Schutten and H. Moustafa, Physica 32:1766 (1966).
21. P.M. Stier and C.F. Barnett, Phys. Rev. 103:896 (1956).
22. T. Andersen, E. Bøving, P. Hedegard, O.J. Østgard, J. Phys. B11:1449 (1978).
23. E. Bloemen, H. Winter, F.J. de Heer, R. Fortner, A. Salop, J. Phys.B11:4207 (1978).

PROTON-IMPACT EXCITATION OF THE 2p-SHELL OF SODIUM ATOMS

O. Schöller and J.S. Briggs
Fakultät für Physik
Universität Freiburg

We consider the scattering process

$$p + Na(2p^6 3s\ ^2S) \xrightarrow{I} p + Na^*(2p^5 3s^2\ ^2P)$$
$$\downarrow II$$
$$p + Na^+(2p^6\ ^1S) + e^-$$

for which the total cross-section[1] and alignment parameter[2,3] have been measured. The angular distribution of autoionization electrons of the decay process II in the above reaction is described by

$$I(\theta) = I_o\ \{1 + A_2 P_2(\cos\theta)\}.$$

The coefficient A_2 of the Legendre polynomial $P_2(\cos\theta)$ is called the alignment parameter. A_2 describes the different excitation of magnetic substates of the Na(2p)-shell during the collision process I. It is related to the corresponding partial cross sections by

$$A_2 = \frac{\sigma_o - \sigma_1}{\sigma_o + 2\sigma_1}$$

where σ_o is the total cross section for the excitation of the ground state to the configuration $Na(2p^5 3s^2\ ^2P)$ with projection quantum number M=0 and σ_1 denotes the cross section of the state with $|M| = 1$. These cross sections are obtained by solving the time dependent Schrödinger equation

$$i \frac{\partial}{\partial t} |\psi(t)\rangle = H(t)|\psi(t)\rangle \qquad (1)$$

where $|\psi(t)\rangle$ is chosen as a superposition of a finite number of Na configurations $|\chi_k\rangle$:

$$|\psi(t)\rangle = \sum_{k=0} c_k(t)|\chi_k\rangle. \qquad (2)$$

Assuming that excitation processes dominate over ionization and charge transfer processes the Hamiltonian H(t) can be split into two parts. The first describes the Na atom and is denoted by the Hamiltonian h_o, the second is the interaction between the proton (projectile) and the target electrons:

$$H(t) = h_o + v(\underline{R}(t)). \qquad (3)$$

The Hamiltonian h_o is approximated by an effective one particle operator

$$h_o = \sum_{i=1}^{11} (t_i + v_i^{eff}). \qquad (3a)$$

The interaction potential $v(\underline{R}(t))$ is written as

$$v(\underline{R}(t)) = -\sum_{i=1}^{11} \frac{1}{|\underline{r}_i - \underline{R}(t)|},$$

where $\underline{r}_i$ is the position of the i-th Na electron. For the collisions of interest, the internuclear vector $\underline{R}(t)$ can be adequately represented by a straight line trajectory with impact parameter $\underline{b}$ and relative velocity $\underline{v}$:

$$\underline{R}(t) = \underline{b} + \underline{v} \cdot t.$$

The resulting system of coupled equations

$$i\,\dot{c}_k(t) = \sum_{\ell=0}^{A} \langle\chi_k|v(\underline{R}(t))|\chi_\ell\rangle c_\ell(t) \qquad (4)$$

has been solved in the following approximations:

i) The term scheme of Na (Fig. 1) suggests that both the initial and the final Na - state are polarized during the scattering process I. We describe the polarization of the ground state by including the excited states $Na(2p^6 3p\ ^2P_{M=0,\pm1})$ in the expansion (2). The polarisation of the final states $Na(2p^5 3s^2\ ^2P_{M=0,\pm1})$ is treated by adding the configurations

$$Na(2p^5(3s3p\ ^1P)\ ^2S,\ ^2P_{M=0,\pm1},\ ^2D_{M=0,\pm1,\pm2}) \text{ and}$$

$$Na(2p^5(3s3p\ ^3P)\ ^2S,\ ^2P_{M=0\pm1},\ ^2D_{M=0,\pm1,\pm2}).$$

Inserting these configurations into expansion (2), one obtains a close coupling problem involving A=25 states.

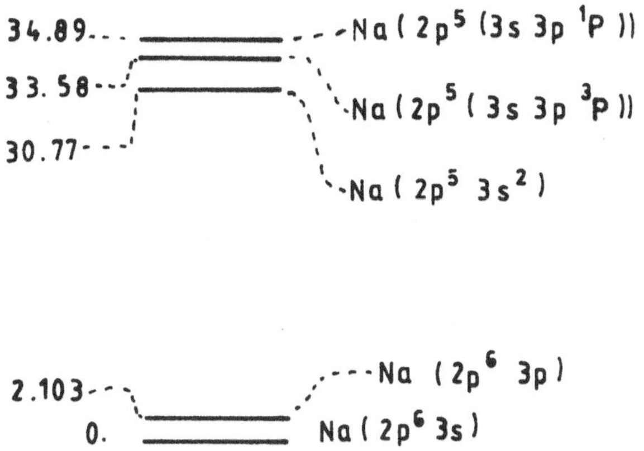

Fig. 1. Term scheme of Na
The energy differences are given in eV

ii) Retaining only the initial and final states of process I, equation (4) is solved with A=4 basis states. This equation is equivalent to a one hole problem in the $(2p^6 3s^2)$ closed-shell configuration. This approximation does not take into account any polarization effects.
iii) In first order perturbation theory the solution of ii) gives cross-sections identical to those obtained by Theodosiou et al.[2] using the plane-wave Born approximation (PWBA). In this case we only regard the direct excitation to the excited Na states from the ground state. Back-coupling and any coupling amongst excited states are ignored. In figure 2 the total cross-section for excitation of the $(2p^5\ 3s^2)\ ^2P$ state calculated in the three approximations is compared with the (somewhat sparse) experimental data. As may be expected for fast collisions ($v/v_{2p} > 1$) the Born approximation agrees with the close coupling results. However in slow collisions the strong influence of back-coupling (approximation ii) and polarisation (approximation i) results in a considerable reduction in the first Born cross-section (not expected to be valid for slow collisions) and essential agreement of the close-coupling results with experiment.

Figure 3 shows the alignment parameter A_2 calculated from the partial cross-sections. Again for fast collisions the first Born approximation agrees with the close-coupling results and with experiment. Curiously enough at lower velocities, experiment shows a deep minimum in the alignment which is reproduced by the Born approximation but not by the more accurate multi-state calculations. In view of the disagreement of the first Born result with experimental total cross - sections in this velocity region, the agreement with the alignment parameter must be considered fortuitous. Nevertheless the

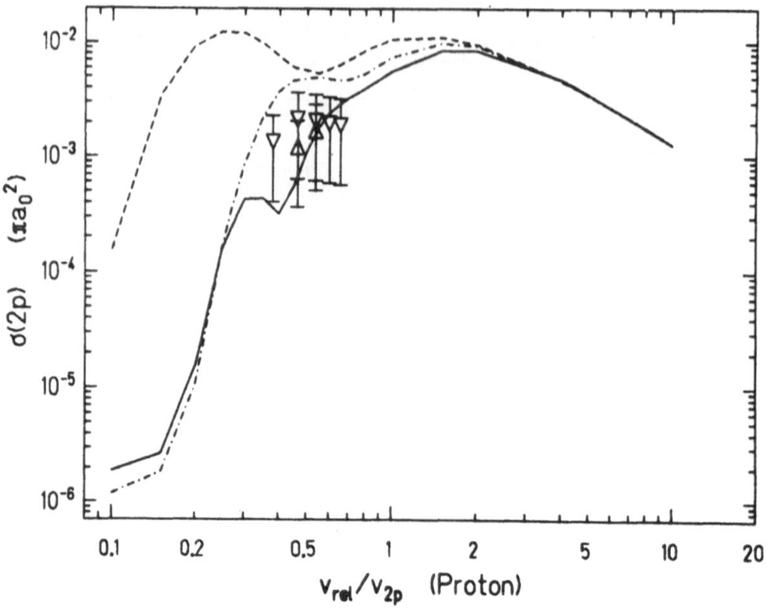

Fig. 2. Total cross section for Na(2p) excitation
Theory: —: 25 state close coupling calculation (approximation i)).
-·-: 4 state close coupling calculation (approximation ii)).
---: Born approximation, identical to the results of Theodosiou et al.[2] and Ziem et al.[3]
Experiment: Δ,∇: Mehlhorn[1]

discrepancy between experiment and close-coupling results is hard to explain. In view of the correspondence between this theory and the total cross-section, the origin of the discrepancy must lie in some mechanism which redistributes probability amongst the magnetic sub-levels without itself absorbing significant probability from the excited state. The only likely candidate appears to be capture, either real or virtual, into projectile states, which mechanism has been ignored in the present calculations.

The important role played by the strong polarisation of the target leads to a significant probability for the dipole-allowed $3s \rightarrow 3p$ transition. The corresponding cross-section can be extracted from approximation (i) and is shown in fig. 4. in comparison with the results of Bell and Skinner [4] and Mathur et al.[5] Bell and Skinner performed a close coupling and a PWBA calculation. Mathur et al. performed a PWBA calculation. The differences in PWBA results are due to the use of different wave functions. Experimental results for this process are not known to us.

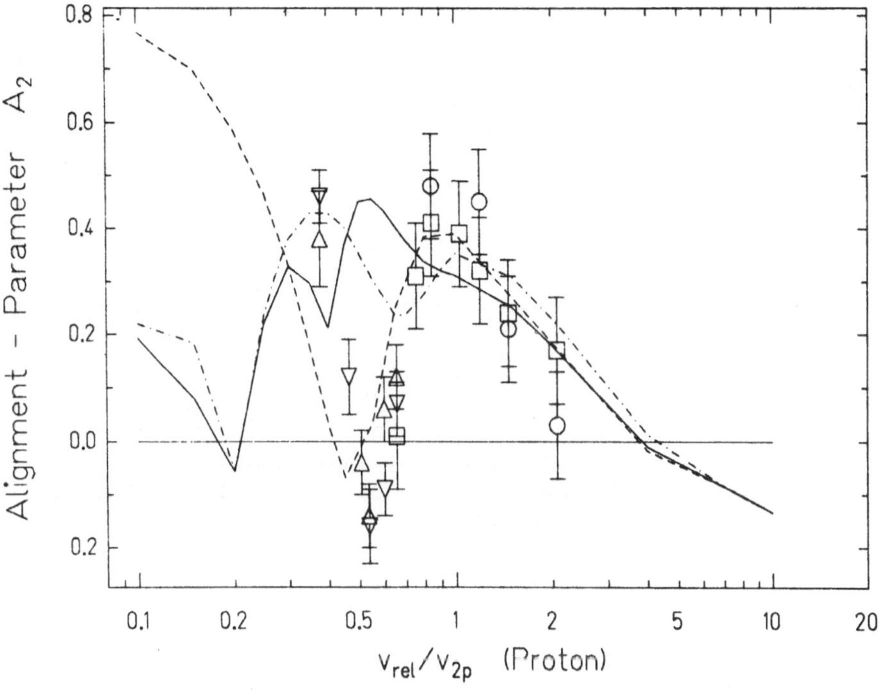

Fig. 3: Alignment parameter A_2 of Na(2p) - shell
Theory: see Figure 2
Experiment: ∇: Mehlhorn[1]
Δ: Theodosiou et al.[2]
O: Ziem et al.[3]
□: Ruckteschler[6]

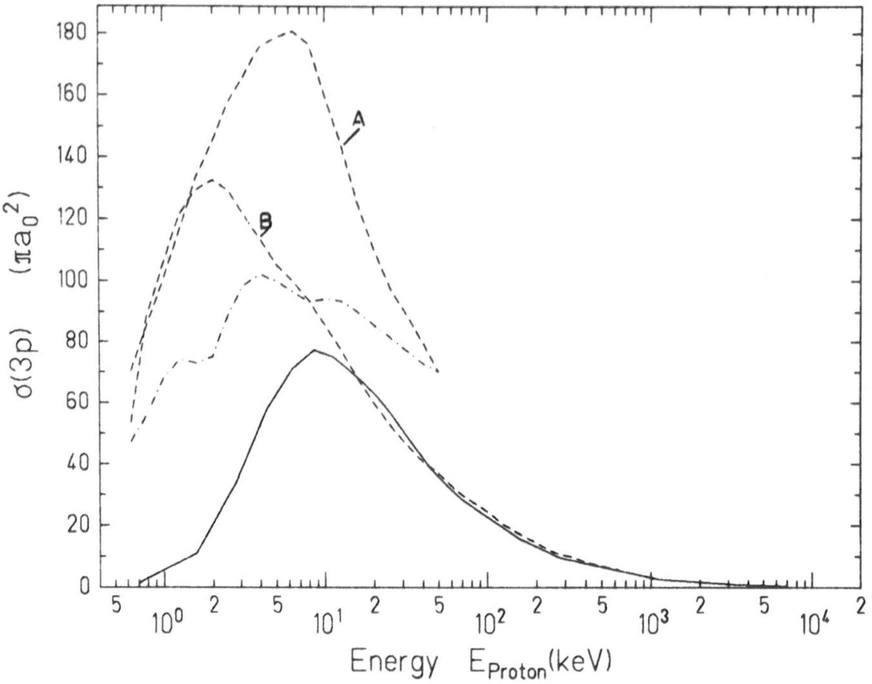

Fig. 4: Total cross section for Na(3p) excitation
 -: present calculation, approximation i)
 -·-: close coupling calculation, Bell and Skinner[4]
 ---: first Born approximation
 curve A: Bell and Skinner[4]
 curve B: Mathur et al.[5]

In summary, it appears that the basis set employed can give a good description of total cross-sections for sodium excitation but serious discrepancies still exist between theory and experiment for the alignment parameter. The study has also shown that the close agreement of the first Born approximation with the alignment parameter for relatively slow collisions is fortuitous.

REFERENCES

1. W. Mehlhorn, private communication (1984)
2. C.E. Theodosiou, E. Breuckmann, B. Breuckmann, W. Mehlhorn, J. Phys. B12 L689 (1979)
3. P. Ziem, R. Morgenstern, J. Phys. B13 L299 (1980)
4. R.J. Bell, B.G. Skinner, Proc. Phys. Soc. 80 404 (1962)
5. K.C. Mathur, A.N. Tripathi, S.K. Joshi, Phys. Letters 35A 139 (1971)
6. R. Ruckteschler, Diplom thesis, Universität Freiburg (1981)

CALCULATIONS OF ION-INDUCED K-SHELL IONIZATION

David J. Land

Nuclear Branch
Naval Surface Weapons Center
Silver Spring, Maryland 20910 USA

INTRODUCTION

Studies of the excitation and ionization of the inner shells of atoms by heavy projectiles (p, He) have been ongoing both theoretically and experimentally for many years. Early calculations for the total cross section for K- and L-shell ionization were based upon the plane-wave Born approximation (PWBA).[1] The results of such calculations agree well with experimental data at sufficiently high projectile velocities but show marked discrepencies of two orders of magnitude or more at low velocities. Calculations based upon the semiclassical approximation (SCA),[2] in which the hyperbolic Coulomb trajectory of the projectile is taken into account, give reasonable predictions for the impact parameter dependence of the ionization probability. Considerable progress in correlating theoretical predictions with experimental results was achieved with the introduction of the idea that the binding energy of the target electron is increased[3,4] and the wave function of the electron is polarized[4,5] by the presence of the positively-charged projectile in the vicinity of the electron. These effects, based upon perturbed-stationary-state (PSS) theory,[4] along with several other physical effects, have been synthesized by Brandt and coworkers[6] into what is known as ECPSSR theory: E, the energy loss of the projectile in exciting the target electron; C, Coulomb deflection and retardation; PSS, binding energy and wave function distortion; and R, relativistic effects of the target electron. Except at very low projectile velocities, this model agrees well with experiment for most known cases involving asymmetric collisions in which $Z_1 \ll Z_2$, where Z_1 and Z_2 are the atomic numbers of the projectile and target atom, respectively.

The picture given above might indicate a fairly complete and successful theoretical description of inner-shell ionization. However, calculations presented in this paper call such a conclusion into question, and suggest that this description is not complete and additional physical mechanisms are involved. We have compared the results of calculations with experimental data for systems with increasing Z_1 and approximately the same Z_2. Systematic discrepencies are found, increasing with increasing projectile energy and, more significantly, with increasing projectile atomic number. Also, we have compared results of our calculations with results from accurate, nonperturbative solutions of the time-dependent Schrödinger equation with the same interaction Hamiltonian.[7,8] These comparisons suggest that the present calculations yield correct trends.

The method of calculation used here is strongly guided by PSS theory[4] and takes as a starting point the expansion of the wave function $\phi(t)$ in a set of time-dependent basis states, chosen to approximate the time development of each state in question. These calculations are restricted to the intermediate velocity region, defined by $\xi \approx 1$, where ξ is the ratio of the adiabatic radius to the K-shell radius. In this velocity region, energy-loss effects, E, and relativistic effects, R, are quite small and Coulomb trajectory effects, C, which are accounted for here approximately, are not large. Thus, these calculations probe the ability of the present method to account adequately for higher-order effects in perturbation theory in the solutions of the Schrödinger equation. This method implements rigorously the ideas embodied in the PSS part of the ECPSSR model,[6] without cutoff parameters for the binding and polarization effects and with a more realistic atomic wave function rather than a simple harmonic oscillator description of the target electron for the polarization effect.

OUTLINE OF THE CALCULATIONAL METHOD

In the semiclassical approximation the behavior of an atomic state in response to the perturbing potential of an incident charged projectile of atomic number Z_1 is governed by the time-dependent Schrödinger equation,

$$i\, \partial\phi/\partial t = H(t)\, \phi,$$

where the Hamiltonian is given by

$$H(t) = H_a + V(t) = H_a - \frac{Z_1}{|\underline{r} - \underline{R}(t)|}; \qquad (1)$$

H_a is the unperturbed Hamiltonian of the target atom, $\underline{r}$ is the electron coordinate, and $\underline{R}(t)$ describes the prescribed trajectory of the projectile. Atomic units are used throughout ($\hbar = e = m_e = 1$).

The wave function $\psi(t)$ for an atomic state of interest (the K shell here) is usually expanded in terms of the eigenfunctions of the unperturbed Hamiltonian, H_a. In the present approach, however, $\psi(t)$ is expanded in terms of a set of time-dependent, perturbed, basis states, $u_n(t)$:

$$\psi(t) = \sum_n a_n(t)\, u_n(t).$$

The states $u_n(t)$ are, in principle, arbitrary, subject to the conditions that the set is complete and that, at $t = \pm\infty$, they are eigenfunctions of the atomic Hamiltonian H_a, and hence are orthogonal at $t = \pm\infty$. At finite times they need not be orthogonal.

The time-dependent basis states, $u_n(t)$, are chosen by the following considerations. Physically, each state of the system evolves in time from its initial configuration to its final one under the action of the Hamiltonian H(t). In the spirit of perturbed-stationary-state theory, each basis state can be chosen to approximate the behavior of the corresponding physical state. For adiabatic changes in H(t) (very low projectile velocities), the basis states are selected to be eigenfunctions of the total time-dependent Hamiltonian, H(t).[4] However, for higher projectile velocities, the dynamic response of the electrons needs to be taken into account. Land and Simons[9] have suggested a model for doing this. For the projectile velocities of interest here, it follows from this model that only the K shell of the atom displays any effects of adiabatic behavior; that is, all other atomic shells can be described by the unperturbed eigenfunctions of the atomic Hamiltonian, H_a. We assume this is the case in the following.

As a description of the K-shell electron a spherically symmetric wave function[3] of the form,

$$u_i(r;R) = \frac{1}{\sqrt{4\pi}} Z_{eff}^{3/2}(R)\; 2\, e^{-Z_{eff}(R) r},$$

is often assumed, where $Z_{eff}(R)$, the effective nuclear charge, is a parameter determined from a variational principle which minimizes the ground-state energy of the Hamiltonian of Eq. (1). More recently, a more general wave function, which exhibits a polarization of the electron cloud, has been proposed by Basbas and Land:[10]

$$u_i(r;R) = N(R)[e^{-Z_1|\underline{r}-\underline{R}|} + f(R)]\, \frac{1}{\sqrt{4\pi}} Z_2^{3/2}\; 2\, e^{-Z_2 r}. \qquad (2)$$

Here the parameter f(R) is determined by the same variational principle and N(R) is chosen to give the usual unit normalization. These two wave functions produce very similar ionization probabilities for impact parameters smaller than the K-shell radius but the second form allows for the continued attraction of the electron at large impact parameters.

As noted above, for the velocities considered in this paper only the K shell displays any significant effects of adiabaticity. For this special case the general equation for the amplitude of the final state, b_f, which takes into account possible effects arising from the use of a nonorthogonal set of basis states, has been shown[9] to reduce to the usual expression in the SCA:

$$b_f = \frac{1}{i} \int_{-\infty}^{\infty} dt \, \exp\{i \int_0^t dt' \, [H_{ff} - H_{ii}(t')]\} \, V_{fi}(t), \qquad (3)$$

where $V_{fi}(t)$ is the matrix element of the interaction potential $V(t)$ evaluated with respect to the final continuum state and the initial state, $u_i(r;R)$, of Eq. (2). H_{ff} is the final-state electron energy and $H_{ii}(t')$ is the matrix element of the Hamiltonian, Eq. (1), taken with respect to the initial state, Eq. (2). The appearance of $H_{ii}(t)$, rather than the binding energy of the initial state, is a reflection of the distortion approximation, according to which a phase transformation on the amplitudes for each of the individual states is made to remove diagonal terms from the equation for that state. There are no effects arising from the nonorthogonality of the initial state with the other states of the atom, which by themselves are unperturbed and, hence, orthogonal. The unperturbed nature of the other states corresponds to the physical situation in which the separated-atom binding algorithm is strictly correct; there is no response by the other electrons and hence no additional change in the binding energy of the K-shell electron.

COMPARISONS WITH DATA AND DISCUSSION

Calculations of the total cross section for K-shell ionization have been performed for three systems for which experimental data are available: p and ^{3}He in Ti[9,11,12] and ^{6}Li in Cr.[13] The target atoms, Ti (Z_2 = 22) and Cr (Z_2 = 24), are sufficiently close in atomic number that these systems probe directly the effect of changing only projectile charge, Z_1.

The expression for the probability amplitude, b_f, given in Eq. (3) is basic to the calculations reported below. In the evaluation of b_f a straightline trajectory characterized by effective values of impact parameter and velocity is used. As suggested in Ref. 4, the impact parameter is taken at d_{min}, the distance of closest approach. The K-shell binding energy, $H_{ii}(t)$, and initial-state wave function, $u(r;R)$, are also evaluated at d_{min}. The effective projectile velocity is taken from the average velocity,[14] $v_{ave} = (v + v_d)/2$, where v is the initial projectile velocity and v_d is the projectile velocity at d_{min}. Screened (Merzbacher-Lewis) hydrogenic wave functions[1] are used to account for outer screening; the Slater approximation, $Z_2 \rightarrow Z_2 - 0.3$, is used to describe the inner screening. For the initial state the polarized wave function of Eq. (2) is employed. For the final state, unperturbed wave functions are used; these are expanded in partial waves and s, p, and d wave are kept. This general prescription is designated by PSS/1P. There are no free parameters available to fit the data.

The results of the calculations are shown in Fig. 1 as the ratio of the experimental values to the corresponding PSS/1P theoretical results. They are shown as a function of the scaled velocity, ξ, the ratio of the adiabatic radius to the K-shell radius,

$$\xi = r_{ad}/r_K = (hv/E_B)/r_K,$$

where E_B is the K-shell binding energy, r_K, the K-shell radius, is defined by a_0/Z_{2K}, a_0 is the Bohr radius, and $Z_{2K} = Z_2 - 0.3$. The energy range in the figure corresponds roughly to 0.5 to 2.5 MeV/amu. For incident protons these values are about 10% below unity at the lowest energies shown, rising quickly through unity and slowly

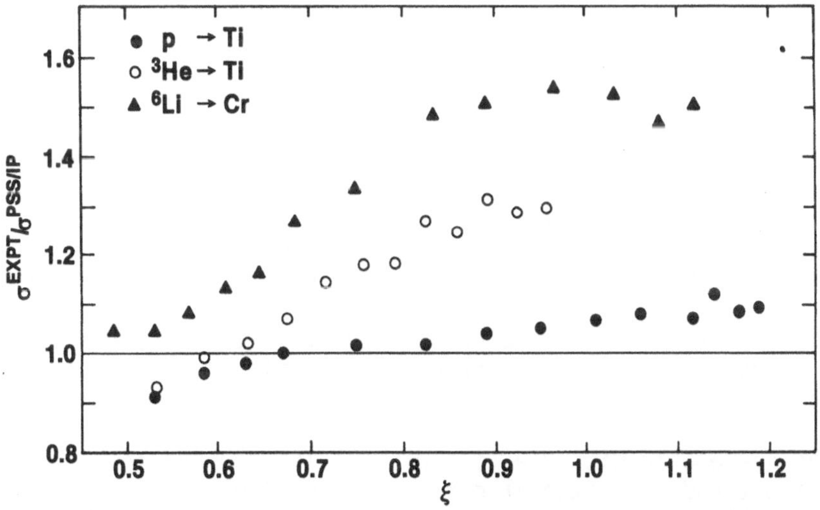

Fig. 1 Ratio of the experimental values to the PSS/1P theoretical values for the K-shell ionization cross sections for protons and ^{3}He in titanium and ^{6}Li in chromium as a function of the scaled velocity ξ. The proton data are from Ref. 11, the ^{3}He from Ref. 9 and 12, and the ^{6}Li data from Ref. 13.

to 10% above unity at the highest energies. For ^{3}He the ratios again begin at 10% below unity and rise steadily to 30% above unity. For ^{6}Li they are near unity at the lowest energy but rise to about 50% above unity and appear to level off. Thus a systematic discrepency with experimental data as a function of projectile atomic number is seen in these comparisons and this increases with Z_1.

If meaningful conclusions are to be drawn from these comparisons, the question must be addressed of how accurate are these calculations. The method employs, basically, lowest-order perturbation theory, but introduces higher-order effects by removing diagonal terms from the basic equations governing the time dependence of the amplitudes (distortion approximation) and using perturbed wave functions to account for the distortion of the initial state wave function. In order to assess the accuracy of the method, comparisons should be made with calculations which solve the same Schrödinger equation but which are themselves exact in some sense.

Such a method has been developed in a series of papers by Reading and coworkers.[7,8] A nonperturbative solution of the Schrödinger equation is obtained through a U-matrix approach operating in a truncated, L^2 basis space. Numerical calculations have been reported for only a few systems, but one of these is protons in Ti. The present results agree with those reported in Ref. 7 to within 8% at 1.0 MeV and within 4% from 1.5 to 2.5 MeV. However, protons as projectiles do not stress details of the solution too severely and, for a more meaningful comparison, systems involving projectiles having atomic number larger than unity should be considered. Results have recently been reported for the ratio of the total cross section for K-shell ionization in Al for alphas to four times the corresponding cross section for protons.[8] The results are shown in Fig. 2. Also shown are the theoretical predictions of CPSS (ECPSSR, without energy loss and relativistic corrections). The differences between the calculations of Reading et al. and the results of CPSS are noteworthy. The PSS/1P calculations agree with those of CPSS at low energies, but they fall below these values at higher energies and cross those of Reading at an energy corresponding to a value of ξ of 0.9. While the shapes of these two curves are different at these higher energies, both fall below CPSS and this suggests the correct trend, overall, of the PSS/1P calculations relative to those of Reading et al.

The systematic discrepencies with Z_1 between the results of calculations and experimental data, coupled with guidance from more accurate numerical calculations discussed above, suggest that the model for direct Coulomb ionization of an inner shell needs additional physical mechanisms to complete the theoretical description. With respect to the present calculation the integral for the final-state amplitude, b_f, needs to be evaluated accurately to account for the variation of $H_{ii}(t')$ with t'. This change[9] should tend to bring the PSS/1P calculations into closer agreement with

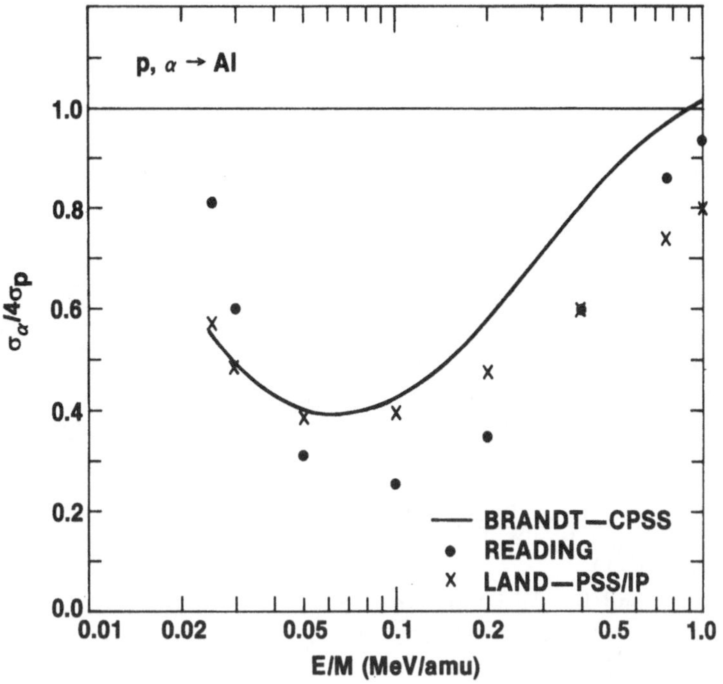

Fig. 2 Ratio of the K-shell ionization cross section for incident alpha particles to four times that for incident protons in aluminum as a function of energy per mass. The solid curve represents results of the CPSS model, the circles those of the U-matrix approach of Reading et al.,[8] and the crosses those discussed in the present paper and denoted by PSS/1P.

those of Reading et al.[8] as shown in Fig. 2. With respect to possible additional mechanisms, we recall a suggestion of Ford et al.[7] These authors noted that electron capture cross sections theoretically should rise with Z_1 as Z_1^5. On the other hand, the experimental values for electron capture are not, in some cases, as large as expected theoretically. Therefore, the idea arose that electron capture may play a larger role in the vacancy production of the K shell than previously thought, and that the projectile loses the captured electron in interaction with electrons in the outer atomic shells. This mechanism is under study by Reading and coworkers.[15]

ACKNOWLEDGEMENTS

The author wishes to thank Dr. B. Raith for making available the ^{6}Li data prior to publication and Dr. D. G. Simons for a careful reading of this manuscript. This work has been supported by the Independent Research Program of the Naval Surface Weapons Center.

REFERENCES

1. E. Merzbacher and H. W. Lewis, X-ray production by heavy charged particles, in: "Handbuch der Physik," vol. 34, pp. 166 - 192, S. Flügge, ed., Springer-Verlag, Berlin (1958).
2. J. Bang and J. M. Hansteen, Mat. Fys. Medd. Dan. Vidensk. Selsk. :31, no. 13 (1959).
3. W. Brandt, R. Laubert, and I. Sellin, Phys. Rev. 151: 56 (1966).
4. G. Basbas, W. Brandt, and R. H. Ritchie, Phys. Rev. A 7: 1971 (1973).
5. G. Basbas, W. Brandt, and R. Laubert, Phys. Rev. A 7: 983 (1973); 17: 1655 (1978).
6. W. Brandt and G. Lapicki, Phys. Rev. A 23: 1717 (1981).
7. A. L. Ford, E. Fitchard, and J. F. Reading, Phys. Rev. A 16: 133 (1977).
8. J. F. Reading, A. L. Ford, J. S. Smith, J. Alexander, and R. L. Becker, Nucl. Instr. and Meth. 192: 266 (1984).
9. D. J. Land and D. G. Simons, Nucl. Instr. and Meth. 232 (B4): 239 (1984).
10. G. Basbas and D. J. Land, in: "Proceedings of the US-Japan seminar on charge states and dynamic screening of swift ions in solids," Honolulu (1982); D. J. Land and G. Basbas, IEEE Trans. Nucl. Sci. NS-30: 1103 (1983).
11. M. D. Brown, D. G. Simons, D. J. Land, and J. G. Brennan, Phys. Rev. A 25: 2935 (1982); IEEE Trans. Nucl. Sci. NS-30: 957 (1983).
12. D. G. Simons and D. J. Land, unpublished.
13. B. Raith, S. Divoux, and B. Gonsior, to be published; B. Raith, Bull. Amer. Phys. Soc. 29: 1091 (1984) and private communication.
14. L. Kocbach, Phys. Norv. 8: 187 (1976).
15. J. F. Reading, private communication.

LONG-RANGE PROCESSES IN Na^{+}-Na(3p) COLLISIONS AND DEPENDENCE ON INITIAL 3p EXCITATION

R. J. Allan,[a] A. Bähring[b] and J. Hanssen[c]

[a] F. B. Chemie der Universität Kaiserslautern
D-6750 Kaiserslautern, F. R. G.

[b] Institut für Molekülphysik, Freie Universität Berlin
Arnimallee 14, D-1000 Berlin 33, F. R. G

[c] Laboratoire des collisiones atomiques, Université de Bordeaux I, 33405 Talence, France
Équipe de rēcherche du CNRS n° 260

INTRODUCTION

There is at present much interest, both experimentally and theoretically, in studying internal details of collision processes - effectively following the collision by looking at state-selected differential cross sections. This is true for all types of atomic interactions from heavy atoms and ions (with which we work) to electron scattering from atoms and molecules and electron ejection in electromagnetic fields (which may be due to other approaching particles). To gain some knowledge of the evolution of a collision system one looks at an angular distribution (which can classically be mapped onto impact parameters through a deflection function) at a variety of kinetic energies. Further information is obtained if we select the observed or prepare the initial atomic states in a known way relative to the collision plane rather than averaging over the quantum numbers. This can be done by measuring Stokes' parameters of the flourescence light emitted in the collision region, or by laser exciting before the interaction and tells us the average shape of the atomic orbitals in the collision.

$Na(3p)+Na^+$ COLLISIONS

In the experimental work of Bähring et al.[1,2] the latter approach is used. Na^+ ions and laser produced Na(3p) atoms collide in crossed beam geometry which allows high resolution measurements of the differential cross sections (DCS) for quenching to Na(3s) (process I) and excitation to Na(3d) (process II). The results show that the cross sections are sharply peaked at an angle β_m and that the product $E_k\Theta_m$ is roughly constant in the range of collision energies $20 \leq E_k \leq 100$eV used. With linearly polarised light the target atom can be produced in a known mixture of σ, π^+ and π^- substates relative to the collision plane. Dependence of the maximal value of the DCS on the angle of polarisation β is measured and shows a smooth sinusoidal behaviour.

Inspection of the adiabatic potential curves for the Na_2^+ system (figure 1) shows four regions relevant to understanding these measurements. The incoming population first evolves among the quasi-degenerate 3p levels leading to a so-called "lockin" process[3] and crossing of $2^2\Sigma_u$ and $1^2\Pi_u$ at $22.5a_0$. The potential curves then separate and crossings take place between the $1^2\Pi_u(3p)$ and $1^2\Sigma_u(3s)$ states at $4.9a_0$

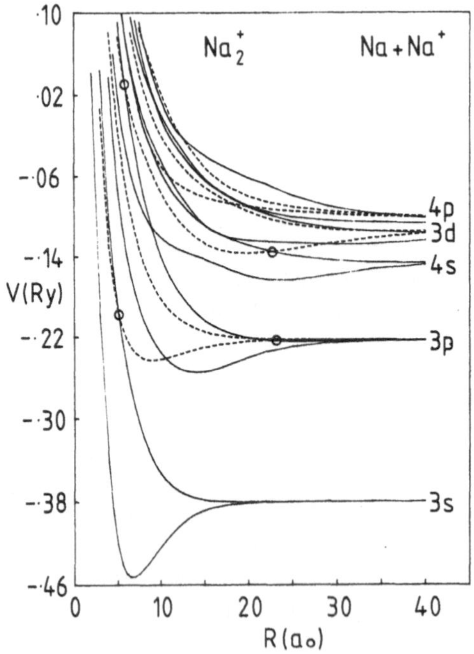

Fig. 1. Adiabatic electronic energy curves for Na_2^+.

and between the $2^2\Sigma_u(3p)$ and $2^2\Pi_u(3d)$ states at $5.56a_0$. Some contribution to the 4s channel due to a further crossing between the $2^2\Pi_u$ and $3^2\Sigma_u$ curves is expected: this is found however to be negligible in both the experiment and the calculations due to the differing impact-parameters of importance for the two processes.

In order to use the experiment to test theoretical assumptions of the scattering process we have performed calculations to compare with the measured quantities. We use accurate molecular energies[2] and inter-state coupling[4] in a quasi-classical impact-parameter method[4,5] to model the dynamical process. The electronic motion is incorporated explicitly through a common translation factor (CTF)[6]. We can thus observe the evolution of the initially prepared coherent mixture of $3p\sigma$ and $3p\pi$ as the distance between Na^+ and Na decreases. In the space-fixed reference frame the elastic and exchange probabilities become different from their asymptotic values at a distance R_L due to the increasing interaction strength[3]. For $R < R_L$ a body-fixed reference frame is most suitable to describe the collision.

In figure 2 we show the DCS for processes I and II. These were calculated for the initial conditions leading to a maximum value of the peaks at E_k=47.5eV (full curves). The experimental values are normalised once only to give the correct height of the 3p-3s peak. There is quite good agreement although the experimental peaks are somewhat broader. A further calculation in which the coupling for $R \gtrsim 12a_0$ (principally the long-range mixing in the 3p manifold) is neglected is shown by the dashed lines. Long-range mixing just leads

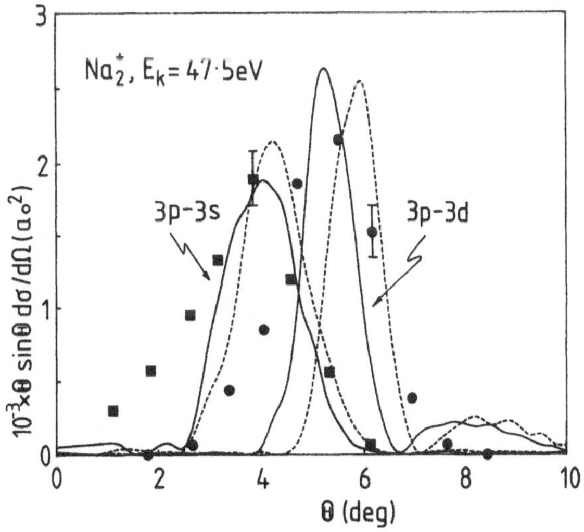

Fig. 2. Differential cross sections in Na_2^+ at E_k=47.5eV. Theoretical predictions are compared to the experimental results for direct 3p-3s and 3p-3d transitions ■ and ●, respectively.

to a shift in the differential pattern. This is all that would be seen if no state-selection (i.e. no coherence) were used. At this energy the transitions are therefore mainly governed by localised rotational coupling mechanisms occuring for critical impact parameters (IP) of 4.83 and $4.91a_o$. It is indeed possible to reproduce the dashed curves using simple two-state semiclassical methods[7]. These do not contain the alignment information which the full curves do. We now look at this information in more detail.

ORBITAL ALIGNMENT

For linearly polarised light we can rotate the E-vector of the laser in the collision plane so that principally σ and π^+ states are excited, or out of the plane to excite σ and π^-. Following the analysis of Hertel and Fischer[8] there is also an incoherent contribution from the third state. Choosing a reference frame S with $\hat{z}\equiv\hat{E}$ the maximum excitation is 1:2.5:1 for the three p states in the basis $l=1$, $m_l=-1,0,1$ respectively. The density matrix $\underline{\underline{\rho}}$ is diagonal in this frame:

$$\rho^n_m = \frac{1}{4.5}\begin{bmatrix}1 & 0 & 0\\ 0 & 2.5 & 0\\ 0 & 0 & 1\end{bmatrix} \quad ;m_l=n-2,\ m_l'=m-2 \qquad (1$$

To find the initial conditions for the collision we must rotate this density matrix into the asymptotic collision frame S' for which z'=v, the initial velocity direction, and y' is parallel to the orbital angular momentum. Defining[9]

$$\mathcal{D}^h_k(\alpha',\beta',\gamma') = e^{i(h-2)\alpha'} d^{(1)}_{h-2,k-2}(\beta') e^{i(k-2)\gamma'} \qquad (2$$

$$\rho'^n_m = \mathcal{D}^n_k(\alpha',\beta',\gamma')\,\rho^k_l\,\bar{\mathcal{D}}^l_m(\alpha',\beta',\gamma') \qquad (3$$

where a bar signifies the inverse matrix. The Euler angles (α',β',γ') to bring S into S' are taken to be $(0,\beta,0)$ if $\underline{E}$ is in the $\hat{z}'.\hat{x}'$ plane and $(0,\beta,90^0)$ if in the $\hat{z}'.\hat{y}'$ plane. A further small correction of $\beta_{corr} = \sin^{-1}(b/R_o)$ is introduced since we start numerical integration of the IP equations at $R_o \approx 70a_o$ rather than ∞. The measured and calculated results are shown in figure 3. The upper curves are for the 3p-3s transition and the lower for 3p-3d. With $\underline{E}$ in the $\hat{x}'.\hat{z}'$ plane the experimental points are marked with symbol $\Diamond$ and for $\underline{E}$ in the $\hat{z}'.\hat{y}'$ plane with o.

This figure essentially contains values of the first and second Stokes' parameters for a single scattering angle[10] (which leads to the peak in fig.2 for each transition). The curves are sinusoidal and may be characterised by three parameters - the transition probability, the

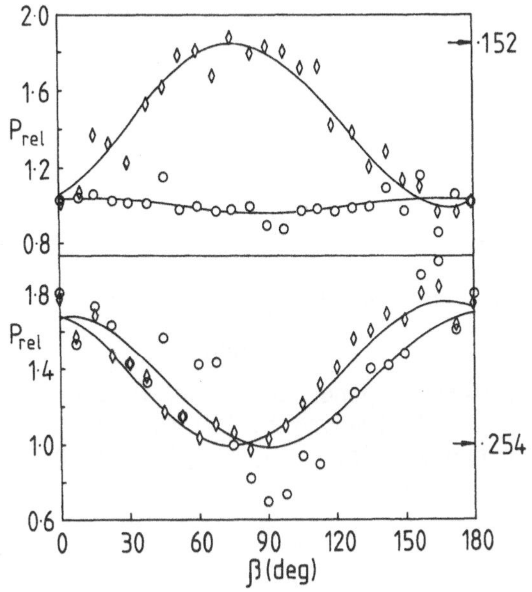

Fig. 3. Comparison of transition probabilities for processes I and II, E_k=47.5eV, b=4.38 and 5.91a_o respectively for different initial conditions (see text).

initial degree of alignment (2.5 in eqn.1) and the shift in β_m from 90° which leads to an extremum in the pattern. Experimentally the first two are confounded unless separate "collisionless" determination of the initial beam properties is carried out. From theoretical considerations we suggest that the actual alignment is around 1:1.8:1 rather than the ideal 1:2.5:1.

Of course both experimentally and numerically a great deal more information is available since we can vary the impact parameter b (or equivalently scattering angle Θ) and kinetic energy E_k. We show a selection of this data related to figure 3 in figures 4 and 5. In the first of these the angle β_m is shown; the energy is kept constant and b is varied. The transitions I and II are represented by the symbols o and + respectively. The full line is the result of a "half-collision" calculation in which we take just the incoming populations of the $1^2\Pi_u$ and $2^2\Sigma_u$ states around $10a_o$ without the interaction at short range - this is more accurate for the higher probability (II) because the second crossing in the outward wing of the collision is ignored. The energy variation of β_m with the same symbols is shown in fig.5 for impact parameters of $4.5a_o$ (I) and $5.0a_o$ (II).

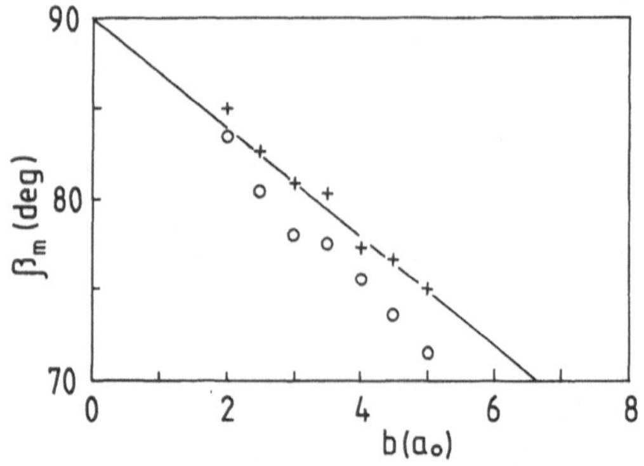

Fig. 4. Impact parameter dependence of the initial polarisation angle β_m near 90° leading to the maximum or minimum in the transition probability I (o) and II (+) for an in-plane E-vector, E_k=47.5eV.

ORIENTED ORBITALS

So far we have looked only at information related to alignment of the orbitals. The effect of orientation can be studied if we use initially right (RHC) or left (LHC) circularly polarised light. Measuring the scattered intensities I for the two cases we can write the third Stokes' parameter for the collision

$$P_3 = (I_{RHC} - I_{LHC}) / (I_{RHC} + I_{LHC}) \qquad (4$$

Experimentally measured values for the inelastic processes I and II lie near P_3= +0.3 respectively and have little energy dependence[1,2]. Using perturbation theory[3]

$$P_3 = 2a_\sigma a_\pi \sin(\eta_\pi - \eta_\sigma + \chi) / (a_\sigma^2 + a_\pi^2) \qquad (5$$

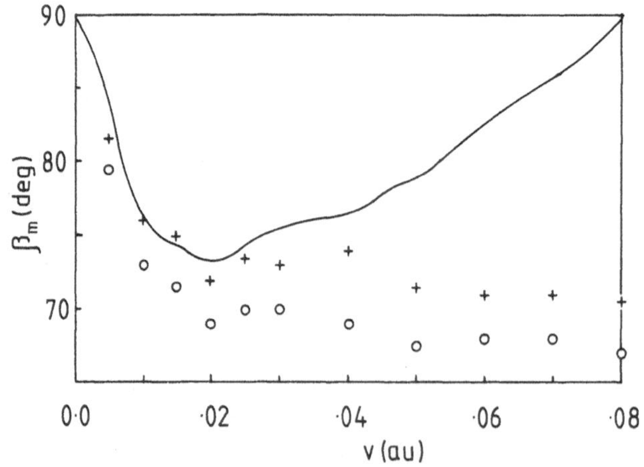

Fig. 5. Energy dependence of β_m (see fig.4). Impact parameters are fixed at $4.5a_o$ (I) and $5.0a_o$ (II).

where a_σ^2 is the amount of flux arriving in the channel (for instance $1^2\Pi_u$)leading to a transition (3p-3s) from the initial $3p\sigma_u$ state, and a_π^2 is the flux coming from $3p\pi_u$. For process I we can think of a_σ^2 as being from a Landau-Zener type transition with a pseudo-crossing at R_L (which changes what we see at the transition point from what we had initially). If a_σ^2 is small the Stueckelberg correction term[11] is $\chi \approx \pi/4$ and since R_L is large in the experiments

$$R_L = b / \cos(\beta_m) \tag{6}$$

there is little phase difference along the collision paths for $R_L \leq R < \infty$ and $\eta_\sigma - \eta_\pi \ll \pi/4$. This leads to

$$P_3^{\,I} \approx 2a_\sigma \sin(\pi/4) \approx \sqrt{2}\cos(\beta_\sigma) \tag{7}$$

for 3p-3s and likewise

$$P_3^{\,II} \approx -\sqrt{2}\cos(\beta_\pi) \quad \text{for 3p-3d.} \tag{8}$$

This simple theory seems to yield quite reasonable results but again neglects short-range effects and transitions occuring on the outgoing wing of the collision. We have therefore evaluated P_3 numerically for transitions I and II with a full collision. The initial density matrix in S for RHC and LHC light is now[8]

$$\underline{\underline{\rho}} = \begin{bmatrix} 1 & 0 & 0 \\ 0 & 0 & 0 \\ 0 & 0 & 0 \end{bmatrix} \quad \text{or} \quad \begin{bmatrix} 0 & 0 & 0 \\ 0 & 0 & 0 \\ 0 & 0 & 1 \end{bmatrix} \tag{9}$$

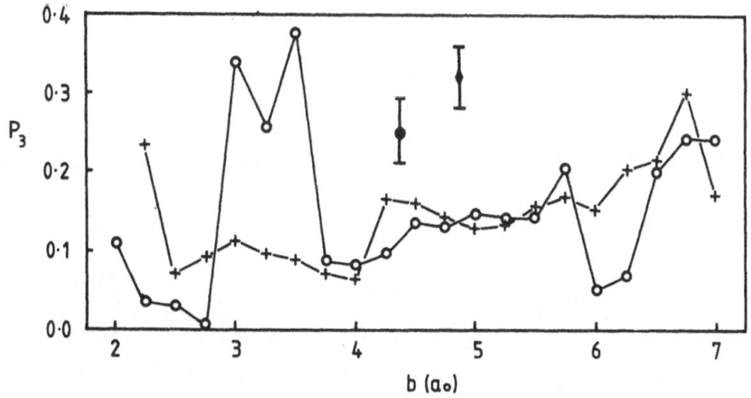

Fig 6. Stokes' parameter $/P_3/$ at E_k=45eV for the transitions I and II. With a straight collision path marked by o and + respectively and with a curved path by ● and ♦ .

The results for $|P_3|$ as a function of b at E_k=45eV follow the rather irregular curves of figure 6. Results are lower than the experimental values in the important range $4 \leqslant b \leqslant 5a_0$, but do have the correct sign.

We are however particularly interested in differences due to different initial conditions and we have therefore used energy-conserving classical trajectories for the separate LHC and RHC polarisations. The values of $|P_3|$ obtained are also shown in the figure. No qualitative change is observed. We thus conclude that the present theoretical predictions for $|P_3|$ lie below the measured values for these inelastic transitions. As a function of increasing energy E_k, $|P_3|$ decreases in the calculation.

Whilst there are clearly some quantitative differences between theory and experiment we feel that the main effects can be understood. We note that without the translation factor in this calculation the parameters investigated would simply not converge with increasing integration radius R_0. The reasonable results obtained are therefore encouraging and we consider that the CTF method can be more widely applied.

ACKNOWLEDGEMENTS

We acknowledge financial support from the Deutsche Forschungsgemeinschaft SFB 91 (RJA) and SFB 161 (AB), and the Centre Nationale de Récherche Scientifique, Équipe n° 260 (JH). RJA also received NATO support to attend the Advanced Study Institute in Palermo.

REFERENCES

1. A.Bähring, I. V. Hertel, E. Meyer and H. Schmidt, Z. für Physik 312:293 (1983)
2. A. Bähring, I. V. Hertel, E. Meyer, W. Meyer, N. Spies and H. Schmidt, J. Phys. B to be published (1984)
3. J. Grosser, J. Phys. B 14:1449 (1981)
4. R. J. Allan, J. Hanssen and A. Salin, J. Phys. B to be published (1984)
5. C. Gaussorgues, R. D. Piacentini and A. Salin, Computer Phys. Comm. 10:223 (1975)
6. L. F. Errea, L. Mendez and A. Riera, J. Phys. B 15:101 (1982)
7. R. J. Allan and H. J. Korsch, Z. für Physik to be published (1984)
8. I. V. Hertel and A. Fischer, Z. für Physik 304:103 (1982)
9. A. R. Edmonds, 1974 "Angular Momentum in Quantum Mechanics," 2nd revised edn. Princeton University Press, Princeton.
10. R. Hippler, 1984 Invited lecture, this volume
11. M. S. Child, 1976 "Molecular Collision Theory," Academic Press, London.

SLOW ION ATOM COLLISIONS IN THE PRESENCE OF A STRONG MAGNETIC FIELD

Uwe Wille

Bereich Kern- und Strahlenphysik
Hahn-Meitner-Institut für Kernforschung Berlin
D-1000 Berlin 39, West Germany

INTRODUCTION

Considerable interest has been devoted in the past few years to the theoretical study of atomic processes proceeding in the presence of strong magnetic fields. The motivation for this interest derives mainly from the occurrence of such fields in the vicinity of white dwarf stars[1,2], with estimated field strengths B ranging from 10^2 T to 10^5 T, and of neutron stars[3,4], with B ranging between 10^7 T and 10^9 T.

Until now, the main subject of investigation has been the structure of the hydrogen atom embedded in a static, uniform magnetic field. Accurate calculations of energy levels and wavefunctions for the "magnetically dressed" hydrogen atom have been reported (see, e.g., Refs. 5 and 6, and references cited therein) for field strengths up to about 10^9 T. A number of studies have dealt with the properties of the hydrogen molecular ion H_2^+ in a strong magnetic field (see, e.g., Refs. 7 and 8).

Collision processes in the presence of a strong magnetic field have been studied in a few cases as yet. Calculations on photoionization and bremsstrahlung processes as well as on electron impact excitation and ionization have been summarized in Refs. 9 and 10. In the realm of ion atom collisions, the first investigation has been that of Bivona et al.[11,12] on resonant charge transfer in H^+-H and Rb^+-Rb collisions. Bivona et al. have calculated transfer probabilities and total cross sections by employing the impact parameter method in conjunction with an atomic two-state expansion. Very recently, Grosdanov and McDowell[13] have applied the classical-trajectory Monte Carlo method in a calculation of cross sections for charge transfer in He^{2+}-H collisions.

In the present work, we consider slow (near-adiabatic) ion atom collisions proceeding in the presence of a strong magnetic field. The treatment of this problem within the impact parameter formulation of atomic-scattering theory requires as a first step the calculation of "magnetically dressed", stationary one-electron molecular orbitals (MO) at arbitrary internuclear distance and arbitrary orientation of the magnetic field with respect to the internuclear line. These MO may be used subsequently in an expansion of the full, time-dependent scattering wavefunction.

STATIONARY MOLECULAR ORBITALS IN A STRONG MAGNETIC FIELD

We choose coordinates x,y,z such that the z-axis coincides with the internuclear line of the collision system and that the y-axis is perpendicular to both the internuclear line and the direction of the static, uniform magnetic field $\vec{B}$ (cf. Fig. 1). The field direction is inclined with respect to the z-axis by an angle Θ, i.e., the vector $\vec{B}$ has components $B_x = B \sin\Theta$, $B_y = 0$, $B_z = B \cos\Theta$. In the symmetric gauge, $\vec{A} = 1/2\ (\vec{B}\times\vec{r})$, the components of the vector potential $\vec{A}$ are given by $A_x = -B/2 \cos\Theta\ y$, $A_y = B/2\ (\cos\Theta\ x - \sin\Theta\ z)$, $A_z = B/2 \sin\Theta\ y$. Atomic units ($e = m_e = \hbar = 1$) will be used in the following. In these units the value of the "critical" field strength $B_o = 2.35 \times 10^5$ T at which the oscillator energy in the magnetic field equals the Rydberg energy is given by the reciprocal of the fine-structure constant, i.e., $B_o = 137.04$.

The Hamiltonian describing the motion of an electron exposed to the field of two nuclei with charge numbers Z_1, Z_2 and to the magnetic

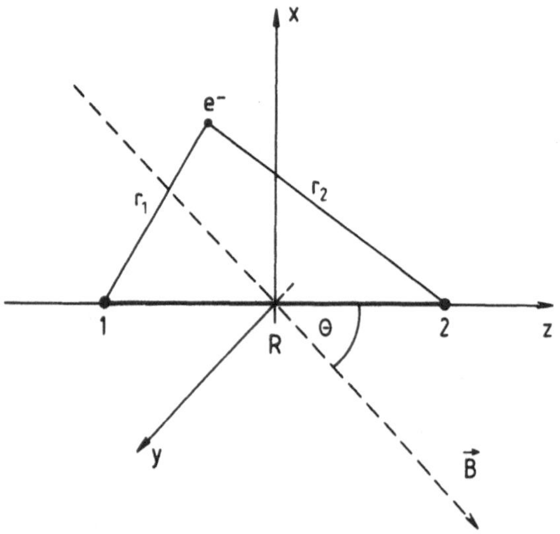

Fig. 1. Coordinates used in the calculation of "magnetically dressed" one-electron molecular orbitals (R = internuclear distance).

field $\vec{B}$ reads

$$H = \frac{1}{2}\{-i\vec{\nabla} + \frac{1}{c}\vec{A}\}^2 - \{\frac{Z_1}{r_1} + \frac{Z_2}{r_2}\}$$

$$\equiv -\frac{1}{2}\vec{\nabla}^2 - \{\frac{Z_1}{r_1} + \frac{Z_2}{r_2}\} \qquad (1)$$

$$+ \frac{B}{2c}\{\cos\Theta\, \ell_z + \sin\Theta\, \ell_x\}$$

$$+ \frac{B^2}{8c^2}\{\rho^2 - \sin^2\Theta\,(x^2 - z^2) - \sin 2\Theta\, xz\} ,$$

where c = 137.04 a.u. is the velocity of light, and $\rho^2 \equiv x^2 + y^2$. The operators ℓ_x and ℓ_z are the components of the electronic orbital angular momentum along the respective coordinate axes. An obvious property of the Hamiltonian (1) is that, for $\Theta \neq 0$, the angular momentum component ℓ_z (quantum number $\underline{m}$) is no longer a conserved quantity. The terms proportional to $\sin\Theta\, \ell_x$ and $\sin 2\Theta\, xz$ couple orbitals differing in $\underline{m}$ by one unit, while the term proportional to $\sin^2\Theta\, x^2$ couples orbitals differing in $\underline{m}$ by two units.

The complete diagonalization of H at arbitrary values of R, Θ, and B clearly constitutes a formidable numerical task. The presence of the magnetic terms removes the separability of the two-center Coulomb problem. The non-separability in conjunction with the non-conservation of ℓ_z entails the necessity to expand the stationary wavefunction in a very large set of suitably chosen basis functions. Moreover, if the stationary solutions are to be used in time-dependent calculations of transition probabilities and cross sections, the diagonalization of H is to be performed for a large manifold of values of the parameters R and Θ.

In order to make a first assessment of the effect of a strong magnetic field on stationary MO, we confine ourselves here to considering the case where the field is parallel to the internuclear line, i.e., $\Theta = 0$. The Hamiltonian $H_0 \equiv H(\Theta=0)$ reads

$$H_0 = -\frac{1}{2}\vec{\nabla}^2 - \{\frac{Z_1}{r_1} + \frac{Z_2}{r_2}\} + \frac{B}{2c}\ell_z + \frac{B^2}{8c^2}\rho^2 . \qquad (2)$$

For diagonalizing H_0 at arbitrary field strength B, a basis set is desirable that contains functions which are adapted to the two-center Coulomb problem as well as functions which are capable of accommodating the limiting case of very large field strength (Landau-type functions). As a compromise we choose basis functions of the old-fashioned Hylleraas type[14]. Expressed in terms of prolate spheroidal coordinates $\lambda = (r_1 + r_2)/R$, $\mu = (r_1 - r_2)/R$, ϕ = azimuthal angle about the z-axis, these functions read

$$\psi_{nl}^{m}(\lambda,\mu,\phi)$$
$$= (2\pi)^{-1/2}(\lambda^2 - 1)^{m/2} \exp(-x/2)\; L_n^m(x)\; P_\ell^m(\mu)\; \exp(im\phi) \quad , \tag{3}$$

where $m \geqslant 0$, $x \equiv (\lambda - 1)/a$, and a is an adjustable parameter. The functions L_n^m ($n = 0,1,\ldots,n_{max}$) and P_ℓ^m ($\ell = m,m+1,\ldots,\ell_{max}$) are generalized Laguerre polynomials and associated Legendre functions, respectively. The Hylleraas functions have proven useful in the solution of non-separable one-electron two-center problems encountered in the theory of inner-shell excitation in slow ion atom collisions[15]. In that case, the non-separability of the Schrödinger equation is caused by the presence of screening terms in the two-center potential. We expect the Hylleraas basis to be well suited for the diagonalization of the Hamiltonian (2) as long as the magnetic interaction quadratic in the field strength B is smaller than or at most of the same order of magnitude as the Coulomb interaction.

The numerical calculation of the MO energies and wavefunctions corresponding to the Hamiltonian (2) is easily accomplished by using an appropriately modified version of a computer code devised for the diagonalization of screened two-center Hamiltonians[15]. In fact, the modified code readily allows the diagonalization of one-electron Hamiltonians more general than H_O, in which the electrostatic potential is given by an effective potential of the form

$$V^{eff}(r_1,r_2;R) = -\frac{Z_1}{r_1}\chi_1(r_1;R) - \frac{Z_2}{r_2}\chi_2(r_2;R) \tag{4}$$

with analytically defined screening functions χ_1, χ_2.

As a first application, we have computed the energy eigenvalues of H_O as function of internuclear distance R for the case $Z_1 = Z_2 = 1$ ("magnetically dressed H_2^+ quasimolecule"). The number of basis functions of the type (3) included in the calculations varied from about 130 for small and intermediate internuclear distances (R < 10 a.u.) to about 175 for R = 100 a.u. Checks on the accuracy of the calculated energies have been performed by comparing them, for large R-values, to the energies (corrected for Stark effect) of the "magnetically dressed" hydrogen atom given in Refs. 5 and 6.

Energy curves of the lowest orbitals of the H_2^+ quasimolecule for $B = B_O$ as well as the corresponding curves for B = 0 are displayed in Fig 2. The energy scale in this figure is chosen such that $\varepsilon = 0$ corresponds to the ionization threshold in the zero-field case. The thresholds for σ, $\pi(m = -1)$, and $\pi(m = +1)$ orbitals in the presence of a field with $B = B_O$ are located at $\varepsilon = 0.5$ a.u., 0.5 a.u., and 1.5 a.u., respectively.

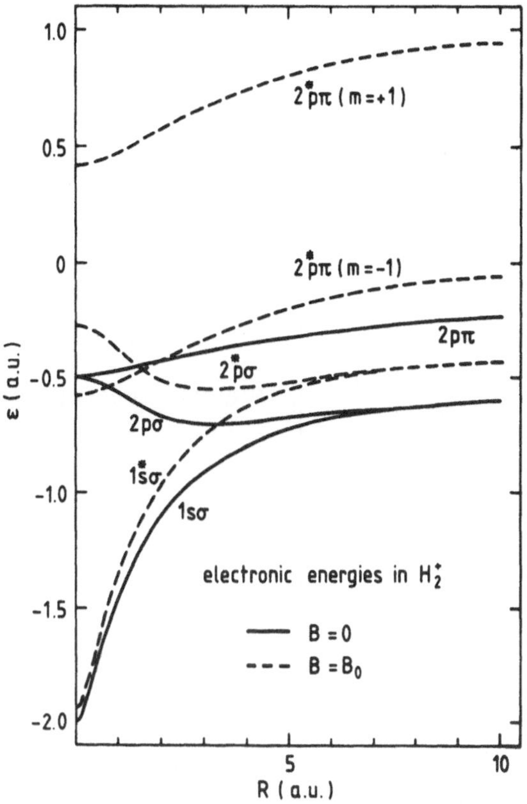

Fig. 2. Energy curves of the lowest orbitals of the H_2^+ quasimolecule for B = O and $B = B_o$ (Θ = O).

It is seen from Fig. 2 that the effect of the quadratic Zeeman term (i.e., the term in H_o that is quadratic in B) on the MO energies increases with decreasing binding energy of the zero-field MO. An interesting feature of the "dressed" energy curves is the occurrence of a real crossing of the $2\overset{*}{p}\sigma$ and $2\overset{*}{p}\pi$(m = -1) curves at R = 1.5 a.u. ("dressed" MO are marked here and in the following by an asterisk). In the zero-field case, the corresponding MO become degenerate only at R = O. The appearance of a $2\overset{*}{p}\sigma$ - $2\overset{*}{p}\pi$(m = -1) crossing at non-zero internuclear distance may possibly give rise to an enhancement of the probability for rotationally induced 2p excitation in H^+-H collisions at very small impact parameters.

RESONANT CHARGE TRANSFER IN SLOW H^+-H COLLISIONS IN THE PRESENCE OF A STRONG MAGNETIC FIELD

We have used the $1\overset{*}{s}\sigma$ and $2\overset{*}{p}\sigma$ MO energies in a calculation of probabilities and total cross sections for resonant charge transfer in slow H^+-H collisions. In the molecular two-state approximation, the transfer probability P(b,v) at impact parameter b and collision

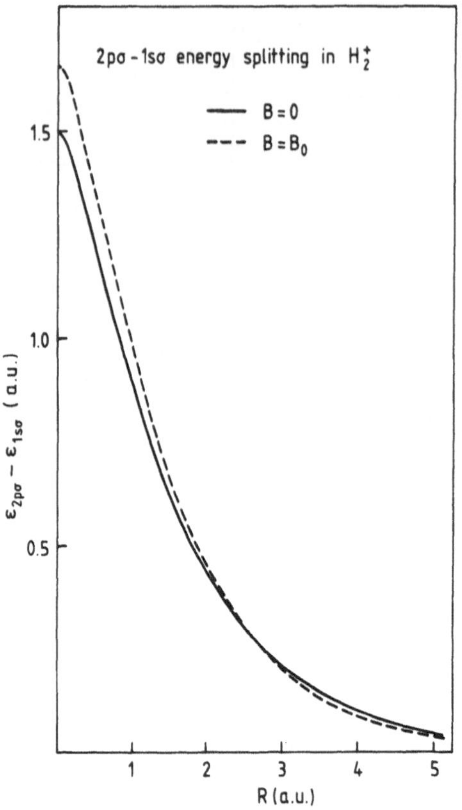

Fig. 3. Energy splitting between the 2pσ and 1sσ MO for B = O and B = B_o.

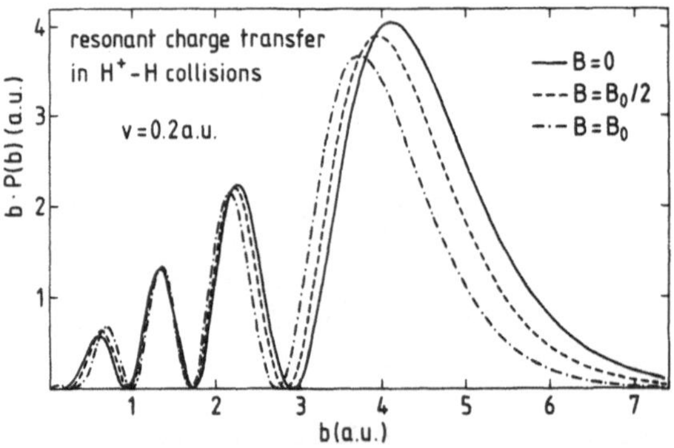

Fig. 4. Impact parameter dependence of the (linearly weighted) probability for resonant charge transfer in H^+-H collisions at a collision velocity of 0.2 a.u. for different values of the magnetic field strength.

velocity $\underline{v}$ is given[16] by

$$P(b,v) = \sin^2\left(\int_{-\infty}^{+\infty} dt \, \{\varepsilon_{2p\sigma}^{*}(R(t)) - \varepsilon_{1s\sigma}^{*}(R(t))\}\right) \qquad (5)$$

where the internuclear distance $R \equiv R(t;b,v)$ is now a given function of time. The R-dependence of the splitting between the $2p\sigma^*$ and $1s\sigma^*$ energies, which decides the phase difference between the MO that accrues during the course of the collision, is shown in Fig. 3.

Figure 4 displays the b-dependence of $b \cdot P(b)$, calculated for a straight-line internuclear trajectory at $v = 0.2$ a.u. The principal effect of the magnetic field on the transfer probability is a shift of the position of the main maximum towards smaller b-values, which monotonically increases with increasing field strength B. Since the main maximum is located in the range 3 a.u. $< b <$ 6 a.u., the shift is caused by the small decrease the $2p\sigma^*$-$1s\sigma^*$ energy splitting experiences in the corresponding R-range when the field strength increases (cf. Fig. 3).

Total cross sections for resonant charge transfer in slow H^+-H collisions as function of collision velocity are shown in Fig. 5. The monotonic decrease of the cross section (at fixed $\underline{v}$) with increasing field strength B reflects the behavior of the corresponding P(b) curves (cf. Fig. 4). As compared to the zero-field case, the average reduction of the cross section is about 10% for $B = B_o/2$ and about 20% for $B = B_o$.

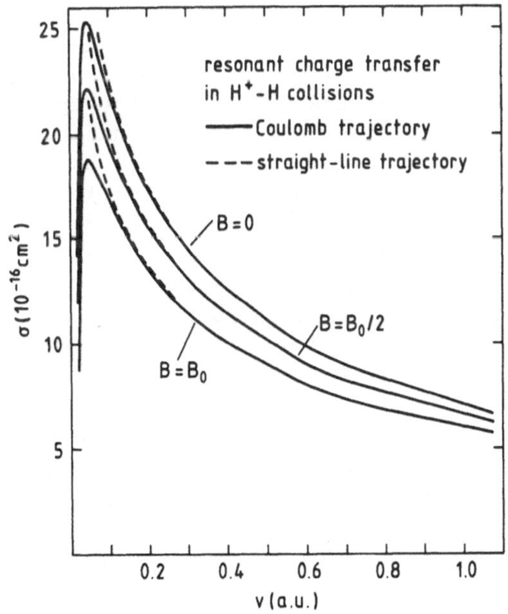

Fig. 5. Total cross section for resonant charge transfer in H^+-H collisions as function of collision velocity, calculated for different values of the magnetic field strength.

The calculations of Bivona et al.[11,12] on resonant charge transfer in H^+-H collisions, which employed an atomic two-state approximation, also showed a monotonic decrease of the cross section with increasing field strength. This decrease, however, appears to be considerably faster than the decrease observed in our results. At $v = 0.25$ a.u., for example, Bivona et al. have found a reduction of the cross section by about 35% when B changes from zero to $B_o/2$. They ascribe this reduction mainly to the appearance of a B-dependent relative phase between the initial and final atomic wavefunctions. This phase arises from the fact that the two wavefunctions are defined with respect to coordinate frames which are centered about different origins. No such phase appears in a molecular treatment since in this case all wavefunctions are defined with respect to a common frame of coordinates.

The classical-trajectory Monte Carlo calculations of Grosdanov and McDowell[13] on He^{2+}-H collisions in the velocity range $v \gtrsim 1$ a.u. have given an increase of the charge transfer cross section by about 20% when B changes from zero to $B_o/2$. The authors attribute this increase to enhanced electron capture into highly excited, field-distorted orbitals. It will be interesting, of course, to apply the classical-trajectory Monte Carlo method also to the case studied in the present work.

In assessing the reliability of the transfer probabilities and cross sections shown in Figs. 4 and 5, one has to consider the effect of the Θ-dependent terms in the full Hamiltonian (1). Assuming the internuclear line to be initially aligned along the field direction and taking a straight-line trajectory to describe the internuclear motion, one has $\sin\Theta = b/R$ and $\sin 2\Theta = 2\,b\,(R^2 - b^2)^{1/2}/R^2$. This shows that the quadratic terms proportional to $\sin^2\Theta$ and $\sin 2\Theta$ have average values much smaller than the average value of ρ^2, and hence are expected to have small influence only on the $2\overset{*}{p}\sigma$-$1\overset{*}{s}\sigma$ energy splitting. The crucial term is obviously the linear term proportional to $\sin\Theta\ l_x$, which will induce, in particular, a mixing of the $2\overset{*}{p}\sigma$ and $2\overset{*}{p}\pi\,(m = -1)$ MO. However, the effect of this mixing will be smaller than what one expects from looking merely at the energy curves of Fig. 2. At non-zero b-values, the linear Zeeman term proportional to $\cos\Theta\ \ell_z$ shifts the $2\overset{*}{p}\pi\,(m = -1)$ curve upwards. For b-values of about 4 a.u., where the transfer probability calculated by diagonalizing H_o maximizes, this shift results in a $2\overset{*}{p}\sigma - 2\overset{*}{p}\pi\,(m = -1)$ energy splitting of at least 0.5 a.u. for all R-values down to $R = b$.

The diagonalization of the full Hamiltonian (1) in the space spanned by the $2\overset{*}{p}\sigma$ and $2\overset{*}{p}\pi\,(m = -1)$ eigenstates of H_o will lead to a depression of the energy curve that originates from the $2\overset{*}{p}\sigma$ curve. This depression will enhance the field-induced reduction of the $2\overset{*}{p}\sigma$-$1\overset{*}{s}\sigma$ splitting in the range $R > 3$ a.u. Accordingly we expect from including the Θ-dependent terms in H into the calculation a further reduction of the cross section for resonant charge transfer.

CONCLUSIONS

The calculations presented in this paper constitute the very first step in a systematic treatment of slow ion atom collisions proceeding in the presence of a strong magnetic field. Future extensions and refinements will have to consider (i) the complete solution of the stationary one-electron problem for two Coulomb centers plus a strong magnetic field, (ii) the evaluation of dynamic couplings among "dressed" MO and the solution of time-dependent coupled-state equations, (iii) the treatment of processes involving outer shells where the effect of the magnetic field is comparatively larger than in the inner shells. The extension of the present calculations into the range $B >> B_0$ will, of course, require for the solution of the stationary problem a basis set that includes Landau-type orbitals.

ACKNOWLEDGMENT

The author is indebted to Professor M.R.C. McDowell for communicating results prior to publication.

REFERENCES

1. R.H. Garstang, Rep. Prog. Phys. 40, 105 (1977).
2. R.H. Garstang, J. Physique 43, C2-19 (1982).
3. J. Trümper, W. Pietsch, C. Reppin, B. Sacco, E. Kendziorra, and R. Staubert, Ann. NY Acad. Sci. 302, 538 (1977).
4. J. Trümper, W. Pietsch, C. Reppin, W. Voges, R. Staubert, and E. Kendziorra, Astrophys. J. 219, L105 (1978).
5. H. Friedrich, Phys. Rev. A26, 1827 (1982).
6. W. Rösner, G. Wunner, H. Herold, and H. Ruder, J. Phys. B: At. Mol. Phys. 17, 29 (1984).
7. J.M. Peek and J. Katriel, Phys. Rev. A21, 413 (1980).
8. G. Wunner, H. Herold, and H. Ruder, Phys. Lett. 88A, 344 (1982).
9. G. Wunner and H. Ruder, J. Physique 43, C2-137 (1982).
10. M.R.C. McDowell, J. Physique 43, C2-387 (1982).
11. S. Bivona, B. Spagnolo, and G. Ferrante, in Electronic and Atomic Collisions (Abstracts of XIII ICPEAC, Berlin 1983), eds. J. Eichler et al. (North-Holland, Amsterdam 1983), p. 692.
12. S. Bivona, B. Spagnolo, and G. Ferrante, J. Phys. B: At. Mol. Phys. 17, 1093 (1984).
13. T. Grosdanov and M.R.C. McDowell, Preprint (1984).
14. E. Hylleraas, Z. Physik 71, 739 (1931).
15. U. Wille and R. Hippler, Physics Reports, to be published.
16. H.S.W. Massey and R.A. Smith, Proc. Roy. Soc. (London) A142, 142 (1933).

CALCULATION OF CHARGE TRANSFER CROSS SECTION FOR ELECTRON CAPTURE COLLISIONS BY N^{2+} IONS IN NEUTRAL ATOMS

Sachchidanand Sharma

Bright Star University of Technology, Brega
PO Box 5858, Ajdabia
Libya

Permanent Address:
R D & D J College (Bhagalpur University)
Munger 811201
India

ABSTRACT

Bates and Moiseiwitsch method is used to calculate the charge transfer cross section for N^{2+} - He and Ne collisions at fixed impact energy (1.70 keV) and the calculated cross sections are compared with the measured charge transfer cross sections based on the measurements of energy loss spectra in electron capture processes.

INTRODUCTION

Electron transfer collisions between multiply charged ions and atoms of the type

$$A^{n+} + B \rightarrow A^{(n-1)+} + B^{+} + \Delta E \qquad (1)$$

represent a very important class of elementary collision processes and have been of considerable application in the field of plasmas, gaseous discharge, astrophysics, chemistry of upper atmosphere etc.

At low velocity $v \leqslant 1$ au, such collisions may take place very effectively in reactions of moderate exothermicity through pseudo-crossing of the adiabatic potential energy curves describing the initial and final molecular systems at internuclear separation R_x au given by

$$R_x \simeq (n-1)/\Delta E \qquad (2)$$

where ΔE is the energy defect for the reactions channel. So, selective capture may take place into a limited number of excited states of the $A^{(n-1)+}$ product ion.

From the difference between the energies, before and after electron capture collisions, of the projectile is obtained the energy loss/gain spectra [1,2,3,4]. Energy loss spectra of the product ions are relevant to the understanding of avoided-crossing and non-crossing mechanism, also to the application to the production and diagnosis of state selected ion beam. If E_1 and E_2 are the kinetic energies of A^{n+} and $A^{(n-1)+}$ ions, the energy defect ΔE corresponding to a particular collision channel characterising the initial state of A^{n+} and the final states of $A^{(n-1)+}$ and B^+ ions is given by

$$\Delta E \simeq E_1 - E_2 \qquad (3)$$

provided $\Delta E/E_1 \ll 1$ and the scattering angles are small.

In this communication we report on the calculation of cross sections for electron capture into particular channel using the methods of Bates and Moiseiwitsch and its comparison with the relative yield of the N^+ product for the reaction

$$N^{2+} + X \rightarrow N^+ + X^+ + \Delta E \qquad (4)$$

where X = He or Ne, using the principle of energy loss spectroscopy [1,2,3].

THEORY

The concept of pseudo-crossing of potential energy curves is utilised to construct an approximation within the framework of the two-state approximation. This method is not restricted to any particular velocity range, although it is most commonly used in the low velocity regime. The explanation of the Landau-Zener Formula [6,7] here is based on the articles by Bates[8], Moiseiwitsch[9] and Mapleton[10].

$$(A + e)_n + B \rightarrow A + (B + e)_m \qquad (5)$$

In the absence of interaction, the left and right hand members of equation (5) can be imagined to form two quasi-molecules, and Figure 1 shows the associated potential energy surfaces in the absence of interaction (the curves containing the dashed lines) between the two systems. In this hypothetical situation the curves cross at an internuclear distance $R = R_x$. If there is an interaction, these curves usually do not cross provided that the interaction matrix element does not vanish. In the region of pseudo-crossing R_x, the difference in the two potential energies is smallest, and a transition is possible, or electronic capture may occur. This transition

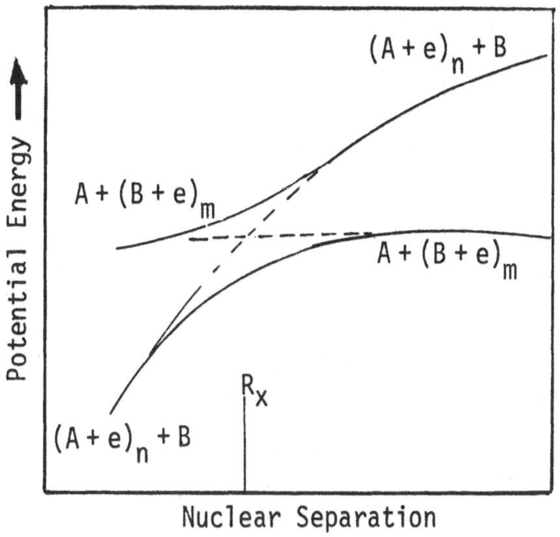

Fig. 1. Associated potential energy surfaces between the two systems of equation (5).

probability is given by

$$p_x = \exp(-w) \tag{6}$$

where

$$w = 2\pi/\hbar \; v(R_x) \cdot h^2_{nm}(R_x) \; / \; (h'_{mm} - h'_{mm}) \tag{7}$$

Here primes indicate differentiation with respect to $\underline{Z}$

$$\underline{Z} = \underline{R} \cdot \underline{v} \tag{8}$$

where $\underline{R}$ is the position vector of the nuclei and $\underline{v}$ the relative velocity of the nuclei, which is taken to be constant and to equal the value at the crossing (v_x) for non adiabatic collisions.

Thus p_x is the probability that in traversing the crossing the system remains on the h_{nn} (R) curve, and ($1-p_x$) is the probability that a jump from the h_{nn}(R) curve to h_{mm}(R) curve occurs. Since in actual collision a crossing must be traversed twice, the total probability that system which initially approaches the state n should finally recede in state m is

$$p = 2\,p_x\,(1 - p_x) \tag{9}$$

It is worth noting that Landau-Zener formula is based on the assumption that it is only in the immediate neighbourhood of the crossing point that there is an appreciable chance of a transition from one potential energy curve to the other.

Bates and Moiseiwitsch[2] have shown that the cross section for collision reaction of type (1) may be expressed as

$$\sigma = 4\pi R_x^2\, p_x\, I(\eta) \tag{10}$$

with $$I(\eta) = \int_1^{\alpha} \exp(-\eta x)\ [1 - \exp(-\eta x)]\ x^{-3}\ dx \tag{11}$$

$$\eta = 247\ (\mu/E_i)^{\frac{1}{2}}\ (n-1)\ (H_{12}(R_x)/\Delta E)^2 \tag{12}$$

Here,

E_i = initial kinetic energy of relative motion (eV)

μ = reduced mass of the whole system (amu)

ΔE = energy defect (eV)

n = initial charge on the projectile

$H_{12}(R_x)$ = interaction matrix element at the crossing point.

Moiseiwitsch[9] has calculated the integral I (η) and finds that it increases until the maximum value of 0.414 is reached when I (η) = 0.113; after that it decreases at higher balues of η. Figure 2 shows the variation of I (η) with log η^{-2}

Bates and Moiseiwitsch[5] have also calculated the energy separation $H_{12}(R_x)$ using the initial and final wave functions of the active electron. Apart from the analytical and computational methods, Olsol et al[11] have parametrized the dependence of the empirical H_{12} upon R_x, which is

$$H_{12}\ (R_x) = A \exp\ (-R_x/c) \tag{13}$$

Where $$A = I_1^{\frac{1}{2}}\ I_2^{\frac{1}{2}}\ R_x/c \tag{14}$$

$$c = 2/(\alpha+r) \tag{15}$$

$\alpha^2/2 = I_1$ = the effective ionization potential of the electron which is transferred from the reactants

$\gamma^2/2 = I_2$ = the effective ionization potential of this electron in its product state

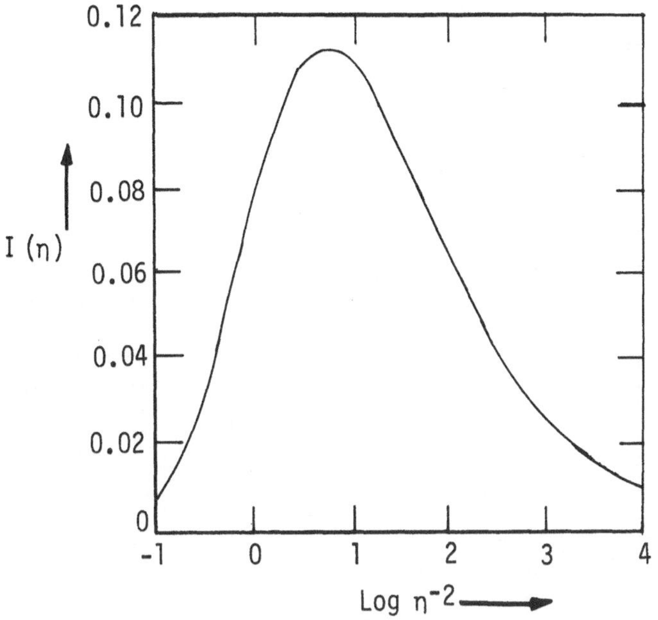

Fig. 2. Variation of $I(\eta)$ with $\log \eta^{-2}$

Here in equations for the interaction matrix element all quantities are in atomic units. The formula permits rapid but crude estimates of $H_{12}(R_x)$.

RESULTS

For easy identification of the observed channels [1,2,3,4] in energy loss spectroscopy measurements the scheme Sharma[12] has been adopted. The designation I is used to indicate the ground state of the incident N^{2+} ions, and II, III etc, any metastable state present while α, β, γ etc, represent the ground and higher states of N^{+} product ions. The ground and higher states of the X^{+} target product ions are designated by X, A, B etc. No special representation for the target has been assigned because it is supposed to be always in the ground state at room temperature. Bracketed design-

TABLE 1. Calculated cross section and measured yield for main collision channels in energy change spectra for one electron capture by N^{2+} in He and Ne at 1.70 keV impact energy.

Projectile	Target	Projectile Product	Target Product	ΔE (ev)	R_x (au)	Designation	H_{12} (au)	σ_{cal} (A^{02})	Projectile Product Yield (arb. Units) S1	L3
$N^{2+}(2p)\,^2P$	$He\,^1S$	$N^{+}(2p^2)\,^1S$	$He^{+}\,^2S$	0.95	28.64	IγX	1.20×10^{-16}	-	1.5	-
		$N^{+}(2p^2)\,^1D$		3.12	8.73	IβX	5.52×10^{-5}	4.79	7.8	1.7
		$N^{+}(2p^2)\,^3P$		5.01	5.43	IαX	3.59×10^{-3}	7.41	9.8	6.2
$N^{2+}(2p^2)\,^4P$		$N^{+}(2p^3)\,^5S$		6.24	4.36	IIδX	0.013	21.01	6	0.5
		$N^{+}(2p^2)\,^1S$		8.05	3.38	(IIγX)	0.038	12.91	-	-
		$N^{+}(2p^2)\,^1D$		10.20	2.67	(IIβX)	0.086	1.12	-	-
		$N^{+}(2p^2)\,^3P$		12.01	2.30	IIαX	0.129	-	-	-
$N^{2+}(2p)\,^2P$	$Ne\,^1S$	$N^{+}(2p^3)\,^5S$	$Ne^{+}\,^2P$	2.20	12.36	(IδX)	6.95×10^{-7}	-	8.3	-
		$N^{+}(2p^2)\,^1S$		4.00	6.86	IγX	7.58×10^{-4}	5.92	12	12.1%
		$N^{+}(2p^2)\,^1D$		6.15	4.56	IβX	0.011	22.98	16	71.6%
		$N^{+}(2p^2)\,^3P$		8.05	3.62	IαX	0.032	19.42	11.5	16.6%
$N^{2+}(2p^2)\,^4P$		$N^{+}(2p^3)\,^5S$		9.29	2.93	IIδX	0.068	0.65	7.4	-
		$N^{+}(2p^2)\,^1S$		11.09	2.45	(IIγX)	0.108	0.108	4.8	-

ations represent the cases of Wigner[13,14] - spin violation. The Wigner spin conservation rule requires that the total spin angular momentum of a pair of atoms or molecules does not change in course of collision. The resultant total spin quantum number is

$$S = |\underline{S}| = (s_1 + s_2), \ldots\ldots\ldots\ldots\ldots |(s_1 - s_2)| \qquad (16)$$

Table 1 shows the collision channels associated with energy charge spectra for one electron capture by N^{2+} ions in ground state (^2P) and metastable state (^4P) in He and Ne at 1.70 keV impact energy. Experimental results[1,3] shows the presence of metastable state in the primary beam of N^{2+} beam.

In N^{2+} - He collision system calculated cross section is maximum for IIδX channel with energy defect of 6.24 eV but measured yield for this channel is minimum may be because of small contamination of N^{2+} beam by the metastable state. Experiments show that in this case energy change spectrum is dominated by the IαX channel leading to ground state N^+ (^3P) ions which has a pseudo-crossing at R_x = 5.43 au. The IβX channel which involves a crossing at larger internuclear separation of 8.73 au also makes a contribution. The small contribution due to IγX channel has been observed by Sharma et al[1] and Sato and Moore[2] at 1.6 keV. The calculated cross section is negligible.

In N^{2+} - Ne collision system calculated cross section is maximum for IβX channel which involves a crossing at an internuclear separation R_x = 4.56 au. In the energy change spectra Sharma et al[1], Sato and Moore and Lennon et al[3] have found IβX as the dominant channel. Sharma et al[1] have also observed channels due to presence of metastable ions in the primary beam.

CONCLUSIONS

For the cases investigated, single electron capture has been found to take place through channels of moderate exothermicity. Channels involving curve crossing at R_x in the range 3 - 8 au appear to be the most effective. Most of the channels observed are in accord with the Wigner total spin conservation rule. Difference between the calculated cross section and measured yield in case of N^{2+} - He collision may be understood by measuring energy loss spectra for N^{2+} ions in metastable state in He.

ACKNOWLEDGEMENT

The author is indebted to Professor J B Hasted, Birkbeck College, University of London, England, for his interest in this work. The author is grateful to the Bright Star University of Technology, Brega, Libya and the NATO ASI for financial support.

REFERENCES

1. Sharma S, Awad G L, Hasted J B and Mathur D, J Phys B 12, L163 (1979)
2. Sato Y and Moore J H, Phys Rev A 19, 495 (1979)
3. Lennon M, McCullough R W and Gilbody H B, Phys B 16, 2191 (1983)
4. Kamber E Y, Mathur D and Hasted J B, J Phys B 15, 263 (1982)
5. Bates D R and Moiseiwitsch B L, Proc Phys Soc (London) A 67, 805 (1954)
6. Landau L D, Z Phys USSR 1, 88 (1932)
7. Zener C, Proc Phys Soc (London) A 137, 692 (1932)
8. Bates D R, Proc Roy Soc (London) A 257, 22 (1960)
9. Moiseiwitsch B L, "Meteors" Spl Suppl 2 to J Atmosph Terr Phys (1955)
10. Mapleton R A, Theory of Charge Exchange, Wiley Interscience, New York (1972)
11. Olson R E, Smith F T and Bauer E, Applied Optic 10, 1848 (1971)
12. Sharma S, PhD Thesis, University of London (1979)
13. Wigner E, Nachr Akad Wiss Göttingen, Math Physik, K1, IIa, 375 (1927)
14. Massey H S W and Burhop E H S, Electronic and Ionic Impact Phenomena, pg 427,522, Oxford Univ Press London (1952)
15. Moore C E, Atomic Energy Level NBS Circular No 467, 1, US Govt Printing Office (1971)

K-SHELL IONIZATION IN RELATIVISTIC HEAVY ION COLLISIONS*

U. Becker, S.R. Valluri, N. Grün and W. Scheid

Institut für Theoretische Physik der
Justus-Liebig-Universität
Giessen, West-Germany

Atomic collision experiments with relativistic heavy ions are already feasible today. Total K-shell ionization cross section measurements have been reported by Anholt et al.[1], who scattered ^{12}C projectiles at E_{lab}= 250 MeV/amu on different target atoms. Recently, Anholt[2] and collaborators have measured K-vacancy and L X-ray production cross sections of Uranium in collisions with relativistic (82 to 670 MeV/amu) projectiles from Ne to U.

In order to study the ionization of strongly bound K-shell electrons (U-target) by the electromagnetic fields of relativistically moving projectiles, we apply the semiclassical approximation (SCA) and first order perturbation theory in the calculation of the K-shell ionization probabilities[3]. Using the usual methods in the SCA[4] we obtain the following differential ionization probabilities from a filled shell, where the summation runs over the magnetic quantum numbers μ_i and μ_f of the initial and final substates, respectively,

$$\frac{dI_b^{fi}}{dE_f} = \sum_{\mu_i,\mu_f} |a_{fi}|^2, \tag{1}$$

where the transition amplitudes are given by

* Work supported by the GSI (Darmstadt) and BMFT

$$a_{fi} = \frac{8\pi i Z_p e^2}{\hbar v} \sum_{\ell,m} i^{\ell-m} \int_0^\infty ds \frac{s}{s^2-q^2\beta^2} \cdot B_{\ell m}(b,q,s) \cdot$$
$$\cdot \langle \psi_f | (1-\beta\alpha_3) j_\ell(sr) Y_{\ell m}(\hat{r}) | \psi_i \rangle \qquad (2)$$

and $q = (E_f - E_i)/\hbar v$, $\beta = \frac{v}{c}$.

The quantities $B_{\ell m}$ are the straight-line path factors[4]:

$$B_{\ell m}(b,q,s) = \Theta(s-q) Y_{\ell m}(\arccos(q/s), 0) J_m(b(s^2-q^2)^{1/2}) \qquad (3)$$

In this formalism the nuclear motion of the projectile with charge $Z_p e$ is classically described by a straight-line path in z-direction with velocity v and impact parameter b. The target atom is set at the origin of the coordinate system and recoil effects are disregarded. The retarded potential of the projectile is taken in Lorentz-gauge in the rest frame of the target nucleus. The initial and final electronic states are described by hydrogenic Dirac-wave functions. In order to calculate the matrix elements in Eq.(2), one separates the radial and angular integration using the standard representation of Dirac-wave functions[5]. The angular integrals are carried out by angular momentum algebra[3]. Also the radial integrals can be carried out analytically[6,7,8] and expressed in finite sums over hypergeometric functions[3]. Finally the integration over s in Eq.(2) is done numerically by quadrature techniques.

Within this formalism we calculated the ionization probabilities for the K-shell of Uranium occupied by two electrons (charge number of the target Z=92). The results are presented in figures 1-4. Since the probabilities are proportional to the square of the projectile charge Z_p (see Eq.(2)), we scale the probabilities in units of Z_p^2. Fig.1 presents the ionization probabilities for a transition into the $s_{1/2}$- and $p_{3/2}$-positive continuum for β=0.99 ($\Gamma=(1-\beta^2)^{-1/2}=7.1$) as function of the impact parameter (full curves). In order to study the contributions of the retarded Coulomb and spin-dependent potentials to the transition amplitudes, which are given by the two matrix elements in Eq.(2) ($\sim 1/r, \sim -\beta\alpha_3/r$), we have calculated the ionization probabilities without the $\beta\alpha_3$-term separately. These probabilities are shown by the dashed curves in Fig. 1 and demonstrate the importance of the spin-dependent contributions, which interfere destructively with those of the Coulomb potential. We em-

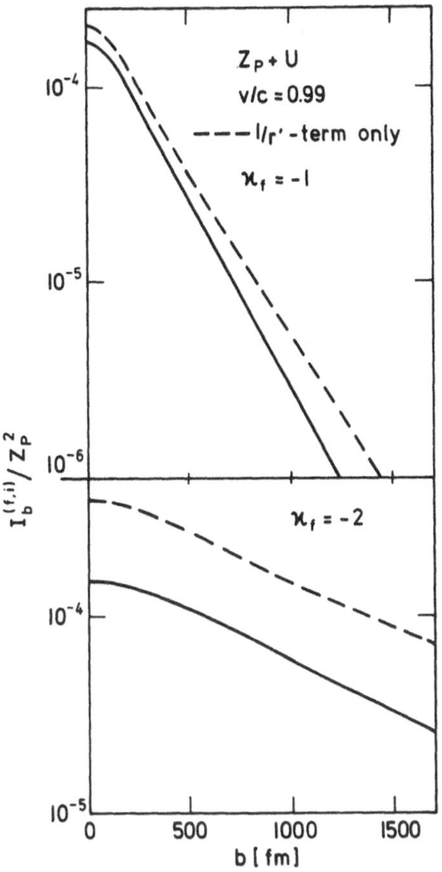

Fig.1: K-shell ionization probabilities for the transition to the $s_{1/2}$ states (κ_f=-1, upper part of the figure) and $p_{3/2}$ states (κ_f=-2, lower part of the figure) of the positive continuum as function of the impact parameter b for the collision system Z_p+U at β=0.99 (Z_p=charge of the projectile). The full curves represent the total probabilities. The dashed curves are calculated without the $\beta\alpha_3$-term in Eq.(2).

phasize that the retarded Coulomb and spin-dependent potentials in Eq.(2) are written in Lorentz-gauge. Therefore, our conclusions about these transition potentials are referred to the case of Lorentz-gauge. A valuable discussion of problems occuring in using the Lorentz- or Coulomb-gauges is given by Amundsen and Aashamar[4].

Fig.2a shows the contributions of the transitions to the various continua for zero impact parameter as a function of the incident energy or relative velocity. For β<0.9 the dominant transitions are the ones to the $s_{1/2}(\kappa_f$=-1)- and $p_{3/2}(\kappa_f$=-2)-continua. The transition

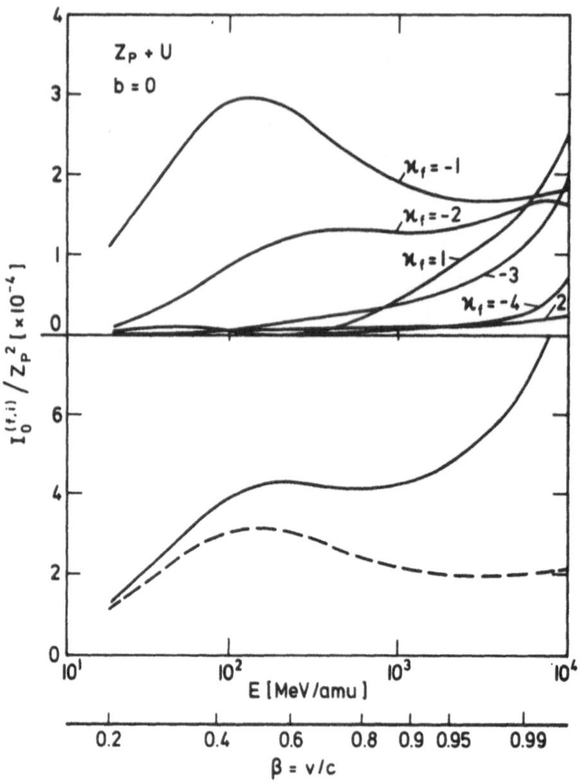

Fig.2: The K-shell ionization probabilities for the collision system Z_p+U at zero impact parameter as a function of the energy or velocity of the projectile.

a) Probabilities for transitions to various continua denoted by κ_f.

b) Total probability summed over κ_f (full curve). The dashed curve shows the contribution of the monopole Coulomb term.

probability to the $s_{1/2}$-continuum has a maximum at $\beta \simeq 0.5$. This maximum can also be recognized in Fig.2b which shows the ionization probability summed over all final states for zero impact parameter. The dashed curve in this figure represents the contributions of the monopole term of the Coulomb potential without spin-dependent contributions. This partial probability has a nearly constant value for energies E_{lab}=1-100 GeV/amu, whereas the total probability increases in the considered range of projectile velocities.

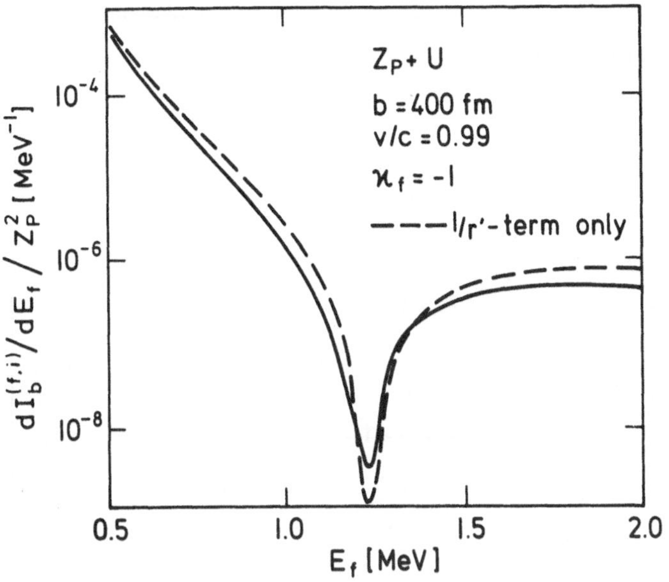

Fig.3: The differential K-shell ionization probabilities for the transition to the positive $s_{1/2}$-continuum ($\kappa_f=-1$) for the collision system Z_p+U at b=400 fm and $\beta=0.99$ as function of E_f (full curve). The dashed curve is calculated without the $\beta\alpha_3$-term.

A closer inspection for certain ranges of the impact parameters and relative velocities reveals interesting features in the differential ionization probabilities. In Figs.3 and 4 we present two types of structures. Fig.3 shows the differential ionization probability to the $s_{1/2}$-continuum ($\kappa_f=-1$) for b=400 fm and $\beta=0.99$. In this case the contributions from the spin-dependent term are small, as one recognizes by the small differences between the solid and dashed curves. The minimum of the curves at E_f=1.22 MeV, arises due to a cancellation in the integration over s:

$$\int_0^\infty \frac{sds}{(s^2-(q\beta)^2)} B_{00}(b,q,s) \langle \psi_f | j_0(sr) Y_{00}(\hat{r}) | \psi_i \rangle. \qquad (4)$$

The differential ionization probability to the $p_{1/2}$-continuum ($\kappa_f=1$) is shown in Fig.4 for zero impact parameter and $\beta=0.6$. Up to E_f=2 MeV this probability has minima at E_f=0.65 and 1.98 MeV. The minima are caused by a destructive interference of the two terms in the matrix element of Eq.(2). If the spin-dependent contributions are neglected, a smooth differential probability is obtained as shown by the dashed curve in Fig.4.

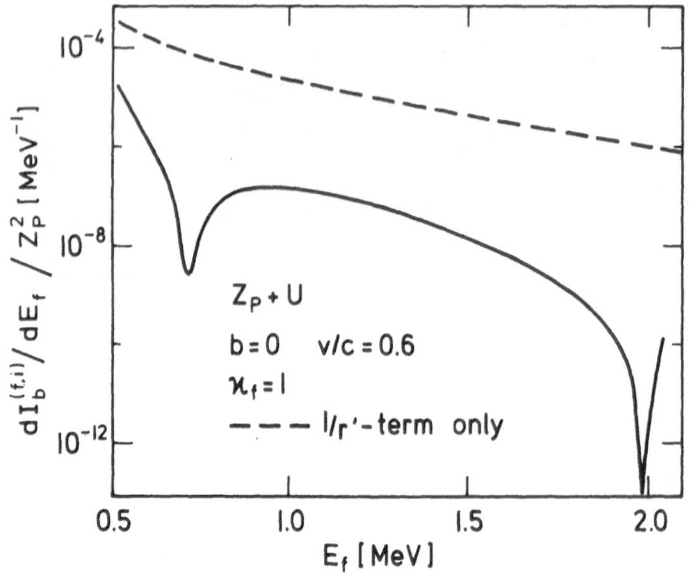

Fig.4: The differential K-shell ionization probabilities for the transition to the positive $p_{1/2}$-continuum (κ_f=1) for the collision system Z_p+U at zero impact parameter and β=0.6 as function of E_f (full curve). The dashed curve is calculated without the $\beta\alpha_3$-term.

If the differential probabilities are summed over all final states for fixed impact parameter, the total differential probabilities are almost smooth, because the partial probabilities having structures are small and give only a very fine undulation in the total probabilities.

As an application of our method to nonrelativistic collisions, we calculated the K-shell ionization probabilities for the scattering of 15 MeV protons and 50 MeV α-particles on ^{208}Pb for fixed impact parameters[9]. In order to take the Coulomb deflection of the projectile in an approximate manner into account, the straight-line trajectories have been modified according to the procedure of Kocbach[10]. In these calculations we used an experimental binding energy of 88 keV for the K-shell electron. The resulting probabilities are compared with those obtained by Trautmann and Baur[11], who used the SCA with Rutherford trajectories and relativistic Hartree-Fock-Slater wave functions, and with the experimental data of van Dijk et al.[12]. As shown in Table 1, there is agreement between our calculations, the calculations of Trautmann and Baur[11] and the experimental data.

Table 1: K-shell ionization probabilities for the collision of 15 MeV protons and 50 MeV alpha-particles on ^{208}Pb for impact parameters 38 fm and 23 fm, respectively.

method	K-shell ionization probability ($\times 10^{-4}$)	
	$p+^{208}Pb$	$\alpha+^{208}Pb$
experiment (van Dijk et al.[12])	1.56±0.1	5.3±0.4
theory (Trautmann and Baur[11])	1.73	5.68
present work	1.71	5.61

From our calculations we conclude, that for relative velocities v/c>0.5 (E_{lab}>150 MeV/amu) the spin-dependent interaction becomes increasingly important and leads to significant effects in the ionization probabilities. Total ionization cross sections are needed for a direct comparison with the experiments of Anholt[2]; but then an additional integration over the impact parameter has to be performed. As pointed out by Amundsen and Aashamar[4], the total cross sections would result identically to those obtained by PWBA calculations[13]. However, the PWBA calculations are carried out up to now only with semi-relativistic wave function, which are less valid for heavy target atoms as Pb or U. PWBA calculations of total cross sections with fully relativistic Dirac-wave functions are in progress.

References

1. R. Anholt, J. Ioannou-Yannou, H. Bowman, E. Rauscher, S. Nagamiya and J.O. Rasmussen, Phys.Lett. 59A: 429 (1977)
2. R. Anholt, W.E. Meyerhof, C. Stoller, E. Morenzoni, S.A. Andriamonje, J.D. Molitoris, O.K. Baker, D.H. H. Hoffmann, H. Bowman, J.S. Xu, Z.Z. Xu, K. Frankel, D. Murphy, K. Crowe and J.O. Rasmussen, Atomic Collisions with Relativistic Heavy Ions I: Target Inner-shell Ionization, Preprint submitted to Phys. Rev. A (1984).
3. S.R. Valluri, U. Becker, N. Grün and W. Scheid, J. Phys.B: At.Mol.Phys. 17 (1984)

4. P.A. Amundsen and K. Aashamar, J.Phys.B: At.Mol.Phys. 14: 4047 (1981)
5. E.M. Rose, "Relativistic Electron Theory", Wiley, New York (1961)
6. P.A. Amundsen, J.Phys.B: At.Mol.Phys. 11: 3197 (1978)
7. D.H. Jakubaßa-Amundsen, Phys.Scr. 26: 319 (1982)
8. D. Trautmann, G. Baur and F. Rösel, J.Phys.B: At.Mol. Phys. 16: 3005 (1983)
9. S.R. Valluri, U. Becker, N. Grün and W. Scheid, to be published in J.Phys.B: At.Mol.Phys.
10. L. Kocbach, Physica Norwegica 8: 187 (1976)
11. D. Trautmann and G. Baur, priv. communic. as quoted by van Dijk et al.[12]
12. J.H. van Dijk, H.W. Wilschut, A.G. Drentje and A. van der Woude, Z.Phys. A314: 1 (1983)
13. R. Anholt, Phys.Rev. A19: 1004 (1979)

EXCITATION OF HE I TRIPLET LEVELS BY ION IMPACT

A. S. Aynacioglu and G. von Oppen

Institut fuer Strahlungs- und Kernphysik
Technische Universitaet Berlin
Rondellstr. 5, D-1000 Berlin 37

INTRODUCTION

According to Wigner's spin conservation rule[1,2] impact excitation of He I triplet states is forbidden in collisions of bare nuclei with He atoms. Other projectiles containing electrons usually may induce intercombination transitions from the 1^1S_o ground state to triplet states via exchange of electrons with opposite spins. The cross sections of these exchange processes are usually largest where the velocity v_P of the projectile is of the order of v_B = 1 a.u. and fall off very rapidly with increasing energy.

Both the validity of Wigner's rule and the cross sections of electron transfer in simple collision systems, e.g. in the three electron systems H-He, H_2^+-He or He^+-He, are of basic interest in atomic collision physics. Accordingly, collisional excitation of He atoms by simple atomic and ionic projectiles has been studied extensively[3,4,5]. At high energies where $v_P \gg v_B$, the experimental results are in general accord with theoretical expectations. Not only for proton impact, triplet lines are not observed, but also for molecular hydrogen ions and hydrogen atoms, the excitation cross sections of triplet levels become vanishingly small. Furthermore, in the singlet system, the optically allowed $1s^2\ ^1S$ - $1snp\ ^1P$ transitions are predominantly induced.[6,7]

The situation is different at medium energies where v_P v_B. Collisions involving electron exchange become probable and orbital angular momenta 1 ≠ 1 are likely to be transferred during the excitation process. As a consequence, excitation of triplet states does occur as expected whenever the projectile contains electrons. However, excitation of triplet lines has also been observed in experiments on p-He collisions.[4] Usually, these excitation processes are attributed to the occurence of multiple collisions of the projectiles. The large cross sections for exchange processes favor the formation of hydrogen atoms which may be responsible for the excitation of triplet states in a subsequent collision.

However, there is another mechanism allowing the population of triplet states in a single p-He collision, which has not yet been investigated thoroughly. Because at medium energies, also angular momenta $l > 1$ are likely to be transferred to the He atom, a considerable amount of atoms may be excited to 1snf states or to states with even higher angular momenta. These states are known to deviate strongly from the Russel-Saunders coupling case. Only S, P and D levels of He I can be considered as pure singlet or triplet levels. The mixing coefficients[8a,b] of these levels can be assumed to be $\Omega^2 \lesssim 0.01$ %. On the contrary, the mixing coefficients of 1snl levels with angular momenta $l \geqslant 3$ have been calculated[8a,b] to be of the order of 1*. Consequently, in addition to electron exchange, the following three-step excitation mechanism can contribute to the excitation of He I triplet levels:

(i) During the collision the He atom is excited to a pure singlet state of a 1snl configuration with $l \geqslant 3$. This state is usually not an eigenstate of the atom.
(ii) The excited state evolves according to the relevant effective hamiltonian. Owing to spin-orbit coupling, there results a mixed state with singlet and triplet components.

(iii) The excited state decays into energetically lower triplet levels, e.g. into the 3^3D level.

Since the relevant cascade transitions are in the infrared, these contributions are usually not analyzed and may lead to misinterpretation if not properly taken into account. In the case of excitation of Helium by bombardment with bare nuclei, these contributions simulate a violation of Wigner's rule. Actually, the results obtained by Myers et al.[10] for excitation of the 3^3D levels in He^{++}-He collisions may have been caused by cascade processes via $l \geqslant 3$ levels. In case of excitation with other projectiles, both electron exchange and cascade processes may contribute significantly to the excitation of He triplet levels, in particular when the lowest ^3L levels, i.e. 2^3S, 2^3P and 3^3D are considered. Therefore, both contributions should be observed separately before comparing experimental and theoretical data. A separation is even

---

* For the $l = 3$ levels the mixing coefficients are probably somewhat lower. For the $1s4f$ multiplet calculated values are $\Omega^2 = (0.4335)^2$ [8a] and $(0.593)^2$ [8b], whereas Aynacioglu et al.[9] deduced from measured Stark constants $\Omega^2 = 0.8$.

more worthwile because the cross sections of both processes are expected to exhibit a similar energy dependence and thus may influence measurements over a wide energy range.

In the following we describe experimental techniques allowing such a separation. First experiments have been performed on the excitation of the 3^3D level of He I by hydrogenic ions accelerated to 5 - 20 keV. For detection we observed the 588 nm line corresponding to the transition $1s3d^3D \rightarrow 1s2p^3P$. Two techniques have been applied to identify the cascade contribution: (i) magnetic depolarization and (ii) electric demixing of singlet and triplet states.

By investigating the magnetic depolarization of impact radiation one may decompose its polarization fraction into one component resulting from direct excitation of the 3^3D level and another one resulting from cascade excitation.[11] By applying electric fields, it becomes possible to analyze also the intensity of the impact radiation. Electric fields of a few kV/cm are sufficient to reduce the singlet-triplet mixing in $l \geqslant 3$ levels considerably. Without singlet-triplet mixing, triplet levels can be excited only by collisions involving electron exchange.

EXPERIMENTAL CONDITIONS

The measurements have been performed by crossing a mass selected beam of hydrogenic ions (H^+, H_2^+, H_3^+ or D^+, D_2^+, D_3^+) at energies of some 10 keV with a Helium atomic beam (fig. 1). In the

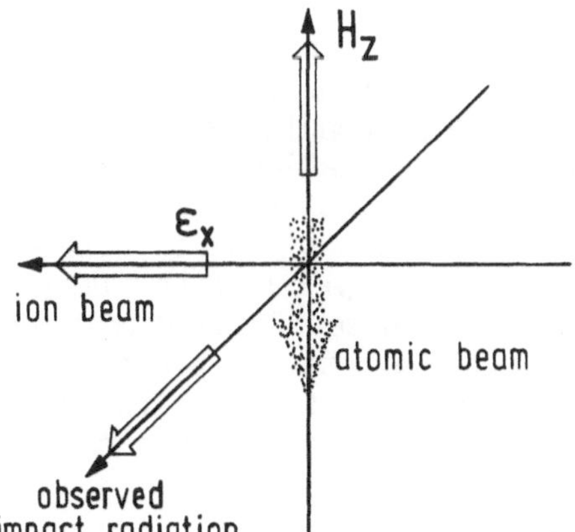

Fig. 1: Scheme of the experimental set up

crossing region, a magnetic field $H_z \lesssim 10$ mT perpendicular to the ion beam could be applied using a pair of Helmholtz coils. Parallel to the ion beam an electric field $E_x \lesssim 5$ kV/cm could be produced between a pair of cylindrical electrodes. The field E_x changes the energy of the ion beam in the target region. However no significant changes of the observed signals occured when switching from accelerating to decelerating fields. Thus the assumption may be justified that the energy change is of minor importance. The impact radiation emitted into a direction perpendicular to the ion beam and to the magnetic field has been observed with a photomultiplier after wave-length selection with an interference filter or with a grating monochromator. During the experiments, care had to be taken to avoid charge exchange collisions between the ions and the target and residual gas atoms. Fig. 2 shows impact radiation spectra induced by a 15 keV deuteron beam. According to Wigner's rule, no excitation of He I triplet lines is expected. Instead, experimentally we observed not only the singlet transitions (3^1S-2^1P) and (3^1D-2^1P) but also the corresponding triplet transitions, with considerable intensity. However, while the intensity ratio $I(3^3S-2^3P)/I(3^1S-2^1P)$ decreases to zero with decreasing intensity of the He atomic beam, the intensity ratio $I(3^3D-2^3P)/I(3^1D-2^1P)$ remains fairly constant, and even for the lowest atomic beam intensity, there is still a significant intensity of the triplet line.

The decrease of the first intensity ratio makes it obvious that at the higher atomic beam intensities a considerable fraction of the deuteron beam is transformed into neutral D atoms by charge exchange collisions. These atoms excite He triplet states in a secondary collision. However, owing to smaller cross sections, this excitation process seems to be less important with regard to the ($^3D-^3P$)-line. Rather, the constancy of the second intensity ratio indicates that the 3^3D level is indeed excited by deuterons.

MAGNETIC DEPOLARIZATION

Magnetic depolarization techniques have been used recently to investigate ion-impact excitation of He I singlet levels.[11,12] A magnetic field is applied perpendicular to the ion beam and the intensity $I_\parallel$ of the impact radiation polarized parallel to the ion beam is measured as a function of the magnetic-field strength. The amplitude of the depolarization signal is related to the polarization fraction $P = (I_\parallel - I_\perp)/(I_\parallel + I_\perp)$ of the impact radiation at zero-magnetic field.[11] The shape of the depolarization signal depends sensitively on the excitation process. In the special case of direct excitation of a singlet level, the depolarization signal of the decay line is Lorentzian shaped and has a half width (FWHM)

$$(1) \qquad \Delta H_{1/2} = \frac{\hbar}{g_J \mu_B \tau}$$

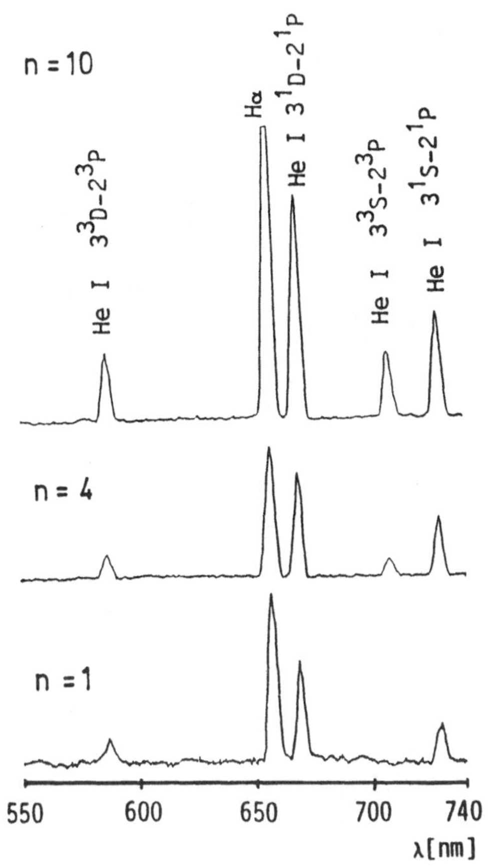

Fig. 2: Spectra of impact radiation induced by 15 keV $D^+ \rightarrow$ He collisions measured at different atomic beam densities n [arb.units.]

where τ is the radiative lifetime of the level and g_J its Landé factor (μ_B Bohr magneton, $\hbar$ Planck's constant). In the case of triplet levels, depolarization signals have a somewhat more complicated structure, even if only direct excitation is involved. Firstly, for unresolved fine structure multiplets, there are several upper levels with different g_J values.

Secondly, the shape of the signal may be affected by magnetic decoupling of angular momenta and by level crossing signals at

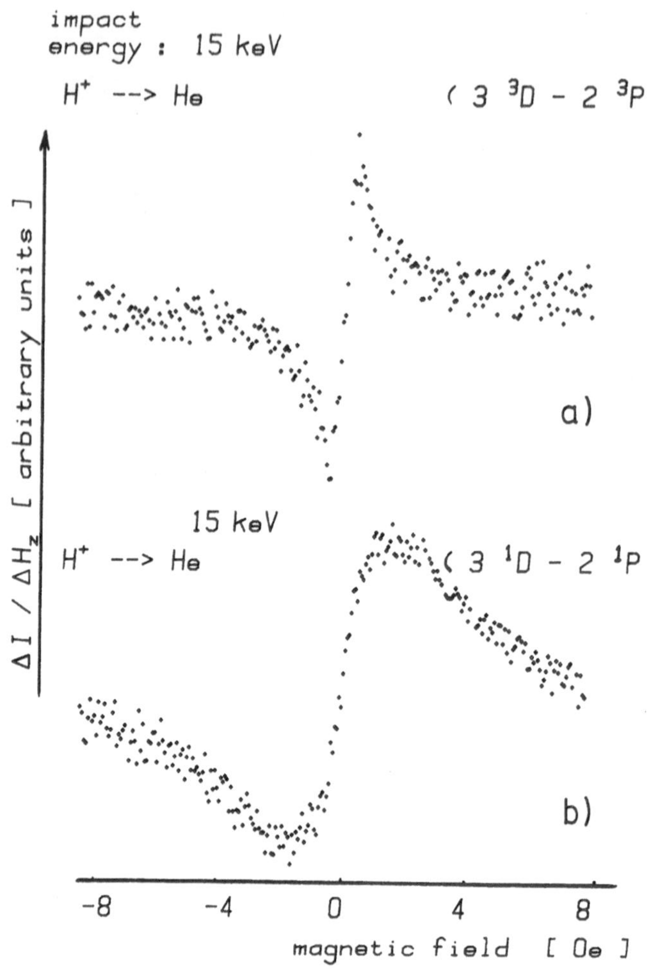

Fig. 3: Magnetic depolarization signals of the 588 nm (above) and the 668 nm line (below) excited by 15 keV proton impact (differential signals due to magnetic field modulation and lock-in detection)

non-zero magnetic fields. Nevertheless, magnetic depolarization signals arising from direct excitation can usually be fitted approximately to a Lorentz curve with a half width determined by (1) using an appropriate average g_J value. Significant deviations from the Lorentz shape arise from cascade excitation through levels with radiative lifetimes $\tau_{casc} \gg \tau$.[11]

In the following we discuss magnetic depolarization measurements on the $\lambda(3^3D-2^3P)$ = 588 nm line. Cascade excitation is expected to take place mainly via levels with $l \geqslant 3$. The radiative lifetimes $\tau(l \geqslant 3) \geqslant \tau(1s4f)$ = 74(2) ns [9] exceed the lifetime $\tau(3^3D) \approx$ 13 ns by more than a factor of 5. Owing to this

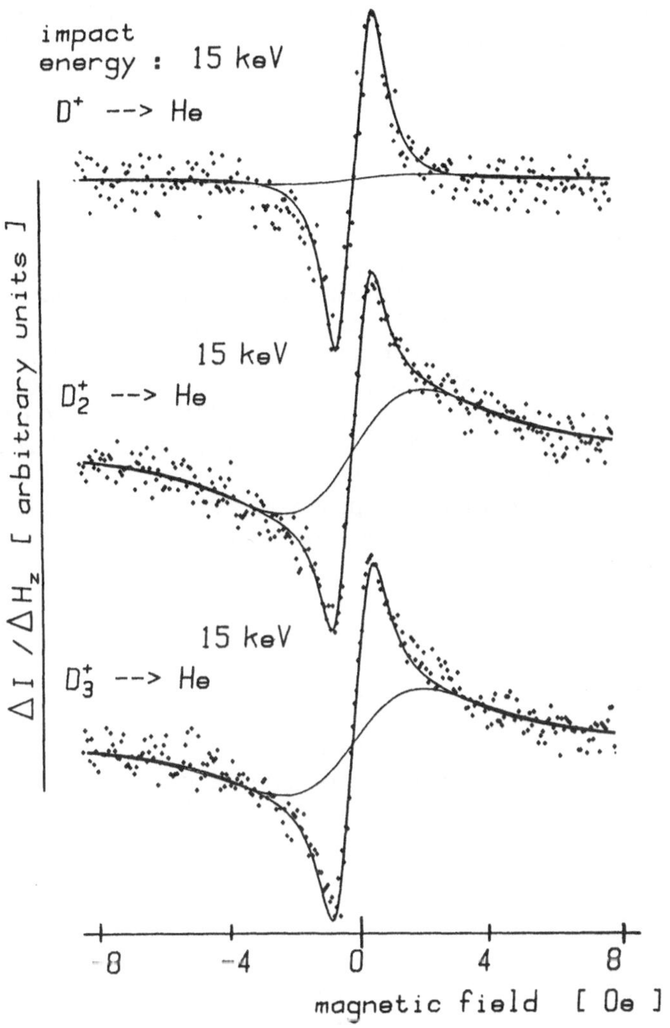

Fig. 4: Magnetic depolarization signals of the 588 nm (3^3D - 2^3P) line of He I for deuteron, D_2^+ and D_3^+ impact. Solid lines: decomposition of the signals into a broad component (direct excitation) and a narrow component (cascade excitation) obtained using a least squares fitting routine

difference, the observed depolarization signals of the 588 nm line (fig. 3 and 4) can be decomposed well into a broad component related to direct excitation and a narrow component related to cascade excitation.

Depolarization signals of the 588 nm line observed after proton or deuteron excitation (fig. 3a) can be fit within the limits of error by a curve representing a pure cascade signal. The shape of these signals is in strong contrast to depolarization signals (fig. 3b) of the corresponding singlet line at 668 nm which is mainly excited directly. The pure shape of the 588 nm signals strongly supports the expectation that, in agreement with Wigner's rule, the 3^3D level is only excited via cascade through $l \geq 3$ levels.

In fig. 4 depolarization signals obtained with atomic and molecular hydrogen ions are compared. In contrast to the signals observed with atomic ions, the signals obtained for molecular-ion impact have both a broad and a narrow component, which indicates that both direct and cascade excitation contribute to the polarization fraction of the 588 nm line. With regard to a comparison of experimental data with theoretical results for electron exchange processes, it is highly desirable to analyze not only the polarization fraction but also the intensity of the impact radiation with respect to the excitation process. Such an analysis can be performed by electric-field demixing techniques discussed in the following section.

ELECTRIC-FIELD DEMIXING

By applying electric fields of some kV/cm to He atoms, the singlet-triplet mixing in $l \geq 3$ levels can be removed. In particular, we consider the 1s4f multiplet, which is expected to be the most important one with respect to cascade excitation. At zero-electric field there are mixed eigenstates for the $1s4f\,^1F_3$ and the $1s4f\,^3F_3$ level, whereas the substates of the $1s4f\,^3F_{2,4}$ levels are pure triplet states. The strong singlet-triplet mixing in $l \geq 3$ levels is due to the fact that in these levels the value of the exchange integral is comparable or even smaller than the spin-orbit coupling. The mixing can be reduced by increasing the exchange integral. Such an increase can be accomplished by applying an electric field that couples $l \geq 3$ states with $l \leq 2$ states, the latter having exchange integrals larger by several orders of magnitude than those of the former.

For a more detailed discussion we refer to Fig. 5, which shows the Stark splitting of the 1s4f multiplet in electric fields of a few kV/cm. The Stark shifts of the 1s4f sublevels are almost exclusively caused by the interaction with the neighbouring 1s4d levels. The energy spacings $\Delta E_1 \approx 5.4$ cm^{-1} and $\Delta E_3 \approx 7.4$ cm^{-1} between the 1s4f multiplet and the 4^1D and 4^3D, respectively, are significantly different owing to the relatively much larger exchange integral of the 1s4d configuration. As a consequence, the $1s4f\,^1F$ states are shifted more than the $1s4f\,^3F$ states. At electric-field strengths of about 2 kV/cm, the energy splitting between corresponding singlet and triplet 1s4f states exceeds the splitting caused by spin-orbit coupling. Accordingly, the singlet-triplet mixing decreases when the electric field is increased significantly beyond 1 kV/cm. The demixing starts at even lower electric fields for the energetically higher $l > 3$ levels.

Actually, only 1snl substates with Zeeman quantum numbers $|m_l| \leq 2$ are demixed. Substates with $|m_l| \geq 3$ do not interact with any substates of $l \leq 2$ levels. Therefore, even with an electric field, some cascade excitation of the 3^3D level is still possible, in particular via the $|m_l| = 3$ substates of the 1s4f multiplet. Nevertheless, the excitation cross section should be reduced by about a factor 2/7 by applying an electric field of several kV/cm.

Fig. 6 shows some preliminary experimental results. The intensity of the 588 nm line excited by 15 keV-deuteron impact has been observed at different strengths of the electric field parallel to the ion beam. As expected, the intensity starts to fall off

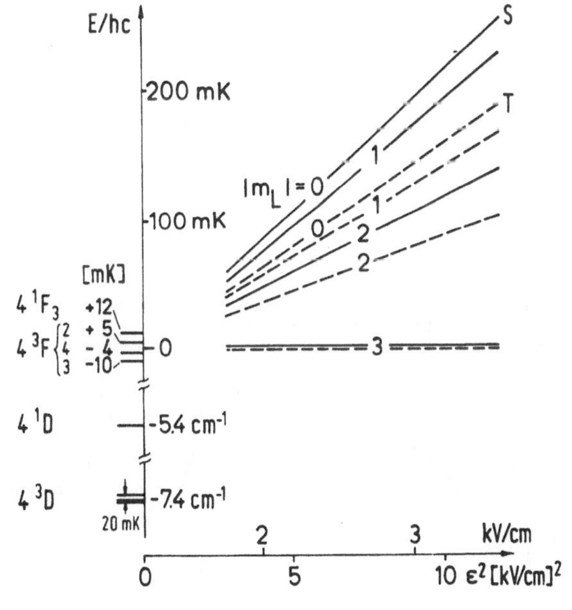

Fig. 5: Electric field splitting scheme of the 1s4f multiplet, 1 mK = 10^{-3} cm^{-1}; (solid lines: singlet substates; dashed lines: triplet substates)

at about 2 kV/cm and decreases to about 1/3 of its initial value. This result is in agreement with the above discussion and once more supports the assumption that the 3^3D level is excited only via cascades through $l \geq 3$ levels.

The electric-field demixing technique has so far not been applied to impact excitation with molecular hydrogen ions. A quantitative evaluation of the signals should in this case enable a decomposition of the excitation cross section. We note that the demixing technique actually does not decompose the cross section into direct and cascade contributions. Rather this technique allows a decomposition into contributions of processes involving electron exchange, regardless of whether the 3^3D level is excited directly or via cascades, and into those contributions based on postcollisional mixing of singlet and triplet states.

CONCLUDING REMARKS

The experiments performed so far demonstrate that both techniques, magnetic depolarization and electric-field demixing are valuable means for analyzing the excitation of He I triplet states and may be helpful for carrying out more accurate investigations of electron exchange processes. In the case of excitation of He I triplet states by impact of bare nuclei, our measurements - though a quantitative evaluation has still to be performed - demonstrate already that the observed excitation processes are not in contradiction to Wigner's spin conservation rule. These processes are, conversely, enabled by the strong singlet-triplet mixing of $l \geq 3$ levels.

A corresponding influence of this singlet-triplet mixing on thermal collisions between He atoms has been well known since the

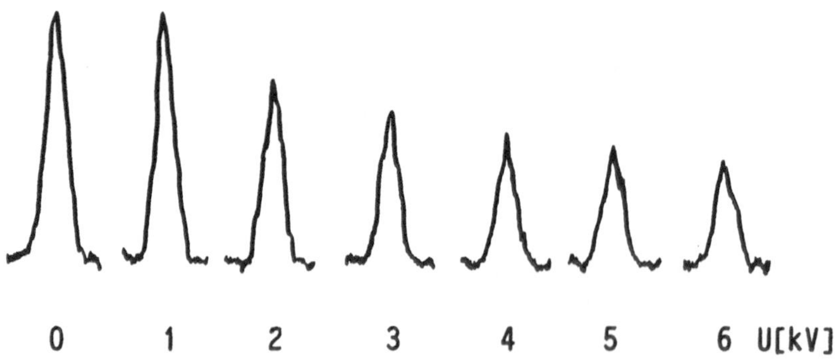

Fig. 6: Intensity of the impact radiation $(1s3d^3D \rightarrow 1s2p^3P)$=588nm at different strengths of an axially applied electric field E_x for 15 keV D^+-He collisions (E_x[kV/cm] ≈ U[kV]/1.5)

1960's.[2] As already observed by Lees and Skinner[13] in 1932, excitation energy of 1snp $^1P_1$ states of He I can be transferred in thermal He*-He collisions to the 3^3D level. The apparent violation of Wigner's rule by this process has puzzled physicists for a long time. The process has finally been explained by St. John and Fowler[14] who pointed out that primarily the excitation energy is transferred to 1snf levels which decay into singlet as well as triplet 1snd levels.

References

1. E.Wigner, Nach. Akad. Wiss. Goettingen, 375 (1927)
2. H.S.W.Massey, E.H.S.Burhop, H.B.Gilbody, Electronic and Ionic Impact Phenomena Vol. III p.1756 2nd Ed., Oxford, At the Clarendon Press 1971
3. H.S.W.Massey, E.H.S.Burhop, H.B.Gilbody, Electronic and Ionic Impact Phenomena, Vol.IV,p.3048ff, 2nd Ed., Oxford At the Claredon Press 1974
4. E.W.Thomas, Excitation in Heavy Particle Collisions, Wiley Intersience, 1972 New York
5. J.T.Park, Adv.At.Mol.Phys. 19, 67 (1983)
6. D.Hasselkamp, R.Hippler, A.Scharmann, K.-H.Schartner, Z.Phys. 248, 254 (1971)
7. R.Hippler, K.-H.Schartner, J.Phys.B 7, 618 (1974)
8. a) R.M.Parish, R.W.Mires, Phys.Rev.A4, 2145 (1971)
   b) R.K.van den Eynde, G.Wiebes, Th.Niemeyer, Physica 59, 401 (1972);
9. A.S.Aynacioglu, G.v.Oppen, W.-D.Perschmann, D.Szostak, Z.Phys.A 303, 97 (1981)
10. G.D.Myers, J.G.Ambrose, P.B.James, J.J.Leventhal, Phys.Rev.A18, 85 (1978)
11. G.v.Oppen: Comments At.Mol.Phys. 15, 87 (1984)
12. A.S.Aynacioglu, G.v.Oppen, W.-D.Perschmann, D.Szostak, J.Phys.B 14, 2611 (1981)
13. J.H.Lees, H.W.B.Skinner, Proc.Roy.Soc.A 137, 186 (1932)
14. R.M.StJohn, R.G.Fowler, Phys.Rev. 122, 1813 (1961)

NONADIABATIC BEHAVIOUR IN ATOMIC AND MOLECULAR PROCESSES: THE RIDGE EFFECT IN MODE TRANSITIONS

Vincenzo Aquilanti, Simonetta Cavalli and Gaia Grossi

Dipartimento di Chimica dell'Università

06100 Perugia, Italy

ABSTRACT

Nonadiabatic effects in atomic and molecular processes are often localized, in the potential energy surface, along ridges which separate basins corresponding to alternative modes. A uniform semiclassical discussion of these phenomena is presented for the simplest models for transitions between modes, the physical pendulum and the Hénon-Heiles potential of coupled oscillators.

I. INTRODUCTION

In a time dependent picture of atomic and molecular processes, Massey's criterion[1] individuates a source for nonadiabatic behaviour in the matching between the frequencies associated with the collision time and those which are characteristic of a given system. Within a time independent picture, it has been shown by recent investigations[2-5] that a typical source of nonadiabatic behaviour can be localized in the potential energy surface along the ridges which separate basins corresponding to alternative modes available to the system.

As stressed by Fano,[2] several phenomena of atomic physics (interelectronic correlation, magnetic perturbation of Rydberg states) are to be explained by studying the role played by the ridges of the potential. Such a ridge effect plays a role also in elementary chemical reactions[3] and in general in processes which involve transitions between localized modes.[4] The description of these effects is best formulated within the adiabatic approximation: this formulation inevitably fails at potential ridges, where nonadiabatic

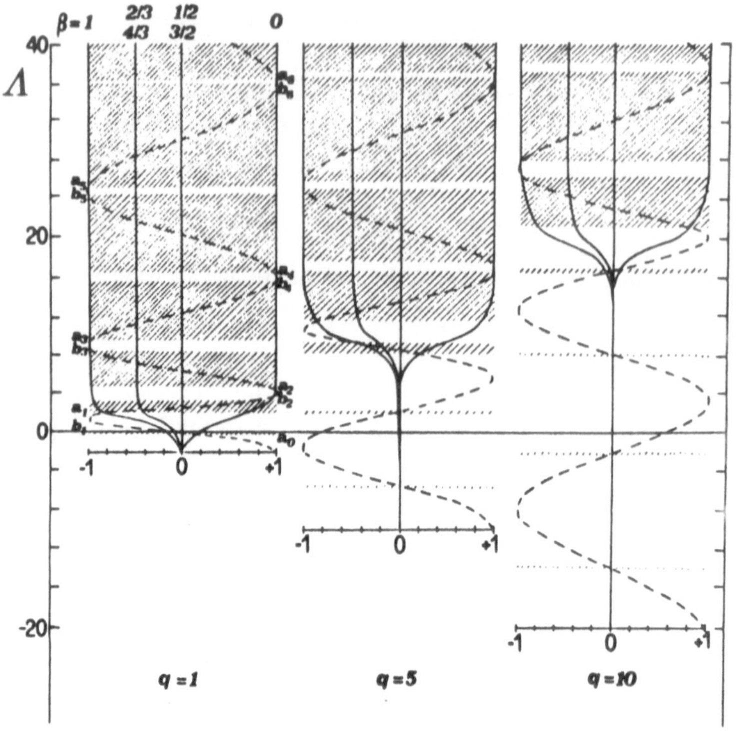

Fig. 1. The quantization rule, Eq. (1) for three values of the parameter q. Continuous lines are the tunnel term (right-hand side of Eq.(1)) for various β values, as a function of Λ. Eigenvalues are given by their intersections with the dashed lines, which represent the function cos σ. Allowed ranges for eigenvalues are hatched.

effects have to be taken into account explicitly.

The simplest model[5] for transitions between modes is that of the physical pendulum,[5] and the transitions show up when its physical characteristics are assumed to be varied slowly (adiabatically). The physical pendulum (Section II.) is mathematically equivalent, within time independent quantum mechanics, to Mathieu equation. An analysis of this equation, which is essentially Schrödinger equation, for a sinusoidal potential, allows a discussion of many interesting features of phenomena associated with transitions between modes, as described by the most familiar model for coupled oscillators, the Hénon-Heiles potential[4] (Section III.). Further applications to the theory of elementary chemical reactions are briefly listed in Section IV.

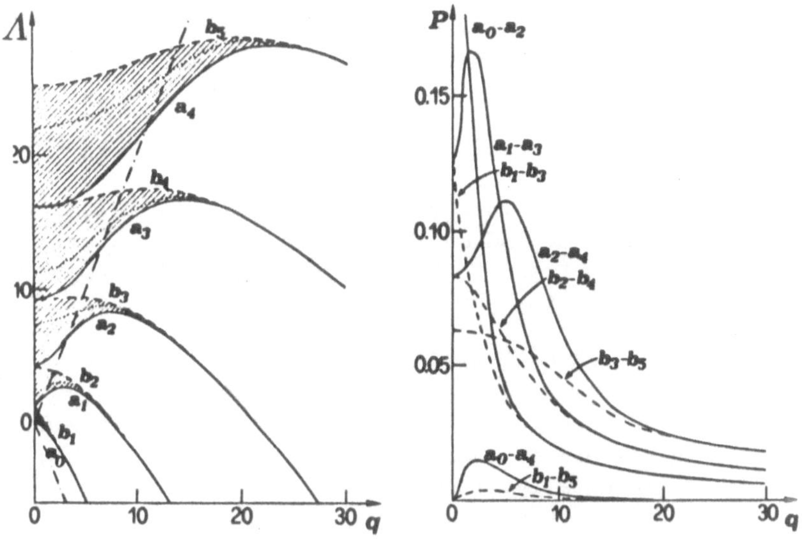

Fig. 2. Some eigenvalues of the Mathieu equation as a function of the parameter q are reported on the left, toghether with the ridge line 2q and the valley bottom line -2q (dash-dotted). Eigenvalues corresponding to $\beta = 2/3$ and 4/3 are shown by dotted lines. Allowed regions for eigenvalues are hatched. The right-hand side shows elements of the P matrix as a function of q.

II. UNIFORM ASYMPTOTIC ANALYSIS OF MATHIEU EQUATION

In a recnt investigation[5] a uniform asymptotic analysis has been presented of the Mathieu equation, whose physical significance is that of the description of the pendular motion in quantum mechanics. The two modes of the pendulum can be taken as models for elementary chemical reactions, for normal versus local molecular vibrations and for several phenomena of atomic and molecular physics which involve transitions between modes at a potential ridge. The two modes are designated vibrating or librating, and rotating or precessing at energies, respectively, lower and higher than the maximum in the potential. In the context of recent investigations of highly excited molecules the two modes would correspond to local and normal vibrations, respectively. The two modes are sharply separated by a trajectory (the separatrix) corresponding to the maximum in the potential (the ridge in our applications). As is often the case in quantum mechanics, transitions between modes are smoother.

We follow here our recent semiclassical analysis[5] (further detailed information has been presented recently)[6] For a sinusoidal potential of N-fold periodicity, the wavefunctions are conveniently characterized by a parameter $\beta = 2k/N$, where $k = 0,1,\ldots,N-1$ and the eigenvalues can be obtained semiclassically by the following quantization rule:

$$\cos \sigma(\Lambda,q) = \{1 + \exp(-2\eta\ (\Lambda,q))\}^{-\frac{1}{2}} \cos \pi\beta \qquad (1)$$

where $\sigma(\Lambda,q)$ is the phase integral for the well, and $\eta(\Lambda,q)$ is a tunnel integral. Both depend on the parameters Λ and q, which are related to the energy E and to the maximum height of the potential V_o. In particular for the physical pendulum Λ is

$$\Lambda = 8ml^2E/\hbar^2, \text{ and } q = 4ml^2V_o/\hbar^2,$$

where m is the mass and l the lenght of the pendulum. (The definition of β and Eq.(1) are slightly misprinted in Ref.5). The quantization rule is graphically illustrated in Fig. 1, which demonstrates the usefulness of this approach for exhibiting most qualitative feature of the transition between modes as a function of the potential maximum V_o: it is semiclassical because it is asymptotic as Planck's constant tends to zero, and it is uniform because it provides a description of the transition between modes, which cannot be obtained by conventional perturbation methods. Several further discussions and applications can be found in Refs. 5 and 6, which also give a quantitative assessment of the semiclassical quantization rule, Eq.(1).

The dependence of eigenvalues as a function of q is shown in Fig. 2, toghether with some computed elements of the matrix P(q), which we found useful to introduce[5] as a measure of the nonadiabatic coupling between eigenvalues. The localization of maxima in the matrix for nonadiabatic coupling along the ridge line 2q shows definitely that nonadiabatic effects are most important there.

III. THE RIDGE EFFECT FOR COUPLED OSCILLATORS

The previous analysis of the pendular motion can be immediately applied to provide a semiclassical discussion of a two dimensional (Hénon-Heiles) model for coupled oscillators,[4] conveniently written in polar coordinates

$$V(\rho,\theta) = \frac{1}{2}\rho^2 + \frac{\lambda\rho^3}{3}\cos 3\theta \qquad (2)$$

We refer to a previous paper[4] for a description of the model and of its relevance for a semiclassical discussion of regular and irregular modes in classical and quantum mechanics.[7,8] By connecting the polar variable ρ and the parameter q through

$$q = \frac{4}{27} \lambda \rho^5 \tag{3}$$

the adiabatic potential energy curves $\varepsilon_n(\rho)$ are obtained by Mathieu eigenvalues $\Lambda_n(q)$ according to the formula

$$\varepsilon_n(\rho) = \frac{9}{8\rho^2} \Lambda_n(q) + \frac{1}{2} \rho^2 \quad . \tag{4}$$

The elements of the nonadiabatic coupling matrix are likewise related:

$$P_{nm}(\rho) = \frac{dq}{d\rho} P_{nm}(q) . \tag{5}$$

Figs. 3, 4 are obtained by these formulas.

A useful aspect of this approach is to provide a classification scheme for levels. When $\lambda = 0$ (the simple isotropic oscillator) a good quantum number exists, and it is designated by $\pm\ell$ in Refs. 7 and 8. For finite λ, the potential belongs to the C_{3v} symmetry group, representations A_1, A_2 and E. Mathieu functions ce_{2n} and se_{2n} ,under the C_{3v} symmetry operations, behave respectively as A_1 and A_2, and their eigenvalues are labelled as A_{2n} and B_{2n+2}, where $n = 0,1,2,...$ (Figs. 1 and 2). They are periodic by π, and correspond to $\beta = 0$. The E representation is induced by Mathieu function of fractional order $ce_{2n+\beta}$ and $se_{2n+\beta}$, and the corresponding doubly degenerate eigenvalues will be designated as $\Lambda_{2n+\beta}$. From the quantization formula (1), we have that for this symmetry β can assume the values 2/3 and 4/3, and therefore, in order that the proper boundary conditions are satisfied, the functions will have periodicities 3π and $3\pi/2$. Therefore, the levels supported by each adiabatic curve will conveniently be labelled both by the proper index of corresponding Mathieu functions $2n+\beta$, and by a progressive number $v = 0,1,2,...$. The Mathieu index is related to ℓ by $|\ell| = 3n + \beta/2$, and in the $\lambda = 0$ limit the energy levels are given by

$$\varepsilon(v,2n+\beta) = 2v + 1 + |\ell| .$$

As illustrated in Fig. 3 failures of the adiabatic picture, as measured by the elements of the matrix for nonadiabatic coupling P, occur at ridge. The correlation between regular modes of classical investigation and the quantum mechanical states which are localized above the ridge has been already pointed out.[4] In our picture, quantum mechanical delocalization of the wavefunction is a process which is favoured by coupling between adiabatic eigenvalues in the proximity of the ridge, where a sequence of avoided crossings shows up, and corresponds to maxima in the P matrix. These features can be discussed within the familiar apparatus of curve crossing theory, and a striking similarity is apparent between these aspects and the theory of level perturbation for diatomic molecules.[9] Actually, our current experience suggests that the semiclassical techniques introduced in such a context, are also extremely fruitful here.

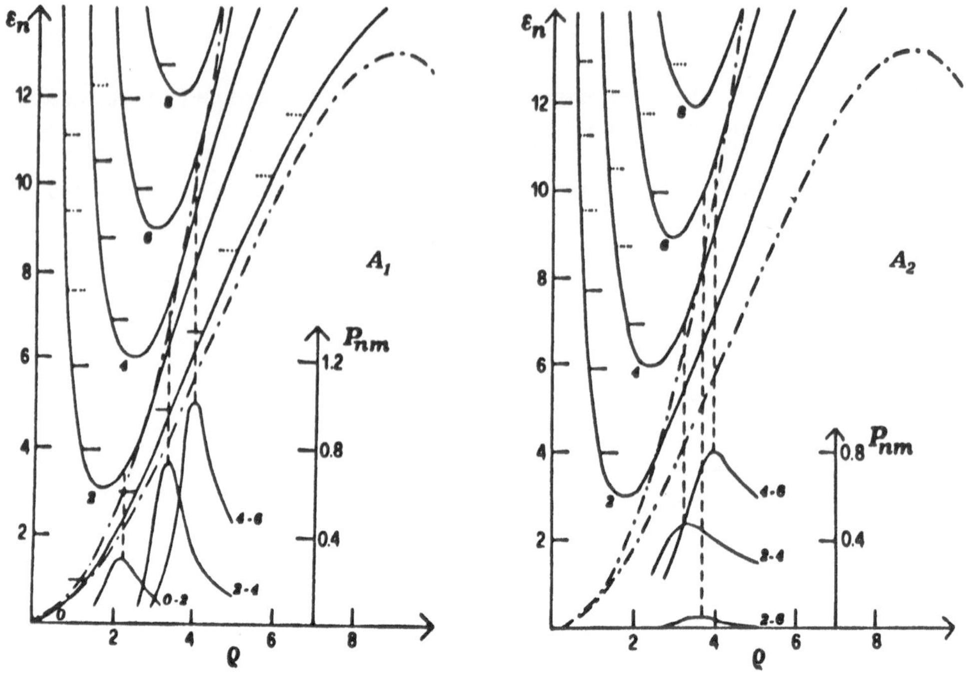

Fig. 3. For the Hénon-Heiles potential with $\lambda = 80^{-\frac{1}{2}}$ (Eq.(2): upper broken curve, ridge profile ($\theta = 0$, $2\pi/3$, $4\pi/3$); lower broken curve, valley bottom profile ($\theta = \pi/3$, π, $5\pi/3$)), adiabatic potential curves $\varepsilon_n(\rho)$ (Eq.(4)) and corresponding nonadiabatic coupling matrix elements $P_{nm}(\rho)$ (Eq.(5)) as a function of radial coordinate ρ for A_1 and A_2 symmetry. Positions of levels indicated by continuous segments for those identified as quasiperiodic (Ref.8) and by dotted segments for those not identified as quasiperiodic. For further details, see Ref. 4.

Fig. 4, which represents results for the E symmetry, focuses the attention to a particularly interesting type of avoided crossings due to the interaction between almost degenerate levels, which the adiabatic curves support. This phenomenon leads to much more pronounced delocalization of the wavefunctions, because of the strong mutual perturbation of the levels, and is strongly dependent on the parameter λ, which measures the strength of the coupling between the oscillators. Therefore it has a relevance to extended discussions[10] of the role of avoided crossings as a function of the parameter λ. In the present approach such avoided crossings, formerly individuated as a road to quantum chaos by Percival,[11] are seen to arise when, because of the increasing importance of anharmonicity for levels with high v quantum numbers, high v levels of lower curves enter into accidental resonance with low v levels of

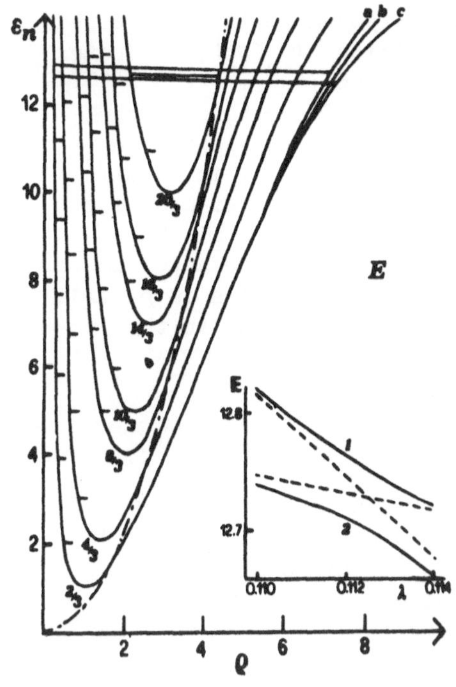

Fig. 4.
Adiabatic curves $\varepsilon_n(\rho)$, Eq.(4), for the E symmetry of Hénon-Heiles potential (Eq.(2)) for λ close to $80^{-\frac{1}{2}} = 0.1118$. Slight changes in λ affect mainly the large ρ region: for example, the curves labelled as a, b and c show how the 2/3 state varies for $\lambda = 0.110, 0.112, 0.114$. The corresponding $v = 7$ level varies as in inset, and thus would cross the $v = 2$ level of the 20/3 state, practically unaffected by a change in λ (see Ref. 10) (dashed curves): actually, the crossing is avoided and the levels behave as the continuous curves 1 and 2.

upper curves. The phenomenon is familiar in spectroscopy leading to strong level repulsion.[9] For the model considered here, this phenomenon happens once in the neighborhood of $\lambda = 80^{-\frac{1}{2}} = 0.118$ (Fig. 4).

IV. APPLICATIONS TO CHEMICAL REACTION THEORY

The previous analysis is of interest not only for general mode transition problems, but also for the discussion of specific problems in the theory of elementary chemical reactions.

For bimolecular reactions, an extensive analysis for $A+BA \rightarrow AB+A$ collinear process has been carried out.[3] Here, the ridge separates the valleys of reactants and products from the saddle of the intermediate complex, which "sits astride it" (see also Ref. 2). The effect of masses on the reaction probabilities and on resonances has been studied, and the role of nonadiabatic behaviour at ridge has been assessed. In particular, it has been found that adiabatic behaviour is propiciated by small skewing angle (low mass of the transferred atom), while diabatic behaviour along the ridge is important for large skewing angles (large mass of the transferred atom). The extension of these ideas to the general three dimensional case is of great interest.

For unimolecular reaction theory, the potential considered in

Section III. and given by Eq.(2) can be extended to large ρ values to provide a model for molecular decomposition. In fact, the potential (2) has three symmetric saddles of height $(6\lambda^2)^{-1}$ at $\rho = \lambda^{-1}$ and therefore levels of energy higher than saddle belong to the dissociation continuum. Actually, all the levels considered in Figs. 3 and 4 are quasibound: Refs. 12 and 13 discuss in detail these points, in a search for mode specificities in unimolecular decay. As will be illustrated elsewhere, a discussion of this phenomenon can be carried out by extending the analysis presented above, and it is found that a mechanism for mode specificities is provided by considering the role played by nonadiabatic effects in the neighborhood of the potential ridge.

REFERENCES

1. H. S. W. Massey, Rep. Progr. Phys. 12:248 (1949).
2. U. Fano, Phys. Rev. A22:2660 (1980); Rep. Progr. Phys. 46:97 (1983); in: "Atomic Physics", vol. 8, I. Lindgren, A. Rosen, and S. Svenberg, eds., Plenum Press, New York (1983); in: "Electronic and Atomic Collisions", J. Eichler, I. W. Hertel, and N. Stolterfoht, eds., North Holland, Amsterdam (1984).
3. V. Aquilanti, G. Grossi and A. Laganà, Chem. Phys. Letters 93: 174 (1982); V. Aquilanti, S. Cavalli and A. Laganà, Chem. Phys. 93:179 (1982); V. Aquilanti, S. Cavalli, G. Grossi and A. Laganà, J. Mol. Struct. 93:319 (1983); 107:95 (1984).
4. V. Aquilanti, S. Cavalli and G. Grossi, in: "Chaotic Behaviour in Quantum Systems", G. Casati,ed., Plenum Press, New York (1984).
5. V. Aquilanti, S. Cavalli and G. Grossi, Chem. Phys. Letters 110:43 (1984).
6. J. N. L. Connor, T. Uzer, R. A. Marcus and A. D. Smith, J. Chem. Phys. 80:5095 (1984).
7. D. W. Noid, M. L. Koszykowski and R. A. Marcus, Ann. Rev. Phys. Chem. 32:267 (1981).
8. G. Hose and H. S. Taylor, J. Chem. Phys. 76:5356 (1982); Chem. Phys. 84:375 (1984).
9. M. S. Child, J. Mol. Spectry. 53:280 (1974); in: "Semiclassical Methods in Molecular Scattering and Spectroscopy", M. S. Child, ed., Reidel, Dordrecht (1980).
10. D. W. Noid, M. L. Koszykowski, M. Tabor and R. A. Marcus, J. Chem. Phys. 72:6169 (1980); see also T. Uzer, D. W. Noid and R. A. Marcus, J. Chem. Phys. 79:4412 (1983), and references therein.
11. I. C. Percival, Advan. Chem. Phys. 36:1 (1977), and references therein.
12. B. A. Waite and W. H. Miller, J. Chem. Phys. 74:3910 (1981); see also B. A. Waite, S. K. Gray and W. H. Miller, J. Chem. Phys. 78:259 (1983), and references therein.
13. Y. Y. Bai, G. Hose, C. W. McCurdy and H. S. Taylor, Chem. Phys. Letters 99:342 (1983).

PHOTON CORRELATIONS, ATOMIC STATE CORRELATIONS AND STATISTICAL DISTRIBUTIONS

L. Moorman, J. van Eck, H.G.M. Heideman and G. Nienhuis

Fysisch Laboratorium
Rijksuniversiteit Utrecht
3584 CC Utrecht, The Netherlands

Abstract: First we will describe our experiment where the simultaneous excitation of two colliding helium atoms to various substate combinations is studied by detecting two emitted photons in coincidence. Next we will give the minimal symmetry properties of these two photon correlations. Furthermore the connection between photon correlations and the fundamental collision amplitudes will be treated and finally we will try to interpret the measurements in the context of statistical distributions.

INTRODUCTION

We measured photon-photon angular distributions resulting from the simultaneous excitation of two colliding helium atoms to the 2^1P state, i.e.

$$\vec{He} + He \rightarrow \vec{He}^*(2^1P) + He^*(2^1P) \rightarrow$$
$$\rightarrow \vec{He}(1^1S) + He(1^1S) + p_1 + p_2 \qquad (1)$$

where the fast projectile is indicated by an arrow and the target atoms have thermal velocities. The angular correlation between the emitted photons p_1 and p_2, which are measured in coincidence, provides information on the excitation of the different magnetic substate combinations of two simultaneously excited atoms. For instance the probabilities to find the system after the collision in the state $|M_1M_2\rangle$, where one atom is in the substate $|JM\rangle = |1M_1\rangle$ and the other in the substate $|1M_2\rangle$. Since the projectile beam has axial symmetry about the beam axis and we do not detect the direction of the scattered particle our collision ensemble will be axially symmetric about the beam axis. Further it will be mirror symmetric with respect to any plane through the beam axis. Finally there is the so called strong symmetry which means that an interchange of the two final substates leaves the amplitude the same if the initial ground state is non-degenerate, which is the case in our experiment. These symmetries reduce the number of independent parameters of the density matrix from 81 complex parameters to four real partial cross sections and one complex coherence integral.

Apart from the total cross section corresponding to the trace of the density matrix (G) we define relative partial cross sections:

$$\alpha = \frac{\langle 00|G|00\rangle}{Tr(G)} \qquad \beta = \frac{\langle 10|G|10\rangle}{Tr(G)}$$

$$\lambda = \frac{\langle 1\text{-}1|G|1\text{-}1\rangle}{Tr(G)} \qquad \mu = \frac{\langle 11|G|11\rangle}{Tr(G)} \tag{2}$$

and the coherence integral:

$$\chi = \frac{\langle 1\text{-}1|G|00\rangle}{Tr(G)}$$

where the brackets contain the projections of the angular momenta of the two atoms onto the beam axis just before they emit their photon.We measured these parameters as can be found in [1] and [3].

In this article we want to answer the question to what extend the angular correlations between the two coincident photons (photon correlations) necessarily indicate certain types of correlations between the excitations of the various magnetic substates of the two simultaneously excited atoms (atomic state correlations). Finally we will compare the measured collision amplitudes in the high and low energy limit with certain types of statistial distributions. This comparison may give hints for the best choice of the basis states in which the final-state distribution is almost random. The best agreement found is that between the low-energy experimental parameters and the statistical distribution including strong symmetry where every substate has an equal statisical weight.

A. Photon Correlations

Let us first look at the photon distribution originating from single atom excitation. The number of photons with polarisation $\vec{\varepsilon}$ emitted by a unit volume per unit of time in a unit solid angle is conveniently given in a polarisation matrix description:

$$N(\vec{\varepsilon}) = \vec{\varepsilon}^{\,+}\cdot\vec{\vec{C}}\cdot\vec{\varepsilon} = \sum_{kq} c_{kq}\, \vec{\varepsilon}^{\,+}\cdot\vec{\vec{S}}_{kq}^{\,+}\cdot\vec{\varepsilon} \tag{3}$$

where $\vec{\vec{C}}$ is the polarisation matrix and $\{\vec{\vec{S}}_{kq}\}$ is the orthonormal complete set of nine cartesian 3×3 matrices [2].

Of course it is impossible to conclude from this polarisation dependence some correlation between two separately emitted photons. Eq. (2) gives the photon distribution in case of a large number of single-photon detections. Dividing (3) by the trace of $\vec{C}$ gives the probability for an arbitrary detected photon to be in polarisation state $\vec{\varepsilon}$.

$$P(\vec{\varepsilon}) = \frac{N(\vec{\varepsilon})}{Tr(\vec{C})} = \sum_{kq} \frac{c_{kq}}{Tr(\vec{C})} \vec{\varepsilon}^{+}.\vec{S}^{+}_{kq}.\vec{\varepsilon} \qquad (4)$$

Assuming that there will be no correlation between two photons emitted within time interval τ and by an interaction volume ΔV the probability for such an event where one photon has polarisation $\vec{\varepsilon}_1$ and the other $\vec{\varepsilon}_2$ is then given by the product of the separate probabilities:

$$P(\vec{\varepsilon}_1,\vec{\varepsilon}_2) = P(\vec{\varepsilon}_1)P(\vec{\varepsilon}_2) \qquad (5)$$

and for the coincidence rate within τ originating from ΔV we have to include the factor $\tau\Delta V$:

$$N(\vec{\varepsilon}_1,\vec{\varepsilon}_2,\tau) = \vec{\varepsilon}_1^{+}.\vec{C}.\vec{\varepsilon}_1 \ \vec{\varepsilon}_2^{+}.\vec{C}\,\vec{\varepsilon}_2 \ \tau\Delta V$$

$$= \sum_{kq} \sum_{k'q'} c_{kq}\, c_{k'q'}\, \tau\Delta V \ \vec{\varepsilon}_1^{+}.\vec{S}^{+}_{kq}.\vec{\varepsilon}_1 \ \vec{\varepsilon}_2^{+}.\vec{S}^{+}_{k'q'}.\vec{\varepsilon}_2$$

$$= \sum_{kq} \sum_{k'q'} c(kq,k'q') \ \vec{\varepsilon}_1^{+}.\vec{S}^{+}_{kq}.\vec{\varepsilon}_1 \ \vec{\varepsilon}_2^{+}.\vec{S}^{+}_{k'q'}.\vec{\varepsilon}_2 \qquad (6)$$

It thus seems that the absence of correlations of the two photons corresponds to factorisation of the multipole components:

$$c(kq,k'q') = c_{kq}\, c_{k'q'}\, \tau\Delta V \qquad (7)$$

If the scattered atom is not detected the collision plane is not fixed and one can show that only the state multipoles with (kq) = (0,0) or (2,0) can be non-zero, corresponding with the multipoles c_{00} and c_{20} in the above case. The angular radiation pattern is given by the sum over two linear polarisation directions transversal to the direction of motion of the photon:

$$I(\vec{n}) = \sum_{\vec{\varepsilon}=\vec{\varepsilon}_1,\vec{\varepsilon}_2} \vec{\varepsilon}^{+}.\vec{C}.\vec{\varepsilon} = Tr(\vec{C}) - \vec{n}.\vec{C}.\vec{n}$$

$$= \left(\frac{2}{\sqrt{3}} c_{00} - \frac{1}{\sqrt{6}} c_{20}\right) + \sqrt{\frac{3}{2}}\, c_{20}(\cos(\theta))^2 \qquad (8)$$

where θ is the polar angle between beam direction and photon direction. Therefore if we measure the angular distribution of two photons originating from two different collisions with single excitation without polarisation analysis we can only find c(00,00), c(20,00) = c(00,20) and c(20,20) to be non-zero. The radiation would be isotropic (and thus the polarisation zero) in case c(00,00) ≠ 0 and all other c(kq,k'q') = 0.

Now that we have identified the absence of correlation between two photons with factorising multipoles as in (7) we want to look more carefully to our experimental case where two photons originating from one collision with simultaneous double atom excitation are detected. The polarisation distribution is

$$N_c(\vec{\varepsilon},\vec{\varepsilon}\,') = \sum_{kqk'q'} c(kq,k'q')\; \vec{\varepsilon}^{\,+}\cdot\vec{S}^{+}_{kq}\cdot\vec{\varepsilon}\;\; \vec{\varepsilon}\,'^{+}\cdot\vec{S}^{+}_{k'q'}\cdot\vec{\varepsilon}\,' \tag{9}$$

and the angular coincidence radiation pattern (compare to (8)) is given by

$$I_c(\vec{n},\vec{\tilde{n}}) = \sum_{\vec{\varepsilon}=\vec{\varepsilon}_1,\vec{\varepsilon}_2}\;\sum_{\vec{\varepsilon}\,'=\vec{\varepsilon}\,'_1,\vec{\varepsilon}\,'_2} N(\vec{\varepsilon},\vec{\varepsilon}\,')$$

$$= A + B\sin^2(\theta) + C\sin^2(\tilde{\theta}) + D\sin^2(\theta)\sin^2(\tilde{\theta})$$

$$+ E\sin(2\theta)\sin(2\tilde{\theta})\cos(\phi-\tilde{\phi}) + F\sin^2(\theta)\sin^2(\tilde{\theta})\cos(2(\phi-\tilde{\phi})) \tag{10}$$

where A, B, C, D are linear combinations of c(00,00) c(20,00) and c(20,20) and:

$$E = -\tfrac{1}{2}c(21,2-1)$$

$$F = \tfrac{1}{2}c(22,2-2) \tag{11}$$

and $\vec{n}$ and $\vec{\tilde{n}}$ are the directions of the two photon detectors with polar and azimuthal angles (θ,ϕ) and $(\tilde{\theta},\tilde{\phi})$, respectively.

At first sight it seems remarkable that in our experiment the multipoles c(21,2-1) and c(22,2-2) also contribute to the anisotropy of the angular distribution. The reason is that here the axial symmetry around the projectile beam does not require q = q' = 0 anymore, but only q + q' = 0. An even more striking feature of eq. (9) is the observation, contrary to the case of single atom excitation, that we cannot have an isotropic polarisation distribution in the case of simultaneous double atom excitation i.e. with c(00,00) ≠ 0 and all other c(kq,k'q') = 0. This can be proven as follows:
One can show that the following proportionalities exist between the multipole components and linear combinations of the density matrix elements:

$$
\begin{aligned}
c(00,00) &\sim \{\alpha + 4\beta + 2\lambda + 2\mu\} \\
c(20,00) &\sim \{\alpha + \beta - \lambda - \mu\} \\
c(20,20) &\sim \{2\alpha - 4\beta + \lambda + \mu\} \\
c(21,2\text{-}1) &\sim \{\mathrm{Re}(\chi) - \beta\} \\
c(22,2\text{-}2) &\sim \{\lambda\}
\end{aligned}
\tag{12}
$$

$$
\begin{aligned}
c(10,10) &\sim \{\mu - \lambda\} \\
c(11,1\text{-}1) &\sim \{\mathrm{Re}(\chi) + \beta\} \\
c(21,1\text{-}1) &\sim \{-i\ \mathrm{Im}(\chi)\}
\end{aligned}
$$

The isotropic polarisation requirement leads to an inconsistent set of eight equations with six unknowns. This means that there always has to be a minimum correlation between the photons. This can be traced back to the strong symmetry because the fully isotropic density matrix obeys axial and planar symmetry but not the strong symmetry. The last three proportionalities are not observable in our experiment, but they are if circular polarisation is measured.

Now we want to investigate for our experimental case the possibility of finding an isotropic angular distribution of coincident photons. At first instance therefore we do not look at the last three multipoles of eq. (12). Normalising with c(00,00) and setting the next four multipoles of (12) equal to zero we find a solvable set. The solutions are:

$$
\begin{aligned}
\alpha = \beta = \mathrm{Re}(\chi) &= \frac{1}{9} \\
\lambda &= 0 \\
\mu &= \frac{2}{9}
\end{aligned}
\tag{13}
$$

And the consequence for the last three multipoles is:

$$
\begin{aligned}
c(10,10) &\sim \{\tfrac{2}{9}\} \\
c(11,1\text{-}1) &\sim \{\tfrac{2}{9}\} \\
c(21,1\text{-}1) &\sim \{-i\ \mathrm{Im}(\chi)\}
\end{aligned}
\tag{14}
$$

This situation can be interpreted as if all of the necessary anisotropy of the polarisation distribution has turned into the multipoles governing the circular polarisation properties of the field.

B. Atomic State Correlation

It seems natural trying to explain the minimum photon correlations given above somehow in terms of correlations between the excitation of the various atomic substates of the two simultaneously excited atoms. Let us discuss the possibility of uncorrelated simultaneous atom excitation in the density operator formalism.

First we consider collisions between atoms A and B in which one of the two atoms is excited. The fundamental theoretical quantity describing the excitation process is the scattering amplitude $g(M\vec{v} \leftarrow \vec{v}_o)$ with $\vec{v}$ the relative velocity after the excitation and $\vec{v}_o$ the initial relative velocity. The ground state is assumed to be non-degenerate and if A is excited $|M^A\rangle$ denotes the magnetic substate of the excited state A^* with total angular momentum J. The density operator becomes:

$$G_A = \frac{n_A n_B}{\Gamma_A} \int d\hat{v} \sum_{M_1^A M_2^A} |M_1^A\rangle g_A(M_1^A\vec{v} \leftarrow \vec{v}_o) v\, g_A^*(M_2^A\vec{v} \leftarrow \vec{v}_o) \langle M_2^A| \tag{15}$$

where n_A, n_B are the densities of A and B atoms respectively and Γ_A is the decay rate of the state A^*. Interchanging A and B in this formula we get the density operator for the case that B is excited during the collision.

The creation of pairs (A^*,B) and (A,B^*), excited within a time interval τ and in a reaction volume ΔV, is now described by the product density operator:

$$\begin{aligned} G_{AB}^{\text{uncorr.}} &= G_A\, G_B\, \tau\Delta V \\ &= \tau\Delta V \frac{n_A^2 n_B^2}{\Gamma_A \Gamma_B} \sum_{M_1^A M_1^B} \sum_{M_2^A M_2^B} |M_1^A M_1^B\rangle \int d\hat{v} \int d\hat{w}\, g_A(\vec{v}) g_B(\vec{w}) v \\ &\qquad w\, g_B^*(\vec{w}) g_A^*(\vec{v})\, \langle M_2^A M_2^B| \end{aligned} \tag{16}$$

Comparing this density operator with the density operator for simultaneous excitation of both the target and the projectile:

$$G_{AB} = \frac{n_A n_B}{(\Gamma_A + \Gamma_B)} \sum_{M_1^A M_1^B} \sum_{M_2^A M_2^B} |M_1^A M_1^B\rangle \int d\hat{v}\, f(M_1^A M_1^B\, \vec{v} \leftarrow \vec{v}_o) v\, f^*(M_2^A M_2^B\, \vec{v} \leftarrow \vec{v}_o)\, \langle M_2^A M_2^B| \tag{17}$$

we see that if we replace the products of g's in the following way:

$$g_A(M^A,\vec{v})\ g_B(M^B,\vec{w}) = \alpha\ f(M^AM^B,\vec{v})\ \delta(\hat{v} - \hat{w}) \tag{18}$$

with $\alpha = \left[\frac{\Gamma_A\Gamma_B}{(\Gamma_A + \Gamma_B)} \frac{1}{n_A n_B\ w\ \tau\Delta V} \right]^{\frac{1}{2}}$

the equations for the true and accidental coincidences become identical. Notice that if $\hat{v} = \hat{w}$ it follows on kinematical grounds that $\vec{v} = \vec{w}$. The proportionality constant α has to be added just for dimensional reasons.

Equation (18) says that in our experiment we are not dealing with a product ensemble of completely uncorrelated excitations but with an ensemble where the elements consist of two collisions with identical final relative velocity. So we see that for the true simultaneous excitation there is a minimum correlation which can not be abolished. This minimum correlation must of course also be reflected in the polarisation distribution of the two photons resulting from simultaneous excitation.

An intuitive way of explaining the minimum correlation is the following. Although coincident photon pairs resulting from simultaneous excitations in different collisions may be emitted with respect to different scattering planes (as we do not detect the scattered atoms) the two photons resulting from simultaneous excitation in one collision are emitted with respect to one and the same scattering plane. So both photons of any particular pair resulting from simultaneous excitation carry information about the same scattering plane and therefore must in some way be correlated. Connected to this notice is the remark below (12) on the strong symmetry.

Now we will derive a strict inequality for density matrix elements and some weaker inequalities, which hold for uncorrelated simultaneous double atom excitations. First we note that an element of the density matrix can be written as a unitary inner product:

$$\langle M_1M_2|G_{AB}|M_1'M_2'\rangle = a\ \langle f_d(M_1M_2).f_d(M_1'M_2')\rangle \tag{19}$$

with $a = \frac{n_A n_B}{(\Gamma_A + \Gamma_B)}$

in which the inner product is defined by

$$\langle f.g\rangle \equiv \int f(\vec{v})\ g^*(\vec{v})v\ d\hat{v} \tag{20}$$

The diagonal elements correspond to the relative partial cross sections (RPCS):

$$\sigma(M_1M_2) = b\ \langle f_d(M_1M_2).f_d(M_1M_2)\rangle \tag{21}$$

with $b = \left[\sum_{M_1M_2} \langle f_d(M_1M_2).f_d(M_1M_2)\rangle\right]^{-1}$

For an inner product as defined in (20) the Schwarz inequality applies:

$$|\langle f.g\rangle|^2 \leq \langle f.f\rangle\langle g.g\rangle \tag{22}$$

With this inequality we can derive strict inequalities between any non-diagonal element of the density matrix and the corresponding diagonal elements:

$$|\langle f(M_1M_2).f(M_1'M_2')\rangle|^2 \leq \sigma(M_1M_2)\ \sigma(M_1'M_2') \tag{23}$$

for example:

$$|\chi|^2 = b^2|\langle 1-1|G|00\rangle|^2 \leq \lambda\alpha \tag{24}$$

and thus certainly: $$|\mathrm{Re}(\chi)|^2 \leq \alpha(\lambda + \mu) \tag{25}$$

This can be checked in the experiment.

Apart from these strict inequalities we can derive, by using a Schwarz inequality, some weaker relations between mutually different diagonal elements. These relations are weaker because they only need to hold for uncorrelated excitation. We split the double atom excitation amplitude $f_d(M_1,M_2,\vec{v})$ in a product of uncorrelated single excitation amplitudes $f_s(M_1,\vec{v})$ and $f_s(M_2,\vec{v})$ and a correlation term as follows:

$$\begin{aligned} f_d(M_1M_2,\vec{v}) = &\{f_s(M_1\vec{v})\ f_s(M_2\vec{v})\}_{uncorr.} \\ &+ \{f_d(M_1M_2,\vec{v}) - f_s(M_1\vec{v})\ f_s(M_2\vec{v})\}_{corr.} \end{aligned} \tag{26}$$

The uncorrelated approximation will be defined by taking only the first term of the right hand side of (26). With the help of the Schwarz inequality we find for any diagonal element in the uncorrelated approximation

$$\begin{aligned} \sigma^2(M_1M_2) &= b^2|\langle f_1f_2.f_1f_2\rangle|^2 = b^2|\langle f_1f_1^*.f_2f_2^*\rangle|^2 = \\ (\text{S ineq}) &\leq b^2\langle f_1^2.f_1^2\rangle\langle f_2^2.f_2^2\rangle = \sigma(M_1M_1)\ \sigma(M_2M_2) \end{aligned} \tag{27}$$

Explicitly: $(M_1M_2) = (01)$ gives $\beta^2 \leq \alpha\mu \leq \alpha(\lambda + \mu)$ (28)

$(M_1M_2) = (1-1)$ gives $\lambda^2 \leq \mu^2$

From fig. 1 it is evident that above 1 keV certainly $\beta^2 > \alpha(\lambda + \mu)$ so that there the excitation process must give rise to correlations between the atoms.

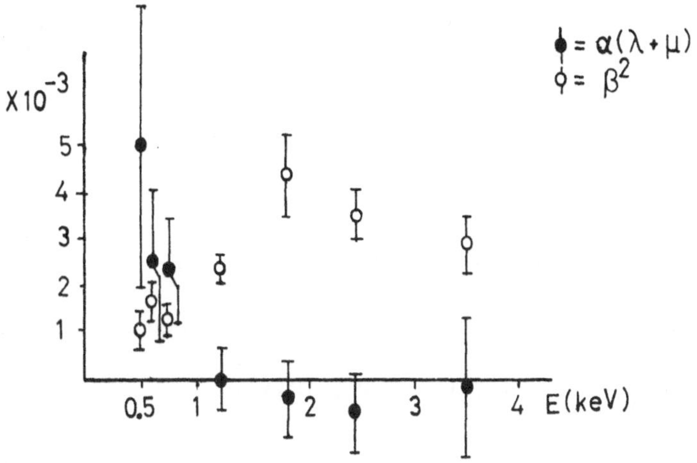

Fig. 1. Experimental evidence of atom correlations.

Unfortunately this does not prove that these correlations really originate from within the dynamically accessible space. By this we mean that in our experiment we cannot determine whether this correlation comes from (18) (i.e. from the fact that their relative velocities after the collision with respect to their collision partner are the same) or from a real dynamical correlation between $|M_1\rangle$ and $|M_2\rangle$.

C. Statistical Distributions

In the foregoing sections we have handled the problem of correlations extensively. However, we did not try to calculate the collision parameters (α, β, λ, μ and Re(χ)). In the following we want to do so for the case where nothing "exceptional" happens. By this we mean the situation where all basis states in which the state of the system is expanded have equal probability and there is no coherence between them. We call this a statistical situation. This definition leaves open the exact choice of the basis set. The aim is trying to find a statistical distribution, which fits the high or low energy data of our experiment (table 1).
Let us consider two possibilities:

1. We choose direct products of the final magnetic substates as a base. This means we require for all M_1 and M_2 an uncorrelated excitation as defined before by the factorisation:

$$f(M_1M_2,\vec{v}) = g_1(M_1\vec{v})\ g_2(M_2\vec{v}) \tag{29}$$

and equal cross sections for all substates of both atoms, i.e.:

$$\text{for all } i,\ j \text{ and } M_1M_2\text{: } |g_i(M_1)| = |g_j(M_2)| \tag{30}$$

With these two conditions we can rewrite the amplitude in the following form:

$$f(M_1M_2,\vec{v}) = f(\vec{v})e^{i\{\phi_1(M_1\vec{v}) + \phi_2(M_2\vec{v})\}} \tag{31}$$

where only the phase functions ϕ_1 and ϕ_2 depend on M_1 and M_2 and $f(\vec{v})$ is a real function.
For a diagonal element of the density matrix we find

$$\langle M_1M_2|G|M_1M_2\rangle = a\int d\hat{v}\ v\ f^2(\vec{v}) = \text{constant} = G$$

So: $$\alpha = \beta = \lambda = \mu = \frac{1}{9} \tag{32}$$

and the coherence integral satisfies $Re(\chi) \leqslant \frac{1}{9}$ in agreement with Schwarz's inequality (24) (see the fourth column of table 1).

2. We choose the magnetic substates of the total orbital angular momentum L_t as a base. Thus we should require

$$\langle L_tM_t|G|L_t'M_t'\rangle = G_o\ \delta_{L_tL_t'}\ \delta_{M_tM_t'} \tag{33}$$

However, Nienhuis [2] proved from the strong symmetry (interchange of the two identical atoms in case the ground state is non-degenerate) that no state can be present with L_t = odd. This would alter the expected statistical situation into

$$\langle L_tM_t|G|L_t'M_t'\rangle = G_o'\ \delta_{L_tL_t'}\ \delta_{M_tM_t'}\ (\delta_{L_t0} + \delta_{L_t2}) \tag{34}$$

Therefore we make the more general "weighted statistical" assumption:

$$\langle L_tM_t|G|L_t'M_t'\rangle = G_{L_t}\ \delta_{L_tL_t'}\ \delta_{M_tM_t'} \tag{35}$$

From this we find $Tr(G) = G_0 + 3G_1 + 5G_2$ and the relationship with the measured parameters is given by

$$\frac{\langle M_1 M_2 | G | M_1' M_2' \rangle}{Tr(G)} = \sum_{L_t=0,1,2} \langle 1M_1 1M_2 | L_t(M_1 + M_2)\rangle \; g_{L_t} \langle L_t(M_1 + M_2) | 1M_1' 1M_2' \rangle \tag{36}$$

Where we have defined normalised weight parameters for excitation to different total angular momentum states

$$g_{L_t} = \frac{G_{L_t}}{\sum_{L_t'} (2L_t' + 1) G_{L_t'}} \tag{37}$$

Explicitly (36) becomes:

$$\begin{aligned}
\alpha &= \tfrac{1}{3} g_0 + \qquad + \tfrac{2}{3} g_2 \\
\beta &= \qquad \tfrac{1}{2} g_1 + \tfrac{1}{2} g_2 \\
\lambda &= \tfrac{1}{3} g_0 + \tfrac{1}{2} g_1 + \tfrac{1}{6} g_2 \\
\mu &= \qquad\qquad g_2 \\
Re(\chi) &= -\tfrac{1}{3} g_0 + \qquad + \tfrac{1}{3} g_2
\end{aligned} \tag{38}$$

Where:

$$g_0 + 3g_1 + 5g_2 = 1 \tag{39}$$

Let us now consider some different statistical models, which means in this case: particular choices for the g-weights. We will motivate these choices and compare their consequences for the various collision parameters with our experimental results (see table 1).

a1) For a uniform distribution over the different total angular momenta in the case where the strong symmetry would not apply, for example if the initial state would be degenerate or the particles not identical, we expect the situation where $g_0 = 1/3$, $g_1 = 1/9$ and $g_2 = 1/15$. Notice from the fifth column of table 1 that $Re(\chi)$ becomes significantly negative.

a2) Giving all magnetic substates an equal weight we expect $g_0 = g_1 = g_2 = 1/9$ and it can be verified easily from (38) that we get the same as eq. (32) (column 4 of table 1) with $Re(\chi) = 0$.

b) If we include strong symmetry caused by the non-degeneracy of the initial ground state for identical atoms we know that in our experiment no total angular momentum $L_t = 1$ can be present in the final state. This means contrary to the foregoing:

$$g_1 = 0 \tag{40}$$

b1) Again taking a uniform distribution but now over the physically accessible total angular momenta we get $g_0 = 1/2$; $g_2 = 1/10$. From table 1 we see comparing these results with the low energy data that $(\lambda + \mu) = 17/60$ is much too large and $Re(\chi)$ is again far too negative for this statistical distribution to apply.

b2) Giving all magnetic substates an equal weight, which means for instance that the $|L_tM_t\rangle = |00\rangle$ state will have the same weight as the $|20\rangle$ state, we find $g_0 = g_2 = 1/6$. According to table 1 this situation with $(\lambda + \mu) = 1/4$ and $Re(\chi) = 0$ is the best candidate of the statistical models considered in this article for the low energy limit.

Table 1. Comparision of the parameters α, β, λ, μ and $Re(\chi)$ as calculated in the various statistical models, with our experimental data in the low ($E_{kin} \downarrow$) and high ($E_{kin} \uparrow$) energy regime.

	Experiment		equal weights	$\|L_tM_t\rangle$		
	0.5 keV	3.5 keV		a1) Excl.	b) Incl. Strong Sym.	
	$E_{kin} \downarrow$	$E_{kin} \uparrow$	$\|M_1M_2\rangle$	Strong sym.	b1	b2
α	$\frac{3}{10}$	0	$\frac{1}{9}$	$\frac{7}{45}$	$\frac{7}{30}$	$\frac{1}{6}$
β	$\approx \frac{1}{10}$	$\approx \frac{3}{20}$	$\frac{1}{9}$	$\frac{4}{45}$	$\frac{1}{20}$	$\frac{1}{12}$
λ } $(\lambda + \mu)$	$\approx \frac{1}{10}$	$\approx \frac{1}{5}$	$\frac{2}{9}$	$\frac{8}{45}$	$\frac{11}{60}$	$\frac{1}{12}$
μ				$\frac{3}{45}$	$\frac{1}{10}$	$\frac{1}{6}$
$Re(\chi)$	$\frac{1}{10}$	$\frac{1}{20}$	$< \frac{1}{9}$	$-\frac{4}{45}$	$-\frac{2}{15}$	0

ACKNOWLEDGEMENT

This work was performed as part of the research programme of the "Stichting voor Fundamenteel Onderzoek der Materie" (FOM) with financial support from the "Nederlandse Organisatie voor Zuiver Wetenschappelijk Onderzoek" (ZWO).

REFERENCES

[1] L. Moorman, K.P.J. Linnartz, J. van Eck and H.G.M. Heideman, XIII ICPEAC contr. paper 338 (1983).

[2] G. Nienhuis, J. Phys. B. 17:587 (1984).

[3] G.J.N.E. de Vlieger, H.G.M. Heideman, J. van Eck and G. Nienhuis, J. Phys. B: At. Mol. Phys. 15:L345 (1982).

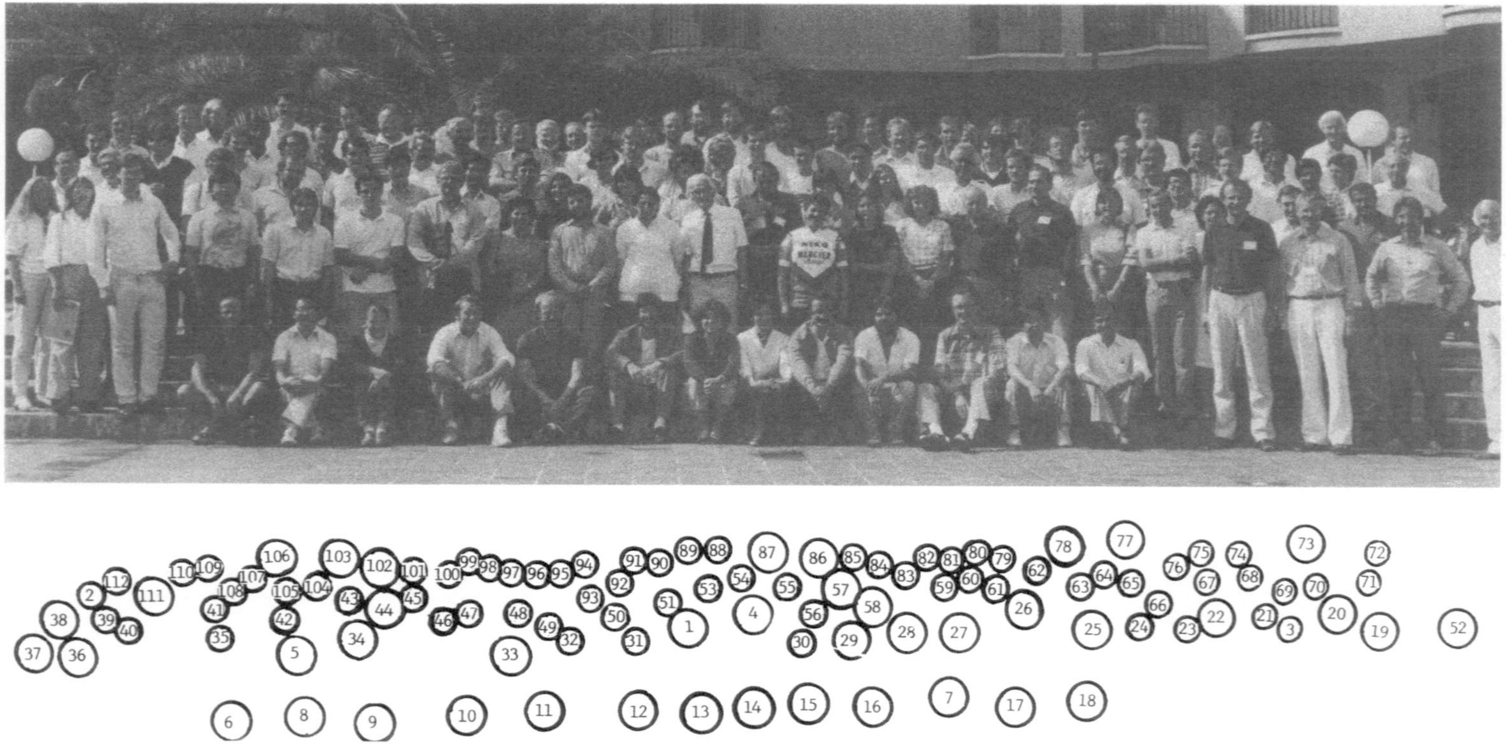

Participants photographed at the A S I on "Fundamental Processes in Atomic Collisions Physics", 10-21 September, 1984, Santa Flavia, Italy .

1. Sir David Bates
2. G. Stefani
3. H. J. Beyer
4. H. A. Silim
5. W. Perrie
6. H. Hamdy
7. A. J. Duncan
8. J. Cai
9. A. Lisini
10. J. S. Briggs
11. R. Hippler
12. E. Fiordilino
13. M. Pehlivan
14. F. Tepehan
15. G. Tepehan
16. A. S. Aynacioglu
17. L. H. Kim
18. C. V. Sheth
19. J. Hinze
20. E. Yurtsever
21. H. Kelly
22. U. Heinzmann
23. C. Cisneros
24. V. E. Ficocelli
25. S. Cavalli
26. H. O. Lutz
27. U. Fano
28. S. J. Ward
29. S. Gozzini
30. A. C. Azeuedo e Souza
31. R. Danielle
32. S. Basile
33. S. Nuzzo Zarcone
34. L. Ulivi
35. S. Watanabe
36. A. Henne
37. A. Toepfer
38. V. Aquilanti
39. D. J. Land
40. J. F. McCann
41. R. M. Dreizler
42. M. Schulz
43. C. D. Lin
44. M. T. Lee
45. K. Blum
46. U. Wille
47. G. Shen
48. M. Zarcone
49. M. Caggegi
50. M. F. V.De Mota Furtado
51. S. Sharma
52. H. Kleinpoppen
53. W. Harbich
54. N. J. Mason
55. L. F. Forrest
56. I. Muriel
57.
58. F. Keller
59. H. G. M. Keidemann
60. L. Avaldi
61. P. J. M. van der Burght
62. U. Bafile
63. U. Becker
64. C. A. Quarles
65. J. C. Alexander
66. P. Decleva
67. A. Wolcke
68. B. Dai
69.
70. M. Allegrini
71. A. Weingartshofer
72. M. Faubel
73. P. G. Burke
74. D. W. Mueller
75. R. B. Kay
76. H. Merz
77. K. J. Ross
78. I. C. Lyon
79. P. F. O'Mahony
80. not ASI participant
81. O. Schöller
82. J. Ganz
83. C. Gabbanini
84. M. A. Bennett
85. R. J. Allan
86. A. Schwab
87. K. Floeder
88. M. H. Kelley
89. G. Basbas
90. H. Schwier
91. H. Reihl
92. H. J. Humpert
93. T. M. Reeves
94. W. J. Humphries
95. E. Merzbacher
96. M. Wilson
97. H. Genz
98. L. Moorman
99. B. Lohmann
100. C. Szmytkowski
101. M. Cavagnero
102. J. F. Reading
103. C. Heckenhamp
104. P. Lambropoulos
105. X. Tang
106. W. Jitschin
107. C. Mandal
108. W. Sandner
109. H. J. Lüdde
110. J. T. Broad
111. A. Palma
112. W. Stich

AUTHOR INDEX

SUBJECT INDEX